Pedagogical Use of Color

The colors that you see in the illustrations of this text are u... improve clarity and understanding. Many figures with three-dimensional perspectives are airbrushed in various colors to make them look as realistic as possible.

Color has been used in various parts of the book to identify specific physical quantities. The following schemes have been adopted.

Chapters 1–10: Motion

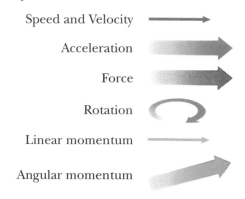

Chapters 17–19: Light and Optical Devices

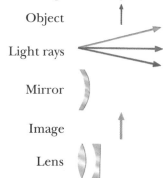

Chapters 20–27: Electricity and Magnetism

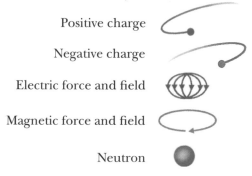

www.brookscole.com

www.brookscole.com is the World Wide Web site for Brooks/Cole and is your direct source to dozens of online resources.

At *www.brookscole.com* you can find out about supplements, demonstration software, and student resources. You can also send e-mail to many of our authors and preview new publications and exciting new technologies.

www.brookscole.com
Changing the way the world learns®

About the Authors

LARRY D. KIRKPATRICK

Larry Kirkpatrick *has always been a teacher; he just didn't know it. After receiving a B.S. in physics from Washington State University and a Ph.D. in experimental high-energy physics from MIT, he began his academic career at the University of Washington as a typical faculty member. However, he found that he was spending more and more time in the classroom and less and less time in the laboratory. Finally, he decided that he would get a position teaching physics full time or he would quit physics and use his computer skills to make lots of money. Fortunately, Montana State University hired him to teach physics. He served eight years as academic director of the U.S. Physics Team that competes in the International Physics Olympiad each summer and has also served as President of the American Association of Physics Teachers. He retired in 2002 so that he can concentrate on teaching, writing, ranching, and playing golf.*

GREGORY E. FRANCIS

Greg Francis *is first and foremost a teacher. As an undergraduate at Brigham Young University, he taught recitation sections normally reserved for graduate students. Later as a graduate student studying plasma physics at MIT, he regularly found opportunities to teach classes normally reserved for research faculty. After finishing his doctorate in 1987, he served as a postdoctoral fellow at Lawrence Livermore National Laboratories. Although his day job gave him the opportunity to work with world-class scientists on exciting problems, he found that he really preferred his night job, teaching physics classes at the local community college. In 1990 Greg joined the Physics Education Research Group at the University of Washington–Seattle, learning the "science" of effective physics teaching. Since 1992 Greg has continued to experiment with active learning approaches in large introductory classes at Montana State University where he is currently Professor of Physics.*

PHYSICS
A WORLD VIEW

SIXTH EDITION

LARRY D. KIRKPATRICK
MONTANA STATE UNIVERSITY

GREGORY E. FRANCIS
MONTANA STATE UNIVERSITY

THOMSON
BROOKS/COLE

Australia • Canada • Mexico • Singapore • Spain
United Kingdom • United States

THOMSON

BROOKS/COLE

Physics: A World View, **Sixth Edition**
Larry D. Kirkpatrick / Gregory E. Francis

Physics Acquisition Editor: *Chris Hall*
Development Editor: *Peter McGahey*
Editorial Assistant: *Jessica Jacobs*
Technology Project Manager: *Sam Subity*
Marketing Manager: *Mark Santee*
Marketing Assistant: *Michele Colella*
Managing Marketing Communications Manager:
  Bryan Vann
Project Manager, Editorial Production: *Teri Hyde*
Art Director: *Lee Friedman*
Print Buyer: *Doreen Suruki*

Permissions Editor: *Sarah Harkrader, Bob Kauser*
Production Service: *G&S Book Services*
Text Designer: *Norman Baugher*
Photo Researcher: *Dena Digilio-Betz*
Copy Editor: *Mary Ann Short*
Illustrator: *Rolin Graphics*
Cover Designer: *Irene Morris*
Cover Image: *Alan Kearney/Getty Images*
Compositor: *G&S Book Services*
Cover and Text Printer: *Quebecor World—Dubuque*

© 2007 Thomson Brooks/Cole, a part of The Thomson Corporation. Thomson, the Star logo, and Brooks/Cole are trademarks used herein under license.

ALL RIGHTS RESERVED. No part of this work covered by the copyright hereon may be reproduced or used in any form or by any means—graphic, electronic, or mechanical, including photocopying, recording, taping, web distribution, information storage and retrieval systems, or in any other manner—without the written permission of the publisher.

Printed in the United States of America
1 2 3 4 5 6 7 10 09 08 07 06

For more information about our products, contact us at:
Thomson Learning Academic Resource Center
1-800-423-0563
For permission to use material from this text or product, submit a request online at http://www.thomsonrights.com.
Any additional questions about permissions can be submitted by email to thomsonrights@thomson.com.

Thomson Wadsworth
10 Davis Drive
Belmont, CA 94002-3098
USA

ExamView® and ExamView Pro® are registered trademarks of FSCreations, Inc. Windows is a registered trademark of the Microsoft Corporation used herein under license. Macintosh and Power Macintosh are registered trademarks of Apple Computer, Inc. Used herein under license.

Library of Congress Control Number: 2005926218

ISBN 0-495-01088-X

WE DEDICATE THIS BOOK TO:

our children

Jennifer   *Joshua*
Monica   *Heidelinde*
Peter   *J. Christian*
     Matthew

who in their very different ways have taught us how to view the world again,

and our wives

Karen   *Sandra*

who have supported us in our efforts to explain the physicists' world view to our students.

Brief Contents

Preface xxi

1 A World View 1
2 Describing Motion 14
3 Explaining Motion 33
4 Motions in Space 56
5 Gravity 74

Interlude **The Discovery of Invariants** 93

6 Momentum 96
7 Energy 111
8 Rotation 134

Interlude **Universality of Motion** 152

9 Classical Relativity 154
10 Einstein's Relativity 176

Interlude **The Search for Atoms** 204

11 Structure of Matter 207
12 States of Matter 227
13 Thermal Energy 247
14 Available Energy 268

Interlude **Waves—Something Else That Moves** 286

15 Vibrations and Waves 288
16 Sound and Music 313

Interlude **The Mystery of Light** 333

17 Light 336
18 Refraction of Light 358
19 A Model for Light 382

Interlude   **An Electrical and Magnetic World**   401

    20   Electricity   403

    21   Electric Current   426

    22   Electromagnetism   446

Interlude   **The Story of the Quantum**   471

    23   The Early Atom   473

    24   The Modern Atom   498

Interlude   **The Subatomic World**   522

    25   The Nucleus   524

    26   Nuclear Energy   548

    27   Elementary Particles   571

    28   Frontiers   590

          Appendix   *Nobel Laureates in Physics*   *A-1*

          Answers to Most Odd-Numbered Questions and Exercises   *A-5*

          Credits   *C-1*

          Glossary   *G-1*

          Index   *I-1*

Contents

Preface xxi

Chapter 1 — A World View 1

First Grade 2
On Building a World View 2
Bode's Law 5
Measurements 6
Sizes: Large and Small 9
Summary 12

Chapter 2 — Describing Motion 14

Average Speed 15
Images of Speed 16
Instantaneous Speed 18
Speed with Direction 19
Acceleration 21
A First Look at Falling Objects 23
Free Fall: Making a Rule of Nature 25
Starting with an Initial Velocity 27
A Subtle Point 27
Summary 28
Fastest and Slowest 19
Galileo: Immoderate Genius 24

Chapter 3 — Explaining Motion 33

An Early Explanation 34
The Beginnings of Our Modern Explanation 35
Newton's First Law 36
Adding Vectors 38
Newton's Second Law 40
Mass and Weight 43
Weight 44
Free-Body Diagrams 44
Free Fall Revisited 45
Galileo versus Aristotle 46
Friction 46

Newton's Third Law 47
Summary 51
Newton: Diversified Brilliance 37
Terminal Speeds 48

Chapter 4 **Motions in Space 56**

Circular Motion 57
Acceleration Revisited 58
Acceleration in Circular Motion 60
Projectile Motion 60
Launching an Apple into Orbit 66
Rotational Motion 66
Summary 69
Banking Corners 62
Floating in Defiance of Gravity 68

Chapter 5 **Gravity 74**

The Concept of Gravity 75
Newton's Gravity 76
The Law of Universal Gravitation 79
The Value of G 80
Gravity near Earth's Surface 81
Satellites 83
Tides 84
How Far Does Gravity Reach? 86
The Field Concept 87
Summary 88
Kepler: Music of the Spheres 76
How Much Do You Weigh? 82

Interlude **The Discovery of Invariants 93**

Chapter 6 **Momentum 96**

Linear Momentum 97
Changing an Object's Momentum 97
Conservation of Linear Momentum 99
Collisions 101
Investigating Accidents 103
Airplanes, Balloons, and Rockets 104
Summary 106
Landing the Hard Way: No Parachute! 99
Noether: The Grammar of Physics 105

Chapter 7 **Energy 111**

What Is Energy? 112
Energy of Motion 113
Conservation of Kinetic Energy 114
Changing Kinetic Energy 115
Forces That Do No Work 116
Gravitational Potential Energy 118
Conservation of Mechanical Energy 119
Roller Coasters 121
Other Forms of Energy 122
Is Conservation of Energy a Hoax? 125
Power 127
Summary 128

Stopping Distances for Cars 117
Exponential Growth 124
Human Power 127

Chapter 8 **Rotation 134**

Rotational Motion 135
Torque 135
Rotational Inertia 138
Center of Mass 139
Stability 141
Rotational Kinetic Energy 142
Angular Momentum 142
Conservation of Angular Momentum 143
Angular Momentum: A Vector 144
Summary 146

Interlude **Universality of Motion 152**

Chapter 9 **Classical Relativity 154**

A Reference System 155
Motions Viewed in Different Reference Systems 156
Comparing Velocities 157
Accelerating Reference Systems 158
Realistic Inertial Forces 160
Centrifugal Forces 163
Earth: A Nearly Inertial System 164
Noninertial Effects of Earth's Motion 166
Summary 170

Living in Zero G 162
Planetary Cyclones 169

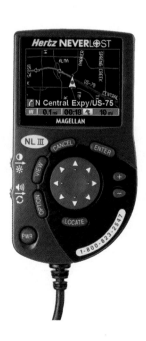

Chapter 10 — Einstein's Relativity 176

- The First Postulate 177
- Searching for the Medium of Light 178
- The Second Postulate 178
- Simultaneous Events 179
- Synchronizing Clocks 181
- Time Varies 185
- Experimental Evidence for Time Dilation 186
- Length Contraction 187
- Spacetime 190
- Relativistic Laws of Motion 191
- General Relativity 192
- Warped Spacetime 195
- Summary 197
- **The Twin Paradox 188**
- **Einstein: Person of the Century 191**
- **Global Positioning System (GPS) 193**
- **Black Holes 196**

Interlude — The Search for Atoms 204

Chapter 11 — Structure of Matter 207

- Building Models 208
- Early Chemistry 209
- Chemical Evidence of Atoms 211
- Masses and Sizes of Atoms 212
- The Ideal Gas Model 214
- Pressure 215
- Atomic Speeds and Temperature 216
- Temperature 218
- The Ideal Gas Law 220
- Summary 222
- **Evaporative Cooling 221**

Chapter 12 — States of Matter 227

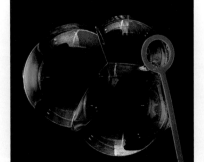

- Atoms 228
- Density 228
- Solids 230
- Liquids 231
- Gases 232
- Plasmas 233
- Pressure 233
- Sink and Float 236

Bernoulli's Effect 238
Summary 241
Density Extremes 229
Solid Liquids and Liquid Solids 235
How Fatty Are You? 239
The Curve Ball 240

Chapter 13 Thermal Energy 247

The Nature of Heat 248
Mechanical Work and Heat 249
Temperature Revisited 250
Heat, Temperature, and Internal Energy 251
Absolute Zero 252
Specific Heat 252
Change of State 255
Conduction 256
Convection 258
Radiation 259
Wind Chill 260
Thermal Expansion 262
Summary 263
Joule: A New View of Energy 251
Freezing Lakes 263

Chapter 14 Available Energy 268

Heat Engines 269
Ideal Heat Engines 271
Perpetual-Motion Machines 272
Real Engines 273
Refrigerators 274
Order and Disorder 275
Entropy 278
Decreasing Entropy 279
Entropy and Our Energy Crisis 281
Summary 281
Arrow of Time 279
Quality of Energy 280

Interlude **Waves—Something Else That Moves 286**

Chapter 15 Vibrations and Waves 288

Simple Vibrations 289
The Pendulum 291

Clocks 293
Resonance 293
Waves: Vibrations That Move 295
One-Dimensional Waves 297
Superposition 299
Periodic Waves 300
Standing Waves 302
Interference 304
Diffraction 306
Summary 307
Tacoma Narrows Bridge 296
Probing the Earth 299

Chapter 16 Sound and Music 313

Sound 314
Speed of Sound 315
Hearing Sounds 316
The Recipe of Sounds 317
Stringed Instruments 319
Wind Instruments 321
Percussion Instruments 323
Beats 324
Doppler Effect 325
Shock Waves 328
Summary 328
Animal Hearing 318
Loudest and Softest Sounds 322
Breaking the Sound Barrier 327

Interlude The Mystery of Light 333

Chapter 17 Light 336

Shadows 337
Pinhole Cameras 339
Reflections 340
Flat Mirrors 341
Multiple Reflections 342
Curved Mirrors 343
Images Produced by Mirrors 345
Locating the Images 346
Speed of Light 348
Color 350
Summary 353

Eclipses 338

Retroreflectors 344

Chapter 18 Refraction of Light 358

Index of Refraction 359

Total Internal Reflection 361

Atmospheric Refraction 362

Dispersion 363

Rainbows 364

Halos 366

Lenses 367

Images Produced by Lenses 368

Cameras 371

Our Eyes 371

Magnifiers 374

Telescopes 375

Summary 377

Mirages 364

Eyeglasses 373

The Hubble Space Telescope 376

Chapter 19 A Model for Light 382

Reflection 383

Refraction 384

Interference 385

Diffraction 387

Thin Films 390

Polarization 392

Looking Ahead 396

Summary 396

Diffraction Limits 389

Holography 394

Interlude An Electrical and Magnetic World 401

Chapter 20 Electricity 403

Electrical Properties 404

Two Kinds of Charge 405

Conservation of Charge 406

Induced Attractions 407

The Electroscope 409

The Electric Force 411

Electricity and Gravity 413

The Electric Field 414
Electric Field Lines 416
Electric Potential 418
Summary 420
Franklin: The American Newton 407
Lightning 419

Chapter 21 Electric Current 426

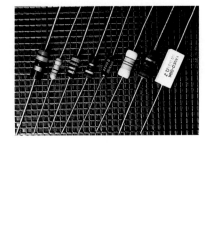

An Accidental Discovery 427
Batteries 428
Pathways 429
A Water Model 430
Resistance 431
The Danger of Electricity 433
A Model for Electric Current 433
A Model for Voltage 436
Electric Power 439
Summary 441
The Real Cost of Electricity 438

Chapter 22 Electromagnetism 446

Magnets 447
Electric Currents and Magnetism 449
Making Magnets 450
The Ampere 451
The Magnetic Earth 453
Charged Particles in Magnetic Fields 454
Magnetism and Electric Currents 455
Transformers 457
Generators and Motors 458
A Question of Symmetry 460
Electromagnetic Waves 461
Radio and TV 464
Summary 466
Superconductivity 452
"Wireless" Battery Charger 459
Maxwell: Unifying the Electromagnetic Spectrum 463
Stereo Broadcasts 465

Interlude The Story of the Quantum 471

Chapter 23 The Early Atom 473

Periodic Properties 474

Atomic Spectra   474
Cathode Rays   477
The Discovery of the Electron   478
Thomson's Model   479
Rutherford's Model   479
Radiating Objects   482
The Photoelectric Effect   484
Bohr's Model   486
Atomic Spectra Explained   488
The Periodic Table   491
X Rays   492
Summary   493
Rutherford: At the Crest of the Wave   481
Planck: Founder of Quantum Mechanics   484
Bohr: Creating the Atomic World   490

Chapter 24 The Modern Atom   498

Successes and Failures   499
De Broglie's Waves   499
Waves and Particles   502
Probability Waves   505
A Particle in a Box   506
The Quantum-Mechanical Atom   507
The Exclusion Principle and the Periodic Table   509
The Uncertainty Principle   511
The Complementarity Principle   513
Determinism   514
Lasers   515
Summary   517
Seeing Atoms   501
Psychedelic Colors   509

Interlude The Subatomic World   522

Chapter 25 The Nucleus   524

The Discovery of Radioactivity   525
Types of Radiation   526
The Nucleus   528
The Discovery of Neutrons   529
Isotopes   529
The Alchemists' Dream   530
Radioactive Decay   533
Radioactive Clocks   535

Radiation and Matter 537
Biological Effects of Radiation 538
Radiation around Us 539
Radiation Detectors 542
Summary 544
Curie: Eight Tons of Ore 527
Smoke Detectors 536
Radon 540

Chapter 26 **Nuclear Energy 548**

Nuclear Probes 549
Accelerators 549
The Nuclear Glue 550
Nuclear Binding Energy 552
Stability 554
Nuclear Fission 555
Chain Reactions 557
Nuclear Reactors 560
Breeding Fuel 562
Fusion Reactors 564
Solar Power 565
Summary 566
Goeppert-Mayer: Magic Numbers 556
Fermi: A Man for All Seasons 558
Meitner: A Physicist Who Never Lost Her Humanity 561
Natural Nuclear Reactors 563

Chapter 27 **Elementary Particles 571**

Antimatter 572
The Puzzle of Beta Decay 575
Exchange Forces 576
Exchange Particles 578
The Elementary Particle Zoo 579
Conservation Laws 581
Quarks 583
Gluons and Color 585
Summary 586
Feynman: Surely You're Joking, Mr. Feynman 577

Chapter 28 **Frontiers 590**

Gravitational Waves 591
Unified Theories 593
Cosmology 594

Cosmic Background Radiation 596
Dark Matter and Dark Energy 597
Neutrinos 598
Quarks, the Universe, and Love 600
The Search Goes On 601

Appendix   Nobel Laureates in Physics   A-1
Answers to Most Odd-Numbered Questions and Exercises   A-5
Credits   C-1
Glossary   G-1
Index   I-1

Preface

This textbook is intended for a conceptual course in introductory physics for students majoring in fields other than science, mathematics, or engineering. It will work very well in courses for future teachers.

Writing this book has been an exercise in translation. We have attempted to take the logic, vocabulary, and values of physics and communicate them in an entirely different language. A good job of translating requires careful attention to both languages, that of the physicist and that of the student. In some areas the physics is so abstract that it took creative bridges to span the gulf between the languages. We are indebted to the many students who shared their confusions with us and wrestled with the clarity of our translations. We are equally indebted to the many physicists who shared our search for the proper word or metaphor that comes closest to capturing the abstract, elusive idea.

Mathematics is the structural foundation for all of the physics world view. As stated previously, this textbook translates most of the ideas into longer, less tightly structured sentences. Still, the mathematics holds much of the beauty and power of physics, and we want to offer a glimpse of this for students whose mathematical background is adequate. Therefore, the more mathematical presentations within the textbook have been placed in boxes labeled *Working It Out* to make the textbook friendlier to those students in courses that do not include this material. These boxes allow the students to skip over the more mathematical material without loss of continuity in the conceptual development of the physics ideas.

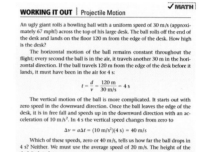

Math presentations are placed in *Working It Out* boxes that can be skipped over in those courses that do not include mathematics.

We have also written a mathematical supplement, *Problem Solving to Accompany Physics: A World View,* that delves deeper into the mathematical structure of the physics world view. The presentations in *Problem Solving* follow those in the textbook and those sections that have extended discussions in the supplement are indicated by a math icon, making it easy to integrate additional mathematics into the course. This supplement can be bundled with the textbook.

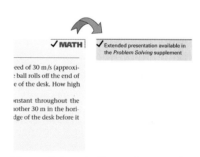

The math icon indicates that complementary mathematical material is in *Problem Solving*.

Objectives

The main objective of this physics textbook is to provide non-science-oriented students with a clear and logical presentation of some of the basic concepts and principles of physics in an appropriate language. Our overriding concern has been to choose topics and ideas for students who will only be taking this single course in physics. We continually reminded ourselves that this may be our one chance to describe the way physicists look at the world and test their ideas. We chose topics that convey the essence of the physics world view. As an example of this concern, we have placed more modern physics—specifically the theories of relativity—in the first half of the book rather than toward the end, as in most traditional textbooks. We also describe the historical development of quantum physics carefully in order to show why various atomic models—models that make common sense—fail to explain the experimental evidence. At the same time, we have attempted to motivate students through practical examples that demonstrate the role of physics in other disciplines and in their everyday lives.

Coverage

The topics covered in this book are the fundamental topics in classical and modern physics. The book is divided into nine parts. Interludes set the theme for the sections that follow.

- Part I (Chapters 1–5) opens with an introduction to the physicists' world view and then deals with the fundamentals of motion, including Newton's three laws of motion. This part ends with a careful look at gravity, our most familiar force.
- Part II (Chapters 6–8) reexamines motion through an investigation of three fundamental conservation laws: momentum, energy, and angular momentum.
- Part III (Chapters 9–10) explores the concepts involved in classical, special, and general relativity.
- In the beginning of Part IV (Chapters 11–14), we set the stage for expanding our understanding of energy by investigating the structure of matter, first macroscopically, then microscopically. This part ends with a study of thermodynamics, including heat, temperature, internal energy, heat engines, and entropy.
- Part V (Chapters 15–16) develops the basic properties of wave phenomena and applies them to a study of sound and music. It also gives the reader a background that will be helpful in understanding much of quantum physics.
- Part VI (Chapters 17–19) covers the study of light and optics, starting with the general question of the basic nature of light, covering interesting applications, and ending with consequences of the wave nature of light.
- Part VII (Chapters 20–22) covers the basic concepts in electricity and magnetism, including a careful examination of simple circuits and the nature of electromagnetic waves.
- In Part VIII (Chapters 23–24), we develop the story of the quantum, starting with the discovery of the electron and ending with quantum physics.
- The final section of the textbook, Part IX (Chapters 25–28), takes the student deeper into the study of the structure of matter by looking at the nucleus and eventually the fundamental particles. It ends with a look at some of the frontiers of physics.

New to the Sixth Edition

After soliciting comments from physics teachers and students, we carefully considered each suggestion and used many of them in reworking the entire textbook. We simplified explanations of some phenomena; updated developing areas, such as elementary particles and cosmology; and added new explanatory material.

- We also reexamined each drawing for clarity and to see if it accomplished its intended functions. Several dozen were modified or entirely redrawn to remove ambiguities or to make the meaning more transparent. Many of the photographs were changed to improve their functionality.
- We carefully reexamined each of the conceptual questions and exercises.
- We added 173 new questions and deleted 72 old questions for a net increase of 101 questions. In addition, we changed 127 questions, either by rewording them or by changing the numerical values. We also added 54 new exercises and deleted 12 of the old ones for a net gain of 42 exercises. Twelve of the exercises were reworded and the numerical values were changed in an additional 148.

In this revision we paid special attention to the three chapters on electricity, magnetism, and electromagnetic waves (Chapters 20–22). In addition to changes throughout these chapters to improve the clarity of the descriptions and explanations, we made a number of more extensive changes.

- In Chapter 20, we made substantial changes in Conservation of Charge, The Electroscope, The Electric Force, Electricity and Gravity, land Electric Potential sections. We introduced a new section on electric field lines and added two new in-text questions and answers.
- We changed the name of Chapter 21 to the more conventional Electric Current. The section on batteries and bulbs was extensively rewritten based on our experiences using this as the inquiry component of the course we are teaching at Montana State University. The ideas have been reordered in a more logical way and the simple current model for circuits containing batteries and bulbs has been expanded to include voltage considerations. The two new sections are A Model for Electric Current and A Model for Voltage. We also introduced Kirchhoff's junction and loop rules.
- We added a new feature box "Wireless" Battery Charger in Chapter 22 as a practical application of electromagnetism. We made a larger number of changes to the conceptual questions and exercises in all three chapters.

We are most excited about our use of personal response systems in our classes. Students use hand-held "clickers" to respond to questions posed on PowerPoint® slides. As others have already discovered, we found that students are much more active learners when they must respond to in-class questions or activities. We have also found graphs of the student responses useful in making mid-class changes in the direction of our presentations. We have developed an extensive collection of clicker slides that are available for adopters of our textbook. For more information, see the listing for JoinIn™ on TurningPoint® in the Supporting Materials for the Instructor section a few pages further along in this Preface.

Author Larry Kirkpatrick has prepared class-tested questions for JoinIn™ on TurningPoint® suitable for use with a variety of "clicker" electronic response systems.

Students are finding web-based support for textbooks increasingly important. In order to better teach students using this book and help them determine which of the many web tools they should devote time to exploring, we have created a PhysicsNow site to complement the new edition. Brooks/Cole has seen great acceptance of this system which uses diagnostic quizzes and personal learning plans to direct student learning to those resources that will help them most. We have tailored this system to the revised edition. Additional details about PhysicsNow are available in the Supporting Materials for the Student section later in the Preface.

Because we realize that the traditional hardback textbook may not fit every classroom's needs or every student's budget, we have included the new edition of *Physics: A World View* in Thomson's digital library TextChoice which allows you to create a customized version of the book that may include locally written material. See the Custom Option section later in the Preface for additional details.

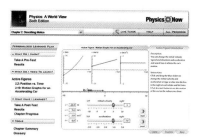

The easy-to-use, browser-based format of PhysicsNow allows users to check their knowledge and target those resources that will most help their study.

Thematic Paths and Teaching Options

A big part of the flow of any course rests on decisions made by the instructor about what to cover. For this reason the textbook contains more material than can be covered in an introductory course in one term. It is possible to take many routes through the material, depending on your interests and the interests of your students. To illustrate some possibilities, we have compiled seven different paths that can be used for semester-long courses. Each thematic path uses about one-half of the material presented in the textbook. These seven different teaching options (or thematic paths) might be called Physical Science, Electricity and Magnetism, Optics, Energy, Vibrations and Waves, Relativity, and Elementary Particles.

- *Physical Science* emphasizes those topics that are basic to both physics and chemistry. After studying motion and the concepts of momentum and energy, this option delves into the structure and states of matter, heat and thermodynamics, the basic properties of waves, and ends up with atomic physics.
- In the *Electricity and Magnetism* option, we begin with motion and the concepts of momentum and energy, skip to the chapter on waves, and then to the three chapters on electricity, magnetism, and electromagnetism. We conclude with the two chapters on atomic physics.
- The *Optics* option also begins with motion and waves. It then covers most of the three chapters on light and finishes with atomic physics.
- The *Energy* course of study begins with motion, momentum, and mechanical energy. It then covers the two chapters on thermal energy and thermodynamics. After the chapter on waves, the course skips to the three chapters on electricity and magnetism, including electromagnetic waves. The course ends with a study of the nucleus and nuclear energy.
- *Vibrations and Waves* covers motion and energy and then concentrates on the wave properties of music, light, electromagnetism, and the quantum-mechanical atom.
- The *Relativity* option yields a very different course. After a study of the basics of motion, momentum, and energy, the course includes the two chapters on classical, special, and general relativity. There are many ways to complete this course; we favor finishing with some of the properties of light.
- For those interested in the search for the ultimate building blocks of the Universe, we suggest the *Elementary Particles* emphasis. The study of motion is followed by the chapter on the structure of matter. The amount of classical and special relativity will depend on the time you have. We then study waves and the wave aspect of light before moving on to selected topics in electricity. The main part of this course is the chapters on atomic and nuclear physics with elementary particles as the capstone.

We have detailed these seven theme-based options in tabular form in the *Instructor's Resource Manual* (ISBN 0-495-01090-1) available to qualifying instructors from your Thomson Brooks/Cole representative or by request at www.physics.brookscole.com. If teaching one of these themes, consider creating a custom version of *Physics: A World View* at www.TextChoice.com. More information about this service is available in the Custom Options section later in the Preface.

Features

Most instructors would agree that the textbook selected for a course should be the student's major "guide" for understanding and learning the subject matter. Furthermore, a textbook should be written and presented to make the material accessible and easier to teach, not harder. With these points in mind, we have included many pedagogical features to enhance the usefulness of our textbook for both you and your students.

Organization The textbook contains a story line about the development of the current physics world view. It is divided into nine parts: the fundamentals of motion and gravity; the conservation laws of momentum and energy; the theories of relativity; the structure of matter, including heat and thermodynamics;

wave phenomena and sound; light and optics; electricity and magnetism; the story of the quantum; and the nucleus, fundamental particles, and frontiers. Each part includes an overview of the subject matter to be covered in that part and some historical perspectives.

Style We have written the book in a style that is clear, logical, and succinct, in order to facilitate students' comprehension. The writing style is somewhat informal and relaxed, which we hope students will find appealing and enjoyable to read. New terms are carefully defined, and we have tried to avoid jargon.

Mathematical Level The mathematical level in the textbook has been kept to a minimum, with some limited use of algebra and geometry. Equations are presented in words as well as in symbols. For those desiring a higher mathematical presentation, we have written a mathematical supplement, *Problem Solving to Accompany Physics: A World View,* which parallels the topics in the textbook and presents additional mathematical aspects of the topic. A math icon in the textbook indicates that supplemental material is available in *Problem Solving.* The textbook is available shrink-wrapped with this ancillary.

We have not shied away from using numbers where they assist in developing a more complete understanding of a concept. On the other hand, we have rounded off the values of physical constants to help simplify the discussion. For example, we use 10 (meters per second) per second for the acceleration due to gravity, except when discussing the law of universal gravitation, where the additional accuracy is needed to understand its development.

Working It Out The more mathematical presentations have been placed into *Working It Out* boxes. These 28 boxed features allow those students following a more mathematical track to find such information easily, while allowing others to skip over this mathematical material without loss of continuity. See, for example, "Acceleration," on page 23, or "Conservation of Kinetic Energy," on page 115.

Flawed Reasoning Questions and Answers These 49 questions are based on research in physics education and highlight topics in which students are known to have misconceptions and/or difficulties grasping the concepts. They are written in a more casual style to make students more comfortable confronting their misconceptions and difficulties. As an example, see the Flawed Reasoning question and answer on page 139.

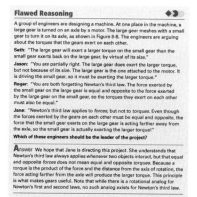

Flawed Reasoning boxes highlight and correct typical misconceptions.

Illustrations The large number of figures, diagrams, photographs, and tables enhance the readability and effectiveness of the text material. Full color is used to add clarity to the artwork and to make it realistic. For example, vectors are color coded for each physical quantity. Three-dimensional effects are produced with the use of color and airbrushed areas, where appropriate. Many of the illustrations show the development of a phenomenon over time as a series of "snapshots." To illustrate the flow of time, we have added a clock icon in these drawings. The color photographs have been carefully selected, and their accompanying captions serve as an added instructional tool. A complete description of the pedagogical use of color appears at the front of the book.

Chapter Opening Questions and Answers Each chapter begins with an inquiry that is answered at the end of the chapter. These focus the student's attention on an important aspect of each chapter.

In-Text Questions and Answers Questions to stimulate thinking are given at key spots throughout each chapter and are set off by a colored screen. These

185 questions allow students to immediately test their comprehension of the concepts discussed. The answers immediately follow the questions. Most questions could also serve as a basis for initiating classroom discussions.

Conceptual Questions and Exercises An extensive set of conceptual questions and exercises is included at the end of each chapter, with a total of 1575 questions and 610 exercises. Almost all questions and exercises are presented in pairs, meaning that each odd-numbered question or exercise has a similar even-numbered one immediately following it. This arrangement allows the student to have one question or exercise with an answer in the back of the textbook and a very similar one without an answer. The pairing also allows you to discuss one exercise in class and assign its "partner" for homework. We have also included a small number of challenging questions and exercises, which are indicated by an icon ▲ printed next to the number. Answers to the odd-numbered questions and exercises are given at the end of the book. Answers to all questions and solutions to all exercises are included in the *Instructor's Resource Manual*.

Physics on Your Own Each chapter contains several projects and simple experiments for students to do on their own. The 88 projects have been designed to require a minimum of apparatus and to illustrate the concepts presented in the text. These illustrate the experimental aspects of physics and the application of physics to our everyday lives.

Special Topic Boxes Almost all chapters include optional special topic boxes to expose students to various practical and interesting applications of physical principles. Some of the 45 special topics include mirages, liquid crystal displays, fluorescent colors, holograms, gravity waves, radon, superconductivity, natural nuclear reactors, and "wireless" battery chargers.

Biographical Sketches Besides the historical perspectives provided in the 8 interludes between the major parts of the textbook, we have added 16 short biographies of important scientists throughout the textbook to give a greater historical emphasis without interrupting the development of the physics concepts.

Important Concepts Important statements and equations are highlighted in several ways for easy reference and review.

- Important principles are boxed for easy reference.
- Marginal notes are used to highlight important statements, equations, and concepts in the text.
- Each chapter ends with a summary reviewing the important concepts of the chapter and the key terms.

Units The international system of units (SI) is used throughout the text. The U.S. customary system of units (conventional system) is used to a limited extent in the chapters on mechanics, heat, and thermodynamics to help the student develop a better feeling for the sizes of the SI units.

Appendix An appendix at the end of the text lists the Nobel laureates in physics.

Endpapers Tables of physical data and other useful information, including fundamental constants, the periodic table of the elements, conversion factors, the Greek alphabet, and standard abbreviations of units, appear on the endpapers. In addition, the front endpaper includes the color code for all figures and diagrams.

TextChoice Custom Options for *Physics: A World View* Create the perfect text to match your syllabus! Realizing that not all instructors cover all material from their text, we are proud to announce our innovative new custom publishing program, TextChoice (www.textchoice.com). This extensive digital library lets you customize learning materials on your own computer by previewing and assembling content from a growing list of Thomson titles including *Physics: A World View,* Sixth Edition. Search for content by course name, keyword, author, title, ISBN, and other categories. You can add your own course notes, supplements, lecture outlines, and other materials to the beginning or end of any chapter as well as arrange text chapters in any order or eliminate chapters that you don't cover in the course. Within 48 hours of saving your project and submitting your order, a consultant will call you with a quote and answer any questions you may have. Once your project is finalized, Thomson Custom Solutions will print the product and ship it to your bookstore.

If you prefer to publish your original learning materials as a standalone book or supplement, contact Thomson Custom Solutions (www.thomsoncustom.com, *or* 1-800-355-9983).

Supporting Materials for the Student

Problem Solving to Accompany Physics: A World View (ISBN 0-495-01093-6) This mathematical supplement, keyed to the textbook, develops some of the numerical aspects of this course that can be addressed with simple algebra and geometry. Sections in the textbook that have an extended, parallel presentation in *Problem Solving* are indicted by a math icon to the right of the section title. The supplement contains extended mathematical discussion for those sections and additional numerical end-of-chapter problems with odd-numbered answers in an appendix.

This supplement is available shrink-wrapped with the textbook (bundle ISBN 0-495-15448-2).

PhysicsNow Website (http://physics.brookscole.com/kf6e) Most new copies of this book come with a free pass code to PhysicsNow, an assessment-centered system that helps students determine their unique study needs and directs them to web-based resources to enhance their conceptual understanding. A Pre-Test creates a Personal Learning Plan that directs students to relevant text sections and end-of-chapter questions, web-based tutorials, and other media assets based on their specific needs. A Chapter Quiz follows the learning plan to help measure comprehension. Select a chapter and log on using your pass code at http://physics.brookscole.com/kf6e.

Pass codes may be purchased independent of the book by following the login link on the student companion site at above web address.

Companion Website (http://physics.brookscole.com/kf6e) The companion website includes study aids such as flash cards, chapter outlines, and learning objectives for each chapter.

Supporting Materials for the Instructor

Qualifying instructors should contact your local Thomson Brooks/Cole representative for review copies. Visit http://physics.brookscole.com to locate your representative or request desk copies online.

ExamView® Computerized Test Bank (ISBN 0-495-01095-2) ExamView® allows instructors to create, deliver, and customize tests and study guides in minutes using questions from the print Test Bank. Both a Quick Test Wizard and an Online Test Wizard provide step by step guidance through the process of creating tests.

Instructor's Resource Manual **(ISBN 0-495-01090-1)** The *Instructor's Resource Manual* is written by the authors and contains answers and solutions to all questions and exercises in the textbook and all problems in the *Problem Solving* supplement. Teaching tips and information about integrating demonstrations into your lecture are provided for each chapter and course outlines help determine which sections to include in themed courses that do not plan to teach every chapter.

JoinIn™ on TurningPoint® for Electronic Response Systems (ISBN 0-495-01099-5) Thomson Brooks/Cole now offers class-tested JoinIn content for electronic response systems written by the authors of *Physics: A World View*. These instant in-class questions are correlated to each major concept and allow you to transform your classroom and assess your students' progress as you lecture. Our exclusive agreement to offer TurningPoint software lets you pose questions and display students' answers seamlessly within your own Microsoft PowerPoint slides in conjunction with the "clicker" hardware of your choice. Enhance how your students interact with you, your lecture, and each other. Contact your local Thomson Brooks/Cole representative to learn more.

The easy-to-use interface of the Multimedia Manager CD allows access to an entire library of text art and tables, full sets of PowerPoint slides written by the text author, and other tools.

Multimedia Manager Instructor CD-ROM (ISBN 0-495-01096-0) The Multimedia Manager is a digital library and presentation tool that provides art, photos, and tables from the text in JPEG and PowerPoint electronic formats. These electronic files can be used to make transparencies and are easily exported into other software packages. This enhanced CD-ROM also contains features that assist in easily preparing lectures for the new edition. Text author Larry Kirkpatrick's full sets of PowerPoint slides for each chapter are suitable for use with electronic response systems, and slides can easily be deleted or new ones added to customize the sets for your course without a great time investment. Animations and movies are provided to supplement lectures and electronic files of various print supplements are included.

Test Bank (ISBN 0-495-01091-X) To help you prepare quizzes and exams, the Test Bank provides over 1800 multiple-choice questions covering every chapter in the textbook. Answers to all questions are included in the Test Bank.

Transparency Acetates (ISBN 0-495-01092-8) The conceptual physics student often has difficulty gleaning all the information presented in physics diagrams. Therefore, a collection of 100 full-color transparency acetates of important figures from the book is available to qualified adopters.

WebAssign WebAssign is the most-utilized homework management system in physics. Designed by physicists for physicists, this system is a trusted companion to your teaching. To preview the content from *Physics: A World View* in WebAssign visit www.webassign.net or contact your Thomson Brooks/Cole representative for more information.

Physics Demonstration Video Clips Professor Clint Sprott of the University of Wisconsin–Madison, has prepared approximately two hours of video clips demonstrating a wide variety of his simple in-class physics experiments for the introductory physics course from projectile motion and Newton's laws to sound and optics. Clips are available at the book's companion website (http://physics.brookscole.com/kf6e). The *Instructor's Resource Manual* gives helpful hints about integrating the clips into your lecture and replicating the demonstrations.

Acknowledgments

Physicists and physics teachers who gave freely of their time to explore the many options of explaining the physics world view with a minimum of mathematics include our colleagues Jeff Adams, John Carlsten, William Hiscock, Robert Swenson, and George Tuthill from Montana State University, as well as the late Arnold Arons (University of Washington), Larry Gould (University of Hartford), and Bob Weinberg (Temple University). We appreciate the special efforts of Montana State University photography graduate David Rogers for many of the photographs used in the textbook. We would also like to thank the many students who have studied from this textbook and provided us with very valuable feedback.

We are also very grateful to Gerry Wheeler, who almost 30 years ago suggested to Larry Kirkpatrick that they write a textbook. Neither Larry nor Gerry could have written a textbook by himself, but together they produced a textbook that has become a best-seller. In the process of understanding the physics, interacting with students to learn how to present physics to a nontechnical audience, and discussing how best to capture the excitement of classroom teaching in a textbook, they became much better physics teachers . . . and lifelong friends. Gerry is the Executive Director of the National Science Teachers Association. NSTA is very fortunate to have his creative mind and his extraordinary ability to work with people for the betterment of science education. We wish him continued success.

The following reviewers were very helpful in producing the current revision:

ELENA BOROVITSKAYA, Temple University
MILTON W. COLE, Pennsylvania State University
MARTIN HACKWORTH, Idaho State University
DOUG BRADLEY-HUTCHISON, Sinclair Community College
LOIS BREUR KRAUSE, Clemson University
ERNEST MA, Montclair State University
PROMOD R. PRATAP, University of North Carolina at Greensboro
INA P. ROBERTSON, University of Kansas
DANIEL STUMP, Michigan State University
D. BRIAN THOMPSON, University of North Alabama
MATTHEW M. WAITE, West Chester University of Pennsylvania
BONNIE L. WYLO, Eastern Michigan University

The reviewers for the fifth edition were:

JAMES ARRISON, Villanova University
ORVILLE DAY, East Carolina University
DAVID DEMUTH, University of Minnesota, Crookston
TINA FANETTI, Western Iowa Technical Community College
DOUGLAS S. HAMILTON, University of Connecticut
TERESA L. LARKIN, American University
ROBERT S. PANVINI, Vanderbilt University
MICHAEL ROTH, University of Northern Iowa
LAWRENCE WEINSTEIN, Old Dominion University
ROBERT A. WILSON, San Bernardino Valley College

The following reviewers offered many helpful suggestions for the fourth edition:

James Arrison, Villanova University
Larry Browning, South Dakota State University
Michael Cree, University of Waikato, New Zealand
Orville Day, East Carolina University
David Donnelly, Sam Houston State University
H. James Harmon, Oklahoma State University
Christian Iliadis, University of North Carolina–Chapel Hill
Richard McCorkle, University of Rhode Island
Irina Nelson, Salt Lake Community College
Robert Packard, Baylor University
Sokrates Pantelides, Vanderbilt University
Alistair Steyn-Ross, University of Waikato, New Zealand
Lara Wilcocks, University of Waikato, New Zealand
Robert Wilson, San Bernardino Valley College
Robert Zbikowski, Hibbing Community College

The following reviewers provided valuable input for the third edition:

Philip Baringer, University of Kansas
Louis Cadwell, Providence College
Jorge Cossio, North Miami–Dade Community College
Gary DeLeo, Lehigh University
Mark Miksic, Queens College
Thomas Smith, Mount San Antonio College
George Smoot, University of California at Berkeley
Jan Yarrison-Rice, Miami University of Ohio

Reviewers who played an important role in the development of the second edition of the textbook include:

Jeffrey Collier, Bismarck State Community College
Leroy Dubeck, Temple University
John R. Dunning, Jr., Sonoma State University
Joseph Hamilton, Vanderbilt University
Roger Herman, Pennsylvania State University
Robert Lieberman, Cornell University
Richard Lindsay, Western Washington University
Robert A. Luke, Boise State University
Allen Miller, Syracuse University
John Mudie, Modesto Junior College
Steven Robinson, Colorado State University
Carl Rosenzweig, Syracuse University
James Watson, Ball State University

Reviewers of the first edition, who helped shape the content and structure of the current edition, include:

John C. Abele, Lewis & Clark College
David Buckley, East Stroudsburg University
Robert Carr, California State University, Los Angeles
Art Champagne, Princeton University
David E. Clark, University of Maine
Robert Cole, University of Southern California

JOHN R. DUNNING, Jr., Sonoma State University
ABBAS FARIDI, Orange Coast College
SIMON GEORGE, California State University, Long Beach
PATRICK C. GIBBONS, Washington University
ROBERT E. GIBBS, Eastern Washington University
TOM J. GRAY, Kansas State University
ROGER HANSON, University of Northern Iowa
STEVEN HOFFMASTER, Gonzaga University
MICHAEL HONES, Villanova University
SARDARI L. KHANNA, York College
JEAN P. KRISH, University of Michigan
DAVID A. KRUEGER, Colorado State University
LEON R. LEONARDO, El Camino College
ROBERT A. LUKE, Boise State University
PETER MCINTYRE, Texas A&M University
PAUL NACHMAN, New Mexico State University
VAN E. NEIE, Purdue University
BARTON PALATNICK, California State Polytechnic University, Pomona
CECIL G. SHUGART, Memphis State University
LEONARD STORM, Eastern Illinois University
PAUL VARLASHKIN, East Carolina University
LEONARD WALL, California State Polytechnic University, San Luis Obispo
JAMES WATSON, Ball State University
ROBERT WILSON, San Bernardino Valley College
JOHN M. YELTON, University of Florida

The final chapter on *frontiers* poses unique challenges as the topics are truly on the frontiers. We especially want to thank Jeff Adams, Neil Cornish, William Hiscock (all Montana State University), the late Robert S. Panvini (Vanderbilt University), and Chris Waltham (University of British Columbia) for contributing and updating essays and assisting us to understand these topics.

We would like to thank our emeritus colleague Pierce Mullen for carefully checking the historical accuracy of the textbook, for writing all but two of the biographical sketches, and for providing many insights into the history of physics.

The current edition continues to benefit from the efforts of our colleague Jeff Adams, who spent many hours revising old conceptual questions and exercises and designing many very innovative and thought-provoking new ones. Thanks also to Natalia Dashkina (Millersville University) and Tracianne Neilsen (Brigham Young University), who painstakingly checked the textual material of all chapters and interludes for accuracy, and to Sytil Murphy for her very careful work on the comprehensive index.

Finally, we would like to thank the staff at Brooks/Cole Publishing for their professionalism, enthusiasm, and generous support: David Harris, Publisher; Michelle Julet, Editor-in-Chief; Chris Hall, Acquisitions Editor; Sam Subity, Technology Project Manager; Mark Santee, Marketing Manager; Teri Hyde, Editorial Production Project Manager. We also thank Leah McAleer and Jamie Armstrong at G&S Book Services, who were the production coordinators. This book would not be of this high quality without the help of those who worked very closely with us; Peter McGahey for his diligent and careful work as Developmental Editor and for keeping us on schedule, George Kelvin for his beautiful illustrations, Dena Digilio-Betz for finding excellent photographs, and Mary Ann Short for her careful work in editing the manuscript.

After giving serious consideration to each of the reviewers' suggestions, we made the final decisions and therefore accept the responsibility for any errors, omissions, and confusions that might remain in the textbook. We would, of course, appreciate receiving any comments that you might have. Send comments and suggestions to Greg Francis, Physics, Montana State University, Bozeman, MT 59717-3840 or via email at francis@physics.montana.edu.

Larry D. Kirkpatrick
Gregory E. Francis
Montana State University
January 2006

1 A World View

The Coma Cluster contains more than 1000 galaxies.

This photograph shows a cluster of galaxies. Each galaxy, a collection of billions of stars, appears small because it is very far from Earth. How far is it to the farthest galaxy that we may ever observe?

(See page 12 for the answer to this question.)

Physics⚛Now™ Test your understanding of this chapter by logging into PhysicsNow at **http://physics.brookscole.com/kf6e**, selecting the chapter, and clicking on the "Take a Pre-Test" link.

PHYSICS is the study of the material world. It is a search for patterns, or rules, for the behavior of objects in the Universe. This search covers the entire range of material objects, from the smallest known particles—millions of millions of times smaller than a marble—to astronomical objects—millions of millions of times bigger than our Sun. The search also covers the entire span of time, from the primordial fireball to the ultimate fate of the Universe. Within this vast realm of space and time, the searchers have one goal: to comprehend the course of events in the whole world—to create a world view.

First Grade

This course could be one of the most challenging experiences that you will ever have—except for first grade. But then you were too young to notice.

What happened in first grade?* Well, you learned to read, and that was really difficult. You first had to learn the names of all those weird little squiggles. You had to learn to tell a *b* from a *d* from a *p*. Even though they looked so much alike, you did it, and it even seemed like fun. Then you learned the sounds each letter represented, and that was not easy because the capitals looked different but made the same sound and some letters could have more than one sound.

Then one day your teacher put some letters on the board: first, the letter *C*, and you all knew it could make a Kuh or Suh sound; then the letter *A*, which had lots of possibilities; finally, a *T*, which luckily had only one sound. You tried out several combinations including Kuh-AAH-Tuh. Then suddenly someone shouted out in triumph, "That isn't kuh-aah-tuh! It's a small furry animal with a long skinny tail that says 'meow.'" And your world was never the same again. When your car paused at an eight-sided red sign, you sounded out *stop* and understood how the drivers knew what to do. You saw the words *ice cream* on the front of a store and knew you wanted to go in.

If this book works, you will become aware of a whole world you never noticed before. You will never walk down a street, ride in a car, or look in a mirror without involuntarily seeing an extra dimension. There are times when you will have to memorize what symbols mean—just as in first grade. There will be times when you will confuse things that seem as much alike as *b*, *d*, and *p* once did, until you suddenly see how different they are. And there will be times when you will look at a combination of events and equations helplessly reciting Kuh-AAH-Tuh in total frustration. This has happened to all of us. But then the moment of insight will come, and you will see whole new images fitting together. You will see the *C-A-T* and will experience fully, and consciously, the exhilaration you felt in first grade.

So, welcome to one of the most challenging (and rewarding) courses you have ever taken in your life. If you work at it and let it happen, this experience will change your world view forever.

On Building a World View

The term *world view* has a fairly elastic meaning. When we think about world views, the interpretations can stretch from the philosophic to the poetic. In physics the world view is a shared set of ideas that represents the current explanations of how the material world operates. These include some rather common constructs, like gravity and mass, as well as strange-sounding ones, like quarks and black holes.

The physics world view is a dynamic one. Ideas are constantly being proposed, debated, and tested against the material world. Some survive the scrutiny

---

*Adapted from an article by Barbara Wolff, "An Introduction to Physics—Find the CAT." *The Physics Teacher* 27 (1989): 427.

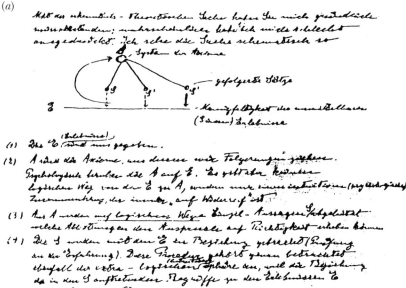
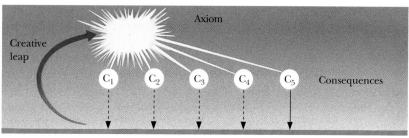

Figure 1-1 (a) Reproduction of a diagram from one of Albert Einstein's letters. (b) Our interpretation of this diagram.

of the community of physicists; some don't. The inclusion of new ideas often forces the rejection of previously accepted ones. Some firmly accepted ideas in the world view are very difficult to discard; in the long run, however, experimentation wins out over personal biases. The model of the atom as a miniature Solar System was reluctantly given up because the experimental facts just didn't support it; it was replaced by a mathematical model that's difficult to visualize.

A few years before his death, Albert Einstein described the process of science to a lifelong friend, Maurice Solovine. Solovine was not a scientist but apparently enjoyed discussing science with his famous friend. In one of their last exchanges, Solovine wrote that he had trouble understanding a certain passage in one of Einstein's essays. The next week Einstein wrote back, carefully explaining his view of the process of science. Figure 1-1(a) is a reproduction of a diagram from this letter. Figure 1-1(b) shows our interpretation of the diagram.

In his text Einstein explains the diagram. The lower horizontal line represents the real world. The curved line on the left signifies the creative leap a scientist makes in attempting to explain some phenomena. The leap is intuitive, and although it may be very insightful, it is not scientific. The scientific process begins, Einstein explains, when the scientist takes the idea, or axiom, and develops consequences based on it. These consequences are illustrated by a number of smaller circles connected to the axiom with lines. A very powerful axiom has a large number of consequences. The final task in the process of science is to test these consequences against the material world. In Einstein's drawing, these tests are vertical arrows returning to the world. If there is no match between the predicted consequences and the real world, the idea is scientifically worthless. If

The study of physics began mostly in an attempt to understand the motions of planets, stars, and other bodies far from Earth. Shown are (a) Galileo's original telescope, which enabled him to discover the four largest moons orbiting Jupiter, and (b) a modern reflecting telescope.

there is a match, there is hope that the idea has merit. With the publication of the idea, the community of scientists is brought into the process, and the original work is often modified.

Ideas for which there are simple physical models are easier to understand and accept. There is comfort in picturing electrons and protons as tiny balls, but those physicists who yearned for the electron to be a miniature billiard ball never got their wish. Cosmologist Sir Hermann Bondi, commenting on doing physics in realms beyond the range of direct human experience, said, "We should be surprised that the gas molecules behave so much like billiard balls and not surprised that electrons don't." This advice, though correct, may leave one feeling like Alice, in *Through the Looking Glass*, listening to the White Queen say that one should be able to "believe six impossible things before breakfast."

Although some of the ideas in physics might appear to be contrary to common sense, they do in fact make sense. Normally, things that make sense—that don't violate your intuition or common sense—are those that fit into your past experience. Common sense is a personal world view. Like the physics world view, your common sense is built on a large experimental base. The difference between what makes sense to you and what makes sense to a physicist is in part due to different ranges of experience. Where our observations are limited by the range of human sensations, the physicist has instruments that bring ultrahuman sensations into consideration. Our twinkling star is the physicist's window into the Universe.

So, without common sense to guide them about whether to accept a new idea, how do physicists decide which ones to adopt? Acceptance is based on whether the idea works, how well it fits into the world view, and if it is better than the old explanations. Although the most basic criteria for accepting an idea are that it agrees with the results of past experiments and successfully predicts the outcome of future experiments, acceptance is a human activity. Because it is a human activity, it has subjective aspects. The phrases "how well it fits" and "if it is better" imply opinions. Ideas have appeal, some more than others.

If an idea is very general, having many consequences in the Einsteinian sense, it can replace many separate ideas. It is regarded as more fundamental and thus more appealing. It is possible to construct a different explanation for each observation. For example, a scheme could be created to explain the disappearance of water from an open container, and another, unrelated idea could be employed to account for the fluidity of water, and so on. An idea about the structure of liquids (not just water) that could be used to explain these phenomena and many others would be a highly valued replacement for the collection of separate ideas.

The simplicity of an idea also influences opinions about its worth. If more than one construct is proposed to explain the same phenomena and if they all predict the experimental results equally well, the most appealing idea is the simplest one. Although elaborate (Rube Goldberg) constructions are cute in cartoons, they hold very little value in the building of a physics world view.

Incredible as an idea may be initially, physicists seem to become more and more comfortable with it the longer it remains in their world view. Most physicists are comfortable with the relativistic notions of slowed-down time and warped space; when the ideas were first introduced, however, they caused quite a stir. As more and more experimental results support an idea, it gains stature and becomes a more established part of our beliefs. But, even if an idea becomes very familiar and comfortable, it is still tentative. Experimental results can never prove an idea; they can only disprove it. If the predictions are borne out, the best that can be claimed is, "So far, so good."

Our goal in writing this book is to help you view the world differently. We describe the building of a physics world view and share with you some of the results. We start in areas that are familiar, where your common sense serves you

well, and chart a course through less-traveled areas. Although there is no end to this journey, the book must stop. We hope it stops at a new place, a place you have never been before.

We also hope that you find joy in the process similar to that expressed by Isaac Newton:

> I do not know what I may appear to the world, but to myself I seem to have been only like a boy playing on the sea-shore, and diverting myself in now and then finding a smoother pebble or a prettier shell than ordinary, whilst the great ocean of truth lay all undiscovered before me.

Bode's Law

In 1776 Titus of Wittenberg developed a numerical rule that gave the relative sizes of the orbits of the planets. This rule was popularized by Bode and is now known as Bode's law. But is this rule a law of physics?

The rule can be developed through the following process: The first and second numbers in the first column of Table 1-1 are 0 and 3. All subsequent numbers are double the preceding one. The second column is obtained by adding 4 to each of the numbers in the first column. The third column is obtained by dividing the entry in the second column by 10. These are the numbers in Bode's law. Because all the radii of the orbits are measured relative to Earth's, the

Table 1-1 | Bode's Law

		Bode's Law	Modern Value	Planet
0	4	0.4	0.39	Mercury
3	7	0.7	0.72	Venus
6	10	1.0	1.00	Earth
12	16	1.6	1.52	Mars
24	28	2.8		
48	52	5.2	5.20	Jupiter
96	100	10.0	9.54	Saturn
192	196	19.6		
384	388	38.8		

value for Earth (the "third rock from the Sun") must be 1. The radius of Earth's orbit, the mean distance between Earth and the Sun, is known as an astronomical unit, or AU.

The rule tells us that the radius of Mercury's orbit must be 0.4 times that of Earth's, or 0.4 astronomical unit. This was in agreement with the value known in 1776 and is close to the modern value of 0.39 astronomical unit. Likewise, the value of 0.7 astronomical unit for Venus agreed with the value at the time the law was developed and is close to the modern value of 0.72 astronomical unit. The modern values for Mars and Saturn differ a bit from the rule, but the rule was consistent with the data known at the time the rule was proposed.

A scientific law must also be testable—that is, it must make predictions that can be tested. Notice that there were no planets known in 1776 that matched the values in the fifth, eighth, or ninth rows. These rows can be viewed as predictions for the orbital radii of planets that had not yet been discovered. Uranus, discovered in 1781, has an orbit with a mean radius of 19.2 astronomical units, near the predicted value. So far, so good. The first of the asteroids, known as Ceres, was discovered in 1801, and its orbit had a mean radius of 2.8 astronomical units, in agreement with the prediction. Does this mean that Bode's law is a physical law of nature?

Neptune, discovered in 1846, has an orbit with a mean radius of 30.1 astronomical units; Pluto, discovered in 1930, has an orbit with a mean radius of 39.4 astronomical units. Clearly, Bode's law does not work for the two outer planets. Because it failed to predict these radii, Bode's law must be discarded.

However, Bode's law was never a candidate for a physical law. It never had any scientific basis; it was simply a way of remembering the values for the known radii. To this day, we know of no way of predicting the radii of the planetary orbits in the recently discovered planetary systems around nearby stars. Furthermore, we believe that each planetary system is unique.

We are not saying that numerical rules such as Bode's law have no role in science. As we will see numerous times in our study of physics, discovering patterns in nature may provide the first steps in developing physical laws.

Question What are the three criteria required for an idea to become a physics law?

Answer (1) It must account for the known data. (2) It must make predictions that can be tested. (3) It must have a scientific basis.

> ✓ Extended presentation available in the *Problem Solving* supplement

Measurements ✓ MATH

If someone offered to sell a bar of gold for $200, you would immediately ask, "How large is the bar?" The size of the bar obviously determines whether it is a good buy. A similar problem existed in the early days of commerce. Even when there were standard units of measure, they were not the same from time to time and region to region. Later, several standardized systems of measurement were developed.

The two dominant systems are the U.S. customary system, based on the foot, pound, and second, and the metric system, based on the meter, kilogram, and second. Thomas Jefferson advocated that the United States adopt the metric system, but his advice was not taken. As a result, most people in the United States do not use the metric system. It is used, however, by the scientific community and those who work on such things as cars. England and Canada have now officially changed to the metric system. The United States is the only major country not to have made the change.

There are obvious advantages in having the entire world use a single system.

It avoids the cumbersome task of converting from one system to another and aids in worldwide commerce. The disadvantages for the countries that must change are the expense of converting the machinery, signposts, and standards; the maintenance of dual systems for a time; and the abandonment of the familiar units for new ones for which people do not have intuitive feelings.

The metric system has advantages over the U.S. customary system and was the system chosen in 1960 by the General Conference on Weights and Measures. The official version is known as **Le Système International d'Unités** and is abbreviated SI.

One problem of the U.S. customary system of measurement is illustrated by Table 1-2, in which we have listed a sample of units used to measure lengths. Some of these units are historic; others have been developed for use in specialized areas. Many of them may be unfamiliar. Few of us have any idea how long a fathom is, let alone how many inches there are in a fathom. Even among the more familiar units, converting from one unit to another is often a difficult task. For instance, to determine how many inches there are in a mile, we must make the following computation:

$$1 \text{ mile} = (5280 \text{ ft})\left[\frac{12 \text{ in.}}{1 \text{ ft}}\right] = 63{,}360 \text{ in.}$$

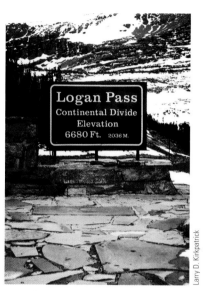

Some highway signs give the elevation in feet and meters—in this case 6680 feet and 2036 meters.

The metric system eliminates the confusion of having many different unfamiliar units and the difficulty of converting from one size unit to another. It does this by adopting a single standard unit for each basic measurement and a series of prefixes that make the unit larger or smaller by factors of 10. For instance, the basic unit for measuring length is the **meter** (m), which is a little longer than a yard. This unit is inconveniently small for measuring distances between cities, so road signs everywhere else in the world display kilometers (km) as a unit of length. The prefix **kilo** means one thousand and indicates that the kilometer is equal to 1000 meters. The kilometer is about $\frac{5}{8}$ mile, so a distance of 50 miles will appear as 80 kilometers. Speed limits appear as 100 kilometers per hour rather than 65 miles per hour.

◄ kilo = 1000

Smaller distances are measured in such units as the centimeter (cm). The prefix **centi** means one-hundredth. It takes 100 centimeters to equal 1 meter. The other prefixes are given in Table 1-3 along with their abbreviations and various forms of their numerical values. These units are pronounced with the accent on the prefix. Note also that the terms *billion* and *trillion* do not have the same meanings in all countries. In this textbook, 1 billion is 1,000,000,000, and 1 trillion is 1000 billion.

◄ centi = $\frac{1}{100}$

Table 1-2 | Partial List of Length Measures

angstrom	micron
astronomical unit	mil
barleycorn	mile
cubit	nautical mile
fathom	pace
fermi	palm
foot	parsec
furlong	pica
hand	point
inch	rod
league	stadium
light-year	thumbnail
meter	yard

Table 1-3 | The Metric Prefixes

Prefix	Symbol		Value	
tera	T	trillion	10^{12}	1,000,000,000,000
giga	G	billion	10^{9}	1,000,000,000
mega	M	million	10^{6}	1,000,000
kilo	k	thousand	10^{3}	1,000
		one	10^{0}	1
centi	c	hundredth	10^{-2}	0.01
milli	m	thousandth	10^{-3}	0.001
micro	μ	millionth	10^{-6}	0.000 001
nano	n	billionth	10^{-9}	0.000 000 001
pico	p	trillionth	10^{-12}	0.000 000 000 001
femto	f	quadrillionth	10^{-15}	0.000 000 000 000 001

Because all the prefixes are multiples of 10, conversions between units are done by moving the decimal point—that is, multiplying and dividing by 10s. For instance, because **milli** means one-thousandth, we can convert from meters to millimeters (mm) by multiplying by 1000. There are 5670 millimeters in 5.67 meters.

▶ milli = $\frac{1}{1000}$

Question What are 10^{-12} boo, 10^{-3} pede, and 10^{12} dactyl?

Answer A picoboo (peek-a-boo), a millipede, and a teradactyl (pterodactyl).

In its purest form, the SI system allows no other units for length. However, other units that have grown up historically will remain in use for some time (and maybe forever). For instance, the terms *micron* and *fermi* continue to be used for micrometer and femtometer. A common unit of length on the astronomical scale is the light-year, the distance light travels in 1 year. Although it is not an SI unit, it is a naturally occurring unit of length on this scale and will continue to be used. On the other hand, the angstrom (10^{-10} meter) has been very popular but is now being replaced by the nanometer (10^{-9} meter).

The metric system also differs from the U.S. customary system in that mass is considered the primary unit and weight (force) the secondary unit. In the U.S. customary system, the situation is reversed. (The distinction between mass and weight is explained in Chapter 3.) The basic unit in the U.S. customary system is the pound, but the basic unit in the SI system is the **kilogram** (kg). (It may seem strange to use the kilogram rather than the gram [g] as the basic unit, but the gram was deemed to be too small for the basic unit.) The weight of 1 kilogram is 2.2 pounds. A U.S. nickel has a mass that is very close to 5 grams. The term *megagram* (1000 kilograms) is not often used in the metric system. This is known as a metric ton and has a weight equal to a long ton (2200 pounds).

Since the invention of the metric system in 1791, many unsuccessful attempts have been made to change the time system to a decimal basis so that time units would also be multiples of 10. They have all failed. The SI system has the same units of time as the U.S. customary system.

Because the metric system is the primary one used in science, we should gain some familiarity with it. On the other hand, our principal goal is to establish connections between your commonsense world view and the physics world view. Learning the metric system at the same time as beginning your study of physics is complicated because you need to develop a feeling for the new units as well as for the scientific ideas. As a compromise, we will predominantly use the metric system, but we will give the approximate English equivalents in parentheses when it is useful.

Sizes: Large and Small

Imagine taking a photograph of three children lying on a blanket in their backyard.* Assume that the camera is located directly above the children and that the scene captured on the film is 1 meter wide and 1 meter tall. One meter, a little more than 3 feet, is the scale of children. In fact, this is also the scale of adults because factors of 2, 3, or even 5 don't matter when we are talking about the approximate sizes of objects.

If we now move the camera 10 times as far away from the children, the film will capture a scene that is 10 meters wide by 10 meters tall. The new scene has 100 times the area and could include approximately 300 children. If we once again move the camera 10 times as far away, the new scene will be 100 meters on a side, about the length of a football or soccer field. At yet another 10 times as far away, our scene would be 1000 meters, or 1 kilometer (0.62 mile), on a side and could include the children's neighborhood.

As we continue increasing our distance from the children by additional factors of 10, the scenes captured by the camera will get larger and larger. In fact, they will get to be so large that it will be very difficult to keep track of the number of zeros in the length and width of the photograph. Therefore, we use the **powers-of-ten notation**, which displays the number of zeros in these numbers. In mathematics the notation 10^2 means 10×10, which is equal to 100. Similarly, 10^3 means $10 \times 10 \times 10 = 1000$. The superscript is called an *exponent* and is equal to the number of 10s that are multiplied together. The exponent is also equal to the numbers of zeros in these numbers. (Note that $10^0 = 1$.)

Positive values for exponents indicate that the numbers are large. Negative exponents indicate small numbers. For instance, $10^{-1} = \frac{1}{10^1} = 0.1$ and $10^{-2} = \frac{1}{10} \times \frac{1}{10} = \frac{1}{10^2} = 0.01$. The minus sign indicates that the power of 10 is in the denominator; that is, it is to be divided into 1. The **order of magnitude** for a quantity is its value rounded off to the nearest power of ten.

At a scale of 10^6 meters (620 miles), our scene could encompass all but the largest states, as shown in the LandSat photograph in Figure 1-2. At 10^7 meters,

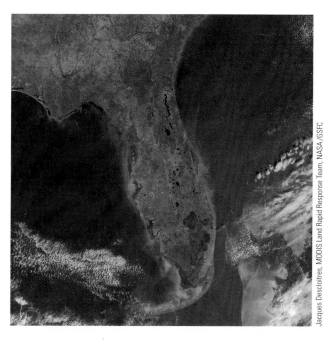

Figure 1-2 This photograph of the southeastern United States shows a scene that is 10^6 meters on a side.

*Based on Philip Morrison, Phylis Morrison, and the Office of Charles and Ray Eames, *Powers of Ten* (Redding, Conn.: Scientific American Books, 1982).

Figure 1-3 Earth spills over the edges of the photograph at a scale of 10^7 meters.

Figure 1-4 The Milky Way Galaxy has a scale of 10^{21} meters.

most of Earth would fit into our scene, as shown in Figure 1-3. The Moon's orbit would be included at a scale of 10^9 meters, and the entire Solar System would be included at a scale of 10^{13} meters. As we continue outward, the photographs look much the same, a field of stars of varying brightness. At a scale of 10^{21} meters, our scene shows the Milky Way Galaxy (Figure 1-4), a collection of billions of stars. Continuing outward, we observe the Milky Way shrink in size and many other galaxies enter the picture. At even larger distance, the galaxies look like stars. (See the chapter-opening photograph.) At a scale of 10^{26} meters, we approach the edge of the Universe that is visible from Earth.

Let's now return to our children and imagine looking at smaller and smaller scales, each time decreasing the size of our photograph by a factor of 10. With a photograph that is $\frac{1}{10}$ meter on a side (a scale of 10^{-1} meter), our photograph could include a child's hand. At a scale of 10^{-2} meter = 1 centimeter (a little less than $\frac{1}{2}$ inch), we might see only a child's fingernail. The thickness of the fingernail is approximately 10^{-3} meter = 1 millimeter. The red blood cells shown in Figure 1-5 are approximately 10^{-5} meter across. Blood capillaries are only a little

Figure 1-5 A scanning electron micrograph of a red blood cell. Red blood cells have diameters of approximately 10^{-5} meter. Notice the white blood cell in the upper right corner.

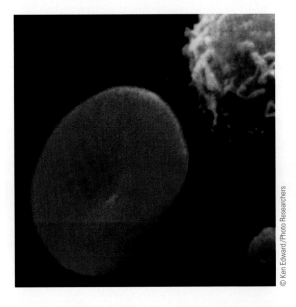

larger than the red blood cells. The smallest living cells are 100 times smaller at a scale of 10^{-7} meter. The atoms making up the molecules in these cells have diameters on the order of 10^{-10} meter. On a scale approximately 10,000 times smaller, we find the nuclei of these atoms; at a scale of 10^{-15} meter, we find the protons and neutrons that make up the nuclei of atoms. Even these protons and neutrons are made up of quarks, which we will study in Chapter 27.

The size of the visible Universe is an incredible 10^{41} times the size of protons and neutrons. Physicists study the material world at both extremes of the size scale and all the way in-between.

WORKING IT OUT | Powers of Ten

In this book we study objects and events that go far beyond the normal human scale of objects and events. When we look at phenomena on very large and very small scales, the sizes of the numbers quickly get out of hand. For instance, the approximate radii of the visible Universe and a proton are

radius of visible Universe = 140,000,000,000,000,000,000,000,000 m
radius of proton = 0.000 000 000 000 001 2 m

These numbers are very difficult to read, write, and manipulate mathematically. It is even easy to make errors in counting the zeros unless they are grouped in threes, as we have done.

Using the powers-of-ten notation, the radius of the visible Universe is written as 1.4×10^{26} m. This indicates that the number in front is to be multiplied by ten 26 times to get the actual number. Because multiplication by ten just moves the decimal point one position to the *right*, the superscript 26 indicates the total number of places the decimal point must be moved. By convention the number out in front is usually written so that it has a value between 1 and 10. You should check that moving the decimal point 26 places to the right in the number 1.4 gives the value in the first paragraph.

The radius of the proton is written as 1.2×10^{-15} m. The number in front must be divided by ten 15 times. Or equivalently, the number can be obtained by moving the decimal point 15 positions to the *left*.

Sometimes you may see a number written with only the power of 10. If the usual number preceding a power of 10 is missing, it is assumed to be 1; that is, $10^5 = 1 \times 10^5$. Similarly, if the exponent is missing, it is assumed to be zero; that is, $4 = 4 \times 10^0$.

The greatest power of using this notation comes when you have to multiply or divide very large or small numbers. For *multiplication*, multiply the two numbers in front and *add* the exponents. For example,

$$R_{Universe} \times R_{proton} = (1.4 \times 10^{26} \text{ m}) \times (1.2 \times 10^{-15} \text{ m})$$
$$= (1.4 \times 1.2) \times 10^{26+(-15)} \text{ m} \times \text{m}$$
$$= 1.7 \times 10^{11} \text{ m}^2$$

This number represents an area that has the width of a proton and a length that extends from Earth to the edge of the visible Universe.

For *division*, you divide the two numbers in front and *subtract* the exponent in the denominator from that in the numerator. For example,

$$\frac{R_{Universe}}{R_{proton}} = \frac{1.4 \times 10^{26} \text{ m}}{1.2 \times 10^{-15} \text{ m}} = \frac{1.4}{1.2} \times 10^{26-(-15)} \frac{\text{m}}{\text{m}}$$
$$= 1.2 \times 10^{41}$$

This number gives the relative size of the visible Universe and the proton—that is, how many times larger the visible Universe is than a proton.

Summary

This course could be one of the most challenging experiences that you've had since first grade. At the same time, it could be one of your most rewarding experiences, one that could change your view of the world around you.

The physics world view is a shared set of ideas that represents the current explanations of how the material world operates. It is a dynamic view with new ideas being proposed, debated, and tested. For a new idea to be accepted, it must (1) agree with the existing data, (2) make predictions that can be tested, and (3) have a scientific basis.

The measurement system used in science (and most of the world) is the metric, or SI, system, which is based on the meter, kilogram, and second. Larger and smaller units are obtained through the use of prefixes.

The sizes of objects studied in physics range from the entire Universe (at a scale of 10^{26} meters) down to neutrons and protons (at a scale of 10^{-15} meter).

Chapter 1 Revisited

The farthest galaxies that we may ever observe lie near the edge of the visible Universe, at a distance of approximately 10^{26} meters from Earth.

KEY TERMS

centi: A prefix meaning $\frac{1}{100}$. A centimeter is $\frac{1}{100}$ meter.

kilo: A prefix meaning 1000. A kilometer is 1000 meters.

kilogram: The SI unit of mass. A kilogram of material weighs about 2.2 pounds on Earth.

meter: The SI unit of length equal to 39.37 inches, or 1.094 yards.

milli: A prefix meaning $\frac{1}{1000}$. A millimeter is $\frac{1}{1000}$ meter.

order of magnitude: The value of a quantity rounded off to the nearest power of ten.

powers-of-ten notation: A method of writing numbers in which a number between 1 and 10 is multiplied by 10 raised to a power.

Système International d'Unités: The French name for the metric system, or International System (SI), of units.

CONCEPTUAL QUESTIONS

Most questions and exercises are paired so that most odd-numbered questions and exercises are followed by a similar even-numbered one. Short answers for most odd-numbered questions and exercises are provided at the back of this textbook. More-challenging questions are indicated by a ▲.

1. Compare and contrast the physics world view with your own personal world view.
2. What is a physics world view?
3. Why is Bode's law, giving the sizes of the orbits of the planets, not considered to be a physical law?
4. Why should you be suspicious of a book titled *The Theory of Everything*?
5. What are the criteria for accepting a theory as a physical law?
6. Which of the criteria for a physical law are not satisfied by Bode's law, which gives the sizes of the orbits of the planets?
7. What role does the prestige of the scientist play in accepting a theory as a physical law?
8. Would you be more likely to accept a scientific theory proposed by a professor at a major university or one developed by the handyman down the street?
9. Which major countries (if any) have not adopted the metric system as the primary system of measurement?
10. What are the advantages and disadvantages of adopting the metric system?
11. What is the height of a typical person, in centimeters?
12. What is the length of a typical newborn baby, in centimeters?
13. What is the typical height of a bedroom ceiling, in meters?
14. What is the length of a full-sized bed, in meters?
15. What is the typical mass of a 6-foot-tall male, in kilograms?
16. What is the typical mass of a 5-foot-6-inch-tall female, in kilograms?
17. What is 10^{-2} nel?
18. What are 10^{12} nosaurs?
19. What is the order of magnitude for the world's population?
20. What is the order of magnitude for the distance across the United States?

EXERCISES

1. How many seconds are there in 1 day?
2. How many seconds are there in 1 year?
3. How long is a 100-m dash in yards if 1 m = 1.094 yd?
4. If 1 in. = 2.54 cm, how tall (in centimeters) is a 6-ft basketball player?
5. How many inches are there in 1 m?
6. How many inches are there in 1 km?
7. Write each of the following numbers in powers-of-ten notation:
    a. 68,200 m
    b. 0.000 000 000 456 g
8. Write each of the following numbers in powers-of-ten notation:
    a. 8,448,000,000 in.
    b. 0.001 48 mm
9. Write each of the following numbers as ordinary numbers:
    a. 3.48×10^3 s
    b. 1.11×10^{-5} kg
10. Write each of the following numbers as ordinary numbers:
    a. 4.72×10^5 ft
    b. 2.73×10^{-3} s
11. Complete the following computations:
    a. $(2.7 \times 10^{-3}) \times (2.3 \times 10^4)$
    b. $\dfrac{9.6 \times 10^6}{3.0 \times 10^{-3}}$
12. Complete the following computations:
    a. $(4.2 \times 10^7) \times (5.2 \times 10^4)$
    b. $\dfrac{4.4 \times 10^5}{5.4 \times 10^2}$
13. Approximately how much larger is the orbit of Pluto than the orbit of the Moon?
14. Approximately how much larger is a child's fingernail than one of the protons in the fingernail?

2 | Describing Motion

The blurring of the train's image shows that it is moving.

The blurred image of the train clearly shows that it's moving. To appear blurred in the photograph, the train had to be in different places during the time the shutter was open. If we know how fast the train is moving, can we determine how long it will take to reach its destination?

See page 28 for the answer to this question.

A property common to everything in the Universe is change. Some things are big, some are small; some are red, some have no color at all; some are rigid, some are fluid; but they all are changing. In fact, change is so important that the fundamental concept of time would be meaningless without it.

Change even occurs where seemingly there is none. Water evaporating, colors fading, flowers growing, and stars evolving are all examples of changes that are beyond our casual observations. Also beyond our sensations is the fact that these changes are a result of the motion of material, often at the submicroscopic level. Because change—and thus motion—is so pervasive, we begin our exploration of the ideas of physics with a study of motion.

Within our commonsense world view, we generally group all motions together, simply observing that an object is moving or that it is not moving. Actually, there is an extraordinary diversity of motion, ranging from the very simple to the extremely complicated. Fortunately, the complex motions—ones more common in our everyday experiences—can be understood as combinations of simpler ones. For example, Earth's motion is a combination of a daily rotation about its axis and an annual revolution around the Sun. Or, closer to home, the motion of a football can be treated as a combination of a vertical rise and fall, a horizontal movement, and a spinning about an axis.

We therefore begin our discussion of motion by trying to describe and understand the simplest kinds of motion. This will yield a conceptual framework within our world view from which even the most complicated motions, such as those associated with a hurricane or with a turbulent waterfall, can be understood.

Average Speed

Imagine driving home from school. For simplicity, assume that you can drive home in a straight line. Normally, you might describe this trip in terms of the time it takes. If pressed for a more detailed account, you would probably give the distance or the actual route taken, adding points of interest along the way. For our purposes we need to develop a more precise description of motion.

First, we note that your position continually changes as you drive. Second, we observe that it takes time to make the trip. These two fundamental notions—space and time—are at the core of our concept of motion. Furthermore, different positions along the trip can be matched with different times.

One relationship between space and time can be illustrated by answering the question, "How fast were you going?" Actually, there are two ways to answer this question; one way looks at the total trip, whereas the other considers the moment-by-moment details of the trip. For the total-trip description, we use the concept of an **average speed**, which is defined as the total distance traveled divided by the time it took to cover this distance.

We can write this relationship more efficiently by using symbols as abbreviations:

$$\bar{s} = \frac{d}{t}$$

where $\bar{s}$ is the average speed, d is the distance traveled, and t is the time taken for the trip. A bar is often used over a symbol to indicate its average value.

This ratio of distance over time gives the average rate at which the car's position changes. Speed is a quantitative measure of how rapidly the change takes place. The definition of average speed states a particular relationship between the concepts of space and time. If any two of the three quantities are known, the third is determined.

A humorous story about a small-time country farmer illustrates this relationship. The farmer was visited by his big-time cousin. Anxious to make a good impression, the host spent the morning showing his cousin around his small

Physics Now™ Test your understanding of this chapter by logging into PhysicsNow at http://physics.brookscole.com/kf6e, selecting the chapter, and clicking on the "Take a Pre-Test" link.

The motion of a football is a combination of three simpler motions.

Havasu Creek on the Colorado River in Grand Canyon National Park.

◄

$$\text{average speed} = \frac{\text{distance traveled}}{\text{time taken}}$$

Mile markers and your wristwatch can be used to calculate your average speed.

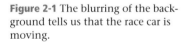
Extended presentation available in the *Problem Solving* supplement

farm. At lunch the cousin could not resist the urge to brag that on his ranch he could get into his car in the early morning and drive until sunset and he would still be on his property. The country farmer thought for a moment and then said, "I had a car like that once."

To measure speed, we need a device for measuring distance, such as a ruler, and one for measuring time, such as a clock. Most highways have mile markers along the side of the road so that maintenance and law enforcement officials can accurately find certain locations. These mile markers and your wristwatch give you all the information you need to determine average speeds.

Assuming that we begin "thinking metric," speeds have units such as meters per second (m/s) or kilometers per hour (km/h). A person walks about $1\frac{1}{2}$ meters per second, and a car traveling at 70 miles per hour is going approximately 113 kilometers per hour.

PHYSICS | ON YOUR OWN

Estimate the average speed of an everyday object such as a falling leaf, a falling snowflake, or a wave traveling from one end of your bathtub to the other.

Images of Speed   ✓ MATH

From the earliest cave drawings to modern time-lapse photography, it has been a part of human nature to try to represent our experiences. Artists, as well as scientists, have devised many ways of illustrating motion. A blurred painting or photograph such as that in Figure 2-1 is one way to "see" motion. One difference between the artist and the scientist is that the scientist uses the representations to analyze the motion.

A clever image of motion that also provides a way of measuring the speed of an object is the multiple-exposure photograph. These photographs are made in a totally dark room with a stroboscope (usually just called a "strobe") and a camera with an open shutter. A strobe is a light source that flashes at a constant, controllable rate. The duration of each flash is very short (about 10 millionths of a second), producing a still image of the moving object.

If the strobe flashes 10 times per second, the resulting photograph will show the position of the object at time intervals of $\frac{1}{10}$ second. Thus, we can "freeze" the motion of the object into a sequence of individual events and use this representation to measure its average speed within each time interval.

As an example of measuring average speed, let's determine the average speed of the puck in Figure 2-2. The puck travels from a position near the 4-centimeter mark to one near the 76-centimeter mark, a total distance of 72 centimeters. Be-

Figure 2-1 The blurring of the background tells us that the race car is moving.

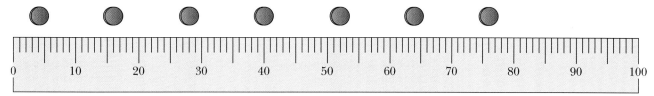

Figure 2-2 A strobe photograph of a moving puck shows its position at different times.

cause there are seven images, there are six intervals and the total time taken is six times the time between flashes—that is, 0.6 second. Therefore, the average speed is

$$\bar{s} = \frac{d}{t} = \frac{72 \text{ cm}}{0.6 \text{ s}} = 120 \text{ cm/s}$$

We can also determine the average speed of the puck between each pair of adjacent flashes. Allowing for the uncertainties in reading the values of the positions of the puck, the average speed for each time interval is the same as the overall average. Therefore, the puck was traveling at a constant speed of 120 centimeters per second.

Suppose you live 40 miles from school and it takes you 2 hours to drive home. Your average speed during the trip is

$$\bar{s} = \frac{d}{t} = \frac{40 \text{ miles}}{2 \text{ h}} = 20 \frac{\text{miles}}{\text{h}}$$

This means that, on the average, you travel a distance of 20 miles during each hour of travel. This answer is read "20 miles per hour" and is often written as 20 miles/hour or, abbreviated, as 20 mph. It is important to include the units with your answer. A speed of "20" does not make any sense. It could be 20 miles per hour or 20 inches per year, very different average speeds.

Actually, you probably weren't moving at 20 mph during much of your trip. At times you may have been stopped at traffic lights; at other times you may have traveled at 50 mph. The use of average speed disregards the details of the trip. Despite this, the concept of average speed is a useful notion.

Question What is the average speed of an airplane that flies 3000 miles in 6 hours?

Answer Using our definition for average speed, we have

$$\bar{s} = \frac{d}{t} = \frac{3000 \text{ miles}}{6 \text{ hours}} = 500 \text{ mph}$$

WORKING IT OUT | Average Speed

If you know the average speed, you can determine other information about the motion. For instance, you can obtain the time needed for a trip. Suppose you plan to drive a distance of 60 miles with the cruise control set at 50 mph. How long will the trip take?

Without consciously doing any calculation, you probably know that the answer is a little more than 1 h. How do you get a more precise answer? You divide the distance traveled by the average speed. For our example we obtain

$$t = \frac{d}{\bar{s}} = \frac{60 \text{ miles}}{50 \text{ miles/h}} = 1.2 \text{ h}$$

You can also calculate how far you could drive if you traveled with a specified average speed for a specified time. Suppose, for example, you plan to maintain an average speed of 50 mph on an upcoming trip. How far can you travel if you drive an 8-h day?

$$d = \bar{s}t = \left(50 \frac{\text{miles}}{\text{h}}\right)(8 \text{ h}) = 400 \text{ miles}$$

Therefore, you would expect to drive 400 miles each day.

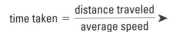

▸ time taken = distance traveled / average speed

▸ distance traveled = average speed × time taken

Figure 2-3 A speedometer tells you the car's instantaneous speed.

Did the bullet have a speed at the instant this picture was taken?

▸ Instantaneous speed is the average speed over a very small time interval.

Instantaneous Speed

The notion of average speed is limited in most cases. Even something as simple as your trip home from school is a much richer motion than our concept of average speed indicates. For example, it doesn't distinguish the parts of your trip when you were stopped waiting for a traffic light to change from those parts when you were exceeding the speed limit. The simple question, "How fast were you going as you passed Third and Vine?" is not answered by knowing the average speed.

To answer the question, "How fast were you going at a specific point or at a specific time?" we need to consider a new concept known as the instantaneous speed. This more complete description of motion tells us how fast you were traveling at any instant during your trip. Because this is the function of your car's speedometer (Figure 2-3), the idea is not new to you, although its precise definition might be new.

Actually, the definitions of average and instantaneous speeds are quite similar. They differ only in the size of the time interval involved. If we want to know how fast you are going at a given instant, we must study the motion during a very small time interval. The **instantaneous speed** is equal to the average speed over a time interval that is very, very small.

As a first approximation in measuring the instantaneous speed, we could measure how far your car traveled during $\frac{1}{10}$ of a second and calculate the average speed for this time interval. With precise equipment we could determine the average speeds during time intervals of $\frac{1}{100}$ of a second, $\frac{1}{1000}$ of a second, or an even smaller interval. How small an interval do we need? For practical purposes, we need a time interval that is small enough that the average speed doesn't change very much if we use an even smaller time interval. It is the instantaneous speed rather than the average speed that plays an important role in the analysis of nearly all realistic motions.

Fastest and Slowest

Britain's Kelly Holmes celebrates as she crosses the finish line to win the women's 1500-meter final at the Athens 2004 Olympic Games.

The fastest speed in the Universe is the speed of light, 300 million meters each second (186,000 miles per second), and the slowest speed is, of course, zero. Between those extremes is a vast range of speeds. With the exception of a few, very small subatomic particles that have been catapulted through huge electric voltages or released in nuclear reactions, most things move at speeds close to the slow end of the range.

The fastest large objects are planets moving at speeds up to 107,000 mph. Our own Earth orbits the Sun at 67,000 mph. (Even this admittedly high speed is 10,000 times slower than the speed of light.) Closer to home, but still in space, the *Apollo* spacecraft returned to Earth traveling at 24,800 mph, and the space shuttles orbit Earth at 17,500 mph.

Our people-carrying machines have a wide range of speeds, from supersonic airplanes with a record speed of 2193 mph (the Lockheed SR-71A Blackbird) to moving stairways that approximate fast walking speeds of 4 mph. In-between we have record speeds set by rocket-powered sleds on rails at 6453 mph, passenger planes (the Concorde) at 1450 mph, jet-powered cars at 763 mph, race cars at 410 mph, magnetically levitated trains at 343 mph, motorcycles at 191 mph (special motorcycles have traveled as fast as 323 mph), and bicycles at 167 mph (while drafting behind a vehicle).

When we give up our machines, we slow down considerably. The fastest recorded human speed is Asafa Powell's 100-meter dash in 9.77 seconds, or about 22.9 mph. The fastest time recorded by a female sprinter is 10.49 seconds (21.3 mph) by Florence Griffith Joyner. (In the time it took either sprinter to run the race, a beam of light could go to the Moon, bounce off a mirror, and return to Earth with 7 seconds to spare!) As the distances get longer, human speeds slow: The 1-mile record is held by Hicham El Guerrouj in a time of 3 minutes and 43.13 seconds, which corresponds to a little more than 16 mph. The record pace for a marathon is a little more than 12 mph.

Other animals range from slow (three-toed sloths that creep at 0.07 mph, giant turtles that lumber along at 0.23 mph, and sea otters that swim at 6 mph) to very fast (killer whales and sailfish, which swim at 35 and 68 mph, respectively, and cheetahs, reported to run up to 68 mph). In 1973 Secretariat set the record for the Kentucky Derby by running $1\frac{1}{4}$ miles in 1 minute and 59.2 seconds, for an average speed of almost 38 mph. The streamlined peregrine falcon diving for its prey has been clocked at 217 mph.

Nobody knows the slowest-moving object. A good candidate for a natural motion is a continent drifting at 1 centimeter per year, or 0.7 billionth of a mile per hour. A few years ago, a machine was built for testing stress corrosion that moves at a million millionths of a millimeter per minute, or 37 billion billionths of a mile per hour. At this rate it would take about 2 billion years to move 1 meter!

Speed with Direction ✓ MATH

We have made a lot of progress in attempting to accurately represent motion. However, as we develop the rules for explaining (and thus predicting) the behavior of objects in the next chapter, we will need to go further. Objects do more than speed up and slow down. They can also change direction, sometimes keeping the same speed but, at other times, changing both their speed and their direction. Either the average speed or the instantaneous speed tells us how fast an object is moving, but neither tells us the direction of motion. If we are discussing a vacation trip, direction doesn't seem important; you obviously know in which direction you're going. However, we are trying to develop rules of motion for all situations, and the direction is as important as the speed. You can get a sense for this by remembering situations in which there is an abrupt change in direction; for example, maybe a car you were riding in swerved sharply. The squeal of the tires and your own body's reaction are clues that new factors are involved when an object changes direction.

In the physics world view, we combine speed and direction into a single concept called **velocity**. When we talk of an object's instantaneous velocity, we give the instantaneous speed (for example, 15 mph) and just add the direction (north, to the left, or 30 degrees above the horizontal). The speed is known as the **magnitude** of the velocity; it gives its size. We use the symbol v to represent the magnitude of the instantaneous velocity.

Quantities that have both a size and a direction are called **vectors**. Vectors do not obey the normal rules of arithmetic. We will study the rules for combining vector quantities in Chapter 3. For now, it is only important to realize that the direction of the motion can be as important as the speed.

There is another important difference between average speed and average velocity besides direction. The average speed is defined as the distance traveled divided by the time taken, whereas the **average velocity** is defined as the **displacement** divided by the time taken. Displacement is a vector quantity; its magnitude is the straight-line distance between the initial and final locations of the object, and its direction is from the initial location to the final location.

The magnitude of the displacement is the same as the distance traveled for motion along a straight line in a single direction. The magnitude of the displacement and the distance traveled differ when the motion retraces part of its straight-line path or takes place in more than one dimension. For instance, assume that you travel a distance of 10 kilometers due west along a straight stretch of road, turn around, and travel 5 kilometers due east along the same road. You have traveled a distance of 15 kilometers, but your displacement is 5 kilometers west; that is, your ending location is 5 kilometers to the west of your starting location. If the trip takes 1 hour, your average speed is 15 kilometers per hour, and your average velocity is 5 kilometers per hour west.

> **Q**uestion A car travels due north a distance of 50 kilometers, turns around, and returns to the starting place along the same route. What distance did the car travel, and what was its displacement?

> **A**nswer The car travels a distance of 100 kilometers, but it has a displacement of zero because it returns to its starting place.

The magnitude of the average velocity of an object is the change in position divided by the time taken to make the change:

$$\bar{v} = \frac{\Delta x}{\Delta t}$$

▶ average velocity = change in position / time taken

We have used x to represent the position of the object. The symbol Δx is called "delta ex." The delta symbol, Δ, is used to represent a change in a quantity. Thus, Δx represents the change in position—the displacement—and must not be thought of as the product of Δ and x. To calculate the displacement, we subtract the position of the object at the beginning of the time interval from its position at the end. For example, if a car travels from milepost 120 to milepost 180, the displacement is 180 miles − 120 miles = 60 miles. Notice we have also written the time taken as Δt to indicate that it is an interval of time rather than an instant of time.

For the rest of this chapter, we will deal with objects traveling in only one dimension. They might be going left and right, east and west, or up and down, but not turning corners. Notice that by doing this we eliminate motions as simple as a home run. The payoff is that we learn to manipulate the new concepts before tackling the many realistic but more difficult situations.

◀ Velocity equals speed with a direction

Flawed Reasoning

A puzzled student claims, "If the average speed over an interval of time is given by $\Delta x/\Delta t$, the instantaneous speed must be given by the instantaneous position divided by the instantaneous time, or x/t." **What is wrong with this reasoning?**

Answer The instantaneous speed is the average speed taken over a very small time interval. Although they may be very small, we still divide the distance traveled by the time interval required to travel this distance. Notice that using x/t can give very nonsensical answers. For example, what if you were at milepost 678 at 2 seconds after midnight?

Acceleration

Because the velocities of many things are not constant, we need a way of describing how velocity changes. We now define a new concept, called *acceleration*, which describes the rate at which velocity changes. The magnitude of the **average acceleration** of an object is the change in its velocity divided by the time it takes to make that change:

$$\bar{a} = \frac{\Delta v}{\Delta t}$$

◀ average acceleration = change in velocity / time taken

As we did with speed, we can speak either of the average acceleration or the instantaneous acceleration, depending on the size of the time interval.

The units of acceleration are a bit more complicated than those of speed and velocity. Remember that the units of velocity are distance divided by time: for example, miles per hour or meters per second. Because acceleration is the change in velocity divided by the time interval, its units are (distance per time) per time—for example, (kilometers per hour) per second or (meters per second) per second.

The concept of acceleration is probably familiar to you. We talk about one car having "better acceleration" than another. This usually means that it can obtain a high speed in a shorter time. For instance, a Dodge Grand Caravan can accelerate from 0 to 60 miles per hour in 11.3 seconds, a Ford Taurus requires 8.7 seconds, and a Chevrolet Corvette requires only 4.8 seconds.

Question Which car has the largest average acceleration?

Answer The Corvette has the largest average acceleration because it reaches 60 mph in the shortest time interval.

Another way of becoming more familiar with acceleration is by experiencing it. For example, when an elevator begins to move up (or down) rapidly, the sensation you get in your stomach is due to the elevator (and you) quickly changing speed. Astronauts feel this when the space shuttle blasts off from its launch pad. Exciting examples of the same effect can be achieved on a roller coaster. In fact, amusement parks can be thought of as places where people pay money to experience the effects of acceleration.

In contrast, you don't feel motion when you're traveling in a straight line at a constant speed—that is, motion with zero acceleration. The motion you do feel when riding in a car on a straight highway is due to small vibrations of the car. (These vibrations are tiny changes in direction or small accelerations of the car caused by bumps in the road.)

Amusement parks sell the thrill of acceleration.

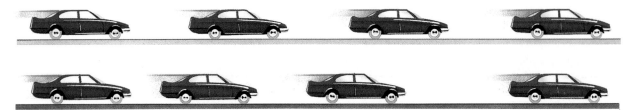

Figure 2-4 Strobe drawings of two cars. Which car is accelerating and which is traveling at a constant speed?

If you are not accelerating but rather are viewing a moving object, how can you tell whether the object is accelerating? One way is to take a strobe photograph of its motion. Figure 2-4 is a drawing of two such pictures. Which of the two corresponds to the car accelerating?

If you answered car (b), you have a qualitative understanding of acceleration. Car (a) travels the same distance during each time interval and therefore is traveling at a constant speed. Car (b) travels farther during each successive time interval; it is accelerating.

Even if the car were slowing down, it would be accelerating. (We don't usually use the word *deceleration* in physics because the word *acceleration* includes slowing down as well as speeding up.) In this case the distance traveled during successive time intervals would be shorter. Acceleration refers to any change in speed or direction—that is, to any change in velocity.

Question You see two cars side by side as they exit a tunnel. The red car has a speed of 40 meters per second and an acceleration of 20 (meters per second) per second. The blue car has a speed of 20 meters per second and an acceleration of 40 (meters per second) per second. At the instant they leave the tunnel, which car is passing the other?

Answer The car with the largest instantaneous speed will travel the farthest down the road in the next small interval of time. Therefore, the red car is passing the blue car as they exit the tunnel. The blue car will have the greatest change on the speedometer in the next small interval of time and will eventually overtake and pass the red car.

Acceleration is a vector quantity. When the acceleration is in the same direction as the velocity, the speed of the object is increasing. When the acceleration and the velocity point in opposite directions, the object is slowing down. The idea of an acceleration having a direction might seem a little abstract and, perhaps, unnecessary. A car's velocity obviously has a direction and probably seems easier to comprehend. However, as we continue our study of acceleration, we will see many examples in which the direction of the acceleration has physical consequences. The discussion of accelerations due only to a change in the direction of the velocity appears in Chapter 4.

Question What is an example from everyday life of something that is slowing with an acceleration vector pointing upward?

Answer If this "something" is slowing, its velocity vector must be pointing in the opposite direction of its acceleration vector, or downward. Therefore, we are looking for something that is moving toward the ground and slowing down. This could be a diver, right after she hits the water, or a parachutist right after the chute is opened. Try to think of another example.

WORKING IT OUT | Acceleration

Consider a car traveling along a straight highway at 40 mph that speeds up to 60 mph during a time interval of 20 s. What is the car's average acceleration?

Using the symbols v_i and v_f to represent the initial and final velocities, we have

$$\bar{a} = \frac{\Delta v}{\Delta t} = \frac{v_f - v_i}{t_f - t_i} = \frac{60 \text{ mph} - 40 \text{ mph}}{20 \text{ s}} = \frac{20 \text{ mph}}{20 \text{ s}} = 1 \text{ mph/s}$$

The car accelerates at 1 (mph) per second; that is, during each second, its speed increases by 1 mph.

If, on the other hand, the car made this change in velocity in 10 s, our new calculation would yield an average acceleration of 2 mph/s. These calculations illustrate that acceleration is more than just a change in velocity; it is a measure of the rate at which the velocity changes.

A First Look at Falling Objects

With these few ideas, we can now look at a common motion: a ball falling near Earth's surface. How does the ball fall? Does it fall faster and faster until it hits the ground? Or does it reach a certain speed and then remain at that speed for the duration of its fall? Does the rate at which it falls depend on its weight? For example, would a cannonball fall faster than a feather?

Questions about this rather simple motion have fascinated scientists since at least the time of Aristotle (4th century BC). They turned out to be quite difficult to answer. In fact, modern answers to these questions were not given until early in the 17th century.

Until that time the accepted answers were those attributed to Aristotle. Every motion required a mover (object) and a goal toward which the object moved. An object falling from a height to its natural resting place upon an immobile Earth—its goal—would travel with a speed determined by its weight divided by the resistance of the medium through which it traveled. Heavier objects would naturally travel faster than light ones. Much heavier objects would presumably fall even more rapidly. A cannonball would fall more slowly in molasses than in air. A feather would flutter down in air, but would float in molasses. In Aristotle's view, all change was motion. Falling bodies were just one example of change—a change of place. Changes of temperature, color, or texture were other examples of motion.

The advantage of Aristotle's system was that he dealt with concrete, observable situations that we encounter every day. This advantage led to serious and rewarding discussions over the next 1500 years. Great scientists like Galileo carefully analyzed Aristotle's ideas, which included a prediction that a 10-pound rock should fall significantly faster than a 1-pound rock.

You can perform an equivalent experiment to test the theory. Hold a heavy book (a physics text is quite appropriate) and a piece of paper at equal heights above the floor and drop them simultaneously. Which falls faster? Now repeat the experiment, but this time wad the paper into a tight ball. How do the results differ?

In the first case, the book fell much faster than the paper. This is in qualitative agreement with Aristotle's claim. But the second case certainly disagrees with Aristotle. Using the Aristotelian rule to predict the fall of the paper and book conflicts with reality. If the book is significantly heavier than the paper, according to Aristotle, the book should drop at a speed significantly faster than the

GALILEO | Immoderate Genius

On February 15, 1564, Galileo Galilei was born in Pisa, Italy, into the family of a Florentine cloth merchant. As a boy he was schooled in Latin, Greek, and the humanities at the local monastery until his family relocated to Florence, where his father assumed the boy's education in mathematics and music. The young Galileo mastered the lute while learning advanced mathematics, physics, and astronomy.

At age 17 Galileo returned to the town of his birth to study medicine. Before completing his training, however, Galileo withdrew from medical school due to conflicts with his mentors caused by his curiosities in the sciences. At age 25 Galileo enlisted the aid of his father's friends in academia and received an appointment as professor of mathematics at the University of Pisa. Other prestigious appointments followed as Galileo engaged in a variety of scientific pursuits.

Galileo Galilei

Galileo studied time, motion, floating bodies, the nature of heat, and the construction of telescopes and microscopes. In 1610 he achieved fame for improving upon previous telescope designs; his allowed magnification of the heavens up to 32 times. Galileo observed features of the Moon, sunspots, and some planets, including Venus in its various phases. He discovered four of Jupiter's moons and showed that the Milky Way consists of an enormous number of stars.

Galileo's many observations supported the Copernican Sun-centered (heliocentric) theory of the Solar System. In this belief, Galileo directly contradicted the Roman Catholic Church dogma of the day, which held that Earth was the stationary center of the Universe; such a model as proposed by Aristotle was supported by biblical references and was held sacred. To deny this view was considered heresy, a crime that carried the highest penalty of being burned at the stake. In 1615 the Church began its investigation of Galileo's beliefs; in 1633, after publishing *Dialogue Concerning the Two World Systems*, he was taken to Rome, charged with heresy, and sentenced to life imprisonment. Later, the sentence was commuted to house arrest at his villa at Arcetri.

In his last years, Galileo reflected upon his life in a manuscript titled *Discourses and Mathematical Discoveries Concerning Two New Sciences*, which was smuggled out of Italy and published in Holland in 1638. The two new sciences were dynamics and the strength of materials. After completion of this work, Galileo became blind and died while still under house arrest in 1642.

Legend has it that Galileo dropped balls from this leaning tower in Pisa while developing his ideas about free fall.

Sources: S. J. Broderick, *Galileo: The Man, His Work, His Misfortunes* (New York: Harper & Row, 1964); AIP Niels Bohr Library; Stillman Drake, *Galileo at Work: His Scientific Biography* (Chicago: University of Chicago Press, 1978).

---

paper. This means that the book would hit the floor well before the paper! Clearly, Aristotle was wrong.

Our experiment with the book and paper might lead you to believe that in the absence of any resistance, as in a vacuum, two objects would fall side by side, independent of their weights. This means that a cannonball and a feather would fall together in a vacuum. This was the opinion of Galileo Galilei, an Italian physicist of the 17th century.

Galileo is often called the founder of modern science, due as much to his style of building a physics world view as to his particular contributions. His style was characterized by a strong desire to verify his theories with measurements; that is, he performed experiments to check his ideas. His goal was simple: to find rules of nature—often in the form of equations—that expressed the results of his investigations.

His work led to some new ideas about motion. He developed the concepts and mathematical language necessary to describe motion. For example, he invented the concept of acceleration.

Although Galileo was unable to directly test his ideas about free fall, he did suggest the *thought* experiment illustrated in Figure 2-5. Imagine dropping three

A hammer and feather dropped on the Moon hit the ground at the same time because there is no air.

identical objects simultaneously from the same height. The Aristotelians would agree that the three would fall side by side. Now imagine repeating the experiment but with two of the objects close to each other. Nothing significant has changed, so there will again be a three-way tie. Finally, consider the situation where the two are touching. Because there was a tie before, there will be no dragging of one on the other, and again there will be a tie. But if the two are touching, they can be considered to be a single object that is twice as big. Consequently, big and small objects fall at the same rate!

A common present-day test of this idea is to drop two objects, such as a coin and a feather (a cannonball is somewhat impractical), inside a plastic tube from which the air has been removed. In a vacuum the race always ends in a tie. Astronauts conducted an ultramodern demonstration of this on the Moon. Because there is no atmosphere on the Moon, a hammer and a feather fell at the same rate.

PHYSICS | **ON YOUR OWN**

Check Galileo's statement that two objects of different weights fall at the same rate. Simultaneously drop balls out of a third- or fourth-floor window. Does the air cause the balls to fall at different rates? Make an estimate of the effects of air resistance by dropping two balls of approximately the same weight but different sizes, such as a solid rubber ball and a tennis ball.

Free Fall: Making a Rule of Nature

If we can find a rule for the motion of an object in free fall, the rule will be equally valid for all objects, heavy or light. Although we will obtain a rule that strictly speaking is valid only in a vacuum, it will be useful in many other situations—whenever the effects of air resistance can be ignored.

The quantitative measurement of a free-falling object was impossible with the technology of Galileo's time. Things simply happened too fast. To get around this, he studied a motion very similar to free fall, that of a ball starting from rest and rolling down a ramp. As the ramp is inclined more and more, the speed of the ball increases. When the ramp is vertical, the ball falls freely. Galileo hoped (correctly) that he would be able to ignore the complications of rolling and be able to deduce the rule for free fall before the ramp became too steep and the ball's motion too fast to measure.

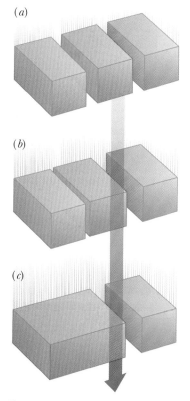

Figure 2-5 Galileo's thought experiment. All objects fall at the same rate in a vacuum.

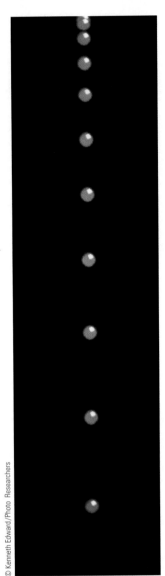

Figure 2-6 This strobe photograph of a falling ball shows that the ball has a constant acceleration.

Experimenting at a single ramp angle, Galileo discovered that the ball traveled with a constant acceleration. After establishing this rule at a single angle, he increased the tilt of the ramp and started over again. At steeper angles he found what you might expect intuitively: the ball traveled down the ramp in less time. The exciting discovery, however, was that at each new ramp angle he discovered the *same* rule: the acceleration was constant. This relationship held *independently* of the ramp angle, a fact that was crucial to the success of his proposal to extrapolate these results to the free-fall situation. He correctly concluded that the ball would obey the same relationship if the ramp was vertical.

Using modern techniques we are able to take a strobe photograph (Figure 2-6) of a falling ball and "see" its motion. The ball is clearly not moving at a constant speed. We can tell this by noting that the distances between the images are continually increasing. As Galileo showed, the ball falls with a constant acceleration.

Constant acceleration means that the speed changes by the same amount during each second. If, for example, we find that the ball's speed changed by a certain amount during 1 second at the beginning of the flight, then it would change by the same amount during any other 1 second of its flight.

We now know that for the case of the vertical ramp (free fall) the acceleration is about 9.8 (meters per second) per second (32 [feet per second] per second). This value is known as the *acceleration due to gravity* and varies slightly from place to place on Earth's surface. Students at Montana State University in Bozeman have determined that the value of the acceleration due to gravity in the basement of the old physics building is 9.800 97 (meters per second) per second. For convenience in calculations we will usually round off this value to 10 (meters per second) per second.

At any time during the fall, if we know its speed and its acceleration, we can calculate how fast the ball will be moving 1 second later. Assume that the ball is traveling 40 meters per second with an acceleration of 10 (meters per second) per second. This acceleration means that the speed will change by 10 meters per second during each second. Because the ball is speeding up, 1 second later the ball will be traveling with a speed of 50 meters per second. One second after that, the speed will be 60 meters per second, and so on.

WORKING IT OUT | Should You Jump?

You are standing at the top of a waterfall looking down at a deep pool of water below. Your friends think it is safe to jump, but you are worried that you might be too high for comfort. You pick up a rock and drop it into the pool. You count "one-one thousand, two-one thousand, three-one thousand" and find that it takes about 3 s for the rock to fall.

How fast would you be going just before you hit the water?

Any falling object, a rock or a person, speeds up by 10 m/s for every second that goes by (if we neglect air resistance). If you drop with an initial speed of zero, you would be traveling 10 m/s after the first second, 20 m/s after the second second, and 30 m/s (nearly 70 mph!) right before you hit the water. We suggest that you point your toes.

How tall is the waterfall?

The rock hit the water going 30 m/s, but clearly the rock did not travel this fast for the entire fall. The rock's initial speed was zero and increased uniformly to 30 m/s. The rock's average speed during the fall was 15 m/s (halfway between zero and 30 m/s), and the rock had this average speed for 3 s. That means that, on average, the rock fell 15 m every second for 3 s, for a total of 45 m.

As a final example of free fall, consider a thrill-seeking skydiver who jumps from a plane and decides not to pull the parachute cord until 30 seconds have elapsed. Our diver accelerates at a rate of 10 meters per second each second. Thus, at the end of $\frac{1}{2}$ minute (assuming our skydiver can resist pulling the cord for that long), the speed will be about 300 meters per second. That is about 675 mph! Actually, as we will see in Chapter 3, this description of free fall is quite inaccurate when the effect of air resistance becomes important.

Air resistance prevents the skydiver from having a constant acceleration.

> PHYSICS | **ON YOUR OWN**
>
> Hold a dollar bill so that it is vertical. Have your friend hold his thumb and index finger on each side of the middle of the bill. Tell him that he can keep the dollar if he can catch it when you let go.
>
> To catch the bill, he must be able to react within 0.13 second. Very few people can do this. However, if he is lucky and anticipates your move, you lose. If you let him hold his fingers near the bottom of the bill, he has 0.18 second to react. Many people can do this.

Starting with an Initial Velocity

What happens if the object is already in motion when we start our observations? Suppose, for example, a ball is thrown vertically upward. Our experience tells us that it will slow, stop, and then fall. Examining strobe photographs of objects thrown vertically upward shows that the behavior of the rising object is symmetrical to that of the same object falling. The speed changes by 10 meters per second during each second. In fact, the strobe photograph in Figure 2-6 could just as well have been a photograph of a ball rising. (Of course, what goes up must come down; in taking the photograph, we would have to close the camera's shutter before the ball started back down.)

We can use the symmetry between motion vertically upward and downward in answering the following question: if you throw a ball vertically upward with an initial speed of 20 meters per second, how long will it take to reach its maximum height? Ignoring air resistance, we know that the ball slows down by 10 meters per second during each second. Therefore, at the end of 1 second, it will be going 10 meters per second. At the end of 2 seconds, it will have an instantaneous speed of zero. Therefore, it takes the ball 2 seconds to reach the top of its path.

Question If you could throw the ball with a vertical speed of 40 meters per second, how long will it take to reach its maximum height?

Answer It will take 4 seconds to reach its maximum height.

A Subtle Point

Let's pause at the end of this chapter to emphasize the fact that Galileo used experiment and reasoning to discover a pattern in nature. He could discern the motion of objects subject only to the pull of Earth's gravity. Using simple mathematics and the rule that he formulated, you can calculate the outcome of future experiments. In a limited but very real way, you can predict the future. Predictions based on this rule are not the crystal-ball type popularized in science fiction, but they represent a very real accomplishment. The discovery of patterns and the creation of rules of nature are central in physicists' attempts to build a world view.

PhysicsNow™ Assess your understanding of this chapter's topics with sample tests and other resources found by logging into PhysicsNow at http://physics.brookscole.com/kf6e.

Summary

We began building a physics world view with the study of motion because motion is a dominant characteristic of the Universe. We can obtain data about the motion of objects from strobe photographs. The average speed $\bar{s}$ of an object is the distance d it travels divided by the time t it takes to travel this distance, $\bar{s} = d/t$. The units for speed are distance divided by time, such as meters per second or kilometers per hour.

Instantaneous speed is equal to the average speed taken over a very small time interval. Speed in a given direction is known as velocity, a vector quantity.

Displacement is a vector quantity giving the straight-line distance and direction from an initial position to a final position. Average velocity is the change in position—displacement—divided by the time taken, $\bar{v} = \Delta x/\Delta t$.

Acceleration is the change in velocity divided by the time it takes to make the change, $\bar{a} = \Delta v/\Delta t$. Acceleration is a vector. The units for acceleration are equal to those of speed divided by time such as (meters per second) per second or (kilometers per hour) per second.

Galileo reasoned that all objects fall at the same rate in the absence of any air resistance. Furthermore, he discovered that these free-falling objects fall with a constant acceleration of about 10 (meters per second) per second.

Chapter 2 Revisited

We can determine how long it will take a train to reach its destination if we know its average speed and how far it has to go. The travel time is equal to the distance divided by the average speed.

KEY TERMS

average acceleration: The change in velocity divided by the time it takes to make the change, $\bar{a} = \Delta v/\Delta t$. Measured in units such as (meters per second) per second. An acceleration can result from a change in speed, a change in direction, or both.

average speed: The distance traveled divided by the time taken, $\bar{s} = d/t$. Measured in units such as meters per second or miles per hour.

average velocity: The change in position—displacement—divided by the time taken, $\bar{v} = \Delta x/\Delta t$.

displacement: A vector quantity giving the straight-line distance and direction from an initial position to a final position.

instantaneous speed: The average speed for a very small time interval. The magnitude of the instantaneous velocity.

magnitude: The size of a vector quantity. For example, speed is the magnitude of the velocity.

vector: A quantity with a magnitude and a direction. Examples are displacement, velocity, and acceleration.

velocity: A vector quantity that includes the speed and direction of an object.

CONCEPTUAL QUESTIONS

Most questions and exercises are paired so that most odd-numbered questions and exercises are followed by a similar even-numbered one. Short answers for most odd-numbered questions and exercises are provided at the back of this textbook. More-challenging questions are indicated by a ▲.

1. Describe the motion depicted in the following strobe drawing.

    • • • • • •   •   •   •   •   •   •

2. Describe the motion of the pucks in the strobe photographs. Assume the pucks move from left to right and do *not* retrace their paths.

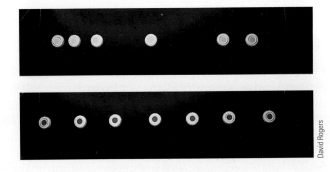

3. Where does the ball shown in the following strobe drawing have the slowest speed?

• • • • •   •   •   •   •   •   •   •

4. Where is the speed the fastest in the following strobe drawing?

• • • • • • • • •   •   •   •   •   •

5. Sketch a strobe drawing for the following description of a caterpillar moving along a straight branch. The caterpillar begins from rest and slowly accelerates to a constant speed. It then slows down to a slower constant speed. Finally it gets tired and stops for a rest.

6. A car is driving along a straight highway at a constant speed when it hits a mud puddle, slowing it down. After the puddle, the driver speeds up until he is going faster than before hitting the puddle and then sets the cruise control. Make a strobe drawing for this motion.

7. Draw a strobe photograph for a sprinter running the 100-yard dash. Represent the sprinter's motion from the firing of the starting gun until she stops after passing the finish line.

8. An ice climber falls from a frozen waterfall into a large snowdrift and gradually comes to rest. Draw a strobe diagram of the climber's motion from the moment he falls until he comes to rest.

9. Which (if either) has the greater average speed: a truck that travels from milepost 83 to milepost 90 in 10 minutes, or a car that travels from milepost 122 to milepost 130 in 10 minutes?

10. Which (if either) has the greater average speed: a car that travels from milepost 35 to milepost 41 in 6 minutes, or one that travels from milepost 68 to milepost 71 in 3 minutes?

11. You are driving down the road, with the cruise control set to 45 mph. You see a rabbit on the road, hit your brakes, and bring your car to rest. Is your average speed while braking greater than, equal to, or less than 45 mph?

12. In Aesop's fable of the tortoise and the hare, the "faster" hare loses the race to the slow and steady tortoise. During the race, which animal has the greater average speed?

13. Pat and Chris both travel from Los Angeles to New York along the same route. Pat rides a bicycle while Chris drives a fancy sports car. Unfortunately, Chris's car breaks down in Salt Lake City for more than a week, causing the two to arrive in New York at exactly the same time. Compare the average speeds of the two travelers.

14. A book falls off a shelf and lands on the floor. Which is greater, the book's average speed or its instantaneous speed right before it lands?

15. For the following strobe drawing, compare the instantaneous speeds at points C and D to the average speed for the time interval between C and D.

A B   C   D   E
••   •   •   •

16. For the following strobe drawing, compare the instantaneous speeds at points C and D to the average speed for the time interval between C and D.

A   B   C   D E
•   •   •   ••

17. How might you estimate your speed if the speedometer in your car is broken?

18. Why is it *not* correct to say that time is more important than distance in determining speed?

19. A truck driver averages 92 kilometers per hour between 2 p.m. and 6 p.m. Can you determine the speed of the truck at 4 p.m.?

20. An ancient marathoner covered the first 20 miles of the race in 4 hours. Can you determine how fast he was running when he passed the 10-mile marker?

21. Which of the following can be used to measure an average speed: stopwatch, odometer, or speedometer? An instantaneous speed?

22. What are the units of the physical properties measured by a stopwatch, an odometer, and a speedometer?

23. What is the essential difference between speed and velocity?

24. If you are told that a car is traveling 65 mph east, are you being given the car's speed or its velocity?

25. In the following strobe drawings, which object (if either) has the greater acceleration?

26. The following strobe drawings represent the motions of two cars, a and b. During which interval of the motion of car a is the average speed of car a approximately equal to the average speed of car b?

27. Which of the following (if any) could *not* be considered an "accelerator" in an automobile—gas pedal, brake pedal, or steering wheel?

28. In what sense can the brakes on your bicycle be considered an "accelerator"?

29. Assume that an airplane accelerates from 550 mph to 555 mph, a car accelerates from 50 mph to 59 mph, and a bicycle accelerates from 0 to 8 mph. If all three vehicles accomplish these changes in the same length of time, which one (if any) has the largest acceleration?

30. If an Acura Integra accelerates from 0 to 60 mph in 4 seconds and a Dodge Stealth accelerates from 20 mph to 75 mph in 4 seconds, which one has the larger acceleration?

31. A motorcycle travels down a straight highway with uniform speed of 35 mph. A sports car starts from rest and accelerates at 10 mph/s. Which will be moving faster after 3 seconds?

32. A Dodge Caravan has a speed of 60 mph and an acceleration of 1 (mph) per second. A Ford Taurus has a speed of 55 mph and an acceleration of 2 (mph) per second. Which car has the higher speed after 10 seconds have elapsed?

33. Carlos and Andrea are driving down the same road in the same direction, with Carlos behind Andrea. Carlos is slowing down and Andrea is speeding up, yet the distance between their cars is getting smaller. Give an example to show how this could happen.

34. Mary and Nathan are driving on a freeway in the same direction. At exactly noon, the cars are side by side. Mary is traveling at constant speed and Nathan is speeding up, yet Mary is passing Nathan. Explain how this could happen.
35. When we say that light objects and heavy objects fall at the same rate, what assumption(s) are we making?
36. Free fall near the surface of the Moon can be described as motion with a constant _____.
37. You are standing on a high cliff above the ocean. You drop a pebble, and it strikes the water 4 seconds later. Ignoring the effects of air resistance, how fast was the pebble traveling just before striking the water?
38. You throw a ball straight up in the air. The instant after leaving your hand the ball's speed is 30 meters per second. Ignoring the effects of air resistance, predict how fast the ball will be traveling 1 second later.
39. What happens to the acceleration of a ball in free fall if the ball's mass is doubled?
40. Two balls have the same size but are made from different materials: one from rubber and the other from steel. How do their accelerations compare after they are dropped?
41. You are bouncing on a trampoline while holding a bowling ball. As your feet leave the trampoline, you let go of the bowling ball. Do you rise to a higher, the same, or a lower height than if you had held on to the bowling ball?
42. You are bouncing on a trampoline while holding a bowling ball. As your feet leave the trampoline, you let go of the bowling ball. When you reach your maximum height, is the bowling ball above, beside, or below you?
43. A penny and a feather are placed inside a long cylinder, and the air is pumped out. When the cylinder is inverted, which hits the bottom first—the penny or the feather?
44. The Moon is a good place to study free fall because it has no atmosphere. An astronaut on the Moon simultaneously dropped a hammer and a feather from the same height. Which one hit the ground first?
45. How did the ideas of Galileo and Aristotle differ concerning the motion of a freely falling object?
46. A sheet of paper and a book fall at different rates unless the paper is wadded up into a ball, as shown in the figure. How would Galileo and Aristotle account for this?

47. A student decides to test Aristotle's and Galileo's ideas about free fall by simultaneously dropping a 20-pound ball and a 1-pound ball from the top of a grain elevator. The two balls have the same size and shape. What actually happens?
48. A table-tennis ball and a golf ball have approximately the same size but very different masses. Which hits the ground first if you drop them simultaneously from a tall building? Do not neglect the effects of air resistance.

49. A table-tennis ball and a marble are dropped side by side from the top of the biology building. Which ball has the greater acceleration? Do not ignore the effects of air resistance.
50. A table-tennis ball and a marble are both thrown straight up in the air at the same initial speed. Which ball has the greater acceleration? Do not ignore the effects of air resistance.
51. A hard rubber ball is bounced on the floor. Compare the ball's acceleration on the way down to its acceleration on its way back up.
52. How (if at all) does the acceleration of a cylinder rolling up a ramp differ from that of one that is rolling down the ramp?
▲ 53. If we ignore air resistance, the acceleration of an object that is falling downward is constant. How do you suppose the acceleration would change if we do *not* ignore air resistance? Explain your reasoning.
54. If we do not neglect air resistance, during which of the first 5 seconds of free fall does a ball's speed change the most?
55. A rubber ball is thrown straight up into the air with an initial speed of 20 m/s. If we do not neglect air resistance as the ball moves upward, is the acceleration of the ball greater than, equal to, or less than the acceleration due to gravity?
56. As the ball in the previous questions returns to the ground, is the acceleration of the ball greater than, equal to, or less than the acceleration due to gravity?
▲ 57. A cart starts from rest and rolls down a ramp with constant acceleration. The cart's average speed is given by the length of the ramp divided by the time required for the trip. For part of the trip down the ramp the cart travels slower than this average speed, and for part of the trip the cart travels faster than this average speed. Does the cart reach the average speed when it is halfway down the

ramp or when half the time has elapsed? Explain your choice.

▲ 58. A cart starts from rest and rolls down a ramp with constant acceleration. At some point in time the cart's instantaneous speed is equal to its average speed. At this instant in time, is the cart less than halfway, exactly halfway, or more than halfway down the ramp? Explain your choice.

EXERCISES

1. The top speed of the Blackbird is 2193 mph. Given that 1 mile = 1.61 km, what is this speed in km/h?
2. Top professional pitchers can throw fastballs at speeds of 100 mph. Given that 1 mph = 0.447 m/s, what is this speed in meters per second?
3. At exactly noon, you pass mile marker 50 in your car. At 2:30 p.m. you pull into a rest stop at mile marker 215. What was your average speed during this time?
4. To be eligible to enter the Boston Marathon, a race that covers a distance of 26.2 miles, a runner must be able to finish in less than 3 h. What minimum average speed must be maintained to accomplish this?
5. In 1993 Sue Ellen Trapp broke the U.S. women's record for a 24-h run by covering a distance of 145.3 miles. What was her average speed?
6. The 10,000-m run world record is 26 min 22.75 s. What was the runner's average speed in m/s?
7. How far can a bus travel in 8 h at an average speed of 70 mph?
8. At an average speed of 10 m/s, how many kilometers can a cyclist travel in an 8-h day?
▲ 9. Your plan was to be on the road by 9 a.m., but you did not leave the garage until 10 a.m. You then drove with the cruise control set at 75 mph until stopping at noon. What was your average speed over the time interval from 9 a.m. to noon?
▲ 10. Starting at 9 a.m., you hike for 3 h at an average speed of 4 mph. You stop for lunch from noon until 2 p.m. What is your average speed over the interval from 9 a.m. to 2 p.m.?
11. If a cheetah runs at 30 m/s, how long will it take a cheetah to run a 100-m dash? How does this compare with human times?
12. How many hours would be required to make a 4400-km trip across the United States if you average 90 km/h?
13. If a runner can average 4 mph, can he complete a 100-mile ultramarathon in less than 24 h?
14. At an average speed of 125 mph, how long would it take a race car to complete a 500-mile race?
15. If a Chevrolet Corvette can accelerate from 0 to 60 mph in 4.8 s, what is the car's average acceleration in mph/s?
16. One of the new hybrid cars is stopped for a traffic light. When the light turns green, the driver accelerates to a speed of 25 mph in a time of 6 s. What is the car's average acceleration?
17. A car speeds up from 40 mph to 70 mph to pass a truck. If this requires 6 s, what is the average acceleration of the car?
18. The world's record for top fuel dragsters is 4.477 s to travel $\frac{1}{4}$ mile from a standing start. The dragster was traveling 332.75 mph at the end of the quarter mile. What was the dragster's average acceleration? What was its average speed?
19. A rock climber drops a piton. If the piton passes you with a speed of 7 m/s, how fast will the piton be traveling 2 s later?
20. A roofer drops a nail that hits the ground traveling at 23 m/s. How fast was the nail traveling 2 s before it hit the ground?
21. A child traveling 5 m/s on a sled passes her younger brother. If her average acceleration on the sledding hill is 2 m/s^2, how fast is she traveling when she passes her older brother 4 s later?
22. You throw a ball straight up at 30 m/s. How many seconds elapse before it is traveling downward at 30 m/s?
23. You accidentally drop a watch from the roof of a six-story building. While picking up the watch, you notice that it stopped 2 s after it was dropped. How tall is the building?

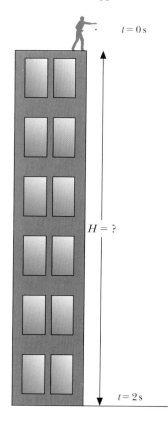

24. A rock is dropped into an abandoned mine, and a splash is heard 4 s later. Assuming that it takes a negligible time for the sound to travel up the mineshaft, determine the depth of the shaft and how fast the rock was falling when it hit the water.
25. A ball is dropped from a height of 45 m. Construct a table showing the height of the ball and its speed at the end of each second until just before the ball hits the ground.
26. A ball is fired vertically upward at a speed of 30 m/s. Construct a table showing the height of the ball and its velocity at the end of each second until just before the ball hits the ground.
▲ 27. You decide to launch a ball vertically so that a friend located 20 m above you can catch it. What is the minimum launch speed you can use? How long after the ball is launched will your friend catch it?
▲ 28. A dummy is fired vertically upward from a cannon with a speed of 40 m/s. How long is the dummy in the air? What is the dummy's maximum height?

3 | Explaining Motion

Air resistance slows the descent of the U.S. Army Parachute Team, the Golden Knights.

An object moving through a fluid must contend with resistance to its motion. For instance, bike racers streamline their bikes and clothing to minimize the air resistance. The effects of air resistance are very noticeable when skydivers open their parachutes. The acceleration is not constant; in fact, at some point the acceleration becomes zero. What happens to the speed of the falling object at this time?

See page 51 for the answer to this question.

PhysicsNow™ Test your understanding of this chapter by logging into PhysicsNow at **http://physics.brookscole.com/kf6e**, selecting the chapter, and clicking on the "Take a Pre-Test" link.

WHAT is meant by explaining motion? Don't motions just happen? These questions can probably be debated by philosophers for hours. In the physics world view, to explain something means to create a scheme, or model, that can predict the outcome of experiments. These experiments don't have to be elaborate; they can be as simple as throwing a baseball or looking at a rainbow.

Physicists try to create a set of ideas that explains how the world might work. Notice the word *might*; we have no proof that the ideas are correct or unique. There may be equally good (or better) schemes yet to be discovered.

Is it reasonable to expect that rules of nature exist? Motions appear to be reproducible; that is, if we start out with the same conditions and do the same thing to an object, we get the same resulting motion. The same motion occurs regardless of (1) when the experiment is done—the results on Monday match those on Tuesday—and (2) whether the experiment is done in San Francisco or Philadelphia. (Imagine the consequences if this were not true, and the "experiment" was throwing a baseball. Baseball teams would need a different pitcher for every day of the week and for every ballpark!) This reproducibility is a necessary condition for even attempting to search for a set of rules that nature obeys. Einstein reflected this idea when he commented that he believed God was cunning but not malicious. His point was that although the rules of nature might be difficult to find, as we search for them we realize that they do not change.

An Early Explanation

Aristotle developed an explanation of motion that lasted for nearly 2000 years. Many of Aristotle's ideas seemed common sense—they were based on our most common experiences.

Aristotle believed that the world was composed of four elements: earth, water, air, and fire. These four were the building blocks of the material world. Each substance was a particular combination of these four elements. If this seems naive to you, consider our modern world view. We take chemical elements, each with its own special attributes, and combine them to form compounds that have quite different attributes. For example, we take hydrogen, a very explosive gas, and oxygen, the element required for combustion, and combine them to form water that we use to fight fires!

Each Aristotelian element had its own natural place in the hierarchy of the Universe. Earth, the heaviest, belonged to the lowest position. Water was next, then air and fire. Aristotle reasoned that if any of these were out of its hierarchical position, its natural motion would be to return. These natural motions occurred in straight lines, toward or away from the center of Earth.

It is interesting to note that if you try to test this part of Aristotle's world view, it works! Put some water and earth (dirt) in a glass and wait. Watch as each element settles into its natural place (Figure 3-1).

For other than natural motions, Aristotle would probably challenge you to think of your own experiences. To move something you have to make an effort. Even though we have developed machines to make the effort for us, we still agree that an effort must be made and that after the effort is stopped the object comes to rest.

Although this seems reasonable, there are problems with the Aristotelian explanation. For example, objects don't stop immediately. An arrow continues to fly even after it loses contact with the bowstring. The Aristotelian explanation invokes an interaction between the arrow and the air. As the arrow moves through the air, it creates a partial vacuum behind it. The air, rushing in behind the arrow to fill the void, pushes on the arrow and causes the continued motion. This explanation, however, predicts that motion without an effort is impossible in a vacuum, while seeming to imply that in air it is perpetual.

Figure 3-1 In the Aristotelian world view, water rises while earth falls.

The Beginnings of Our Modern Explanation

What is the motion of an object when there is nothing external trying to change its motion? One might guess, in agreement with Aristotle, that the only natural motion of an object is to return to Earth; otherwise, it has no motion—it remains at rest. Medieval thinkers agreed that objects have this tendency not to move.

Let's do a simple experiment. Give this book a brief push across a table or desk. Although the book starts in a straight line at some particular speed, it quickly slows and stops. It seems natural for an object to come to rest and remain at rest.

Remaining at rest is a natural state. However, there is another state of motion that is just as natural but not nearly as obvious. Suppose you were to repeat this book-pushing experiment on a surface covered with ice. The book would travel a much greater distance before coming to rest. Our explanation of the difference in these two results is that the ice is slicker than the desktop. Different surfaces interact with the book with different strengths. The book's interaction with the ice is less than that with the wood. Can you predict what would happen to the book if the surface were perfectly slick? The book would not slow down at all; it would continue in a straight line at a constant speed forever. Stated differently, when the interaction is reduced to zero, the book's motion is constant.

Thus, it seems that a natural motion is one in which the speed and direction are constant. Note that this statement covers the object whether it is at rest or in motion. An interaction with an external agent is required to cause an object to change its velocity. If left alone, it would naturally continue in its initial direction with its initial speed.

Galileo reached this same conclusion by deducing the outcome of a thought experiment. He thought about the motion of a perfectly round ball placed on a tilted surface free of "all external and accidental obstacles." He noted—presumably from the same experiences we have all had—that a ball rolling down a slope speeds up (Figure 3-2[a]). Conversely, if the ball rolls up the slope, it naturally slows down. The ball experiences an interaction on the falling slope that speeds it up and an interaction on the rising slope that slows it down.

Now, Galileo asked himself, what would happen to the ball if it were placed on a level surface? Nothing. Because the surface does not slope, the ball would neither speed up nor slow down (Figure 3-2[b]). The ball would continue its motion forever.

It is important to remember that this is another of Galileo's thought experiments and not an account of an actual experiment. He assumed that there were no resistive interactions between the ball and the surface. There was no friction. By doing this he was able to strip motion of its earthly aspects and focus on its essential features.

Galileo was the first to suggest that constant-speed, straight-line motion was just as natural as at-rest motion. This property of remaining at rest or continuing to move in a straight line at a constant speed is known as **inertia**.

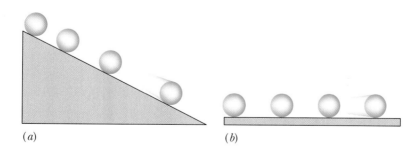

Figure 3-2 Galileo's thought experiment. (a) On a tilted surface, a ball's speed changes. (b) On a level surface, a ball's speed and direction are constant.

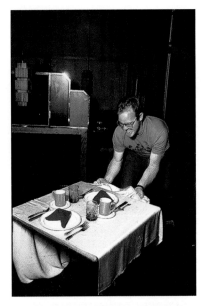

Figure 3-3 The tablecloth can be pulled from under the dishes because of their inertia.

The common use of the word *inertia* usually refers to an emotional state, one of feeling sluggish. Often when people say something has a lot of inertia, they are referring to their difficulty in getting it moving. (Sometimes they are talking about themselves.) As you build your physics world view, it is important to distinguish between the everyday uses of words and the usage within physics.

You already have a vast set of experiences that are directly related to inertia as it is used here. If something is at rest, it takes an interaction of some kind to get it moving. A magician uses the inertial property of cups and saucers when abruptly pulling the tablecloth from beneath them. If the tablecloth is smooth enough, the interaction with the cloth will be small, and the dishes will remain (nearly) at rest. Figure 3-3 shows the before and after photographs of this trick.

PHYSICS | ON YOUR OWN

Impress your friends by pulling a tablecloth out from under some dishes. To increase your chances of avoiding a disaster, you should do the following: Use a smooth, hemless cloth about the size of a pillowcase. Don't be timid; pull the cloth quickly in a downward direction across the straight edge of the table. Choose some dishes that are stable (and cheap!).

Another example of how inertia can be fascinating is the circus strongman who boasts of his strength by asking someone to hit him on the head with a sledgehammer. The strongman always does this demonstration holding a big, heavy block on his head. The inertia of the block is large enough that the blow from the sledgehammer does little to move it. The happy strongman only has to be strong enough to hold up the block!

But there is more to inertia than getting things moving. If something is already moving, it is difficult to slow it down or speed it up. An example is drying your wet hands by shaking them. When you stop your hands abruptly, the water continues to move and leaves your hands. In a similar way, seat belts counteract your body's inertial tendency to continue forward at a constant speed when the car suddenly stops.

However, all objects do not have the same inertia. For example, imagine trying to stop a baseball and a cannonball, each of which is moving at 150 kilometers per hour (about the speed a major-league pitcher throws a baseball). The cannonball has more inertia and, as you can guess, requires a much larger effort to stop it. Conversely, if you were the pitcher trying to throw them, you would find it much harder to get the cannonball moving.

Question   What is the main difference between the everyday usage of the word *inertia* and its use in physics?

Answer   The physics usage also includes the idea that objects tend to keep moving.

Although Galileo did not fully explain motion, he did take the first important step and, by doing so, radically changed the way we view the motion of objects. His work profoundly influenced Isaac Newton, the originator of our present-day rules of motion.

Newton's First Law

Isaac Newton, an Englishman, was born a few months after Galileo's death. Although he is probably best known for his work on gravitation, his most profound contribution to our modern world view is his three laws of motion. Like

NEWTON | Diversified Brilliance

On December 25, 1642, Isaac Newton was born in Woolsthorpe in Lincolnshire, England. Newton's father—a country gentleman in Lincolnshire—had died three months before he was born, leaving him to be reared by his mother. When he was 3 years old, his mother married a local pastor and moved a few miles away, leaving young Newton in the care of the housekeeper.

At the age 14, Newton returned home from school at the request of his mother to help work on the family farm. He proved to be not much of a farmer, however, and spent most of his time reading. Times when he could be alone, Newton would amuse himself by building model windmills powered by mice, water clocks, sundials, and kites carrying fiery lanterns, which frightened the country folk. A local schoolmaster recognized Newton's abilities and helped him enter Trinity College at Cambridge at age 18, where he would receive his bachelor of arts degree 4 years later, in 1665.

Later that same year, the bubonic plague raged through the English countryside and, consequently, the university was closed. Newton returned to Woolsthorpe, and the next 18 months proved to be his most productive. It was during this interlude that Newton developed his theories and ideas about optics, celestial mechanics, calculus, the laws of motion, and his famous law of gravity. After the Great Plague, Newton returned to Cambridge, where he was appointed professor of mathematics at the age of 26. From here Newton went on to develop a reflecting telescope, one that utilized a mirror to collect light instead of a lens, which was used in earlier models. Newton also published his most notable book—

Isaac Newton

with the help of Edmund Halley—titled *Principia Mathematica Philosophiae Naturalis* (*Mathematical Principles of Natural Philosophy*) in 1687.

In 1701 Newton was appointed master of the mint, and in 1703 he was elected president of the Royal Society, a position he retained until his death. Newton's honors did not end there, however; in 1705 Queen Anne knighted him in recognition of his many accomplishments, forever changing his name to Sir Isaac Newton. This was the first knighthood given for scientific achievement.

Sir Isaac Newton's life was not all discoveries and honors. In fact, he spent a great deal of his later life quarreling with fellow scientists. Robert Hooke accused Newton of stealing some of his ideas about gravity and light. Newton also fought bitterly with Gottfried Leibniz, a German mathematician who claimed to have developed calculus first, and Christiaan Huygens, who worked independently on the wave theory of light. In 1727 Sir Isaac Newton became seriously ill, and on March 20 one of the greatest physicists of all time died. He was accorded a state funeral and interred in the nave of Westminster Abbey—a high and rare honor for a commoner.

Sources: Adapted from R. A. Serway and J. S. Faughn, *College Physics* (Philadelphia: Saunders, 1992). Also, AIP Niels Bohr Library; F. E. Manuel, *A Portrait of Isaac Newton* (Cambridge, Mass.: Harvard University Press, 1968); Richard S. Westfall, *Never at Rest: A Biography of Isaac Newton* (Cambridge, UK: Cambridge University Press, 1980).

Galileo, Newton was interested in the interactions that occur while an object is in motion rather than in its final destination. He formulated Galileo's observations into what is now called **Newton's first law of motion**. It is also referred to as the **law of inertia**.

> The velocity of an object remains constant unless an unbalanced force acts on the object.

◀ Newton's first law

For the velocity of an object to remain constant, its speed and its direction must both remain constant. Note that this law applies to the special case of an object at rest: an object at rest remains at rest unless acted on by an unbalanced force.

The first law incorporates Galileo's idea of inertia and introduces a new concept, **force**. In the Newtonian world view, the book sliding across the table slows down and stops because there is a force (called friction) that opposes the motion. Similarly, a falling rock speeds up because there is a force (called gravity) continually changing its speed. In short, there is no acceleration unless there is a net, or unbalanced, force.

All of us have an intuitive understanding of forces; casually speaking, a force is a push or a pull. But it should be noted that the concept of force is a human construct. Because we have grown up with forces as a part of our personal world view, most of us feel quite comfortable with them. But we don't actually *see* forces. We see objects behave in a certain way, and we infer that a force is present. In fact, alternative world views have been developed that do not include the

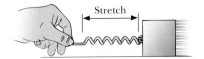

Figure 3-4 The stretch of the spring is a measure of the applied force.

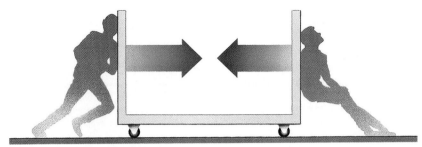

Figure 3-5 Equal forces acting in opposite directions cancel, and the cart does not accelerate.

concept of force. This concept, however, has greatly aided the process of building a physics world view.

Although the concept of force includes much more than our intuitive ideas of push and pull, we use these thoughts as a beginning. A force can be defined in terms of the observed behavior of objects. For example, a force measurer (Figure 3-4) constructed with rubber bands or springs would allow us to quantify our observations of force by measuring the amount of stretch using some arbitrary scale. We will have more to say about these devices after we learn about Newton's second law.

Another important characteristic of forces is that they are directional, meaning that the direction of the force is as important as its size. Different results are produced by forces of the same size when applied in different directions. Imagine a skater coasting across the ice. A force in the direction of the original motion increases the skater's speed. A force applied in the opposite direction slows the skater. So we need to incorporate this difference into our understanding of motion. As you might guess from the discussion in Chapter 2, we will do this by treating forces as vectors.

Remember that Newton's first law refers to the *unbalanced* force. In many situations there is more than one force on an object. There is an unbalanced force only if the sum of the forces is *not* zero. When two forces of equal size act along a straight line but in opposite directions, they cancel each other. In this case the forces tend to stretch or compress the object, but the unbalanced, or *net*, force is zero. The "helpers" in Figure 3-5 could each be exerting a very large force, but if the forces are equal in size and opposite in direction, there is no unbalanced force on the cart.

The converse of this also holds. When we observe an object with no acceleration, we infer that there is no unbalanced force on that object. If you see a car moving at a constant speed on a level, straight highway, you infer that the frictional forces balance the driving forces. This is not to say that there are no forces acting on the car, because there are many. The crucial point is that the sum of all of these forces is zero; there is no unbalanced, or net, force acting on the car.

Question   What is the net force acting on an airplane in level flight flying at 500 mph due east?

Answer   Because the speed and direction are constant, there is no acceleration, and the net force must be zero.

✓ Extended presentation available in the *Problem Solving* supplement

Adding Vectors

Mathematicians have developed rules for combining vector quantities such as displacements, velocities, accelerations, or forces. We can represent any vector by an arrow; its length represents the magnitude of the quantity, and its direction represents the direction of the quantity. To complete this representation, we assign a convenient scale to our drawing. For instance, we let 1 centimeter in the drawing in Figure 3-6 represent a distance of 20 meters on the ground. Then an arrow that is 4 centimeters long represents a displacement of 80 meters in the direction the arrow points.

Question How many meters would a 10-centimeter arrow represent?

Answer Because each centimeter represents 20 meters, a 10-centimeter arrow represents 200 meters.

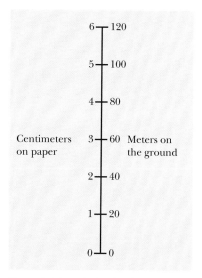

Figure 3-6 A scale for displacement vectors. One centimeter on paper represents 20 meters on the ground.

In texts, vector quantities are represented by boldface symbols (such as **x**) and in handwritten materials with an arrow over the symbol (such as $\vec{x}$). The size, or magnitude, of the vector quantity is represented by an italic symbol. Therefore, a force is written as **F**, and its magnitude is written as F.

We can combine vectors using a graphical method and the scale shown in Figure 3-6. Let's assume that you walk a distance of 80 meters due north. This displacement is represented by an arrow 4 centimeters long pointing straight up the page. Then you continue walking due north for another 60 meters. This displacement is represented by an arrow 3 centimeters long pointing straight up the page. Your total displacement is 140 meters north. Notice in Figure 3-7(a) that we can represent this graphically by drawing the second arrow starting at the head of the first arrow, just as your second displacement started at the end of the first displacement. The sum of the two arrows is the arrow drawn from the tail of the first arrow to the head of the second. In this case, the arrow representing the sum is 7 centimeters long and points north, representing a displacement of 140 meters north.

Now let's say you walk 80 meters due north, turn around, and walk 60 meters due south along the same path. What is your displacement? You are 20 meters north of your starting place, so your displacement is 20 meters north. This is shown graphically in Figure 3-7(b).

The third time, you walk 80 meters due north, turn to the right, and walk 60 meters due east. To find your displacement on the ground, you could put stakes at your beginning and ending locations, measure the distance between

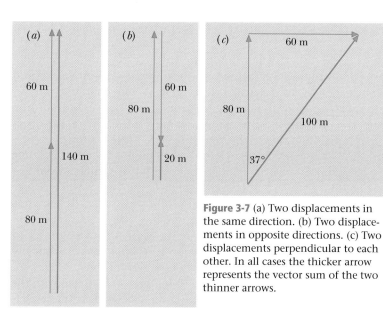

Figure 3-7 (a) Two displacements in the same direction. (b) Two displacements in opposite directions. (c) Two displacements perpendicular to each other. In all cases the thicker arrow represents the vector sum of the two thinner arrows.

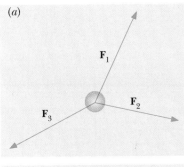
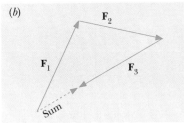
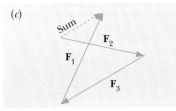

Figure 3-8 The three forces acting on the ball (a) can be added to find the net force (b). The order in which the forces are added (c) does not matter.

the stakes, and use a compass to get the direction from the beginning stake to the ending stake. Graphically, we draw a 4-centimeter arrow pointing up the page. We then draw a 3-centimeter arrow to the right, as shown in Figure 3-7(c). The displacement is the arrow drawn from the tail of the first arrow to the head of the second arrow, as shown. To get the displacement, we measure the length of the arrow. We get 5 centimeters, telling us that the distance is 100 meters. We then use a protractor to measure the angle indicated, obtaining 37 degrees. Thus, we obtain a displacement of 100 meters at 37 degrees east of north.

This method of adding vectors is easily generalized to more than two vectors. After the first arrow is drawn, each succeeding arrow is drawn beginning at the head of the previous arrow. The arrow drawn from the tail of the first arrow to the head of the last arrow represents the vector sum. The order of the arrows does not affect the final answer.

When more than one force acts on an object, we can find the net force acting on the object by adding all of the forces using the method just described. Consider the three forces acting on the ball shown in Figure 3-8(a). To add these forces, we move F_2 (without changing its direction) and place its tail on the head of F_1, as shown in Figure 3-8(b). We then place the tail of F_3 on the head of F_2. The sum of these three forces is the arrow from the tail of F_1 to the head of F_3. The size of the force is determined by the scale used in the drawing, and the direction is determined with a protractor.

Newton's Second Law

Newton's first law tells us what happens when there is no net force acting on an object: the speed and direction don't change. If there is a net force, the object accelerates and thus changes its velocity. Newton's second law describes the relationship between a net force and the resulting acceleration.

Our development of the second law will present some simple experiments that illustrate this relationship before we state it formally. Assume that we have a collection of identical springs (the force measurers mentioned earlier), a collection of objects, and all the necessary equipment to measure accelerations. Further assume that if we stretch a spring by a fixed amount and maintain this stretch, the spring exerts a constant force.

Because the second law describes the net force, we need a situation in which the frictional forces are so small that they can be disregarded and any other forces are balanced so that the forces that we apply are the only ones affecting the acceleration. A horizontal air-hockey table is a good experimental surface. The hockey puck rides on a cushion of air, so it experiences very small frictional forces.

If we pull a hockey puck with a spring stretched by a certain amount and maintain the direction and amount of force even while the puck moves, we find that the puck experiences a constant acceleration in the direction of the force. After doing this many times with differing amounts of stretch, we conclude that a constant net force produces a constant acceleration. Furthermore, the direction of the acceleration is always in the same direction as the net force.

Let's now compare the results we obtained when pulling the puck with one spring with what happens when we pull the puck with two springs. When two springs are pulling side by side, as shown in Figure 3-9, the force is twice as large as that of the single spring. If we stretch each of the two springs by the same amount as before, we find that the two springs produce twice the acceleration. If we use three springs, they produce three times the acceleration, and so on. In general, we find that the acceleration of an object is **proportional** to the net force acting on it. This relationship will be part of the second law.

But this is not the entire story. Imagine pushing on a cannonball with the same force that is used on the hockey puck. Intuition tells you that the acceleration of the cannonball will be smaller. If pressed for a reason, you might re-

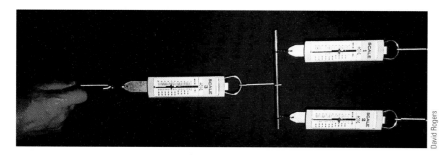

Figure 3-9 Two springs pulling side by side exert twice the force of one spring, as indicated by the scale readings.

spond, "Because there is more 'stuff' in a cannonball," or "It weighs more." We will soon see that, although the term *weight* is not technically correct, this intuition leads in the correct direction.

We build on this intuition by investigating how the acceleration of an object depends on the amount of matter in the object. The **mass** of an object is a measure of the amount of matter in the object. We assume that masses combine in the simplest possible way: the masses add. Therefore, the combined mass of two identical objects is twice the mass of one of them.

Again, we use a spring to pull on one of the hockey pucks and record its acceleration. We then look at the acceleration of two pucks (somehow tied together) pulled by a single spring. If the spring is stretched by the same amount as before, the acceleration is one-half the original. Likewise, one spring pulling on three pucks yields one-third the acceleration, and so on. Mass and acceleration are **inversely proportional**, where *inversely* indicates that the changes in the two values are opposite each other. If the mass is *increased* by a certain multiple, the acceleration produced by the force is *reduced* by the same multiple.

Notice that the more mass an object has, the more force it takes to produce a given acceleration. This means that the object has more inertia. We therefore take an object's mass to be the measure of the amount of inertia the object possesses.

Newton put the two preceding ideas together into one of the most important physical laws of nature ever proposed. This law states that the acceleration of an object is equal to the net force on the object divided by its inertial mass and can be written symbolically as

$$\mathbf{a} = \frac{\mathbf{F}_{net}}{m}$$

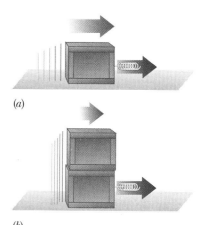

(a)

(b)

The acceleration of two masses pulled by identical springs is one-half as large as that for a single mass.

◂ acceleration = $\dfrac{\text{net force}}{\text{mass}}$

where we have written the acceleration and the net force as vectors to emphasize that they always point in the same direction.

Any such mathematical equation can be rearranged using algebra. **Newton's second law of motion** is more commonly written as

$$\mathbf{F}_{net} = m\mathbf{a}$$

◂ net force = mass × acceleration

> The net force on an object is equal to its mass times its acceleration and points in the direction of the acceleration.

◂ Newton's second law

The second law describes a specific relationship among three quantities: net force, mass, and acceleration. Although we have a prescription for determining the numerical value of an acceleration, we have not yet done this for the other two quantities. We have a choice to make. We can choose a standard spring stretched by a specific amount as our definition of 1 unit of force, or we can take a certain amount of matter and define it as 1 unit of mass, or we could even choose the two units independently.

Historically, a certain amount of matter was chosen as a mass standard. It was assigned the value of 1 **kilogram** (kg). One liter of water (a little more than a quart) has a mass of 1 kilogram. The value of the unit force is then defined in terms of the observed acceleration of this standard mass. The force needed to accelerate a 1-kilogram mass at 1 (meter per second) per second is called 1 **newton** (N), in honor of Isaac Newton. The gravitational force on a very small apple is about 1 newton.

In the United States, a commonly used unit of force is the *pound* (lb). The unit of mass, a *slug*, is used so seldom that you may never have heard of it. One pound is the force required to accelerate a mass of 1 slug at 1 (foot per second) per second.

Once again we have found a pattern in nature. We can use Newton's second law to predict the motion of objects before we actually do the experiment.

WORKING IT OUT | The Second Law ✓ MATH

What is the net force needed to accelerate a 5-kg object at 3 m/s²? Applying the second law, we have

$$F_{net} = ma = (5 \text{ kg})(3 \text{ m/s}^2) = 15 \text{ kg} \cdot \text{m/s}^2 = 15 \text{ N}$$

In using any rule of nature, we must use a consistent set of units. The units are an integral part of the rules of nature. In the preceding case, when accelerations are measured in (meters per second) per second, the masses must be in kilograms and the forces in newtons. The combination $\text{kg} \cdot \text{m/s}^2$ is equal to a newton.

Question Suppose that in this situation you discovered that there is a 5-N force of friction opposing the motion. How large is the applied force acting on the object?

Answer The net force is the vector sum of the applied force and the frictional force. To obtain a net force of 15 N, the applied force must be 20 N. That is, 20 N in the forward direction plus 5 N in the backward direction gives a sum of 15 N in the forward direction.

We can use the second law to ask other questions. For example, what acceleration would be produced by a 2-N net force acting on the 5-kg object? Rearranging the second law and putting in the values of mass and force, we get

$$a = \frac{F_{net}}{m} = \frac{2 \text{ N}}{5 \text{ kg}} = \frac{2 \text{ kg} \cdot \text{m/s}^2}{5 \text{ kg}} = 0.4 \text{ m/s}^2$$

Question A crate falls from a helicopter and lands on a very deep snowdrift. The snow slows the crate and eventually brings it to a stop. During the time that the crate is moving downward through the snow, is the magnitude of the upward force exerted on the crate by the snow greater than, equal to, or less than the magnitude of the gravitational force acting downward on the crate?

Answer Because the crate is moving downward, its velocity is pointing down. Because the crate is losing speed, its acceleration must be pointing in the opposite direction—that is, up. The net force always points in the same direction as the acceleration. Therefore, the force acting upward on the crate must be larger than the force acting downward. Thus, the snow exerts the greater force.

Mass and Weight

Mass is often confused with weight. Part of the confusion lies in the fact that mass and weight are proportional to each other; doubling the value of one doubles the value of the other. In addition, the differences don't come up in our everyday experience. In the physics world view, however, the differences are profound and thus important to understanding motion.

We measure our **weight** by how much we can compress a calibrated spring, such as that in a bathroom scale. We compress the spring because Earth is attracting us; we are being pulled downward. Our weight depends on the strength of this gravitational attraction. If we were on the Moon, our weight would be less because the Moon's gravitational force on us would be less.

Our mass, however, is *not* dependent on our location in the Universe. It is a constant property that depends only on how much there is of us. If we were far, far away from any planet or other celestial body, we would be weightless but not massless. And, because we are not massless, the force required to accelerate us is still given by Newton's second law.

The idea of weightlessness fascinates science fiction writers. Some of them, however, confuse the concepts of mass and weight. Contrary to some fictional accounts of weightlessness, the laws of motion still hold in these situations. For example, suppose that while far out in space, where all gravitational forces are negligible, you float across your spacecraft and collide with a wall. You won't just bounce off, feeling no pain. The wall provides a force to slow and reverse your motion. Newton's second law tells us that this force depends only on your mass and the acceleration you experience, both of which are the same as here on Earth. If the force can break bones on Earth, it can do the same in the spacecraft. Being weightless does not mean that you are massless. Similarly, imagine a huge truck in outer space "hanging" from a spring scale. Although the scale would read zero, if you tried to kick the truck, you would find that it resisted moving.

The marketplace is another place where mass and weight are often confused. We talk about buying "a pound of butter." A pound is a unit of weight and is determined by how much the butter stretches a spring in the scales. It is a measure of the gravitational attraction and varies slightly from place to place. The shopper doesn't really care about the weight of the butter but is interested in purchasing a certain amount of butter; the important thing is its mass. Stores using spring scales calibrate them with standard masses to compensate for the value of the local gravitational attraction. In the rest of the world, the units on spring scales are usually mass units to reflect the fact that you are buying a certain mass of the product.

This confusion between mass and weight will not be resolved by switching over to the metric system. Most people will probably still refer to the standard masses as "weights" and the process of determining the amount of butter as "weighing." What is really meant by saying "the butter weighs 1 kilogram" is that the amount of butter has a weight that is equal to the weight of 1 kilogram. Because that is quite a mouthful, people will probably refer to the weight of the butter as being 1 kilogram.

A 1-kilogram mass near Earth's surface has a weight of 9.8 newtons, or about 2.2 pounds. Therefore, a "pound" of butter has a weight of 4.5 newtons and a mass a little less than $\frac{1}{2}$ kilogram. When the United States is fully converted to the metric system, butter will most likely be purchased by the $\frac{1}{2}$ kilogram, as is done presently in most of the world.

Although the distinctions between mass and weight are not important in the marketplace, the spacecraft example demonstrates that we have to be careful when discussing these concepts in physics.

Shoppers use supermarket scales to determine the masses of the produce.

Weight

The force causing an object to accelerate toward Earth's surface is just the gravitational force. We often call this force the weight of the object. (We will refine this definition in Chapter 9).

In the idealized situation of no air resistance described in Chapter 2, we concluded that all objects near Earth's surface fall at a constant acceleration. Let's represent the acceleration due to gravity by the symbol **g**, where we've used a vector to indicate both the size and direction. If we replace the net force $\mathbf{F}_{net}$ by the weight **W** and the acceleration **a** by the acceleration due to gravity **g** in Newton's second law, we obtain

$$\mathbf{W} = m\mathbf{g}$$

weight = mass × acceleration due to gravity ▶

This is just a mathematical way of saying that the weight of an object is proportional to its mass and directed downward.

WORKING IT OUT | Weight ✓ MATH

As a numerical example, let's calculate the weight of a child with a mass of 25 kg:

$$W = mg = (25 \text{ kg})(10 \text{ m/s}^2) = 250 \text{ N}$$

Therefore, the child has a weight of 250 N (about 55 lb).

Question What is the weight of a wrestler who has a mass of 120 kg?

Answer 1200 N.

This process can be reversed to obtain the mass of a dog that has a weight of 150 N:

$$m = \frac{W}{g} = \frac{150 \text{ N}}{10 \text{ m/s}^2} = 15 \text{ kg}$$

Free-Body Diagrams ✓ MATH

Imagine that you are pulling your little sister on a sled and that the sled is speeding up. There are many forces acting on the sled. The rope is exerting a tension on the sled, pulling it forward. Earth is pulling down on the sled with a gravitational force. The snow is pushing up on the sled with a force commonly called a *normal force*. (*Normal* means perpendicular, and this force acts perpendicular to the surface between the sled and the snow.) Your sister is pushing down on the sled with a normal force, and the snow is resisting your efforts with a frictional force that acts parallel to the surface of the snow.

Which of these forces do we use in Newton's second law to find the acceleration of the sled? Would it be the force of the rope? Would it be the largest of the forces? No. It is the net force. The net force is the vector sum of all the forces acting *on the sled*. It is important, therefore, to correctly identify *all* the forces acting on an object when analyzing its motion.

We identify the forces by drawing a *free-body diagram*. As the name suggests, we isolate, or free, the object in question (in this case the sled) from everything else. We represent that object by a dot. We then draw all the forces acting on the object with each tail starting on the dot. We label each vector to indicate what type of force it represents—**W** for a gravitational force, **N** for a normal force, **f** for a frictional force, and **T** for a tension force (a pull exerted by a string or a rope). Because every force is an interaction between two objects (things you can touch, taste, and smell), it is also useful to include two subscripts for each force label, one to indicate which object is exerting the force and the second to indicate which object is being acted on. For example, the tension force exerted by the rope on the sled would be labeled $\mathbf{T}_{rope, sled}$. If you are not able to identify the object that is exerting a force, you should consider the possibility that the force does not exist.

Take a moment and draw a free-body diagram for the sled described above. When you are finished, compare your diagram to Figure 3-10. All the second subscripts should be "sled" because only forces acting on the sled appear on this free-body diagram.

Remember that your diagram should be consistent with the acceleration of the sled. We know that the sled is speeding up. This means that the sled's acceleration must point in the same direction as its velocity. If the sled is moving to the right and speeding up, the acceleration must also point to the right. If the acceleration is to the right, the sum of the forces acting to the right must be larger than the sum of the forces acting to the left. Because the sled is not accelerating upward or downward, the sum of the forces acting upward must balance the sum of the forces acting downward.

Drawing a free-body diagram should always be the first step in solving a problem involving Newton's second law. The time required to draw the diagram is seldom wasted, because most real problems are too complicated to correctly answer without first drawing a diagram. We are reminded of the story of the woodcutter sawing down a tree. A passing hiker asks, "How long have you been sawing down that tree?" and receives the reply, "Nearly three hours!" The hiker then asks, "Why is it taking so long?" and is told, "My saw is very dull." "Why don't you sharpen your saw?" asks the hiker. "I am too busy sawing down this tree."

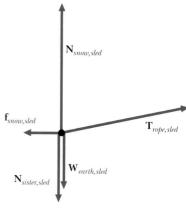

Figure 3-10 The free-body diagram for the sled.

Flawed Reasoning

You are analyzing a problem in which two forces act horizontally on an object. A 20-newton force pulls to the right and a 5-newton force pulls to the left. Your classmate asserts that the net force is 20 newtons because that is the dominant force that is acting. **What is wrong with this assertion?**

Answer The net force is the vector sum of all forces acting on the object. In this case there are 15 more newtons pulling to the right than to the left. The net force would therefore be 15 newtons to the right.

Free Fall Revisited

Objects falling on Earth don't fall in a vacuum but through air, which offers a resistive force to the motion. Thus, in realistic situations, a falling object has two forces acting on it simultaneously: the weight acting downward and the air resistance acting upward. Among other factors, the force due to air resistance depends on the speed of the object. The greater the speed, the greater the air resistance. You can experience this by sticking your hand out a car window as the car's speed increases.

With these facts in mind, consider the downward motion of a falling rock. Initially, it falls at a low speed, and the air resistance is small. There is a net force downward that is equal to the weight of the rock minus the force of the air resistance (Figure 3-11[a]). Because there is a net force, the rock accelerates, thus increasing its speed. As the rock speeds up, however, its weight remains constant

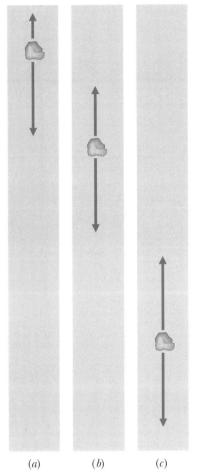

Figure 3-11 As the speed of a falling rock increases (a and b), the force of the air resistance increases until it equals (c) the weight of the rock.

while the air resistance increases (Figure 3-11[b]). Thus, the net force and the acceleration decrease. The rock continues to speed up but at a decreasing rate. Eventually, the rock reaches a speed for which the air resistance equals the weight (Figure 3-11[c]). There is no longer a net force acting on the rock, and it stops accelerating—its speed remains constant. This maximum speed is called the **terminal speed** of the object.

The terminal speeds of different objects are not necessarily equal. Even if the shape and size of two objects are identical—thus having identical frictional forces—the objects have different terminal speeds if they have different masses. In all cases, though, the object continues to accelerate until the frictional force is the same size as its weight. The value of the terminal speed is determined by a combination of many factors: the size, shape, and weight of the object, as well as the properties of the medium. A BB has a much larger terminal speed than a feather, primarily because the feather's shape creates a resistive force that quickly becomes comparable to its weight.

Let's look again at the skydivers discussed in the last chapter. Assuming no air resistance, we calculated that the skydivers would be falling at a speed of 1080 kilometers per hour (675 mph) after only 30 seconds. As skydivers know, the maximum speed they can obtain near sea level is a little over 300 kilometers per hour (190 mph). Obviously, air resistance is responsible for this difference. Skydivers also know that their terminal speed can be altered by changing their shape—falling feet first or spread-eagled—because the larger the surface of the object facing into the wind, the greater the air resistance.

Galileo versus Aristotle

Recall that in Chapter 2 we discussed the views of Aristotle and Galileo on the subject of falling objects, and came down firmly on the side of Galileo. Now it seems that we are agreeing with Aristotle! If a falling object reaches a terminal speed whose value is determined by the object's weight and its interaction with the medium, wasn't Aristotle correct? It might seem so.

The motion of a bowling ball dropped from a great height shows that each man correctly described a part of the ball's motion. Initially, before the air resistance becomes significant, the ball exhibits constant acceleration, as hypothesized by Galileo. As the air resistance grows, the acceleration is no longer constant but decreases to zero. From that point on, the object travels at a constant speed, as described by Aristotle.

Each person described different extremes of the motion: Galileo, the extreme of negligible air resistance; Aristotle, the extreme of maximum air resistance. One might, naively, suggest that we determine how much of the motion is accelerated and how much is at a constant speed. We could then award a physics prize to the person whose explanation holds for the longest time. Aristotle would win; we only need to drop the object from higher and higher positions, making the constant-speed portion of the fall as large as desired.

But building a physics world view doesn't always progress by choosing on such a basis. Galileo has fared better in the eyes of science historians because his idealization stripped away the nonessentials of falling motion and thus uncovered the more fundamental behavior of motion. Galileo therefore paved the way for Newton's work, which explains the entire motion of a free-falling object (including Aristotle's observations). As long as all the forces acting on an object are known, the resulting acceleration can be calculated.

Friction

Newton's insight can be turned around; rather than predicting the motion from the forces, we can use an object's motion to tell us something about the forces acting on it. Imagine pushing horizontally on a large wooden crate (Figure 3-12[a]).

At first you don't push hard enough to move the crate. If it doesn't move, there is no acceleration and, according to Newton's second law, there can be no net force on the crate. This means that there must be at least one other force canceling out the push. This other force is the force of friction exerted on the crate by the floor. As long as the crate does not move, the frictional force must be equal in size and opposite in direction to your applied force. This force is called **static friction**, to distinguish it from the frictional force that occurs when the crate moves.

This static frictional force seems a bit mysterious. Because it is equal to the force you exert, the frictional force is small if you push with a small force. But if you push with a large force, the frictional force is large (Figure 3-12[b]). It is a force that opposes the applied force and ceases to exist when the applied force is removed. The static frictional force can have any value from zero up to a maximum value determined by the surfaces and the weight of the crate. Notice that the behavior of the static frictional force is very similar to the force exerted by a spring.

If your applied force exceeds the maximum static frictional force, the crate accelerates in the direction of your applied force. Although the crate is now sliding, there is still a frictional force (Figure 3-12[c]). The value of this **kinetic friction** is less than the maximum value of the static frictional force. Unlike air resistance, *kinetic friction has a constant value, independent of the speed of the object.*

It is important to understand the difference between static and kinetic friction when making an emergency stop in an automobile. Because you want to stop the car as quickly as possible, you want to have the maximum frictional force with the road. This occurs when the tires are rolling because the surface of the tire is not sliding along the surface of the road, and it is the larger static friction that is important. Therefore, you should not brake so hard that the tires skid. The same thing occurs when a car takes a corner too fast. Once the tires start to skid, the frictional force is reduced, making it difficult to recover from the skid. If you've ever been in one of these unfortunate situations, you may recall how fast the car slides once it starts to skid.

PHYSICS | ON YOUR OWN

With a simple wooden block and a long rubber band, you can verify the behavior of static and kinetic friction. Connect the rubber band to the block with a thumbtack, and slowly pull on the block. The stretch of the band provides a visual indication of the force you are applying. If the block does not move, the static force is equal but opposite in direction to the force of the rubber band. Continue to increase your pull. What happens?

Repeat the experiment, with the block sliding across the table at a constant speed. How does the stretch of the rubber band now compare with its maximum stretch in the static situation?

A significant advance in the automotive industry is based on the fact that static friction is greater than kinetic friction. Antilock brakes are a computer-controlled braking system that keeps the wheels from skidding, thus maximizing the frictional forces. Sensors monitor how fast the wheels are rotating and continuously feed the data to an onboard computer. The computer controls the braking by repeatedly applying and releasing pressure to the brake pads. Without antilock braking, a driver who jams on the brakes, hoping to avoid danger, causes the wheels to lock, which often results in a loss of control and an increase in the stopping distance.

Newton's Third Law   ✓ MATH

There is still one more Newtonian law to consider. Imagine that you are playing tennis and have just hit a ball. The racket exerts a force on the ball that causes it to accelerate. The high-speed photograph in Figure 3-13 shows that the strings

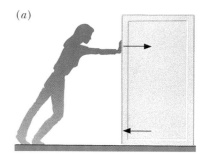
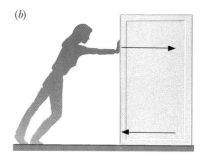
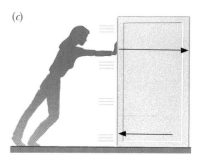

Figure 3-12 The static frictional force is equal and opposite to the applied force if the crate does not accelerate. The applied force can be small (a) or large (b) as long as it doesn't cause the crate to move. (c) The kinetic frictional force has a constant value independent of the speed.

Figure 3-13 A high-speed photograph illustrating Newton's third law. The ball exerts a force on the strings, and the strings exert an equal and opposite force on the ball.

Terminal Speeds

Downhill skiers gain speed by reducing their air resistance and the resistance of the skis with the snow.

How fast something can move depends on the forces that retard its motion as well as those that propel it forward. The primary retarding forces are frictional forces, often due to the medium through which the object moves. A 100-meter race in 3 feet of water would produce times far slower than the current record of 9.77 seconds in air. But, even in air, there is resistance to motion. A clean, waxed car has a measurable increase in gas mileage over a dirty car.

When the retarding forces equal the propelling forces, there is no net force on the object, and the object stops accelerating; it reaches a constant speed known as the terminal speed. Minimizing the retarding forces increases the terminal speed.

Streamlining an object minimizes its air resistance. In 1980 Steve McKinney set the world unpowered land-speed record by paying a great deal of attention to minimizing air resistance and friction. He skied down a 40-degree slope at slightly more than 200 kilometers per hour (125 mph!). The current downhill skiing record is held by Phillippe Goitschel at 251 kilometers per hour (156 mph). The woman's record is held by Karine Dubouchet at 242 kilometers per hour (151 mph). The peregrine falcon, already streamlined, dives for prey at speeds up to 350 kilometers per hour.

The effect of minimizing air resistance was convincingly demonstrated by U.S. Air Force Captain Joseph Kittinger when he jumped from a balloon at 31,330 meters and attained a speed of more than 1006 kilometers per hour (625 mph) after falling approximately 4000 meters. At this speed he nearly broke the sound barrier! He was then slowed down as the atmosphere became denser.

---

of the tennis racket are pushed back at the same time the ball is flattened. The ball is squashed by the force of the racket *on the ball*; at the same time, the racket strings are stretched by the force of the ball *on the racket*. At the same time the racket is exerting a force on the ball, the ball is exerting an opposite force on the racket.

If you wish to pursue this point further, find a friend who will help with a simple experiment. Give your friend a shove. At the same time you are pushing, you will feel a force being exerted on you, regardless of whether your friend pushes back. If you can, try this wearing ice skates or in-line skates; it will be even more dramatic.

Let's carry this one step further. Lean on a wall. Notice the force of the wall pushing back on you. If this force did not exist, you would fall.

It could have been these kinds of experiences that led Newton to his third law of motion. He realized that there is no way to push something without being pushed yourself. For every force there is always an equal and opposite force. The two forces act on different objects, are the same size, and act in opposite directions. Formally, we state **Newton's third law of motion** as follows:

> Newton's third law ▶ If an object exerts a force on a second object, the second object exerts an equal force back on the first object.

Because Newton referred to these forces as action and reaction, they are often known as an action–reaction pair. However, because the two forces are equivalent, it doesn't matter which one is called the action and which the reaction. Another statement of the third law might be that for every action there is an equal and opposite reaction.

Forces always occur in pairs. In Newton's words, "If you press a stone with your finger, the finger is also pressed by the stone." These forces *never act on the same body*. When you press the stone with your finger, you exert a force *on the stone*. The reaction force is *acting on you*.

Consider a ball with a weight of 10 newtons falling freely toward Earth's surface. Ignoring air resistance, there is only one force acting on the ball; Earth's gravity is pulling it downward with a force of 10 newtons. What is the second force in the action–reaction pair? The first force is the force of Earth acting on the ball and can be labeled $\mathbf{W}_{eb}$ (where the subscripts *eb* remind us that this is the force of *Earth on ball*). The second force involves the objects in the reverse order and is written $\mathbf{W}_{be}$ for *ball on Earth* (Figure 3-14). Therefore, Newton's third law tells us that the ball must be exerting an upward force on Earth of 10 newtons. Although your common sense may tell you that Earth must exert a larger force because it is so much larger, this is not true. No matter what the origin of the forces, Newton's third law tells us that the forces must be equal in size and opposite in direction.

But if this is true, why doesn't Earth accelerate toward the ball? It does, but if we put the values into Newton's second law, we find that Earth's mass is so large that its acceleration is minuscule. Earth does accelerate, but we don't notice it.

A very important point concerning third-law forces is that one of the forces acts on the ball while the other acts on Earth; one causes the ball to accelerate, and the other causes Earth to accelerate. Because the two forces act on different objects, the two forces cannot cancel; if these are the only two forces acting, both objects accelerate. Third-law forces *never* appear on the same free-body diagram.

Without the third law, paradoxical events would occur in the Newtonian world view. Consider, for example, why a person doesn't fall through the floor. There is a gravitational force pulling the person down. If this were the only force acting on the person, according to Newton's second law, the person would accelerate downward through the floor. Because the person is not accelerating, the net force on the person must be zero. Thus, the question arises, "What is the force that balances the downward gravitational force?"

Newton's third law provides the answer. Earth attracts the person with a force that we can label $\mathbf{W}_{ep}$, as shown in Figure 3-15. The person pushes down on the floor with a force that we label $\mathbf{N}_{pf}$. According to the third law, the floor pushes upward on the person with a force $\mathbf{N}_{fp}$ that is equal in size and opposite in direction to $\mathbf{N}_{pf}$. Therefore, there are two forces acting on the person: the person's weight $\mathbf{W}_{ep}$ and the upward force of the floor $\mathbf{N}_{fp}$. Although these two forces are equal in size and act in opposite directions, they are not third-law forces; they both act on the person. They are equal and opposite because the person is not accelerating; therefore, by Newton's second law, the net force on the person must be zero. Note that the third-law force associated with the person's weight $\mathbf{W}_{ep}$ is the gravitational force of the person acting on Earth $\mathbf{W}_{pe}$.

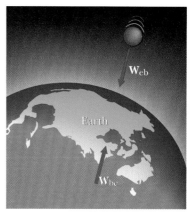

Figure 3-14 Earth exerts a force $\mathbf{W}_{eb}$ on the ball. According to Newton's third law, the ball exerts an equal and opposite force $\mathbf{W}_{be}$ on Earth.

Question   A branch exerts an upward force on an apple in a tree. What is the third-law companion to this force?

Answer   It is the downward force of the apple on the branch. Note that it is not the downward force of gravity on the apple. Although the gravitational force is equal and opposite to the upward force on the apple, both forces act on the apple and they cannot be action–reaction forces.

When you fire a rifle, it recoils. Why? As explained by the third law, when the rifle exerts a forward force on the bullet (by virtue of an explosion), the bullet simultaneously exerts an equal force on the rifle but in the backward direction. But why doesn't the rifle accelerate as much as the bullet? The force of the rifle on the bullet produces a large acceleration because the mass of the bullet is

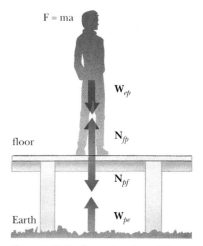

Figure 3-15 Earth exerts a force $\mathbf{W}_{ep}$ on the person, which causes the person to exert a force $\mathbf{N}_{pf}$ on the floor. By Newton's third law, the floor exerts an equal and opposite force $\mathbf{N}_{fp}$ on the person. Although $\mathbf{W}_{ep}$ and $\mathbf{N}_{fp}$ are equal and opposite, they are not an action–reaction pair; they both act on the person.

Figure 3-16 As the man walks to the left, he exerts a force on the boat that causes the boat to move to the right.

small. The force of the bullet on the rifle is the same size but produces a small acceleration because the mass of the rifle is large.

Even the common act of walking is possible only because of third-law forces. In walking you must have a force exerted on you in the direction of your acceleration. And yet the force you produce is clearly in the opposite direction. The solution to this apparent paradox lies in the third law. As you start to walk, you exert a force against the floor (down and backward); the floor therefore exerts a force back, causing you to go forward (and up a little). If there is any sand or loose earth where you walk, you can see that it has been pushed *back*. If you want a clearer demonstration of the fact that you push backward against the floor, try walking on a skateboard or in a rowboat (Figure 3-16). But be careful!

PHYSICS | **ON YOUR OWN**

Determine your weight on a bathroom scale placed in an elevator when the elevator is stopped, as it accelerates upward, as it travels between floors at a constant speed, and as it stops. What do you expect to happen? Explain your reasoning in each case.

Flawed Reasoning

Let's reconsider the crate that fell from the helicopter into a deep snowdrift. Three students are discussing which force is bigger, the force exerted by the snow upward on the bottom of the crate or the force exerted downward by the bottom of the crate on the snow.

Jennifer: "The crate must be pushing down on the snow more than the snow is pushing up on the crate because the crate is still moving down through the snow."

Monica: "The snow must be pushing up on the crate harder than the crate is pushing down on the snow because the crate is slowing down."

Peter: "The two forces, crate on snow and snow on crate, are part of the same interaction. They must always be equal in magnitude and opposite in direction by Newton's third law."

With which student do you agree?

Answer   Third-law forces always involve the same players. If A pushes on B, then B pushes back on A. When we refer to the two forces as crate on snow and snow on crate, it becomes obvious that these are third-law forces. They must always be equal in magnitude and opposite in direction.

Summary

According to Newton's first law of motion, an object at rest remains at rest, and an object in motion remains in motion with a constant velocity unless a net outside force acts on the object. The net force is determined by adding all the forces acting on an object according to the rules for combining vector quantities. The converse of Newton's first law is also true: if an object has a constant velocity (including the case of zero velocity), the unbalanced, or net, force acting on the object must be zero.

If there is a net force, the object accelerates with a value given by Newton's second law of motion, $\mathbf{a} = \mathbf{F}_{net}/m$. The direction of the acceleration is always the same as the net force. In the metric system, the unit of mass is the kilogram and the unit of force is the newton.

Newton's second law of motion can be used to study frictional forces. Static friction can range from zero to a maximum value that depends on the force pushing the surfaces together and the nature of these surfaces. Kinetic friction has a constant value less than the maximum static value. A special kind of friction is air resistance, which varies with the speed of the object. As the air resistance acting on a falling object becomes equal to the object's weight, the acceleration goes to zero, and the object falls at its terminal speed.

The weight of an object close to Earth's surface is given by $\mathbf{W} = m\mathbf{g}$, where $\mathbf{g}$ is the acceleration due to gravity, about 10 (meters per second) per second downward. Weight, a force, should not be confused with mass.

There is no such thing as an isolated force. All forces occur in pairs that are equal in size and opposite in direction. As you stand on the floor, you exert a downward force *on the floor*. According to Newton's third law, the floor must exert an upward force *on you* of the same size. These two forces do not cancel because they act on different objects—one on you and one on the floor. The other force acting on you is the force of gravity.

PhysicsNow™ Assess your understanding of this chapter's topics with sample tests and other resources found by logging into PhysicsNow at http://physics.brookscole.com/kf6e.

Chapter 3 Revisited

The acceleration of an object in free fall becomes smaller and smaller because the force due to the air resistance increases as the speed of the object increases, becoming closer and closer in magnitude to the gravitational force pulling the object down. Therefore, the net force continually decreases. When the net, or total, force becomes zero, the acceleration also becomes zero and the speed assumes a constant value. This speed is known as the terminal speed.

KEY TERMS

force: A push or a pull. Measured by the acceleration it produces on a standard, isolated object, $\mathbf{F}_{net} = m\mathbf{a}$. Measured in newtons.

inertia: An object's resistance to a change in its velocity.

inversely proportional: A relationship in which two quantities have a constant product. If one quantity increases by a certain factor, the other decreases by the same factor.

kilogram: The standard international system (SI) unit of mass. A kilogram of material weighs about 2.2 pounds on Earth.

kinetic friction: The frictional force between two surfaces in relative motion. This force does not depend very much on the relative speed.

law of inertia: Newton's first law of motion.

mass: A measure of the quantity of matter in an object. The mass determines an object's inertia. Measured in kilograms.

motion, Newton's first law of: The velocity of an object remains constant unless an unbalanced force acts on the object.

motion, Newton's second law of: $\mathbf{F}_{net} = m\mathbf{a}$. The net force on an object is equal to its mass times its acceleration. The net force and the acceleration are vectors that always point in the same direction.

motion, Newton's third law of: If an object exerts a force on a second object, the second object exerts an equal force on the first object.

newton: The SI unit of force. A net force of 1 newton accelerates a mass of 1 kilogram at a rate of 1 (meter per second) per second.

proportional: A relationship in which two quantities have a constant ratio. If one quantity increases by a certain factor, the other increases by the same factor.

static friction: The frictional force between two surfaces at rest relative to each other. This force is equal and opposite to the net applied force if the force is not large enough to make the object accelerate.

terminal speed: The speed obtained in free fall when the upward force of air resistance is equal to the downward force of gravity.

weight: $\mathbf{W} = m\mathbf{g}$. The force of gravitational attraction of Earth for an object. This definition is modified in Chapter 9 for accelerating systems such as elevators and spacecraft.

CONCEPTUAL QUESTIONS

1. Assume you drop a bag of snacks while riding in an airplane that is flying due west at 800 kilometers per hour. Will the bag fall straight down, or angle toward the front or back of the airplane? Explain your reasoning.
2. The room you are sitting in is currently moving at about 400 meters per second as a result of Earth spinning about its axis. The walls of the room are attached to Earth but, if your keys fall out of your pocket, they are not. Why do the keys *not* appear to fly back toward the west wall?
3. Assume that you are pushing a car across a level parking lot. When you stop pushing, the car comes to a stop. Does this violate Newton's first law? Why?
4. If you give this book a shove so that it moves across a tabletop, it slows and comes to a stop. How can you reconcile this observation with Newton's first law?
5. How does the net force on the first subway car compare with that on the last car if the subway train has a constant velocity?
6. What can you say about the forces acting on a motorcycle that is traveling at a constant speed down a straight stretch of highway?
7. Why does a tassel hanging from the rearview mirror appear to swing forward as you apply the brakes?
8. When dogs finish swimming, they often shake themselves to dry off. What is the physics behind this?
9. Assume that you're not wearing your seat belt and the car stops suddenly. Why would your head hit the windshield?
10. Modern automobiles are required to have headrests to protect your neck during collisions. For what type of collision are these headrests most effective?
11. Why does a blacksmith use an anvil when hammering a horseshoe?
12. You find that every time you pound a steak on your kitchen counter, the bottles fall out of the spice rack hanging on your wall. To solve the problem, you buy a large oak cutting board, which you place on the counter under the steak. Why does this help?
13. In everyday use, *inertia* means that something is hard to get moving. Is this the only meaning it has in physics? If not, what other meaning does it have?
14. How would you determine if two objects have the same inertia?
15. When a number of different forces act on an object, is the net force necessarily in the same direction as one of the individual forces? Why?
16. You are analyzing a problem in which two forces act on an object. A 200-newton force pulls to the right and a 40-newton force pulls to the left. Your classmate asserts that the net force is 200 newtons because that is the dominant force that is acting. What is wrong with this assertion?
17. Forces of 40 newtons and 90 newtons act on an object. What are the minimum and maximum values for the sum of these two forces?
18. Two ropes are being used to pull a car out of a ditch. Each rope exerts a force of 700 newtons on the car. Is it possible for the sum of these two forces to have a magnitude of 1000 newtons? Explain your reasoning.
19. You apply a 75-newton force to pull a child's wagon across the floor at constant speed. If you increase your pull to 80 newtons, will the wagon speed up to some new constant speed or will it continue to speed up indefinitely? Explain your reasoning.
20. You push a crate full of books across the floor at a constant speed of 0.5 meter per second. You then remove some of the books and push exactly the same as you did before. How does the crate's motion differ, if at all?
21. If the net force on a boat is directed due east, what is the direction of the acceleration of the boat? Would your answer change if the boat had a velocity due north but the net force still acted to the east?
22. If the net force on a hot-air balloon is directed vertically upward, what is the direction of the acceleration of the balloon? What would be the direction of the acceleration if the balloon were being blown westward (with the net force still acting vertically upward)?
23. You are riding an elevator from your 10th-floor apartment to the parking garage in the basement. As you approach the garage, the elevator begins to slow. What is the direction of the net force on you?
24. You are riding an elevator from the parking garage in the basement to the 10th floor of an apartment building. As

you approach your floor, the elevator begins to slow. What is the direction of the net force on you?

25. If you double the net horizontal force applied to a wagon, what happens to the wagon's acceleration?
26. What happens to the acceleration of a rocket if the net force on it is cut in half?
27. A car can accelerate at 2 (meters per second) per second when towing an identical car. What will its acceleration be if the towrope breaks?
28. How does the net force on the first subway car compare with that on the last subway car if a subway train has a constant acceleration? Assume that the subway cars are identical.
29. When an astronaut walks on the Moon, is either her mass or her weight the same as on Earth? Explain.
30. If you buy a bag of pretzels labeled 0.1 kilogram, are you buying the pretzels by mass or by weight?
31. How does the weight of a can of pop compare with the weight of a six-pack of the same pop?
32. What happens to the weight of an object if you take it from Earth to the Moon, where the acceleration due to gravity is one-sixth as large?
33. A skier is slowing down as she skis over level ground. Draw a free-body diagram for the skier.
34. A car on a level section of highway is speeding up to pass a truck. Draw a free-body diagram for the car.
35. Under what conditions will a golf ball and a table-tennis ball that are dropped simultaneously from the same height reach the ground at the same time?
36. If a golf ball and a table-tennis ball are simultaneously dropped from the same height, they do not reach the ground at the same time. How would Aristotle explain this? How would Galileo?
37. A marble dropped into a bottle of liquid soap quickly reaches a terminal speed. Draw a free-body diagram for the marble just before it hits the bottom of the bottle. What is the acceleration of the marble at this time?
38. Draw a free-body diagram for a parachutist who has reached terminal speed. What is his acceleration?
39. Sara is taking the high-speed elevator, which travels at a constant speed of 5 meters per second, to the 43rd floor of a high-rise building. Sam is stuck making the same trip in the freight elevator, which travels at a constant speed of only 1.5 meters per second. Compare the net forces on Sara and Sam.
40. Pat and Chris are pushing identical crates across a rough floor. Pat's crate is moving at a constant 1 meter per second while Chris's crate is moving at a constant 2 meters per second. Compare the net forces on the two crates.
41. A friend falsely claims, "Newton's first law doesn't work if there is any friction." How would you correct this claim?
42. One of your classmates falsely asserts, "Newton's second law only works when there are no frictional forces." How would you correct this assertion?
43. You are applying a 400-newton force to a freezer full of chocolate chip ice cream in an attempt to move it across the basement. It will not budge. Is the frictional force exerted by the floor on the freezer greater than, equal to, or less than 400 newtons?
44. You find that you must push with a force of 10 newtons to keep a jar of cold cream sliding at constant speed across your bathroom counter. With the jar at rest, you apply a force of 11 newtons. Is it possible that the jar will stay at rest? Explain.
45. What force is required to pull a dog in a wagon along a level sidewalk, as in the figure, at a constant speed if the frictional force is 250 newtons?

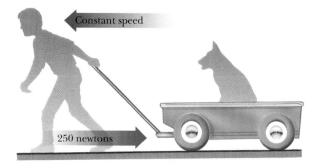

46. A skateboarder is boarding down a steep ramp at constant terminal speed. What are the size and direction of the net force on the boarder?
47. You are driving along the freeway at 75 mph when a bug splats on your windshield. Compare the force of the bug on the windshield with the force of the windshield on the bug.
48. You leap from a bridge with a bungee cord tied around your ankles. As you approach the river below, the bungee cord begins to stretch and you begin to slow down. Which is greater (if either), the force of the cord on your ankles to slow you or the force of your ankles on the cord to stretch it? Explain.
49. Is the force that the Sun exerts on Earth bigger, smaller, or the same size as the force that Earth exerts on the Sun? Explain your reasoning.
50. Assume that you are riding on a merry-go-round. How does the force the merry-go-round exerts on you compare with the force you exert on the merry-go-round? Explain your reasoning.
51. What is the net force on an apple that weighs 4 newtons when you hold it at rest?
52. Suppose you are holding an apple that weighs 4 newtons. What is the net force on the apple just after you drop it?
53. Why do the cannons aboard pirates' ships roll backward when they are fired?
54. Why does a tennis racket slow down when it hits a ball?

55. Describe the force or forces that allow you to walk across a room.
▲ 56. We often say that the engine supplies the forces that propel a car. This is an oversimplification. What are the forces that actually move the car?
57. A ball with a weight of 40 newtons is falling freely toward the surface of the Moon. What force does this ball exert on the Moon?
58. The figure shows a ball hanging by a string from the ceiling. Identify the action–reaction pairs in this drawing. Argue that the magnitudes of all the forces are the same.

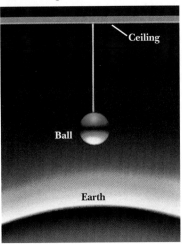

▲ 59. If the force exerted by a horse on a cart is equal and opposite to the force exerted by the cart on the horse, as required by Newton's third law, how does the horse manage to move the cart?
60. Gary reads about Newton's third law while sitting in a room with a single closed door. He reasons that if he applies a force to the door there will be an equal and opposite force that will cancel his pull and he will never be able to escape. He bemoans, "Why did I ever take physics?" What is wrong with Gary's reasoning?
61. A soft-drink can sits at rest on a table. Which of Newton's laws explains why the upward force of the table acting on the can is equal and opposite to Earth's gravitational force pulling down on the can?
62. A mouse sits at rest on a mouse pad. Which force does Newton's third law tell us is equal and opposite to the gravitational force acting on the mouse?

EXERCISES

1. Find the size of the net force produced by a 6-N and an 8-N force in each of the following arrangements:
    a. The forces act in the same direction.
    b. The forces act in opposite directions.
    c. The forces act at right angles to each other.
2. Find the size of the net force produced by a 5-N and a 12-N force in each of the following arrangements:
    a. The forces act in the same direction.
    b. The forces act in opposite directions.
    c. The forces act at right angles to each other.
3. Two horizontal forces act on a wagon, 550 N forward and 300 N backward. What force is needed to produce a net force of zero?
4. Three forces act on an object. A 4-N force acts due east and a 3-N force acts due north. If the net force on the object is zero, what is the magnitude of the third force?
5. What is the acceleration of a 600-kg buffalo if the net force on the buffalo is 1800 N?
6. What is the acceleration of a 2000-kg car if the net force on the car is 4000 N?
7. A 30-06 bullet has a mass of 0.010 kg. If the average force on the bullet is 9000 N, what is the bullet's average acceleration?
8. The net horizontal force on a 60,000-kg railroad boxcar is 6000 N. What is the acceleration of the boxcar?
9. What net force is needed to accelerate a 60-kg ice skater at 3 m/s²?
10. If a sled with a mass of 20 kg is to accelerate at 5 m/s², what net force is needed?
11. If a 6-kg sledge hammer has a weight of 10 N on the Moon, what is its acceleration when it is dropped?
12. A salesperson claims a 1200-kg car has an average acceleration of 3 m/s² from a standing start to 100 km/h. What average net force is required to do this?
13. If a skydiver has a net force of 300 N and an acceleration of 4 m/s², what is the mass of the skydiver?
14. A child on roller skates undergoes an acceleration of 0.4 m/s² due to a horizontal net force of 24 N. What is the mass of the child?
15. A 0.5-kg ball has been thrown vertically upward. If we ignore the air resistance, what are the direction and size of each force acting on the ball while it is traveling upward?
16. A 1-kg ball is thrown straight up in the air. What is the net force acting on the ball when it reaches its maximum height? What is the ball's acceleration at this point?
17. Skip Parsec, intrepid space explorer, travels to a new planet and finds that he weighs only 320 N. If his mass is 80 kg, what is the acceleration due to gravity on this planet?
18. A fully equipped astronaut weighs 1500 N on the surface of Earth. If the astronaut has a weight of 555 N standing on the surface of Mars, what is the acceleration due to gravity on Mars?
▲ 19. A crate has a mass of 24 kg. What applied force is required to produce an acceleration of 3 m/s² if the frictional force is known to be 90 N?
▲ 20. A rope is used to pull a 10-kg block across the floor with

an acceleration of 3 m/s². If the frictional force acting on the block is 50 N, what is the tension in the rope?

▲ 21. If a pull of 210 N accelerates a 40-kg child on ice skates at a rate of 5 m/s², what is the frictional force acting on the skates?

▲ 22. If you stand on a spring scale in your bathroom at home, it reads 600 N, which means your mass is 60 kg. If instead you stand on the scale while accelerating at 2 m/s² upward in an elevator, how many newtons would it read?

23. Terry and Chris pull hand over hand on opposite ends of a rope while standing on a frictionless frozen pond. Terry's mass is 75 kg and Chris's mass is 50 kg. If Terry's acceleration is 2 m/s², what is Chris's acceleration?

▲ 24. A mother of mass 50 kg and her daughter of mass 25 kg are ice-skating. They face each other, and the mother pushes on the daughter such that the daughter's acceleration is 2 m/s². What is the force exerted by the mother on the daughter? What is the force exerted by the daughter on the mother? What is the mother's acceleration?

4 | Motions in Space

Newton's first law states that an object naturally travels in a straight line. Yet in nearly every motion we observe, the objects execute much more complicated motions. Skiers fly through the air, and cars maneuver tight corners. What causes these motions? And what happens to you when you're part of the motion—say, as a passenger in a car rounding a corner?

See page 69 for the answer to this question.

The motion of the ski jumper is a combination of a horizontal motion with constant speed and a vertical motion with constant acceleration.

So far, we have restricted our discussion to straight-line, or one-dimensional, motion. Most motions, however, take place in more than one dimension—most commonly in three-dimensional space.

Going from one dimension to two or three dimensions is less difficult than you might anticipate because all motions can be divided into separate motions in each of the three dimensions. A knuckleball's motion can be thought of as the sum of three separate motions: up/down, left/right, and near/far (along the line between the pitcher and catcher). This separation means that we can apply the laws that were developed for one dimension to many common motions in space.

Our first consideration will be to look at two-dimensional motion—that is, motion confined to a flat surface. Adding this second dimension allows us to study a variety of motions, from the path of a football to Earth's annual journey around the Sun.

Circular Motion

The motion of Earth around the Sun, the Moon around Earth, a race car around a circular track, and a ball swinging on the end of a string are examples of circular (or nearly circular) motion. Each of these occurs in a two-dimensional plane.

As we look at these motions, we start with the simplest situation, motion in which the speed of the object remains constant. In the one-dimensional cases we studied in the previous chapter, constant speed implied the absence of a net force. This is not the case in two dimensions. An object moving along a circular path at a constant speed must have a net force acting on it.

To see this, imagine whirling a ball on the end of a string in a circle above your head, as in Figure 4-1. Because you have to pull on the string to do this, the string must be exerting a force on the ball. And because a string can exert a force only along its length, this force must act toward the center of the circle. It has the special name **centripetal force**, which means *center-seeking* force. What force is pulling outward on the ball? What object could be producing this force? There is no outward force. According to Newton's first law, the ball will travel in a straight line unless acted on by a net force. The inward force of the string is required to pull the ball inward along the circular path.

It is important to distinguish between the adjectives centri*petal* (center-seeking) and centri*fugal* (center-fleeing). The word *centripetal* is rarely used in our everyday language. The word *centrifugal* is much more common and often mistakenly substituted for *centripetal*. The force we are discussing, the centripetal force, is directed toward the center of the circle. We discuss centrifugal effects in Chapter 9.

If you cut the string, the ball no longer moves in a circle but flies off in a straight line (if we ignore the force of gravity). It continues to travel in the direction of its velocity at the time the string was cut due to its inertia. The ball's path (Figure 4-2) is perpendicular to the string and tangent to the circular path. Without applied forces the ball exhibits straight-line motion, as described by Newton's first law.

A car racing on an unbanked, circular track (again, at a constant speed) doesn't have a string anchoring it to the center of the track. The centripetal force, in this case, is provided by the frictional force of the track acting on the car's tires *toward the center* of the track. An oil slick on the track would greatly reduce the frictional force, resulting in a potentially disastrous situation.

Question Assuming the frictional force goes to zero when the car hits the oil slick, describe its motion.

Answer The car would slide in a straight line in the direction it was traveling at the time it hit the oil slick. Because there is no net force acting on the car, the car maintains a constant velocity.

PhysicsNow™ Test your understanding of this chapter by logging into PhysicsNow at **http://physics.brookscole.com/kf6e**, selecting the chapter, and clicking on the "Take a Pre-Test" link.

Figure 4-1 The force acting inward along the string keeps the ball in a circular path.

◄ a centripetal force is a center-seeking force

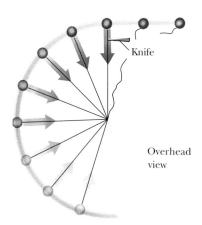

Figure 4-2 When the string is cut, the ball travels in a straight line tangent to the circular path.

Figure 4-3 As the race car turns the corner, it momentarily moves along a circular path.

Let's reexamine Newton's first law of motion. It says that an object will go in a *straight line* unless acted on by a net force. Therefore, any time an object changes direction, there must be a net force acting on it. Newton's first law—the law of inertia—applies to direction as well as speed. Conversely, if the object is traveling in a straight line, there can be no net force acting to either side.

This discussion can be generalized to an even larger class of motions. Any slight change in direction can be thought of as momentarily traveling along an arc of a circle, as shown in Figure 4-3. This requires that a force act toward the inside of the turn. For this reason it is dangerous to drive your car on icy, curved roads—you need friction (traction) to turn the corners.

As mentioned in Chapter 3, the law of inertia applies to all objects everywhere. If a spacecraft were far from all massive bodies (where gravitational forces can be ignored), it would coast in a straight line. The only time NASA would need to turn on the spacecraft's engines would be to make midcourse corrections to change its direction or speed.

✓ Extended presentation available in the *Problem Solving* supplement

Acceleration Revisited   ✓ MATH

We have established that an object moving in a circle with constant speed must have a net force acting on it. As a consequence, according to Newton's second law, the object must be accelerating. This acceleration is equal to the change in velocity divided by the time it takes to make the change. In this case the velocity changes direction without changing size. If this seems a little bizarre—that an object whose speed remains constant is accelerating—remember that a vector such as velocity can change in two ways.

Like the force vector, the velocity vector can be represented by an arrow, as shown in Figure 4-4. The length of the arrow in this case tells us the speed (for example, 1 centimeter = 5 meters per second), and the direction of the arrow is the direction of motion. The velocity vectors corresponding to circular motion at a constant speed are tangent to the circular path and have equal lengths. Although they have equal lengths, they are different vectors because their directions are different.

Figure 4-4 The velocity vector for circular motion with constant speed changes direction but not size.

The change in velocity is determined by subtracting the initial velocity vector from the final one. Suppose the initial velocity of an object is represented by the arrow $\mathbf{v}_i$ and its final velocity by the arrow $\mathbf{v}_f$ (Figure 4-5[a]). Then, we need to calculate $\mathbf{v}_f - \mathbf{v}_i$. We can apply our rule for vector addition from Chapter 3 to evaluate this expression by rewriting it as the sum of the final velocity and the negative of the initial velocity, $\mathbf{v}_f - \mathbf{v}_i = \mathbf{v}_f + (-\mathbf{v}_i)$, where the negative of a vector means that it points in the opposite direction. To subtract the velocities, we turn $\mathbf{v}_i$ around and place its tail on the head of $\mathbf{v}_f$ (making certain not to change their original orientations or lengths). The change in velocity $\Delta \mathbf{v}$ is also a vector and is represented by an arrow drawn from the tail of $\mathbf{v}_f$ to the head of $-\mathbf{v}_i$, as shown in Figure 4-5(b).

You can verify that this procedure for subtracting vectors is correct. Because $\Delta \mathbf{v}$ is the change in the velocity, adding it to the initial velocity $\mathbf{v}_i$ (using the rules of vector addition developed earlier) yields the final velocity $\mathbf{v}_f$. Figure 4-5(c) illustrates this.

The fact that the acceleration is defined as a vector ($\Delta \mathbf{v}$) divided by a number (the time elapsed) means that acceleration is also a vector. Its direction is the same as that of $\Delta \mathbf{v}$, and its magnitude is obtained by dividing the magnitude of $\Delta \mathbf{v}$ by the elapsed time:

$$\text{acceleration} = \frac{\text{change in velocity}}{\text{time taken}} \qquad\qquad \mathbf{a} = \frac{\Delta \mathbf{v}}{\Delta t}$$

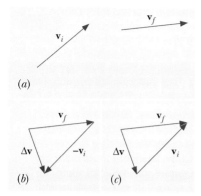
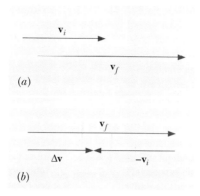

Figure 4-5 (a) Arrows $\mathbf{v}_i$ and $\mathbf{v}_f$ represent initial and final velocities with the same speeds. (b) $\Delta\mathbf{v} = \mathbf{v}_f - \mathbf{v}_i$ is the change in the velocity. (c) The initial velocity plus the change in velocity equals the final velocity.

Figure 4-6 (a) Two parallel velocity vectors. (b) Subtracting the two velocities by adding the negative of the initial velocity $\mathbf{v}_i$ to the final velocity $\mathbf{v}_f$ gives the change in velocity $\Delta\mathbf{v}$.

Unlike the velocity vector, which points in the direction of motion, the direction of the acceleration vector is not intuitive. The safest procedure is to not make an intuitive guess but to formally subtract the velocity vectors to obtain the direction of the acceleration vector.

Let's check this definition for acceleration against the familiar case of straight-line motion, where the object changes speed but not direction. If the object just speeds up, the final velocity vector will be longer than the initial vector but will still point in the same direction. This is shown in Figure 4-6(a). To determine $\Delta\mathbf{v}$, we again add the negative of the initial velocity to the final velocity and draw $\Delta\mathbf{v}$ from the tail of $\mathbf{v}_f$ to the head of $-\mathbf{v}_i$, as shown in Figure 4-6(b). The change in velocity is just the numerical difference in the lengths of the velocity vectors (that is, the difference in the initial and final speeds); it points in the same direction as $\mathbf{v}_f$. Therefore, the acceleration is in the same direction as the velocity, and the speed increases.

Question Which way would the acceleration point if the object were slowing down?

Answer In this case the final velocity vector is the shorter one. Adding the negative of the longer initial velocity to it yields a change in velocity that points backward. Whenever an object slows down, the velocity and the acceleration point in opposite directions.

The new physics student often greets this more complete definition of acceleration with disbelief and bewilderment. "How can you claim an acceleration when the speed stays constant?" is a typical reaction. The reason for the confusion is that the concept of acceleration has a more complex meaning in physics than in everyday language.

Although this difference in meaning may seem strange and perhaps even unnecessary, remember our goal is to describe motion completely. We must include the observed fact that a change in *just* the direction (without a change in speed) requires a net force and, therefore, according to Newton's second law, there must be an acceleration.

Flawed Reasoning

At the county fair Billy finds himself pressed up against the wall of the Rotor, a circular room that spins about a vertical axis. When the room is spinning fast enough, the floor drops from under the people. Two students are arguing about the free-body diagram for Billy when he is at the location shown in Figure 4-7(a).

Isabel: "The gravitational force is pulling down on Billy, and the frictional force of the wall is keeping him from falling. The wall is exerting a normal force inward and the centrifugal force is acting outward, pinning Billy to the wall" (Figure 4-7[b]).

Caitlin: "I agree with you except for the centrifugal force. Billy would travel in a straight-line path if it weren't for the inward centripetal force pushing him into a circular path" (Figure 4-7[c]).

Each student has made a critical error in reasoning. Identify each student's mistake, and draw the correct free-body diagram for Billy.

Answer   Isabel thinks that Billy's free-body diagram should be balanced to show that he is not moving (and hence not accelerating) relative to the wall. Billy is, however, accelerating with respect to the ground. He is moving along a circle at constant speed, so his acceleration must be pointing toward the center of the circle. The free-body diagram should indicate a net force toward the right. There is no centrifugal force.

Caitlin correctly recognizes that Billy is accelerating toward the center of the circle, but believes that an extra centripetal force is needed to cause this acceleration. However, centripetal force is just a name for the net force in the special case of uniform circular motion. Only real forces exerted by real objects should ever appear on free-body diagrams. Only three forces act on Billy: the gravitational force, the frictional force, and the normal force. (See Figure 4-8.) The normal force exerted by the wall on Billy is the unbalanced force and provides the centripetal force causing Billy to travel in a circle.

Acceleration in Circular Motion

In the case of circular motion at a constant speed, we know that the net force acts inward along the circle's radius. We find the **centripetal acceleration** by subtracting two velocity vectors separated by a short time interval. This gives a change in velocity that points close to the center of the circle. As we shorten the time interval, the change in velocity points closer to the center. As the time interval becomes very small, the instantaneous change in velocity points at the center. This agrees with Newton's second law because the instantaneous acceleration must always point in the same direction as the net force.

You can get a feeling for how the centripetal force depends on the parameters of the motion by twirling a ball on a string above your head and noticing how hard you have to pull on the string. If you increase the speed of the ball while keeping the radius the same, you have to pull harder. This tells you that the force increases as the speed increases. You can also increase the radius (lengthen the string) while keeping the speed along the circle the same. Increasing the radius requires less pull on the string. Therefore, the force varies inversely with the radius of the circular path.

Projectile Motion

When something is thrown or launched near Earth's surface, it experiences a constant downward gravitational force. Motion under these conditions is called **projectile motion**. It occurs whenever an object is given some initial velocity and thereafter travels in a trajectory subject only to the force of gravity. Examples of

The hammer thrower provides the centripetal force required to make the hammer go in a circle.

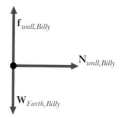

Figure 4-8 The correct free-body diagram for Billy

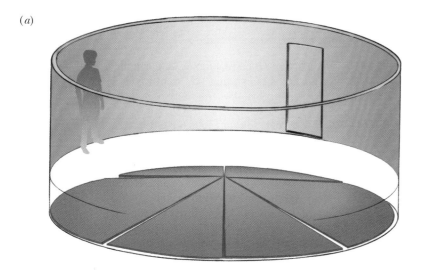

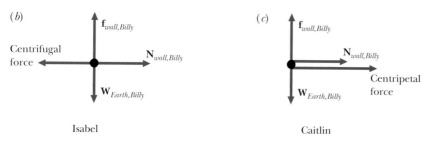

Figure 4-7 (a) Billy riding the Rotor. (b and c) Incorrect free-body diagrams for Billy.

WORKING IT OUT | Centripetal Acceleration ✓ MATH

A more detailed analysis of circular motion with a constant speed tells us that the centripetal acceleration a is equal to the object's speed v squared divided by the radius r of the circle:

$$a = \frac{v^2}{r}$$

◂ centripetal acceleration = speed squared / radius

As an example, we can calculate the centripetal acceleration of a 0.2-kg ball traveling in a circle with a radius of 1 m if it completes one revolution every second. Because this circle has a circumference of $2\pi r = 6.3$ m, the ball has a speed of 6.3 m/s:

$$a = \frac{v^2}{r} = \frac{(6.3 \text{ m/s})^2}{1 \text{ m}} = 40 \text{ m/s}^2$$

Because $F_{net} = ma$ is always valid, we can use this result to find the centripetal force required to produce this circular motion:

$$F_{net} = ma = (0.2 \text{ kg})(40 \text{ m/s}^2) = 8.0 \text{ N}$$

Question   If you double the speed of the ball traveling in a circle, what happens to the centripetal acceleration and the centripetal force?

Answer   Both the acceleration and the force quadruple.

Banking Corners

It takes a centripetal force to make a car go around a corner. On a flat parking lot or an unbanked curve, this force comes from the frictional interaction between the tires and the road. The maximum size of this frictional force varies a little bit with the type of tire and the type of surface but depends mostly on whether the surface is dry, wet, or icy. The maximum frictional force does not depend very much on the speed. However, as the speed of the car increases for a given turning radius, the centripetal force provided by friction must also increase. At some speed, the car is moving too fast, and the frictional forces are simply not large enough for the car to make it around the curve. Disaster may result!

There is another way to get a centripetal force. By tilting, or banking, the road, the demands on the frictional forces can be reduced. Let's first consider the case when there is no frictional force. As shown in the figure, the car's weight acts vertically downward, pushing the car against the road. Because there is no frictional force, the road can only exert a force on the car that is perpendicular to the road. (If we assume that the car maintains the same elevation in the curve, the car does not accelerate in the vertical direction. Therefore, the vertical part of the force due to the road must just cancel out the car's weight.) As shown in the figure, the sum of these two forces on the car—the force due to gravity and the force due to the road—is a horizontal force acting toward the inside of the curve. It is this horizontal force that provides the centripetal acceleration that makes the car go around the curve.

For a given bank to the curve, there is a well-determined centripetal force and a corresponding centripetal acceleration. It therefore requires a specific speed to execute circular motion with a radius that matches the curve. This speed is the design speed of the curve and is the speed that does not require any frictional forces between the tires and the road. This is the speed that you want to use if the road is icy. If your speed is higher, the circle your car follows will be larger than that of the roadway, and you will skid off the outside of the curve. If your speed is lower than the design speed, the centripetal force will cause your car's path to have too small a radius, and the car will slide off the road to the inside!

The relationship between the design speed and the banking angle is precise and can be calculated. To get a feeling for this, consider a curve with a radius of 100 meters. For a bank angle of 13 degrees, the design speed is found to be 15 meters per second (34 mph). Doubling the design speed to 30 meters per second requires a bank angle of 43 degrees, and tripling the design speed to 45 meters per second (100 mph) requires a bank angle of 65 degrees.

In actuality the centripetal force is provided by the banking of the curve and the force of friction. If the car is traveling faster than the design speed of the curve, the frictional force acts toward the inside of the curve to provide the extra centripetal force needed. However, if the car is traveling slower than the design speed, the frictional force acts outward, reducing the net centripetal force. Thus, friction means that you can safely execute corners at speeds above and below their design speeds. For instance, the corner with the 13-degree bank described in the previous paragraph could be taken at speeds up to 30 meters per second (67 mph).

The centripetal force experienced by a car going around a curve increases as the weight of the car increases. This is fortunate because the larger mass of the car requires a greater centripetal force to make the car go around the corner. Therefore, the design speed of the curve is the same for all cars. Can you imagine what the road signs would be like if this were not true?

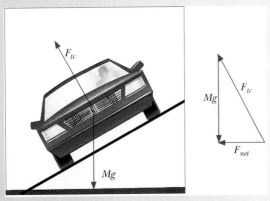

The forces acting on a car going around a banked curve with no friction.

Banking corners allows cars to take curves at high speeds.

projectile motion range from human cannonballs to the forward pass in football if we ignore the effects of air resistance (which we will do throughout the remainder of this chapter).

The study of projectile motion is simplified because the motion can be treated as two mutually independent, perpendicular motions, one horizontal and the other vertical. This reduces a complicated situation to two independent, one-dimensional motions that we already know how to handle.

Even though this division works, its consequences are often difficult to accept. Suppose a bullet is fired horizontally from a pistol and simultaneously another bullet is dropped from the same height. Which bullet hits the ground first?

Many students incorrectly answer, "The dropped bullet hits first." The time it takes to reach the ground is determined by the vertical motion. Although the two bullets have different horizontal speeds, their vertical motions are identical *if we ignore the air resistance*. Therefore, they take the same time to reach the ground. Note that we do not claim that they travel the same distance or with the same speed; clearly the fired bullet has a much greater speed and consequently travels much farther by the time it hits the ground.

This result is convincingly demonstrated in the strobe photograph in Figure 4-9. The ball on the right was fired in a horizontal direction at the same time the ball on the left was dropped. The horizontal lines are included as a visual aid to help you see that the vertical race ends in a tie.

Which bullet will hit the ground first if they are simultaneously released from the same height?

PHYSICS | ON YOUR OWN

Investigate the independence of vertical and horizontal motion using the apparatus shown in Figure 4-10. A headless nail holds the two washers. Hit the stick sharply so that one washer flies off horizontally while the other simply drops. Listen to determine whether they hit the floor simultaneously. Does the result depend on how hard you hit the stick?

Let's look at the motion of the ball thrown in Figure 4-11. Assuming that gravity is the only force acting on the ball, the acceleration of the ball must be in the vertical direction. Therefore, only the vertical speed changes; the horizontal speed remains constant throughout the flight.

How do we describe the vertical motion? The ball takes off with a certain initial vertical speed and has a constant downward acceleration throughout the flight. On the upward part of its flight, the downward acceleration slows the ball's vertical speed by 10 meters per second during each second. At some point the vertical speed becomes zero (point A in Figure 4-11). Then the ball starts its vertical descent, speeding up at 10 (meters per second) each second.

Although the vertical speed is zero at point A, the ball is still moving horizontally. The horizontal spacings between the images are equal, indicating that the horizontal speed is constant along the entire path. This curve also explains why basketball players such as Michael Jordan appear to hang in the air when they drive to the basket for a slam dunk. Note that near the top of the curve the ball travels very little in the vertical direction while it covers a much larger distance horizontally.

To further our understanding, we contrast the motion of a projectile on Earth with its corresponding motion on a mythical planet called Narang, a planet with **N**o **A**ir **R**esistance **A**nd **N**o **G**ravity. We fire a bullet from a horizontal rifle. On Narang the bullet travels in a straight, horizontal line at a constant velocity, because it experiences no force after it leaves the rifle barrel. On Earth the bullet's path differs because Earth's gravity affects its vertical motion. The horizontal motion of the bullet, however, is exactly the same on each planet, as shown in Figure 4-12.

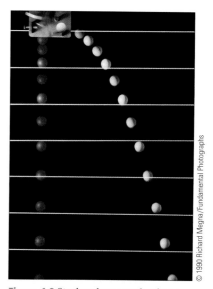

Figure 4-9 Strobe photograph of two balls. The yellow ball was fired horizontally; the red ball was dropped.

Figure 4-10 Which washer will hit the floor first?

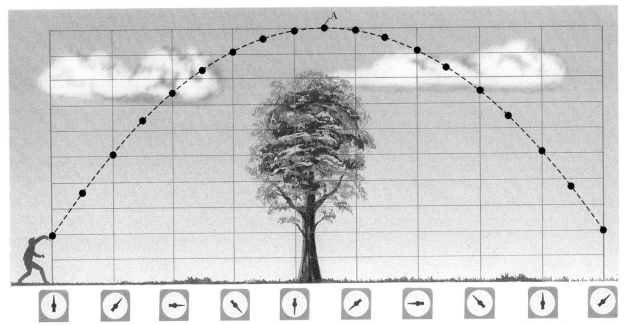

Figure 4-11 Strobe drawing of a thrown ball's motion. Note that the horizontal motion has a constant speed. The clock icon indicates the flow of time.

Michael Jordan appears to defy gravity as he slam dunks the ball.

WORKING IT OUT | Projectile Motion ✓ MATH

An ugly giant rolls a bowling ball with a uniform speed of 30 m/s (approximately 67 mph!) across the top of his large desk. The ball rolls off the end of the desk and lands on the floor 120 m from the edge of the desk. How high is the desk?

The horizontal motion of the ball remains constant throughout the flight; every second the ball is in the air, it travels another 30 m in the horizontal direction. If the ball travels 120 m from the edge of the desk before it lands, it must have been in the air for 4 s:

$$t = \frac{d}{v} = \frac{120 \text{ m}}{30 \text{ m/s}} = 4 \text{ s}$$

The vertical motion of the ball is more complicated. It starts out with zero speed in the downward direction. Once the ball leaves the edge of the desk, it is in free fall and speeds up in the downward direction with an acceleration of 10 m/s². In 4 s the vertical speed changes from zero to

$$\Delta v = a\Delta t = (10 \text{ m/s}^2)(4 \text{ s}) = 40 \text{ m/s}$$

Which of these speeds, zero or 40 m/s, tells us how far the ball drops in 4 s? Neither. We must use the average speed of 20 m/s. The height of the desk is therefore

$$h = \bar{v}t = (20 \text{ m/s})(4 \text{ s}) = 80 \text{ m}$$

Imagine now that we fire the rifle at an upward angle. On Narang the bullet again follows a straight line, but this time its path is inclined upward. On Earth the bullet continuously falls downward from the straight-line path. At any time during the flight, the distance between the two paths is equal to the distance an object would fall during the same elapsed time (Figure 4-13).

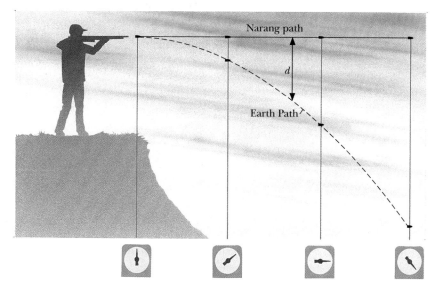

Figure 4-12 Paths of bullets fired horizontally on Narang and Earth. The distance d between the two paths is due to free fall.

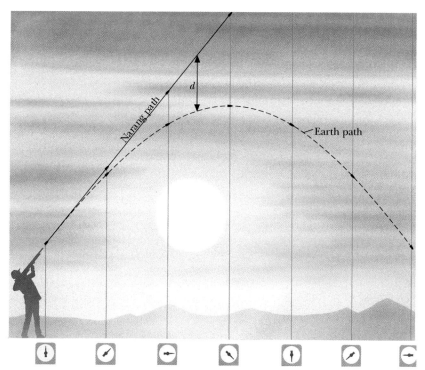

Figure 4-13 Paths of bullets fired at an upward angle on Narang and Earth. The distance d between the two paths is due to free fall.

Question You are hunting gorillas with a rifle that shoots tranquilizer darts. Suddenly you see a gorilla wearing a bright red button hanging from a limb (Figure 4-14). At the instant you pull the trigger, the gorilla lets go of the limb. Where should you have aimed to hit the gorilla's red button?

Answer Directly at the button. The dart falls away from the line of sight at the same rate as the gorilla, and therefore the falling dart will hit the falling gorilla.

Figure 4-14 Tranquilizing a gorilla in a tree. Where should you aim?

Launching an Apple into Orbit

Legend has it that Isaac Newton developed his ideas about gravity after seeing an apple fall. As a tribute to this legend, let's discuss the problem of launching an apple into orbit around Earth. We will of course neglect the effects of air resistance. (Incidentally, our discussion parallels the discussion in Newton's famous book *Principia*; the apple part is ours.)

To accomplish this space-age thought experiment, we need a very high place to stand and a very, very strong arm. For each launch we will throw the apple horizontally. On our first attempt, we throw the apple with an ordinary speed and find that the apple follows a projectile path (Figure 4-15) like the ones we discussed in the previous section.

On our next attempt, imagine that we throw the apple much faster. The apple still falls to the ground, but the path is unlike the first one. If the apple travels very far, Earth's curvature becomes important. The force of gravity points in slightly different directions at the beginning and end of the path.

Normally, we are not aware of the curvature of Earth's surface because Earth is so huge. Over large distances, however, this cannot be ignored. If we imagine that Earth is perfectly smooth—without hills and valleys—and construct a large horizontal plane at our location as shown in Figure 4-16, Earth's surface will be 5 meters below the plane at a distance of 8 kilometers away. (This is about 16 feet at a distance of 5 miles.)

Imagine, then, that on our next attempt to launch the apple, we throw it with a speed of 8 kilometers per second. (This may seem slow, but it is 18,000 mph!) During the first second, the apple drops 5 meters, but so does the surface of Earth. Thus, the apple is still moving horizontally at the end of the first second. The motion during the next second is a repeat of that during the first. And so on. The apple is in orbit.

If air resistance is negligible, the speed remains constant. The only force is the force of gravity. It is a centripetal force acting perpendicular to the instantaneous velocity. By throwing the apple hard enough, we have changed the motion from projectile to circular. Unlike projectile motion our apple will not come down even though it is continually falling. The illustration used by Newton in his discussion of this is reproduced in Figure 4-17.

Rotational Motion

When something is thrown, it usually undergoes another motion. As it moves though space, it often spins or tumbles. We need to ask if it is legitimate to treat these motions as independent of each other.

Figure 4-15 An unsuccessful launch results in projectile motion.

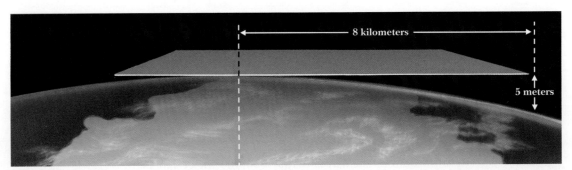

Figure 4-16 A plane touching a spherical Earth is 5 meters above Earth's surface at a distance of 8 kilometers from the point of contact.

Flawed Reasoning

A newspaper report reads in part, "The space shuttle orbits Earth at an altitude of nearly 200 miles and is traveling at a speed of 18,000 mph. The shuttle remains in orbit because the gravitational force pulling it toward Earth is balanced by the centrifugal force (the force of inertia) that is pulling it away from Earth." **Explain why this newspaper should hire a new reporter.**

Answer All forces are exerted by one object on another object. Earth exerts the gravitational force on the shuttle. We have great difficulty, however, finding an object responsible for exerting a centrifugal, or outward, force on the shuttle. This is our first clue that such a force does not exist and, indeed, is not needed. Circular motion requires a net force acting *toward the center* of the circle, and the gravitational force provides this force. There is also no such force as a "force of inertia." Objects travel at constant velocity in the *absence* of a force, not *because* of a force.

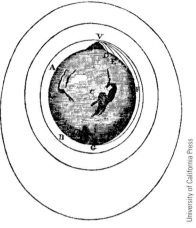

Figure 4-17 This illustration was used in Newton's *Principia* in the discussion of launching an object into orbit around Earth.

The strobe photograph in Figure 4-18 shows the top view of a wrench sliding on a nearly frictionless surface. The wrench is doing two things: it is moving along a path, and it is rotating. Motion along a path is called **translational motion**. Notice that the spot marked with the white dot is moving in a straight line and at a constant speed; it is moving at a constant velocity.

If we mentally shrink any object so that its entire mass is located at a certain point, the translational motion of this new, very compact object would be the same as the original object. Furthermore, if the object is rotating freely, it rotates about this same point. This point is called the **center of mass**. The center of mass of the wrench in Figure 4-18 is marked by the white dot. There is a simple way of finding the center of mass of the wrench. If you place the wrench with the white dot over your finger, the wrench will balance on your finger.

Although realistic motions are a combination of rotational and translational motions, the photograph shows that treating them as two separate motions can reduce the complexity. In other words, we can look at the rotational motion as if the object were not moving along a path and look at the translational motion as if the object were not rotating.

PHYSICS | ON YOUR OWN

Make a small hole in a table-tennis ball so that you can load up one side with glue. After the glue has dried, throw the ball to a friend. Try putting some spin on the ball. What happens and why?

Figure 4-18 Strobe photograph of a wrench sliding across a horizontal surface. Notice that the white dot moves with a constant velocity.

Floating in Defiance of Gravity

Many performers—long jumpers, basketball players, and ballet dancers, to name a few—would like to defy gravity and stay in the air longer. Regardless of their desires, however, the force of gravity is ever present, and the center of mass of the projectile—be it a pebble or a performing artist—follows the parabolic path discussed in this chapter. The only control over the path of the center of mass is in the initial conditions at the moment of launch. A different angle or different launch speed alters the projectile's path. However, once the object is launched, the center of mass follows the predetermined path.

The grand jeté in ballet seems to be a contradiction. In this popular ballet move, dancers execute a running leap across the stage, creating a seemingly floating motion that suspends them in the air longer than gravity should allow (Figure A).

There are two parts to this illusion. First, all objects spend most of the time of flight near the peak of the motion. By counting the images of the ball in Figure 4-11, it is easy to verify that the ball spends half the time in the top one-quarter of the vertical space. During this time, the object is moving mostly horizontally.

The second part of the illusion depends on the dancer's skill. Although the dancer's center of mass follows a parabolic path, a skillful dancer can change the position of the center of mass within her body during the flight. This allows the head and torso to stay at a nearly fixed height for a longer time. As illustrated in Figure B, the movements of the arms and legs raise and lower the location of the center of mass within the body. So, during the beginning and end of the jump, the dancer's arms and legs are pointing downward, keeping the center of mass low. During the middle of the flight, the dancer's arms are up and the legs are outstretched, raising the center of mass relative to the rest of the dancer's body. The total effect is a flattening of the path of the head and torso.

Figure A

Source: Kenneth Laws, *The Physics of Dance* (New York: Schirmer, 1984).

Figure B The dancer raises and lowers the center of mass within the body during the grand jeté.

Summary

Multidimensional motion can be divided into separate motions—translation in each of the three dimensions (up/down, left/right, and near/far). The laws for one-dimensional motion apply to each dimension separately.

An object moving along a circular path at a constant speed must have a net force acting on it. This centripetal force causes the circular motion; without it the object would fly off in a straight line, moving in the direction of its velocity at the time of release. Whenever there is any change in the velocity of an object, it experiences an acceleration and therefore must experience a net force. A net force with a constant magnitude acting perpendicular to the velocity produces circular motion with constant speed.

Projectile motion results from the constant, downward force of gravity. Again, these problems are simplified by the fact that the horizontal and vertical motions are independent. Assuming that air resistance is negligible, the only acceleration is in the vertical direction, the direction of the force of gravity. Therefore, the vertical motion is just that of free fall. The horizontal speed remains constant throughout the flight. For an extended object, there is a single point, the center of mass, that follows the projectile path.

Ignoring air resistance, a projectile launched with a horizontal speed of 8 kilometers per second will go into orbit around Earth. The projectile's speed and altitude remain constant, and it does not return to Earth even though it is continually falling.

Physics Now™ Assess your understanding of this chapter's topics with sample tests and other resources found by logging into PhysicsNow at http://physics.brookscole.com/kf6e.

Chapter 4 Revisited

If an object changes direction, it is accelerating. There must be a net force acting on the object to cause this acceleration. The car turns the corner because the pavement exerts a force on its tires that provides the centripetal force toward the center of the curve. If you are a passenger in the car, your inertia keeps you moving forward in a straight line. The frictional forces between you and the car's seat are usually large enough to provide the force that makes you follow the same curve. You may get any additional needed force from the door. You feel that you've been pushed outward against the door, when in fact the door came to you and is pushing you into the curved path.

KEY TERMS

center of mass: The balance point of an object. This location has the same translational motion as the object would if it were shrunk to a point.

centripetal acceleration: The acceleration of an object toward the center of its circular path. For uniform circular motion, it has a magnitude v^2/r.

centripetal force: The force on an object directed toward the center of its circular path. For uniform circular motion, it has a magnitude mv^2/r.

projectile motion: A type of motion that occurs near Earth's surface when the only force acting on the object is that of gravity.

translational motion: Motion along a path.

CONCEPTUAL QUESTIONS

Important: Ignore the effects of air resistance in the following questions and exercises.

1. A motorcycle drives through a vertical loop-the-loop at constant speed, as shown in the figure. Draw arrows to show the directions of the instantaneous velocity and the net force on the motorcycle at points A, B, and C.

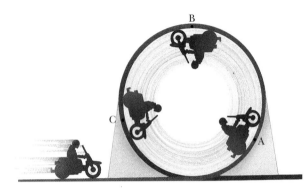

2. The figure shows a racetrack with identical cars at points A, B, and C. The cars are moving clockwise at constant speeds. Draw arrows indicating the direction of the net force on each car and the instantaneous velocity of each car. In what direction would car A travel if there were an oil slick at point A? Why?

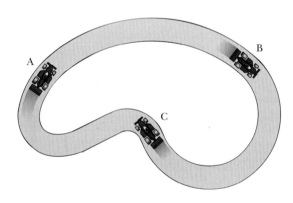

3. What is the force that causes a communications satellite to orbit Earth?
4. What is the force that allows a person on in-line skates to turn a corner? What happens if this force is not strong enough?
5. Consider the motorcycle in the figure for Question 1 when it is at point B. In which directions do the velocity, change in velocity, acceleration, and net force point?
6. A child rides on a carousel. In which direction does each of the following vectors point?
    a. velocity
    b. change in velocity
    c. acceleration
    d. net force
7. How do the velocity vectors differ on opposite sides of the path for uniform circular motion? How are they the same?
8. How do the acceleration vectors differ on opposite sides of the path for uniform circular motion? How are they the same?
9. An object executes circular motion with a constant speed whenever a net force acts perpendicular to the object's velocity. What happens to the speed if the net force is not perpendicular to the velocity?
▲ 10. An object is acted on by a force that always acts perpendicular to the velocity. If the force continually increases in magnitude, does the speed change? Draw a sketch of the path of the object.
11. A water bug is skittering across the surface of a pond. In each case below, the bug's initial and final velocity vectors are shown for a time interval Δt. For each case find the direction of the bug's average acceleration during this interval.

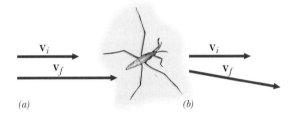

12. A dog is running loose across an open field. In each case below, the dog's initial and final velocity vectors are shown for a time interval Δt. For each case, find the direction of the dog's average acceleration during this interval.

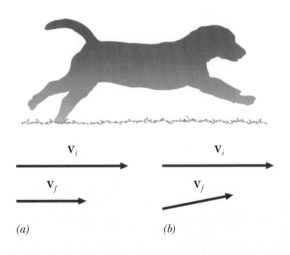

13. Each case below depicts an object's velocity vector and acceleration vector at an instant in time. State whether the object is (i) speeding up, slowing down, or maintain-

ing the same speed and (ii) turning right, turning left, or moving in a straight line.

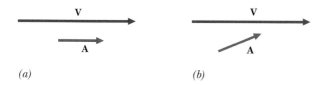

14. Each case below depicts an object's velocity vector and acceleration vector at an instant in time. State whether the object is (i) speeding up, slowing down, or maintaining the same speed and (ii) turning right, turning left, or moving in a straight line.

15. Give examples from everyday life in which an object's acceleration is pointed downward and the object is (a) speeding up and (b) slowing down.
16. You get into an elevator on the eighth floor and hit the button for the lobby on the ground floor. What is the direction of your acceleration right after the elevator starts moving and right before it stops at the ground floor?
17. What force allows you to turn a bicycle while riding on a flat parking lot?
18. Most footraces take place on unbanked tracks. How do the racers turn the corners?
19. A monkey is swinging from tree to tree on vines. At the bottom of a swing, what force provides (or forces provide) the centripetal force required for the monkey to travel along a circular path?
20. A race car is traveling around a banked curve as described in the box "Banking Corners." What force provides (or forces provide) the centripetal force required for the race car to travel along its circular path?
▲ 21. Imagine that you swing a bucket in a vertical circle at constant speed. Will you need to exert more force when the bucket is at the top of the circle or at the bottom? Explain.
▲ 22. A vine is just strong enough to support Tarzan when he is hanging straight down. However, when he tries to swing from tree to tree, the same vine breaks at the bottom of the swing. How could this happen?
23. Earth executes a nearly circular orbit around the Sun. What does this tell you about the speed of Earth along its orbit?
24. According to Newton's third law, Earth exerts a force on the Sun. Does the Sun move in a circular path as Earth goes around it each year? How can this idea be used to determine if nearby stars have planets?
25. You are driving your race car around a circular test track. Which would have a greater effect on the magnitude of your acceleration, doubling your speed or moving to a track with half the radius of curvature? Why?

▲ 26. A figure skater skates a figure-8 pattern with a small circle and a big circle, as shown in the drawing. The big circle has twice the radius of the small circle, and he skates it at twice the speed. Compare the magnitude of his centripetal acceleration on the two circles.

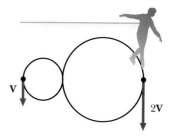

27. A playful astronaut decides to throw rocks on the Moon. What forces act on the rock while it is in the "vacuum"? (We can't say "air"!)
28. A book slides along a frictionless table at a constant velocity and then sails off the edge. Draw a free-body diagram for the book while it is on the table and while it is in the air.
29. A left fielder throws a baseball toward home plate. At the instant the ball reaches its highest point, what are the directions of the ball's velocity, the net force on the ball, and the ball's acceleration?
30. The figure shows the path of a thrown baseball. Draw arrows to indicate the directions of the ball's velocity and acceleration vectors at the three labeled points.

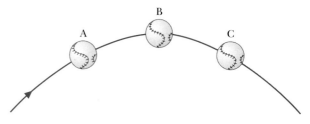

31. A hammer dropped on the surface of the Moon falls with an acceleration of 1.6 (meters per second) per second. Would its acceleration be smaller, larger, or the same if it was thrown horizontally at 6 meters per second? Why?
32. A rock dropped from 3.3 meters above the surface of the Moon requires 2 seconds to reach the ground. Would it require a shorter, a longer, or the same time if the rock was thrown horizontally from this height with a speed of 12 meters per second? Why?
33. Two identical balls roll off the edge of a table. One leaves the table traveling twice the speed of the other. Which ball hits the floor first? Why?
34. Two balls, one of mass 1 kilogram and one of mass 4 kilograms, roll off the edge of a table at the same time traveling at the same speed. Which ball hits the floor first? Why?
35. A physics student reports that upon arrival on planet X, she promptly sets up the "gorilla-shoot" demonstration. She does not realize that the gravity on planet X is stronger than it is on Earth. Will the demonstration work? Explain why or why not.

36. A fearless bicycle rider announces that he will jump the Beaver River Canyon. If he does not use a ramp, but simply launches himself horizontally, is there any way that he can succeed? Why?

37. In football and soccer, it is often desirable to give up some of the distance a kick travels to gain hang time, the time the ball remains in the air. How does the kicker do this?
38. The irons used in golf have faces that make different angles with the shaft of the club. How does this affect the distance traveled and maximum height of the golf ball?

39. If Earth exerts a gravitational force of 5 newtons on the apple while it is in orbit, what force does the apple exert on Earth?
40. Is the size of the gravitational force that Earth exerts on the apple in orbit smaller than, larger than, or the same size as the force the apple exerts on Earth? Why?
41. We know that Earth orbits the Sun in a nearly circular orbit. Draw a free-body diagram for Earth.
42. Draw a free-body diagram for our fictitious apple in orbit near the Earth's surface.
43. A carpenter's square is tossed through the air. As it tumbles, only one point follows a simple parabolic path. In the following picture, which of the four labeled points most likely represents this point? Why?

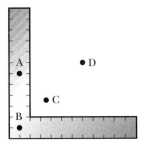

44. A tennis ball and a softball are fastened together by a light, rigid rod as shown. When this arrangement is thrown tumbling through the air, which of the labeled points is most likely to follow a parabolic path? Why?

EXERCISES

1. Find the size and direction of the change in velocity for each of the following initial and final velocities:
   a. 5 m/s west to 10 m/s west
   b. 10 m/s west to 5 m/s west
   c. 5 m/s west to 10 m/s east
2. What is the change in velocity for each of the following initial and final velocities?
   a. 75 km/h right to 100 km/h right
   b. 75 km/h right to 100 km/h left
3. What are the size and direction of the change in velocity if the initial velocity is 30 m/s south and the final velocity is 40 m/s west?
4. What is the change in velocity of a car that is initially traveling west at 50 km/h and then travels 120 km/h toward the north?
5. A migrating bird is initially flying south at 8 m/s. To avoid hitting a high-rise building, the bird veers and changes its velocity to 6 m/s east over a period of 2 s. What is the bird's average acceleration (magnitude and direction) during this 2-s interval?
6. A fox is chasing a bunny. The bunny is initially hopping east at 1 m/s when it first sees the fox. Over the next half second, the bunny changes its velocity to west at 4 m/s and escapes. What was the bunny's average acceleration (magnitude and direction) during this half-second interval?
7. A cyclist turns a corner with a radius of 50 m at a speed of 10 m/s.
   a. What is the cyclist's acceleration?
   b. If the cyclist and cycle have a combined mass of 120 kg, what is the force causing them to turn?
8. A 60-kg person on a merry-go-round is traveling in a circle with a radius of 3 m at a speed of 2 m/s.
   a. What acceleration does the person experience?
   b. What is the net force? How does it compare with the person's weight?
9. Earth orbits once around the Sun every 365 days at an average radius of 1.5×10^{11} m. Earth's mass is 6×10^{24} kg.
   a. How many seconds does it take Earth to orbit the Sun?
   b. What distance does Earth travel in one year?
   c. What is Earth's average centripetal acceleration?
   d. What is the average force that the Sun exerts on Earth?
10. Given that the average distance from Earth to the Moon is 3.8×10^{8} m, that the Moon takes 27 days to orbit Earth, and that the mass of the Moon is 7.4×10^{22} kg, what is the average centripetal acceleration of the Moon and the size of the attractive force between Earth and the Moon?
11. A baseball is hit with a horizontal speed of 45 m/s and a vertical speed of 18 m/s upward. What are these speeds 1 s later?
12. What are the horizontal and vertical speeds of the baseball in the previous exercise 2 s after it is hit?
13. An SUV accidentally drove off a cliff with a horizontal velocity of 40 m/s. Given that it took 5 s for the SUV to hit the ground, how far vertically and horizontally from the top of the cliff did the SUV land?
14. Angel Falls in southeastern Venezuela is the highest uninterrupted waterfall in the world, dropping 979 m (3212 ft). Ignoring air resistance, it would take 14 s for the water to fall from the lip of the falls to the river below. If the water lands 50 m from the base of the vertical cliff, what was its horizontal speed at the top?

15. A tennis ball is hit with a vertical speed of 10 m/s and a horizontal speed of 30 m/s. How long will the ball remain in the air? How far will the ball travel horizontally during this time?
16. If a baseball is hit with a vertical speed of 30 m/s and a horizontal speed of 6 m/s, how long will the ball remain in the air? How far from home plate will it land?
17. Given that the radius of Earth is 6400 km, calculate the acceleration of an apple in orbit just above Earth's surface.
18. The Moon orbits Earth at a distance of 3.8×10^{8} m and a speed of 1 km/s. What is the centripetal acceleration of the Moon?

5 Gravity

This view of Earth greeted the Apollo 11 astronauts as they orbited the Moon.

A popular legend has it that Newton had his most creative thought while watching an apple fall to the ground. He made a huge conceptual leap by equating the motion of the apple to the motion of the Moon and developing the concept of gravity. How far does gravity reach? Do we see any evidence of gravity elsewhere in the Universe?

(See page 89 for the answer to this question.)

IN the physicist's view of the world, there are four fundamental forces: the gravitational, the electromagnetic, the weak, and the strong. We begin our studies of forces with the most familiar force in our everyday lives. Every school-age child knows that objects fall because of gravity. But what is gravity? Saying that it is what makes things fall doesn't tell us much.

Is gravity some material like a fluid or a fog? Or is it something more ethereal? No one knows. Because we have given something a name doesn't mean we understand it completely. We do understand gravity in the sense that we can precisely describe how it affects the motion of objects. For instance, we have already seen how to use the concept of gravity to describe the motion of falling objects. We can do more. By looking carefully at the motions of certain objects, we can develop an equation that describes this attractive force between material objects and explore some of its consequences.

On the other hand, we cannot answer questions like "What is gravity?" or "Why does gravity exist?"

The Concept of Gravity

The concept of gravity hasn't always existed. It was conceived when changes in our world view required a new explanation of why things fell to Earth. When Earth was believed to be flat, gravity wasn't needed. Objects fell because they were seeking their natural places. A stone on the end of a string hung down because of its tendency to return to its natural place. Up and down were absolute directions.

The realization that Earth was spherical required a change in perspective. What happens to those unfortunate people on the other side of Earth who are upside down? But the change in thinking was made without gravity. The center of Earth was at the center of the Universe, and things naturally moved toward this point. Up and down became relative, but the location of the center of the Universe became absolute.

Gravity was also not needed to understand the motion of the heavenly bodies. The earliest successful scheme viewed Earth as the center of the Universe, with each of the celestial bodies going around Earth in circular orbits. Perpetual, circular orbits were considered quite natural for celestial motions; little attention was given to the causes of these motions. Aristotle did not recognize any connections between what he saw as perfect, heavenly motion and imperfect, earthly motion. He stated that circular motion with constant speed was the most perfect of all motions and thus the natural heavenly motion needed no further explanation.

This changed slightly with a new view of heavenly motion by Nicholas Copernicus, a 16th-century Polish scientist and clergyman. He proposed that the planets (including Earth) go around the Sun in circular orbits and that the Moon orbits Earth. This is essentially the scheme taught in schools today. A hint of a concept of gravity appears in Copernicus's work. He believed that the Sun and Moon would attract objects near their surfaces—each would have a local gravity—but he had no concept of that attractive influence spreading throughout space.

A hundred years later, Johannes Kepler, a German mathematician and astronomer, suggested that the planets move because of an interaction between them and the Sun. Kepler also moved us away from the assumption that the planets traveled in circular paths. After many years of trial and error, Kepler correctly deduced that the orbits of the planets were *ellipses*—but ellipses that are close to being circles. Furthermore, the planets do not have constant speeds in their journeys around their elliptical paths but speed up as they approach the Sun and slow down as they move farther away.

Influenced by early, important work on magnetism, Kepler postulated that

KEPLER | Music of the Spheres

Johannes Kepler was born two days after Christmas in 1571. In his horoscope, which he later compiled, he noted that the family spelled their names in a variety of ways, that he was premature at birth, and that he was a sickly child. The village in which he was born is now part of greater Stuttgart in the German state of Baden-Württemberg. His peasant origins could not mask his precocity and gift for mathematics. He was chosen by the local duke to receive a good education and later attended the new Lutheran seminary at Tübingen University. There he became acquainted with the techniques necessary to work on advanced astronomical problems.

Johannes Kepler

Kepler's adult life was centered upon mathematical astronomy. A Lutheran, he spent most of his life in Prague in the employment of a Catholic Holy Roman emperor. He was a committed Copernican and sought to extend the accuracy of that new astronomy. His first work—*Cosmic Mysteries* (1597)—was a somewhat mystical and numerological theory relating distances and times of revolutions of the planets. His generation regarded mathematics as a language that revealed the inner harmony of creation and that celestial motions revealed physical harmony and unity in action. So his search for a "music" of the spheres truly was a search for a mathematical description of God's creation.

His mentor in astronomy, Tycho Brahe, was a Danish master of observational technique and precision. In fact, Brahe's work reached the limit of naked-eye observations. After Brahe's death, Kepler gained access to a wealth of astronomical data on the motions of Mars. This planet was the most perplexing in its apparently irregular movements. Kepler published his book on Mars in 1609 and ushered in an age of "new" astronomy. It is here that his first two laws of motion can be found.

His work was known throughout Europe, and he and Galileo (and even Galileo's father) engaged in a desultory correspondence concerning issues of astronomy. Again, music was frequently discussed in this context.

Kepler found himself engulfed in a series of major European wars, in personal tragedy, and when his mother was accused of witchcraft in Baden-Württemberg, with the law. His great ability continued to shine through in all this turmoil. He developed a technique—logarithms—to speed up calculations, and he continued his astronomical work. He desperately needed money and so wrote astrological predictions for the powerful. He developed the concept of *satellite* and even wrote a little fable on space travel.

Always seeking harmony in the cosmos, he published in 1618 a book similar to his earlier one on cosmic mysteries. In this later work, he laid out his third law, which related planetary orbits, times of revolution, and heliocentric theory. It was this third law upon which Isaac Newton so successfully built his physics. Again, a musical theme emerges: it was titled *Harmonies of the World*.

Johannes Kepler provided impetus to the development of a new physics because his astronomy destroyed the Ferris wheels of perfect circles and the customary ideas of natural and unnatural motions. When he died in 1630, his legacy was secure. He had created a new astronomy and demanded a new concept of physics to support it.

—*Pierce C. Mullen, historian and author*

Sources: Max Caspar, *Kepler*, trans. and ed. C. Doris Hellman (New York: Abelard-Schuman, 1959); Arthur Koestler, *The Watershed: A Biography of Johannes Kepler* (Garden City, N.Y.: Anchor Books, 1960).

---

the planets were magnetically driven along their paths by the Sun. Because this work occurred shortly before the acceptance of the idea of inertia, Kepler did not realize that a force was needed not to drive the planets along their orbits but to cause the orbits to be curved. Kepler postulated an interaction reaching from the Sun to the various planets and driving the planets, but he didn't consider the possibility of any interaction between the planets themselves; the Sun reigned supreme, a metaphor for his god, from which everything else gained strength.

Newton developed our present view of gravity. He started by saying there was nothing special about the rules of nature that he had developed for use on Earth. They should also apply to heavenly motions. As we learned in the previous chapter, anything traveling in a circle must be accelerating. The acceleration must be toward the center of the circle, and the object must therefore have a net force acting on it. Newton went searching for this force.

Newton's Gravity

Creativity often involves bringing together ideas or things from seemingly unrelated areas. After an artist or scholar has done it, the connection often seems obvious to others. Newton made such a synthesis between motions on Earth and motions in the heavens. Legend has it that he made his intellectual leap while contemplating such matters and seeing an apple fall. In the last chapter, we discussed the problem of launching an apple into orbit around Earth. The transition of the Earth-bound apple into heavenly orbits provides an analogy of Newton's intellectual leap.

Newton believed that the laws of motion that worked on Earth's surface should also apply to motion in the heavens. Because the Moon orbits Earth in a nearly circular orbit, it must be accelerating toward Earth. According to the second law, any acceleration requires a force. He believed that if this force could be shut off, the Moon would no longer continue to move along its circular path but would fly off in a straight line like a stone from a sling.

The genius of Newton was in relating the cause of this heavenly motion to earthly events. Newton felt that the Moon's acceleration was due to the force of gravity—the same gravity that caused the apple to fall from the tree. How could he demonstrate this? First, he calculated the acceleration of the Moon. Because the distance to the Moon and the time it took the Moon to make one revolution were already known, he was able to calculate that the Moon accelerated 0.002 72 (meter per second) per second. This is a very small acceleration. In 1 second the Moon moves about 1 kilometer along its orbit but falls only 1.4 millimeters (about $\frac{1}{20}$ inch in 0.6 mile).

In contrast to the Moon's acceleration, the apple has an acceleration of 9.80 (meters per second) per second and falls about 5 meters in its first second of flight. (In earlier discussions we rounded the acceleration off to 10 (meters per second) per second for ease of computation. The small difference is important here.) We can compare these two accelerations by dividing one by the other:

$$\frac{0.002\,72 \text{ m/s}^2}{9.80 \text{ m/s}^2} = \frac{1}{3600}$$

Legend has it that Newton conceived his law of universal gravitation while observing an apple fall from a tree in his yard.

◀ $\dfrac{\text{acceleration of Moon}}{\text{acceleration of apple}}$

Why are these two accelerations so different? The mass of the Moon is certainly much larger than that of the apple. But that doesn't matter. As we saw in Chapter 2, free-falling objects all have the same acceleration independent of their masses.

However, the accelerations of the apple and the Moon weren't equal. Could Newton's idea that both motions were governed by the same gravity be wrong? Or could the rules of motion that he developed on Earth not apply to heavenly motion? Neither. Newton reasoned that the Moon's acceleration is smaller because Earth's gravitational attraction is smaller at larger distances; it is "diluted" by distance.

How did the force decrease with increasing distance? Retracing Newton's reasoning is impossible because he didn't write about how he arrived at his conclusions, but he may have used the following kind of reasoning.

Many things get less intense the farther you are from their source. Imagine a paint gun that can spray paint uniformly in all directions. Suppose the gun is in the center of a sphere of radius 1 meter, and at the end of 1 minute of spraying, the paint on the inside wall of the sphere is 1 millimeter thick. If we repeat the experiment with the same gun but with a sphere that is 2 meters in radius, the paint will be only $\frac{1}{4}$ millimeter thick because a sphere with twice the radius has a surface area that is four times the original (Figure 5-1). If the sphere has three times the radius, the surface is nine times bigger, and the paint is $(\frac{1}{3})^2 = \frac{1}{9}$ as thick. The thickness of the paint decreases as the square of the radius of the sphere increases. This is known as an **inverse-square** relationship. A force reaching into space could be diluted in a similar manner.

Question   If the sphere were 4 meters in radius, how thick would the paint be?

Answer   It would be $(\frac{1}{4})^2 \times 1$ millimeter = $\frac{1}{16}$ millimeter thick.

Figure 5-1 If the sphere's radius is doubled, the sphere's surface area increases by a factor of four.

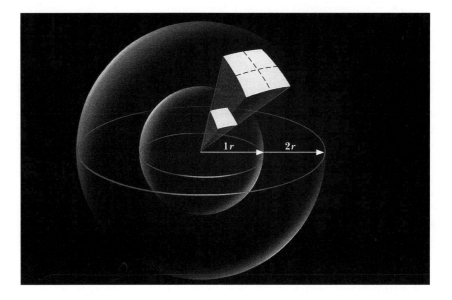

Newton may also have received encouragement for this explanation by working backward from observational results on the motion of the planets that had been developed by Kepler. Kepler found a relationship that connected the orbital periods of the planets with their average distances from the Sun. Using Kepler's results and an expression for the acceleration of an object in circular motion, we can show that the force decreases with the square of the distance. This means that if the distance between the objects is doubled, the force is only one-fourth as strong. If the distance is tripled, the force is one-ninth as strong. And so on. This relationship is shown in Figure 5-2.

What, exactly, does it mean to say that the force is one-ninth as strong? We cannot be referring to the gravitational forces acting on different objects; the gravitational force on an automobile is obviously much larger than the gravitational force on a person. We must compare the gravitational force acting on a single object, such as our apple, at different distances. When the apple is moved three times as far away from Earth's center, the gravitational force on the apple is one-ninth as strong.

The obvious test of the notion of gravity was to see if the relationship between distance and force gave the correct relative accelerations for the apple on Earth and the orbiting Moon. Newton could use this rule and make the comparison. The distance from the center of Earth to the center of the Moon is about 60 times the radius of Earth. The Moon is therefore 60 times farther away from the center of Earth than the apple. Thus, the force at the Moon's location—and its acceleration—should be 60^2, or 3600, times smaller. This is in agreement with the previous calculation. The data available in Newton's time were not as good as those we have used here, but they were good enough to convince him of the validity of his reasoning. Modern measurements yield more precise values and agree that *the gravitational force is inversely proportional to the square of the distance*.

Figure 5-2 The force on a 0.1-kilogram mass at various distances from Earth. Notice that the force decreases as the square of the distance.

Question What happens to the force of gravity if the distance between the two objects is cut in half?

Answer The force becomes four times as strong.

Newton now knew how gravity changed with distance: the force of Earth's gravity exists beyond Earth and gets weaker the farther away you go. But there are other factors. He already knew that the force of gravity depended on the object's mass. His third law of motion said that the force exerted on the Moon by Earth was equal in strength to that exerted on Earth by the Moon; they attracted each other. This symmetry indicated that both masses should be included in the same way. *The gravitational force is proportional to each mass.*

The Law of Universal Gravitation ✓ MATH

✓ Extended presentation available in the *Problem Solving* supplement

Having made the connection between celestial motion and motion near Earth's surface with a force that reaches across empty space and pulls objects to Earth, Newton took another, even bolder, step. He stated that the force of gravity existed between *all* objects, that it was truly a *universal* law of gravitation.

The boldness of this assertion becomes apparent when one realizes that the force between two ordinary-sized objects is extremely small. Clearly, as you walk past a friend, you don't feel a gravitational attraction pulling you together. But that is exactly what Newton was claiming. Any two objects have a force of attraction between them; his rule for gravity is a **law of universal gravitation**.

Putting everything together, we arrive at an equation for the gravitational force:

$$F = G\frac{m_1 m_2}{r^2}$$

◄ law of universal gravitation

where m_1 and m_2 are the masses of the two objects, r is the distance between their centers, and G is a constant that contains information about the strength of the force.

Although Newton arrived at this conclusion when he was 24 years old, he didn't publish his results for more than 20 years. This was partly due to one unsettling aspect of his work. The distance that appears in the relationship is the distance from Earth's center. This means that Earth's mass is assumed to be concentrated at a point at its center. This might seem like a reasonable assumption when considering the force of gravity on the Moon; Earth's size is irrelevant when dealing with these huge distances. But what about the apple on Earth's surface? In this case the apple is attracted by mass that is only a few meters away and mass that is 13,000 kilometers away, as well as all the mass between (Figure 5-3). It seems less intuitive that all this would somehow act like a very compact mass located at Earth's center. But that is just what happens. Newton was eventually able to show mathematically that the sum of the forces due to each cubic meter of Earth is the same as if all of them were concentrated at its center.

This result holds if Earth is spherically symmetric. It doesn't have to have a uniform composition; it need only be composed of a series of spherical shells, each of which has a uniform composition. In fact, 1 cubic meter of material near Earth's center has almost four times the mass of a typical cubic meter of surface material.

Newton applied the laws of motion and the law of universal gravitation extensively to explain the motions of the heavenly bodies. He was able to show that three observational rules developed by Kepler to describe planetary motion were a mathematical consequence of his work. Kepler's rules were the results of years of work reducing observational data to a set of simple patterns.

By the 18th century, scientists were so confident of Newton's work that

Figure 5-3 The sum of all the forces on the apple exerted by all portions of Earth acts as if all the mass were located at Earth's center.

when a newly discovered planet failed to behave "properly," they assumed that there must be other, yet to be discovered, masses causing the deviations. When Uranus was discovered in 1781, a great effort was made to collect additional data on its orbit. By going back to old records, additional times and locations of its orbit were determined. Although the main contribution to Uranus's orbit is the force of the Sun, the other planets also have their effects on Uranus. In this case, however, the calculations still differed from the actual path by a small amount. The deviations were explained in terms of the influence of an unknown planet. This led to the discovery of Neptune in 1846.

This still didn't completely account for the orbits of Uranus and Neptune; a search began for yet another planet. The discovery of Pluto in 1930 still left some discrepancies. Although the search for new planets continues, analysis of the paths of the known planets indicates that any additional planets must be very small or very far away or both.

The Value of G

Even though Newton had an equation for the gravitational force, he couldn't use it to actually calculate the force between two objects; he needed to know the value of the constant G. The way to get this is to measure the force between two known masses separated by a known distance. However, the force between two objects on Earth is so tiny that it couldn't be detected in Newton's time.

It was more than 100 years after the publication of Newton's results before Henry Cavendish, a British scientist, developed a technique that was sensitive enough to measure the force between two masses. Modern measurements yield the value

gravitational constant ▶

$$G = 0.000\,000\,000\,066\,7\,\frac{\text{N} \cdot \text{m}^2}{\text{kg}^2} = 6.67 \times 10^{-11}\,\frac{\text{N} \cdot \text{m}^2}{\text{kg}^2}$$

(See the box "Working It Out" in Chapter 1 for an explanation of this notation.) Putting this value into the equation for the gravitational force tells us that the force between two 1-kilogram masses separated by 1 meter is only 0.000 000 000 066 7 newton. This is minuscule compared with a weight of 9.8 newtons for each mass. The small value of G explains why two friends don't feel their mutual gravitational attraction when standing near each other.

Cavendish referred to his experiment as one that "weighed" Earth, although it would have been more accurate to claim that it "massed" the Earth. His point, though, was important. By measuring the value of G, Cavendish made it possible to accurately determine Earth's mass for the first time. The acceleration of a mass near Earth's surface depends on the value of G and Earth's mass and radius. Because he now knew the values of all but Earth's mass, he could calculate it. Earth's mass is 5.98×10^{24} kilograms; that's about a million million million million times as large as your mass.

Once Earth's mass is known, we can use the law of universal gravitation to calculate the acceleration due to gravity near Earth's surface:

$$g = \frac{F}{m} = \frac{GM_E}{R_E^2}$$

where M_E is Earth's mass and $R_E = 6370$ kilometers is Earth's radius. Plugging in the numerical values yields $g = 9.8$ (meters per second) per second, as expected.

Because Earth orbits the Sun, the Sun's mass can also be calculated with the Cavendish results. In making these computations, it is assumed that the value of G measured on Earth is valid throughout the Solar System. This cannot be proved. On the other hand, there is no evidence to the contrary, and this assumption gives consistent results. Newton made this same claim more than 100 years earlier.

WORKING IT OUT | Gravity

Let's calculate the gravitational force between two friends. To make the calculation of this force easier, we make one unrealistic assumption: we assume that the friends are spheres! This allows us to use the distance between their centers as their separation and still get a reasonable answer. Assuming that the friends have masses of 70 and 86 kg (about 154 and 189 lb, respectively) and are standing 2 m apart, we have

$$F = G\frac{m_1 m_2}{r^2} = \left(6.67 \times 10^{-11}\ \frac{\text{N}\cdot\text{m}^2}{\text{kg}^2}\right)\frac{(70\ \text{kg})(86\ \text{kg})}{(2\ \text{m})^2} = 1.00 \times 10^{-7}\ \text{N}$$

This very tiny force is about one 10-billionth (10^{-10}) of either friend's weight.

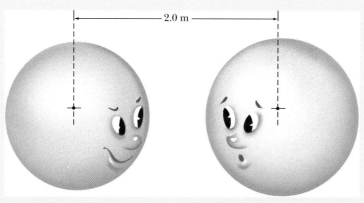

The attraction between these spheroidal friends depends on the distance between their centers.

Flawed Reasoning

In the blockbuster movie *Armageddon*, the heroes land their space shuttle on a Texas-sized comet that is careening toward Earth. They then walk around the comet like construction workers here on Earth. **What is wrong with this picture?**

Answer A comet is not massive enough to provide the gravity required to walk around normally. The astronauts would have to tether themselves to keep from flying away due to the smallest exertion. The astronauts were tethered in the other comet-coming-to-destroy-Earth movie *Deep Impact*.

Gravity near Earth's Surface

In earlier chapters we assumed that the gravitational force on an object was constant near Earth's surface. We were able to do this because the force changes so little over the distances in question. In fact, to assume otherwise would have unnecessarily complicated matters.

Near Earth's surface the gravitational force decreases by one part in a million for every 3 meters (about 10 feet) of gain in elevation. Therefore, an object that weighs 1 newton at Earth's surface would weigh 0.999 999 newton at an elevation of 3 meters. An individual with a mass of 50 kilograms has a weight of 500 newtons (110 pounds) in New York City; this person would weigh about 0.25 newton (1 ounce) less in mile-high Denver.

The variations in the gravitational force result in changes in the acceleration

How Much Do You Weigh?

In the 1600s people didn't imagine that anybody would one day travel to distant planets. Nevertheless, Newton's law of universal gravitation allowed them to predict what they would weigh if they ever found themselves on another planet.

According to Newton's law of universal gravitation, a person's weight on a planet depends on the mass and radius of the planet, as well as the mass of the person. Your weight on Jupiter compared with that on Earth depends on these factors. Assuming that Jupiter is the same size as Earth (it isn't) but that it has 318 times as much mass as Earth (it does) means that you would weigh 318 times as much on this fictitious Jupiter as on Earth. Actually, Jupiter's diameter is 11.2 times that of Earth's. Because the law of universal gravitation contains the radius squared in the denominator, your weight is actually reduced by a factor of 11.2 squared, or 125. Combining these two factors means that you would tip a Jovian bathroom scale at $\frac{318}{125}$, or $2\frac{1}{2}$, times your weight on Earth. You would be lightest on Pluto, weighing only 8% of your Earth weight because Pluto's small mass decreases your weight more than its small radius increases it. Your weight on each of the planets is given in the table.

On the Moon the force of gravity is only $\frac{1}{6}$ of that on Earth. Thus, an astronaut's weight is only $\frac{1}{6}$ of what it would be on Earth. This means that astronauts can jump higher and will fall more slowly, as we have seen from the television images transmitted to Earth during the lunar explorations. Vehicles designed for lunar travel would collapse under their own weight on Earth.

Table 5-01 | Weights on Each of the Planets

Planet	Relative Weight	150-lb Person
Mercury	0.38	57 lb
Venus	0.91	136
Earth	1.00	150
Mars	0.38	57
Jupiter	2.53	380
Saturn	1.07	160
Uranus	0.92	138
Neptune	1.18	177
Pluto	0.08	12

The lunar rover would collapse under its own weight if used on Earth.

Flawed Reasoning

You read in a comic book that the gravity on the Moon is not as strong as on Earth because there is no atmosphere on the Moon. This doesn't seem right, so you do a little research. **What do you find?**

Answer The source of the gravitational attraction is mass, not air. The gravitational attraction on the Moon is less than on Earth because the Moon's mass is so much less than Earth's. Actually, the argument in the comic book is completely backward. The Moon does not have an atmosphere because its gravity is too weak to hold one.

The passengers in this airplane weigh less because of their altitude.

due to gravity. The value of the acceleration—normally symbolized as g—is nearly constant near Earth's surface. As long as one stays near the surface, the distance between the object and Earth's center changes very slightly. If an object is raised 1 kilometer (about $\frac{5}{8}$ mile), the distance changes from 6378 kilometers to 6379 kilometers, and g changes only from 9.800 (meters per second) per second to 9.797 (meters per second) per second.

However, even without a change in elevation, g is not strictly constant from place to place. Earth would need to be composed of spherical shells, with each shell being uniform; this is not the case. Underground salt deposits have less mass per cubic meter and give smaller values of g than average, whereas metal deposits produce larger g values. Therefore, measurements of g can be used to locate large-scale underground ore deposits. By noting variations, geologists can map regions for further exploration.

Question Given the fact that water has less mass per cubic meter than soil and rock, would you expect the value of g to be smaller or larger than average over a lake?

Answer A cubic meter of water would provide less attraction than a cubic meter of soil and rock. Therefore, the value of g would be smaller.

The experimental value of g also varies with latitude because of Earth's rotation on its axis. The value is smallest near the equator and increases toward each pole. We will study this nongravitational effect more carefully in Chapter 9.

Satellites ✓ MATH

Newton's theory also predicts the orbits of artificial satellites orbiting Earth. By knowing how the force changes with distance from Earth, we know what accelerations—and consequently, other orbital characteristics—to expect at different altitudes. For instance, a satellite at a height of 200 kilometers should orbit Earth in 88.5 minutes. This is close to the orbit of the satellite *Vostok 6* that carried the first woman, Valentina Tereskova, into Earth orbit in June 1963. Its orbit varied in height from 170 to 210 kilometers and had a period of a little over 88 minutes.

> **PHYSICS | ON YOUR OWN**
>
> Many satellites have north–south orbits with periods of approximately 90 minutes. Search the night sky near the North Star until you locate one of these satellites moving southward. Estimate the time this satellite spends above the horizon. Why is this time much, much shorter than 45 minutes?

Satellite dishes are aimed at communications satellites in geosynchronous orbits above Earth's equator.

The higher a satellite's orbit, the longer it takes to complete one orbit. The Moon takes 27.3 days; *Vostok 6* took 88 minutes. It is possible to calculate the height that a satellite would need to have a period of 1 day. With this orbit, if the satellite were positioned above the equator, it would appear to remain fixed directly above one spot on Earth—an orbit called geosynchronous. Such geosynchronous satellites have an altitude of 36,000 kilometers, or about $5\frac{1}{2}$ Earth radii, and are useful in establishing worldwide communications networks. Backyard satellite dishes that pick up television signals point to geosynchronous satellites. The first successful geosynchronous satellite was *Syncom II*, launched in July 1963. Some geosynchronous satellites are used to monitor the weather on Earth.

GEOS weather satellites orbit Earth once each day, maintaining fixed locations above the equator.

Question If you could spot a geosynchronous satellite in the sky, how could you distinguish it from a star?

Answer The satellite remains in the same location in the sky, whereas the stars drift westward as the Earth rotates under them.

> PHYSICS | **ON YOUR OWN**
>
> Estimate the locations of the weather satellites based on the views of the United States they present during the evening weather forecast and your knowledge of the possible orbits of geosynchronous satellites.

Any space probe requires the same computations as those done for satellites; NASA's computers calculate the trajectories for space flights using Newton's laws of motion and the law of gravitation. The forces on the spacecraft at any time depend on the positions of the other bodies in the Solar System. These can be calculated with the gravitation equation by inserting the distance to and the mass of each body. The net force produces an acceleration of the spacecraft that changes its velocity. From this the computer calculates a new position for the spacecraft. It also calculates new positions for the other celestial bodies, and the process starts over. In this manner the computer plots the path of the spacecraft through the Solar System.

Tides

Before Newton's work with gravity, no one was able to explain why we have tides. Some things were known. The tides are due to bulges in the surface of Earth's oceans. There are two bulges, one on each side of Earth, as shown in Figure 5-4. The occurrence of tides at a given location is due to Earth's rotation. Imagine for simplicity that the bulges are stationary—pointing in some direction in space—and that Earth is rotating. Each point on Earth passes through both bulges in 24 hours, and we have high tides at these times. Low tides occur halfway between the bulges. So we have two low and two high tides each day.

What wasn't known was why Earth had these bulges. Newton claimed they were due to the Moon's gravity. Earth exerts a gravitational force on the Moon that causes the Moon to orbit it. But the Moon exerts an equal and opposite force on Earth that causes Earth to orbit the Moon. Actually, both Earth and the Moon orbit a common point located between them. This point is the center of mass of the Earth–Moon system. Because Earth is so much more massive than the Moon, the center of mass is much closer to Earth. In fact, its location is inside Earth, as shown in Figure 5-5. Earth's orbital motion would look more like a wobble to somebody viewing its motion from high above the North Pole. But it is an orbit.

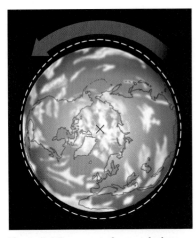

Figure 5-4 Exaggerated ocean bulges. As Earth rotates the bulges appear to move around Earth's surface.

Question Are the forces between Earth and the Moon a Newtonian third-law pair?

Answer Yes, one of the forces is the force of *Earth on Moon*, and the other is the force of the *Moon on Earth*.

Because Earth has an orbital motion, we can use the conclusions developed for the Moon's motion to help us understand Earth's tides. Namely, because we concluded that the Moon is continually falling toward Earth, Earth then is continually falling toward the Moon. This centripetal acceleration toward the Moon is the key to understanding tidal bulges.

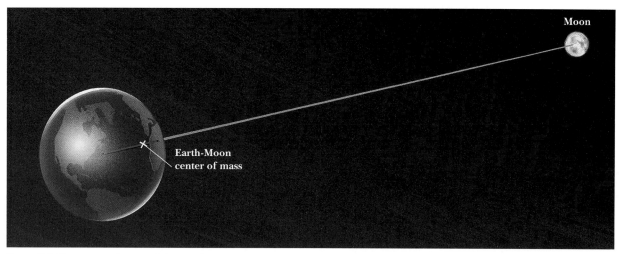

Figure 5-5 The center of mass of the Earth–Moon system is located inside Earth.

Forget momentarily that Earth is moving along its orbit and just consider Earth falling toward the Moon, as in Figure 5-6. This acceleration is the major contributor to the tides. Because the strength of the Moon's gravity gets weaker with increasing distance, the force on different parts of Earth is different. For example, on the side nearest the Moon, 1 kilogram of ocean water feels a stronger force than an equal mass of rock at Earth's center. Similarly, 1 kilogram of material on the far side of Earth feels a smaller force than both the kilogram on the near side and the one at the center.

If there are different-sized forces at different spots on Earth, there are different accelerations for different parts. Parts of Earth outrace other parts in their fall toward the Moon. Material on the side of Earth facing the Moon tries to get ahead, while the material on the other side lags behind. Of course, Earth has internal forces keeping it together that eventually balance these inequalities. But we do end up with a stretched-out Earth.

Although this reasoning accounts for the occurrence of the two high tides each day, it is too simple to get the details right. We observe that high tides do not occur at the same time each day. This happens because the Moon orbits the rotating Earth once a month. The normal time interval between successive high tides is 12 hours and 25 minutes. High tides do not occur when the Moon is overhead but later—as much as 6 hours later. This is due to such factors as the frictional and inertial effects of the water and the variable depth of the ocean.

Although the height difference between low and high tides in the middle of the ocean is only about 1 meter, the shape of the shoreline can greatly amplify the tides. The greatest tides occur in the Bay of Fundy, on the eastern seaboard between Canada and the United States; there the maximum range from low to high tide is 16 meters (54 feet)!

We would also expect to observe solar tides because the Sun also exerts a gravitational pull on Earth and Earth is "falling" toward the Sun. These do occur, but their heights are a little less than one-half those due to the Moon. This value may seem too small, taking into consideration that the Sun's gravitational force on Earth is about 180 times as large as the Moon's. The solar effect is so small because it is the difference in the force from one side of Earth to the other that matters and not the absolute size. The tides due to the planets are even smaller, that of Jupiter being less than one ten-millionth that due to the Sun.

The continents are much more rigid than the oceans. Even so, the land experiences measurable tidal effects. Land areas may rise and fall as much as 23 centimeters (9 inches). Because the entire area moves up and down together, we don't notice this effect.

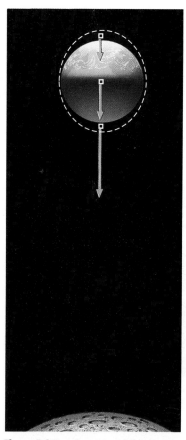

Figure 5-6 Equal masses of Earth experience different gravitational forces due to their different distances from the Moon. The effect is exaggerated in the diagram.

Photographs of the same view in the Minas Basin, Nova Scotia, at low and high tides

> **Question** Is the height of the high tide related to the phase of the Moon? That is, is it higher when the Sun and Moon are on the same side of Earth (new moon), when they are on opposite sides (full moon), or when they are at right angles to each other (first- or third-quarter moon)?
>
> **Answer** The highest high tides and the lowest low tides occur near new and full moons when Earth, the Moon, and the Sun are in a line.

How Far Does Gravity Reach?

The law of gravitation has been thoroughly tested within the Solar System. It accounts for the planets' motions, including their irregularities due to the mutual attraction of all the other planets.

What about tests outside the Solar System? We haven't sent probes out there. We are fortunate, however, because nature has provided us with ready-made probes. Astronomers observe that many stars in our galaxy revolve around a companion star. These binary star systems are the rule rather than the exception. These pairs revolve around each other in exactly the way predicted by Newton's laws.

Occasionally, a star is spotted that appears to be alone yet is moving in an elliptical path. Our faith in Newton's laws is so great that we assume a companion star is there; it is just not visible. Some of these invisible stars have later been detected because of signals they emit other than visible light.

Star clusters provide evidence of gravity at work between stars.

Photographs of star clusters show that the gravitational interaction occurs between stars. In fact, measurements show that all the stars in the Milky Way Galaxy are rotating about a common point under the influence of gravity. This has been used to estimate the total mass of the Galaxy and the number of stars in it. The Milky Way Galaxy is very similar in size and shape to our neighboring galaxy, the Andromeda Galaxy.

Such successes are a remarkable witness to Newton's genius. For more than two centuries, scientists applied his laws of motion and the law of gravitation without discovering any discrepancies. However, some exceptions to the Newtonian world view were eventually discovered. It should not take away from his fame to admit these exceptions. They only occur when we venture very far from the realm of our ordinary senses. In the world of very high velocities and extremely large masses, we must replace Newton's ideas with the theories of special and general relativity (Chapter 10). In the world of the extremely small, we must use the theories of quantum mechanics (Chapter 24). It should be noted, however, that when these newer theories are applied in the realm where Newton's laws work, the new theories give the same results.

The Andromeda Galaxy is similar to our own Milky Way Galaxy.

We also do not know if the value of G varies with time. No such variation has been detected, but a small variation with time could exist. Because the measurements of G are still limited in accuracy, it has been suggested that NASA orbit two satellites about each other. Accurate knowledge of the satellites' masses as well as their orbital data would give a more accurate value for G.

The Field Concept

Implicitly, we have assumed the force between two masses to be the result of some kind of direct interaction—sort of an action-at-a-distance interaction. This type of interaction is a little unsettling because there is no direct pushing or pulling mechanism in the intervening space. Gravitational effects are evident even in situations in which there is a vacuum between the masses.

It is both conceptually and computationally useful to separate the gravitational interaction into two distinct steps using the **field** concept. First, one of the objects modifies, by virtue of its mass, the surrounding space; it produces a **gravitational field** at every point in space. Second, the other object interacts, by virtue of its mass, with this gravitational field to experience the force. The field concept divides the task of determining the force on a mass into two distinct parts: determining the field from the first mass and then calculating the force that this field exerts on the second mass.

◀ gravitational field

If this were the only purpose of the field idea, it would play a minor role in our physics world view. In fact, it probably seems like we are trading one unsettling idea for another. However, as we continue our studies, we will find that the field takes on an identity of its own and is a valuable aid in understanding these and many other phenomena.

By convention the value of the gravitational field at any point in space is equal to the force experienced by a 1-kilogram mass if it were placed at that point. Then the gravitational force on any other object is the product of its mass and the gravitational field at that point. If you hold a 1-kilogram block near Earth's surface, it feels a gravitational force of 10 newtons. Therefore, the gravitational field has a magnitude of 10 newtons per kilogram at this point. If you replace this block with a 5-kilogram block, the gravitational force changes to 50 newtons; this is just the product of (5 kilograms) and (10 newtons per kilogram). We can also see that we do not need to use a 1-kilogram block to find the gravitational field. We could just as easily use the 5-kilogram block and divide the resulting force by the mass of the block: (50 newtons)/(5 kilograms) = 10 newtons per kilogram.

"THAT WRAPS IT UP—THE MASS OF THE UNIVERSE."

Because force is a vector quantity, the gravitational field is a vector field; it has a magnitude and a direction at each point in space. It is often convenient to talk about the gravitational field rather than the gravitational force. The strength of the gravitational force depends on the object being considered, whereas the strength of the gravitational field is independent of the object.

We saw in the last chapter that the gravitational force **W** could be expressed through Newton's second law as

$$\mathbf{W} = m\mathbf{g}$$

where **g** is the acceleration due to gravity, 9.8 (meters per second) per second. Because the gravitational force is equal to the mass times either the gravitational *field* or the gravitational *acceleration*, these two must be numerically the same. Indeed, we use the symbol **g** for both the gravitational field and the acceleration due to gravity. We can also show that newtons per kilogram may be rewritten as (meters per second) per second.

Summary

Although no one knows what gravity is or why it exists, we can accurately describe how gravity affects the motions of objects. The same laws of motion work on Earth and in the heavens. Newton's universal law of gravitation states that a gravitational attraction exists between every pair of objects and is given by

$$F = G\frac{m_1 m_2}{r^2}$$

where m_1 and m_2 are the masses of the two objects, r is the distance between their centers, and G is the gravitational constant. The value of G was first determined by Cavendish and is believed to be constant with time and space.

The higher a satellite's orbit, the longer it takes to complete one orbit. A satellite with a period of 1 day and positioned above the equator would appear to remain fixed in the sky. The Moon, a natural satellite, takes 27.3 days to complete one orbit around Earth.

The force of gravity can be considered constant when the motion occurs over short distances near Earth's surface. However, small variations occur in the

PhysicsNow™ Assess your understanding of this chapter's topics with sample tests and other resources found by logging into PhysicsNow at **http://physics.brookscole.com/kf6e**.

acceleration due to gravity with latitude, elevation, and the types of surface material. At larger distances the force decreases as the square of the distance. Stars in binary systems revolving around each other and the motion of stars within galaxies support this idea.

The value of the gravitational field at any point in space is equal to the force experienced by a 1-kilogram mass placed at that point.

Chapter 5 Revisited

Studying the motions of distant celestial objects, as well as the distributions of mass within multiparticle systems such as globular clusters and galaxies, provides direct evidence that gravity extends throughout the entire visible Universe.

KEY TERMS

field: A region of space that has a number or a vector assigned to every point.

gravitational field: The gravitational force experienced by a 1-kilogram mass placed at a point in space.

inverse-square: A relationship in which a quantity is related to the reciprocal of the square of a second quantity. An example is the law of universal gravitation; the force is inversely proportional to the square of the distance. If the distance is doubled, the force decreases by a factor of four.

law of universal gravitation: All masses exert forces on all other masses. The force F between any two objects is given by $F = Gm_1m_2/r^2$, where G is a universal constant, m_1 and m_2 are the masses of the two objects, and r is the distance between their centers.

CONCEPTUAL QUESTIONS

1. What force (if any) drives the planets along their orbits?
2. What force (if any) causes the planets to execute (nearly) circular orbits?
3. Is the size of the gravitational force that Earth exerts on the Moon smaller than, larger than, or the same size as the force the Moon exerts on Earth? Why?

4. Earth exerts a gravitational force of more than 1 million newtons on the International Space Station. What force does the ISS exert on Earth?
5. How does the average acceleration of the Moon about the Sun compare with that of Earth about the Sun?
6. If an apple were placed in orbit at the same distance from Earth as the Moon, what acceleration would the apple have?
7. What happens to the surface area of a cube when the length of each side is doubled? How does this compare with what happens to the surface area of a sphere when you double its radius?
8. What happens to the volume of a cube if the length of each side is doubled? How does this compare with what happens to the volume of a sphere when you double its radius?
9. A future space traveler, Skip Parsec, lands on the planet MSU3, which has the same mass as Earth but twice the radius. If Skip weighs 800 newtons on Earth's surface, how much does he weigh on MSU3's surface?
▲ 10. Astronaut Skip visits planet MSU8, which is composed of the same materials as Earth but has twice the radius. If Skip weighs 800 newtons on Earth's surface, how much does he weigh on MSU8's surface?
11. Why didn't Newton have to know the mass of the Moon to obtain the law of universal gravitation?
12. Comment on the following statement made by a TV newscaster during an Apollo flight to the Moon: "The spacecraft has now left the gravitational force of Earth."
13. In a parallel universe, there is a planet with the same mass and radius as Earth. However, when an apple is dropped on this planet, it falls with an acceleration of 20 (meters per second) per second. What is the value of the gravitational constant G in this parallel universe?
▲ 14. For simplicity we use 10 (meters per second) per second for the acceleration due to gravity instead of the more accurate 9.8 (meters per second) per second. If Cavendish had made the same approximation, would his estimate for Earth's mass have been too high or too low?

15. If a satellite in a circular orbit above Earth is continually "falling," why doesn't it quickly return to Earth?
16. As a satellite orbits Earth, the gravitational force is constantly pulling the satellite inward. What counters this force?
17. Astronaut Story Musgrave spent a total of 1281 hours, 59 minutes, and 22 seconds in space on his six space shuttle missions. If Story's mass was 80 kilograms, then the gravitational force acting on him in orbit was approximately 730 newtons. Why did he feel weightless?
18. You are standing on a bathroom scale in an elevator when suddenly the cable breaks and the elevator falls freely down the shaft. How does the reading on the scale change from just before to just after the cable breaks? How does the force of gravity that Earth exerts on you change over the same time interval?
19. NASA uses the famous Vomit Comet, a KC-135 cargo plane, to provide astronauts and scientists a simulated zero-gravity environment. The plane flies a series of parabolic arcs, as shown in the figure. Explain why the passengers feel "weightless" when the plane is near the top of its arc.

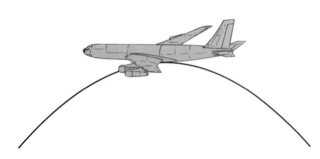

20. You have no doubt seen pictures of the astronauts floating around inside a space shuttle as it orbits some 300 kilometers above Earth's surface. Would the force of gravity Earth exerts on an astronaut be the same, a little less, or much less than the force that the astronaut would experience on Earth's surface? Why?
▲ 21. If Earth were hollow but still had the same mass and radius, would your weight be different? Why?
22. Astronomers believe that when Earth first formed, its composition was uniform. Over time, the heavier materials sank to the middle to create a dense iron core, with the less dense materials toward the outside. How did the value of the acceleration due to gravity at Earth's surface change while this process was occurring?
23. The gravitational force between two books sitting on a table does not cause them to accelerate toward each other because of frictional forces. If these same two books were floating near each other in deep space, they would still not appear to accelerate toward each other. Why not?
▲ 24. Why do we use the form $W = mg$ for the gravitational force on an object near Earth, but the form $F = Gm_1m_2/r^2$ when the object is far from Earth?
25. Skylab caused quite a commotion when it returned to Earth in July 1979. Why would it suddenly return to Earth after it had been in orbit for many years?

26. When the Hubble Space Telescope (HST) was originally launched by the space shuttle *Discovery*, its approximately circular orbit was at an altitude of about 600 kilometers. However, over the next several years, the altitude decreased so that subsequent servicing missions were required to lift the HST back into the higher orbit. What is responsible for the orbital decay?
27. How could we determine the mass of a planet such as Venus, which has no moon?
28. Why can we not determine the mass of the Moon by noting that it orbits Earth in a nearly circular orbit? What can we do to determine the Moon's mass?
29. Would you expect the value of the acceleration due to gravity to be larger or smaller than normal over a large deposit of uranium ore? Why?
30. You are steaming across the Atlantic Ocean on a large cruise ship. What happens to your weight as the ship leaves the deep waters of the North Atlantic and enters the shallow coastal waters of the United States?
31. What changes would occur in the Solar System if the gravitational constant G were slowly getting larger?
32. What do you think would happen to the Moon's orbit if the gravitational attraction between the Moon and Earth were slowly growing stronger?
33. Is it possible for an Earth satellite to remain stationary over Paris? Why or why not?
34. During the Gulf War with Iraq in 1991, a newspaper story reported that American spy satellites were in stationary orbits over Iraq, providing continuous intelligence information. Explain why this is impossible.
▲ 35. Assume that NASA fails in its attempt to put a communications satellite into geosynchronous orbit. If the orbit is too big, what apparent motion will the satellite have as seen from the rotating Earth?
36. Some of NOAA's (National Oceanic and Atmospheric Administration) Earth satellites remain above a single location on Earth. Why don't these geosynchronous satellites fall to Earth under the influence of gravity?

37. Newton's third law says that the gravitational force exerted on Earth by the Moon is equal to that exerted on the Moon by Earth. Why is it, then, that Earth doesn't appear to orbit the Moon?
38. The Sun has a profound influence on Earth's motion. Does Earth influence the Sun's motion? Explain.
39. When the tide is high along the American western seaboard, is the tide in Japan nearer high tide or low tide? Japan is approximately 90 degrees west of San Francisco.
40. When it is high tide off the coast of Ecuador, is the tide off the coast of Indonesia, which is 180 degrees around the globe from Ecuador, nearer high tide or low tide?
41. Jupiter rotates once in 9 hours 50 minutes. How long is it between high tides in its atmosphere?
▲ 42. The Moon is observed to keep the same side facing Earth at all times. If the Moon had oceans, how much time would elapse between its high tides?
43. Which position in the figure corresponds to the new moon? This is when the Moon is above the horizon but cannot be seen because the lit side faces away from Earth. Why are high tides higher than normal during this phase?
44. Which position in the figure corresponds to the full moon, which is when the Moon appears as a fully lit disk? Why are high tides higher than normal during this phase?
45. A classmate asserts that when the Moon is in position b in the figure, the gravitational effects of the Sun and Moon tend to cancel, producing lower than normal high tides. What is wrong with this reasoning?
46. Which positions of the Moon in the figure correspond to the smallest difference between high tide and low tide?
▲ 47. In *The Jupiter Effect*, authors John Gribbin and Stephen Plagemann claim that the additional tidal force produced when all the planets lie along one line might be enough to trigger an earthquake along the San Andreas Fault in California. What do you think about the possibility?
▲ 48. Why would the inertia and friction of water cause the tides to occur after the Moon passes overhead?
49. How does the magnitude of Earth's gravitational field change with increasing distance?
50. Show that the units newton per kilogram are equivalent to the units (meters per second) per second.
51. Is it possible for the gravitational field to point in two different directions at the same location in space? Why or why not?
52. Which of the following are independent of the mass of an object falling freely near Earth's surface: the acceleration of the object, the gravitational force acting on the object, the gravitational force acting on Earth, and the magnitude of the gravitational field?

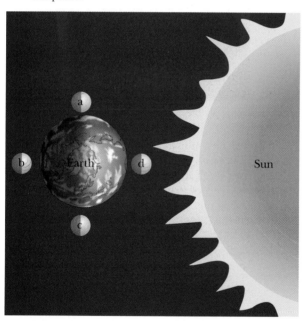

Questions 43–46

EXERCISES

1. Earth's speed in its orbit about the Sun is about 30 km/s. What is Earth's acceleration?
2. The Moon's speed in its orbit is approximately 1 km/s, and the Earth–Moon distance is 380,000 km. Show that these numbers yield an acceleration for the Moon that is very close to that given in the text.
3. What is the acceleration due to gravity at a distance of one Earth radius above Earth's surface?
4. If you were located halfway between Earth and the Moon, what acceleration would you have toward Earth? The Earth-Moon separation is 60 Earth radii. (Ignore the gravitational force of the Moon because it is much less than Earth's.)
5. A solid lead sphere of radius 10 m (about 66 ft across!) has a mass of about 57 million kg. If two of these spheres are floating in deep space with their centers 20 m apart, the gravitational attraction between the spheres is only 540 N (about 120 lb). How large would this gravitational force be if the distance between the centers of the two spheres were tripled?

6. Two spacecraft in outer space attract each other with a force of 20 N. What would the attractive force be if they were one-half as far apart?
7. How would the Sun's gravitational force on Earth change if Earth had twice its present mass? Would Earth's acceleration change?
8. The gravitational force between two very large metal spheres in outer space is 50 N. How large would this force be if the mass of each sphere were cut in half?
9. What is the ratio of the gravitational force on you when you are 6400 km above Earth's surface versus when you are standing on the surface? (Earth's radius is 6400 km.)
10. How does Earth's gravitational force on you differ when you are standing on Earth and when you are riding in a space shuttle 400 km above Earth's surface? (Earth's radius is 6400 km.)
11. A 320-kg satellite experiences a gravitational force of 800 N. What is the radius of the satellite's orbit? What is its altitude?
12. A 600-kg geosynchronous satellite has an orbital radius of 6.6 Earth radii. What gravitational force does Earth exert on the satellite?
13. What is the gravitational force between two 20-kg iron balls separated by a distance of 0.5 m? How does this compare with the weight of either ball?
14. The masses of the Moon and Earth are 7.4×10^{22} kg and 6×10^{24} kg, respectively. The Earth–Moon distance is 3.8×10^{8} m. What is the size of the gravitational force between Earth and the Moon? Does the acceleration of the Moon produced by this force agree with the value given in the text?
15. If an astronaut in full gear has a weight of 1200 N on Earth, how much will the astronaut weigh on the Moon?
16. The acceleration due to gravity on Titan, Saturn's largest moon, is about 1.4 m/s². What would a 50-kg scientific instrument weigh on Titan?
17. Mercury has a radius of about 0.38 Earth radius and a mass of only 0.055 Earth mass. Estimate the acceleration due to gravity on Mercury.
18. Mars has a radius of about 0.53 Earth radius and a mass of only 0.11 Earth mass. Estimate the acceleration due to gravity on Mars.
19. A geosynchronous satellite orbits at a distance from Earth's center of about 6.6 Earth radii and takes 24 h to go around once. What distance (in meters) does the satellite travel in one day? What is its orbital velocity (in m/s)?
20. An 80-kg satellite orbits a distant planet with a radius of 4000 km and a period of 280 min. From the radius and period, you calculate the satellite's acceleration to be 0.56 m/s². What is the gravitational force on the satellite?
21. The radius of Venus's orbit is 0.72 times that of Earth's orbit. How much stronger is the Sun's gravitational field at Venus than at Earth?

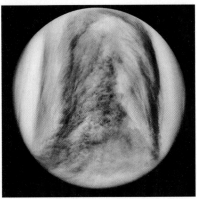

22. By what factor is Earth's gravitational field reduced at a distance of 4 Earth radii from the center of Earth?
23. If Earth shrank until its diameter were only one-half its present size without changing its mass, what would a 1-kg mass weigh at its surface?
24. If Earth expanded to twice its diameter without changing its mass, find the resulting magnitude of the gravitational field.

Interlude

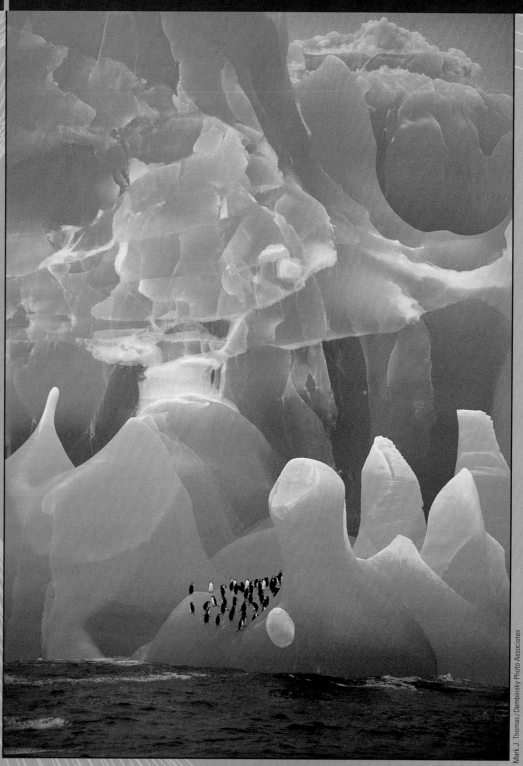

Chinstrip penguins on a rare blue iceberg.

The Discovery of Invariants

One theme of the physics world view has been that change is an essential part of the Universe. In fact, change has been so pervasive in our journey that you might be surprised to learn that some things don't change. A number of quantities are in fact constant, or invariant. For something to be invariant, the numerical value associated with it must be constant, meaning that we obtain the same value at all times. We say that the quantity is conserved. In our everyday language, we often use the word *conserve* to mean that something is saved, or at least used sparingly. In physics when something is conserved, it remains constant; its value does not change.

According to Swiss child psychologist Jean Piaget, an essential part of an individual's development is the establishment of invariants, or conservation rules. That is, as we process sensory data, we build constructs that are relatively permanent components of our personal world views. Most of this building occurs instinctively at an early age. We certainly don't go around consciously searching for invariants.

A striking example of such a learned invariant that occurs at a very early age is the permanence of the physical size of objects. Although the size of the image on your retina gets smaller as an object moves away from you, common sense tells you that the object remains the same size. How do you know this? Piaget would claim that because you have seen objects return to their original sizes many times, you therefore have formed this invariant.

> For something to be an invariant, the numerical value associated with it must be constant, meaning that we obtain the same value at all times.

The concept that objects have a permanent size is so much a part of our common sense that our brain overrides the sensory information it receives. When you perceive objects as being far away through depth perception clues, you automatically compensate for the smaller retinal images. Artists can create nonsensical situations by providing our brain with conflicting information, as in the optical illusion shown in the photographs.

One of the first invariants discovered was the quantity of matter. When a piece of wood burns, it loses most of its mass. This loss is quite obvious if it is burned on an equal-arm balance. As the wood burns, the balance rises continually, indicating a decrease in mass. Where does the lost mass go? Does it actually disappear, or does it simply escape into the air?

Burning wood in a closed container gives different results. Provided there is enough air to allow the wood to burn and the products of the burning—the ashes and smoke—don't escape, the equal-arm balance remains level. Thus, the mass of the closed system does not change.

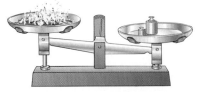

As the wood burns, the left side rises.

Which figure is taller?

Near the end of the 18th century, the chemist Antoine Lavoisier conducted a large number of experiments showing that mass does not change when chemical reactions take place in closed flasks. These investigations led to the generalization that the mass of any closed system is invariant. Lavoisier's generalization is known as a law of nature: the law of **conservation of mass**. (When we study relativity and nuclear physics, we will discover that this law has to be modified. However, for ordinary physical and chemical processes, this law is obeyed to a high degree of accuracy and is very useful in analyzing many processes.)

Other invariants have been discovered. Imagine a moving billiard ball hitting a stationary billiard ball head-on. After the collision, the first ball stops and the second has a velocity that is equal to the original velocity of the first ball. It appears that something is transferred from the first ball to the second. Christian Huygens, a contemporary of Galileo and Newton, suggested that a "quantity of motion" is invariant in this collision. This quantity of motion is transferred from one ball to the other.

Discovering invariants is difficult. A lot of effort has been spent and will continue to be spent searching for them because the rewards are worth it. Their discovery yields powerful generalizations in the physics world view.

In principle, we can use Newton's second law to predict the motion of any object, whether it be a leaf, an airplane, a billiard ball, or even a planet. All we have to do is determine the net force acting on the object and calculate the resulting acceleration. Solving these equations, however, often turns out to be quite difficult. To use the second law, we need to know all the forces acting on the object at all times. This is complicated for most motions because the forces are often not constant in direction or size.

The use of invariants often allows us to bypass the details of the individual interactions. In some cases, such as those involving nuclear forces, it is not a case of trying to avoid these troubling complications, but rather of taking the only possible path; we don't know the forces well enough to apply Newton's laws.

In the next three chapters, we will look at three important invariants—linear momentum, energy, and angular momentum. These three "quantities of motion" allow us to look at the behaviors of objects and systems of objects from an entirely new point of view.

6 | Momentum

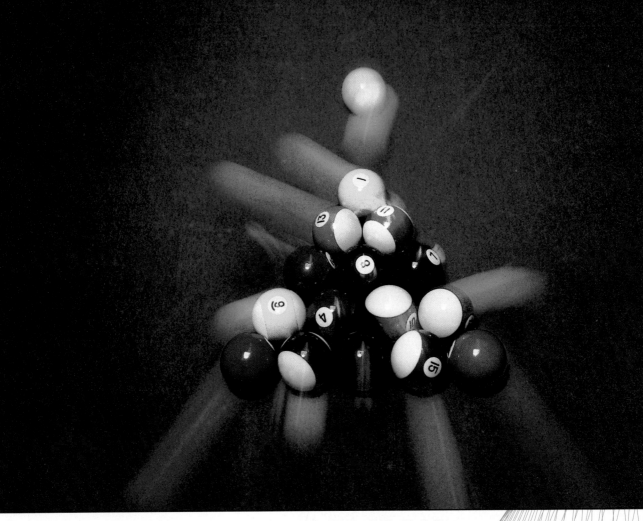

The collisions of billiard balls illustrate the law of conservation of linear momentum.

Collisions send objects scattering in seemingly random directions. Yet nature has a scheme—a rule—that governs the path of each individual particle in collisions and explosions, whether it is a group of billiard balls or colliding galaxies. If the rule predicts the future, can it also tell us about the past?

(See page 106 for the answer to this question.)

We know that it is harder to get a more massive object moving from rest than a less massive object. This is the concept of inertia that we have already included in our world view. We also know intuitively that for two objects moving with the same speed, the more massive object would be the harder one to stop. For example, if a kayak and a battleship were both moving through the harbor at 5 kilometers per hour, you would clearly have an easier time stopping the kayak.

We need to add a new concept to our world view to address the question, "How hard would it be to stop an object?" We call this new concept momentum, and it depends on both the mass of the object and how fast it is moving.

PhysicsNow™ Test your understanding of this chapter by logging into PhysicsNow at http://physics.brookscole.com/kf6e, selecting the chapter, and clicking on the "Take a Pre-Test" link.

The kayak and the ship have different momenta, even when they are moving at the same speed, because of their different masses.

Linear Momentum

The **linear momentum** of an object is defined as the product of its mass and its velocity. Momentum is a vector quantity that has the same direction as the velocity. Using the symbol **p** for momentum, we write the relationship as

$$\mathbf{p} = m\mathbf{v}$$

◀ linear momentum = mass × velocity

The adjective *linear* distinguishes this from another kind of momentum that we will discuss in Chapter 8. Unless there is a possibility of confusion, this adjective is usually omitted.

There is no special unit for momentum as there is for force; the momentum unit is simply that of mass times velocity (or speed)—that is, kilogram-meter per second (kg·m/s).

An object may have a large momentum due to a large mass, a large velocity, or both. A slow battleship and a rocket-propelled race car have large momenta.

Question Which has the greater momentum, an 18-wheeler parked at the curb or a Volkswagen rolling down a hill?

Answer Because the 18-wheeler has zero velocity, its momentum is also zero. Therefore, the VW has the larger momentum as long as it is moving.

The word *momentum* is often used in our everyday language in a much looser sense, but it is still roughly consistent with its meaning in the physics world view; that is, something with a lot of momentum is hard to stop. You have probably heard someone say, "We don't want to lose our momentum!" Coaches are particularly fond of this word.

The catcher is protected from the baseball's momentum.

Changing an Object's Momentum ✓ MATH

✓ Extended presentation available in the *Problem Solving* supplement

The momentum of an object changes if its velocity or mass changes, or both. We can obtain an expression for the amount of change by rewriting Newton's second law ($\mathbf{F}_{net} = m\mathbf{a}$) in a more general form. Actually, Newton's original formulation is closer to the new form. Newton realized that mass as well as velocity could change. His form of the second law says that the net force is equal to the change in the momentum divided by the time required to make this change:

$$\mathbf{F}_{net} = \frac{\Delta(m\mathbf{v})}{\Delta t}$$

◀ net force = change in momentum / time taken

impulse = net force × time = change in momentum ▶

If we now multiply both sides of this equation by the time interval Δ*t*, we get an equation that tells us how to produce a change in momentum:

$$\mathbf{F}_{net}\Delta t = \Delta(m\mathbf{v})$$

This relationship tells us that this change is produced by applying a net force to the object for a certain time interval. The interaction that changes an object's momentum—a force acting for a time interval—is called **impulse**. Impulse is a vector quantity that has the same direction as the net force.

Because impulse is a product of two things, there are many ways to produce a particular change in momentum. For example, two ways of changing an object's momentum by 10 kilogram-meters per second are to exert a net force of 5 newtons on the object for 2 seconds or to exert 100 newtons for 0.1 second. They each produce an impulse of 10 newton-seconds (N·s) and therefore a momentum change of 10 kilogram-meters per second. The units of impulse (newton-seconds) are equivalent to those of momentum (kilogram-meters per second).

Question Which of the following will cause the larger change in the momentum of an object—a force of 2 newtons acting for 10 seconds or a force of 3 newtons acting for 6 seconds?

Answer The larger impulse causes the larger change in the momentum. The first force yields an impulse of (2 newtons)(10 seconds) = 20 newton-seconds; the second, of (3 newtons)(6 seconds) = 18 newton-seconds. Therefore, the first impulse produces the larger change in momentum.

Modern cars employ air bags to protect passengers by increasing their stopping time.

Although the momentum change may be the same, certain effects depend on the particular combination of force and time. Suppose you had to jump from a second-story window. Would you prefer to jump onto a wooden or a concrete surface? Intuitively, you would choose the wooden one. Our commonsense world view tells us that jumping onto a surface that "gives" is better. But why is this so?

You undergo the same change in momentum with either surface; your momentum changes from a high value just before you hit to zero afterward. The difference is in the time needed for the collision to occur. When a surface gives, the collision time is longer. Therefore, the average net force must be correspondingly smaller to produce the same impulse.

Because our bones break when forces are large, the particular combination of force and time interval is important. For a given momentum change, a short collision time could cause large enough forces to break bones. You may break a leg landing on the concrete. On the other hand, the collision time with wood might be large enough to keep the forces in a huge momentum change from doing any damage.

This idea has many applications. Dashboards in cars are covered with foam rubber to increase the collision time during an accident. New cars are built with shock-absorbing bumpers to minimize damage to cars and with air bags to minimize injuries to passengers. The barrels of water or sand in front of highway median strips serve the same purpose. Stunt people are able to leap from amazing heights by falling onto large air bags that increase their collision times on landing. Volleyball players wear knee pads. Small pieces of Styrofoam are used as packing material in shipping boxes to smooth out the bumpy rides.

Even without a soft surface, we have learned how to increase the collision time when jumping. Instead of landing stiff-kneed, we bend our knees immediately upon colliding with the ground. We are then brought to rest gradually rather than abruptly.

A pole-vaulter lands on thick pads to increase the collision time and thus reduce the force.

Landing the Hard Way: No Parachute!

An extreme example of minimizing the effects of momentum change occurred during World War II. A Royal Air Force rear gunner jumped (without a parachute!) from a flaming Lancaster bomber flying at 5500 meters (18,000 feet). He attained a terminal speed (no pun intended) of more than 54 meters per second (120 mph) but survived because his momentum change occurred in a series of small impulses with some branches of a pine tree and a final impulse from 46 centimeters (18 inches) of snow. Because this took a longer time than hitting the ground directly, the forces were reduced. Miraculously, he suffered only scratches and bruises.

The record for surviving a fall without a parachute is held by Vesna Vulovic. She was serving as a hostess on a Yugoslavian DC-9 that blew up at 10,160 meters (33,330 feet) in 1972. She suffered many broken bones and was hospitalized for 18 months after being in a coma for 27 days.

Source: *Guinness Book of Records* (New York: Bantam Books, 1999).

PHYSICS | **ON YOUR OWN**

Play catch with a raw egg. After each successful toss, take one step away from your partner. How do you catch the egg to keep from breaking it?

Conservation of Linear Momentum

Imagine standing on a giant skateboard that is at rest (Figure 6-1[a]). What is the total momentum of you and the skateboard? It must be zero because everything is at rest. Now suppose that you walk on the skateboard. What happens to the skateboard? When you walk in one direction, the skateboard moves in the other direction, as shown in Figure 6-1(b). An analogous thing happens when you fire a rifle: the bullet goes in one direction, and the rifle recoils in the opposite direction.

These situations can be understood even though we don't know the values of the forces—and thus the impulses—involved. We start by assuming that there is no net external force acting on the objects. In particular, we assume that the frictional forces are negligible and that any other external force—such as gravity—is balanced by other forces.

When you walk on the skateboard, there is an interaction. The force you exert on the skateboard is, by Newton's third law, equal and opposite to the force the skateboard exerts on you. The time intervals during which these forces act on you and the skateboard must be the same because there is no way that one can touch the other without also being touched. Because you and the skateboard each experience the same force for the same time interval, you must each experience the same-size impulse and, therefore, the same-size change in momentum.

But impulse and momentum are vectors, so their directions are important. Because the impulses are in opposite directions, the changes in the momenta are also in opposite directions. Thus, your momentum and that of the skateboard still add to zero. In other words, even though you and the skateboard are moving and, individually, have nonzero momenta, the total momentum remains zero. Notice that we arrived at this conclusion without considering the details of the forces involved. It is true for all forces between you and the skateboard.

Question   Suppose the skateboard has half your mass and you walk at a velocity of 1 meter per second to the left. Describe the motion of the skateboard.

Answer   The skateboard must have the same momentum but in the opposite direction. Because it has half the mass, its speed must be twice as much. Therefore, its velocity must be 2 meters per second to the right.

Figure 6-1 (a) Person and skateboard at rest have zero momentum. (b) When the person walks to the right, the board moves to the left, keeping the total momentum zero.

Because the changes in the momenta of the two objects are equal in size and opposite in direction, the value of the total momentum does not change. We say that the total momentum is **conserved**.

We can generalize these findings. Whenever any object is acted on by a force, there must be at least one other object involved. This other object might be in actual contact with the first, or it might be interacting at a distance of 150 million kilometers, but it is there. If we widen our consideration to include all of the interacting objects, we gain a new insight.

Consider the objects as a system. Whenever there is no net force acting on the system from the outside (that is, the system is isolated, or closed), the forces that are involved act only between the objects within the system. As a consequence of Newton's third law, the total momentum of the system remains constant. This generalization is known as the law of **conservation of linear momentum**.

conservation of linear momentum ▶

> The total linear momentum of a system does not change if there is no net external force.

This means that if you add up all of the momenta now and leave for a while, when you return and add the momenta again, you will get the same number even if the objects were bumping and crashing into each other while you were gone. In practice we apply the conservation of momentum to systems where the net external force is zero or the effects of the forces can be neglected.

You experience conservation of momentum firsthand when you try to step from a small boat onto a dock. As you step toward the dock, the boat moves away from the dock, and you may fall into the water. Although the same effect occurs when we disembark from an ocean liner, the large mass of the ocean liner reduces the speed given it by our stepping off. A large mass requires a small change in velocity to undergo the same change in momentum.

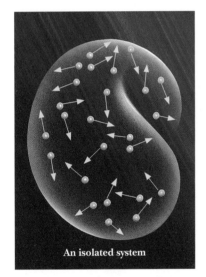

The total linear momentum of an isolated system is conserved. A system is isolated if there's no net external force acting on it.

WORKING IT OUT | Momentum

Let's calculate the recoil of a rifle. A 150-grain bullet for a 30-06 rifle has a mass m of 0.01 kg and a muzzle velocity v of 900 m/s (2000 mph). Therefore, the magnitude of the momentum p of the bullet is

$$p = mv = (0.01 \text{ kg})(900 \text{ m/s}) = 9 \text{ kg} \cdot \text{m/s}$$

Because the total momentum of the bullet and rifle was initially zero, conservation of momentum requires that the rifle recoil with an equal momentum in the opposite direction. If the mass M of the rifle is 4.5 kg, the speed V of its recoil is given by

$$V = \frac{p}{M} = \frac{9 \text{ kg} \cdot \text{m/s}}{4.5 \text{ kg}} = 2 \text{ m/s}$$

If you do not hold the rifle snugly against your shoulder, the rifle will hit your shoulder at this speed (4.5 mph!) and hurt you.

Question Why does holding the rifle snugly reduce the recoil effects?

Answer Holding the rifle snugly increases the recoiling mass (your mass is now added to that of the rifle) and therefore reduces the recoil speed.

Although we don't notice it, the same effect occurs whenever we walk. Our momentum changes; the momentum of something else must therefore change in the opposite direction. The something else is Earth. Because of its enormous mass, Earth's speed need change by only an infinitesimal amount to acquire the necessary momentum change.

Collisions

Interacting objects don't need to be initially at rest for conservation of momentum to be valid. Suppose a ball moving to the left with a certain momentum crashes head-on with an identical ball moving to the right with the same-size momentum. Before the collision, the two momenta are equal in size but opposite in direction, and because they are vectors, they add to zero.

After the collision the balls move apart with equal momenta in opposite directions. Because the masses of the balls are the same, the speeds after the collision are also the same. These speeds depend on the type of ball. The speeds may be almost as large as the original speeds in the case of billiard balls, quite a bit smaller in the case of lead balls, or even zero if the balls are made of soft putty and stick together. In all cases the two momenta are the same size and in opposite directions. The total momentum remains zero.

Flawed Reasoning

A question on the final exam asks, "What do we mean when we claim that the total momentum is conserved during a collision?" The following two answers are given:

Answer 1: Total momentum of the system stays the same before and after the collision.

Answer 2: Total momentum of the system is zero before and after the collision.

Which answer (if either) do you agree with?

Answer While we have considered several examples in which the total momentum of the system is zero, this is not the most general case. The momentum of a system can have any magnitude and any direction before the collision. If momentum is conserved, the momentum of the system always has the same magnitude and direction after the collision. Therefore, answer 1 is correct. This is a very powerful principle because of the word *always*.

A boxcar traveling at 10 meters per second approaches a string of four identical boxcars sitting stationary on the track. The moving boxcar collides and links with the stationary cars, and the five boxcars move off together along the track. What is the final speed of the five cars immediately after the collision?

Conservation of momentum tells us that the total momentum must be the same before and after the collision. Before the collision, one car is moving at 10 meters per second. After the collision, five identical cars are moving with a common final speed. Because the amount of mass that is moving has increased by a factor of five, the speed must decrease by a factor of five. The cars will have a final speed of 2 meters per second. Notice that we did not have to know the mass of each boxcar, only that they all had the same mass.

We can use the conservation of momentum to measure the speed of fast-moving objects. For example, consider determining the speed of an arrow shot from a bow. We first choose a movable, massive target—a wooden block sus-

Figure 6-2 Determining the speed of an arrow using conservation of momentum. The momentum of the block and arrow after the collision is equal to the momentum of the arrow before the collision.

pended by strings. Before the arrow hits the block (Figure 6-2[a]), the total momentum of the system is equal to that of the arrow (the block is at rest). After the arrow is embedded in the block (Figure 6 2[b]), the two move with a smaller, more measurable speed. The final momentum of the block and arrow just after the collision is equal to the initial momentum of the arrow. Knowing the masses, the arrow's initial speed can be determined.

Another example of a small, fast-moving object colliding with a much more massive object is graphically illustrated by one of your brave(?) authors, who lies down on a bed of nails, as shown in Figure 6-3. (This in itself may seem like a remarkable feat. However, your author does not have to be a fakir with mystic powers, because he knows that the weight of his upper body is supported by 500 nails so that each nail has to support only 0.4 pound. It takes approximately 1 pound of force for the nail to break the skin.) Once on the bed of nails, he places a board on his chest and tops that off with a concrete block, as shown in Figure 6-3(a). He then invites a student to break the concrete block with a sledgehammer (Figure 6-3[b]).

This dramatic demonstration illustrates several ideas. The board on the chest spreads out the blow so that the force on any one part of the chest is small. Because it takes time for the hammer to break through the concrete block, the collision time is increased, and the force is therefore decreased even further. Finally, momentum is conserved in the collision, but the much larger mass of your author ensures that the velocity imparted to his body is much less than the velocity of the hammer. This means that his body is only slowly pushed down onto the nails, and the additional force that each nail must exert to stop his body is small. Therefore, your author's back is not perforated, and he wins the admiration of his students without sacrificing his body.

Figure 6-3 Your author demonstrates some physics by letting a student break a concrete block on his chest while lying on a bed of nails.

Investigating Accidents

Accident investigators use conservation of momentum to reconstruct automobile accidents. Newton's laws can't be used to analyze the collision itself because we do not know the detailed forces involved. Conservation of linear momentum, however, tells us that regardless of the details of the crash, the total momentum of the two cars remains the same. The total momentum immediately before the crash must be equal to that immediately after the crash. Because the impact takes place over a very short time, we normally ignore frictional effects with the pavement and treat the collision as if there were no net external forces.

As an example, consider a rear-end collision. Assume that the front car was stopped and the two cars locked bumpers on impact. From an analysis of the length of the skid marks made *after* the collision and the type of surface, the total momentum of the two cars just after the collision can be calculated. (We will see how to do this in Chapter 7; for now, assume we know their total momentum.) Because one car was stationary, the total momentum before the crash must have been due to the moving car. Knowing that the momentum is the product of mass and velocity (*mv*), we can compute the speed of the car just before the collision. We can thus determine whether the driver was speeding.

The initial speeds of these cars can be determined by analyzing the collision.

WORKING IT OUT | A Collision ✓ MATH

Let's use conservation of momentum to analyze this collision. For simplicity assume that each car has a mass of 1000 kg (a metric ton) and that the cars traveled along a straight line. Further assume that we have determined that the speed of the two cars locked together was 10 m/s (about 22 mph) after the crash. The total momentum after the crash was equal to the total mass of the two cars multiplied by their combined speed:

$$p = (m_1 + m_2)v = (1000 \text{ kg} + 1000 \text{ kg})(10 \text{ m/s}) = 20{,}000 \text{ kg} \cdot \text{m/s}$$

But because momentum is conserved, this was also the value before the crash. Before the crash, however, only one car was moving. So if we divide this total momentum by the mass of the moving car, we obtain its speed:

$$v = \frac{p}{m} = \frac{20{,}000 \text{ kg} \cdot \text{m/s}}{1000 \text{ kg}} = 20 \text{ m/s}$$

The car was therefore traveling at 20 m/s (about 45 mph) at the time of the accident.

Question If the stationary car was not stationary but was slowly rolling in the direction of the total momentum, how would the calculated speed of the other car change if the final momentum remains the same?

Answer Because the rolling car accounts for part of the total momentum before the collision, the other car had less initial momentum and therefore a lower speed.

Assuming that the cars stick together after the collision simplifies the analysis but is not required. Conservation of momentum applies to all types of collisions. Even if the two cars do not stick together, the original velocity can be determined if the velocity of each car just after the collision can be determined. The cars do not even have to be going in the same initial direction. If the cars suffer a head-on collision, we must be careful to include the directions of the momenta, but the procedure of equating the total momenta before and after the accident remains the same.

Studying physics in a billiard parlor improves both your physics and your game.

> **Flawed Reasoning**
>
> Two students are arguing about a collision between two gliders on an air track. Glider A hits glider B, which is twice as large and initially stationary.
>
> **Jose:** "I think that glider B will have the largest final speed when glider A's final speed is zero. In this case, glider A gives *all* of its momentum to glider B."
>
> **Shaq:** "You are forgetting that momentum is a vector. If glider A bounces backward during the collision, it experiences a greater change in momentum than if it stops. Glider B must always experience the same change in momentum (but in the opposite direction) as glider A, so it would have a faster final speed in this case."
>
> **With which student (if either) do you agree?**
>
> **Answer** Shaq was paying attention in class. Anyone who has credit cards knows that it is possible to lose more than everything you have. If glider A is initially moving at 3 meters per second in the positive direction and stops, its change in velocity is −3 meters per second. If, on the other hand, the same glider bounces back with a final velocity of −2 meters per second, the change in its velocity is −5 meters per second. Because the change in momentum is just the mass times the change in velocity, bouncing results in the greater change of momentum.

Because momentum is a vector, this procedure can also be used in understanding two-dimensional collisions such as occur when cars collide while traveling at right angles to each other. The total vector momentum must be conserved.

PHYSICS | ON YOUR OWN

> Visit a billiard parlor and try the following momentum experiments. Ask yourself if momentum is conserved in each situation. (Keep things simple by not putting any additional spin on the ball.)
>
> a. A billiard ball strikes a stationary one head-on.
> b. A billiard ball strikes two others placed one behind the other.
> c. A head-on collision occurs between billiard balls with equal but opposite velocities.
> d. A ball with a smaller mass (for instance, a snooker ball) collides head-on with a stationary billiard ball.
> e. A billiard ball collides head-on with a stationary ball with a smaller mass.
> f. A billiard ball experiences a glancing blow with a stationary one so that the two balls no longer travel along a straight line.

Airplanes, Balloons, and Rockets

Conservation of momentum also applies to flight. If we look only at the airplane, momentum is certainly not conserved. It has zero momentum before takeoff, and its momentum changes many times during a flight.

But if we consider the system of the airplane plus the atmosphere, momentum is conserved. In the case of a propeller-driven airplane, the interaction occurs when the propeller pushes against the surrounding air molecules, increasing their momenta in the backward direction. This is accompanied by an equal

In flight a Cessna 172's propeller blades push the air backward to go forward.

NOETHER | The Grammar of Physics

Upon entering Plato's Academy, the novice of science was enjoined, "Let None Enter Here Who Know Not Geometry." Mathematics has always been the language of the physical sciences. In recent centuries science students have sometimes been frustrated by the bewildering array of symbols and manipulations unique to mathematics: to provide the most general statements for a range of phenomena, it has become necessary to state the rules and theories in mathematical terms. So students of science have undertaken increasingly complex studies of mathematics simply to master the basic grammar of science.

For centuries it was nearly impossible for half the human race—women—even to aspire to a life in science or mathematics. Fortunately, that changed dramatically in the 20th century. An example of hard struggle, sheer brilliance, and determination can be found in the life of Amalie Emmy Noether (1882–1935). She came from a family of distinguished parents and siblings. She attended Erlangen University, where her father was a research mathematician, and then Göttingen but was not allowed to matriculate formally. (It was just not done by women in the Germany of that day.) The quality of her work drew high praise, however, and she finally obtained a Ph.D. cum laude from Erlangen in 1907.

She was already attracting attention from the finest minds in Germany. Her work on algebraic invariants sparked an invitation by David Hilbert to teach at Göttingen in 1915. There she supplied some elegant formulations for Albert Einstein's general the-

Amalie Emmy Noether

ory of relativity. But she was never one of the "boys." Her official title was unofficial associate professor. Yet she was a widely lauded, paid instructor in algebra.

Her work on the relationships between the symmetries of space and time and the conservation laws is widely used in modern theoretical physics: If the equations of the theory do not explicitly contain time, energy is conserved. If the theory does not depend on translations in space, linear momentum is conserved. And, likewise, if the theory does not depend on rotations of space, angular momentum is conserved.

When Adolf Hitler came to power in 1933, Noether, like many Jews, was expelled. She migrated to the United States, where she acquired a regular professorship at Bryn Mawr College in Pennsylvania. She was also an associate of Albert Einstein at the Institute of Advanced Study in Princeton.

Her unexpected and untimely death occurred in the course of a routine surgical procedure. Today she is regarded as the finest female mathematician in the history of that discipline.

—Pierce C. Mullen, historian and author

Sources: Auguste Dick, *Emmy Noether, 1882–1935*, trans. H. I. Blocher (Boston: Birkhaeser, 1981). For a general study of women in science, see Merelene F. Rayner-Canham, *A Devotion to Their Science: Pioneer Women of Radioactivity* (Montreal: McGill–Queen's University Press, 1997); and Evelyn Fox Keller, *Reflections on Gender and Science* (New Haven, Conn.: Yale University Press, 1985).

---

change of the airplane's momentum in the forward direction. If we could ignore the air resistance, the airplane would continually gain momentum in the forward direction.

Question Why doesn't the airplane continually gain momentum?

Answer As the airplane pushes its way through the air, it hits air molecules, giving them impulses in the forward direction. This produces impulses on the airplane in the backward direction. In straight, level flight at a constant velocity, the two effects cancel.

Release an inflated balloon, and it takes off across the room. Is this similar to the propeller-driven airplane? No, because the molecules in the atmosphere are not necessary. The air molecules in the balloon rush out, acquiring a change in momentum toward the rear. This is accompanied by an equal change in momentum of the balloon in the forward direction. The air molecules do not need to push on anything; the balloon can fly through a vacuum.

This is also true of rockets and explains why they can be used in space flight. Rockets acquire changes in momentum in the forward direction by expelling gases at very high velocities in the backward direction. By choosing the direction of the expelled gases, the resulting momentum changes can also be used to change the direction of the rocket. An interesting classroom demonstration of this is often done using a modified fire extinguisher as the source of the high-velocity gas, as shown in Figure 6-4.

Figure 6-4 A fire extinguisher provides the impulse for a hallway rocket. Why does the person move backward?

PHYSICS ON YOUR OWN

You can make a popular model rocket (available in toy stores) work by filling it with water and then pumping it up with air. Place a piece of cardboard behind the rocket before you fire it. Notice where the water goes. What happens if you don't use any water?

Jet airplanes lie somewhere between propeller-driven airplanes and rockets. Jet engines take in air from the atmosphere, heat it to high temperatures, and then expel it at high speed out the back of the engine. The fast-moving gases impart a change in momentum to the airplane as they leave the engine. Although the gases do not push on the atmosphere, jet engines require the atmosphere as a source of oxygen for combustion.

PhysicsNow™ Assess your understanding of this chapter's topics with sample tests and other resources found by logging into PhysicsNow at http://physics.brookscole.com/kf6e.

Summary

The momentum of an object changes if its velocity or its mass changes. This change is produced by an impulse, a net force acting on the object for a certain time $\mathbf{F}\Delta t$. Impulse is a vector quantity with the same direction as the force; this is also the direction of the change in momentum. There are many ways of producing a particular change in momentum by changing the strength of the force and the time interval during which it acts.

The momentum of a system is the vector sum of all the momenta of the system's particles. Assuming that there is no net external force acting on the system, the total momentum does not change. This generalization is known as the law of conservation of linear momentum. Conservation of momentum applies to many systems, from balloons to billiard balls.

Chapter 6 Revisited

The rule—called the conservation of linear momentum—is valid in both directions of time. If we know the velocities and masses of all objects at any time, we can back up the equations to see where the objects came from, and we can go forward to see where they are headed.

KEY TERMS

conservation of linear momentum: If the net external force on a system is zero, the total linear momentum of the system does not change.

conserved: This term is used in physics to mean that a number associated with a physical property does not change; it is invariant.

impulse: The product of the force and the time during which it acts, $\mathbf{F}\Delta t$. This vector quantity is equal to the change in momentum.

linear momentum: A vector quantity equal to the product of an object's mass and its velocity, $\mathbf{p} = m\mathbf{v}$.

CONCEPTUAL QUESTIONS

1. What does it mean in physics to say that something is conserved?
2. Under what conditions is mass conserved?
3. Two identical carts with identical speeds collide head-on and stick together. Sydney argues, "Momentum for this system is conserved because the momentum of the first cart cancels the momentum of the second cart to give zero." Toby responds, "No, momentum is conserved because it's zero both before and after the collision." Which student do you agree with, and why?
4. A 2-kilogram cart moving at 6 meters per second hits a stationary 2-kilogram cart. The two move off together at 3 meters per second. Lee contends, "Momentum is conserved in this collision because the momentum of the system has the same value before and after the collision." Jackie counters, "The momentum of the system before

the collision is 12 kilogram-meters per second, not zero, so momentum is not conserved." With which student do you agree, and why?

5. Why are supertankers so hard to stop? To turn?
6. Which has the greater momentum, a parked cement truck or a child on a skateboard moving slowly down the street? Why?
7. State Newton's second law in terms of momentum.
8. State Newton's first law in terms of momentum.
9. How does the padding (or air pockets) in the soles of running shoes reduce the forces on your legs? Explain your answer in terms of impulse and momentum.
10. How does padding dashboards in automobiles make them safer? Explain your answer in terms of impulse and momentum.
11. An astronaut training at the Craters of the Moon in Idaho jumps off a platform in full space gear and hits the surface at 5 meters per second. If later, on the Moon, the astronaut jumps from the landing vehicle and hits the surface at the same speed, will the impulse be larger, smaller, or the same as that on Earth? Why?
12. Why is skiing into a wall of deep powder less hazardous to your health than skiing into a wall of bricks? Assume in both cases that you have the same initial speed and come to a complete stop. Explain your answer in terms of impulse and momentum.
13. Assume that a friend jumps from the roof of a garage and lands on the ground. How will the impulses the ground exerts on your friend compare if the landing is on grass or on concrete?
14. Why does an egg break when it is dropped onto a kitchen tile floor but not when it lands on a living room carpet?
15. A 2-kilogram sack of flour falls off the counter and lands on the floor. Just before hitting the floor, the sack has a speed of 4 meters per second. What impulse (magnitude and direction) does the floor exert on the sack?
16. A 2-kilogram rubber ball falls off a counter and lands on the floor. Just before hitting the floor, the ball has a speed of 4 meters per second. If the ball bounces, is the magnitude of the impulse the floor exerts on the ball less than, equal to, or greater than 8 kilogram-meters per second? Why?
17. Greg and Jeff are walking down the sidewalk when identical flowerpots fall out of a window above. One flowerpot lands on Greg's head and does not bounce while the other lands on Jeff's head and bounces. Which of the flowerpots experiences the greater impulse? Assuming that the collision time is the same for both cases, who ends up with the worse headache? Explain.
18. Two balls are dropped on the floor from the same height. The balls are made of different types of rubber so that one bounces back to nearly the same height while the other does not bounce at all. Assuming both balls have the same mass, which of the balls experiences the greater impulse in colliding with the floor? Why?
19. Explain why the 12-ounce boxing gloves used in amateur fights hurt less than the 6-ounce gloves used in professional fights.
20. You kick a soccer ball 15 meters without hurting your foot much. You then pump the ball up until it is really hard (the extra air does not significantly change the ball's mass) and again kick it 15 meters This time it hurts a lot. Using the concept of impulse, explain why it hurts more in the second case.
21. Two people are playing catch with a ball. Describe the momentum changes that occur for the ball, the people, and Earth. Is momentum conserved at all times?
22. Describe the momentum changes that occur when you dribble a basketball.
23. Which produces the larger change in momentum: a force of 3 newtons acting for 5 seconds or a force of 4 newtons acting for 4 seconds? Explain.
24. Which produces the larger impulse: a force of 3 newtons acting for 3 seconds or a force of 4 newtons acting for 2 seconds? Explain.
25. How can you explain the recoil that occurs when a rifle is fired?
▲ 26. How might you design a rifle that does not recoil?
27. Young Bill loves to fly model rockets. In his current project, however, he worries that once the rocket leaves the launch pad it will have nothing left to push on. To fix this, Bill fastens to the rocket, directly below its engine, an aluminum pie plate that will travel with the rocket. Explain why Bill will be sorely disappointed with the results.
28. A student who recently studied the law of conservation of linear momentum decides to propel a go-cart by having a fan blow on a board as shown in the figure. This idea won't work very well. Why not?

29. While a ball is falling toward the floor, it is continually speeding up and therefore increasing its momentum. Why is this not a violation of the law of conservation of linear momentum?
30. A cue ball hits a stationary eight ball on a pool table. For which of the following systems is there a change in momentum during the collision? Explain why.
    a. the cue ball
    b. the eight ball
    c. both balls
31. Two identical objects moving at the same speed collide with each other as shown in the figure. If the two objects stick together after the collision, will they be moving to the left, to the right, or not at all? Justify your answer using the concept of linear momentum.

32. An object of mass m and an object of mass $3m$, both moving at the same speed, collide with each other as shown in the figure. If the two objects stick together after the collision, will they be moving to the left, to the right, or

not at all? Justify your answer using the concept of linear momentum.

33. An object of mass m and an object of mass $3m$ collide with each other as shown in the figure. The lighter object is initially moving twice as fast as the heavier one. If the two objects stick together after the collision, will they be moving to the left, to the right, or not at all? Justify your answer using the concept of linear momentum.

34. Two identical objects, one moving twice as fast as the other, collide with each other as shown in the figure. If the two objects stick together after the collision, will they be moving to the left, to the right, or not at all? Justify your answer using the concept of linear momentum.

35. Your teacher runs across the front of the classroom with a momentum of 250 kilogram-meters per second and foolishly jumps onto a giant skateboard. The skateboard is initially at rest and has a mass equal to your teacher's. If you ignore friction with the floor, what is the total momentum of your teacher and the skateboard before and after the landing?

36. A friend is standing on a giant skateboard that is initially at rest. If you ignore frictional effects with the floor, what is the momentum of the skateboard if your friend walks to the right with a momentum of 150 kilogram-meters per second? What is the momentum of the skateboard–person system?

37. The figure shows two air-track gliders held together with a string. A spring is tightly compressed between the gliders and is released by burning the string. The mass of the glider on the left is twice that of the glider on the right, and they are initially at rest. What is the total momentum of both gliders after the release?

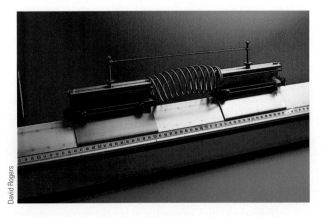

38. If the glider on the right in Question 37 has a speed of 2 meters per second after the release, how fast will the glider on the left be moving?

39. Explain why people who try to jump from rowboats onto docks often end up getting wet.

40. An astronaut in a space shuttle pushes off a wall to float across the room. What effect (if any) does this have on the motion of the shuttle?

41. Sometimes a star "dies" in an enormous explosion known as a supernova. What happens to the total momentum of such a star?

▲ 42. During a Fourth of July celebration, a rocket is launched from the ground and explodes at the top of its arc. If we ignore air resistance, what happens to the total momentum of all of the rocket's fragments?

43. An astronaut is floating in the center of a space station with no translational motion relative to the station. Is it possible for the astronaut to move to the floor? Explain why or why not.

44. During his last trip, Al the Astronaut happened on an enormous bag of gold coins floating in space. He quickly brought his spaceship to a halt, put on his space suit, tied a rope around his waist, and pushed off in the direction of the gold. But problems developed; as he reached the bag of gold, the rope broke. Devise a way of getting Al back to his spaceship before his oxygen runs out. Although Al cares most about his life, the creative problem solver can get Al back alive with money.

45. Two identical objects, one moving north and the other moving east, collide and stick together. If the northbound object is initially moving twice as fast as the eastbound object, which of the indicated paths represents the most likely final motion of the pair? Justify your answer using the concept of linear momentum.

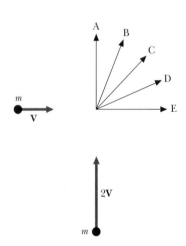

46. Two objects with the same speed, one moving north and the other moving east, collide and stick together. If the northbound object has twice the mass of the eastbound object, which of the indicated paths represents the most likely final motion of the pair? Justify your answer using the concept of linear momentum.

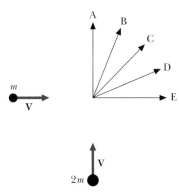

47. Two objects, one moving north and the other moving east, collide and stick together. If the eastbound object has three times the mass and is initially moving half as fast as the northbound object, which of the indicated paths represents the most likely final motion of the pair? Justify your answer using the concept of linear momentum.

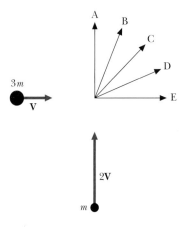

48. Two objects, one moving north and the other moving east, collide and stick together. If the eastbound object has twice the mass and is initially moving half as fast as the northbound object, which of the indicated paths represents the most likely final motion of the pair? Justify your answer using the concept of linear momentum.

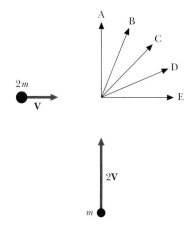

EXERCISES

1. What is the momentum of a 1200-kg sports car traveling down the road at a speed of 30 m/s?
2. Does a defensive end with a mass of 120 kg running at 6 m/s have a larger or smaller momentum than a running back with a mass of 100 kg running at 8 m/s?
3. How fast would you have to throw a baseball ($m = 145$ g) to give it the same momentum as a 10-g bullet traveling at 900 m/s?
4. How fast (in mph) would a person with a mass of 80 kg have to run to have the same momentum as an 18-wheeler ($m = 24,000$ kg) rolling along at 1 mph?
5. What average net force is needed to accelerate a 1500-kg car to a speed of 30 m/s in a time of 8 s?
6. It takes about 30 s for a jet plane to go from rest to the takeoff speed of 100 mph (44.7 m/s). What is the average horizontal force that the seat exerts on the back of a 60-kg passenger during takeoff? How does this force compare to the weight of the passenger?
7. What impulse is needed to stop a 1400-kg car traveling at 25 m/s?
8. A soft rubber ball ($m = 0.5$ kg) was falling vertically at 6 m/s just before it hit the ground and stopped. What was the impulse experienced by the ball? If the ball had bounced, would the impulse have been less than, equal to, or greater than what you calculated?
9. A 1500-kg car has a speed of 30 m/s. If it takes 8 s to stop the car, what are the impulse and the average force acting on the car?
10. A coach is hitting pop flies to the outfielders. If the baseball ($m = 145$ g) stays in contact with the bat for 0.04 s and leaves the bat with a speed of 50 m/s, what is the average force acting on the ball?
11. A very hard rubber ball ($m = 0.6$ kg) is falling vertically at 8 m/s just before it bounces on the floor. The ball rebounds at essentially the same speed. If the collision with the floor lasts 0.04 s, what is the average force exerted by the floor on the ball?
12. A tennis ball ($m = 0.2$ kg) is thrown at a brick wall. It is traveling horizontally at 12 m/s just before hitting the wall and rebounds from the wall at 8 m/s, still traveling

horizontally. The ball is in contact with the wall for 0.04 s. What is the magnitude of the average force of the wall on the ball?

13. A 150-grain 30-06 bullet has a mass of 0.01 kg and a muzzle velocity of 900 m/s. If it takes 1 ms (millisecond) to travel down the barrel, what is the average force acting on the bullet?

14. A 30-06 rifle fires a bullet with a mass of 10 g at a velocity of 800 m/s. If the rifle has a mass of 4 kg, what is its recoil speed?

15. A father ($m = 80$ kg) and son ($m = 40$ kg) are standing facing each other on a frozen pond. The son pushes on the father and finds himself moving backward at 3 m/s after they have separated. How fast will the father be moving?

16. A woman with a mass of 50 kg runs at a speed of 8 m/s and jumps onto a giant skateboard with a mass of 25 kg. What is the combined speed of the woman and the skateboard?

17. A 3-kg ball traveling to the right with a speed of 4 m/s collides with a 4-kg ball traveling to the left with a speed of 3 m/s. What is the total momentum of the two balls before and after the collision?

18. A 4-kg ball traveling to the right with a speed of 4 m/s collides with a 5-kg ball traveling to the left with a speed of 2 m/s. What is the total momentum of the two balls before they collide? After they collide?

19. A 1200-kg car traveling north at 14 m/s is rear-ended by a 2000-kg truck traveling at 25 m/s. What is the total momentum before and after the collision?

20. If the truck and car in Exercise 19 lock bumpers and stick together, what is their speed immediately after the collision?

21. Two identical boxcars ($m = 18,000$ kg) are traveling along the same track but in opposite directions. Both boxcars have a speed of 5 m/s. If the cars collide and link, what will be the final speed of the pair?

22. A boxcar traveling at 10 m/s approaches a string of three identical boxcars sitting stationary on the track. The moving boxcar collides and links with the stationary cars, and the four move off together along the track. What is the final speed of the four cars immediately after the collision?

7 Energy

Windmills are being developed to transform the energy in the wind to electric energy.

Energy is an important commodity. People and countries that know how to obtain and use energy are generally the wealthiest and most powerful. But what is energy? And what does it mean to say that energy is conserved?

(See page 129 for the answer to this question.)

PhysicsNow™ Test your understanding of this chapter by logging into PhysicsNow at **http://physics.brookscole.com/kf6e**, selecting the chapter, and clicking on the "Take a Pre-Test" link.

I N the preceding chapter, we came to understand momentum by considering the effect of a force acting for a certain time. If, instead, we look at an object's motion after the force has acted for a certain distance, we find another quantity that is sometimes invariant, the energy of motion. This is, however, only one form of a more general and much more profound invariant known as energy.

conservation of energy ▶

> The total energy of an isolated system does not change.

Energy is one of the most fundamental and far-reaching concepts in the physics world view. It took nearly 300 years to fully develop the ideas of energy and energy conservation. In view of its importance and popularity, you might think that it would be easy to give a precise definition of energy. Not so.

What Is Energy?

Nobel laureate Richard Feynman, in his *Lectures on Physics*, captures the essential character of energy and its many forms when he discusses the law of conservation of energy:

> There is a certain quantity, which we call energy, that does not change in the manifold changes which nature undergoes. That is a most abstract idea, because it is a mathematical principle; it says that there is a numerical quantity which does not change when something happens. It is not a description of a mechanism, or anything concrete; it is just a strange fact that we can calculate some number and when we finish watching nature go through her tricks and calculate the number again, it is the same. (Something like the bishop on a red square, and after a number of moves—details unknown—it is still on some red square. It is a law of this nature.) Since it is an abstract idea, we shall illustrate the meaning of it by an analogy.
>
> Imagine a child, perhaps "Dennis the Menace," who has blocks which are absolutely indestructible, and cannot be divided into pieces. Each is the same as the other. Let us suppose that he has 28 blocks. His mother puts him with his 28 blocks into a room at the beginning of the day. At the end of the day, being curious, she counts the blocks very carefully, and discovers a phenomenal law—no matter what he does with the blocks, there are always 28 remaining! This continues for a number of days, until one day there are only 27 blocks, but a little investigating shows that there is one under the rug—she must look everywhere to be sure that the number of blocks has not changed. One day, however, the number appears to change—there are only 26 blocks. Careful investigation indicates that the window was open, and upon looking outside, the other two blocks are found. Another day, careful count indicates that there are 30 blocks! This causes considerable consternation, until it is realized that Bruce came to visit, bringing his blocks with him, and he left a few at Dennis' house. After she has disposed of the extra blocks, she closes the window, does not let Bruce in, and then everything is going along all right, until one time she counts and finds only 25 blocks. However, there is a box in the room, a toy box, and the mother goes to open the toy box, but the boy says "No, do not open my toy box," and screams. Mother is not allowed to open the toy box. Being extremely curious, and somewhat ingenious, she invents a scheme! She knows that a block weighs three ounces, so she weighs the box at a time when she sees 28 blocks, and it weighs 16 ounces. The next time she wishes to check, she weighs the box again, subtracts sixteen ounces and divides by three. She discovers the following:
>
> $$\begin{pmatrix} \text{number of} \\ \text{blocks seen} \end{pmatrix} + \frac{(\text{weight of box}) - 16 \text{ ounces}}{3 \text{ ounces}} = \text{constant}$$
>
> There then appear to be some new deviations, but careful study indicates that the dirty water in the bathtub is changing its level. The child is throwing blocks into the water, and she cannot see them because it is so dirty, but she can find out how many blocks are in the water by adding another term to her formula. Since the original height of the water was 6 inches and each block raises the water a quarter of an inch, this new formula would be:

$$\left(\begin{array}{c}\text{number of}\\\text{blocks seen}\end{array}\right) + \frac{(\text{weight of box}) - 16 \text{ ounces}}{3 \text{ ounces}} + \frac{(\text{height of water}) - 6 \text{ inches}}{\frac{1}{4} \text{ inch}} = \text{constant}$$

In the gradual increase in the complexity of her world, she finds a whole series of terms representing ways of calculating how many blocks are in places where she is not allowed to look. As a result, she finds a complex formula, a quantity which *has to be computed*, which always stays the same in her situation....

The analogy has the following points. First, when we are calculating the energy, sometimes some of it leaves the system and goes away, or sometimes some comes in. In order to verify the conservation of energy, we must be careful that we have not put any in or taken any out. Second, the energy has a large number of *different forms*, and there is a formula for each one.... If we total up the formulas for each of these contributions, it will not change except for energy going in or out.

It is important to realize that in physics today, we have no knowledge of what energy is. We do not have a picture that energy comes in little blobs of a definite amount. It is not that way. However, there are formulas for calculating some numerical quantity, and when we add it all together it gives "28"—always the same number. It is an abstract thing in that it does not tell us the mechanisms or the *reasons* for the various formulas.*

Energy of Motion ✓ MATH

✓ Extended presentation available in the *Problem Solving* supplement

The most obvious form of energy is the one an object has because of its motion. We call this quantity of motion the **kinetic energy** of the object. Like momentum, kinetic energy depends on the mass and the motion of the object. But the expression for the kinetic energy is different from that for momentum. The kinetic energy *KE* of an object is

$$KE = \frac{1}{2}mv^2$$

◀ kinetic energy
= $\frac{1}{2}$ mass × speed squared

where the factor of $\frac{1}{2}$ makes the kinetic energy compatible with other forms of energy, which we will study later.

Notice that the kinetic energy of an object increases with the square of its speed. This means that if an object has twice the speed, it has four times the kinetic energy; if it has three times the speed, it has nine times the kinetic energy; and so on.

Question What happens to the kinetic energy of an object if its mass is doubled while its speed remains the same?

Answer Because the kinetic energy is directly proportional to the mass, the kinetic energy doubles.

The units for kinetic energy, and therefore for all types of energy, are kilograms multiplied by (meters per second) squared (kg · m²/s²). This energy unit is called a **joule** (rhymes with tool). Kinetic energy differs from momentum in that it is not a vector quantity. An object has the same kinetic energy regardless of its direction as long as its speed does not change.

A typical textbook dropped from a height of 10 centimeters (about 4 inches) hits the floor with a kinetic energy of about 1 joule (J). The kinetic energy of a 70-kilogram (154-pound) person running at a speed of 8 meters per second is

$$KE = \frac{1}{2}mv^2 = \frac{1}{2}(70 \text{ kg})(8 \text{ m/s})^2 = (35 \text{ kg})(64 \text{ m}^2/\text{s}^2) = 2240 \text{ J}$$

*R. P. Feynman, R. B. Leighton, and M. Sands, *The Feynman Lectures on Physics* (Reading, Mass.: Addison-Wesley, 1963), 1: 4-1 and 4-2.

Conservation of Kinetic Energy

The search for invariants of motion often involved collisions. In fact, early in their development, the concepts of momentum and kinetic energy were often confused. Things became much clearer when these two were recognized as distinct quantities. We have already seen that momentum is conserved during collisions. Under certain, more restrictive conditions, kinetic energy is also conserved.

Consider the collision of a billiard ball with a hard wall. Obviously, the kinetic energy of the ball is not constant. At the instant the ball reverses its direction, its speed is zero and therefore its kinetic energy is zero. As we will see, even if we include the kinetic energy of the wall and Earth, the kinetic energy of the system is not conserved.

However, if we don't concern ourselves with the details of what happens during the collision and look only at the kinetic energy before the collision and after the collision, we find that the kinetic energy is nearly conserved. During the collision the ball and the wall distort, resulting in internal frictional forces that reduce the kinetic energy slightly. But once again we ignore some things to get at the heart of the matter. Let's assume we have "perfect" materials and can ignore these frictional effects. In this case the kinetic energy of the ball after it leaves the wall equals its kinetic energy before it hit the wall. Collisions in which kinetic energy is conserved are known as **elastic** collisions.

Actually, many atomic and subatomic collisions are perfectly elastic. A larger-scale approximation of an elastic collision would be one between air-hockey pucks that have magnets stuck on top so that they repel each other.

PHYSICS | ON YOUR OWN

The apparatus shown in Figure 7-1 may look familiar. It has five balls of equal mass suspended in a line so that they can swing along the direction of the line. If you pull on a ball at one end and release it, the last ball at the other end leaves with the velocity of the original. Knowing that momentum and kinetic energy are conserved, can you predict what will happen if you pull back two balls and release them together? Try this experiment if you have one of these toys.

Figure 7-1 The collision-ball apparatus demonstrates the conservation of momentum and kinetic energy.

Collisions in which kinetic energy is lost are known as **inelastic** collisions. The loss in kinetic energy shows up as other forms of energy, primarily in the form of heat, which we will discuss in Chapter 13. Collisions in which the objects move away with a common velocity are never elastic. We will see this in the billiard ball example presented at the end of this section.

The outcomes of collisions are determined by the conservation of momentum and the extent to which kinetic energy is conserved. We know that the collisions of billiard balls are not perfectly elastic because we hear them collide. (Sound is a form of energy and therefore carries off some of the energy.)

PHYSICS | ON YOUR OWN

Determine the relative elasticities of balls made of various materials by dropping them from a uniform height onto a very hard surface, such as a concrete floor or a thick steel plate. The more elastic the material, the closer the ball will return to its original height. Is a rubber ball more or less elastic than a glass marble? Do your results agree with the everyday use of the word *elastic*?

WORKING IT OUT | Conservation of Kinetic Energy ✓ MATH

Collisions between very hard objects such as billiard balls are nearly elastic. Consider the head-on collision of a moving billiard ball with a stationary billiard ball (Figure 7-2[a]). The moving ball stops, and the stationary one acquires the initial velocity of the moving ball (Figure 7-2[b]). Suppose the mass of each ball is 0.2 kg, and the initial velocity is 4 m/s. The total momenta before and after the collision are

$$p(\text{before}) = m_1 v_1 + m_2 v_2 = (0.2 \text{ kg})(4 \text{ m/s}) + 0 = 0.8 \text{ kg} \cdot \text{m/s}$$

$$p(\text{after}) = m_1 v_1 + m_2 v_2 = 0 + (0.2 \text{ kg})(4 \text{ m/s}) = 0.8 \text{ kg} \cdot \text{m/s}$$

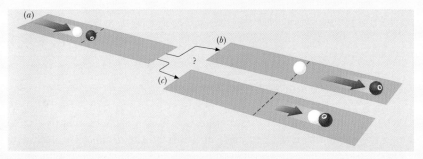

Figure 7-2 (a) A moving billiard ball collides head-on with a stationary one. Which possibility occurs? (b) The moving ball stops, and the stationary ball takes on the initial velocity. (c) The balls move off together with one-half the initial velocity.

Therefore, momentum is conserved. Likewise, the kinetic energies before and after the collision are

$$KE(\text{before}) = \frac{1}{2}m_1 v_1^2 + \frac{1}{2}m_2 v_2^2 = \frac{1}{2}(0.2 \text{ kg})(4 \text{ m/s})^2 + 0 = 1.6 \text{ J}$$

$$KE(\text{after}) = \frac{1}{2}m_1 v_1^2 + \frac{1}{2}m_2 v_2^2 = 0 + \frac{1}{2}(0.2 \text{ kg})(4 \text{ m/s})^2 = 1.6 \text{ J}$$

Therefore, kinetic energy is also conserved.

But this is not the only possibility that conserves momentum. Another is for the two balls to both move in the forward direction, each with one-half of the initial velocity (Figure 7-2[c]). Both possibilities conserve the total momentum of the two-ball system. Yet when we do the experiment, the first possibility is the one that we always observe.

Why does one occur and not the other? The first possibility conserves kinetic energy, but the second does not. To see that the second doesn't conserve kinetic energy, we again calculate the kinetic energy after the collision:

$$KE(\text{after}) = \frac{1}{2}m_1 v_1^2 + \frac{1}{2}m_2 v_2^2 = \frac{1}{2}(0.2 \text{ kg})(2 \text{ m/s})^2 + \frac{1}{2}(0.2 \text{ kg})(2 \text{ m/s})^2 = 0.8 \text{ J}$$

One-half of the kinetic energy is lost in this case.

Changing Kinetic Energy

A cart rolling along a frictionless, horizontal surface has a certain kinetic energy because it is moving. A net force on the cart can change its speed and thus its kinetic energy. If you push in the direction the cart is moving, you increase the cart's kinetic energy. Pushing in the opposite direction slows the cart and decreases its kinetic energy (Figure 7-3).

Figure 7-3 (a) A force in the direction of motion increases the kinetic energy. (b) A force opposite the direction of motion decreases the kinetic energy.

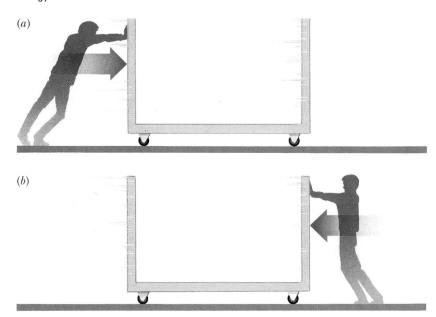

When a force acts for a certain time, the force produces an impulse that changes the momentum of the object. In contrast, the *distance* through which the force acts determines how much the kinetic energy changes. The product of the force F *in the direction of motion* and the distance moved d is known as the **work** W:

work = force × distance moved ▶

$$W = Fd$$

From the definition of work, we conclude that the units of work are newton-meters. If we substitute our previous expression for a newton (kg · m/s^2), we find that a newton-meter (N · m) equals a kg · m^2/s^2, which is the same as the units for energy—that is, a joule. Therefore, the units of work are the same as those of energy. In the U.S. customary system, the units of work are foot-pounds.

Newton's second law and our expressions for acceleration and distance traveled can be combined with the definition of work to show that the work done on an object is equal to the change in its kinetic energy:

work done = change in kinetic energy ▶

$$W = \Delta KE$$

If the net force is in the same direction as the velocity, the work is positive, and the kinetic energy increases. If the net force and velocity are oppositely directed, the work is negative, and the kinetic energy decreases.

Forces That Do No Work

The meaning of *work* in physics is different from the common usage of the word. Commonly, people talk about "playing" when they throw a ball and "working" when they study physics. The physics definition of work is quite precise—work occurs when the product of the force and the distance is nonzero. When you throw a ball, you are actually doing work on the ball; its kinetic energy is increased because you apply a force through a distance. Although you may move pages and pencils as you study physics, the amount of work is quite small.

Similarly, if you hold a suitcase above your head for 30 minutes, you would probably claim it was hard work. According to the physics definition, however, you did not do any work on the suitcase; the 30 minutes of straining and groaning did not change the suitcase's kinetic energy. A table could hold up the suitcase just as well. Your body, however, is doing physiological work because the muscles in your arms do not lock in place but rather twitch in response to nerve

Stopping Distances for Cars

The data for stopping distances given in driver's manuals do not seem to have a simple pattern, other than the faster you drive, the longer it takes to stop. The driver's manual for the state of Montana, for example, states that it takes at least 186 feet to stop a car traveling at 50 mph and only 65 feet for a car that is traveling at 25 mph. Why is the stopping distance one-third as much when the speed is one-half as much? Furthermore, these tables make no mention of the size of the car.

A car's kinetic energy depends on its speed and its mass. When the brakes are applied, the brake pads slow the wheels, which in turn apply a force on the highway. The reaction force of the highway on the car does work on the car, reducing its kinetic energy to zero. To stop the car, the total work done on the car must be equal to its initial kinetic energy.

The frictional force between the tires and the road depends on the car's mass and whether it is rolling or skidding. (Skidding greatly reduces the frictional force.) The force of friction does not change very much for various tire designs or road surfaces providing the roads are not wet or icy. Because both the frictional force and kinetic energy are proportional to the car's mass, the stopping distance is independent of the car's mass. So how far the car travels after the brakes are applied depends almost entirely on its speed.

The distances in driver's manual tables include the distance traveled during approximately 1 second of reaction time in addition to the distance required for the brakes to stop the car. At 50 mph the car is traveling at 74 feet per second, so the distance required to stop the car once the brakes are applied is 186 feet − 74 feet = 112 feet.

Because the car's kinetic energy changes as the square of its speed, a car that is traveling at 25 mph has only one-fourth the kinetic energy of one traveling at 50 mph. Because it requires only one-fourth as much work to stop the car, the force must only act through one-fourth the distance. A car traveling at 25 mph can stop in $\frac{1}{4} \times$ 112 feet = 28 feet. During the 1-second reaction time, the car travels an additional $\frac{1}{2} \times$ 74 feet = 37 feet so that the total stopping distance is 28 feet + 37 feet = 65 feet. The stopping distances for other speeds are given in the table.

Stopping Distances for Automobiles Traveling at Selected Speeds

Speed		Stopping Distance		
(mph)	(ft/s)	Reaction (ft)	Braking (ft)	Total (ft)
10	15	15	5	20
20	29	29	18	47
30	44	44	40	84
40	59	59	72	131
50	74	74	112	186
60	88	88	161	249
70	103	103	220	323
80	117	117	287	404
90	132	132	363	495
100	147	147	448	595

impulses. This work shows up as heat (as evidenced by your sweating) rather than as a change in the kinetic energy of the suitcase.

There are other situations in which a net force does not change an object's kinetic energy. If the force is applied in a direction perpendicular to its motion, the velocity of the object changes, but its speed doesn't. Therefore, the kinetic energy does not change. The definition of work takes this into account by stating that it is only the force that acts along the direction of motion that can do work.

Often, a force is neither parallel nor perpendicular to the displacement of an object. Because force is a vector, we can think of it as having two components, one that is parallel and one that is perpendicular to the motion as illustrated in Figure 7-4. The parallel component does work, but the perpendicular one does not do any work.

Consider an air-hockey puck moving in a circle on the end of a string attached to the center of the table shown in Figure 7-5. Because the speed is constant, the kinetic energy is also constant. The force of gravity is balanced by the upward force of the table. These vertical forces cancel and do no work. The tension that the string exerts on the puck is not canceled but does no work because it always acts perpendicular to the direction of motion.

If Earth's orbit were a circle with the Sun at the center, the gravitational force the Sun exerts on Earth would do no work, and Earth would have a constant kinetic energy and therefore a constant speed. However, because the orbit is an el-

It takes no work to hold a cheerleader in the air.

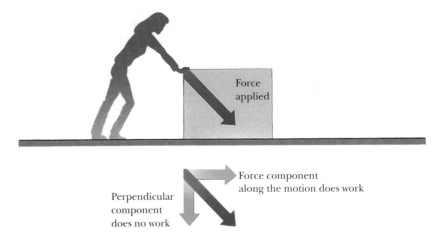

Figure 7-4 Any force can be replaced by two perpendicular component forces. Only the component along the direction of motion does work on the box.

Figure 7-5 When the force is perpendicular to the velocity, the force does no work.

lipse, the force is not always perpendicular to the direction of motion (Figure 7-6) and therefore does work on Earth. During one-half of each orbit, a small component of the force acts in the direction of motion, increasing Earth's kinetic energy and speed. During the other half of each orbit, the component is opposite the direction of motion, and Earth's kinetic energy and speed decrease.

Gravitational Potential Energy

When a ball is thrown vertically upward, it has a certain amount of kinetic energy that disappears as it rises. At the top of its flight, it has no kinetic energy, but as it falls, the kinetic energy reappears. If we believe that energy is an invariant, we must be missing one or more forms of energy.

The loss and subsequent reappearance of the ball's kinetic energy can be understood by examining the work done on the ball. As the ball rises, the force of gravity performs negative work on the ball, reducing its kinetic energy until it reaches zero. On the way back down, the force of gravity increases the ball's kinetic energy by the same amount it lost on the way up. Rather than simply saying that the kinetic energy temporarily disappears, we can retain the idea of the conservation of energy by defining a new form of energy. Kinetic energy is then transformed into this new form and later transformed back. This new energy is called **gravitational potential energy**.

We have some clues about the expression for this new energy. Its change must also be given by the work done by the force of gravity, and it must increase when the kinetic energy decreases, and vice versa. Therefore, we define the gravitational potential energy of an object at a height h above some zero level as equal to the work done by the force of gravity on the object as it falls to height zero. The gravitational potential energy GPE of an object near Earth's surface is then given by

$$GPE = mgh$$

▶ gravitational potential energy = force of gravity × height

Figure 7-6 The gravitational force of the Sun does work on Earth whenever the force is not perpendicular to Earth's velocity. The elliptical nature of the orbit has been exaggerated to show the parallel component.

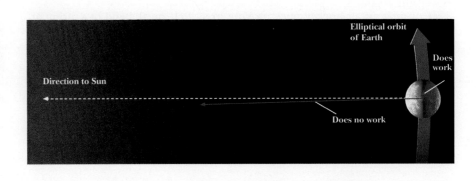

As an example, we can calculate the gravitational potential energy of a 6-kilogram ball located 0.5 meter above the level that we choose to call zero:

$$GPE = mgh = (6 \text{ kg})(10 \text{ m/s}^2)(0.5 \text{ m}) = 30 \text{ J}$$

Notice that only the vertical height is important. Moving an object 100 meters sideways does not change the gravitational potential energy because the force of gravity is perpendicular to the motion and therefore does no work on the object.

Question What is the change in gravitational potential energy of a 50-kilogram person who climbs a flight of stairs with a height of 3 meters and a horizontal extent of 5 meters?

Answer The change in gravitational potential energy is

$$GPE = mgh = (50 \text{ kg})(10 \text{ m/s}^2)(3 \text{ m}) = 1500 \text{ J}$$

The horizontal extent has no effect on the answer.

The amount of gravitational potential energy an object has is a relative quantity. Its value depends on how we define the height—that is, what height we take as the zero value. We can choose to measure the height from any place that is convenient. The only thing that has any physical significance is the *change* in gravitational potential energy. If a ball gains 20 joules of kinetic energy as it falls, it must lose 20 joules of gravitational potential energy. It does not matter how much gravitational potential energy it had at the beginning; it is only the amount lost that has any meaning in physics.

Flawed Reasoning

Bill and Will are calculating the gravitational potential energy of a 5-newton ball held 2 meters above the floor of their classroom.

Bill: "This is easy. Gravitational potential energy is *mgh*, where *mg* is the weight of the ball. We just multiply the 5 newtons by the 2 meters to get the gravitational potential energy of 10 joules."

Will: "You are forgetting that our classroom is on the second floor. We are going to have to find out how high the ball is relative to the ground."

Do you agree with either of these students?

Answer It is only the difference in gravitational potential energy that matters. We can either say that the ball fell from a height of 2 meters to a height of zero, or we can say that it fell from a height of 5 meters (relative to the ground) to a height of 3 meters. Either way, we get the same decrease in gravitational potential energy and the same increase in kinetic energy. Both students would get correct answers. In general, it is usually easiest to take the lowest point in each problem to be the zero for height.

Conservation of Mechanical Energy   **MATH**

The sum of the gravitational potential and kinetic energies is conserved in some situations. This sum is called the **mechanical energy** of the system:

$$ME = KE + GPE = \frac{1}{2}mv^2 + mgh$$

◂ mechanical energy = kinetic energy + gravitational potential energy

When frictional forces can be ignored and the other nongravitational forces do not perform any work, the mechanical energy of the system does not change.

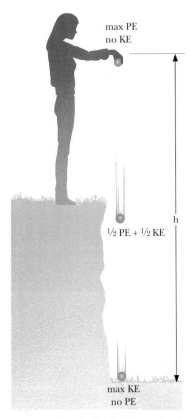

Figure 7-7 As a ball falls, its mechanical energy is conserved; any loss in gravitational potential energy shows up as a gain in kinetic energy.

The simplest example of this circumstance is free fall (Figure 7-7). Any decrease in the gravitational potential energy shows up as an increase in the kinetic energy, and vice versa.

Let's use the conservation of mechanical energy to analyze an idealized situation originally discussed by Galileo. Galileo released a ball that rolled down a ramp, across a horizontal track, and up another ramp, as shown in Figure 7-8. He remarked that the ball always returned to its original height. From an energy point of view, he gave the ball some initial gravitational potential energy by placing it at a certain height above the horizontal ramp. As the ball rolled down the first ramp, its gravitational potential energy was transformed into kinetic energy. Going up the opposite ramp merely reversed this process; the ball's kinetic energy was transformed back into gravitational potential energy. The ball continued to move until its kinetic energy was entirely transformed back to gravitational potential energy, which occurred when it once again reached its original height. This result is independent of the slopes of the ramps.

During the horizontal portion of the ball's trip, the gravitational potential energy remained constant. Hence, the kinetic energy did not change and the speed remained constant, in agreement with Newton's first law.

The pendulum bob shown in Figure 7-9 cyclically gains and loses kinetic and gravitational potential energy. Note that the tension exerted by the string does no work. Therefore, if we ignore frictional forces, the total mechanical energy is conserved. Suppose that at the beginning the bob has zero speed at point A and a gravitational potential energy of 10 joules. (We have chosen the zero for gravitational potential energy to be at the lowest point of the bob's path.) Because the kinetic energy is zero, the total mechanical energy is the same as the gravitational potential energy—that is, 10 joules.

The bob is released. As it swings down, it loses gravitational potential energy and gains kinetic energy. The gravitational potential energy is zero at the lowest point of the swing, and so the mechanical energy is all kinetic and equal to 10 joules. The bob continues to rise up the other side until all the kinetic energy is transformed back to gravitational potential energy. Because the gravitational potential energy depends on the height, the bob must return to its original height. And the motion repeats.

Question Suppose the bob is released at twice the height. What is the maximum kinetic energy?

Answer The initial gravitational potential energy is now twice as big, so the maximum kinetic energy will also be twice as big—that is, 20 joules.

Even when the bob is someplace between the highest and lowest points of the swing, the total mechanical energy is still 10 joules. If at this point we de-

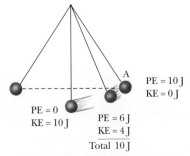

Figure 7-9 The total mechanical energy (kinetic plus gravitational potential) remains equal to 10 joules.

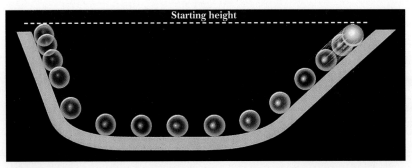

Figure 7-8 Galileo's two-ramp experiment can be analyzed in terms of the conservation of energy.

termine from the height of the bob that it has 6 joules of gravitational potential energy, we can immediately declare that it has 4 joules of kinetic energy. Because we know the expression for the kinetic energy, we can calculate the speed of the bob at this point.

PHYSICS | ON YOUR OWN

Construct a simple pendulum. If you let it swing freely, it returns to its original height. If you place a pencil in the path of the string as shown in Figure 7-10, does the bob still swing to the original height? Do your results depend on the vertical height of the pencil's position? Try placing the pencil near the lowest point of the swing.

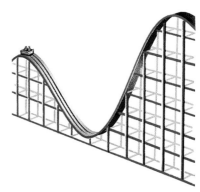

Figure 7-10 How high will the pendulum bob swing?

Roller Coasters

Imagine trying to determine the speed of a roller coaster inside a thrilling loop-the-loop as it traverses the track's hills and valleys. Determining the speed at any spot using Newton's second law is difficult because the forces are continually changing magnitude and direction. We can, however, use conservation of mechanical energy to determine the speed of an object without knowing the details of the net forces acting on it, providing we can ignore the frictional forces. If we know the mass, speed, and height of the roller coaster at some spot, we can calculate its mechanical energy. Now determining its speed at any other spot is greatly simplified. The height gives us the gravitational potential energy. Subtracting this from the total mechanical energy yields the kinetic energy from which we can obtain the roller coaster's speed. Notice that we don't need to worry about all the energy transformations that occurred earlier in the ride.

Suppose the roller-coaster ride was designed like the one in Figure 7-11, and upon reaching the top of the lower hill, your car almost comes to rest. Assuming that there are no frictional forces to worry about (not true in real situations), is there any possibility that you can get over the higher hill? The answer is no. Your gravitational potential energy at the top of the hill is nearly equal to the mechanical energy. This energy is not enough to get you over the higher hill. You will gain speed and thus kinetic energy as you coast down the hill, but as you

Figure 7-11 The car does not have enough gravitational potential energy to coast over the higher hill.

The mechanical energy of a roller coaster on a loop-the-loop is conserved if the frictional forces can be neglected.

start up the other hill, you will find that you cannot exceed the height of the original hill.

Question Is there any way that the roller-coaster car can make it over the second hill when starting on top of the first hill?

Answer Yes. If your car has some kinetic energy at the top of the first hill, it might be possible to make it over the second hill. You would need enough kinetic energy to equal or exceed the extra gravitational potential energy required to climb the second hill.

Our use of the conservation of mechanical energy is limited because we ignore losses due to frictional forces, and in many cases this is not realistic. These frictional forces do work on the car and thus drain away some of the car's mechanical energy. If we know the magnitude of these frictional forces, however, and the distances through which they act, we can calculate the energy transformed to other forms and allow for the loss of mechanical energy.

Where does the energy go that is thus drained away? The answer to this question is still ahead of us. To maintain the notion that energy is an invariant, we will have to search for other forms of energy.

Flawed Reasoning

The following question appears on the final exam: "Three bears are throwing identical rocks from a bridge to the river below. Papa Bear throws his rock upward at an angle of 30 degrees above the horizontal. Mama Bear throws hers horizontally. Baby Bear throws the rock at an angle of 30 degrees below the horizontal. Assuming that all three bears throw with the same speed, which rock will be traveling fastest when it hits the water?" Three students meet after the exam and discuss their answers.

Emma: "Baby Bear's rock will be going the fastest because it starts with a downward component of velocity."

Hector: "But Papa Bear's rock will stay in the air the longest, so it will have more time to speed up. I think his rock will be traveling the fastest."

M'Lynn: "Papa Bear's rock does stay in the air longer, but part of that time it is moving upward and slowing down. I think Mama Bear's rock will be traveling fastest when it hits the water because it is in the air longer than Baby Bear's and it is speeding up all of the time."

With which student (if any) do you agree?

Answer All three students are wrong. They are making an easy problem much too difficult by ignoring the power of the energy approach to problem solving. Because each of the three rocks started with the same kinetic energy (same speed) and the same gravitational potential energy (same height), they must all end up with the same final kinetic energy before hitting the water. All three rocks must therefore hit the water with the same speed. Note that the three rocks will not hit the water at the same time, with the same velocity, or at the same distance from the bridge. However, the energy method will not give us this information.

Other Forms of Energy

We can identify other places where kinetic energy is temporarily stored. For example, a moving ball can compress a spring and lose its kinetic energy (Figure 7-12). While the spring is compressed, it stores energy much like the ball in

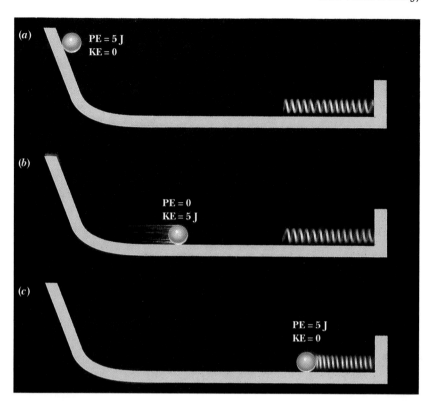

Figure 7-12 The gravitational potential energy (a) is converted to kinetic energy (b), which is then converted to elastic potential energy of the spring (c).

the gravitational situation. If we latch the spring while it is in the compressed state, we can store its energy indefinitely as elastic potential energy. Releasing it at some future date will transform the spring's elastic potential energy back into the ball's kinetic energy.

Question When a ball is hung from a vertical spring, it stretches the spring. As it drops, it loses gravitational potential energy, but this does not all show up as kinetic energy. What happens to the gravitational potential energy?

Answer The gravitational potential energy is converted to kinetic energy and to elastic potential energy of the spring. At the bottom, it is all elastic potential energy.

A thrilling example of this is bungee jumping. A nylon rope is securely fastened to the ankles of the jumper, who then dives headfirst from a very high platform. As the jumper falls, gravitational potential energy is converted to kinetic energy. As the rope tightens and stretches, both the kinetic energy and some additional gravitational potential energy are converted to elastic potential energy. When the rope reaches its maximum stretch, the jumper bounces back up into the air because much of the elastic potential energy is converted back into kinetic and gravitational potential energy. After several bounces the bungee jumper is lowered to the ground. Notice that if there were no loss of mechanical energy, the bungee jumper would bounce forever!

Many objects or materials when distorted by a force hold some elastic potential energy as a result of the distortions. A floor gives when we jump on it. Our kinetic energy on impact is transformed, in part, to elastic potential energy of the floor. As the floor springs back to its original shape, we regain some of the kinetic energy. In these cases some of the mechanical energy (maybe most) is lost to the dissipating effects of the distortion.

We have described a gravitational potential energy that is associated with the gravitational force. Other potential energies are associated with other forces

As the bungee jumper falls, gravitational potential energy is converted to kinetic energy.

Exponential Growth

America's energy usage has grown at a rate of about 5% per year for the past few decades. At first glance this sounds relatively harmless, but don't be fooled. Anything that grows in proportion to its current size gets out of control. This kind of growth is called *exponential growth*, and whether it's a bank account, energy usage, or population, it follows the same pattern. This pattern can be illustrated with a simple story of a creative mathematician.

Legend has it that a mathematician in ancient India invented the game of chess. The ruler of India was so pleased that the mathematician was allowed to choose his own reward. This clever fellow shunned the obvious bounties of gold and jewels and instead requested grains of wheat. His plan was to put one grain on the first square of the chessboard and then double the number of grains on each successive square, so there would be two grains of wheat for the second square, four grains for the third square, and so on. You can determine the number on each square with a simple calculator. Just start with the first grain and keep multiplying by two, 64 times.

At the end of the first row, there are 128 grains on the 8th square (Figure A). But that's not very many. By the end of the second row, the mathematician has only enough wheat to make a few loaves of bread. But what happens if we continue this pattern of doubling? The photo in Figure B

Figure B The 64th square has 9 billion billion grains of wheat. The photograph is of course only symbolic.

Figure A The 8th square has 128 grains of wheat.

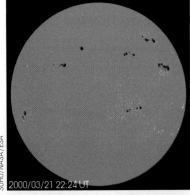

The Sun is a great storehouse of nuclear potential energy.

in nature. The elastic potential energy in a spring is due to electromagnetic (electric and magnetic) forces. There are also other forms of electromagnetic potential energy. Chemical energy is really just a potential energy associated with the electromagnetic force. We will see in Chapter 26 that nuclear potential energy is associated with the nuclear force.

The transformation of these various forms of potential energy to kinetic energy is what powers our civilization. The gravitational potential energy of the water behind dams powers hydroelectric plants; that of a weight runs grandfather clocks. Most of the energy we use every day is the result of releasing the chemical potential energy of fossil fuels. Nuclear power plants are designed to release nuclear potential energy. Nuclear potential energy is the ultimate source of the energy we receive from the Sun.

We have defined a variety of potential energies to explain the temporary losses of energy. However, this explanation doesn't apply to situations that include friction. To understand the physics of friction, consider the following. We start a box sliding across the floor with a given amount of kinetic energy. As it slows and comes to a halt, its kinetic energy decreases and finally reaches zero. We might imagine that this energy was stored in some form of "frictional poten-

is only symbolic, because the number of grains on the 64th square is about 9 billion billion! And the grand total is twice this! This is roughly 500 times the amount of wheat grown in the entire world in a year's time and probably more wheat than has been grown during the entire history of humankind.

The pattern is very important. The doubling process starts out very slowly—a gentle, rather innocent, growth that eventually gets totally out of hand. The legend ends with the mathematician being beheaded around the 35th square when the granaries of India were exhausted.

There are many other examples of exponential growth. Consider the compound interest that you earn in a savings account. Suppose the bank offers a guaranteed interest rate of 5% per year. The growth pattern is the same as for the chess example. Your money will double after a certain number of years and double again each time this number of years elapses. The *doubling time* is obtained by dividing 70 by the percentage rate. For your savings account this would be 70/(5%/year), or 14 years. If you put $1 in the bank today, it will double every 14 years. In 100 years your ancestors will collect $128. If they choose to leave the money in the bank for their ancestors—say, for another 400 years—your initial $1 becomes $64 billion! (Bankers are not worried about 500-year-old accounts, but they are well aware of this pattern and have rules about inactive accounts.)

Historically, the growth of the world's population has been roughly exponential. According to the estimate by the United Nations, the world's population reached 6 billion in October 1999 and was increasing at a rate of 1.3% per year. Most people are not worried, because this seems like a very low rate of increase. However, if this rate were to continue, the world's population would double in 55 years to a total of 12 billion by 2054. Few people believe that the world's population will actually reach this number by the middle of the century because the rate of increase has been getting smaller and may be as low as 0.5% (a doubling time of 140 years) by 2050. If this occurs, the world's population will be somewhere between 9 and 10 billion in 2050.

Exponential growth can have very serious consequences because the big changes come on very fast. We can look at another example to illustrate the nature of the problem. Bacteria are rather curious in that they multiply by dividing. Assume that you have a colony of bacteria in which each bacterium divides exactly in two at the end of each minute. Therefore, the doubling time is 1 minute. Let's further assume that the bottle the bacteria are in will be exactly full at the end of 1 hour; that is, it will take 60 doublings to fill the bottle. When will the bottle be half-filled? In 59 minutes. If you were a bacterium, when would you realize that you were running out of space? Because you're reading this essay, assume that at 55 minutes you've recognized that the growth is exponential and that there will be a major problem. At this time the bottle is only 3% full. How successful would you be in convincing your fellow bacteria that there's only 5 minutes left?

Imagine that during this debate an adversary argues that with proper government funding the colony can find more space, and in fact they do. After a great deal of expense and effort, they locate three new bottles. The resources have quadrupled. How much longer can the growth continue? The answer is only 2 more minutes!

The point of this discussion is to understand our current usage of energy. Clearly, we cannot let our usage grow exponentially. Although there are sizable variations in the estimates of the world's nonrenewable energy resources, all estimates are rather inconsequential if exponential growth continues. Does it really make a difference if we have underestimated these resources by a factor of two, or even four? No. If we continue to increase our use of these resources by even a small percentage each year, they will quickly be exhausted. During the next doubling time, we will use as much energy as has been used by humankind up to the present.

Source: A. A. Bartlett, "Forgotten Fundamentals of the Energy Crisis," *The Physics Teacher* 46 (1978): 876.

---

tial energy." If this were the case, we could somehow release this frictional potential energy, and the box would move across the floor, continually gaining kinetic energy. This does not happen. As we will discuss in Chapter 13, when frictional forces act, some of the energy changes form and appears as thermal energy.

Is Conservation of Energy a Hoax?

It may seem strange that whenever we discover a situation in which energy does not appear to be conserved, we invent a new form of energy. How can the conservation law have any validity if we keep modifying it whenever it appears to be violated? In fact, the whole procedure would be worthless if it were not internally consistent, meaning that the total amount of energy stays the same no matter what sequence of changes takes place. This idea of internal consistency places a rather strong constraint on the law of conservation of energy.

Imagine a rotating wheel that has 10 joules of kinetic energy. If we extract this energy by stopping the wheel with a brake, we end up with 10 joules of thermal energy, as shown in Figure 7-13(a). Instead imagine that we convert the 10 joules of kinetic energy to electric energy by turning a generator to charge a

The gravitational potential energy of the water behind Hoover Dam is converted to electric energy.

battery, then use the battery's chemical energy to produce the electricity to run a heater (Figure 7-13[b]). We still get 10 joules of thermal energy. This result is reassuring. If the result were different, we would not have a valid scientific principle. Conservation of energy is a useful concept because the mathematical expressions for the various kinds of energy are independent of the specific situations in which they occur.

Conservation of energy is one of the most powerful principles in the physics world view, partly because of its generality; its only restriction is that the system be isolated, or closed. Conservation of energy can be applied to a wide range of problems in physics and in our everyday world.

Figure 7-13 The two different paths for the transformation of the kinetic energy of the wheel yield the same amount of thermal energy.

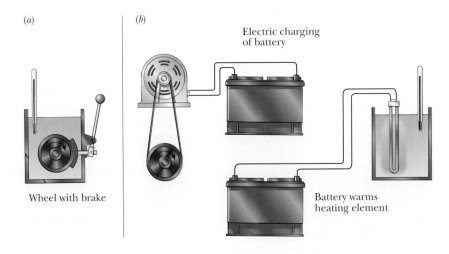

Human Power

A human can generate 1500 watts (2 horsepower) for very short periods of time, such as in weightlifting. The maximum average human power for an 8-hour day is more like 75 watts (0.1 horsepower). Each person in a room generates thermal energy equivalent to that of a 75-watt lightbulb. That's one of the reasons why crowded rooms warm up!

Achieving human-powered flight has been a dream for centuries. People failed because it has been difficult to create enough aerodynamic lift with the power output that's humanly possible. In 1979 an American team led by Paul MacCready designed, built, and flew a "winged bicycle" that convincingly demonstrated that human-powered flight is possible. Professional cyclist Bryan Allen crossed the English Channel in the *Gossamer Albatross*. During this flight he generated an average power output of **190 watts** (0.25 horsepower).

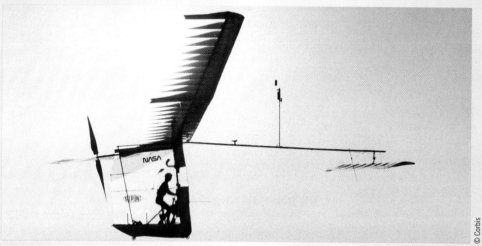

The human-powered aircraft *Gossamer Albatross*, which successfully flew across the English Channel in 1979

Power

In previous chapters we discussed how various quantities change with time. For example, speed is the change of position with time, and acceleration is the change of velocity with time. The change of energy with time is called **power**. Power P is equal to the amount of energy converted from one form to another ΔE divided by the time Δt during which this conversion takes place:

$$P = \frac{\Delta E}{\Delta t}$$

◂ power = energy converted / time taken

Power is measured in units of joules per second, a metric unit known as a **watt** (W). One watt of power would raise a 1-kilogram mass (with a weight of 10 newtons) a height of 0.1 meter each second.

The English unit for electric power is the watt, but a different English unit is used for mechanical power. A horsepower is defined as 550 foot-pounds per second. This definition was proposed by the Scottish inventor James Watt because he found that an average strong horse could produce 550 foot-pounds of work during an entire working day. One horsepower is equal to 746 watts.

We get electric energy from our local power company. But power is not an amount of energy. The energy transformed during a period of time is given by the power multiplied by the time this power is expended. Power companies usually bill us for the amount of energy we use, not the rate of consumption.

WORKING IT OUT | Power ✓ MATH

A compact car traveling at 27 m/s (60 mph) on a level highway experiences a frictional force of about 300 N due to the air resistance and the friction of the tires with the road. Therefore, the car must obtain enough energy by burning gasoline to compensate for the work done by the frictional forces each second:

$$\Delta E = W = Fd = (300 \text{ N})(27 \text{ m}) = 8100 \text{ J}$$

$$P = \frac{\Delta E}{\Delta t} = \frac{8100 \text{ J}}{1 \text{ s}} = 8100 \text{ W}$$

This means that the power needed is 8100 W, or 8.1 kW. This is equivalent to a little less than 11 horsepower.

How much electric energy does a motor running at 1000 W for 8 h require?

$$\Delta E = P\Delta t = (1000 \text{ W})(8 \text{ h}) = 8000 \text{ Wh}$$

This is usually written as 8 kilowatt-hours (kWh). Although this doesn't look like an energy unit, it is—(energy/time) × time = energy.

The energy used by the motor in 1 h is

$$\Delta E = P\Delta t = (1000 \text{ W})(1 \text{ h}) = (1000 \text{ J/s})(3600 \text{ s}) = 3,600,000 \text{ J}$$

In other words, 1 kWh = 3.6 million J.

Question How much energy is required to leave a 75-watt yard light on for 8 hours?

Answer $\Delta E = P\Delta t = (75 \text{ watts})(8 \text{ hours}) = 600$ watt-hours = 0.6 kilowatt-hour.

PhysicsNow™ Assess your understanding of this chapter's topics with sample tests and other resources found by logging into PhysicsNow at http://physics.brookscole.com/kf6e.

Summary

Energy is an abstract quantity that is conserved whenever a system is closed. Independent of the kinds of transformations that take place within a closed system, the total amount of energy remains the same.

Kinetic energy is the energy of motion and is defined to be one-half the mass times the speed squared, $KE = \frac{1}{2}mv^2$. If the kinetic energy before a collision is the same as that after the collision, the collision conserves kinetic energy and is said to be elastic. Kinetic energy is transformed into other forms of energy in inelastic collisions.

Work is equal to the product of the force in the direction of motion and the distance traveled—that is, $W = Fd$. If the force is perpendicular to the displacement of the object or if the object does not move, no work is done by the force. The change in kinetic energy of an object is equal to the work done on the object.

The gravitational potential energy is equal to the work done by the force of gravity when an object falls through a height h—that is, $GPE = mgh$. The location for the zero value of gravitational potential energy is arbitrary because only the change in gravitational potential energy has any physical meaning. If gravity is the only force that does work on an object, the total mechanical energy (kinetic plus gravitational potential) is conserved. Therefore, any loss in gravitational potential energy shows up as a gain in the kinetic energy, and vice versa.

Other forms of potential energy can be associated with the electromagnetic and nuclear forces. However, a potential energy cannot be associated with the frictional force. This force transforms mechanical energy into thermal energy.

Power is the rate at which energy is transformed from one form into another, $P = \Delta E/\Delta t$. A kilowatt-hour is an energy unit because it is power multiplied by time.

Chapter 7 Revisited

We really don't know what energy *is*, but we know the many forms it takes and we have an accounting system for determining the amount of energy in a system. When we say that energy is conserved, we are acknowledging that although energy changes from one form into other forms, the total amount of energy in the system stays the same. Because of this, we can keep track of the energy.

KEY TERMS

elastic: A collision or interaction in which kinetic energy is conserved.

gravitational potential energy: The work that would be done by the force of gravity if an object fell from a particular point in space to the location assigned the value of zero, $GPE = mgh$.

inelastic: A collision or interaction in which kinetic energy is not conserved.

joule: The SI unit of energy equal to 1 newton acting through a distance of 1 meter.

kinetic energy: The energy of motion, $KE = \frac{1}{2}mv^2$, where *m* is the object's mass and *v* is its speed.

mechanical energy: The sum of the kinetic energy and various potential energies, which may include the gravitational and the elastic potential energies.

power: The rate at which energy is converted from one form to another, $P = \Delta E/\Delta t$. Measured in joules per second, or watts.

watt: The SI unit of power, 1 joule per second.

work: The product of the force along the direction of motion and the distance moved, $W = Fd$. Measured in energy units, joules.

CONCEPTUAL QUESTIONS

1. Two identical cars traveling at the same speed collide head-on and come to rest in a mangled heap. At first glance it appears that energy is not conserved in this collision. However, like Dennis's mother in Richard Feynman's story at the beginning of the chapter, we find the energy "hidden" in many different forms. The initial kinetic energy is transformed into sound energy, thermal energy, and deformation energy. Where does the initial momentum of the system hide?

2. Energy is always conserved for an isolated system. However, this chapter begins by stating that the energy of motion is only sometimes invariant, even for an isolated system. How can total energy be conserved while energy of motion is not?

3. You have been asked to analyze a collision at a traffic intersection. Will you be better off to begin your analysis using conservation of momentum or conservation of kinetic energy? Why?

4. A sports car with a mass of 1200 kilograms travels down the road with a speed of 20 meters per second. Why can't we say that its momentum is smaller than its kinetic energy?

5. If a system has zero kinetic energy, does it necessarily have zero momentum? Give an example to illustrate your answer.

6. If a system has zero momentum, does it necessarily have zero kinetic energy? Give an example to illustrate your answer.

7. Which has the greater kinetic energy, a supertanker berthed at a pier or a motorboat pulling a water skier? Why?

8. Two pickup trucks with the same mass are driving on the freeway. If the Chevy has twice the speed of the Ford, does the Chevy have twice as much kinetic energy as the Ford? Explain your answer.

9. Assume that a minivan has a mass of 2000 kilograms and a sports car has a mass of 1000 kilograms. If both vehicles are traveling at the same speed, which vehicle has the higher kinetic energy? Why?

10. If the sports car in Question 9 has twice the speed of the minivan, which vehicle has the higher kinetic energy? Why?

11. A silver Camry is driving on the freeway at a constant 70 mph. Another Camry, identical but white, is on the on-ramp and is speeding up at a rate of 5 mph per second. Compare their kinetic energies at the instant the white Camry reaches 70 mph.

12. A jet is circling above the Salt Lake City airport at constant speed and elevation. How does the jet's kinetic energy change, if at all, as it circles? How does the jet's momentum change, if at all, as it circles?
13. What will happen if you pull two balls from the same side of the collision-ball apparatus in Figure 7-1 and let them go?
14. What will happen if the end balls of the collision-ball apparatus in Figure 7-1 are pulled out the same distances and let go?
15. A bowler lifts a bowling ball from the floor and places it on a rack. If you know the weight of the ball, what else must you know to calculate the work he does on the ball?

16. Bill's job is to lift bags of flour and place them in the back of a truck, which is parked right next to him. Sally is loading the same bags of flour into a similar truck that is located 10 meters away. Sally wants a raise because she says that she is doing more work than Bill. Does the physics definition of work support her claim?
17. An airplane is flying due south when it experiences a wind gust that exerts a force on the airplane acting due north. Will the kinetic energy of the airplane initially increase, decrease, or stay the same? Explain.
18. A bowling ball is rolling directly north along a smooth floor. Using a hammer, you tap the ball such that the force is directed east. How does the tap affect the ball's kinetic energy and its momentum?
▲ 19. In tryouts for the national bobsled team, each competing team pushes a sled along a level, smooth surface for 5 meters. One team brings a sled that is much lighter than all the others. Assuming that each team pushes with the same net force, compare the kinetic energy of the light sled to that of the others after 5 meters. Compare the momentum of the light sled to that of the others after 5 meters. (*Hint*: Think about the times involved.)
▲ 20. Suppose the rules were changed in Question 19 so that the teams pushed for a fixed time of 5 seconds rather than a fixed distance of 5 meters. Compare the momentum of the light sled to that of the others after 5 seconds. Compare the kinetic energy of the light sled to that of the others after 5 seconds. (*Hint*: Think about the distances involved.)
21. The tractor of an 18-wheeler performs work on its trailer when the truck is traveling along a level highway with a constant velocity. Why doesn't the trailer continually gain kinetic energy—that is, continually speed up?
22. The Chandra X-ray satellite orbits the Earth in a highly elliptical orbit, as shown in the figure. The force that Earth exerts on the satellite is always directed toward Earth. Is the satellite's kinetic energy increasing, decreasing, or staying the same at each of the points indicated? Explain your reasoning. (*Note*: The velocity vectors on the figure are *not* drawn to scale.)

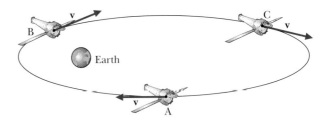

23. Two forces are used to move a block 2 meters across a level surface as shown. Is the work done by force A greater than, equal to, or less than the work done by force B? (*Note*: The force vectors are drawn to scale.)

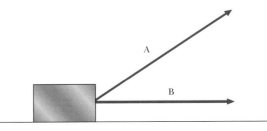

24. Two forces are used to move a block 2 meters across a level surface as shown. Is the work done by force A greater than, equal to, or less than the work done by force B? (*Note*: The force vectors are drawn to scale.)

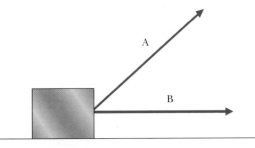

25. Two cars have different masses but the same kinetic energies. If the same frictional force is used to stop each of the cars, which car, if either, will stop in the shorter distance?
▲ 26. Two cars have different masses but the same linear momenta. If the same frictional force is used to stop each of the cars, which car, if either, will stop in the shorter distance? (*Hint*: Which car has the larger kinetic energy?)
▲ 27. We can use Newton's third law to demonstrate that the momentum lost by one object is gained by another. Can you also do this for kinetic energy? Explain why or why not.
28. Is it possible to change an object's momentum without changing its kinetic energy? What about the reverse situation?
29. Which of the following, if either, does more work: a force of 3 newtons acting through a distance of 3 meters or a force of 4 newtons acting through a distance of 2 meters?

30. Which of the following, if either, produces the larger change in the kinetic energy: a force of 5 newtons acting through a distance of 3 meters or a force of 4 newtons acting through a distance of 4 meters?
31. On a test, the physics teacher asks, "What is the gravitational potential energy of a 10-newton ball resting on a shelf 2 meters above the floor?" Jamie got no points for responding that the answer was zero. What argument could Jamie use to convince the teacher that zero could be the right answer?
32. As the firefighter in the picture slides down the pole, he initially speeds up to some terminal velocity, which he maintains until reaching the bottom. Gravitational potential energy is constantly decreasing during this process. What happens to the energy?

33. The kinetic energy of a free-falling ball is not conserved. Why is this not a violation of the law of conservation of mechanical energy?
34. Which of the following is conserved as a ball falls freely in a vacuum: the ball's kinetic energy, gravitational potential energy, momentum, or mechanical energy?
35. At which point in the swing of an ideal pendulum (ignoring friction) is the gravitational potential energy at its maximum? At which point is the kinetic energy at its maximum?
36. As an ideal pendulum (ignoring friction) swings from the bottom to the top of its arc, the string is always exerting a force on the ball. Why then is the gravitational potential energy at the top not greater than the kinetic energy at the bottom?
37. If we do not ignore frictional forces, what can you say about the height to which a pendulum bob swings on consecutive swings?
38. A block of wood, released from rest, loses 100 joules of gravitational potential energy as it slides down a ramp. If it has 90 joules of kinetic energy at the bottom of the ramp, what can you conclude?
39. Describe the energy transformations that occur as a satellite orbits Earth in a highly elliptical orbit.
40. Imagine a giant catapult that could hurl a spaceship to the Moon. Describe the energy transformations that would take place on such a journey.
41. Describe the energy changes that take place when you dribble a basketball.

42. An elephant, an ant, and a professor jump from a lecture table. Assuming no frictional losses, which of the following could be said about their motion just before they hit the floor?
    a. They all have the same kinetic energy.
    b. They all started with the same gravitational potential energy.
    c. They will all experience the same force on stopping.
    d. They all have the same speed.
43. Magnets mounted on top of air-hockey pucks allow the pucks to "collide" without touching each other. Describe the energy transformations that take place when one puck collides head-on with another.
44. An air-hockey puck is fastened to the table with a spring so that it oscillates back and forth on top of the table. Describe the energy transformations that take place. How would your description change if the puck were suspended from the ceiling by the spring?
45. Mountain highways often have emergency ramps for truckers whose brakes fail. Why are these covered with soft dirt or sand rather than pavement?

46. Athletes will sometimes run along the beach to increase the effect of their workouts. Why is running on soft sand so tiring?
47. What happens to the chemical potential energy in the batteries used to power electric socks?
48. Describe the energy changes that take place when you stop your car using the brakes.
49. Why can we not associate a potential energy with the frictional force as we did with the gravitational force?
50. A physics textbook is launched up a rough incline with a kinetic energy of 200 joules. When the book comes momentarily to rest near the top of the incline, it has gained 180 joules of gravitational potential energy. How much kinetic energy will it have when it returns to the launch point?
51. The winch on your pickup truck is rated at 600 watts. Is it possible to do more than 600 joules of work with this winch? Explain.
52. A company advertises a new battery, which it claims is twice as powerful as anything else on the market. If you were to put this new battery in your flashlight, would you expect the light to be brighter? Would you expect it to last longer?
53. When you get your power bill, you are charged for the number of kilowatt-hours that you have used. Is kilowatt-hour a unit of power or a unit of energy?
54. Valerie is able to do 1200 joules of work in 10 seconds. Brett is able to do 5000 joules of work in 50 seconds. Who is more powerful?
55. Which of the following is an energy unit: newton, kilowatt, kilogram-meter per second, or kilowatt-hour?
56. Which of the following is not a unit of energy: joule, newton-meter, kilowatt-hour, or watt?

EXERCISES

1. What is the kinetic energy of an 1800-kg pickup truck traveling down the road with a speed of 20 m/s?
2. What is the kinetic energy of an 84-kg sprinter running at 10 m/s?
3. In reviewing his lab book, a physics student finds the following description of a collision: "A 4-kg air-hockey puck with an initial speed of 6 m/s to the right collided head-on with a 1-kg puck moving to the left at the same speed. After the collision, both pucks traveled to the right, the 4-kg puck at 3 m/s and the 1-kg puck at 12 m/s." Is momentum conserved in this description? Is kinetic energy conserved in this description? Could this collision actually have taken place as described?
4. In reviewing her lab book, a physics student finds the following description of a collision: "A 4-kg air-hockey puck with an initial speed of 6 m/s to the right collided head-on with a 1-kg puck moving to the left at the same speed. After the collision, both pucks traveled to the right, the 4-kg puck at 2 m/s and the 1-kg puck at 10 m/s." Is momentum conserved in this description? Is kinetic energy conserved in this description? Could this collision actually have taken place as described?
5. A 4-kg toy car with a speed of 5 m/s collides head-on with a stationary 1-kg car. After the collision, the cars are locked together with a speed of 4 m/s. How much kinetic energy is lost in the collision?
6. A 3-kg toy car with a speed of 6 m/s collides head-on with a 2-kg car traveling in the opposite direction with a speed of 4 m/s. If the cars are locked together after the collision with a speed of 2 m/s, how much kinetic energy is lost?
7. A 0.5-kg air-hockey puck is initially at rest. What will its kinetic energy be after a net force of 0.7 N acts on it for a distance of 3 m?
8. A 40-N block lifted straight upward by a hand applying a force of 40 N has an initial kinetic energy of 12 J. If the block is lifted 0.5 m, how much work does the hand do? What is the block's final kinetic energy?
9. A radio-controlled car increases its kinetic energy from 5 J to 27 J over a distance of 2 m. What was the average net force on the car during this interval?
10. A toy car has a kinetic energy of 12 J. What is its kinetic energy after a frictional force of 0.8 N has acted on it for 4 m?
11. Earth, which orbits the Sun in an elliptical path, reaches its closest point to the Sun on about January 4 each year. Will the work done by the gravitational force of the Sun on Earth be positive, negative, or zero over the next 6 months? Over the next year?
12. How much work is performed by the gravitational force on a satellite in near-Earth orbit during one revolution?
13. How much work does a 55-kg person do against gravity in walking up a trail that gains 720 m in elevation?
14. A woman with a mass of 65 kg climbs a set of stairs that are 3 m high. How much gravitational potential energy does she gain?

15. A baseball (mass = 145 g) is thrown straight upward with kinetic energy 8.7 J. When the ball has risen 6 m, find (a) the work done by gravity, (b) the ball's kinetic energy, and (c) the ball's speed.

16. What is the gravitational potential energy of a ball with a weight of 50 N when it is sitting on a shelf 1.5 m above the floor? What assumption do you need to make to get your answer?

17. If a 0.5-kg ball is dropped from a height of 6 m, what is its kinetic energy when it hits the ground?

18. A 2-kg block is released from rest at the top of a 20-m-long frictionless ramp that is 4 m high. At the same time, an identical block is released next to the ramp so that it drops straight down the same 4 m. What are the values for each of the following for the blocks just before they reach ground level? Which quantities are the same for the two blocks?
    a. gravitational potential energy
    b. kinetic energy
    c. speed
    d. momentum

19. A 1200-kg frictionless roller coaster starts from rest at a height of 24 m. What is its kinetic energy when it goes over a hill that is 12 m high?

20. You reach out the second-story window that is 5 m above the sidewalk and throw a 0.1-kg ball straight upward with 6 J of kinetic energy.
    a. What is the ball's gravitational potential energy when it is released?
    b. What is the ball's gravitational potential energy just before hitting the sidewalk?
    c. What is the ball's kinetic energy just before hitting the sidewalk?
    d. How would the answer to part (c) change if the ball had initially been thrown straight down with 6 J of kinetic energy?

21. What average power does a weightlifter need to lift 300 lb a distance of 4 ft in 0.8 s?

22. If an 80-kg sprinter can accelerate from a standing start to a speed of 10 m/s in 3 s, what average power is generated?

23. If a CD player uses electricity at a rate of 15 W, how much energy does it use during an 8-h day?

24. If a hair dryer is rated at 800 W, how much energy does it require in 3 min?

8 | Rotation

Helicopters are very useful in fighting forest fires.

Objects in translational motion have momentum. Objects in rotational motion also have momentum. A net force is required to change an object's linear momentum. What causes a change in an object's rotational momentum? For instance, why does a helicopter have a small rotor at the back?

(See page 146 for the answer to this question.)

A well-thrown football follows a projectile path while spinning around its long axis. The motion of the football is easy to analyze because the spin does not affect the path and the path does not affect the spin. That is, the rotational motion about the axis is independent of the motion through the air.

In this chapter we examine rotational motion without examining any accompanying translational motion.

Rotational Motion

The rules for rotational motion have many analogies with translational motion. The distance traveled is no longer measured in ordinary distance units like feet or meters, but in an angular measure such as degrees, revolutions, or radians. Just as feet and meters can be converted to each other, these angular measures are related by simple conversion factors. There are 360 degrees in a complete circle, so 1 degree is equal to $\frac{1}{360} \cong 0.0028$ revolutions. You may not have encountered the radian; it is defined as the angle for which the arc length along the circle is equal to the radius of the circle. The drawing in Figure 8-1 shows the size of the radian. There are $2\pi \cong 6.28$ radians in a complete circle, so a radian is a little larger than 57 degrees.

Just as translational speed v is the displacement (change in position) divided by the time required, **rotational speed** ω is the angular displacement (change in angular position) divided by the time required. The units used to express this measurement could be radians per second or revolutions per minute (rpm). A modern compact disc (CD) spins at variable rates between 500 and 200 revolutions per minute as it plays from the inside track to the outside track.

Rotational acceleration α is a measure of the rate at which the rotational speed changes. The change in the rotational speed is equal to the final rotational speed minus the initial rotational speed. If a CD contains 60 minutes of music, its average rotational acceleration is

$$\alpha = \frac{\Delta \omega}{\Delta t} = \frac{200 \text{ rpm} - 500 \text{ rpm}}{60 \text{ min}} = -5 \frac{\text{rpm}}{\text{min}}$$

Because the CD is slowing, this change is negative. In this example the CD's rotational speed decreases by 5 revolutions per minute for each minute of music.

Like translational velocity and acceleration, **rotational velocity** and rotational acceleration are vectors. The assignment of directions to these rotational vectors is not as obvious as it is for their translational counterparts. The only directions associated with a rotating body that are not continually changing are the directions of the axis of rotation. Therefore, we assign the direction of the rotational velocity to be along the axis of rotation. By convention, if you curl the fingers of your *right hand* along the direction of rotation as shown in Figure 8-2, your thumb points along the axis in the direction of the rotational velocity.

The direction of the rotational acceleration is also along the axis of rotation. If the acceleration causes the object to speed up, the direction of the acceleration is the same as that of the velocity. If the acceleration causes the object to slow down, the acceleration points in the direction opposite to the velocity.

Torque

✓ MATH

Newton's first law has the same form for rotational motion as it does for translational motion. The first law says that in the absence of a net external interaction the natural motion is one in which the rotational velocity remains constant. If the object is not rotating, it continues to not rotate. If it is rotating, it continues to rotate with the same rotational velocity. This can be seen with the thrown wrench in Figure 4-18. If you imagine riding along with the wrench, you see that

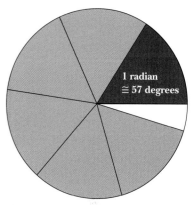

Figure 8-1 There are a little more than $6\frac{1}{4}$ radians in a complete circle.

Figure 8-2 If you curl the fingers of your right hand along the direction of motion, your thumb points along the axis in the direction of the rotational velocity.

✓ Extended presentation available in the *Problem Solving* supplement

the wrench rotates about the white dot by the same amount between successive flashes; that is, its rotational speed is constant.

A change in the rotational speed can only occur when there is a net external interaction on the object. This interaction involves forces, but unlike translational motion, the locations at which the forces act are as important as their sizes and directions. The same force can produce different effects depending on where and in which direction it is applied.

You can experiment with these ideas by exerting different forces on a door to your room. If you push directly toward the hinges or pull directly away from the hinges, the door does not rotate. Rotations only occur when a horizontal force is applied in any other direction; the largest occurs when the force is perpendicular to the face of the door.

Even when you apply the force in the perpendicular direction, you get different results depending on where you push. Try opening the door by pushing at different distances from the hinges. The largest effect occurs when the force is applied farthest from the hinges. That's why doorknobs are put there!

PHYSICS | ON YOUR OWN

A young child can easily beat you in a game of door "push"-of-war if you carefully choose the points where each of you pushes. Discover where you should choose the points.

The rotational analog of force is called **torque** and combines the effects of the force on the door and the distance from the hinges. If you push on the doorknob, the doorknob moves along a circular path with a radius equal to the distance from the hinge to the doorknob, as shown in Figure 8-3. If we restrict ourselves to the case when the force is perpendicular to the radius, the magnitude of the torque τ is equal to the radius r multiplied by the force F:

▶ torque = radius × force

$$\tau = rF$$

Although we will find that torque is a vector that lies along the axis of rotation, it is easier for us to describe a torque by the effect it has in rotating an object that is initially stationary. If the object rotates clockwise, we say that the torque is clockwise.

The development of the concept of torque allows us to restate Newton's first law for rotation:

▶ Newton's first law for rotation

The angular velocity of an object remains constant unless acted on by an unbalanced torque.

For a vector to remain constant, both its magnitude and its direction must remain constant.

Because torque is equal to a product, we can see why the same applied force can produce different torques on an object. The torque is increased if the force is applied farther from the axis of rotation. This fact is useful if you have to loosen a stubborn nut. The biggest torque occurs when you push or pull the wrench at the spot farthest from the nut. We can make the distance longer (for the really stubborn nuts) by slipping a pipe over the wrench.

Imagine that you have a flat tire and one of the nuts is stuck tight. Suppose further that your wrench is 0.3 meter long. How much torque could you apply by stepping on the end of the wrench?

If you weigh 500 newtons (110 pounds), the maximum torque you could apply would be 150 newton-meters:

$$\tau = rF = (0.3 \text{ m})(500 \text{ N}) = 150 \text{ N} \cdot \text{m}$$

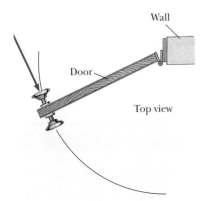

Figure 8-3 The torque exerted on the door is equal to the product of the distance from the hinge and the force applied to the doorknob.

Question Suppose that this is not enough but you found a pipe in your car that could be slipped over the wrench, tripling its effective length. What torque could you now apply?

Answer Because the torque is a product of the force and the distance, you would get a torque that is three times as large as the original, or 450 newton-meters.

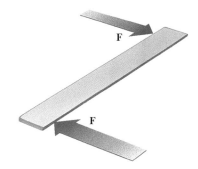

Figure 8-4 Two equal but opposite forces can produce a rotational acceleration if they do not act along the same line.

When there is more than one applied force, situations can arise when the net force is zero but the net torque is not zero. In other words, a pair of forces that produces no translational acceleration can still produce rotational acceleration. The two forces on the board in Figure 8-4 are equal in size and opposite in direction, but they do not act along the same line. The stick accelerates clockwise because the torques about the center of the board are nonzero and act in the same rotational direction.

Figure 8-5 shows two girls on a seesaw. Each girl's weight multiplied by her distance from the pivot point gives the torque that she applies to the board. If the torques are equal in magnitude and opposite in direction, there will be no rotational acceleration. Of course, if this was all that happened, playing on a seesaw would be dull. The seesaw's motion alternates between two rotations—first in one direction, then in the other. The momentary torque that makes the transition from one direction to the other is provided when a child pushes off the ground with her feet.

Two people with quite different weights are still able to balance the seesaw. The lighter person sits farther from the pivot point, as shown in Figure 8-6. This equalizes the torques. The extra distance compensates for the reduced force due to the smaller weight.

Figure 8-5 If a seesaw is balanced, the torques exerted by the two children are equal.

Figure 8-6 The heavier person sits closer to the pivot point to equalize the torques.

Figure 8-7 (a) Two masses taped near the ends of a meter stick have a large rotational inertia. (b) When the masses are moved closer to the center, the rotational inertia is considerably less.

Rotational Inertia

The rotational acceleration depends on the object as well as the net torque. If you push on the door to a bank vault and a door in your house, you get different rotational accelerations. In the translational case, the same net force produces different accelerations for different inertial masses. In the rotational case, the same net torque produces different rotational accelerations, but now the acceleration depends on more than the object's mass; the distribution of the mass is also important.

Consider a dumbbell arrangement of a meter stick with a $\frac{1}{2}$-kilogram mass taped on each end. Holding the dumbbell at its center and rotating it back and forth demonstrates convincingly that a large torque is required to give it a substantial rotational acceleration (Figure 8-7[a]). This is completely analogous to the translational inertial properties we have encountered. The rotational analog of inertia is **rotational inertia**.

Changing the arrangement gives different results. If the two masses are moved closer to the center of the meter stick, it is much easier to start and stop the rotation (Figure 8-7[b]). The dumbbell has less rotational inertia even though no mass was removed. Simply changing the distribution of the mass changed the rotational inertia; it is larger the farther the mass is located from the point of rotation.

Newton's second law for rotational motion is analogous to that for translational motion, $F_{net} = ma$, where the net torque τ replaces the net force F, the rotational inertia I replaces the mass m, and the rotational acceleration α replaces the translational acceleration a:

$$\tau_{net} = I\alpha$$

▶ net torque = rotational inertia × rotational acceleration

▶ Newton's second law for rotation

The net torque on an object is equal to its rotational inertia times its rotational acceleration.

Just as translational acceleration must always point in the same direction as the net force causing it, rotational acceleration must always point in the same direction as the net torque. Therefore, the net torque must lie along the axis of rotation.

Losing one's balance on the high wire amounts to gaining a rotation off the wire. Tightrope walkers increase their rotational inertia by carrying long poles. Their increased rotational inertia helps them maintain their balance by allowing them more time to react. We naturally do something like this when we try to keep our balance. Picture yourself walking on a railroad track. Where are your arms?

Extended arms help children maintain their balance on an abandoned railroad track.

Flawed Reasoning

A group of engineers are designing a machine. At one place in the machine, a large gear is turned on an axle by a motor. The large gear meshes with a small gear to turn it on its axle, as shown in Figure 8-8. The engineers are arguing about the torques that the gears exert on each other.

Seth: "The large gear will exert a larger torque on the small gear than the small gear exerts back on the large gear, by virtual of its size."

Jason: "You are partially right. The large gear does exert the larger torque, but not because of its size. The large gear is the one attached to the motor. It is driving the small gear, so it must be exerting the larger torque."

Roger: "You are both forgetting Newton's third law. The force exerted by the small gear on the large gear is equal and opposite to the force exerted by the large gear on the small gear, so the torques they exert on each other must also be equal."

Jane: "Newton's third law applies to *forces*, but not to *torques*. Even though the forces exerted by the gears on each other must be equal and opposite, the force that the small gear exerts on the large gear is acting farther away from the axle, so the small gear is actually exerting the larger torque!"

Which of these engineers should be the leader of the project?

Answer We hope that Jane is directing this project. She understands that Newton's third law always applies whenever two objects interact, but that equal and opposite *forces* does not mean equal and opposite *torques*. Because a torque is the product of the force and the distance from the axis of rotation, the force acting farther from the axle will produce the larger torque. This principle is what makes gears useful. Note that while there is a rotational analog for Newton's first and second laws, no such analog exists for Newton's third law.

Figure 8-8 Which gear exerts the larger torque?

Center of Mass

If we mentally shrink any object so that its entire mass is located at a certain point, the translational motion of this new, very compact object would be the same as that of the original object. Furthermore, if the original object is rotating freely, it rotates about this same point. This point is called the **center of mass**.

The center-of-mass concept is also useful for examining the effect of gravity on objects. Rather than dealing with an incredibly large number of forces acting on the object, we treat the object as if the total force (that is, its weight) acts at the center of mass. By doing this, we can account for the translational and rotational motions of the object.

Now we need a way of finding the center of mass. This could be determined by a mathematical averaging procedure that considers the distribution of the object's mass. A certain amount of mass on one side of the object is balanced, or averaged, with some mass on the other side. But there are easier ways.

Finding the center of mass for a regularly shaped object is fairly simple. The symmetry of the object tells us that the center of mass must be at the geometric center of the object. It is interesting to note that there does not have to be any mass at that spot; a hollow tennis ball's center of mass is still at its geometric center.

Question Where would you expect the center of mass of a donut to be?

Answer Because the donut is approximately symmetric, its center of mass is near the center of the hole.

Locating the center of mass of an irregularly shaped object is a little more difficult. However, because the weight can be considered to act at the center of mass, we can locate it with a simple experiment. Hang the object from some point along its surface so that it is free to swing, as in Figure 8-9(a). The object will come to rest in a position where there is no net torque on it. At this position the weight acts along a vertical line through the support point. Therefore, the center of mass is located someplace on this vertical line. Now suspend the object from another point, establishing a second line. Because the center of mass must lie on both lines, it must be at the intersection of the two lines (Figure 8-9[b]).

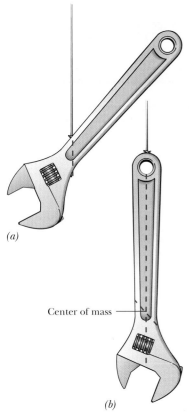

(a)

Center of mass

(b)

Figure 8-9 The center of mass of the wrench is located at the intersection of the vertical lines obtained by hanging the wrench from two or more places.

Flawed Reasoning

Roger finds the center of mass of a baseball bat by balancing the bat on his finger. He then saws the bat in two at the location of the center of mass. He expected the masses of the two pieces to be identical because the average location of the mass must have half the mass on one side and half the mass on the other. But when he held the two pieces, one was obviously heavier than the other. **What mistake did Roger make in his reasoning?**

Answer The center of mass of the bat is not the average location of its mass. It is a *weighted average*, like the calculation of your GPA. Because the bat balances at the center of mass, the torque exerted by the weight of the fat end about this pivot must balance the torque exerted by the skinny end about this pivot. Because the mass in the fat end is located closer to the pivot point, the gravitational force acting on it must be greater. The fat end therefore weighs more than the skinny end.

PHYSICS | ON YOUR OWN

Guess the location of the center of your home state. You can check your guess by taping a map of your state onto a piece of cardboard. After cutting around the edges of the state, suspend it from several points, as shown in Figure 8-10, to locate its center.

Figure 8-10 The center of the state of Vermont is located by hanging it from two points.

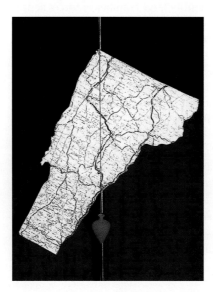

Stability

We can extend our ideas about rotating objects to see why some things tip over easily and others are quite stable. Picture a child making a tall tower out of toy blocks. Much to the child's delight, the tower always tips over. But why does this happen? Clearly there are taller structures in the world than this child's tower.

We answer this by looking at the stability of a one-block tower. In Figure 8-11(a) the left side of the block is slightly above the table. If let go in this position, the block's weight (acting at the center of mass) provides a counterclockwise torque about the right edge. The force of the table on the block acts through this edge but produces no torque because it acts at the pivot point. Thus, the net torque is counterclockwise, and the block falls back to its original position.

In Figure 8-11(b) the block is tilted far enough that the weight acts to the right of the pivot point. The weight produces a clockwise torque, and the block falls over. The block tips over whenever its center of mass is beyond the edge of the base.

As the child's toy tower gets taller and its center of mass gets higher, the amount the tower has to sway before the center of mass passes beyond the base gets smaller. We can make the tower more stable by keeping it short, widening its base, or both.

If you get bumped while standing with your feet close together, you begin to fall over. To stop this, you quickly spread your feet and increase your support base. Car manufacturers promote superwide wheelbases because this innovation makes the car more stable.

Tightrope walking is difficult because the support base (the wire's thickness) is so small. A slight lean to the left or right puts the center of mass past the support point and creates a torque. The torque produces a rotation in the same direction as the initial lean, making the situation worse. Such a situation is known as **unstable equilibrium**.

The most stable arrangement occurs when the center of mass is below the support point, as in Figure 8-12. As the center of mass sways left or right, the torque that is created rotates the object back to the original orientation. This situation is known as **stable equilibrium**.

PHYSICS | ON YOUR OWN

Cut a piece of plywood into the shape shown in Figure 8-13. Try hooking the skyhook over your fingers as indicated. Try this again after you have inserted a stiff belt into the slot as shown. Where is the center of mass of the skyhook and belt?

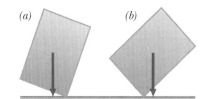

Figure 8-11 (a) When the center of mass is above the base, the block returns to its upright position. (b) When the center of mass is beyond the base, the block topples over.

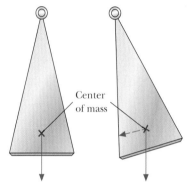

Figure 8-12 Stable equilibrium occurs when the center of mass is located below the point of suspension.

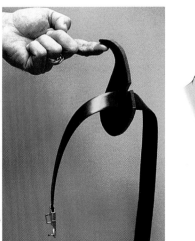
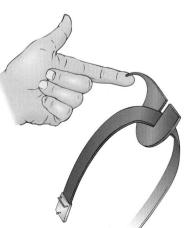

Figure 8-13 The center of mass of the skyhook and belt is located directly below the pivot point.

Rotational Kinetic Energy

✓ MATH

If we drop a yo-yo to the floor, it speeds up as gravitational potential energy is converted to kinetic energy. If, instead, we hold on to the string while the yo-yo drops, the yo-yo does not speed up as quickly. Only part of the lost gravitational potential energy has been converted to translational kinetic energy of the center of mass. Because the total energy is conserved, the rest must have been converted to a new form of energy. This new form is associated with the spinning motion of the yo-yo *about* its center of mass and is called **rotational kinetic energy**.

The rotational kinetic energy of the yo-yo can be calculated by treating the yo-yo as if it were millions of connected pieces. If we use our formula for linear kinetic energy, $KE = \frac{1}{2}mv^2$, for each of these pieces and add all of the contributions, we find that the total takes a very familiar form. Whereas the linear kinetic energy depends on the linear inertia (mass) and the square of the linear speed, the rotational kinetic energy depends on the rotational inertia I and the square of the rotational speed ω:

▶ rotational kinetic energy

$$KE_{rotation} = \frac{1}{2}I\omega^2$$

As for translational kinetic energy, this quantity is not a vector; there is no direction associated with rotational kinetic energy. Because this is an energy, the units for rotational kinetic energy are joules.

Just as the rotational kinetic energy of the yo-yo can be converted to other forms of energy, the rotational kinetic energy stored in flywheels can be used to make automobiles or buses more fuel efficient. As the vehicle slows to a stop, the translational kinetic energy can be used to spin up a heavy flywheel. Then, when the light turns green, this spinning flywheel can be used to accelerate the car. We will find in the next three sections that building a car with a single flywheel could have disastrous effects. In practice we need to use two identical flywheels spinning in opposite directions at the same speed.

Angular Momentum

There is another kind of momentum in which an object orbiting a point has a *rotational* quantity of motion that is different from linear momentum. This new quantity is called **angular momentum** and is represented by the letter L. The magnitude of the angular momentum in this example is equal to the object's linear momentum multiplied by the radius r of its circular path:

▶ angular momentum = linear momentum × radius

$$L = mvr$$

A spinning object also has angular momentum because it is really just a large collection of tiny particles, each of which is revolving around the same axis. The total angular momentum of a spinning object is just the sum of the individual angular momenta of the individual particles. We find that the angular momentum of the spinning object is equal to the product of its rotational inertia I and its rotational speed ω:

▶ angular momentum = rotational inertia × rotational speed

$$L = I\omega$$

which is analogous to the expression for linear momentum, $p = mv$, where the angular momentum L replaces the linear momentum p, the rotational inertia I replaces the mass m, and the rotational velocity ω replaces the translational velocity v.

Earth has both types of angular momenta: the angular momentum due to its annual revolution around the Sun and that due to its daily rotation on its axis.

Conservation of Angular Momentum

The angular momentum of a system does not change under certain circumstances. The law of **conservation of angular momentum** is analogous to the conservation law for linear momentum. The difference is that the interaction that changes the angular momentum is a torque rather than a force.

> If the net external torque on a system is zero, the total angular momentum of the system does not change.

◂ conservation of angular momentum

Note that the net external force need not be zero for angular momentum to be conserved. There can be a net external force acting on the system as long as the force does not produce a torque. This is the case for projectile motion because the force of gravity can be considered to act at the object's center of mass. Therefore, even though a thrown baton follows a projectile path, it continues to spin with the same angular momentum around its center of mass. There is no net torque on the baton.

There are some interesting situations in which the angular momentum of a spinning object is conserved but the object changes its rotational speed. Near the end of a performance, many ice skaters go into a spin. The spin usually starts out slowly and then gets faster and faster. This might appear to be a violation of the law of conservation of angular momentum but is in fact a beautiful example of its validity.

Angular momentum is the product of the rotational inertia and the rotational speed and, in the absence of a net torque, remains constant. Therefore, if the rotational inertia decreases, the rotational speed must increase. This is exactly what happens. The skater usually begins with arms extended. As the arms are drawn in toward the body, the rotational inertia of the body decreases because the mass of the arms is now closer to the axis of rotation. This requires that the rotational speed increase. To slow the spin, the skater reverses the procedure by extending the arms to increase the rotational inertia.

PHYSICS | ON YOUR OWN

> Sit on a rotating stool, start yourself spinning slowly with your arms extended, and draw your arms in toward your body. Why does your rotational speed increase? You can increase the effect by holding a large mass in each hand.

The same principle applies to the flips and twists of gymnasts and springboard divers. The rate of rotation and hence the number of somersaults that can be completed depends on the rotational inertia of the body as well as the angular momentum and height generated during the take-off. The drawings in Figure 8-14 give the relative values of the rotational inertia for the tuck, pike, and layout positions. The more compact tuck has the smallest rotational inertia and therefore has the fastest rotational speed.

Question Imagine you are executing a running front somersault when you suddenly realize that you are not turning fast enough to make it around to your feet. What can you do?

Answer You can tighten your tuck to reduce your rotational inertia. Because angular momentum is conserved, you will rotate faster.

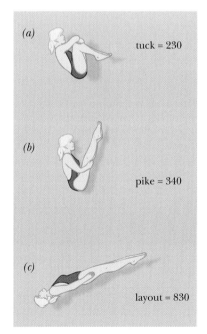

Figure 8-14 Relative values of the rotational inertia for the (a) tuck, (b) pike, and (c) layout positions.

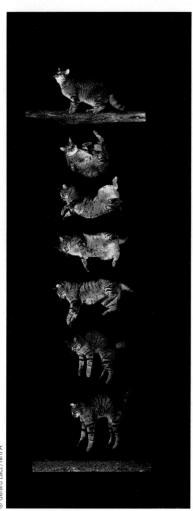

Figure 8-15 Strobe photos of a falling cat

Cats have the amazing ability to land on their feet regardless of their initial orientation. Modern strobe photographs (Figure 8-15) have shown that the cat does not acquire a rotation by kicking off. The cat's initial angular momentum is zero. Because the force of gravity acts through the cat's center of mass, it produces no torque, and the angular momentum remains zero.

The cat rotates by turning the front and hind ends of its body in different directions. The entire cat has zero angular momentum as long as the angular momenta of the two parts are equal and opposite. Even though these angular momenta are the same size, the amount of rotation can be different because it depends on the rotational inertia of that part of the body. The cat adjusts the rotational inertia by retracting and extending its legs.

As Earth moves along its orbit, it is continually being attracted toward the Sun. Because the gravitational force always acts toward the Sun, there is no net torque affecting Earth's motion around the Sun. Therefore, Earth's orbital angular momentum must be conserved. As Kepler discovered, Earth's orbit about the Sun is not a circle but an ellipse. Thus, Earth is not always the same distance away from the Sun. This means that when Earth is closer to the Sun, its speed must be faster to keep the angular momentum constant. Similarly, Earth's speed must be slower when it is farther away from the Sun.

The Solar System began as a huge cloud of gas and dust that had a very small rotation as part of its overall motion around the center of the Galaxy. As it collapsed under its mutual gravitational attraction, it rotated faster and faster in agreement with conservation of angular momentum. This explains why the planets all revolve around the Sun in the same direction and why the rotation of the Sun itself is also in this direction.

Angular Momentum: A Vector

Like linear momentum, angular momentum is a vector quantity. The conservation of a vector quantity means that both the magnitude and direction are constant. There are some interesting consequences of conserving the direction of angular momentum. The direction of the angular momentum is the same as that of the rotational velocity; that is, it lies along the axis of rotation.

One important application of this principle is the use of a gyroscope for guidance in airplanes and spacecraft. A gyroscope is simply a disk that is rotating rapidly about an axle. The axle is mounted so that the mounting can be rotated in any direction without exerting a torque on the rotating disk (Figure 8-16).

Figure 8-16 A spinning gyroscope maintains its direction in space even when its mount is rotated.

Once the gyroscope is rotating, the axle maintains its direction in space no matter what the orientation of the spacecraft.

A couple of students who were studying angular momentum decided to play a practical joke on a classmate. They mounted a heavy flywheel in an old suitcase and gave it a large rotational speed. They then asked a classmate to carry the suitcase into another room. When the classmate turned a corner, the bottom of the suitcase quickly rose, almost spraining his arm! What happened? The suitcase did not follow the classmate around the corner because the large angular momentum of the flywheel resisted any change in its orientation. Not only did it resist any change in its orientation, it turned in a different, and unexpected, direction.

Question Assume that you are at the North Pole holding a rapidly spinning gyroscope that has its angular momentum vector pointing straight up. Which way will the gyroscope point if you transport it to the South Pole without exerting any torques on it?

Answer It will point toward the ground. Remember this is the same direction (directly toward the North Star) as before. You have changed your orientation because your feet must point toward the center of the spherical Earth.

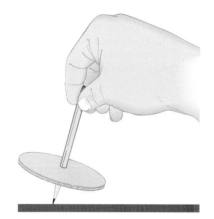

Figure 8-17 Can you make the pencil stand on its point without holding on to it?

PHYSICS | **ON YOUR OWN**

Push a pencil through the center of a circular piece of cardboard so that it looks like Figure 8-17. Can you make the pencil stand on its point without holding on to it? Try spinning it. Why does this work? What happens if you try this without the cardboard? Explain your observations.

A spinning top has angular momentum, but it is not usually constant. When the top's center of mass is not directly over the tip, the gravitational force exerts a torque on the top. If the top were not spinning, this torque would cause it to simply topple over. But when it is spinning, the torque produces a change in the angular momentum that causes the angular momentum to change its direction, not its magnitude. The spin axis of the top (and hence its angular momentum) traces out a cone, as shown in Figure 8-18. We say that the top *precesses*. The friction of the top's contact point with the table produces another torque. This torque reduces the magnitude of the angular momentum and eventually causes the top to slow down and topple over.

A similar situation exists with Earth. Because Earth's shape is irregular, the gravitational forces of the Sun and Moon on Earth produce torques on the spinning Earth. These torques cause Earth's spin axis to precess. This precession is very slow but does cause the direction of our North Pole to sweep out a big cone in the sky once every 25,780 years (Figure 8-19). Thus, Polaris (the Pole Star) is not always the North Star.

Figure 8-18 A top precesses due to the torque produced by the top's weight.

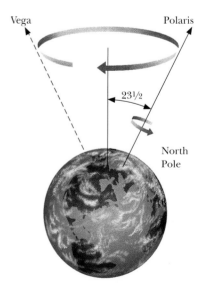

Figure 8-19 The precession of Earth's axis causes the North Pole to follow a circular path among the stars.

Summary

Objects can rotate or revolve around some axis, and this can happen whether they have a fixed or moving axis. The rotational and translational motions are independent of each other. The rotation of a free body takes place about its center of mass.

The rules for rotational motion are similar to the rules for translational motion. The displacement is a change in angular position, rotational speeds are angular displacements divided by time, and rotational accelerations are changes in rotational speeds divided by time. Newton's first and second laws for rotational motion have the same form as the laws for translational motion. A change in rotational speed occurs only when there is a net torque on the object. The torque τ is equal to the radius r multiplied by the perpendicular force F, or $\tau = rF$, and has units of newton-meters. Rotational inertia depends on the distance of the mass from the axis of rotation as well as on the mass itself.

The stability of an object depends on the torques produced by its weight (acting at the center of mass) and on the supporting forces.

For an object orbiting a point, its angular momentum is defined as the product of its linear momentum and the radius of its circular path, $L = mvr$. For a spinning object, its angular momentum is the product of its rotational inertia and its rotational speed, $L = I\omega$.

The angular momentum of a system is conserved if no net external torque acts on the system. External forces may act on the system as long as these forces do not produce a net torque. Even though angular momentum is conserved, the rotational speed can change if the rotational inertia changes.

The conservation of a vector quantity means that both its magnitude and direction are constant. Similarly, a change in angular momentum can be a change in magnitude or direction or both. Conservation of angular momentum can be used to analyze problems ranging from the motion of tops to that of gymnasts and cats.

Chapter 8 Revisited

The helicopter's engine must exert a torque on the rotor to turn the blades. In turn, the rotor exerts a torque on the helicopter in the opposite direction. If acting alone this torque would cause the helicopter to rotate about a vertical axis, gaining unwanted angular momentum. The small rotor produces a torque in the opposite direction to prevent this. In straight, level flight, the net torque is zero, and therefore the angular momentum remains zero. On a larger helicopter, the twin rotors turn in opposite directions so that the helicopter's total angular momentum is zero.

KEY TERMS

angular momentum: A quantity giving the rotational momentum. For an object orbiting a point, it is the product of the linear momentum and the radius of the path, $L = mvr$. For a spinning object, it is the product of the rotational inertia and the rotational speed, $L = I\omega$.

center of mass: The balance point of an object. This location has the same translational motion as the object would if it were shrunk to a point.

conservation of angular momentum: If the net external torque on a system is zero, the total angular momentum of the system does not change.

rotational acceleration: The change in rotational speed divided by the time it takes to make the change.

rotational inertia: The property of an object that measures its resistance to a change in its rotational speed.

rotational kinetic energy: Kinetic energy associated with the rotation of a body, $KE = \frac{1}{2}I\omega^2$.

rotational speed: The angle of rotation or revolution divided by the time taken. Measured in units such as degrees per second or revolutions per minute.

rotational velocity: A vector quantity that includes the rotational speed and the direction of the axis of rotation.

stable equilibrium: An equilibrium position or orientation to which an object returns after being slightly displaced.

torque: The rotational analog of force. It is equal to the radius multiplied by the force perpendicular to the radius, $\tau = rF$.

unstable equilibrium: An equilibrium position or orientation from which an object leaves after being slightly displaced.

CONCEPTUAL QUESTIONS

1. A figure skater is spinning with her arms held straight out. Which has greater rotational speed, her shoulders or her fingertips? Why?
2. Who has the greater rotational speed, a person living on the equator or one living in New York City?
3. Drew and Blake are riding on a merry-go-round at the country fair. Drew is riding near the center while Blake is near the outside. Compare their rotational accelerations.
4. You are looking down on a merry-go-round and observe that it is rotating clockwise. What is the direction of the merry-go-round's rotational velocity? If the merry-go-round is slowing down, what is the direction of its rotational acceleration?
5. What is the direction of the rotational velocity of Earth?
6. Earth's rotational speed is slowing due to the tidal influences of the Sun and Moon. What is the direction of Earth's rotational acceleration?
7. What do we call an object's resistance to a change in its rotational velocity?
8. What is needed to change the rotational velocity of an object?
9. Future space stations will rotate to produce artificial gravity. What torque (if any) is needed to keep the space stations rotating?
10. A flywheel with a large rotational inertia is often attached to the drive shaft of automobile engines. What purpose does the flywheel serve?
11. If the object shown in the figure is fixed but free to rotate about point A, which force will produce the larger torque? Why?

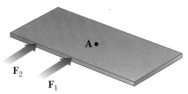

Questions 11 and 12

12. If the object shown in the figure is not fixed and point A is the object's center of mass, which force will produce translational motion without rotation?
13. Use the concept of torque to explain how a claw hammer is used to pull nails.
14. Apply the concept of torque to explain how a wheelbarrow allows you to transport a heavy load with a lifting force much less than the weight of the load.
15. You are a window washer and, rather than use the fancy lift, you decide just to lean a ladder up against a large plate glass window. Use the concept of torque to explain why the likelihood of breaking the glass increases the higher you climb up the ladder.
16. A 10-speed bicycle has five gears on the rear wheel. When the bicycle is in first gear, is the chain on the gear with the largest radius or the smallest radius? Use the concept of torque to explain your answer.
17. Sam and Kelly are carrying an office desk. Sam has to exert a much greater force than Kelly does to keep the desk level. Is the desk's center of mass closer to Sam or to Kelly? Use the concept of torque to justify your answer.
18. Dana and Loren are carrying a steel girder. As shown below, Dana is holding the girder at the end while Loren is not. Who is exerting the greater force on the girder? Justify your answer using the concept of torque.

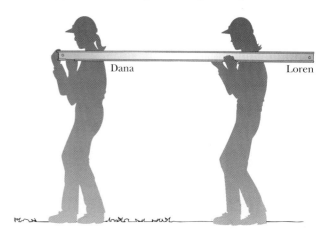

19. Two flywheels have the same mass but one has a radius twice that of the other. If both flywheels are spinning about their axes at the same rate, which one would be harder to stop? Why?
20. Which would be harder to rotate about its center, a 12-foot-long, 2″ × 4″ board or a 6-foot-long, 4″ × 4″ board? Why?
21. Does an object's rotational inertia increase or decrease with an increase in mass? Does it increase or decrease as the mass is moved closer to the axis of rotation?
22. Would you have a larger rotational inertia in the tuck, pike, or layout positions? Why? (See Figure 8-14 if you are not familiar with these diving positions.)
23. A solid sphere and a solid cylinder are made of the same material. If they have the same mass and radius, which one has the smaller rotational inertia about its center? Why?

24. If a solid disk and a hoop have the same mass and radius, which would have the smaller rotational inertia about its center of mass? Why?
25. Earth spins about its own axis once every 23 hours and 56 minutes. Why has this rate changed very little since the time of Isaac Newton?
26. Find an everyday example that clearly illustrates the meaning of Newton's second law for rotation.
▲ 27. How would you determine the center of mass of an automobile?
28. In the text we found the center of mass at the intersection of two lines. If you suspend the object from a third point, this line passes through the intersection of the first two. Why?
29. Where is the center of mass of the figurine resting on the pedestal in the figure?

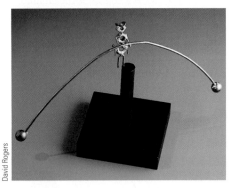

30. A spoon and a fork can be suspended beyond the edge of a glass by using a flat toothpick, as shown in the figure. Where is the center of mass of the spoon–fork combination?

31. Using diagrams, show why an empty ice cream cone is more stable when it is placed upside down rather than on its tip.
32. The Leaning Tower of Pisa is stable even though it is tilted significantly from the vertical. If a greedy developer decided to add three more stories to this historical landmark, it might well topple over. Use the concept of stable equilibrium to explain this.
33. A marble is resting in a round bowl. Is the marble in stable or unstable equilibrium? Why?

34. The Pacific Science Center in Seattle, Washington, has an exhibit in which patrons can ride a bicycle around a narrow track high above the ground. A large concrete block hangs beneath the bike on a long bar that is fastened rigidly to the bike's frame. Why is this exhibit safe?

35. If you stand with your back against the wall and try to bend over and touch your toes, you will invariably tip over. Use the concept of equilibrium to explain why.
36. If you stand facing a wall with your toes touching the wall, you cannot raise yourself up on your tip toes. Use the concept of equilibrium to explain why.
37. It is possible (and quite likely) for a high jumper's center of mass to pass *under* the bar while the jumper passes *over* the bar as shown in the figure. Explain how this is possible.

38. A solid cylinder and a hoop have the same mass and radius. If both are rotating with the same speed, which will have the largest rotational kinetic energy?
39. Suppose you race the solid cylinder and the hoop from Question 38 down a ramp. Use the concept of rotational kinetic energy to argue that the solid cylinder will reach the bottom of the ramp first.
40. If you are asked to design a 50-pound flywheel for use in a new car, would you want to concentrate the mass as close as possible to the axis of rotation or as far away as possible? Explain your reasoning.
41. Why does a small helicopter have a rotor on its tail?

42. Why does a helicopter with two sets of rotors not need a rotor on the tail?

43. In which of the following positions would a diver have the smallest rotational inertia for performing a front somersault: tuck, pike, or layout? Why? (See Figure 8-14 if you are not familiar with these diving positions.)
44. Why is it possible for a high diver to execute more front somersaults in the tuck position than in the layout position?
45. Why do figure skaters spin faster when they pull in their arms?

46. An astronaut "floating" in a space shuttle has an initial rotational motion but no initial translational motion relative to the shuttle. Why does the astronaut continue to rotate?

47. A cat that is held upside down and dropped with no initial angular momentum manages to land on its feet. Does the cat need to acquire any angular momentum to do this? Explain your reasoning.
▲ 48. A common early maneuver learned on a trampoline is to land sitting with your feet pointing in one direction and to then reverse directions on the bounce to land with your feet pointing in the opposite direction. With practice it is possible to turn either direction on command *after* leaving the mat. How is this possible?
▲ 49. If you look down the inside of the barrel of a rifle, you see long spiral grooves. When the bullet travels down the barrel, these grooves cause the bullet to spin. Why would we want the bullet to spin?
▲ 50. A billiard ball without spin hits perpendicular to the cushion and bounces back perpendicular to the cushion. However, if the ball is spinning about the vertical, it bounces off to one side and spins at a slower rate. What is the force that causes the change (a) in the ball's linear momentum? (b) In its angular momentum?
▲ 51. Figure 6-4 shows a fire extinguisher being used to propel a person in a straight line. A fire extinguisher could also be used to turn a merry-go-round if a person were to sit on one of the horses and fire the extinguisher backward. As it speeds up, the merry-go-round acquires angular momentum. Use the concept of torque to account for this change in angular momentum.
52. An air-hockey puck is whirling on the end of a string that passes through a small hole in the center of the table as shown in the figure. What happens to the speed of the puck as the string is slowly pulled down through the hole?

53. A gyroscope that points horizontally at the North Pole is transported to the South Pole while it continues to spin. Which way does it point? Explain.
54. A gyroscope is oriented so that it points toward the North Star when it is in Seattle. If it is carried to the equator while it continues to spin, which way will it point, and why?
55. Some people have proposed powering cars by extracting energy from large rotating flywheels mounted in the car. There is a problem with this suggestion if only one flywheel is used. What is the problem and how can it be remedied?
56. The film crew of *Candid Camera* replaces a person's briefcase with an identical one that contains a mounted and spinning flywheel. Explain what happens when the person tries to carry the briefcase around a corner.
57. A teacher sits on a stool that is free to rotate. She holds a rotating flywheel with the axis vertical such that the wheel spins clockwise when viewed from above. She turns the wheel completely over, and she begins to spin on the stool. Does she spin clockwise or counterclockwise? Justify your answer using the concept of conservation of angular momentum.
58. The rotational axis of Earth's spin is tilted $23\frac{1}{2}$ degrees relative to Earth's axis of revolution about the Sun. The North Pole is tilted toward the Sun on June 22. Which pole is tilted toward the Sun on December 22?
59. If you stand outside all night and watch the stars, they all appear to move except one. Which star appears to remain stationary, and why?
60. Why is the star Polaris not always located directly above Earth's geographic North Pole?
61. You jump onto a merry-go-round that is rotating clockwise as viewed from above. Find the direction of (a) the merry-go-round's angular momentum, (b) the merry-go-round's change in angular momentum, (c) your change in angular momentum, and (d) the change in angular momentum for the system made up of you and the merry-go-round. Assume the merry-go-round has very good bearings.
62. As you walk from the center of a merry-go-round toward the outer edge, the merry-go-round slows. Is angular momentum conserved? Explain why or why not.
▲ 63. Bicycle wheels are normally about 2 feet in diameter. Imagine that a bicycle was constructed with 4-inch-diameter wheels but was otherwise the same size as a normal bicycle. How would the small wheels affect the bicycle's stability. Justify your answer using the concept of angular momentum.
▲ 64. A salesman at the local motorcycle dealership tells you about the new ultralight motorcycle tires that have just come on the market. These tires are the same size as normal tires but have only one-tenth the mass. How would the light wheels affect the motorcycle's stability? Justify your answer using the concept of angular momentum.

EXERCISES

1. If the beaters on a food mixer make 1000 revolutions in 5 min, what is the average rotational speed of the beaters? Express your answer in both revolutions per minute and revolutions per second.
2. If a CD makes 1500 revolutions in 5 min, what is the CD's average rotational speed?
3. What is the rotational speed of the hand on a clock that measures the minutes?
4. What is the rotational speed of the hand on a clock that measures the seconds?
5. If it takes 3 s for a modern player to stop a DVD with a rotational speed of 7490 rpm, what is the DVD's average rotational acceleration?
6. A variable-speed drill, initially turning at 400 rpm, speeds up to 1000 rpm in a time of 0.5 s. What is its average rotational acceleration?
7. What torque does a 140-N salmon exert on a 1.5-m-long fishing pole if the pole is horizontal and the salmon is out of the water?
8. You are holding a 5-kg dumbbell straight out at arm's length. Assuming your arm is 0.70 m long, what torque is the dumbbell exerting on your shoulder?
9. A pirate with a mass of 90 kg stands on the end of a plank that extends 2 m beyond the gunwale. What torque does he exert on the plank?
10. Robin is standing terrified at the end of a diving board that is high above the water. If Robin has a mass of 65 kg and is standing 1.5 m from the board's pivot point, what torque is Robin exerting on the board?
11. Two children with masses of 20 kg and 30 kg are sitting on a balanced seesaw. If the lighter child is sitting 3 m from the center, where is the heavier child sitting?
12. A child with a mass of 20 kg sits at a distance of 2 m from the pivot point of a seesaw. Where should a 14-kg child sit to balance the seesaw?
13. Is the system shown in the figure balanced? If not, which end will fall? Explain your reasoning.

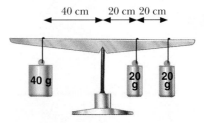

14. What mass would you hang on the right side of the system in the figure to balance it—that is, to make the clockwise and counterclockwise torques equal?

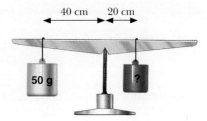

15. A child with a mass of 50 kg is riding on a merry-go-round. If the child has a speed of 3 m/s and is located 2 m from the center of the merry-go-round, what is the child's angular momentum?
16. A 1600-kg car is traveling at 20 m/s around a curve with a radius of 120 m. What is the angular momentum of the car?
▲ 17. Which has the larger angular momentum about the Sun, Mars or Earth? The radius, speed, and mass of Mars are 1.5, 0.8, and 0.11 times those of Earth, respectively.
▲ 18. Mercury follows an elliptical orbit that takes it as close as 46 million km to the Sun and as far as 70 million km from the Sun. At both of these locations, Mercury's velocity makes a right angle to the direction to the Sun. If Mercury's speed is 38 km/s when it is farthest from the Sun, how fast is it moving when it is closest to the Sun?

Interlude

Why do the riders not fall out of the roller-coaster car as it executes a loop-the-loop?

Universality of Motion

A set of rules—Newton's laws—correctly describes the motion of ordinary objects. These rules are the foundation of the classical physics world view, and they match our commonsense notions about the behavior of the material world. Are Newton's laws valid for all situations and for all regions of the Universe?

As we developed these rules, we presumed that all observers were standing still on Earth's surface. In Chapter 2, when describing the speed of a ball dropped from a height, we simply said, "The ball is falling at 30 meters per second." It was unnecessarily cumbersome at that time to say, "The ball is falling at 30 meters per second *relative to Earth's surface*." Is this extra qualification ever needed? To answer this question, we need to ask how the motion would appear to somebody moving relative to Earth's surface. We need to ask whether that person—say, somebody riding in a train or standing in a free-falling elevator—would develop the same rules of motion. If the rules are different, it is possible that the laws of motion are not universal. The Universe might have an enormous number of different rules—one set for each different point of view. On the other hand, if we can deduce that these laws are universal, the payoffs in terms of our world view are large. In science, the fewer the rules, the more beautiful the world view. When a rule is universal, such as Newton's law of gravitation (Chapter 5), we believe it is a more fundamental aspect of nature.

As we try to establish the universality of the laws of motion, we find that the price for this new level of understanding is a restructuring of our ideas about the concepts of space and time. The person primarily responsible for this restructuring was Albert Einstein, who, in addition to his stature among physicists, captured the popular imagination as no other scientist ever has. The mere mention of his name conjures up such images as time as the fourth dimension, people growing older slowly, and warped space. His popularity was due to the seemingly bizarre ideas that he brought to our world view. In fact, Einstein didn't like the publicity. He wished to be left alone in the solitude of his work. But his ideas were too shocking not to create a stir.

In 1905 this quiet man was a clerk in a Swiss patent office. During that year he published four papers and his doctoral thesis; two were on the subject we now call the special theory of relativity. Interestingly, it was work in another area of physics that resulted in his being awarded the Nobel Prize in 1921. His ideas revolutionized the physics world view and propelled him into the center of scientific activity for the next half century. How this person—who could not get a university teaching job when he graduated—created such revolutionary ideas is a fascinating story in itself.

Some people may feel that Einstein's ideas have little or no connection with reality—that they are a fantasy-based creation resulting from some mathematical trickery. This perception couldn't be further from the truth.

> Although the ideas of relativity had their beginnings in a realm of the physical world beyond our everyday experience, they produced profound changes in the very foundations of our world view. But the ideas of relativity didn't start with Einstein. As early as Galileo, questions were being asked about the absolute nature of position, speed, and acceleration.

Although the ideas of relativity had their beginnings in a realm of the physical world beyond our everyday experience, they produced profound changes in the very foundations of our world view. But the ideas of relativity didn't start with Einstein. As early as Galileo, questions were being asked about the absolute nature of position, speed, and acceleration.

9 Classical Relativity

Everybody has been told that Earth rotates on its axis once each day, and yet it appears that the Sun, Moon, and stars all go around Earth. What evidence do we have to support the idea that it is Earth that is really moving?

(See page 171 for the answer to this question.)

The apparent motion of the stars is due to Earth's rotation.

A Reference System

Do observers moving relative to each other agree on the description of the motion of an object? Most of us feel that they would not. Consider, for example, the situation in which one observer is unfortunate enough to be in a free-falling elevator and the other is standing safely on the fifth floor. How do the two observers describe the motion of an apple that is dropped by the observer in the elevator?

The observer in the elevator sees the apple suspended in midair (Figure 9-1[a]). It has no speed and no acceleration. The observer on the fifth floor sees the apple falling freely under the influence of gravity. It has a constant downward acceleration and therefore is continually gaining speed (Figure 9-1[b]).

Is there something fundamentally different about these descriptions, or are the differences just cosmetic? And most important, do the differences mean that the validity of Newton's laws of motion is in question? Are the laws valid for the observer in the elevator? If not, the consequences for our physics world view could be serious.

Physics⦿Now™ Test your understanding of this chapter by logging into PhysicsNow at **http://physics.brookscole.com/kf6e**, selecting the chapter, and clicking on the "Take a Pre-Test" link.

(a)

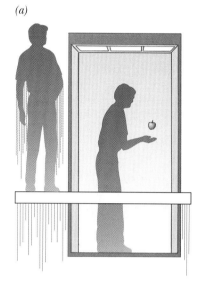

A Reference System

We see motion when something moves relative to other things. Imagine sitting in an airplane that is in straight, level flight at a constant speed. As far as the activities inside the plane are concerned, you don't think of your seat as moving. From your point of view, the seat remains in the same spot relative to everything else in the plane.

The phrase *point of view* is too general. Because all motion is viewed relative to other objects, we need to agree on a set of objects that are not moving relative to each other and that can therefore be used as the basis for detecting and describing motion. This collection of objects is called a **reference system**.

One common reference system is Earth. It consists of such things as the houses, trees, and roads that we see every day. This reference system appears to be stationary. In fact, we are so convinced that it is stationary that we occasionally get tricked. If, while you sit in a car waiting for a traffic light to change, the car next to you moves forward, you occasionally experience a momentary sensation that your car is rolling backward. This illusion occurs because you expect your car to be moving and everything outside the car to be stationary.

It doesn't matter if you are in a moving car or sitting in your kitchen; both are good reference systems. Consider your room as your reference system. To describe the motion of an object in the room, you measure its instantaneous position with respect to some objects in the room and record the corresponding time with a clock. This probably seems reasonable and quite obvious. But complications—and interesting effects—arise when the same motion is described from

(b)

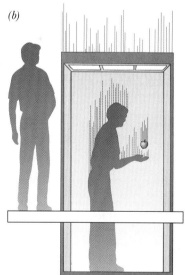

Figure 9-1 The dropped apple appears suspended in midair (a) as viewed by the elevator passenger and (b) as falling freely by the observer standing on the floor.

Which car is moving?

two different reference systems. We begin by studying these interesting effects in classical relativity.

Motions Viewed in Different Reference Systems

Imagine that you are standing next to a tree and some friends ride past you in a van, as shown in Figure 9-2. Suppose that the van is moving at a very high, constant velocity relative to you and that you have the ability to see inside the van.

One of your friends drops a ball. What does the ball's motion look like? When your friends describe the motion, they refer to the walls and floor of the van. They see the ball fall straight down and hit the van's floor directly below where it was released (Figure 9-3).

You describe the motion of the ball in terms of the ground and trees. Before the ball is released, you see it moving horizontally with the same velocity as your friends. Afterward, the ball has a constant horizontal component of velocity, but the vertical component increases uniformly. That is, you see the ball follow the projectile path shown in Figure 9-4.

The ball's path looks quite different when viewed in different reference systems. Galileo asked if observers could decide whose description was "correct." He concluded that they couldn't. In fact, each observer's description was correct. We can understand this by looking at the explanations that you and your friends give for the ball's motion.

We begin by examining the horizontal motion. Your friends, observing that the ball doesn't move horizontally, conclude that the net horizontal force is zero. On the other hand, you do see a horizontal velocity. But because it is constant, you also conclude that the net horizontal force is zero.

What about vertical forces? Your friends see the ball exhibit free fall with an acceleration of 10 (meters per second) per second. The vertical component of the

Figure 9-2 Your friends move by you at a very high, constant velocity and drop a ball.

Figure 9-3 From your friends' point of view, they are at rest and see you moving. In their system they see the ball fall vertically.

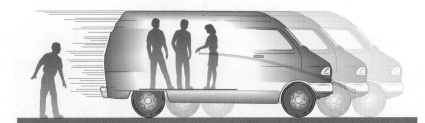

Figure 9-4 From the ground you see the ball follow a projectile path.

projectile motion that you observe is also free-fall motion with the same acceleration. Each of you concludes that there is the same net constant force acting downward.

Although you disagree with your friends' description of the ball's path, you agree on the acceleration and the forces involved. Any experiments that you do in your reference system will yield the same accelerations and the same forces that your friends find in their system. In both cases the laws of motion explain the observed motion.

We define an **inertial reference system** as one in which Newton's first law (the law of inertia) is valid. Each system above was assumed to be an inertial reference system. In fact, any reference system that has a constant velocity relative to an inertial system is also an inertial system.

◀ inertial reference system

The principle that the laws of motion are the same for any two inertial reference systems is called the **Galilean principle of relativity**. Galileo stated that if one were in the hold of a ship moving at a constant velocity, there would be no experiment this person could perform that would detect the motion. This means that there is no way to determine which of the two inertial reference systems is "really" at rest. There seems to be no such thing in our Universe as an absolute motion in space; all motion is relative.

The laws of physics are the same in all inertial reference systems.

◀ principle of relativity

The principle of relativity says that the laws of motion are the same for your friends in the van as they are for you. A very important consequence is that the conservation laws for mass, energy, and momentum are valid in the van system as well as the Earth system. If your friends say that momentum is conserved in a collision, you will agree momentum is conserved even though you do not agree on the values for the velocities or momenta of each object.

PHYSICS | ON YOUR OWN

Conduct some physics experiments the next time you ride in a car, train, bus, or airplane. During the constant velocity portions of the trip, try tossing a ball into the air, playing catch with a friend, using a plumb line or a carpenter's level, observing the surface of a beverage, using an equal-arm balance, or weighing something. Do you notice any differences in the results of these experiments and those performed in your home?

Comparing Velocities   ✓ MATH

✓ Extended presentation available in the *Problem Solving* supplement

Is there any way that you and your friends in the van can reconcile the different velocities that you have measured? Yes. Although you each see different velocities, you can at least agree that each person's observations make sense within their respective reference system. When you measure the velocity of the ball moving in the van, the value you get is equal to the *vector* sum of the van's velocity (measured in your system) and the ball's velocity (measured relative to the van).

Suppose your friends roll the ball on the floor at 2 meters per second due east and the van is moving with a velocity of 3 meters per second due east relative to your system. In this case the vectors point in the same direction, so you simply add the speeds to obtain 5 meters per second due east, as shown in Figure 9-5(a). If, instead, the ball rolls due west at 2 meters per second, you measure the ball's velocity to be 1 meter per second due east (Figure 9-5[b]).

Figure 9-5 The velocity of the ball relative to the ground is the vector sum of its velocity relative to the van and the van's velocity relative to the ground.

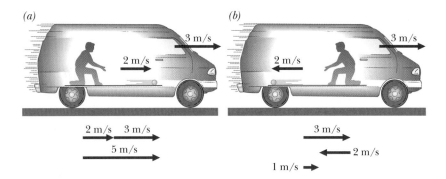

Question What do you observe for the velocity of the ball if it is rolling eastward at 2 meters per second while the van is moving westward at 4 meters per second?

Answer The ball is moving 2 meters per second westward.

Although this rule works well for speeds up to millions of kilometers per hour, it fails for speeds near the speed of light, about 300,000 kilometers per second (186,000 miles per second). This is certainly not a speed that we encounter in our everyday activities. The fantastic, almost unbelievable, effects that occur at speeds approaching that of light are the subject of our next chapter.

Flawed Reasoning

Why is the following statement wrong? "If energy is conserved, it must have the same value in every inertial reference system."

Answer Kinetic energy is given by $\frac{1}{2}mv^2$. This formula depends on speed, so it must yield different values in different inertial systems. Take the example of a person on a moving train dropping a 1-kilogram ball from a height of 2 meters above the floor. In the system of the train, the ball initially has 20 joules of gravitational potential energy (relative to the floor) and no kinetic energy for a total of 20 joules. An observer on the ground, however, sees the ball initially moving with the same speed as the train, say, 30 meters per second. This observer agrees that the ball initially has 20 joules of gravitational potential energy but finds that the initial kinetic energy is 450 joules for a total of 470 joules. Conservation of energy simply means that just before the ball hits the floor, the person in the train will still calculate the total energy to be 20 joules, and the observer on the ground will still calculate the total energy to be 470 joules.

Accelerating Reference Systems

Let's expand our discussion of your friends in the van. This time, suppose their system has a constant forward acceleration relative to your reference system. Your friends find that the ball doesn't land directly beneath where it was released but falls toward the back of the van, as shown in Figure 9-6. In your reference system, however, the path looks the same as before. It is still a projectile path with a horizontal velocity equal to the ball's velocity at the moment it was released. The ball stops accelerating horizontally when it is released, but your friends continue to accelerate. Thus, the ball falls behind.

Figure 9-6 In an accelerating system, your friends see the ball fall toward the back of the van.

Question Where would the ball land if the van was slowing down?

Answer It would land forward of the release point because the ball continues moving with the horizontal velocity it had when released, whereas the van is slowing down.

As before, the descriptions of the ball's motion are different. But what about the explanations? Your explanation of the motion—the forces involved, the constant horizontal velocity, and the constant vertical acceleration—doesn't change. But your friends' explanation does change; the law of inertia does not seem to work anymore. The ball moves off with a horizontal acceleration. In their reference system, they would have to apply a horizontal force to make an object fall vertically, a contradiction of the law of inertia. Such an accelerating system is called a **noninertial reference system**.

There are two ways for your friends to explain the motion. First, they can abandon Newton's laws of motion. This is a radical move requiring a different formulation of these laws for each type of noninertial situation. This is intuitively unacceptable in our search for universal rules of nature.

Second, they can keep Newton's laws by assuming that a horizontal force is acting on the ball. But this would indeed be strange; there would be a horizontal force in addition to the usual vertical gravitational force. This also poses problems. In inertial reference systems, we can explain all large-scale motion in terms of gravitational, electric, or magnetic forces. The origin of this new force is unknown; furthermore, its size and direction depend on the acceleration of the system. We know, from your inertial reference system, that the strange new force your friends seem to experience is due entirely to their accelerated motion. Forces that arise in accelerating reference systems are called **inertial forces**.

If inertial forces seem like a way of getting around the fact that Newton's laws don't work in accelerated reference systems, you are right. These forces do not exist; they are invented to preserve the Newtonian world view in reference systems where it does not apply. In fact, another common label for these forces is *fictitious forces*.

If you are in the accelerating system, these fictitious forces seem real. We have all felt the effect of being in a noninertial system. If your car suddenly changes its velocity—speeding up, slowing down, or changing direction—you feel pushed in the direction opposite the acceleration. When the car speeds up rapidly, we often say that we are being pushed back into the seat.

Question What is the direction and cause of the fictitious force you experience when you suddenly apply the brakes in your car?

Answer Assuming that you are moving forward, the inertial force acts in the forward direction, "throwing" you toward the dashboard. It arises because of the car's acceleration in the backward direction due to the braking.

PHYSICS | **ON YOUR OWN**

Fasten a helium-filled balloon on a string to the seat in your car. Observe which way it floats when the car speeds up, moves at a constant velocity, and slows down. Describe the fictitious inertial forces in each case.

Realistic Inertial Forces ✓ MATH

If you were in a windowless room that suddenly started accelerating relative to an inertial reference system, you would know something happened. You would feel a new force. Of course, in this windowless room, you wouldn't have any visual clues to tell you that you were accelerating; you would only know that some strange force was pushing in a certain direction.

This strange force would seem very real. If you had force measurers set up in the room, they would all agree with your sensations. This experience would be rather bizarre; things initially at rest would not stay at rest. Vases, chairs, and even people would need to be fastened down securely, or they would move.

This situation occurs whenever we are in a noninertial reference system. Imagine riding in an elevator accelerating upward from Earth's surface. You would experience an inertial force opposite the acceleration in addition to the gravitational force. In this case the inertial force would be in the same direction as gravity, and you would feel heavier. You can even measure the change by standing on a bathroom scale.

If the elevator stands still or moves with a constant velocity, a bathroom scale indicates your normal weight (Figure 9-7[a]). Because your acceleration is zero, the net force on you must also be zero. This means that the force $\mathbf{F}_s$ exerted on you by the scale must balance the gravitational force $\mathbf{F}_g$ exerted on you by Earth, and therefore $\mathbf{F}_s$ is equal to and opposite $\mathbf{F}_g$. Because the size of the gravitational force is equal to the mass m times the acceleration due to gravity g, we sometimes say that you experience a force of 1 g.

If the elevator accelerates upward (Figure 9-7[b]), you must experience a net upward force as viewed from the ground. Because the gravitational force does not change, the force $\mathbf{F}_s$ exerted on you by the scale must be larger than the gravi-

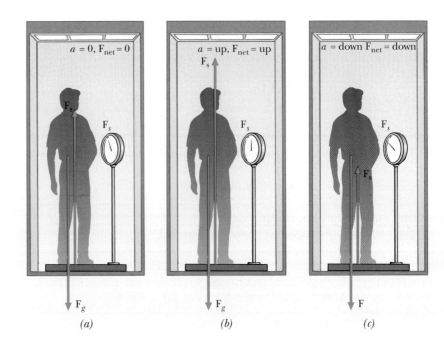

Figure 9-7 The weight of the elevator passenger $\mathbf{F}_s$ registered on the dial depends on the acceleration and the force of gravity. (a) $\mathbf{F}_s$ is equal to the force of gravity $\mathbf{F}_g$ when the elevator has no acceleration; (b) $\mathbf{F}_s$ is larger than $\mathbf{F}_g$ when the elevator accelerates upward; and (c) $\mathbf{F}_s$ is smaller than $\mathbf{F}_g$ when the elevator accelerates downward.

tational force $\mathbf{F}_g$. This change in force would register as a heavier reading on the scale. You would also experience the effects on your body. Your stomach "sinks" and you feel heavier.

If the upward acceleration is equal to that of gravity, the net upward force on you must have a magnitude equal to $\mathbf{F}_g$. Therefore, the scale must exert a force equal to twice $\mathbf{F}_g$, and the reading shows this. You experience a force of 2 g's and feel twice as heavy. Astronauts experience maximum forces of 3 g's during launches of a space shuttle. During launches of the Apollo missions to the Moon, the astronauts experienced up to 6 g's. When pilots eject from jet fighters, the forces approach 20 g's for very short times.

Figure 9-7(c) shows the situation as seen from the ground when the elevator accelerates in the downward direction. In the elevator the inertial force is upward and subtracts from the gravitational force. You feel lighter; we often say that our stomachs are "up in our throats."

Question   If you are traveling upward in the elevator and slowing down to stop at a floor, will the scale read heavier or lighter?

Answer   Because you are slowing while traveling upward, the acceleration is downward and therefore the inertial force is upward and the scale will read lighter.

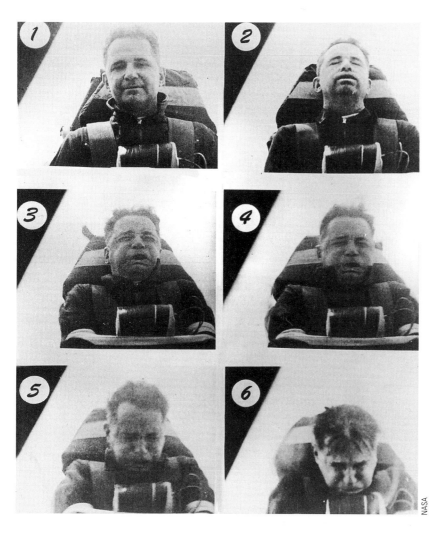

This sequence of photographs taken during the experiments before the first space flight shows the effects of inertial forces during large accelerations.

Living in Zero G

Astronauts in space stations orbiting Earth experience weightlessness—"zero g." They float about the station and can do gymnastic maneuvers involving a dozen somersaults and twists. They can release objects and have them stay in place suspended in the air. This happens because the astronauts (and other "floating" objects) are in orbit about Earth just like the space station. Even when the astronauts leave the space station to go for a space walk, the effect is the same: they float along with the space station.

Of course, when the astronauts try to move a massive object, they still experience the universality of Newton's second law; being weightless does not mean being massless.

Although they experience the sensation of being weightless, the gravitational force on them is definitely *not* zero; at an altitude of a few hundred kilometers, the gravitational force is approximately 10% lower than at Earth's surface. In the accelerating, noninertial reference system of the space station, the gravitational force and the inertial force cancel each other, producing the sensation of weightlessness. Any experiment they could perform inside the space station, however, yields the same result; gravity appears to have been turned off.

Although a pleasant experience for a while, living in zero g can create problems over long periods because our bodies have evolved in a gravitational field. Astronauts report puffiness in the face, presumably from body fluids not being held down by gravity. Scientists also report that changes occur in astronauts' hearts due to the lower stress levels—sort of the reverse of exercise. For longer periods of living in zero g, it is expected that bone growth may be impaired.

Flight Engineer Susan Helms and Mission Commander Yury Usachev aboard the U.S. Laboratory/Destiny module of the International Space Station in April 2001.

All these issues will need to be addressed in the next decade as the United States and Russia develop plans to send astronauts and cosmonauts to Mars. Such a trip will require several years, much longer than the record 439 days for living in zero g held by the Russians.

Astronauts working in the Spacelab science module in *Atlantis*'s cargo bay.

If the downward acceleration is equal to that of gravity, you feel *weightless*. You and the elevator are both accelerating downward with the rate of acceleration due to gravity. The bathroom scale does not exert any force on you. You appear to be floating in the elevator, a situation sometimes referred to as "zero g."

If somehow your elevator accelerates in a sideways direction, the extra force is like the one your friends felt in the van; the inertial force is horizontal and opposite the acceleration. During the takeoff of a commercial jet airplane, passengers typically experience horizontal accelerations of $\frac{1}{4}$ g.

PHYSICS | **ON YOUR OWN**

Try some of the experiments mentioned in the first Physics on Your Own in this chapter during an elevator ride. Can you tell from these experiments whether you are moving up or down? Accelerating up or down?

PHYSICS | **ON YOUR OWN**

Fasten a cork to the inside of a lid to a quart jar with a string that is approximately three-fourths the height of the jar, as shown in Figure 9-8. Fill the jar with water, put the lid on tight, and invert the jar. Which way does the cork swing when the jar is accelerated? How do you explain your observations?

Figure 9-8 The floating cork can be used as a detector of inertial forces.

Centrifugal Forces ✓ MATH

A rotating reference system—such as a merry-go-round—is also noninertial. If you are on the merry-go-round, you feel a force directed outward. This fictitious force is the opposite of the centripetal force we discussed in Chapter 4 and is called the **centrifugal force**. It is present only when the system is rotating. As soon as the ride is over, the centrifugal force disappears.

Consider the carnival ride the Rotor, which spins you in a huge cylinder (Figure 9-9[a]). As the cylinder spins, you feel the centrifugal force pressing you against the wall. When the cylinder reaches a large enough rotational speed, the floor drops out from under you. You don't fall, however, because the centrifugal force pushing you against the wall increases the frictional force with the wall enough to prevent you from sliding down the wall. If you try to "raise" your arms away from the wall, you feel the force pulling them back to the wall.

Somebody looking into the cylinder from outside (an inertial system) sees

(a)

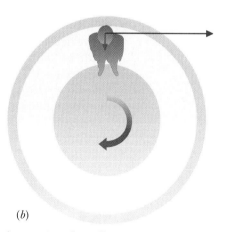
(b)

Figure 9-9 Although the people in the Rotor feel forces pushing them against the wall, an inertial observer says that the wall must push on the people to make them go in a circle.

the situation shown in Figure 9-9(b); the only force is the centripetal one acting inward. Your body is simply trying to go in a straight line, and the wall is exerting an inward-directed force on you, causing you to go in the circular path. This real force causes the increased frictional force.

> **Flawed Reasoning**
>
> You are riding in the Rotor at the state fair, as shown in Figure 9-9. A friend explains that two equal and opposite forces are acting on you, a centripetal force inward and a centrifugal force outward. Your friend further explains that these forces are third-law forces. **Are there some things that you should not learn from your friend?**
>
> **Answer** Third-law forces never act on the same object, so these two "forces" cannot form a third-law pair. In the inertial system, there is only one force acting on you, the centripetal force exerted by the wall on your back. This force causes you to accelerate in a circle. The third-law companion to this force is the push your back exerts on the wall. In your noninertial frame you are at rest, so you invent a fictitious force acting outward to balance the centripetal force. This outward "force" is not a real force.

An artificial gravity in a space station can be created by rotating the station. A person in the station would see objects "fall" to the floor and trees grow "up." If the space station had a radius of 1 kilometer, a rotation of about once every minute would produce an acceleration of 1 g near the rim. Again, viewed from a nearby inertial system, the objects don't fall, they merely try to go in straight lines. Living in this space station would have interesting consequences. For example, climbing to the axis of rotation would result in "gravity" being turned off.

Question What is the net force on someone standing on the floor of the rotating space station as viewed from their reference system?

Answer The net force would be zero because the person is at rest relative to the floor. The pilot of an approaching spaceship would see a net centripetal force acting on the person in the space station.

The International Space Station as seen from space shuttle *Endeavour* in June 2002.

Earth: A Nearly Inertial System

Earth is moving. This is probably part of your commonsense world view because you have heard it so often. But what evidence do you have to support this statement? To be sure that you are really a member of the moving-Earth society, point in the direction that Earth is moving right now. This isn't easy to do.

We do not feel our massive Earth move, and it seems more likely that it is motionless. But in fact it is moving at a very high speed. A person on the equator travels about 1700 kilometers per hour due to Earth's rotation. The speed due to Earth's orbit around the Sun is even larger, 107,000 kilometers per hour (67,000 mph)!

What was it that led us to accept this idea that Earth is moving? If we look at the Sun, Moon, and stars, we can agree that *something* is moving. The question is this: are the heavenly bodies moving and is Earth at rest, or are the heavenly bodies at rest and Earth is moving? The Greeks believed that the motion was due to the heavenly bodies traveling around a fixed Earth located in the center of the Universe. This scheme is called the **geocentric model**.

They assumed that the stars were fixed on the surface of a huge celestial sphere with Earth at its center (Figure 9-10). This sphere rotated on an axis through the North and South Poles, making one complete revolution every 24 hours. You can easily verify that this model describes the motion of the stars by observing them during a few clear nights.

The Sun, Moon, and planets were assumed to orbit Earth in circular paths at constant speeds. When this theory did not result in a model that could accurately predict the positions of these heavenly bodies, the Greek astronomers developed an elaborate scheme of bodies moving around circles that were in turn moving on other circles, and so on. Although this geocentric model was fairly complicated, it described most of the motions in the heavens.

This brief summary doesn't do justice to the ingenious astronomical picture developed by the Greeks. The detailed model of heavenly motion developed by Ptolemy in AD 150 resulted in a world view that was accepted for 1500 years. Ptolemy's theory was so widely accepted because it predicted the positions of the Sun, Moon, planets, and stars accurately enough for most practical purposes. It was also very comforting for philosophic and religious reasons. It accorded well with Aristotle's view of Earth's central position in the Universe and humankind's correspondingly central place in the divine scheme of things.

In the 16th century, a Polish scientist named Copernicus examined technical aspects of this Greek legacy and found them wanting. In 1543 his powerful and revolutionary astronomy offered an alternative view: Earth rotated about an axis once every 24 hours while revolving about the Sun once a year. Only the Moon remained as a satellite of Earth; the planets were assumed to orbit the Sun. Because his proposal put the Sun in the center of the Universe, it is called the **heliocentric model** (Figure 9-11).

Figure 9-10 The geocentric view of the Universe has Earth at its center.

Question   Which way does Earth rotate, toward the east or west?

Answer   Earth rotates toward the east, making the stars appear to move to the west.

How does one choose between two competing views? One criterion—simplicity—doesn't help here. Although Copernicus's basic model was simpler to visualize than Ptolemy's, it required about the same mathematical complexity to

Figure 9-11 In the heliocentric model of the Solar System, the planets orbit the Sun.

Figure 9-12 The position of the finger changes relative to the background when viewed by the other eye.

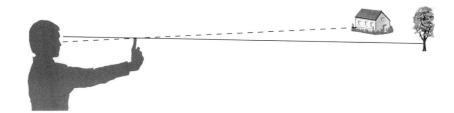

achieve the same degree of accuracy in predicting the positions of the heavenly bodies.

A second criterion is whether one model can explain more than the other. Here Copernicus was the clear winner. His model predicted the order and relative distances of the planets, explained why Mercury and Venus were always observed near the Sun, and included some of the details of planetary motion in a more natural way. It would seem that the Copernican model should have quickly replaced the older Ptolemaic model.

But the Copernican model appeared to fail in one crucial prediction. Copernicus's model meant that Earth would orbit the Sun in a huge circle. Therefore, observers on Earth would view the stars from vastly different positions during Earth's annual journey around the Sun. These different positions would provide different perspectives of the stars, and thus they should be observed to shift their positions relative to each other on an annual basis. This shift in position is called *parallax*.

You can demonstrate parallax to yourself with the simple experiment shown in Figure 9-12. Hold a finger in front of your face and look at a distant scene with your left eye only. Now look at the same scene with only your right eye. Because your eyes are not in the same spot, the two views are not the same. You see a shifting of your finger relative to the distant scene. Notice also that this effect is more noticeable when your finger is close to your face.

Unfortunately for Copernicus, the stars did not exhibit parallax. Undaunted by the null results, Copernicus countered that the stars were so far away that Earth's orbit about the Sun was but a point compared with the distances to the stars. Instruments were too crude to measure this effect. Although his counterclaim was a possible explanation, the lack of observable parallax was a strong argument against his model and delayed its acceptance. It is interesting to note that the biggest stellar parallax is so small that it was not observed until 1838—300 years later.

There was another problem with Copernicus's model. Copernicus developed these ideas before Galileo's time and did not have the benefit of Galileo's work on inertia or inertial reference systems. Because it was not known that all inertial reference systems are equivalent, most people ridiculed the idea that Earth could be moving: after all, one would argue, if a bird were to leave its perch to catch a worm on the ground, Earth would leave the bird far behind! For these reasons the ideas of Copernicus were not accepted for a long time. In fact, 90 years later Galileo was being censured for his heretical stance that Earth does indeed move.

One of the reasons that it took thousands of years to accept Earth's motion is that Earth is very nearly an inertial reference system. Were Earth's motion undergoing large accelerations, the effects would have been indisputable. Even though the inertial forces are very small, they do provide evidence of Earth's motion.

Noninertial Effects of Earth's Motion

A convincing demonstration of Earth's rotation was given by French physicist J. B. L. Foucault around the middle of the 19th century. He showed that the plane of swing of a pendulum appears to rotate. Foucault's demonstration is very popu-

Copernicus's critics argued that if Earth were moving, birds would be left behind.

A Foucault pendulum shows that Earth rotates.

lar in science museums; almost every one has a large pendulum with a sign saying that it shows Earth's rotation. But how does this show that Earth is rotating?

First, we must ask what would be observed in an inertial system. In the inertial system, the only forces on the swinging bob are the tension in the string and the pull of gravity; both of these act in the plane of swing. So in an inertial system, there is no reason for the plane to change its orientation.

The noninertial explanation is simplest with a Foucault pendulum on the North Pole, as shown in Figure 9-13. The plane of the pendulum rotates once every 24 hours; that is, if you start it swinging along a line on the ground, some time later the pendulum will swing along a line at a slight angle to the original line. In 12 hours it will be along the original line again (the pendulum's plane is halfway through its rotation). Finally, after 24 hours the pendulum will once again be realigned with the original line. If you lie on your back under the pendulum and observe its motion with respect to the distant stars, you would see that the plane of the pendulum remains fixed relative to them. It is Earth that is rotating.

Top views

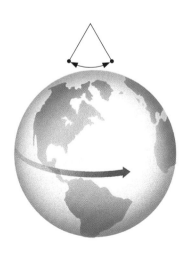
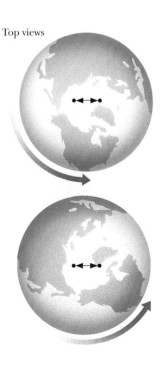

Figure 9-13 A Foucault pendulum at the North Pole appears to rotate relative to the ground once in 24 hours.

You can experience the effects of the Coriolis force by playing catch while riding on a merry-go-round.

Air moving poleward from the Equator is traveling east faster than the land beneath it and veers to the east (turns right in the Northern Hemisphere and left in the Southern Hemisphere).

Air moving toward the Equator is traveling east slower than the land beneath it and veers to the west (turns right in the Northern Hemisphere and left in the Southern Hemisphere).

Coriolis effect. Air moving north or south is deflected by Earth's rotation.

At more temperate latitudes, the plane of a Foucault pendulum requires longer times to complete one rotation. The time increases continuously from the pole to the equator, with the time becoming infinite at the equator; that is, the plane does not rotate.

The weight of a person on Earth is affected by Earth's rotation. A person on the equator is traveling along a circular path, but a person on the North Pole is not. The person on the North Pole feels the force of gravity; the person on the equator feels the force of gravity plus the fictitious centrifugal force. The effect of this centrifugal force is small; it is only one-third of 1% of the gravitational force. That means if we transported a 1-newton object from the North Pole to the equator, its weight would be 0.997 newton.

Another inertial force in a rotating system, known as the **Coriolis force**, is the force you feel when you move along a radius of the rotating system. If, for example, you were to walk from the center of a merry-go-round to its edge, you would feel a force pushing you in the direction opposite the rotation.

From the ground system, the explanation is straightforward. A point on the outer edge of a rotating merry-go-round has a larger speed than a point closer to the center because it must travel a larger distance during each rotation. As you walk toward the outer edge, the floor of the merry-go-round moves faster and faster. Your inertial tendency is to keep the same speed relative to the ground system. The merry-go-round moves out from under you, giving you the sensation of being pulled in the opposite direction. If you move inward toward the center of the merry-go-round, the direction of this inertial force is reversed. The Coriolis force is more complicated than the centrifugal force in that it depends on the velocity of the object in the noninertial system as well as the acceleration of the system.

Question If you drop a ball from a great height, it will experience a Coriolis force. Will the ball be deflected to the east or the west?

Answer This situation is analogous to walking toward the center of the merry-go-round. Therefore, the ball will be deflected in the direction of Earth's rotation—that is, to the east.

The Coriolis force acts on anything moving along Earth's surface and deflects it toward the right in the Northern Hemisphere and toward the left in the Southern Hemisphere. British sailors experienced this reversal during World War I. During a naval battle near the Falkland Islands (50 degrees south latitude), they noticed that their shells were landing about 100 meters to the left of the German ships. The Coriolis corrections that were built into their sights were correct for the Northern Hemisphere but were in the wrong direction for the Southern Hemisphere!

PHYSICS | ON YOUR OWN

> You can experience the Coriolis force by playing catch with a friend while riding on a merry-go-round. Notice how the ball appears to curve in the air when you expect it to go straight.

The Coriolis force also causes large, flowing air masses in the Northern Hemisphere to be deflected to the right. As the air flows in toward a low-pressure region, it is deflected to the right. The result is that hurricanes in the Northern Hemisphere rotate counterclockwise as viewed from above. The circulation pattern is reversed for hurricanes in the Southern Hemisphere and for high-pressure regions in the Northern Hemisphere. Figure 9-14 shows a hurricane in the Northern Hemisphere as seen from one of NASA's satellites. Folklore has it that the

Planetary Cyclones

The atmospheres of the gaseous planets—Jupiter, Saturn, Uranus, and Neptune—are very unlike Earth's atmosphere. The atmospheres are composed primarily of hydrogen molecules with a much smaller amount of helium. All other gases comprise less than 1% of the atmospheres. Yet the colors provided by these gases (for instance, the clouds on Jupiter and Saturn are composed of crystals of frozen ammonia and those on Uranus and Neptune are composed of frozen methane) give us some visual clues about the effects of the Coriolis force on the large-scale motions in these planetary atmospheres.

The most famous cyclone in the Solar System is the Great Red Spot on Jupiter, which was first observed more than 300 years ago. It is a giant, reddish oval that is about 26,000 kilometers across the long dimensions—about the size of two Earths side by side. Because the Great Red Spot is located in Jupiter's southern hemisphere, it might be expected to rotate in the clockwise direction. However, it is observed to rotate *counterclockwise* with a period of 6 days.

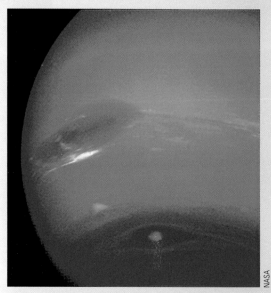

Voyager 2 discovered this Great Dark Spot on Neptune during its flyby in the fall of 1989.

Therefore, the Coriolis effect tells us that the Great Red Spot must be a high-pressure storm rather than the low-pressure regions typical of hurricanes and cyclones on Earth. Jupiter also has three white ovals that were first observed in 1938 and have diameters about 10,000 kilometers across.

When *Voyager 2* flew by Neptune in the fall of 1989, planetary scientists were surprised and pleased to observe a Great Dark Spot. It is located in Neptune's southern hemisphere, is about 10,000 kilometers across, and rotates counterclockwise with a period of 17 days. *Voyager* also observed a few small storms on the order of 5000 kilometers across in Saturn's atmosphere but none in Uranus's atmosphere.

Although no one knows the origins of these planetary storms, scientists can explain their long lives. Hurricanes on Earth die out rather quickly when they travel across land areas. Although each of these planets has a rocky core, the cores are relatively small compared with the planet's size. The resultant thickness of the atmospheres contributes to the long lifetimes of the storms. Another factor is size. Larger storms are more stable and last longer.

The Great Red Spot on Jupiter is a high-pressure cyclonic storm that has lasted for at least 300 years.

Coriolis force causes toilets and bathtubs to drain counterclockwise in the Northern Hemisphere, but its effects on this scale are so small that other effects dominate.

Even if Earth was not rotating, it would still not be an inertial reference system. Although Earth's orbital velocity is very large, the change in its velocity is small. The acceleration due to its orbit around the Sun is about one-sixth that of its daily rotation on its axis. In addition, the Solar System orbits the center of the Milky Way Galaxy once every 250 million years with an average speed of 1 million kilometers per hour. The associated inertial forces are about 100 million times as small as those due to rotation. The Milky Way Galaxy has an acceleration within the local group of galaxies, and so on.

Figure 9-14 Hurricanes in the Northern Hemisphere turn counterclockwise, as seen from above. This image of Hurricane Fran was taken from GEOS-8 less than seven hours before the eye went ashore at Cape Fear, North Carolina.

Earth is located in one of the spiral arms of the Milky Way Galaxy.

Physics Now™ Assess your understanding of this chapter's topics with sample tests and other resources found by logging into PhysicsNow at **http://physics.brookscole.com/kf6e**.

In terms of our daily lives, Earth is very nearly an inertial reference system. Any system that is moving at a constant velocity relative to its surface is, for most practical purposes, an inertial reference system.

Summary

All motion is viewed relative to some reference system, the most common being Earth. An inertial reference system is one in which the law of inertia (Newton's first law) is valid. Any reference system that has a constant velocity relative to an inertial reference system is also an inertial reference system.

The Galilean principle of relativity states that the laws of motion are the same for any two inertial reference systems. Observers moving relative to each other report different descriptions for the motion of an object, but the objects obey the same laws of motion regardless of reference system.

Observers in different reference systems can reconcile the different velocities they obtain for an object by adding the relative velocity of the reference systems to that of the object. However, this procedure breaks down for speeds near that of light.

In a reference system accelerating relative to an inertial reference system, the law of inertia does not work without the introduction of fictitious forces that are due entirely to the accelerated motion. Centrifugal and Coriolis forces arise in rotating reference systems and are examples of inertial forces. Earth is a noninertial reference system, but its accelerations are so small that we often consider Earth to be an inertial reference system.

Chapter 9 Revisited

The most direct evidence is provided by the Foucault pendulum. The plane in which the pendulum swings stays fixed relative to the distant stars and rotates relative to the ground, demonstrating that Earth is rotating. The effect of the Coriolis force on storm systems is also evident. This force occurs only in a rotating system and causes hurricanes to rotate in opposite directions on either side of the equator.

KEY TERMS

centrifugal force: A fictitious force arising when a reference system rotates (or changes direction). It points away from the center, in the direction opposite the centripetal acceleration.

Coriolis force: A fictitious force that occurs in rotating reference systems. It is responsible for the direction of winds in hurricanes.

Galilean principle of relativity: The laws of motion are the same in all inertial reference systems.

geocentric model: A model of the Universe with Earth at its center.

heliocentric model: A model of the Universe with the Sun at its center.

inertial force: A fictitious force that arises in accelerating (noninertial) reference systems. Examples are the centrifugal and Coriolis forces.

inertial reference system: Any reference system in which the law of inertia (Newton's first law of motion) is valid.

noninertial reference system: Any reference system in which the law of inertia (Newton's first law of motion) is not valid. An accelerating reference system is noninertial.

reference system: A collection of objects not moving relative to each other that can be used to describe the motion of other objects. See inertial and noninertial reference systems.

CONCEPTUAL QUESTIONS

1. Newton's first law states, "Every object remains at rest or in motion in a straight line at constant speed unless acted on by an unbalanced force." Is this "law" true in all reference frames? Explain.

2. Newton's third law states, "If an object exerts a force on a second object, the second object exerts an equal force back on the first object." Is this "law" true in all reference frames? Explain.

3. *Alice in Wonderland* begins with Alice falling down a deep, deep rabbit hole. As she falls, she notices that the hole is lined with shelves and grabs a jar of orange marmalade. Upon discovering that the jar is empty, she tries to set it back on a shelf—a difficult task while she is falling. She is afraid to drop the jar because it might hit somebody on the head. What would really happen to the jar if Alice had dropped it? Describe its motion from Alice's reference system and from the reference system of someone sitting on a shelf on the hole's wall.

▲ 4. Imagine riding in a glass-walled elevator that goes up the outside of a tall building at a constant speed of 20 meters per second. As you pass a window washer, he throws a ball upward at a speed of 20 meters per second. Assume, furthermore, that you drop a ball out a window at the same instant.
   a. Describe the motion of each ball from the point of view of the window washer.
   b. Describe the motion of each ball as you perceive it from the reference system of the elevator.

5. You wake up in a windowless room on a train, which rides along particularly smooth, straight tracks. Imagine that you have a collection of objects and measuring devices in your room. What experiment could you do to determine whether the train is stopped at a station or moving horizontally at a constant velocity?

6. Assume that you are riding in a windowless train on perfectly smooth, straight tracks. Imagine that you have a

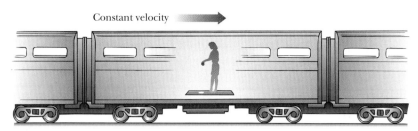

Questions 7–12 A train is traveling along a straight, horizontal track at a constant velocity of 50 kilometers per hour. An observer in the train holds a ball directly over a white spot on the floor of the train and drops it.

collection of objects and measuring devices in the train. What experiment could you do to determine whether the train is moving horizontally at a constant velocity or is speeding up?

7. The woman riding the train in the figure drops a ball directly above a white spot on the floor. Where will the ball land relative to the white spot? Explain.
8. How would the woman in the figure describe the ball's horizontal velocity while the ball is falling? Would an observer on the ground standing next to the tracks agree? Explain.
9. What would the woman in the figure say about the horizontal forces acting on the ball as it falls? Would an observer on the ground standing next to the tracks agree? Explain.
10. What value would the woman in the figure obtain for the acceleration of the ball as it falls? Would an observer on the ground standing next to the tracks obtain the same value? Explain.
11. Would the woman and the observer on the ground in the figure agree on the ball's kinetic energy just before it leaves her hand? Would they agree on the *change* in kinetic energy of the ball from the moment it leaves her hand until just before it hits the floor? Explain.
12. Would the woman and the observer on the ground in the figure agree on the ball's momentum just before it leaves her hand? Would they agree on the change in momentum of the ball from the moment it leaves her hand until just before it hits the floor? Explain.
13. Gary is riding on a flatbed railway car, which is moving along a straight track at a constant 20 meters per second. Applying a 600-newton force, Gary is trying in vain to push a large block toward the front of the car. His two friends, Cindy and Mitch, are watching from beside the track. Would Gary and Cindy agree on the value of the block's kinetic energy at the instant Gary passes his friends? Would they agree on the *change* in the block's kinetic energy in the next second? Explain.

▲ 14. Mitch, in Question 13, has just returned from physics class where he was studying about work. Mitch argues that, from his frame of reference, Gary applies a 600-newton force for a distance of 20 meters in 1 second and therefore does 12,000 joules of work on the block. He wonders why the block does not appear to speed up as a result of this work. What is the flaw in Mitch's reasoning?
15. Assume that you are driving down a straight road at a constant speed. A small ball is tied to a string hanging from the rearview mirror. Which way will the ball swing when you apply the brakes? Explain your reasoning.
▲ 16. Assume that you are driving down a straight road at a constant speed. A helium-filled balloon is tied to a string that is pinned to the front seat. Which way will the balloon swing when you apply the brakes? Explain your reasoning.
17. If the train in the figure is traveling to the right, is it speeding up or slowing down? What if it is traveling to the left?

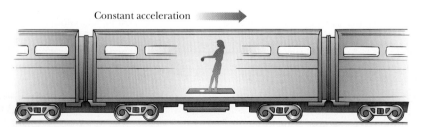

Questions 17–24 A train is traveling along a straight, horizontal track with a constant acceleration as indicated. An observer in the train holds a ball directly over a white spot on the floor of the train. At the instant the speed is 50 kilometers per hour, she drops the ball.

18. If all the curtains in the train in the figure were closed, what experiment (if any) could the woman perform to determine whether the train was traveling to the right or to the left? Explain.
19. The woman riding the train in the figure drops a ball directly above a white spot on the floor. Where will the ball land relative to the white spot? Why doesn't it matter whether the train is moving to the right or to the left?
▲ 20. The woman in the train in the figure observes that the ball falls in a straight line that is slanted away from her. Is the magnitude of the ball's acceleration along this line greater than, equal to, or less than the usual acceleration due to gravity? Explain.
21. How would the woman in the figure describe the ball's horizontal velocity both just after releasing the ball and just before it strikes the floor of the train? Explain.
22. What would an observer on the ground obtain for the horizontal speed of the ball in the figure right after the ball is released and right before it hits the floor of the train? Explain.
23. What would the woman in the figure say about the horizontal forces acting on the ball as it falls? Would an observer on the ground standing next to the tracks agree? Explain.
24. Draw the free-body diagram for the falling ball in the figure from the reference frame of the woman in the train. For each force on the diagram, state, if possible, the object responsible for the force. Repeat this from the reference frame of the observer on the ground.
25. A 180-pound person takes a ride in the elevator that goes up the side of the Space Needle in Seattle. Much to the amusement of the other passengers, this person stands on a bathroom scale during the ride. During the time the elevator is accelerating upward, is the reading on the scale greater than, equal to, or less than 180 pounds? Explain.

26. During the time the elevator in Question 25 is moving upward with a constant speed, is the reading on the scale greater than, equal to, or less than 180 pounds? Explain.
27. Assume you are standing on a bathroom scale while an elevator slows down to stop at the top floor. Will the reading on the scale be greater than, equal to, or less than the reading after the elevator stops? Why?
28. The elevator in Question 27 now starts downward to return to the ground floor. Will the reading on the scale be greater than, equal to, or less than the reading with the elevator stopped? Why?
29. If a child weighs 200 newtons standing at rest on Earth, would she weigh more, less, or the same if she were in a spaceship accelerating at 10 (meters per second) per second in a region of space far from any celestial objects? Why?
30. The child in Question 29 enters a circular orbit at constant speed around a distant planet. The spacecraft's centripetal acceleration is 10 (meters per second) per second. Would she weigh less than, equal to, or more than 200 newtons? Why?
31. If you were allowed to leave your tray down while your DC-9 accelerates for take off, why would objects slide off the tray?
32. What happens to the surface of a drink if you hold the drink while your Boeing 777 accelerates down the runway?
▲ 33. Assume that a meter-stick balance is balanced with a 20-gram mass at 40 centimeters from the center and a 40-gram mass at 20 centimeters from the center. Will it remain balanced if it is in an elevator accelerating downward? Explain your reasoning.
▲ 34. Assume that you weigh a book on an equal-arm balance while an elevator is stopped at the ground floor. Would you get the same result if the elevator were accelerating upward? Explain your reasoning.
35. You wake up in a windowless room on a train, traveling along particularly smooth, horizontal tracks. You don't know in which direction the train is moving, but you are carrying a compass. You place a ball in the center of the floor and observe as it rolls east. If the train is moving west, is it speeding up, slowing down, or turning with constant speed? (If turning, state right or left.) What if it is moving east? Explain.

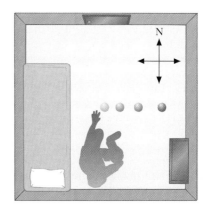

36. Consider the train in Question 35. If the train is moving north, is it speeding up, slowing down, or turning with constant speed? (If turning, state right or left.) What if it is moving south? Explain.
37. In an inertial reference system, we define *up* as the direction opposite the gravitational force. In a noninertial reference system, *up* is defined as the direction opposite the vector sum of the gravitational force and any inertial

forces. Which direction is up in each of the following cases?
  a. An elevator accelerates downward with an acceleration smaller than that of free fall.
  b. An elevator accelerates upward with an acceleration larger than that of free fall.
  c. An elevator accelerates downward with an acceleration larger than that of free fall.

▲ 38. Using the definition of *up* in Question 37, which direction is up for each of the following situations?
  a. A child rides near the outer edge of a merry-go-round.
  b. A train's dining car going around a curve turns to the right.
  c. A skier skies down a hill with virtually no friction.

39. Which direction is up for astronauts orbiting Earth in a space shuttle?

40. What happened to the astronauts' sense of up and down as the *Apollo* spacecraft passed the point in space where the gravitational forces of Earth and the Moon are equal? Explain.

41. You and a friend are rolling marbles across a horizontal table in the back of a moving van traveling along a straight section of an interstate highway. You roll the marbles toward the side of the van. What can you say about the velocity and acceleration of the van if you observe the marbles (a) head straight for the wall? (b) Curve toward the front of the truck?

42. A ball is thrown vertically upward from the center of a moving railroad flatcar. Where, relative to the center of the car, does the ball land in each of the following cases?
  a. The flatcar moves at a constant velocity.
  b. The velocity of the flatcar increases.
  c. The velocity of the flatcar decreases.
  d. The flatcar travels to the right in a circle at constant speed.

43. You fill a bucket half full of water and swing it in a vertical circle. When the bucket is at the top of its arc, the bucket is upside down but the water does not spill on your head. What direction is up for the water? Explain.

44. Would it be possible to take a drink at the top of a loop-the-loop on a roller coaster? Explain.

45. For a science project, a student plants some bean seeds in water and lets them grow in containers fastened near the outer edge of a merry-go-round that is continually turning. Draw a side view of the experiment showing the direction the plants will grow.

46. A student carries a ball on a string onto the rotating cylinder ride shown in Figure 9-9. With the ride in operation, she holds her hand straight in front of her and lets the ball hang by the string. Using a side view, draw a free-body diagram for the ball from the student's reference frame. For each force on the diagram, state, if possible, the object responsible for the force.

47. Why does the mud fly off the tires of a pickup traveling down an interstate?

48. The Red Cross uses centrifuges to separate the various components of donated blood. The centrifugal force causes the denser component (the red blood cells) to go to the bottom of the test tube. If there were a dial on the wall of the lab that allowed the "local gravity" to be increased to any value, would the centrifuge still be required? Why or why not?

▲ 49. Copernicus had difficulty convincing his peers of the validity of his heliocentric model because if Earth were moving around the Sun, stellar parallax should have been observed, which it wasn't. If Earth's orbital radius about the Sun were magically doubled, would this make stellar parallax easier or harder to observe? Explain.

50. Assuming that Earth is a perfect sphere and that the gravitational field has a constant magnitude at all points on the surface, would your weight at the equator be greater, smaller, or the same as at the North Pole?

51. Would a Foucault pendulum rotate at the equator? Explain your reasoning.

52. If you were to set up a Foucault pendulum at the South Pole, would it appear to rotate clockwise or counterclockwise when viewed from above? Why?

53. Why are there no hurricanes on the equator?

54. In preparation for hunting season, you practice at a shooting range in which the targets are located straight to the south. You find that you must aim slightly to the right of the target to account for the Coriolis force. Are you in the Northern or the Southern Hemisphere? If you are out hunting and shoot to the north, do you have to aim slightly to the right or slightly to the left to ensure a direct hit? Explain your reasoning.

▲ 55. It is known that Earth is bigger around the equator than around the poles. How does this equatorial bulge support the idea that Earth is rotating?

EXERCISES

1. A spring gun fires a ball horizontally at 15 m/s. It is mounted on a flatcar moving in a straight line at 25 m/s. Relative to the ground, what is the horizontal speed of the ball when the gun is aimed (a) forward? (b) Backward?
2. An aircraft carrier is moving to the north at a constant 25 mph on a windless day. A plane requires a speed relative to the air of 125 mph to take off. How fast must the plane be traveling relative to the deck of the aircraft carrier to take off if the plane is headed (a) north? (b) South?
3. A child can throw a ball at a speed of 50 mph. If the child is riding in a bus traveling at 20 mph, what is the speed of the ball relative to the ground if the ball is thrown (a) forward? (b) Backward?
4. A transport plane with a large rear-facing cargo door flies at a constant horizontal speed of 400 mph. A major-league baseball pitcher hurls his best fastball, which he throws at 95 mph, out the rear door of the plane. Describe what the motion of the baseball would look like to an observer on the ground.
5. What would an observer measure for the magnitude and direction of the free-fall acceleration in an elevator near the surface of Earth if the elevator (a) accelerates downward at 5 m/s^2? (b) Accelerates downward at 15 m/s^2?
6. An observer measures the free-fall acceleration in an elevator near the surface of Earth. What would the magnitude and direction be if the elevator (a) accelerates upward at 2 m/s^2? (b) Travels upward with a constant speed of 2 m/s?
7. A person riding a train at a constant speed of 30 m/s drops a 2-kg backpack from a height of 1.25 m. The fall requires half a second and the backpack acquires a vertical velocity of 5 m/s. Find the initial kinetic energy, the final kinetic energy, and the change in kinetic energy from the reference system of an observer on the train.
▲ 8. Consider the falling backpack described in Exercise 7 from the reference system of an observer standing along the side of the track. Find the initial kinetic energy, the final kinetic energy, and the change in kinetic energy. How do the changes in kinetic energy compare in the two cases?
9. What is the maximum total force exerted on a 50-kg astronaut by her seat during the launch of a space shuttle?
10. What would be the maximum total force exerted on a 90-kg fighter pilot when ejecting from an aircraft?
11. A child weighs 300 N standing on Earth. What is the weight of the child in an elevator accelerating upward at 0.2 g?
12. An elevator is moving downward and slowing down with an acceleration of 0.1 g. If a person who weighs 800 N when at rest on Earth steps on a bathroom scale in this elevator, what will the scale read?
13. An 8-kg monkey rides on a bathroom scale in an elevator that is accelerating upward at $\frac{1}{4}$ g. What does the scale read?
14. What does the scale read if a 5-kg cat lies on a bathroom scale in an elevator accelerating downward at 0.2 g?
15. A room is being accelerated through space at 3 m/s^2 relative to the "fixed stars." It is far away from any massive objects. If a man weighs 800 N when he is at rest on Earth, how much will he weigh in the room?
16. A woman with a weight of 700 N on Earth is in a spacecraft accelerating through space a long way from any massive objects. If the acceleration is 4 m/s^2, what is her weight in the ship?
▲ 17. A cylindrical space station with a 40-m radius is rotating so that points on the walls have speeds of 20 m/s. What is the acceleration due to this artificial gravity at the walls?
▲ 18. What is the centrifugal acceleration on the equator of Mars given that it has a radius of 3400 km and a rotational period of 24.6 h? How does this compare with the acceleration due to gravity on Mars of 3.7 m/s^2?

10 Einstein's Relativity

At various times in our lives, we have all had impressions of the passage of time. An hour in a dentist's chair seems much longer than 2 hours watching a good movie. But what is time? If we develop a foolproof way of measuring time, will all observers in the Universe accept our measurements?

(See page 198 for the answer to this question.)

Time and space are central in the theories of special and general relativity, and they take on new roles.

WHEN observers in different inertial reference systems describe the same events, their reports don't match. In the framework of classical relativity, they disagree in their descriptions of the paths and on the values of an object's velocity, momentum, and kinetic energy. On the other hand, they agree on relative positions, lengths, time intervals, accelerations, masses, and forces. Even the laws of motion and the conservation laws are the same.

We never asked, or even thought to ask, whether some of these were actually the same for all reference systems or whether we had just assumed them to be the same. In classical relativity we assumed that the concepts of length, time, and mass were the same. But are they really the same?

Albert Einstein asked this question. He reexamined the process of describing events from different reference systems with an emphasis on the concepts of space and time. This led to the development of the **special theory of relativity**.

Einstein arrived at the special theory of relativity by setting forth two postulates, or conditions, that were assumed to be true. He then examined the effects of these postulates on our basic concepts of space and time. The predictions of special relativity were then compared with actual experimental measurements. The theory had to agree with nature to have any validity.

The First Postulate

The **first postulate of special relativity** is related to the question of whether there exists an absolute space—some signpost in the Universe from which all motion can be regarded as absolute. This postulate says that there is no absolute space; any inertial reference system is just as good as any other. Einstein's first postulate is a reaffirmation of the Galilean principle of relativity.

Physics Now™ Test your understanding of this chapter by logging into PhysicsNow at **http://physics.brookscole.com/kf6e**, selecting the chapter, and clicking on the "Take a Pre-Test" link.

Albert Einstein (1879–1955).

The laws of physics are the same in all inertial systems.

◀ first postulate of special relativity

As we discussed in the previous chapter, Galileo argued that a traveler in the hold of a ship moving with a constant velocity could not conduct experiments that would determine whether the ship was moving or at rest. However, the Galilean principle of relativity came into question near the end of the 19th century. A theory by Scottish physicist James Clerk Maxwell describing the behavior of electromagnetic waves, such as light and radio, yielded unexpected results.

In Newton's laws, reference systems moving at constant velocities are equivalent to each other. If, however, one system accelerates relative to another, the systems are not equivalent. Because Newton's laws depend on acceleration and not on velocity, acceleration of a reference system can be detected, but its velocity cannot.

In Maxwell's theory, however, the *velocity* of the electromagnetic waves appears in the equations rather than their acceleration. According to the classical ideas, the appearance of a velocity indicated that inertial systems were not equivalent. In principle you could merely turn on a flashlight and measure the speed of light to determine how your reference system was moving.

Maxwell's equations and the Galilean principle of relativity were apparently in conflict. It seemed that the physics world view could not accommodate both. During his studies, Einstein developed a firm belief that the principle of relativity must be a fundamental part of any physical theory. At the same time, he wasn't ready to abandon Maxwell's new ideas about light. He felt that the conflict could be resolved and that both the principle of relativity and Maxwell's equations could be retained.

Meanwhile, others were pursuing different options. If there were an absolute reference system in the Universe, it should be possible to find it. The key seemed to lie in the behavior of light.

Figure 10-1 A race between two light beams in perpendicular directions was supposed to detect the hypothesized ether. In this experiment a light beam from the source S is split by the partially silvered mirror P and travels two different paths to the telescope T.

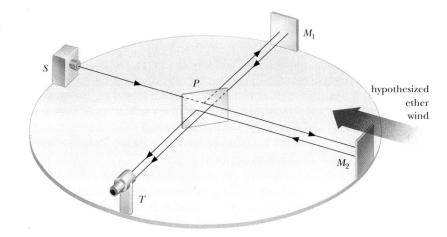

✓ Extended presentation available in the *Problem Solving* supplement

Searching for the Medium of Light ✓ MATH

Although it was well established in the late 1800s that light was a wave phenomenon, nobody knew what substance was waving. Sound waves move by vibrating the air, ocean waves vibrate water, and waves on a rope vibrate the rope. What did light vibrate? It was assumed that there must be some medium through which light traveled. This medium was called the **ether**.

But if space were filled with such a medium, it should be detectable. From their knowledge of the behavior of other waves, scientists were convinced that the ether had to be fairly rigid. Therefore, as Earth passed through this ether in its annual journey around the Sun, it should be slowed by friction. However, no such slowing was detected. How could the ether be rigid and yet so intangible that Earth could pass through it without slowing?

Two American physicists, A. A. Michelson and E. W. Morley, tried to detect the ether with an experiment that raced two light beams in perpendicular directions, as shown in Figure 10-1. They reasoned that Earth's annual motion around the Sun should create an ether "wind" on Earth much like a moving car creates a wind for the passengers. This ether wind would affect the speed of light differently along the two paths, and the race would not end in a tie. They calculated that their experiment was sensitive enough to measure a speed relative to the ether as small as one-hundredth of Earth's orbital speed. Although the experiment was conducted at many times of the year and in many different orientations, the results were always the same—every race ended in a tie!

Not finding the ether wind with such a straightforward experiment was shocking. Physicists were receiving conflicting information. First, there *must* be an ether wind; second, there *must not* be an ether wind. The problem was in the first message: light does not require a medium. It can travel through a vacuum. It is a wave that doesn't wave anything.

The Second Postulate

It is difficult (if not impossible) to re-create a creative process. Although Einstein mentioned the failure to find the ether in his 1905 paper, years later he indicated that his primary motivation in formulating the **second postulate of special relativity** was his deep belief in the principle of relativity. He could reconcile the apparent contradiction between the principle of relativity and Maxwell's equations with his second postulate because it eliminated the possibility of using the speed of light to distinguish between inertial reference systems.

The speed of light in a vacuum is a constant regardless of the speed of the source or the speed of the observer.

◄ second postulate of special relativity

At first glance the second postulate might seem like a rather innocent statement. But consider the situation of your friends in the van from the previous chapter. We agreed that the velocity of an object measured relative to the ground was different from the velocity measured relative to the van—the difference depended on how fast the van was moving relative to the ground. Einstein's second postulate says that this doesn't happen with light.

If your friends move toward us and turn on a flashlight, we might expect that we would measure the speed of light to be greater than that from a flashlight on the ground. We find, however, that we get the same speed. It doesn't matter that the flashlight is moving relative to us. Even if we move very rapidly toward the flashlight, the results would be the same. Regardless of any relative motion, any measurement of the speed of light yields the same value: 300,000 kilometers per second (186,000 miles per second).

Question If we were communicating with an alien spaceship approaching Earth at 20% of the speed of light, at what speed would we receive their signals and at what speed would they receive ours?

Answer Because radio and light behave the same, it would not matter which type of signal we used. In either case the second postulate tells us that both observers would receive the signals at the speed of light, not at 120% of this speed, as would be predicted intuitively.

Simultaneous Events

When Einstein's two postulates are applied to rather simple measurements, unexpected consequences occur. Consider the question of determining whether two events took place at the same time. How would we know, for example, whether two explosions happened simultaneously? We all have an intuitive feeling about this and don't usually even think to question it. Einstein cautioned that we must not simply accept this intuitive feeling. We should look very carefully at how we determine the validity of such statements.

To determine the simultaneity of two events, we must receive some type of signal indicating that each event occurred. To be specific let's determine if two paint cans exploded at the same time. If the two cans are in the same place, as in Figure 10-2, we can agree that they exploded simultaneously if the light signals from the two explosions arrived together. The signals traveled the same distance, and their simultaneous arrival means that the explosions occurred at the same time. The simultaneity of events at a single location does not present a problem.

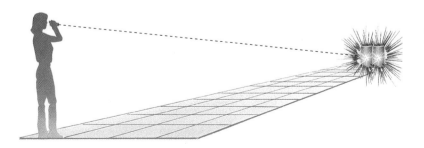

Figure 10-2 The simultaneity of events at a single location presents no problem.

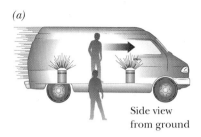

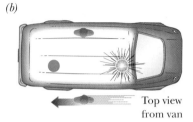

Figure 10-3 (a) The paint cans explode when they are equal distances from each observer. While the ground observer claims the events were simultaneous, the observer in the van (b) claims the can on the right exploded first.

If the paint cans are not in the same place, we have to be more careful. The signals could arrive together even though the explosions occurred at different times. The results depend on the distances to each explosion. Clearly, the easiest case occurs when the observer is the same distance from each event; then the simultaneous arrival of the signals indicates that the events occurred simultaneously.

Einstein had no quarrel with this method of determining simultaneity. His concern was whether *all* observers would agree on the simultaneity; he concluded that they wouldn't. He claimed that two observers moving relative to each other at a constant velocity cannot agree on whether two events happen at the same time. This statement probably seems incredible. You might say, "How can two people see the same physical events and disagree on their simultaneity? They really did happen at the same time . . . didn't they?"

Einstein would say that there is no such thing as universal agreement about simultaneity. To understand this, let's return to your friends in the van. Assume that the paint cans are located equal distances to the right and left of one of your friends and that the van is moving with constant velocity to the right relative to you on the ground. Assume that, at the moment when you were also equal distances from the paint cans, the cans exploded, as shown in Figure 10-3(a). You know that they exploded simultaneously because the signals arrived at your eyes simultaneously and you can verify that you are equal distances from the paint marks on the ground.

How would this apply to one of your friends in the van? During the time it took the signals to reach him, he approached the right-hand signal and receded from the left-hand one, as shown in Figure 10-3(b). The signal from the right-hand explosion therefore reached his eyes before the left-hand one. Both of you agree that the arrival of the signals at his eyes was not simultaneous.

Question   Why do you both agree that the signals arrived at your friend's eyes at different times?

Answer   Different observers agree on the simultaneity (or nonsimultaneity) of events at a single location—in this case, at your friend's eyes.

Your friend concludes that the explosions were not simultaneous. He reports, "I'm standing here in the middle of the van. I can tell by the paint marks on the floor of the van that the explosions happened equal distances from me, but the signals did not reach my eyes simultaneously. Clearly, the one that reached me first came from the explosion that happened first."

"Well," you might counter, "I understand why you think that. You were moving and that's why you reached a different conclusion."

"I'm not moving!" retorts your friend.

According to the first postulate, his motion is no more certain than yours. From his point of view, he is standing still, and you are moving. There is nothing either of you can do to determine who is *really* moving. From his point of view, *you* falsely concluded that the events were simultaneous because *you* moved to the left and thus shortened the distance that the left-hand signal traveled to your eyes.

How do we get ourselves out of this predicament? Einstein concluded that we don't. You and your friend are both correct. You each believe that you have the correct answer and that the other is moving and therefore has been fooled. There is no way to resolve the conflict other than to admit that simultaneity is relative.

▸ simultaneity is relative

Flawed Reasoning

Your friend says, "I don't see why special relativity is so special. It is obvious that two events can be simultaneous for one person and not for another. For example, if I am camped exactly halfway between volcano A and volcano B and hear them both erupt at the exactly the same time, someone camped closer to volcano A would hear it erupt first. The difference is just the time delay for the sound to travel the greater distance." Your friend is just not seeing the big picture. **What is the error in your friend's thinking?**

Answer Special relativity predicts that events that are simultaneous in one reference system may not be simultaneous in another. This is much more fundamental than time delays due to signal speed. The camper closer to volcano A could calculate the time required for the sound to travel from each volcano and determine that they must have erupted at the same time. This camper would agree with your friend about the simultaneity of the events, even though she heard the eruptions at different times. An observer flying past the volcanoes in a spaceship, however, would do the same calculation and determine that they did not erupt simultaneously.

Synchronizing Clocks

Because our concepts of motion are very fundamental to the physics world view, disagreements in simultaneity could result in a radical revision. For example, when we discussed the motion of a ball and talked about the ball being 20 meters above the ground at a time of 3 seconds, we were claiming that the ball was at this position at the same instant that the hands on the clock indicated 3 seconds. That is, the two events occurred simultaneously.

Remember that disagreements only occur when the events are at different locations. Rather than trying to determine distant events with a single clock located near you, you could set up a series of clocks distributed throughout space. Then each event could be recorded on a clock at that location, and there would be no problem with the simultaneity of the event and the clock reading.

However, for this to work, all the clocks must be synchronized. But how do we know that they are synchronized? Even if they are synchronized in one reference system, will they be synchronized in all inertial reference systems?

To answer these questions, we need to examine the process of synchronizing clocks in different places. It might be tempting to suggest that we follow the procedure used for years in war movies. The soldiers rendezvous to synchronize their wristwatches and then disperse. Clearly, this method worked quite satisfactorily for them, but we have no guarantee that the watches remain synchronized. We don't know, for example, whether the motion of a clock affects its timekeeping ability.

One way of synchronizing separated clocks is illustrated in the sequence of strobe drawings in Figure 10-4. A flashbulb is mounted on top of a pole located midway between the two clocks. Initially, the clocks are preset to the same time and are not running. They are designed to start when light signals hit photocells mounted on their roofs. After the flash (a), the light expands in a sphere centered on the top of the pole (b and c). The light signals are detected as they arrive at each clock (d), starting the clocks simultaneously. The two clocks are now synchronized (e).

Let's now attempt to synchronize clocks in two different inertial systems. We assume that each system has the same setup as that used in Figure 10-4 and that the clocks are located along a line parallel to the direction of *relative* motion.

Figure 10-4 Strobe drawings illustrating a method of synchronizing two separated clocks in a single inertial system.

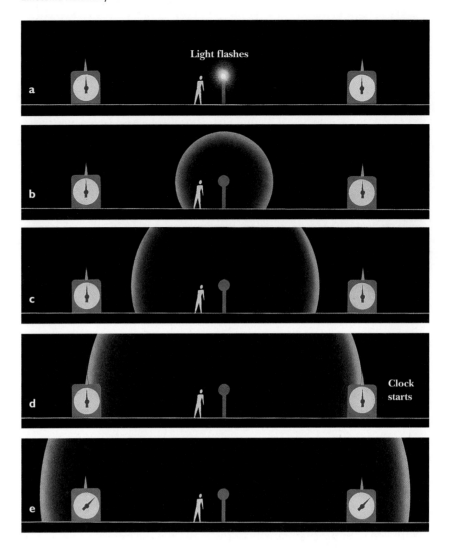

Pretend you are located in the lower system of Figure 10-5 and see the upper system moving to the right with constant velocity. The flashbulb goes off as the two poles meet (b). You see the light expanding in a sphere about the pole in your inertial system (c). (It does not matter which bulb flashes, or even if both flash, because you measure the speed of light to be a constant independent of the motion of the source.) Because you see the upper system moving to the right, the left-hand clock moves toward the light signal and starts first (d). The two clocks in your system start simultaneously (e). Notice, however, that it takes some additional time for the light signal to catch up with the right-hand clock (h) in the upper system, and it starts after the left-hand clock.

You report that your clocks are synchronized but that the clocks in the other system are not synchronized. Because the upper system was moving to the right during the time the light signal was en route, the clock on the left moved toward the signal, while the one on the right moved away from it. You observe that the light traveled a shorter distance to the left-hand clock and it was therefore started before the right-hand one.

What would the observer viewing the events from the other inertial system say? Let's repeat the analysis, assuming that you are now in the upper system. From this point of view, you observe the lower system moving to the left, as shown in Figure 10-6. Once again you see the light signal expand in a sphere centered on the top of the pole in your system (c). The two clocks in your sys-

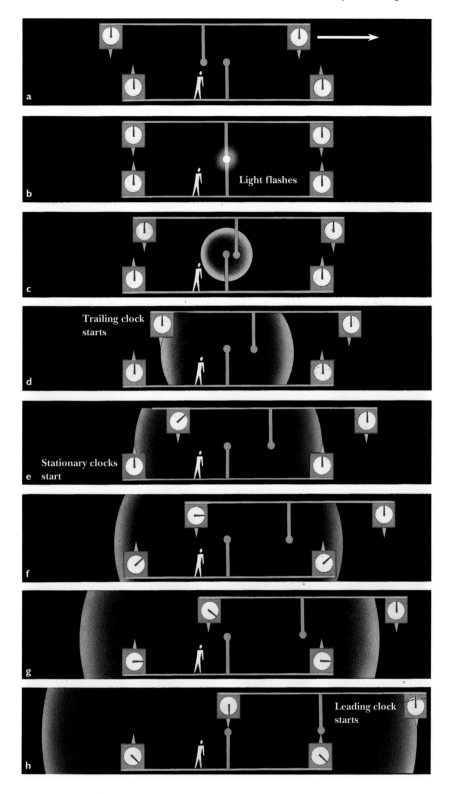

Figure 10-5 An attempt to synchronize clocks in two inertial systems as viewed by an observer in the lower system.

tem start simultaneously (e). You see the right-hand clock in the lower system approach the light signal and start first (d). Only later is the left-hand clock in the lower system started (h). You conclude that the clocks in your system are synchronized, but the clocks in the lower system are not synchronized.

All observers conclude that the clocks in their own reference system are synchronized and the clocks in all other reference systems are not synchronized.

Figure 10-6 The same attempt to synchronize the clocks shown in Figure 10-5 but as viewed by an observer in the upper system.

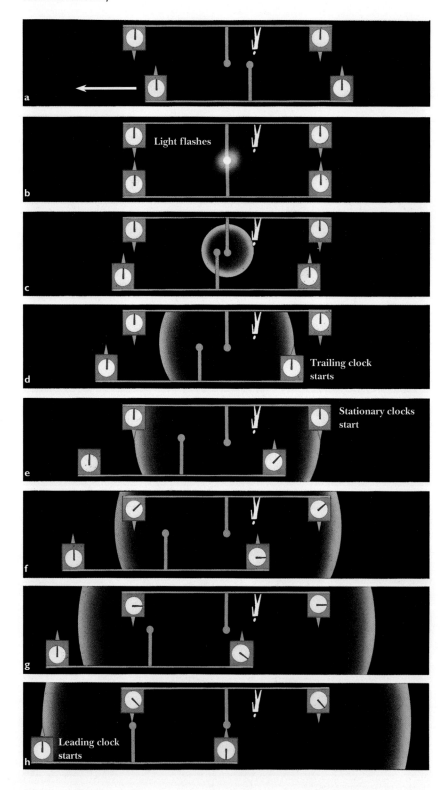

This conflict cannot be resolved. The first postulate says that the two inertial reference systems are equivalent; no experiments can be performed to determine which observer is "really" at rest.

The equivalence can be made more apparent by noting that in each case it was the trailing clock that moved toward the light signal and thus started early. We can summarize the situation by observing that the *trailing clocks lead*.

trailing clocks lead ➤

Question A conductor on board a fast-moving train verifies that all clocks in the train are synchronized. What do observers on the ground say about this?

Answer They find that the clock in the caboose is ahead of the clock in the engine.

Time Varies

Can observers in different inertial systems agree on the time interval between two events taking place at the same location and measured on the same clock? To examine this question, consider the very unusual but legitimate clock shown in Figure 10-7. Clocks keep time by counting some regular cycle. In this clock the cycle is initiated by firing a flashbulb at the bottom. The light signal travels to the mirror at the top of the cylinder and is reflected downward. The photocell receiving the signal initiates a new cycle by firing the flashbulb again.

Imagine an identical clock in your friends' van. The light that strikes the top mirror and returns to the photocell must be that portion of the flash that left at an angle to the right of the vertical. Therefore, it travels the larger distance shown in Figure 10-8. Because the speed of light is a constant, the time for the round-trip must be longer. The time interval between flashes is longer for the moving clock; that is, time is dilated. The moving clock runs slower.

Your friends' report is different. The light signal in their clock travels straight up and down, whereas the signal in your clock takes the longer path. They claim that your time is dilated. Note that each of you agrees that moving clocks run slower. This equivalence is in agreement with the first postulate.

The implications of time dilation are startling. According to the first postulate, *all* clocks in the moving system must run at the same rate. This statement applies to physical, chemical, and even biological clocks. Thus, pulse rates will be lower, biological aging will be slower, the pitch of musical notes will be lower, and so on. Time itself changes when viewed from different inertial systems.

At first glance it may seem like Einstein discovered the fountain of youth. By traveling at a high velocity, clocks would run slower, and we would live longer. Unfortunately, this isn't the case. Within our own inertial system, everything is normal. Our biological clocks run at their normal pace, and we age normally. Nothing changes.

We should also note that we have not invented a time machine that will

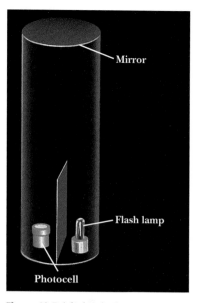

Figure 10-7 A light clock.

◀ moving clocks run slower

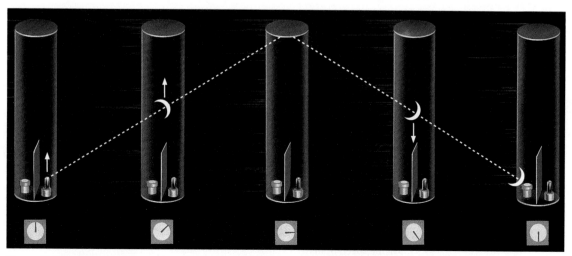

Figure 10-8 The light in the moving clock travels farther, and therefore the clock runs slower.

allow us to go back into history. Although we can make moving clocks run very slowly by giving them very high speeds, we cannot make them run backward. If such were possible, a person moving relative to you could conceivably see your death before your birth! Obviously, this would play havoc with our ideas of cause and effect.

Question Assume that both you and your friends are carrying clocks. If you determine that your friends' clock is running 10% slower, what will your friends say about your clock?

Answer Because the first postulate requires the situations to be symmetric, your friends will observe your clock to be running 10% slower than theirs.

Experimental Evidence for Time Dilation ✓ MATH

The size of the effects predicted by the special theory of relativity increases with speed. The time interval in the moving system is equal to the time interval in the rest system multiplied by an adjustment factor. The relativistic adjustment factor is called gamma (γ) and is given by

▶ relativistic adjustment factor

$$\gamma = \frac{1}{\sqrt{1 - \left(\frac{v}{c}\right)^2}}$$

In this expression, v is the relative speed of the inertial systems and c is the speed of light. Notice that the value of the adjustment factor depends only on the ratio of these speeds and always has a value greater than or equal to 1.

In Table 10-1 we have calculated the values for the adjustment factor for different speeds of the moving system. As you can see from the first two entries in the table, a clock moving at ordinary speeds relative to an observer is slowed by a seemingly negligible amount. For instance, a clock moving at three times the speed of sound would have to travel for 63 centuries before it lost 1 second relative to a clock at rest!

An experiment to detect the slowing of a clock during a transcontinental flight would need to detect differences of a few billionths of a second. However, modern atomic clocks are sensitive to such small time differences. Jet planes, each with several atomic cesium clocks, were flown in opposite directions around Earth. Two experiments—one in 1971 and one in 1977—confirmed the predictions.

Figure 10-9, the graph of the adjustment factor versus speed, shows that the

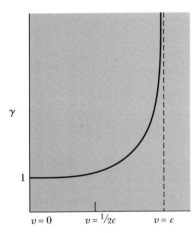

Figure 10-9 A graph of the adjustment factor versus the ratio of the speed of the system or object to the speed of light.

Table 10-1 | The Value of the Adjustment Factor for Various Speeds

Speed	Adjustment Factor
The fastest subsonic jet plane	1.000 000 000 000 6
Three times the speed of sound	1.000 000 000 005
1% the speed of light	1.000 05
10% the speed of light	1.005
25% the speed of light	1.03
50% the speed of light	1.15
80% the speed of light	1.67
99% the speed of light	7.09
99.99% the speed of light	70.7
99.9999% the speed of light	707
The speed of light	Infinite

effects become infinitely large as the speed approaches that of light. An early verification of time dilation at these large speeds involved the behavior of subatomic particles known as *muons*. Muons are created high in our atmosphere by collisions of particles from outer space with air molecules. Time dilation can be tested with these fast-moving muons because they are radioactive; they spontaneously break up into other particles. This radioactive decay provides us with a simple but very accurate clock.

Question Assuming that you could measure the radioactivity of muons flying past you, would you expect the muons to live for a longer or a shorter time due to the relativistic effects?

Answer Because the muons are moving, their radioactive decays should be slowed as viewed from Earth. Therefore, they will live longer than if they were at rest in the laboratory or if you were moving along with them.

Knowing the number of muons present at a high elevation and the characteristics of the radioactive "clocks," one can predict quite accurately, in the absence of any relativistic effects, how many muons should reach Earth's surface before disintegrating. Experiments yielded a much greater number of muons at sea level than predicted. In fact, the number of muons agrees with calculations that assume the decay time for the muons is dilated as predicted.

WORKING IT OUT | Relativistic Times ✓ MATH

Muons at rest in the laboratory have an average lifetime of 2.2 microseconds ($1\ \mu s = 10^{-6}$ s). This average lifetime is measured with a clock that is also at rest in the laboratory.

What is the average lifetime of muons traveling at 99% of the speed of light? According to Table 10-1, the adjustment factor for this speed is 7.09. Let's imagine a clock moving with the muons. This clock is at rest relative to the muons and will measure an average lifetime of 2.2 μs. Observers on the ground know that this clock is running slow. This means that the lifetime of the muons (as measured by clocks on the ground) is longer by an amount determined by the adjustment factor:

$$\tau_{moving} = \gamma \tau_{at\ rest} = (7.09)(2.2\ \mu s) = 15.6\ \mu s$$

Therefore, the average lifetime of the fast-moving muons is 15.6 μs. Note that the shortest time is determined by the clocks *at rest* relative to the muons.

Length Contraction ✓ MATH

The existence of time dilation suggests that space travel to distant galaxies is possible. Although the galaxies are enormous distances from Earth, space explorers traveling fast enough could complete the trips within their (time-dilated) lifetimes.

An examination of such a trip leads us to another startling consequence of Einstein's ideas. Consider a trip to our nearest neighbor star, Proxima Centauri, which is 42 trillion kilometers from Earth. Even light traveling at its incredible speed takes 4.4 years to make the trip. A spaceship capable of traveling at 99% of the speed of light would make the trip in about 4.5 years, according to clocks on Earth. However, according to Table 10-1, clocks inside the spaceship will record that the trip takes one-seventh as long, or about 0.64 year.

The Twin Paradox

The prediction of time dilation is usually greeted with disbelief. Surely people in different inertial systems can stop their experiment, come together, and compare clocks. They should be able to resolve the question of which clock is really running slower. This feeling was ingeniously expressed in a hypothetical situation that led to an apparent paradox, called the *twin paradox*. One twin gets in a spaceship and flies away from Earth. The spaceship travels out to a distant star and returns to Earth.

The twin on Earth observes that clocks in the spaceship run slower than on Earth. Therefore, the twin in the spaceship ages more slowly. The twin should return from the journey at a younger age than the one who stayed at home. Meanwhile, the twin in the spaceship observes that the clocks on Earth are running slower. So the Earth-bound twin will age slower and should be the younger at the reunion. Thus, we have a paradox. How can each twin be younger than the other?

The paradox arises because we assumed that everything was symmetric, that there was no way of deciding who was taking the trip. But the situation is not symmetric. It becomes clear that the twin in the spaceship is taking the trip as soon as the spaceship accelerates to leave or turns around to return. The inertial forces that arise during the acceleration give it away.

It is sometimes thought that the special theory of relativity can only be applied to situations involving inertial systems—that is, where there is no acceleration. This is not true; there are several ways of getting the correct answer within the framework of special relativity. All solutions agree that the twin in the spaceship is younger at the reunion than the twin who stayed on Earth.

Imagine making such a trip. Suppose the journey takes 40 years as measured by clocks on Earth, but only 10 years elapse on the spaceship's clocks. On your return you would find that society's technology and institutions have jumped ahead by 40 years. It is possible that you would return and be younger than your children. The social consequences of this *family* shock could be even more mind boggling than the expected *future* shock that you would experience.

One twin makes a journey to a distant star while the other remains at home. Upon meeting, the twin who made the journey is younger than the twin who stayed at home.

But there is a catch. Both the space travelers and their Earth-bound friends agree that their relative velocity is 99% of the speed of light. How, then, can the space travelers reconcile making this trip in only 0.64 year? The distance to Proxima Centauri must be *contracted*. The amount of contraction is just right to compensate for the time dilation. Our space travelers measure a distance that is one-seventh as long. Measurements of space, like those of time, change with relative motion.

To see why this happens, we follow Einstein's advice and carefully consider how we measure the length of something. If you are at rest relative to a stick, there is no problem measuring its length. You simply measure it with a ruler or mark the position of the two ends on the floor and measure the distance between the marks.

What if the stick is moving? Again, you could mark the floor at each end of the stick as it passes by. But you have to be careful. Clearly, you could get a variety of lengths if you mark one end first and the other end at various times later. To obtain the correct length, you must mark the position of the two ends *simultaneously*—for example, by exploding paint cans at each end.

What will a person at rest relative to the stick (Figure 10-10) think of this measurement? She agrees with your procedure but says that the paint cans didn't explode simultaneously. She claims that the paint can at the front exploded ear-

Figure 10-10 According to the observer on the ground, the paint can at the front of the pole explodes before the can at the back. Therefore, the two paint splashes on the ground are closer together.

lier. By the time the can at the back exploded, the back end of the stick had moved closer to the first mark. Thus, the length you measured is shorter than hers; that is, the length of a moving stick is contracted.

Consider another way to measure the length of the moving stick. You could measure the stick's velocity and record the elapsed time between the passing of the front and back ends. Again, the observer on the moving stick disagrees with your results. She says that your clocks are running slower and the elapsed time is therefore shorter. Once again, your measurement yields a contracted length.

The moving length is equal to the length measured at rest *divided* by the relativistic adjustment factor. It is important to note that this length contraction occurs only along the direction of the relative motion. Lengths along the direction perpendicular to this motion are the same in the two inertial systems.

◄ moving sticks are shorter

WORKING IT OUT | Relativistic Lengths ✓ MATH

Let's assume that a train is traveling along a straight, horizontal track at a constant speed of 80% of the speed of light. What is the size of the adjustment factor?

$$\gamma = \frac{1}{\sqrt{1 - \left(\frac{v}{c}\right)^2}} = \frac{1}{\sqrt{1 - \left(\frac{0.8c}{c}\right)^2}} = \frac{1}{\sqrt{1 - 0.64}} = \frac{5}{3}$$

Mary is a passenger in the train and measures the length of the dining car to be 30 m. Bill is standing on the platform of a railway station. What will Bill obtain for the length of the dining car?

We know that we should either divide or multiply by the adjustment factor, but which do we do? We know that the length measured in a reference system at rest relative to the object (in this case the dining car) must always be the longest. Therefore, Mary's length will be the longer, and we must divide by the adjustment factor to obtain Bill's length:

$$L_{Bill} = \frac{L_{Mary}}{\gamma} = \frac{30 \text{ m}}{\frac{5}{3}} = 18 \text{ m}$$

If Mary measures the length of the railway platform to be 120 m, how long will Bill measure it to be?

Bill is at rest relative to the platform and must therefore obtain a longer length. In this case we must multiply by the adjustment factor:

$$L_{Bill} = \gamma L_{Mary} = \tfrac{5}{3}(120 \text{ m}) = 200 \text{ m}$$

Question Assume that both you and your friends in the van are carrying meter sticks pointing along the direction of relative motion. If your friends measure your stick to be $\frac{1}{2}$ meter long, what length would you measure for your friends' stick?

Answer The first postulate requires that the situations be symmetric. *Each observer says that the other's stick is contracted.* Therefore, you would also measure your friends' stick to be $\frac{1}{2}$ meter long.

Spacetime

Einstein's ideas changed the role time plays in our world view. In the Newtonian world view, we considered motion by looking at the spatial dimensions and looking *independently* at time. Einstein demonstrated that time is not a separate quantity but rather is intimately connected to the spatial dimensions. When space changes, there is a corresponding change in the time. Time is now truly the fourth dimension of **spacetime**.

The theory of special relativity must be self-consistent. All observers must find that events obey the laws of physics. As we have seen, they do not have to agree on their particular measurements, but they must be able to make sense of the events within their own reference system.

A hypothetical situation illustrates this point. Suppose an ingenious student claims that she can fit a 10-meter pole into a 6-meter-long barn (Figure 10-11). Knowing about length contraction, she proposes to propel the pole into the barn at 80% of the speed of light because the adjustment factor is $\frac{5}{3}$, giving a moving length of 6 meters. Just enough to fit into the barn! Of course, the pole will only be in the barn for an instant because it is moving very fast. Our ingenious experimenter plans to prove that the pole was entirely in the barn by closing the front door and simultaneously opening the back door.

Now consider this situation from the point of view of a person riding on the pole (Figure 10-12). The pole is 10 meters long, but the barn is moving and is

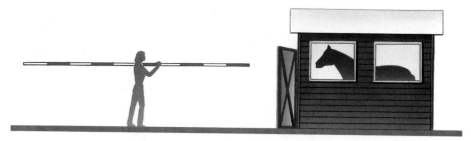

Figure 10-11 The observer on the ground tries to put a 10-meter-long pole in a 6-meter-long barn.

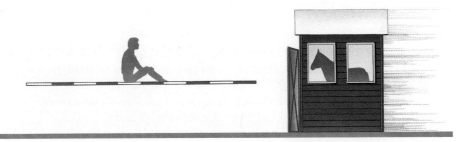

Figure 10-12 The observer on the pole tries to put a 10-meter-long pole in a 3.6-meter-long barn.

EINSTEIN | Person of the Century

Albert Einstein (1879–1955) was a great physicist who devoted his life to peace and humanity. Born in Ulm, Germany, he received his basic education in Munich, studied in Italy and Switzerland, and accepted major scientific posts in Berlin just before World War I. After the National Socialists and Hitler took control of Germany, Einstein found refuge and citizenship in the United States.

In his autobiography, Einstein states that at age 3 he was much taken by the fact that an uncle's compass always pointed north—even in a dark closet. Invisible laws of nature seemed to him to be the route to the most powerful understanding of our being and the Universe.

His heroes were Newton and Maxwell. Their portraits were always displayed in his office. He received a solid if, for him, dull German education. After a year of individual tutoring, he entered the Swiss Federal Technical University and began serious study of physics. He lived during a period of great upheaval in science and felt revolutionary change in the air.

He began to publish early and in 1905 brought out four important papers. Two founded the study of special relativity. Another on the photoelectric effect explained a puzzling phenomenon and earned him the Nobel Prize in 1921. In 1907 he published his now famous equation on energy and mass: $E = mc^2$. His great work on general relativity, the nature of gravitation, appeared in 1915 in the midst of the greatest war in human history. This powerful theory related gravity to a warping of spacetime.

After the war he worked on early quantum theory but could accept it only as a stopgap until a better theory evolved. He was unpopular in Germany because he was so outspoken against the war, because he was a Jew, and because his theories were so revolutionary and unsettling.

Albert Einstein

He chose to settle in the United States because he believed it to be an open, democratic society. A research institute was arranged for him at Princeton University. In America he could work more effectively for a homeland for the oppressed Jews. Ironically, this outspoken pacifist triggered American interest in nuclear power. In 1939 he received word from Niels Bohr (see "Bohr: Creating the Atomic World" in Chapter 23) that German scientists had fissioned uranium. He wrote to President Franklin Roosevelt, alerting him officially that atomic power for military use might be possible. He was never active in nuclear research himself and sought mightily to mitigate conditions in which such terrible weapons might be used.

He also played a significant role in creating the modern state of Israel and was offered the first presidency of that nation (he declined). He also worked for better Jewish–Arab relations. Einstein was, for most of his life, the most famous scientist in the world and was named Person of the Century by *Time* magazine. UNESCO designated 2005 as the World Year of Physics, and many physical societies around the world honored the 100th anniversary of Einstein's remarkable year, coincidentally the 50th anniversary of his death.

A fine writer, Einstein is often the best source to read for those seeking an understanding of his life and work.

—*Pierce C. Mullen, historian and author*

Sources: Abraham Pais, *Subtle Is the Lord: The Science and Life of Albert Einstein* (Oxford: Oxford University Press, 1982); and P. A. Schilpp, *Albert Einstein: Philosopher-Scientist* (New York: Harper, 1949).

---

contracted to 3.6 meters! Clearly, he is not going to agree that the pole was ever entirely in the barn, not even for an instant.

There is no paradox, however. The rider does not agree that the back door was opened at the same time that the front door was closed. Recalling that trailing clocks lead, he says that the back door opened before the front door closed. In fact, careful calculations of this situation verify the consistency. The time interval between these events is just enough to allow the "extra" 6.4 meters to pass through.

Relativistic Laws of Motion ✓ MATH

As we did in our study of the classical ideas of motion, we now expand our considerations beyond describing motion to consider the laws of motion. Many approaches can be taken to develop laws of motion that are consistent with the ideas of special relativity, although we don't have an entirely free hand. The new laws must have a structure that is logical and internally consistent, and the predictions of the laws must agree with the results of experiments. Furthermore, the new formulations must reduce to the older ones (Newton's laws) when the velocities are small because we know Newton's laws work for small velocities.

The form of Newton's second law regarding momentum carries over into special relativity, providing that the expression for the momentum is modified so that $p = \gamma mv$. This is the classical formula multiplied by the adjustment factor. Notice that this expression reduces to the classical one for small speeds be-

The speed of light is the speed limit of the Universe.

▶ relativistic form of Newton's second law

▶ relativistic kinetic energy

▶ mass–energy relationship

What is the acceleration of a ball dropped in an accelerating spaceship?

cause the adjustment factor is very close to 1 in this case. With this modification, the second law is written as

$$F = \frac{\Delta p}{\Delta t} = \frac{\Delta(\gamma m v)}{\Delta t}$$

Careful analysis of symmetric collisions of identical balls in different inertial reference systems demonstrates that conservation of momentum is still valid provided this relativistic form for momentum is used.

A force acting for a time still produces the same impulse and therefore the same change in the relativistic momentum. However, because the adjustment factor increases with speed, the acceleration decreases and goes to zero as the object approaches the speed of light. This means that a material object cannot be accelerated to a speed equal to or greater than that of light. Nothing can go faster than the speed of light.

The law of conservation of energy is also valid in special relativity. If we calculate the work done by a force acting on an object that is initially at rest and equate this work to the kinetic energy of the object as we did in classical physics, we arrive at the expression

$$KE = \gamma m c^2 - m c^2$$

The relativistic kinetic energy of an object is equal to the difference between two terms. The second term mc^2 is the energy of the particle at rest. Therefore, it is known as the **rest–mass energy** E_o. This is the origin of Einstein's famous mass-energy equation

$$E_o = mc^2$$

Because the kinetic energy is the additional energy of the object due to its motion, the first term, $\gamma m c^2$, is identified as the total energy of the particle. This expression tells us that the total energy of an object increases with speed. In fact, because the adjustment factor approaches infinity as the speed approaches that of light, the energy also approaches infinity.

Notice also that even when the object is at rest, it has an amount of energy mc^2 stored in its mass. Mass is another form of energy. Thus, the law of conservation of energy must be modified once more to include a new form of energy, mass energy. This relationship produced a major change in the physics world view. It tells us that mass can be converted into energy and energy can be converted into mass. We will discuss this more fully when we discuss nuclear reactors and the properties of subatomic particles.

The authors of some popular books and articles about special relativity make the statement that mass increases with speed. These authors define a relativistic mass $m = \gamma m_o$, where m_o is called the *rest mass* and is the mass measured in a system at rest relative to the object. This statement does not change any of the mathematics, but it does change the interpretation of some of the expressions. The modern view is that mass is an invariant and that the introduction of a relativistic mass is unnecessary and sometimes leads to errors.

General Relativity

For about a decade after Einstein's publication of his theory of special relativity, he worked on generalizing his ideas. The outcome—his **general theory of relativity**—deals with the roles of acceleration and gravity in our attempts to find our place in the Universe.

Armed with his deep relativistic philosophy, Einstein started by expanding the principle of relativity to include all areas of physics and noninertial, or accelerating, reference systems. Imagine a spaceship very far from any stars that is accelerating at 10 (meters per second) per second. The astronauts feel an inertial force equivalent to the gravitational force they would feel on Earth. If they re-

Global Positioning System (GPS)

If you ever get seriously hurt while hiking in the mountains, two products of modern technology might save your life: you can use a cell phone to call for help, and you can use a GPS receiver to tell the rescuers where to find you.

Beginning in 1969, the U.S. military developed the Global Positioning System to accurately determine locations any place on or near Earth's surface. This is accomplished with 24 satellites arranged in groups of 4 in each of six orbital planes. A worldwide network of ground stations monitors the satellites and uploads time and orbital data. Each satellite sends out digitally coded information that gives the time the information was sent and the location of the satellite at that time.

Assume for the moment that the clock in your GPS receiver is synchronized with the clocks in the GPS satellite. Knowing the speed of light in the atmosphere and the time delay in receiving the time code tells you how far the receiver is located from the satellite. This determines the location of your receiver to be some place on the surface of a sphere centered on the location of the satellite at the time the signal was sent. The reception of a signal from a second satellite narrows the location of your receiver to be along the intersection of two spheres, one centered on each satellite. The signal from a third satellite narrows the location to a single point determined by the intersection of three spheres. In practice the GPS receiver must receive signals from a minimum of four satellites, the fourth being used to determine the difference in times on the clock (usually a crystal clock) in the receiver and GPS time. GPS provides an accuracy of 15 meters in the horizontal plane and 22 meters in the vertical plane.

Large improvements in accuracy can be obtained by using differential GPS. The locations of stationary receivers are very accurately surveyed to determine their locations. Comparison of these locations with locations determined by GPS yields the errors in these positions. A map of the errors can then be broadcast to portable receivers to correct their readings. This results in an accuracy of 3–5 meters. The Wide Area Augmentation System (WAAS) developed for aviation and marine navigation uses a similar technique to obtain an accuracy of less than 3 meters. More sophisticated techniques used in surveying can yield measurements with an accuracy of centimeters.

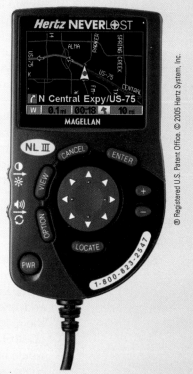

GPS receivers are being installed in automobiles to aid travelers in finding their way.

It is crucial to the operation of the GPS that all satellites use the same time to a very high degree of accuracy. This requires that the effects of both special and general relativity be taken into account. Don't let anyone tell you that relativity doesn't have any consequences in the real world.

---

lease a ball, it falls freely with an acceleration of 10 (meters per second) per second. However, an observer in an inertial system outside the spaceship would give a different explanation: The ball continues in the forward direction with the velocity it had at the time of release. The floor accelerates toward the ball at 10 (meters per second) per second, making it seem as if the ball were falling.

Question What would the astronauts observe if they release two balls with different masses?

Answer The two balls would appear to fall with the same acceleration. This is easiest to see from outside the spaceship. The two balls move side by side with the same velocity while the floor accelerates toward them.

The fact that the astronauts can attribute their motion to gravitational effects is possible only because the mass that appears in Newton's second law (Chapter 3) is identical to the mass that appears in the law of universal gravitation (Chapter 5). It might seem surprising that these two notions of mass are not automatically the same, but recall that they arise in different physical circumstances. Newton's second law gives a relationship between applied force and the resulting acceleration. This idea of mass depends on the inertial properties of mass and is therefore called the **inertial mass**. The law of universal gravitation refers to the strength of the attractive force between two objects. This mass is known as the **gravitational mass**. Experiments have shown that inertial and gravitational masses differ by less than a part per billion.

▶ gravitational mass = inertial mass

As a result of the equality of inertial and gravitational mass, any experiment using material objects would not be able to reveal to the astronauts whether the force is due to the gravitational attraction of a nearby mass or the accelerated motion of their spaceship. Believing that *all* motion is relative, Einstein felt that the astronauts could not make any distinction between the two alternatives. He formalized his belief as the **equivalence principle**.

▶ equivalence principle

> Constant acceleration is completely equivalent to a uniform gravitational field.

Before Einstein, there seemed to be a way for the astronauts to distinguish between gravitational and inertial forces. According to the ideas of that time, the astronauts would have only to shine a flashlight across their windowless ship and observe the path of the light by placing frosted glass at equal intervals across the spaceship, as shown in Figure 10-13. With the ship at rest on a planet, the beam of light would pass straight across the room because the gravitational field would have no effect on it. However, in an accelerating spaceship, the astronauts could see the light bend. While the light travels across the ship, the ship accelerates upward, making the differences in the vertical positions on adjacent screens get larger and larger.

Einstein agreed that light would bend in the accelerating spaceship but disagreed that light would be unaffected in the ship in the presence of gravity. Because he firmly believed in the equivalence principle, he concluded that the astronauts should not be able to tell any difference between the two situations. Einstein was forced to conclude that light is bent by gravity. All experimental measurements are in agreement with this prediction.

Figure 10-13 While the light travels across the spaceship, the spaceship accelerates upward, causing the light to intersect the frosted glass closer and closer to the floor. The path relative to the ship is a parabola just like that of a falling ball that is projected horizontally on Earth.

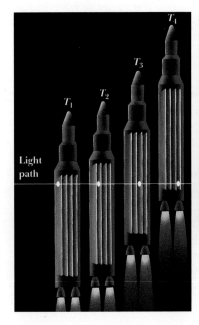
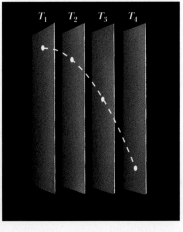

Flawed Reasoning

A space shuttle is in orbit around Earth. One of the astronauts turns on the headlights. He has studied the theory of general relativity and knows that the beam of light will experience the same acceleration as the space shuttle. He quickly glances in the rearview mirror expecting to see the light from his own headlights, which he thinks should orbit Earth just like the space shuttle. **What is wrong with his reasoning?**

Answer The space shuttle and everything in it are falling around Earth in a circular orbit. The space shuttle is moving fast enough that by the time it has fallen 5 meters, it is still the same distance from Earth's surface. The light from the headlights is traveling much faster than the space shuttle, so the light will leave Earth orbit before it drops 5 meters.

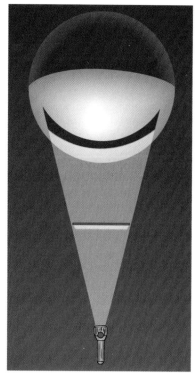

Figure 10-14 The shadow of a straight stick projected onto a sphere is not necessarily straight.

Warped Spacetime

Einstein's work in general relativity also showed that time is altered by a gravitational field. Clocks run slower in a gravitational field. The stronger the gravitational field, the slower the clocks run.

The special theory of relativity has shown us that there is a very intimate relationship between space and time. Because gravity affects time, we should also expect it to affect space.

The Newtonian world view considered space to be flat (Euclidean) and completely independent of matter. Objects naturally travel in straight lines in this space. The addition of matter introduced forces that caused objects to deviate from these natural paths. The matter interacted with the objects but did not affect space. Time was independent of space.

The Einsteinian world view begins with a four-dimensional flat spacetime. The addition of matter warps this spacetime. The matter does not act directly on objects but changes the geometry of space. The objects travel in straight lines in this four-dimensional spacetime. Although these lines are straight in four dimensions, the paths that we view in three-dimensional space are not necessarily straight. This situation is analogous to that in which the shadow of a straight stick is projected onto the surface of a sphere. The shadow is not necessarily straight, as shown in Figure 10-14.

We can see what is meant by replacing the gravitational field by a warped spacetime by considering the following situation. Imagine that you are looking down into a completely darkened room where somebody is rolling bowling balls that glow in the dark along the floor. You record the paths of the balls, and your record looks like Figure 10-15.

What reasons can you give to explain the pattern that emerges from these paths? You might suggest that the balls are attracted to an invisible mass that is

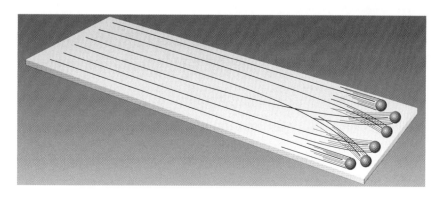

Figure 10-15 The paths of the glowing bowling balls as seen in the dark.

Black Holes

A bizarre astronomical object has dramatically confirmed the ideas that Einstein put forth in his theory of general relativity. Stars are known to collapse due to their own gravitational attraction after their source of fuel is exhausted. Stars more massive than our Sun can collapse to objects so compact that they are only tens of kilometers in diameter. One type of such object is so massive and so small that the increased gravity near the star would prohibit anything—including light—from escaping. This is called a *black hole* because no light can come from this region of space.

Because even light can't escape from a black hole, we have to search for more indirect ways of "seeing" a black hole. The key to finding a black hole is the influence its gravitational field has on nearby objects. The majority of stars in the Universe occur in groups of two or more that are bound together by their mutual gravitational attraction. In some binary star systems, one of the stars is very compact and not visible. The mass of the unseen star can be determined by examining the behavior of the visible companion. The visible star orbits along an elliptical path because the compact star continuously exerts a centripetal force on the visible star.

Current, well-established theories of stellar evolution indicate that of several possibilities for the end stages of stars only black holes can have masses larger than five times the mass of the Sun. Although the determination of the masses of the unseen stars in binary stars is not very accurate, there are several cases known with masses large enough to be black holes.

Current evidence indicates that Cygnus X-1, the first X-ray source observed in the constellation Cygnus, is very compact and has a mass at least 9 times that of the Sun. (The best experimental value is 16 solar masses.) It is almost certainly a black hole. Several other binary star systems also contain good candidates for black holes.

Although light from a black hole cannot escape, light from events taking place near the black hole should be visible. In binary systems, a black hole's powerful gravitational field may capture mass from its companion star. As the mass falls into the black hole, it should emit X rays. Although this is not a black hole "fingerprint," X rays compatible with the existence of black holes have been observed.

The strongest evidence for the existence of black holes is the presence of supermassive, compact objects at the centers of galaxies. Observations of the orbits of stars near the centers of galaxies indicate that the stars are orbiting very massive, yet very small objects. Our own Milky Way Galaxy has a black hole at its center with a mass about 2.5 million times as large as the mass of the Sun. Most galaxies that have been examined closely have a black hole at their center, some with masses greater than a billion solar masses.

Some galaxies are expected to have two large black holes at their centers. These binary black holes orbit each other due to their mutual gravitational attraction and may exist because of the collision and coalescing of galaxies. The detection of gravitational waves (see Chapter 28) emitted by these binary black holes would enhance our understanding of black holes.

The circle indicates the location of Cygnus X-1, which is believed to be a black hole.

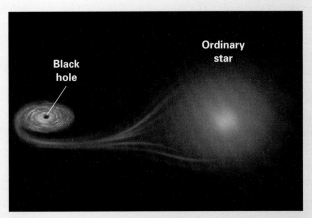

Matter from an ordinary star falling into its companion black hole produces characteristic X rays.

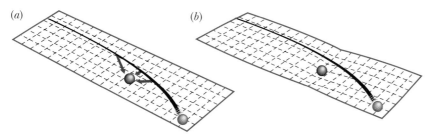

Figure 10-16 Two possible explanations for the paths in Figure 10-15. (a) The bowling balls are attracted by a mass. (b) The mass causes the floor to sag.

fixed in the center of the floor (Figure 10-16[a]). Or you might suggest that the floor has a dip in the center (Figure 10-16[b]).

In his classic book *Flatland*, author and mathematician Edwin Abbott tells a bizarre tale of an inhabitant of a two-dimensional world. This fellow was given the "pleasure" of going to another dimension. Once he went into the third dimension, the shape of his normally flat world became obvious to him. Things that were incomprehensible in Flatland—like looking inside a closed figure— became trivial in the third dimension. When he returned to Flatland, he tried to convince his fellow inhabitants of his newly gained insight. They thought he was insane with his strange talk of "up."

Are we similarly doomed to never understand four-dimensional spacetime? If our space has certain shapes that are obvious in the next dimension, can we deduce them without stepping into that additional dimension? Much as two-dimensional creatures on the surface of a sphere can examine their geometry to determine that their space is not flat, we can examine the geometry of our space to learn about the geometry of spacetime.

Summary

The ideas contained in the special theory of relativity are based on two postulates: (1) the laws of physics are the same in all inertial reference systems, and (2) the speed of light in a vacuum is a constant, regardless of the speed of the source or the observer.

As a consequence of adopting these two postulates, observers in different inertial reference systems cannot agree on the simultaneity of events at different places or on the synchronization of separated clocks. These observers do agree that trailing clocks lead.

Time seems normal in one's own reference system but is dilated when viewed from another reference system. Both observers see the other's clocks running slower. The time interval in the moving system is equal to the interval in the rest system multiplied by the relativistic adjustment factor

$$\gamma = \frac{1}{\sqrt{1 - \left(\dfrac{v}{c}\right)^2}}$$

Observers in two different inertial systems agree that objects in the other's system are shorter along the direction of the relative velocity. However, both observers agree on the relative speed of the two systems.

The conservation laws are valid in special relativity if we introduce the ideas of relativistic momentum and energy and consider mass to be another form of energy. The speed of light is the speed limit of the Universe.

In general relativity, gravitational fields can replace accelerations. The equivalence principle requires that the mass that appears in Newton's second law (in-

Physics Now™ Assess your understanding of this chapter's topics with sample tests and other resources found by logging into PhysicsNow at **http://physics.brookscole.com/kf6e**.

ertial mass) be identical to the mass that appears in the universal law of gravitation (gravitational mass) and that light be bent by gravitational fields.

Space and time form a four-dimensional spacetime that is warped by the presence of matter. Objects travel in straight lines in this four-dimensional spacetime, but the paths that we view in three-dimensional space are not necessarily straight. Time is slowed as the strength of the gravitational field increases.

Chapter 10 Revisited

Time intervals are measured in comparison to some kind of periodic motion. Even though all observers in our own inertial reference system will agree with our foolproof method for measuring time, observers moving relative to us will not accept these time measurements. There is no absolute time; time measurements depend on the particular observer.

KEY TERMS

equivalence principle: Constant acceleration is completely equivalent to a uniform gravitational field.

ether: The hypothesized medium through which light was believed to travel.

first postulate of special relativity: The laws of physics are the same for all inertial reference systems.

general theory of relativity: An extension of the special theory of relativity to include the concept of gravity.

gravitational mass: The property of a particle that determines the strength of its gravitational interaction with other particles. Measured in kilograms.

inertial mass: An object's resistance to a change in its velocity. Measured in kilograms.

rest–mass energy: The energy associated with the mass of a particle. Given by $E_o = mc^2$, where c is the speed of light.

second postulate of special relativity: The speed of light in a vacuum is a constant regardless of the speed of the source or the speed of the observer.

spacetime: A combination of time and three-dimensional space that forms a four-dimensional geometry.

special theory of relativity: A comprehensive theory of space and time that replaces Newtonian mechanics when velocities get very high.

CONCEPTUAL QUESTIONS

1. If you were located in a spaceship traveling with a constant velocity somewhere in the Galaxy, could you devise experiments to determine your speed? If so, what kinds of experiments?
2. If you were on the starship *Enterprise* in a room with no windows, could you devise experiments to determine your acceleration? If so, what kinds of experiments?
3. Why did Maxwell's equations appear to be in conflict with the Galilean principle of relativity?
4. Does the first postulate require that the speed of light in a vacuum be the same in all inertial reference systems?
5. Your friend is driving her 1964 Thunderbird convertible straight toward you at 40 mph. She stands up and throws a baseball forward at 30 mph. How fast do you see the ball approaching you?

6. Your friend from Question 5 finds that hanging fuzzy dice from the rearview mirror allows the car to travel up to 98% of the speed of light. She is driving straight toward you when she turns on her headlights. How fast do you see the light approaching you?
7. An observer in the train in the figure stands in the back of the car. He turns on a light and measures the time it takes for the light to get to the front of the car, bounce off a mirror, and return to him. (Assume that the light is traveling in a vacuum.) Knowing the length of the car, he is able to calculate the speed of light. Will he obtain a speed less than, greater than, or equal to c? Explain.
8. If an observer on the ground uses her own instruments to measure the speed of the light in Question 7, will she obtain a value less than, greater than, or equal to c? Explain.

Questions 7, 8, 13, 14, and 35–40 A train is traveling along a straight, horizontal track at a constant speed that is only slightly less than that of light.

9. According to the special theory of relativity, a twin who makes a long trip at a high speed can return to Earth at a younger age than the twin who remains at home. Is it possible for one twin to return before the other is born? Explain.

10. Suppose that in the situation depicted in Figure 10-3 the observer on the ground saw the rear paint can explode before the front one. For the observer in the van, is it possible that the two explosions occurred (a) simultaneously, (b) in the order as observed from the ground, or (c) in the reverse order as observed from the ground? Explain.

11. Can observers in two different inertial systems agree on the simultaneity of events at a single location?

12. Can observers in different inertial systems agree on the simultaneity of events at different locations?

13. An observer on the ground reports that, as the midpoint of the train in the figure passes her, simultaneous flashes occurred in the engine and caboose. How would an observer in the train describe these same events?

14. An observer in the train in the figure determines that firecrackers go off simultaneously in the engine and in the caboose. How would an observer on the ground describe these same events?

15. As a friend passes you at a very high speed to the right, he explodes a firecracker at each end of his skateboard. These explode simultaneously from his point of view. Which one explodes first from your point of view? How must a third person be moving for her to have observed the other firecracker explode first?

16. Two lights on lampposts flash simultaneously as seen by an observer on the ground. How would you have to be moving to see (a) the right-hand light flash first? (b) The left-hand light flash first?

17. It is possible for observers moving relative to one another to disagree on the order of two events. However, the theory of special relativity preserves cause and effect. If one event caused, or could have caused, the other, then the order of the two events must be preserved for all observers. Two light sources, A and B, are located 186,000 miles apart (the distance light travels in 1 second). An observer at the midpoint between the sources receives a light signal from source A half a second before receiving a signal from B. Is it possible that the light from source A caused source B to flash? Could another observer have seen B flash before A?

18. If the signal from source B in Question 17 was received 2 seconds after the signal from source A, is it possible that the light from source A caused source B to flash? Could another observer have seen B flash before A?

19. Space travelers on the way to colonize a planet orbiting a distant star decide to cook a 3-minute egg. Would a clock on Earth record the cooking time as less than, equal to, or greater than 3 minutes? Why?

20. Skip Parsec ventured into space without taking his watch. Wishing to cook a perfect 3-minute egg on board his fast-moving spaceship, Skip is forced to rely on a clock on Earth. Because Skip missed the day that special relativity was taught at training camp, he cooks his egg for 3 minutes according to the Earth clock. Is his egg undercooked or overcooked?

21. If a musician plays middle C on a clarinet while traveling at 85% of the speed of light in a spaceship, will passengers in the ship hear a lower note, a higher note, or the same note? Why?

22. Superman wants to travel back to his native Krypton for a visit, a distance of 3,000,000,000 kilometers. (It takes light 10,000 seconds to travel this distance.) Superman can hold his breath for only 1000 seconds, but he can travel at any speed less than that of light. Can he make it?

23. In an experiment to measure the lifetime of muons moving through the laboratory, scientists obtained an average value of 8 microseconds before a muon decayed into an electron and two neutrinos. If the muons were at rest in the laboratory, would they have a longer, a shorter, or the same average life? Why?

24. On average, an isolated neutron at rest lives for 17 minutes before it decays. If neutrons are moving relative to you, will you observe that they have a longer, a shorter, or the same average life? Explain.

25. A warning light in the engine of a fast-moving train flashes once each second according to a clock on the train. Will an observer on the ground measure the time between flashes to be greater than, less than, or equal to 1 second? Explain.

26. A warning light on the ground flashes once each second. Will an observer in a fast-moving train measure the time between flashes to be greater than, less than, or equal to 1 second? Why?

27. Peter volunteers to serve on the first mission to visit Alpha Centauri. Even traveling at 80% of the speed of light, the round-trip will take a minimum of 10 years. When Peter returns from the trip, how will his biological age compare with that of his twin brother Paul, who will remain on Earth?

28. Is it physically possible for a 30-year-old college professor to be the natural parent of a 40-year-old student? Would this imply that the child was conceived before the professor was born?

29. What does the special theory of relativity say about the possibility of the event described in the following limerick?

    There was a young lady named Bright
    Who could travel much faster than light.
    She went away one day
    In a relative way
    And returned on the previous night.

30. In *A Connecticut Yankee in King Arthur's Court*, Mark Twain chronicles the adventures of a New England craftsman who in 1879 is suddenly transported back in time to Camelot in the year 528. What does the special theory of relativity say about this possibility? What effect would such a trip have on our beliefs about cause and effect?

31. Suppose you had a row of clocks along a line perpendicular to the direction of relative motion. Would observers in both reference systems agree on the synchronization of these clocks? Explain

32. Two events occur at different locations along a line perpendicular to the direction of relative motion. Will observers in both reference systems agree on the simultaneity of these events? Explain.

33. Muons are created in the upper atmosphere, thousands of meters above sea level. A muon at rest has an average lifetime of only 2.2 microseconds, which would allow it

to travel an average distance of 660 meters before disintegrating. However, most muons created in the upper atmosphere survive to strike Earth. This effect is often explained in terms of time dilation. In this explanation, is the observer in the reference system of Earth or the reference system of the muon? Explain.

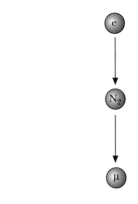

34. An alternative explanation for the survival of muons as described in Question 33 invokes length contraction. In this explanation is the observer in the reference system of Earth or the reference system of the muon? Explain.
35. An observer on the ground and an observer in the train in the figure each measure the distance between two posts located along the tracks. The observer on the ground measures the distance to be 100 meters. Does the observer on the train obtain a measurement that is less than, equal to, or greater than 100 meters? Why?
36. Suppose a ground-based observer in Question 35 measures the distance between the posts at 100 meters. She then places paint cans on the two posts and detonates them simultaneously as the train passes. An observer on the train then measures the distance between the paint splatters on the side of the train. Will his measurement be less than, equal to, or greater than 100 meters? Why?
37. An observer on the ground and an observer in the train in the figure each measure the length of the train. The observer on the train measures the distance to be 400 meters. Does the observer on the ground obtain a measurement that is less than, equal to, or greater than 400 meters? Why?
38. Suppose a train-based observer in Question 37 measures the length of the train to be 400 meters. He then places paint cans on the front and back of the train and detonates them simultaneously. An observer on the ground then measures the distance between the paint splatters left on the tracks. Will her measurement be less than, equal to, or greater than 400 meters? Why?
39. An observer on the ground and an observer in the train in the figure each measure the distance between the rails. Does the observer on the ground obtain a longer, a shorter, or the same distance as the observer in the train? Explain.
40. An observer on the ground and an observer in the train in the figure each measure the width of the train. Does the observer on the ground obtain a longer, a shorter, or the same width as the observer in the train? Explain.
41. An observer on the ground claims that the engine of a rapidly moving train came out of a tunnel at the same time as the caboose entered.
    a. Would an observer in the train agree? If not, which event would the observer say happened first?
    b. According to this observer, which is longer, the train or the tunnel?
    c. Are your answers consistent with each other?
42. An observer in a rapidly moving train claims that the engine came out of the tunnel at the same time as the caboose entered it.
    a. Would an observer on the ground agree? If not, which event would the observer say happened first?
    b. According to this observer, which is longer, the train or the tunnel?
    c. Are your answers consistent with each other?
43. A proton is accelerated from 10% to 99% of the speed of light. If the magnitude of the proton's acceleration is to remain constant during this interval, how does the force exerted on the proton have to change as it speeds up?
44. A constant force acts on a proton and causes it to speed up. Does the magnitude of the proton's acceleration increase, decrease, or remain constant as its speed gets closer and closer to the speed of light?
45. Why is it *not* correct to claim that matter can neither be created nor destroyed?
46. In a nuclear fusion reaction, one deuterium atom (one proton and one neutron) combines with one tritium atom (one proton and two neutrons) to form one helium atom (two protons and two neutrons) plus a free neutron. In this reaction a huge amount of energy is released. Using Einstein's idea of mass-energy equivalence, what can you conclude about the mass of the final products compared with the mass of the initial fuel?

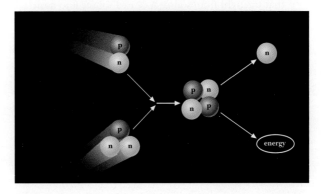

47. An artist is making a metallic statue of Einstein. Does the mass of the statue change as the metal cools? If so, does it get larger or smaller?

48. In view of the fact that clocks run slower and meter sticks are shorter in a moving system, how is it possible for an observer in a moving system to obtain the same speed for light as we do in our system?
49. What types of reference systems does the general theory of relativity address that are specifically excluded from the special theory of relativity?
50. In what way is the special theory of relativity more "special" than the general theory of relativity?
51. The postulates of the special theory of relativity imply that no experiment can distinguish between two reference systems moving at different constant velocities. Does the statement of the general theory of relativity imply that no experiment can distinguish between two reference systems moving with different constant accelerations?
52. Student 1 claims, "General relativity says that there is no experiment I can do in a closed room to tell if my system is accelerating or not accelerating, which means the path of a thrown ball should be the same in both systems." Student 2 counters, "The path of the ball could be different. General relativity only says that you can't tell if it's different because the system is accelerating or because of the presence of some new gravitational force." Which student do you agree with?
53. Jordan and Blake are asked to compare the masses of two objects. Jordan holds one object in each hand and shakes them. Blake holds the objects stationary in each hand. Which student is comparing the gravitational masses of the objects, and which is comparing the inertial masses? Explain your reasoning.
54. You are an astronaut in deep space and you are holding a sledgehammer in one hand and a nail in the other. How could you determine which object has the greater mass? Would you be comparing the gravitational masses or the inertial masses?
55. Imagine a universe in which inertial and gravitational masses are not the same. Specifically, if you double the inertial mass, the gravitational mass increases three times. If you were to drop a hammer and a penny from the same height above the floor, which would hit first? Explain your reasoning.
56. Imagine a universe in which inertial and gravitational masses are not the same. Specifically, if you double the gravitational mass, the inertial mass increases three times. If you were to drop a hammer and a penny from the same height above the floor, which would hit first? Explain your reasoning.
57. Spaceship A is traveling through deep space with twice the acceleration of spaceship B. If the passengers on the two spaceships believe that they are actually sitting on planets with identical masses, which passengers believe their planet has the larger radius? Why?
58. To create "artificial gravity" for inhabitants of a space station located in deep space, the station is rotated, as shown in the figure. If one of the inhabitants were to compare the weight of a ball held near the floor with its weight near the ceiling, which would be greater, and why? Compare this with the case where a ball's weight is measured as it is moved up from Earth's surface.

59. Why do we usually not notice the bending of light?
60. The barrel of a rifle and a laser are both pointed directly toward a target that is 1000 meters away. General relativity says that the bullet and the beam of light from the laser experience the same acceleration, and yet the bullet hits the target well below the beam of light. How can you explain this result?
61. You are asked to predict whether a clock at the North Pole would run faster or slower than one at the equator. You know that because Earth is somewhat flattened at the poles, the clock at the North Pole would be closer to Earth's center. You also know that Earth is spinning, so the clock at the equator has a centripetal acceleration. Will the clock at the North Pole tend to run faster or

slower than the one at the equator due to (a) the Earth's shape and (b) the Earth's spin?

62. We normally describe Earth as orbiting the Sun as a result of the Sun's gravitational attraction of Earth. What alternative explanation does general relativity provide to explain this orbital motion?

EXERCISES

1. The Moon shines by reflecting light from the Sun. The distance from Earth to the Moon is 3.84×10^8 m, and the distance from Earth to the Sun is 1.5×10^{11} m.
    a. How long does it take for light to reach Earth from the Sun?
    b. How long does it take for light to reach Earth from the Moon?
2. If it takes light 4.4 years to reach Earth from the nearest star system, how far is it to the star system?
3. When Venus is closest to Earth, it is approximately 45 million km away. If the radio telescope at Arecibo, Puerto Rico, bounces a radio signal from Venus's surface, how long will it take the radio signal to make the round-trip?

4. How long would it take a radio signal to reach a space probe in orbit about Saturn when Saturn is 1.5×10^{12} m from Earth?
5. What is the size of the adjustment factor for a speed of $0.4c$?
6. What is the size of the adjustment factor for a speed 20% that of light?
7. The average lifetime of a pion moving at 99% the speed of light is measured to be 2.69 nanoseconds (1 ns = 10^{-9} s). What would be the average lifetime of a pion at rest in the laboratory?
8. The average lifetime of isolated neutrons measured at rest relative to the lab is 920 s. What is the average lifetime of neutrons traveling at 80% of the speed of light?
9. An astronaut traveling at 99% of the speed of light waits 4 h (on his watch) after breakfast before eating lunch. To an observer on Earth, how long did the astronaut wait between meals?
10. The ground-based mission doctor for the astronaut in Exercise 9 is concerned that the astronaut is getting out of shape and requires him to exercise. The doctor tells the astronaut to begin pedaling the stationary bicycle and continue until she tells him to stop. She waits for 1 h on her clock. How long does the astronaut have to exercise according to his watch?
11. A ground-based observer measures a rocket ship to have a length of 60 m. If the rocket was traveling at 50% of the speed of light when the measurement was made, what length would the rocket have if brought to rest?
12. A rocket ship is 80 m long when measured at rest. What is its length as measured by an observer who sees the rocket ship moving past at 99% of the speed of light?
13. The conductor of a high-speed train uses a meter stick to measure the length of her train at 200 m while the train is stopped at the station. The train then travels at 80% of the speed of light (this is a super-supersonic train!). If she repeats the measurement on the moving train, what answer will she get?
14. An observer standing beside the tracks in Exercise 13 measures the length of the moving train as it goes by. What value does he get?
15. What is the distance to the nearest star system measured by an observer in a rocket ship traveling to the star system with a speed of $0.95c$? (The distance is 42 trillion km as measured by an observer on Earth.)
16. The pilot of an interstellar spaceship traveling at $0.98c$ determines the diameter of our Milky Way Galaxy to be about 1.2×10^{14} km. What value would an Earth-based observer calculate for the Galaxy's diameter?
17. According to the classical form of Newton's second law, $F \Delta t = \Delta p$, it would require a force of 9.5 N acting for a year to accelerate a 1-kg mass to a speed of $0.9999c$. Us-

ing the relativistic form of Newton's second law, what force is required?
18. Calculate the impulse ($F\,\Delta t$) needed to accelerate a 1-kg mass to 80% of the speed of light, using both the classical and relativistic forms of Newton's second law of motion.
19. How fast would a proton have to be traveling for its kinetic energy to equal its rest–mass energy?
20. By what factor does the total energy of a particle increase when its speed doubles from $0.4c$ to $0.8c$?
21. A spaceship in deep space has a velocity of 200 km/s and an acceleration in the forward direction of 5 m/s². What is the acceleration of a ball relative to the spaceship after it is released in this spaceship?
22. Two spaceships, one red and one blue, are traveling through deep space. The red spaceship has a velocity of 20 m/s and an acceleration of 40 m/s², and the blue spaceship has a velocity of 40 m/s and an acceleration of 20 m/s². In which spaceship do the astronauts experience the greater effective gravitational force?
23. A spacecraft is descending to land on planet Y and slows by 4 m/s every second. The strength of the planet's gravitational field is 7 N/kg. If the passengers in the spacecraft account for the forces they feel in terms of a single gravitational field, how strong would this field have to be?
24. A windowless spaceship is lifting off the surface of planet X with an acceleration of 20 m/s². The strength of the planet's gravitational field is 10 N/kg. If the passengers in the spacecraft account for the forces they feel in terms of a single gravitational field, how strong would this field have to be?
25. If light could somehow continuously travel perpendicular to a gravitational field with a strength of 10 N/kg—the strength at Earth's surface—how far would the light bend in 1 s?
26. How far would light bend due to gravity in traveling across the United States, a distance of approximately 5000 km?
27. The sum of the angles of a triangle drawn on the surface of a sphere is greater than 180°. What is the largest possible sum? What does this triangle look like?
28. The ratio of the circumference to the diameter of a circle drawn on a flat surface is 3.14. What is the value of this ratio if the circle is Earth's equator? In this case, the center of the circle is the North Pole, as shown in the figure.

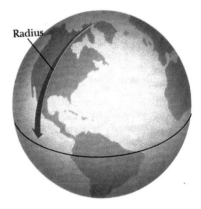

Interlude

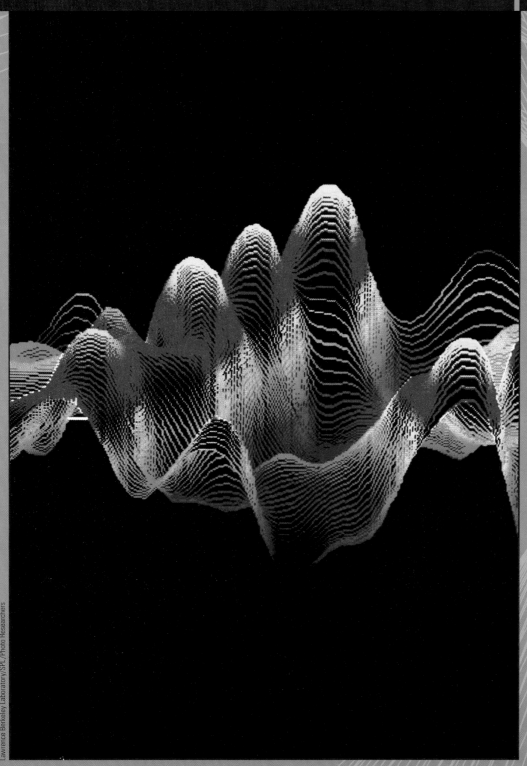

A sample of DNA as seen through a scanning tunneling micrograph.

The Search for Atoms

One of the oldest challenges in building a physics world view is the search for the fundamental building blocks of matter. This search began more than 2000 years ago. The first ideas appeared in writings about the Greek philosopher Leucippus, who lived in the 5th century BC. Leucippus asked a simple question: "If you take a piece of gold and cut it in half and then cut one of the halves in half, and so on, will you always have gold?" We know what happens initially: one piece of gold yields two pieces of gold; either of these pieces yields two more pieces of gold; and so on. But what if you could continue the process indefinitely? Do you think you would eventually reach an end—a place where either you couldn't cut the piece or, if you could, you would no longer have gold?

Leucippus and his student Democritus felt that this process would eventually stop—that gold has a definite elementary building block. Once you get to this level, further cuts either fail or yield something different. These elementary building blocks were (and still are) known as *atoms*—from the Greek word for "indivisible."

We know of these two early atomists through the writings of Aristotle in the 4th century BC. Aristotle disagreed with their atomistic view. He realized that if mat-ter were made of atoms, one needed to ask, "What is between them?" Presumably nothing. But Aristotle felt that a void—pure nothingness—between pieces of matter was philosophically unacceptable. Furthermore, the atomistic view held that the atoms were eternal and in continual motion. Aristotle felt this was foolish and in total conflict with everyday experience. He saw a world where all objects wear out and where their natural motion is one of rest: even if pushed, they quickly slow down and stop when the pushing stops.

Above all, Aristotle believed that everything must have a purpose, a goal toward which it is directed. If these atoms whirled about in empty space, what was their purpose? Atoms seemed to defy the natural view of the universe that there was a goal to creation.

Aristotle's arguments were powerful, and his world view remained relatively unchanged for the next 19 centuries. Galileo's idea of inertia and the possibility of perpetual motion in the absence of friction did not occur until the 17th century. The technology for making a good vacuum—Aristotle's unacceptable void—did not occur until later. But eventually Aristotle's two impossibilities became possible.

Although the evidence of atoms has grown tremen-

> Although the evidence of atoms has grown tremendously during the last 300 years, it might be surprising to learn that the first proof of the existence of atoms didn't come until 1905.

As we view the woman's face from closer and closer, we begin to see the building blocks that form the image. Can we do the same for the building blocks of our material world?

dously during the last 300 years, it might be surprising to learn that the first proof of the existence of atoms didn't come until 1905.

Albert Einstein published a paper showing that atoms in continual, random motion could explain a strange jiggling motion of microscopic objects suspended in fluids observed by a Scottish botanist, Robert Brown, almost 80 years earlier. Not until recently have new electron microscopes been able to show direct evidence of atoms.

Different combinations and arrangements of a relatively small number of atoms make up the diverse material world around us. What are the properties of these solids, liquids, and gases? How are their macroscopic properties, such as mass, volume, temperature, pressure, and elasticity, related to the underlying microscopic properties? We will examine both macroscopic and microscopic properties and their connections in the next two chapters. This study will then allow us to expand our understanding of one of the most fundamental concepts in the physics world view: the conservation of energy.

11 | Structure of Matter

Hot-air balloons are a beautiful illustration of the ideal gas law.

Hot air allows balloonists to enjoy the scenery from a high vantage point. But what do the macroscopic properties of gases—volume, pressure, and temperature—tell us about the existence of atoms and molecules and their microscopic properties of size, mass, and speed?

(See page 222 for the answer to this question.)

PhysicsNow™ Test your understanding of this chapter by logging into PhysicsNow at **http://physics.brookscole.com/kf6e**, selecting the chapter, and clicking on the "Take a Pre-Test" link.

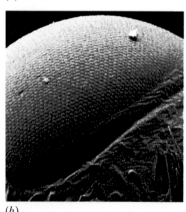

An electron micrograph of (a) a fly's head (27×) and its eye at (b) 122× and (c) 1240×.

WHEN we talk about the properties of objects, we usually think about their bulk, or **macroscopic**, properties such as size, shape, mass, color, surface texture, and temperature. For instance, a gas has mass, occupies a volume, exerts a pressure on its surroundings, and has a temperature. But a gas is composed of particles that have their own characteristics, such as velocity, momentum, and kinetic energy. These are the **microscopic** properties of the gas. It seems reasonable that connections should exist between these macroscopic and microscopic properties. At first glance, you might assume that the macroscopic properties are just the sum, or maybe the average, of the microscopic ones; however, the connections provided by nature are much more interesting.

We could begin our search for the connections between the macroscopic and the microscopic by examining a surface with a conventional microscope and discovering that rather than being smooth, as it appears to the naked eye, the surface has some texture. The most powerful electron microscope shows even more structure than an optical microscope reveals. But, until recently, instruments could not show us the basic underlying structure of things. To gain an understanding of matter at levels beyond which they could observe directly, scientists constructed models of possible microscopic structures to explain their macroscopic properties.

Building Models

By the middle of the 19th century, the body of chemical knowledge had pretty well established the existence of **atoms** as the basic building blocks of matter. All the evidence was indirect but was sufficient to create some idea of how atoms combine to form various substances. This description of matter as being composed of atoms is a model. It is not a model in the sense of a scale model, such as a model railroad or an architectural model of a building, but rather a theory or mental picture.

To illustrate this concept of a model, suppose someone gives you a tin can without a label and asks you to form a mental picture of what might be inside. Suppose you hear and feel something sloshing when you shake the can. You guess that the can contains a liquid. Your model for the contents is that of a liquid, but you do not know if this model actually matches the contents. For example, it is possible (but unlikely) that the can contains an electronic device that imitates sloshing sounds. However, you can use your knowledge of liquids to test your model. You know, for example, that liquids freeze. Therefore, your model predicts that cooling the can would stop the sloshing sounds. In this way you can use your model to help you learn more about the contents. The things you can say, however, are limited. For example, there is no way to determine the color of the liquid or its taste or smell.

Sometimes a model takes the form of an analogy. For example, the flow of electricity is often described in terms of the flow of water through pipes. Similarly, little, solid spheres might be analogous to atoms. We could use the analogy to develop rules for combining these spheres and learn something about how atoms combine to form molecules. However, it is important to distinguish between a collection of atoms that we cannot see and the spheres that we can see and manipulate. Sooner or later the analogy breaks down because electricity is not water and atoms are not tiny spheres.

Sometimes a model takes the form of a mathematical equation. Physicists have developed equations that describe the structure and behavior of atoms. Although mathematical models can be very abstract, they can also be very accurate descriptions of the way nature behaves. Suppose that in examining a tin can you hear a sliding sound followed by a clunk whenever you tilt the can. You might devise an equation for the length of time it takes a certain length rod to slide along the wall before hitting an end. This mathematical model would allow you to predict the time delays for different-length rods.

Regardless of its form, a model should summarize and account for the known data. It must agree with the way nature behaves. And to be useful, it must also be able to make predictions about new situations. The model for the contents of the tin can allows you to predict that liquid will flow out of the can if it is punctured. Our model for the structure of matter must allow us to make predictions that can be tested by experiment. If the predictions are borne out, they strengthen our belief in the model. If they are not borne out, we must modify our model or abandon it and invent a new one.

Early Chemistry

Experimental techniques, as opposed to careful observation, became of increasing interest to scientists around the end of the first millennium. They flourished especially in the Arab-speaking world, where so many protochemical ideas found a home. We now unconsciously use many chemical terms—alcohol, alchemy, alembic, and so on—that derive from Arabic.

Alchemy was the search for the magic formula for perpetual life—the elixir (note again the Arabic)—and for the secret of how to make gold. Alchemists set the stage for our modern atomic world view, but their contributions are often overlooked. We hear only that these strange pseudoscientists spent all their time foolishly trying to turn lead into gold. But they weren't so strange. In the Aristotelian world view that all matter was composed of various portions of four elements—earth, water, fire, and air—and that matter was continuous, it made sense that different materials could be made by changing the proportions of the four elementary substances. Making solutions and heating materials were natural beginnings because water and fire were two of the elements. A cooked egg was not the same as a raw one. Adding fire changed its characteristics. Adding fire to other materials released some air even though the original material gave no hint of containing air. Adding water to sugar changed it from a white solid to a clear liquid. This kind of experimentation—heating, dissolving in water, stirring, blending, grinding, and so on—was the work of the alchemists.

This early 19th-century Japanese woodcut shows an artisan extracting copper from its ore.

But the alchemists' role in the building of our world view was more than just finding ways of manipulating materials. By showing that the common materials were not all that existed, alchemists set the stage for the experimenters who followed. These new experimenters began with a different background of knowledge, a different common sense about the world, and a different expectation of what was possible. They went beyond cataloging ways of manipulating substances and began considering the structures of matter that could be causing these results—they built models. Although it is difficult to pinpoint the transition from alchemy to chemistry, the early chemists were part of the Newtonian age and had a different way of looking at the world. Isaac Newton himself spent many arduous hours in a little wooden laboratory outside his rooms at Trinity College, Cambridge, performing alchemical experiments.

By the latter half of the 17th century, the existence of the four Aristotelian elements was in doubt. They were of little or no help in making sense out of the chemical data that had been accumulated. However, the idea that all matter was composed of some basic building blocks was so appealing that it persisted. This belief fueled the development of our modern atomic model of matter.

The simplest, or most elementary, substances were known as **elements**. These elements could be combined to form more complex substances, the **com-

Table 11-1 | Lavoisier's List of the Elements (1789)

Lavoisier's Name	Modern English Name	Lavoisier's Name	Modern English Name
Lumiere	Light*	Etain	Tin
Calorique	Heat*	Fer	Iron
Oxygene	Oxygen	Manganese	Manganese
Azote	Nitrogen	Mercure	Mercury
Hydrogene	Hydrogen	Molybdene	Molybdenum
Soufre	Sulfur	Nickel	Nickel
Phosphore	Phosphorus	Or	Gold
Carbone	Carbon	Platine	Platinum
Radical muriatique[†]	—	Plomb	Lead
Radical fluorique[†]	—	Tungstene	Tungsten
Radical boracique[†]	—	Zinc	Zinc
Antimoine	Antimony	Chaux	Calcium oxide (lime)[‡]
Argent	Silver	Magnésie	Magnesium oxide[‡]
Arsenic	Arsenic	Baryte	Barium oxide[‡]
Bismuth	Bismuth	Alumina	Aluminum oxide[‡]
Cobalt	Cobalt	Silice	Silicon dioxide (sand)[‡]
Cuivre	Copper		

*Not elements.
[†]No elements with these properties have ever been found.
[‡]These "elements" are really compounds.

The electrolysis of water breaks up the water molecules into hydrogen and oxygen. Notice that the volume of hydrogen collected in the left-hand tube is twice that of the oxygen collected in the right-hand tube.

pounds. It was not until the 1780s that French chemist and physicist Antoine Lavoisier and his contemporaries had enough data to draw up a tentative list of elements (Table 11-1). Something was called an element if it could not be broken down into simpler substances. (The modern periodic table of the elements is printed on the inside front cover of this book.)

A good example of an incorrectly identified element is water. It was not known until the end of the 18th century that water is a compound of the elements hydrogen and oxygen. Hydrogen had been crudely separated during the early 16th century, but oxygen was not discovered until 1774. When a flame is put into a test tube of hydrogen, it "pops." One day while popping hydrogen, an experimenter noticed some clear liquid in the tube. This liquid was water. This was the first hint that water was not an element. The actual decomposition of water was accomplished at the end of the 18th century by a technique known as *electrolysis*, by which an electric current passing through a liquid or molten compound breaks it down into its respective elements.

Question Could these early chemists know for sure that a substance was an element?

Answer They never really knew whether the substance was an element or whether they had not yet figured out how to break it down.

The radically different properties of elements and their compounds are still striking in the modern world view. Hydrogen, a very explosive gas, and oxygen, the element required for all burning, combine to form a compound that is great for putting out fires! Similarly, sodium, a very reactive metal that must be kept in oil to keep it from reacting violently with moisture in the air, combines with chlorine, a very poisonous gas, to form a compound that tastes great on mashed potatoes—common table salt!

Chemical Evidence of Atoms

Another important aspect of elements and compounds was discovered around 1800. Suppose a particular compound is made from two elements and when you combine 10 grams of the first element with 5 grams of the second, you get 12 grams of the compound and have 3 grams of the second element remaining. If you now repeat the experiment, only this time adding 10 grams of each element, you still get 12 grams of the compound, but now have 8 grams of the second element remaining. This result was exciting. It meant that, rather than containing some random mixture of the two elements, the compound had a very definite ratio of their masses. This principle is known as the **law of definite proportions**.

Question   How much of the compound would you get if you added only 1 gram of the second element?

Answer   Because 10 grams of the first element require 2 grams of the second, 1 gram of the second will combine with 5 grams of the first. The total mass of the compound is just the sum of the masses of the two elements, so 6 grams of the compound will be formed.

This law is difficult to explain using the Aristotelian model, which viewed matter as a continuous, smooth substance. In the continuum model, one would expect there to be a range of masses in which the elements could combine to form the compound. An atomic model, on the other hand, provides a simple explanation: the atoms of one element can combine with atoms of another element to form **molecules** of the compound. It may be that one atom of the first element combines with one atom of the second to form one molecule of the compound that contains two atoms. Or it may be that one atom of the first element combines with two atoms of the second element. In any case the ratio of the masses of the combining elements has a definite value. (Note that the value of the ratio doesn't change with differing amounts. The mass ratio of 10 baseballs and 10 basketballs is the same as that of 1 baseball and 1 basketball. It doesn't matter how many balls we have as long as there is 1 baseball for *each* basketball.)

The actual way the elements combined and what caused them to always combine in the same way was unknown. English mathematician and physicist John Dalton hypothesized that the elements might have hooks (Figure 11-1) that control how many of one atom combine with another. Dalton's hooks can be literal or metaphorical; the actual mechanism is not important. The essential point of his model was that different atoms have different capacities for attaching to other atoms. Regardless of the visual model we use, atoms combine in a definite

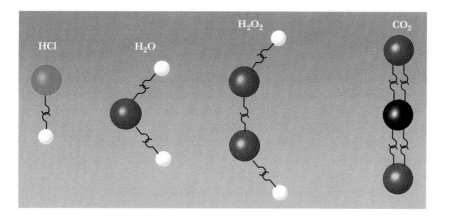

Figure 11-1 Dalton's atomic model has hooks to explain the law of definite proportions.

ratio to form molecules. One atom of chlorine combines with one atom of sodium to form salt. The ratio in salt is always one atom to one atom.

In retrospect it may seem that the law of definite proportions was a minor step and that it should have been obvious once mass measurements were made. However, seeing this relationship was difficult because some processes did not obey this law. For instance, *any* amount of sugar (up to some maximum) dissolves completely in water. One breakthrough came when it was recognized that this process was distinctly different. The sugar–water solution was not a compound with its own set of properties but simply a mixture of the two substances. Mixtures had to be recognized as different from compounds and eliminated from the discussion.

Another complication occurred because some elements can form more than one compound. Carbon atoms, for example, could combine with one or two oxygen atoms to form two compounds with different characteristics. When this happened in the same experiment, the final product was not a pure compound but a mixture of compounds. This result yielded a range of mass ratios and was quite confusing until chemists were able to analyze the compounds separately.

Fortunately, the atomic model makes predictions about situations in which two elements form more than one compound. Imagine for a moment that atoms can be represented by nuts and bolts. Suppose a hypothetical molecule of one compound consists of one nut and one bolt and another consists of two nuts and one bolt. Because there are twice as many nuts for each bolt in the second compound (Figure 11-2), it has a mass ratio of nuts to bolts that is twice that of the first compound. This prediction was confirmed for actual compounds and provided further evidence for the existence of atoms.

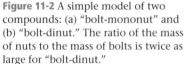

Figure 11-2 A simple model of two compounds: (a) "bolt-mononut" and (b) "bolt-dinut." The ratio of the mass of nuts to the mass of bolts is twice as large for "bolt-dinut."

PHYSICS | ON YOUR OWN

Collect a number of identical bolts and identical nuts. Assemble a number of "bolt-mononuts"; that is, put one nut on each bolt. Take these apart and measure the total mass of the bolts and the total mass of the nuts. What is the ratio of the mass of nuts to that of bolts? Does this ratio depend on how many bolt-mononuts you built?

Repeat this process with "bolt-dinuts"; that is, put two nuts on each bolt. How does the mass ratio for the bolt-dinuts compare with the mass ratio for the bolt-mononuts?

✓ Extended presentation available in the *Problem Solving* supplement

Masses and Sizes of Atoms ✓ MATH

Even with their new information, the 18th-century chemists did not know how many atoms of each type it took to make a specific molecule. Was water composed of one atom of oxygen and one atom of hydrogen, or was it one atom of oxygen and two atoms of hydrogen, or two of oxygen and one of hydrogen? All that was known was that 8 grams of oxygen combined with 1 gram of hydrogen. These early chemists needed to find a way of establishing the relative masses of atoms.

The next piece of evidence was an observation made when gaseous elements were combined; the gases combined in definite *volume* ratios when their temperatures and pressures were the same. This statement was not surprising except that the volume ratios were always simple fractions. For example, 1 liter of hydrogen combines with 1 liter of chlorine (a ratio of 1 to 1), 1 liter of oxygen combines with 2 liters of hydrogen (a ratio of 1 to 2), 1 liter of nitrogen combines with 3 liters of hydrogen (a ratio of 1 to 3), and so on (Figure 11-3).

It was very tempting to propose an equally simple underlying rule to explain these observations. Italian physicist Amedeo Avogadro suggested that under identical conditions each liter of any gas contains the same number of molecules. Al-

Figure 11-3 Gases combine completely to form compounds when the ratios of their volumes are equal to the ratios of small whole numbers. One liter of nitrogen combines completely with 3 liters of hydrogen to form 2 liters of ammonia.

though it took more than 50 years for this hypothesis to be accepted, it was the key to unraveling the question of the number of atoms in molecules.

Question Given that oxygen and hydrogen gases are each composed of molecules with two atoms each, how many atoms of oxygen and hydrogen combine to form water?

Answer The observation that 2 liters of hydrogen gas combine with 1 liter of oxygen means that there are two hydrogen molecules for each oxygen molecule. Therefore, there are four hydrogen atoms for every two oxygen atoms. The simplest case would be for these to form two water molecules with two hydrogen atoms and one oxygen atom in each water molecule. This is confirmed by the observation that 2 liters of water vapor are produced.

Once the number of atoms in each molecule was known, the data on the mass ratios could be used to calculate the relative masses of different atoms. For example, an oxygen atom has about 16 times the mass of a hydrogen atom.

To avoid the use of ratios, a mass scale was invented by choosing a value for one of the elements. An obvious choice was to assign the value of 1 to hydrogen because it is the lightest element. However, setting the value of carbon equal to 12 **atomic mass units** (amu) makes the relative masses of most elements close to whole-number values. These values are known as **atomic masses** and keep the value for hydrogen very close to 1. Even though the values for the actual atomic masses are now known, it is still convenient to use the relative atomic masses.

Question What is the atomic mass of carbon dioxide, a gas formed by combining two oxygen atoms with each carbon atom?

Answer We have 12 atomic mass units for the carbon atom and 16 atomic mass units for each oxygen atom. Therefore, 12 atomic mass units + 32 atomic mass units = 44 atomic mass units.

The problem of determining the masses and diameters of individual atoms required the determination of the number of atoms in a given amount of material. Diffraction experiments, like those described in Chapter 19 but using X rays, determined the distance between individual atoms in solids to be about 10^{-10} meter, one 10-billionth of a meter. If we assume that atoms in a solid can be represented by marbles like those in Figure 11-4, the diameter of an atom is about equal to their spacing. This means that it would take 10 billion atoms to make a line 1 meter long. Stated another way, if we imagine expanding a baseball to the size of Earth, the individual atoms of the ball would only be the size of grapes!

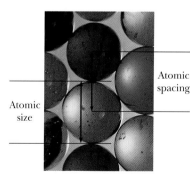

Figure 11-4 Assuming that the atoms are "touching" like marbles, the spacing between their centers is the same as their diameters.

A useful quantity of matter for our purposes is the **mole**. If the mass of the molecule is some number of atomic mass units, 1 mole of the substance is this same number of grams. For example, 1 mole of carbon is 12 grams. Further experiments showed that 1 mole of any substance contained the same number of molecules—namely, 6.02×10^{23} molecules, a number known as **Avogadro's number**. With this number we can calculate the size of the atomic mass unit in terms of kilograms. Because 12 grams of carbon contain Avogadro's number of carbon atoms, the mass of one atom is

$$m_{\text{carbon}} = \frac{12 \text{ g}}{6.02 \times 10^{23} \text{ molecules}} = 2 \times 10^{-23} \text{ g/molecule}$$

Because one carbon atom also has a mass of 12 atomic mass units, we obtain

$$\frac{2 \times 10^{-23} \text{ g}}{12 \text{ amu}} = 1.66 \times 10^{-24} \text{ g/amu}$$

Therefore, 1 atomic mass unit equals 1.66×10^{-27} kilogram, a mass so small that it is very hard to imagine. This is the approximate mass of one hydrogen atom. The most massive atoms are about 260 times this value.

Flawed Reasoning

Two students are arguing after class about gases.

Dominic: "If two gases are both at the same temperature and pressure, then equal volumes will contain equal numbers of atoms. This means that 1 mole of ammonia (NH_3) would take up twice as much volume as 1 mole of nitrogen (N_2) because each ammonia molecule has four atoms and each nitrogen molecule has only two."

Angelina: "No, equal volumes will contain equal numbers of *molecules*, not *atoms*. One mole of ammonia would contain the same number of molecules as 1 mole of nitrogen—namely, Avogadro's number—so they would take up the same volume."

Do you agree with either of these students?

Answer Angelina was paying attention in class. Avogadro found that the number of molecules determined the volume of a gas for a given temperature and pressure. The number of atoms in each molecule does not matter.

The Ideal Gas Model

Many macroscopic properties of materials can be understood from the atomic model of matter. Under many situations the behavior of real gases is very closely approximated by an **ideal gas**. The gas is assumed to be composed of an enormous number of very tiny particles separated by relatively large distances. These particles are assumed to have no internal structure and to be indestructible. They also do not interact with each other except when they collide, and then they undergo elastic collisions much like air-hockey pucks. Although this model may not seem realistic, it follows in the spirit of Galileo in trying to get at essential features. Later we can add the complications of real gases.

For this model to have any validity, it must describe the macroscopic behavior of gases. For instance, we know that gases are easily compressed. This makes sense; the model says that the distance between particles is very much greater than the particle size and they don't interact at a distance. There is, then, a lot of space in the gas, and it should be easily compressed. This aspect of the model also accounts for the low mass-to-volume ratio of gases.

Because a gas completely fills any container and the particles are far from one another, the particles must be in continual motion. Is there any other evidence that the particles are continually moving? We might ask, "Is the air in the room moving even with all the doors and windows closed to eliminate drafts?" The fact that you can detect an open perfume bottle across the room indicates that some of the perfume particles have moved through the air to your nose.

More direct evidence for the motion of particles in matter was observed in 1827 by Scottish botanist Robert Brown. To view pollen under a microscope without it blowing away, Brown mixed the pollen with water. He discovered that the pollen grains were constantly jiggling. Brown initially thought that the pollen might be alive and moving erratically on its own. However, he observed the same kind of motion with inanimate objects as well.

Brownian motion is not restricted to liquids. Observation of smoke under a microscope shows that the smoke particles have the same very erratic motion. This motion never ceases. If the pollen and water are kept in a sealed container and put on a shelf, you would still observe the motion years later. The particles are in continual motion.

It was 78 years before Brownian motion was rigorously explained. Albert Einstein demonstrated mathematically that the erratic motion was due to collisions between water molecules and pollen grains. The number and direction of the collisions occurring at any time is a statistical process. When the collisions on opposite sides have equal impulses, the grain is not accelerated. But when more collisions occur on one side, the pollen experiences an abrupt acceleration that is observed as Brownian motion.

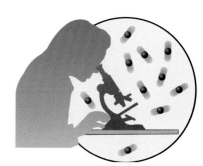

Pollen grains suspended in a liquid exhibit continual, erratic motion known as Brownian motion.

Pressure

✓ MATH

Let's take a look at one of the macroscopic properties of an ideal gas that is a result of the atomic motions. **Pressure** is the force exerted on a surface divided by the area of the surface—that is, the force per unit area:

$$P = \frac{F}{A}$$

◄ pressure = $\frac{\text{force}}{\text{area}}$

This definition is not restricted to gases and liquids. For instance, if a crate weighs 6000 newtons and its bottom surface has an area of 2 square meters, what pressure does it exert on the floor under the crate?

$$P = \frac{F}{A} = \frac{6000 \text{ N}}{2 \text{ m}^2} = 3000 \text{ N/m}^2$$

Therefore, the pressure is 3000 newtons per square meter. The SI unit of pressure (newton per square meter [N/m²]) is called a pascal (Pa). Pressure in the U.S. customary system is often measured in pounds per square inch (psi) or atmospheres (atm), where 1 atmosphere is equal to 101 kilopascals, or 14.7 pounds per square inch.

Question Susan asks politely if it would be all right with you if she pushes on your arm with a force of 5 newtons (about 1 pound). Should you let her?

Answer That depends. If she pushes on your arm with the palm of her hand, you will hardly notice a force of 5 newtons. If, on the other hand, she pushes on your arm with a sharp hatpin, you will definitely notice! The damage to your arm does not depend on the force but on the pressure.

Imagine a cubical container of gas in which particles are continually moving around and colliding with each other and with the walls (Figure 11-5). In each collision with a wall, the particle reverses its direction. Assume a head-on

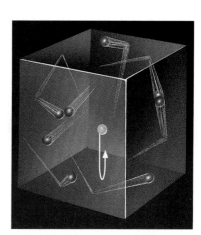

Figure 11-5 Gas particles are continually colliding with each other and with the walls of the container.

collision as in the case of the yellow particle in Figure 11-5. (If the collision is a glancing blow, only the component of the velocity perpendicular to the wall would be reversed.) This means that its momentum is also reversed. The change in momentum means that there must be an impulse on the particle and an equal and opposite impulse on the wall (Chapter 6). Our model assumes that an enormous number of particles strike the wall. The average of an enormous number of impulses produces a steady force on the wall that we experience as the pressure of the gas.

We can use this application of the ideal gas model to make predictions that can be tested. Suppose, for example, that we shrink the volume of the container. This means that the particles have less distance to travel between collisions with the walls and should strike the walls more frequently, increasing the pressure. Therefore, decreasing the volume increases the pressure, provided that the average speeds of the molecules do not change. Similarly, if we increase the number of particles in the container, we expect the pressure to increase because there would be more frequent collisions with the walls.

Atomic Speeds and Temperature

Presumably, the atomic particles making up a gas have a range of speeds due to their collisions with the walls and with each other. The distribution of these speeds can be calculated from the ideal gas model and a connection made with

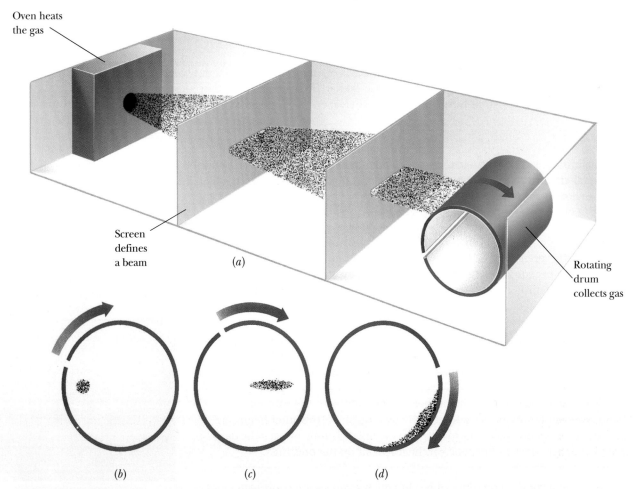

Figure 11-6 (a) Apparatus for measuring the speeds of atomic particles in a gas. (b) A small bunch of particles enters the rotating drum through the narrow slit. (c) The particles spread out as they move across the drum because of their different speeds. (d) The particles are deposited on a sensitive film at locations determined by how much the drum rotates before they arrive at the opposite wall.

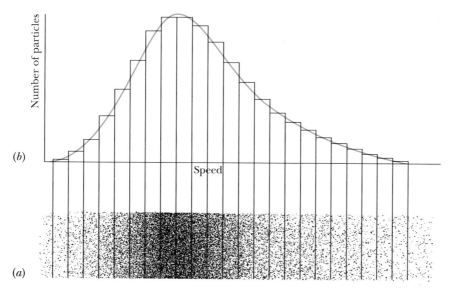

Figure 11-7 (a) The distribution of atomic particles as recorded by the film along the circumference of the drum. (b) The distribution of atomic speeds calculated from this experiment.

temperature. Therefore, a direct measurement of these speeds would provide additional support for the model. This is not an easy task. Imagine trying to measure the speeds of a very large group of invisible particles. One needs to devise a way of starting a race and recording the order of the finishers.

One creative approach led to a successful experiment in 1920. The gas leaves a heated vessel and passes through a series of small openings that select only those particles going in a particular direction (Figure 11-6[a]). Some of these particles enter a small opening in a rapidly rotating drum, as shown in Figure 11-6(b). This arrangement guarantees that a group of particles start across the drum at the same time. Particles with different speeds take different times to cross the drum and arrive on the opposite wall after the drum has rotated by different amounts. The locations of the particles are recorded by a film of sensitive material attached to the inside of the drum (Figure 11-6[d]). A drawing of the film's record is shown in Figure 11-7(a). A graph of the number of particles versus their position along the film is shown in Figure 11-7(b).

Most of the particles have speeds near the average speed, but some move very slowly and some very rapidly. The average speed is typically about 500 meters per second, which means that an average gas particle could travel the length of five football fields in a single second. This high value may seem counter to your experience. If the particles travel with this speed, why does it take several minutes to detect the opening of a perfume bottle on the other side of the room? This delay is due to collisions between the gas particles. On average a gas particle travels a distance of only 0.0002 millimeter before it collides with another gas particle. (This distance is about 1000 particle diameters.) Each particle makes approximately 2 billion collisions per second. During these collisions the particles can radically change directions, resulting in very zigzag paths. So although their average speed is quite fast, it takes them a long time to cross the room because they travel enormous distances.

When the speeds of the gas particles are measured at different temperatures, something interesting is found. As the temperature of the gas increases, the speeds of the particles also increase. The distributions of speeds for three temperatures are given in Figure 11-8. The calculations based on the ideal gas model agree with these results. A relationship can be derived that connects temperature, a macroscopic property, with the average kinetic energy of the gas particles,

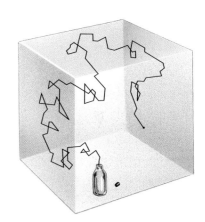

Atomic particles in air travel in zigzag paths because of numerous collisions with air molecules.

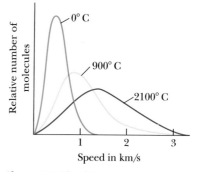

Figure 11-8 The distribution of atomic speeds changes as the temperature of the gas changes.

Figure 11-9 The height of the liquid in Galileo's first thermometer indicated changes in temperature.

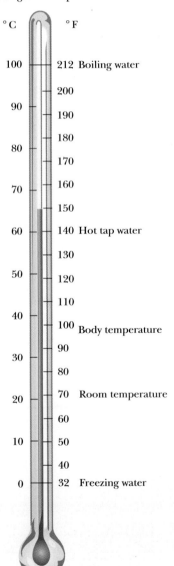

Figure 11-10 A comparison of the Fahrenheit and Celsius temperature scales.

Temperature

We generally associate temperature with our feelings of hot and cold; however, our subjective feelings of hot and cold are not very accurate. Although we can usually say which of two objects is hotter, we can't state just how hot something is. To do this we must be able to assign numbers to various temperatures.

PHYSICS | ON YOUR OWN

Place one hand in a pan of hot water and the other in a pan of ice-cold water. After several minutes, place both hands in a pan of lukewarm water. Do your hands feel like they are at the same temperature? What does this say about using your body as a thermometer?

Assigning numbers to various temperatures turns out to be a difficult task that has occupied some of the greatest scientific minds. Just as it is not possible to define time in a simple way, it is not possible to define temperature in a simple way. In Chapters 12 through 14, we will return to the subject of temperature a number of times with the aim of helping you incorporate the concept of temperature into your world view. We start with a familiar concept, measuring temperature with a thermometer.

Galileo was the first person to develop a thermometer. He observed that some of an object's properties change when its temperature changes. For example, with only a few exceptions, when an object's temperature goes up, it expands. Galileo's thermometer (Figure 11-9) was an inverted flask with a little water in its long neck. As the enclosed air got hotter, it expanded and forced the water down the flask's neck. Conversely, the air contracted on cooling, and the water rose. Galileo completed his thermometer by marking a scale on the neck of the flask.

Unfortunately, the water level also changed when atmospheric pressure changed. The alcohol-in-glass thermometer, which is still popular today, replaced Galileo's thermometer. The column is sealed so that the rise and fall of the alcohol is due to its change in volume and not the atmospheric pressure. The change in height is amplified by adding a bulb to the bottom of the column, as shown in Figure 11-10. When the temperature rises, the larger volume in the bulb expands into the narrow tube, making the expansion much more obvious.

In 1701 Newton proposed a method for standardizing the scales on thermometers. He put the thermometer in a mixture of ice and water, waited for the level of the alcohol to stop changing, and marked this level as zero. He used the temperature of the human body as a second fixed temperature, which he called 12. The scale was then marked off into 12 equal divisions, or degrees.

Shortly after this, German physicist Gabriel Fahrenheit suggested that the zero point correspond to the temperature of a mixture of ice and salt. Because this was the lowest temperature producible in the laboratory at that time, it avoided the use of negative numbers for temperatures. The original 12 degrees were later divided into eighths and renumbered so that body temperature became 96 degrees.

It is important that the fixed temperatures be reliably reproducible in different laboratories. Unfortunately, neither of Fahrenheit's reference temperatures could be reproduced with sufficient accuracy. Therefore, the reference temperatures were changed to those of the freezing and boiling points of pure water at standard atmospheric pressure. To get the best overall agreement with the previous scale, these temperatures were defined to be 32°F and 212°F, respectively. This is how we ended up with such strange numbers on the **Fahrenheit temperature scale**. On this scale, normal body temperature is 98.6°F.

At the time the metric system was adopted, a new temperature scale was defined with the freezing and boiling points as 0°C and 100°C. The name of this centigrade (or 100-point) scale was changed to the **Celsius temperature scale** in 1948 in honor of Swedish astronomer Anders Celsius, who devised the scale. A comparison of the Fahrenheit and Celsius scales is given in Figure 11-10. This figure can be used to convert temperatures from one scale to the other.

Question   What are room temperature (68°F) and body temperature (98.6°F) on the Celsius scale?

Answer   Using Figure 11-10, we see that room temperature is about 20°C and body temperature is about 37°C.

PHYSICS | **ON YOUR OWN**

If you live in an area with a high elevation, measure the boiling point of water. Does it boil at 100°C (212°F)? Does it matter how hard the water boils? What does this say about the possibility of using the boiling point of water as a fixed point for calibrating thermometers?

Assume that we have a quantity of ideal gas in a special container designed to always maintain the pressure of the gas at some constant low value. When the volume of the gas is measured at a variety of temperatures, we obtain the graph shown in Figure 11-11. If the line on the graph is extended down to the left, we find that the volume goes to zero at a temperature of −273°C (−459°F). Although we could not actually do this experiment with a real gas, this very low temperature arises in several theoretical considerations and is the basis for a new, more fundamental temperature scale.

The **Kelvin temperature scale**, also known as the **absolute temperature scale**, has its zero at −273°C and the same-size degree marks as the Celsius scale. The difference between the Celsius and Kelvin scales is that temperatures are 273 degrees higher on the Kelvin scale. Water freezes at 273 K and boils at 373 K. (Notice that the degree symbol is dropped from this scale. The freezing point of water is read "273 kelvin," or "273 kay.") The Kelvin scale is named for British physicist William Thomson, who is more commonly known as Lord Kelvin.

◄ K = C + 273

Question   What is normal body temperature (37°C) on the Kelvin scale?

Answer   Body temperature is 37 + 273 = 310 K.

It would seem that all temperature scales are equivalent and which one we use would be a matter of history and custom. It is true that these scales are equiv-

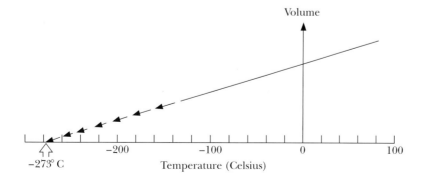

Figure 11-11 When the graph of volume versus temperature of an ideal gas is extrapolated to zero volume, the temperature scale reads −273°C.

alent because conversions can be made between them. However, the absolute temperature scale has a greater simplicity for expressing physical relationships. In particular, the relationship between the volume and temperature of an ideal gas is greatly simplified using absolute temperatures. *The volume of an ideal gas at constant pressure is proportional to the absolute temperature.* This means that if the absolute temperature is doubled while keeping the pressure fixed, the volume of the gas doubles.

The volume of an ideal gas at constant pressure can be used as a thermometer. All we need to do to establish the temperature scale is to measure the volume at one fixed temperature. Of course, thermometers must be made of real gases. But real gases behave like the ideal gas if the pressure is kept low and the temperature is well above the temperature at which the gas liquefies.

T is proportional to KE_{ave}

This new scale also connects the microscopic property of atomic speeds and the macroscopic property of temperature. *The absolute temperature is directly proportional to the average kinetic energy of the gas particles.* This means that if we double the average kinetic energy of the particles, the absolute temperature of a gas doubles. Remember, however, that the average speed of the gas particles does not double, because the kinetic energy depends on the square of the speed (Chapter 7).

> **Flawed Reasoning**
>
> When you wake, the temperature outside is 40°F, but by noon it is 80°F. **Why is it not reasonable to say that the temperature doubled?**
>
> **A**nswer The zero point for the Fahrenheit scale was arbitrarily chosen as the temperature of a mixture of ice and salt. If this zero point had been chosen differently—say, 30 degrees higher—the temperature during the morning would have changed from 10°F to 50°F; an increase of a factor of 5! Clearly, we can attach no physical significance to the doubling of the temperature reading on the Fahrenheit scale (or the Celsius scale). If, on the other hand, the temperature of a gas doubles on the Kelvin scale, we can say that the average kinetic energy of the gas particles has also doubled.

The Ideal Gas Law

The three macroscopic properties of a gas—volume, temperature, and pressure—are related by a relationship known as the **ideal gas law**. This law states that

ideal gas law

$$PV = nRT$$

where P is the pressure, V is the volume, n is the number of moles, T is the *absolute* temperature, and R is a number known as the gas constant.

This relationship is a combination of three experimental relationships that had been discovered to hold for the various pairs of these three macroscopic properties. For example, if we hold the temperature of a quantity of gas constant, we can experimentally determine what happens to the pressure as we compress the gas. Or we can vary the pressure and measure the change in volume. This experimentation leads to a relationship known as Boyle's law, which states that at constant temperature the product of the pressure and the volume is a constant. This is equivalent to saying that they are inversely proportional to each other; as one increases, the other must decrease by the same factor.

In a similar manner, we can investigate the relationship between temperature and volume while holding the pressure constant. The results for a gas at one

Evaporative Cooling

Many of the concepts of the ideal gas model apply to liquids as well as gases. One big difference between liquids and gases is the strength of the attractive forces between the molecules. In our ideal gas model, we assumed that these forces could be neglected until the molecules collided. As a consequence a gas expands to fill its container. However, liquids have a definite volume. It is intermolecular forces that hold the molecules in the liquid together.

If we assume that a model for liquids is similar to that for gases, we can begin to understand the evaporation of liquids. Assume that the kinetic energies of the molecules in liquids have a distribution similar to that in gases and that the average kinetic energy of the particles increases with increasing temperature. The intermolecular forces perform work on molecules that try to escape the liquid, allowing only those with large enough kinetic energies to succeed. Therefore, the molecules that leave the liquid have higher than average kinetic energies. Conservation of energy requires that the average kinetic energy of the molecules left behind is lower, and the liquid is cooled. Water is often carried in canvas bags on desert trips to keep the water cool. The canvas bag is kept wet by water seeping through the canvas. The evaporation of the water from the wet canvas keeps the rest of the water cool.

Evaporative cooling can be demonstrated with a simple experiment. Wrap the end of a thermometer in cotton soaked in room-temperature water and observe the temperature as the water evaporates.

Your body uses evaporative cooling to maintain your body temperature on very hot days or during strenuous exercise. The evaporating sweat cools our bodies. If you didn't sweat, you might die of heat prostration! A more effective way of cooling your body is the alcohol rub used to reduce fevers.

Canvas bags like the one shown here keep water cool, even in a hot desert.

pressure are shown in Figure 11-11. As stated in the previous section, the volume in this case is directly proportional to the absolute temperature.

The third relationship is between temperature and pressure at a constant volume. The pressure in this case is directly proportional to the absolute temperature.

Each of these relationships can be obtained from our model for an ideal gas. For example, let's take a qualitative look at Boyle's law. As we decrease the volume while keeping the temperature the same, the molecules will be moving at the same average speed as before but will now hit the sides more frequently; therefore, the pressure increases in agreement with the statement of Boyle's law.

Question How does the ideal gas model explain the rise in pressure of a gas as its temperature is raised without changing its volume?

Answer Raising the temperature of the gas increases the kinetic energies of the particles. The increased speeds of the particles mean not only that they have larger momenta but also that they hit the walls more frequently.

PhysicsNow™ Assess your understanding of this chapter's topics with sample tests and other resources found by logging into PhysicsNow at http://physics.brookscole.com/kf6e.

Summary

The most elementary substances are elements, which are composed of a large number of very tiny, identical atoms. These atoms combine in definite mass ratios to form molecules according to the law of definite proportions. Observations that gaseous elements combine in definite volume ratios allowed the determination of the relative masses of different atoms. The mass of a carbon atom is set equal to 12 atomic mass units.

Modern experiments show that atoms have diameters of approximately 10^{-10} meter and masses ranging from 1 to more than 260 atomic mass units, where 1 atomic mass unit is equal to 1.66×10^{-27} kilogram.

The Celsius temperature scale is defined with the freezing and boiling points of water as 0°C and 100°C, respectively. The absolute, or Kelvin, temperature scale has its zero point at −273°C and the same-size degree as the Celsius scale.

The ideal gas model assumes that the gas is composed of an enormous number of tiny, indestructible spheres with no internal structure, separated by relatively large distances, and interacting only via elastic collisions. The pressure exerted by a gas is due to the average of the impulses exerted by the gas particles on the walls of the container. Most of the particles have speeds near the average speed, but some move very slowly and some very rapidly. The average speed is typically about 500 meters per second. The absolute temperature of a gas is proportional to the average kinetic energy of the gas particles.

The ideal gas law states the relationship between the pressure, the volume, the number of moles, and the absolute temperature of a gas as $PV = nRT$, where R is the gas constant.

Chapter 11 Revisited

Observation of the motion of smoke particles provided evidence of the existence of atoms and molecules. Knowledge of the volumes of gases that combined chemically to form other gases helped establish the masses of individual atoms and molecules. The ideal gas law tells us that the average kinetic energy of the gas particles is determined by the absolute temperature.

KEY TERMS

absolute temperature scale: The temperature scale with its zero point at absolute zero and degrees equal to those on the Celsius scale. Also called the Kelvin temperature scale.

atom: The smallest unit of an element that has the chemical and physical properties of that element.

atomic mass: The mass of an atom in atomic mass units.

atomic mass unit: One-twelfth of the mass of a carbon atom. Equal to 1.66×10^{-27} kilogram.

Avogadro's number: 6.02×10^{23} molecules, the number of molecules in 1 mole of any substance.

Celsius temperature scale: The temperature scale with the values of 0°C and 100°C for the temperatures of freezing and boiling water, respectively.

compound: A combination of chemical elements that forms a substance with its own properties.

element: Any chemical species that cannot be broken up into other chemical species.

Fahrenheit temperature scale: The temperature scale with the values of 32°F and 212°F for the temperatures of freezing and boiling water, respectively. Its degree is five-ninths of that on the Celsius or Kelvin scales.

ideal gas: An enormous number of very tiny particles separated by relatively large distances. The particles have no internal structure, are indestructible, and do not interact with each other except when they collide; all collisions are elastic.

ideal gas law: $PV = nRT$, where P is the pressure, V is the volume, T is the absolute temperature, n is the number of moles, and R is the gas constant.

Kelvin temperature scale: The temperature scale with its zero point at absolute zero and a degree equal to that on the Celsius scale. Also called the absolute temperature scale.

law of definite proportions: When two or more elements combine to form a compound, the ratios of the masses of the combining elements have fixed values.

macroscopic: The bulk properties of a substance, such as mass, size, pressure, and temperature.

microscopic: Properties not visible to the naked eye, such as atomic speeds or the masses and sizes of atoms.

mole: The amount of a substance that has a mass in grams numerically equal to the mass of its molecules in atomic mass units.

molecule: A combination of two or more atoms.

pressure: The force per unit area of surface. Measured in newtons per square meter, or pascals.

CONCEPTUAL QUESTIONS

1. The two essential elements of a good model are insight and predictive power. Many ancient cultures explained natural phenomena in terms of the actions of their gods. Did these models fail primarily because of lack of insight or lack of predictive power?

2. The two essential elements of a good model are insight and predictive power. Choose a model with which you are familiar and point out how it meets these two criteria.

3. A friend has created a model of how a candy vending machine works. His theory says that a little blue person (LBP) lives inside each vending machine. This person takes your coins and gives you candy in return. Although this LBP theory may not seem reasonable to you, can you suggest ways of disproving it without opening the machine?

4. For most of human history, we believed that Earth was stationary and the Sun and planets orbited Earth (the geocentric model). Beginning about 500 years ago, a second model emerged in which Earth orbits the Sun (the heliocentric model). When we read in the paper that the Sun rose at 6:40 this morning, which of these two models is being used?

5. Your friend notices that a brown can of diet cola floats whereas a green can of lemon–lime soda and a can of orange soda both sink. He postulates a model in which only nonbrown cans of soda sink. To prove his model, he tries a brown can of diet root beer and finds that it floats as expected. Has he proven that his model is correct? In general, can a model ever be proven true?

6. Following the experiments described in Question 5, your friend tries a brown can of nondiet root beer and finds that it sinks. He rightfully discards his original model and proposes an alternative. What would this new model be? Has it been proven correct?

7. Alchemists held a model in which matter was continuous. Atomists showed the fallacy of this model and replaced it with a model in which all objects are made of small, discrete particles. In your day-to-day living, which model do you appeal to more often? Is the most complete model always the most useful?

8. What role did the alchemists play in the development of an atomistic world view?

9. Which of the following are not elements: hydrogen, salt, nitrogen, granite, sodium, chlorine, water?

10. Would you expect carbon monoxide to be an element or a compound? Why?

11. When the element mercury is heated in air, a red powder is formed. Careful measurement shows that the mass of the resulting powder is greater than the mass of the original mercury. Is this powder an element or a compound? How do you account for this additional mass?

12. Give an example that clearly illustrates the meaning of the law of definite proportions.

13. What are the basic differences between mixtures and compounds?

14. Do water and salt form a compound or a mixture?

15. The atomic mass of zinc oxide is 81 atomic mass units. If a molecule of zinc oxide consists of one atom of zinc and one atom of oxygen, what is the atomic mass of zinc?

16. The atomic mass of iron oxide (rust) is 160 atomic mass units. What is the atomic mass of iron if a molecule of iron oxide consists of two atoms of iron and three atoms of oxygen?

17. Silver has an atomic mass of 108. Which, if either, contains more atoms: 1 gram of silver or 1 gram of hydrogen?

18. Silver has an atomic mass of 108. Which, if either, contains more atoms: 1 mole of silver or 1 mole of hydrogen?

19. How does the number of molecules in 1 liter of oxygen compare with the number of molecules in 1 liter of carbon dioxide if they are both at the same temperature and pressure?

20. Oxygen molecules contain two oxygen atoms, and carbon dioxide molecules contain one atom of carbon and two atoms of oxygen. How does the total number of atoms in 1 liter of oxygen compare with the total number of atoms in 1 liter of carbon dioxide if they are both at the same temperature and pressure?

21. The atomic mass of sulfur is 32 atomic mass units. How many grams of sulfur are needed to have an Avogadro's number of sulfur atoms?

22. How many grams of water are needed to have an Avogadro's number of water molecules?

23. If a gas condenses to a liquid, the liquid occupies a much smaller volume than the gas. How does the ideal gas model account for this?

24. The ideal gas model accounts very well for the behavior of gases at standard temperature and pressure. Would the ideal gas model begin to fail for very large pressures or for very small pressures? Explain your answer.

25. A cube and a spherical ball are made of the same material and have the same mass. Which exerts the larger pressure while resting on a floor?

26. You can apply enough force to the head of a pushpin to push it into a plaster wall with your thumb. However, it is not a good idea to try to do this with a needle. Use the

concept of pressure to explain the difference between these two situations.

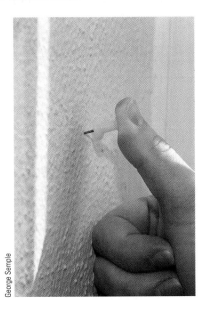

27. If you screw the cap of an empty plastic drinking bottle on tightly while walking in the mountains, why are the sides of the bottle caved in when you return to the valley?
28. Your right rear tire has to support a weight of 3000 newtons. Normally, the contact area of your tire with the road is 200 square centimeters. If the pressure in your tire is suddenly reduced from 32 pounds per square inch to 16 pounds per square inch, what must be the new contact area to support the car?
29. Use the microscopic model of a gas to explain why the pressure in a tire increases as you add more air.
30. As you drive your car down the road, the friction of the rubber with the road causes the air inside the tire to increase in temperature, resulting in an increase in pressure. Use the microscopic model of a gas to explain why the pressure increases.
31. If the average speed of a perfume molecule is 500 meters per second, why does it take several minutes before you smell the perfume from a bottle opened across the room?
32. What happens to the average speed of the molecules of a gas as it is heated?
33. Why does an alcohol-in-glass thermometer have a bulb at the bottom?
34. It is possible to cut the top off an alcohol-in-glass thermometer without any of the alcohol spilling out. However, it will no longer function as a good thermometer. Why not?
35. What conditions must be imposed before the boiling point of water can be used as a fixed temperature?
36. Why is body temperature not a good fixed temperature for establishing a temperature scale?
37. Two students are sick in bed with 2-degree fevers. One has a temperature of 39.0°C; the other, 100.6°F. Which student has the higher fever?
38. Is a sauna at a temperature of 190°F hotter or colder than one at 90°C?
39. At what temperature should you set your new Celsius thermostat so that your hot tub stays at a comfortable 102°F?

40. You move to Canada and find that the thermostat in your home is in Celsius degrees. You normally like your house about 72°F. To what temperature should you set your new thermostat?
41. What is the freezing point of water on the Kelvin scale?
42. Nitrogen boils at 77 K. At what Celsius temperature does it boil?
43. What microscopic property of an ideal gas doubles when the absolute temperature is doubled?
44. What temperature change would be needed to double the average speed of the molecules in an ideal gas?
45. Air is a mixture of several gases, primarily nitrogen and oxygen. Is the average kinetic energy of the nitrogen molecules greater than, equal to, or less than the average kinetic energy of the oxygen molecules?
46. Consider a mixture of helium and neon gases. The atomic masses of helium and neon are 4 atomic mass units and 20 atomic mass units, respectively. Is the average speed of a helium atom greater than, equal to, or less than the average speed of a neon atom?
47. If you heat a gas in a container with a fixed volume, the pressure increases. Use the ideal gas model to explain this.
48. If the volume of an ideal gas is held constant, what happens to the pressure if the absolute temperature is cut in half?
49. What macroscopic property of an ideal gas doubles when the absolute temperature is doubled while the pressure remains constant?
50. If you put a sealed plastic bottle partially filled with hot tea in the refrigerator, the sides of the bottle will cave in as the tea cools. Why?
51. What happens to the temperature of an ideal gas if you reduce its volume by one-half while holding the pressure constant?
52. Why does the pressure inside the tires increase after a car has been driven?
53. Use the microscopic gas model to explain why the pressure of a gas rises as the volume is reduced while the temperature remains constant.
54. If you hold the temperature of an ideal gas constant, what happens to its volume when you triple its pressure?
55. The water in a canvas water bag placed in front of your car when driving across the desert stays cooler than the surrounding air. Explain this in terms of the average ki-

netic energies of the water molecules that leave and those that stay behind.

56. How does an alcohol rub cool your body?
57. Why might hikers get hypothermia during wet weather even when the temperature is above freezing?
58. The temperature of boiling water does not increase even if the heat is turned on high. Use the microscopic model to explain this.

EXERCISES

1. If 1 g of hydrogen combines completely with 8 g of oxygen to form water, how many grams of hydrogen does it take to combine completely with 24 g of oxygen?
2. In ammonia, 14 g of nitrogen combines completely with 3 g of hydrogen. How many grams of nitrogen does it take to combine completely with 18 g of hydrogen?
3. Given that 1 g of hydrogen combines completely with 8 g of oxygen to form water, how many grams of water can you make with 8 g of hydrogen and 16 g of oxygen?
4. Given that 12 g of carbon combines completely with 16 g of oxygen to form carbon monoxide, how many grams of carbon monoxide can be made from 48 g of carbon and 48 g of oxygen?
5. A ham sandwich consists of one slice of ham (10 g) and two slices of bread (25 g each). You have 1 kg of ham and 1 kg of bread. You make as many sandwiches as you can. How many sandwiches did you make? What is the mass of the sandwiches? Which ingredient is left over? What is the mass of the ingredient that is left over?
6. One mole of water molecules consists of 1 mole of oxygen (16 g) and 2 moles of hydrogen (1 g each). You combine 1 kg of oxygen with 1 kg of hydrogen to make water. How many moles of water did you make? What is the mass of the water? What is the mass of the element that is left over?
7. Given that the carbon atom has a mass of 12 amu, how many carbon atoms are there in a diamond with a mass of 1 g?
8. Given that the nitrogen molecule has a mass of 28 amu, how many nitrogen molecules are in 1 g of nitrogen?
9. One liter of water has a mass of 1 kg, and the mass of a water molecule is 18 amu. How many molecules of water are there in 1 L of water?
10. One liter of oxygen has a mass of 1.4 g, and the oxygen molecule has a mass of 32 amu. How many oxygen molecules are there in 1 L of oxygen?
▲ 11. One liter of nitrogen combines with 3 L of hydrogen to form 2 L of ammonia. If the molecules of nitrogen and hydrogen have two atoms each, how many atoms of hydrogen and nitrogen are there in one molecule of ammonia?
▲ 12. One liter of oxygen combines with 1 L of hydrogen to form 1 L of hydrogen peroxide. Given that the molecules of hydrogen and oxygen contain two atoms each, how many atoms are there in one molecule of hydrogen peroxide?
13. About how many atoms would it take to deposit a single layer on a surface with an area of 1 cm^2?
14. About how many atoms would you expect there to be in a cube of material 1 cm on each side?
15. You exert a force of 30 N on the head of a thumbtack. The head of the thumbtack has a radius of 5 mm. What is the pressure on your thumb?
▲ 16. The pressure in each of your car tires is 2.5×10^5 Pa. The mass of your car is 1600 kg. Assuming that each of your tires bears one-quarter of the total load, what is the contact area of each tire with the road?
17. What happens to the volume of 1 L of an ideal gas when the pressure is tripled while the temperature is held fixed?
18. An ideal gas at 27°C is contained in a piston that ensures that its pressure will always be constant. Raising the temperature of the gas causes it to expand. At what temperature will the gas take up twice its original volume?
▲ 19. A helium bottle with a pressure of 100 atm has a volume of 2 L. How many balloons can the bottle fill if each balloon has a volume of 1 L and a pressure of 1.25 atm?
▲ 20. When the temperature of an automobile tire is 20°C, the pressure in the tire reads 32 psi on a tire gauge. (The gauge measures the difference between the pressures in-

side and outside the tire.) What is the pressure when the tire heats up to 40°C while driving? You may assume that the volume of the tire remains the same and that atmospheric pressure is a steady 14 psi.

▲ 21. A volume of 150 cm³ of an ideal gas has an initial temperature of 20°C and an initial pressure of 1 atm. What is the final pressure if the volume is reduced to 100 cm³ and the temperature is raised to 40°C?

▲ 22. An ideal gas has the following initial conditions: $V_i = 500$ cm³, $P_i = 3$ atm, and $T_i = 100$°C. What is its final temperature if the pressure is reduced to 1 atm and the volume expands to 1000 cm³?

12 | States of Matter

Ice crystals.

Many materials exist as solids, liquids, or gases, each with its own characteristics. However, different materials share many characteristics; for instance, many materials form crystalline shapes in the solid form. What does the shape of a crystal tell us about the underlying structure of the material?

(See page 242 for the answer to this question.)

PhysicsNow™ Test your understanding of this chapter by logging into PhysicsNow at http://physics.brookscole.com/kf6e, selecting the chapter, and clicking on the "Take a Pre-Test" link.

ALL matter is composed of approximately 100 different elements. Yet the material world we experience—say, in a walk through the woods—holds a seemingly endless variety of forms. This variety arises from the particular combinations of elements and the structures they form, which can be divided into four basic forms, or states: solid, liquid, gas, and plasma.

Many materials can exist in the solid, liquid, and gaseous states if the forces holding the chemical elements together are strong enough that their melting and vaporization temperatures are lower than their decomposition temperatures. Hydrogen and oxygen in water, for example, are so tightly bonded that water exists in all three states. Sugar, on the other hand, decomposes into its constituent parts before it can turn into a gas.

If we continuously heat a solid, the average kinetic energy of its molecules rises and the temperature of the solid increases. Eventually, the intermolecular bonds break, and the molecules slide over one another (the process called melting) to form a liquid. The next change of state occurs when the substance turns into a gas. In the gaseous state, the molecules have enough kinetic energy to be essentially independent of each other. In a plasma, individual atoms are literally ripped apart into charged ions and electrons, and the subsequent electrical interactions drastically change the resulting substance's behavior.

Atoms

At the end of the previous chapter, we established the evidence for the existence of atoms. It would be natural to ask, "Why stop there?" Maybe atoms are not the end of our search for the fundamental building blocks of matter. In fact, they are not. Atoms have structure, and we will devote two chapters near the end of the book to further developing our understanding of this structure. For now it is useful to know a little about this structure so that we can understand the properties of the states of matter.

A useful model for the structure of an atom for our current purposes is the solar-system model developed early in the 20th century. In this model the atom is seen as consisting of a very tiny central nucleus that contains almost all the atom's mass. This nucleus has a positive electric charge that binds very light, negatively charged electrons to the atom in a way analogous to the Sun's gravitational attraction for the planets. The electrons' orbits define the size of the atom and give it its chemical properties.

The basic force that binds materials together is the electrical attraction between atomic and subatomic particles. As we will see in later chapters, the gravitational force is too weak and the nuclear forces are too short-ranged to have much effect in chemical reactions. We live in an electrical universe when it comes to the states of matter. How these materials form depends on these electric forces. And the form they take determines the properties of the materials.

✓ Extended presentation available in the *Problem Solving* supplement

Density ✓ MATH

One characteristic property of matter is its **density**. Unlike mass and volume, which vary from one object to another, density is an inherent property of the material. A ton of copper and a copper coin have drastically different masses and volumes but identical densities. If you were to find an unknown material and could be assured that it was pure, you could go a long way toward identifying it by measuring its density.

Density is defined as the amount of mass in a standard unit of volume and is expressed in units of kilograms per cubic meter (kg/m^3):

$$\text{density} = \frac{\text{mass}}{\text{volume}}$$

$$D = \frac{M}{V}$$

Density Extremes

Which weighs more: a pound of feathers or a pound of iron? The answer to this junior high school puzzle is of course that they weigh the same. The key difference between the two materials is *density*; although the two weigh the same, they have very different volumes. Density is a comparison of the masses of two substances with the same volumes. Obviously, 1 cubic meter of iron has much more mass than 1 cubic meter of feathers.

The densities of objects vary over a large range. The densities of common Earth materials pale in comparison to those of some astronomical objects. After a star runs out of fuel, its own gravitational attraction causes it to collapse. The collapse stops when the outward forces due to the pressure in the star balance the gravitational forces. The resultant stellar cores can have astronomically large densities. White dwarf stars are the death stage of most stars. They can have masses up to 1.4 times that of our Sun compressed to a size about that of Earth, with resulting densities a million times larger than the density of water. Neutron stars are cores left after a star explodes and can have densities a billion times larger than white dwarfs. A teaspoon of material from a neutron star would weigh a billion tons on Earth!

A very low-density solid, called *silica aerogel*, was created at the Lawrence Livermore National Laboratory in California. This solid is made from silicon dioxide and has a density that is only three times that of air. Because of this very low density, it is sometimes known as "solid smoke." Because silica aerogel is a solid, it holds its shape. In fact, it can support 1600 times its own weight!

Silica aerogel has an extremely low density but still can support 1600 times its own weight.

The density of interstellar space is very much smaller than that of air; there is about one atom per cubic centimeter, resulting in a density about 1 billion-trillionth that of air at 1 atmosphere of pressure, or about 1 trillion-trillionth (10^{-24}) that of water.

For example, an aluminum ingot is 3 meters long, 1 meter wide, and 0.3 meter thick. If it has a mass of 2430 kilograms, what is the density of aluminum? We calculate the volume first and then the density:

$$V = lwh = (3 \text{ m})(1 \text{ m})(0.3 \text{ m}) = 0.9 \text{ m}^3$$

$$D = \frac{M}{V} = \frac{2430 \text{ kg}}{0.9 \text{ m}^3} = 2700 \text{ kg/m}^3$$

Therefore, the density of aluminum is 2700 kilograms per cubic meter. Densities are also often expressed in grams per cubic centimeter. Thus, the density of aluminum is also 2.7 grams per cubic centimeter (g/cm^3). Table 12-1 gives the densities of a number of common materials.

Question Which has the greater density, 1 kilogram of iron or 2 kilograms of iron?

Answer They have the same density; the density of a material does not depend on the amount of material.

The densities of materials range from the small for a gas under normal conditions to the large for the element osmium. One cubic meter of osmium has a mass of 22,480 kilograms (a weight of nearly 50,000 pounds), about 22 times as large as the same volume of water. It is interesting to note that the osmium atom is less massive than a gold atom. Therefore, the higher density of osmium indicates that the osmium atoms must be packed closer together.

The materials that we commonly encounter have densities around the density of water, 1 gram per cubic centimeter. A cubic centimeter is about the vol-

Table 12-1

Densities of Some Common Materials

Material	Density (g/cm^3)
Air*	0.0013
Ice	0.92
Water	1.00
Magnesium	1.75
Aluminum	2.70
Iron	7.86
Copper	8.93
Silver	10.5
Lead	11.3
Mercury	13.6
Uranium	18.7
Gold	19.3
Osmium	22.5

*At 0°C and 1 atm.

ume of a sugar cube. The densities of surface materials on Earth average approximately 2.5 grams per cubic centimeter. The density at Earth's core is about 9 grams per cubic centimeter, making Earth's average density about 5.5 grams per cubic centimeter.

Question If a hollow sphere and a solid sphere are both made of the same amount of iron, which sphere has the greater average density?

Answer The solid sphere has the greater average density because it occupies the smallest volume for a given mass of iron.

PHYSICS | **ON YOUR OWN**

> Determine your density. You can get a pretty good measure of your volume by submerging yourself in a bathtub. Multiply the area of the tub by the difference in the water levels with you in and out of the tub. Your mass can be determined from your weight. (On Earth, 1 kilogram has a weight of 2.2 pounds.)

✓ MATH

WORKING IT OUT | Density

Suppose you find a chunk of material that you cannot identify. You find that the chunk has a mass of 87.5 g and a volume of 50 cm³. What is the material, and what is the mass of a 6-cm³ piece of this material?

We could easily find the mass of 6 cm³, if only we knew the mass of 1 cm³. This is just the density. We can find the density from the measurements made on the original chunk:

$$D = \frac{M}{V} = \frac{87.5 \text{ g}}{50 \text{ cm}^3} = 1.75 \text{ g/cm}^3$$

This density is the same as that of magnesium. Therefore, the material could be magnesium, but we would need to look at other characteristics to be sure.

The 6-cm³ piece has a mass 6 times larger than the mass of 1 cm³:

$$M = DV = (1.75 \text{ g/cm}^3)(6 \text{ cm}^3) = 10.5 \text{ g}$$

Solids

Solids have the greatest variety of properties of the four states of matter. The character of a solid substance is determined by its elemental constituents and its particular structure. This underlying structure depends on the way it was formed. For example, slow cooling often leads to solidification with the atoms in an ordered state known as a **crystal**.

PHYSICS | **ON YOUR OWN**

> Salol, a compound used by pharmacists in medicines, is interesting because its melting point is only a little above room temperature. Melt a pinch of salol in a small glass bowl floating in hot water. As soon as it melts, remove the glass bowl from the hot water and watch the liquid as it solidifies.

Crystals grow in a variety of shapes. Their common property is the orderliness of their atomic arrangements. The orderliness consists of a basic arrangement of atoms that repeats throughout the crystal, analogous to the repeating geometric patterns in some wallpapers.

The microscopic order of the atoms is not always obvious in macroscopic samples. For one thing there are very few perfect crystals; most samples are aggregates of small crystals. However, macroscopic evidence of this underlying structure does exist. A common example in northern climates is a snowflake (Figure 12-1). Its sixfold symmetry is evidence of the structure of ice. Another example is mica (Figure 12-2), a mineral you might find on a hike in the woods. Shining flakes of mica can be seen in many rocks. Larger pieces can be easily separated into thin sheets. The thinness of the sheets seems (at least on the macroscopic scale) to be limitless. It is easy to convince yourself that the atoms in mica are arranged in two-dimensional sheets with relatively strong bonds between atoms within the sheet and much weaker bonds between the sheets.

In contrast to mica, ordinary table salt exhibits a three-dimensional structure of sodium and chlorine atoms. If you dissolve salt in water and let the water slowly evaporate, the salt crystals that form have very obvious cubic structures. If you try to cut a small piece of salt with a razor blade, you find that it doesn't separate into sheets like mica but fractures along planes parallel to its faces. (Salt from a saltshaker displays this same structure, but the grains are usually much smaller. A simple magnifying glass allows you to see the cubic structure.) Precious stones also have planes in their crystalline structure. A gem cutter studies the raw gemstones very carefully before making the cleavages that produce a fine piece of jewelry.

Figure 12-1 The sixfold symmetry exhibited by snowflakes is evidence that ice crystals have hexagonal shapes.

PHYSICS | **ON YOUR OWN**

Examine salt and sugar crystals with a magnifying glass. If you sprinkle salt or sugar crystals on a black piece of paper and carefully separate them, you can see distinct structures that provide clues to the underlying atomic ordering.

Some substances have more than one crystalline structure. A common example is pure carbon. Carbon can form diamond or graphite crystals (Figure 12-3). Diamond is a very hard substance that is treasured for its optical brilliance. Diamond has a three-dimensional structure. Graphite, on the other hand, has a two-dimensional structure like mica, creating sheets of material that are relatively free to move over each other. Because of its slippery nature, graphite is used as a lubricant and as the "lead" in pencils.

Figure 12-2 Samples of mica exhibit a two-dimensional crystalline structure as evidenced by the fact that one can peel thin sheets from the larger crystal.

Liquids

When a solid melts, interatomic bonds break, allowing the atoms or molecules to slide over each other, producing a liquid. **Liquids** fill the shape of the container that holds them, much like the random stacking of a bunch of marbles.

The temperature at which a solid melts varies from material to material simply because the bonding forces are different. Hydrogen is so loosely bound that it becomes a liquid at 14 K. Oxygen and nitrogen—the constituents of the air we breathe—melt at 55 K and 63 K, respectively. The fact that ice doesn't melt until 273 K (0°C) tells us that the bonds between the molecules are relatively strong.

Water is an unusual liquid. Although water is abundant, it is one of only a few liquids that occur at ordinary temperatures on Earth. The bonding between the water molecules is relatively strong, and it requires a high temperature to separate them into the gaseous state.

The intermolecular forces in a liquid create a special "skin" on the surface of

Figure 12-3 Synthetic diamonds and finely divided graphite are two different crystalline forms of carbon.

Figure 12-4 A glass filled with milk beyond the brim is evidence of surface tension.

Figure 12-5 A steel needle floats on water because of the surface tension of the water.

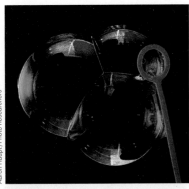

Figure 12-6 Surface tension minimizes the surface area of the soap film forming these bubbles.

the liquid. This can be seen in Figure 12-4, in which a glass has been filled with milk beyond its brim. What is keeping the extra liquid from flowing over the edge?

Imagine two molecules, one on the surface of a liquid and one deeper into the liquid. The molecule beneath the surface experiences attractive forces in all directions because of its neighbors. The molecule on the surface only feels forces from below and to the sides. This imbalance tends to pull the surface molecules back into the liquid.

PHYSICS | ON YOUR OWN

A dry needle with a density much greater than that of water can actually float on the surface of the water, as shown in Figure 12-5. Use tweezers to slowly lower the needle onto the surface.

Surface tension also tries to pull liquids into shapes with the smallest possible surface areas. The shapes of soap bubbles are determined by the surface tension trying to minimize the surface area of the film (Figure 12-6). If there are no external forces, the liquid forms into spherical drops. In fact, letting liquids cool in space has been proposed as a way of making nearly perfect spheres. In the free-fall environment of an orbiting space shuttle, liquid drops are nearly spherical.

Surface tensions vary among liquids. Water, as you might expect, has a relatively high surface tension. If we add soap or oil to the water, its surface tension is reduced, meaning that the water molecules are not as attracted to each other. It is probably reasonable to infer that the new molecules in the solution are somehow shielding the water molecules from each other.

Gases

When the molecules separate totally, a liquid turns into a **gas**. (See Chapter 11 for a discussion of an ideal gas.) The gas occupies a volume about 1000 times as large as that of the liquid. In the gaseous state, the molecules have enough kinetic energy to be essentially independent of each other. A gas fills the container holding it, taking its shape and volume. Because gases are mostly empty space, they are compressible and can be readily mixed with each other.

Gases and liquids have some common properties because they are both "fluids." All fluids are able to flow, some more easily than others. The **viscosity** of a fluid is a measure of the internal friction within the fluid. You can get a qualitative feeling for the viscosity of a fluid by pouring it. Those fluids that pour easily, such as water and gasoline, have low viscosities. Those that pour very slowly, such as molasses, honey, and egg whites, have high viscosities. Glass is a fluid with an extremely high viscosity. In the winter, drivers put lower-viscosity oils in their cars so that the oils will flow better on cold mornings.

The viscosity of a fluid determines its resistance to objects moving through it. A parachutist's safe descent is due to the viscosity of air. Air and water have drastically different viscosities. Imagine running a 100-meter dash in water 1 meter deep!

Question How might you explain the observation that the viscosities of fluids decrease as they are heated?

Answer The increased kinetic energy of the molecules means that the molecules are more independent of each other.

Plasmas

At around 4500°C, all solids have melted. At 6000°C, all liquids have been turned into gases. And at somewhere above 100,000°C, most matter is ionized into the **plasma** state. In the transition between a gas and a plasma, the atoms themselves break apart into electrically charged particles.

Although more rare on Earth than the solid, liquid, and gaseous states, the fourth state of matter, plasma, is actually the most common state of matter in the Universe (more than 99%). Examples of naturally occurring plasmas on Earth include fluorescent lights and neon-type signs. Fluorescent lights consist of a plasma created by a high voltage that strips mercury vapor of some of its electrons. "Neon" signs employ the same mechanism but use a variety of gases to create the different colors.

Perhaps the most beautiful naturally occurring plasma effect is the aurora borealis, or northern lights. Charged particles emitted by the Sun and other stars are trapped in Earth's upper atmosphere to form a plasma known as the Van Allen radiation belts. These plasma particles can interact with atoms of nitrogen and oxygen over both magnetic poles, causing them to emit light as discussed in Chapter 23.

Plasmas are important in nuclear power as well as in the interiors of stars. An important potential energy source for the future is the "burning" of a plasma of hydrogen ions at very high temperatures to create nuclear energy. We will discuss nuclear energy more completely in Chapter 26.

Honey is a very viscous fluid.

Pressure

✓ MATH

A macroscopic property of a fluid—either a gas or a liquid—is its change in pressure with depth. As we saw in Chapter 11, pressure is the force per unit area exerted on a surface, measured in units of newtons per square meter (N/m^2), a unit known as a pascal (Pa).

When a gas or liquid is under the influence of gravity, the weight of the material above a certain point exerts a force downward, creating the pressure at that point. Therefore, the pressure in a fluid varies with depth. You have probably felt this while swimming. As you go deeper, the pressure on your eardrums increases. If you swim horizontally at this depth, you notice that the pressure doesn't change. In fact, there is no change if you rotate your head; the pressure at a given depth in a fluid is the same in all directions.

The aurora borealis results from the interaction of charged particles with air molecules.

Consider the box of fluid shown in Figure 12-7. Because the fluid in the box does not move, the net force on the fluid must be zero. Therefore, the fluid below the box must be exerting an upward force on the bottom of the box that is equal to the weight of the fluid in the box plus the force of the atmosphere on the top of the box. The pressure at the bottom of the box is just this force per unit area.

Our atmosphere is held in a rather strange container, Earth's two-dimensional surface. Gravity holds the atmosphere down so that it doesn't escape. There is no definite top to our atmosphere; it just gets thinner and thinner the higher you go above Earth's surface.

The air pressure at Earth's surface is due to the weight of the column of air above the surface. At sea level the average atmospheric pressure is about 101 kilopascals. This means that a column of air that is 1 square meter in cross section and reaches to the top of the atmosphere weighs 101,000 newtons and has a mass of 10 metric tons. A similar column of air 1 square inch in cross section weighs 14.7 pounds; therefore, atmospheric pressure is also 14.7 pounds per square inch.

The pressure on a scuba diver increases with depth.

We can use these ideas to describe what happens to atmospheric pressure as we go higher and higher. You might think that the pressure drops to one-half the surface value halfway to the "top" of the atmosphere. However, this is not true, because the air near Earth's surface is much denser than that near the top of the atmosphere. This means that there is much less air in the top half compared with the bottom half. Because the pressure at a given altitude depends on the weight of the air above that altitude, the pressure changes more quickly near the surface. In fact, the pressure drops to half at about 5500 meters (18,000 feet) and then drops by half again in the next 5500 meters. This means that commercial airplanes flying at a typical altitude of 36,000 feet experience pressures that are only one-fourth those at the surface.

Like fish living on the ocean floor, we land-lovers are generally unaware of the pressure due to the ocean of air above us. Although the atmospheric pressure at sea level might not seem like much, consider the total force on the surface of your body. A typical human body has approximately 2 square meters (3000 square inches) of surface area. This means that the total force on the body is about 200,000 newtons (20 tons!).

Question Why doesn't the very large force on the surface of your body crush you?

Answer You aren't crushed because the pressure inside your body is the same as the pressure outside. Therefore, the inward force is balanced by the outward force.

An ingenious experiment conducted by a contemporary of Isaac Newton demonstrated the large forces that can be produced by atmospheric pressure. German scientist Otto von Guericke joined two half spheres (Figure 12-8) with just a simple gasket (no clamps or bolts). He then pumped the air from the sphere, creating a partial vacuum. Two teams of eight horses could not pull the hemispheres apart!

In weather reports, atmospheric pressure is often given in units of millimeters or inches of mercury. A typical pressure is 760 millimeters (30 inches) of mercury. Because pressure is a force per unit area, reporting pressure in units of length must seem strange. This scale comes from the historical method of measuring pressure. Early pressure gauges were similar to the simple mercury barometer shown in Figure 12-9. A sealed glass tube is filled with mercury and inverted into a bowl of mercury. After inversion the column of mercury does not pour out into the bowl but maintains a definite height above the pool of mercury. Because the mercury is not flowing, we know that the force due to atmospheric pressure at the bottom of the column is equal to the weight of the mercury column. This means that the atmospheric pressure is the same as the pressure at the bottom of a column of mercury 760 millimeters tall if there is a vacuum above the mercury. Therefore, atmospheric pressure can be characterized by the height of the column of mercury it will support.

Atmospheric pressure also allows you to drink through a straw. As you suck on the straw, you reduce the pressure above the liquid in the straw, allowing the pressure below to push the liquid up. In fact, if you could suck hard enough to produce a perfect vacuum above water, you could use a straw 10 meters (almost 34 feet) long! So although we often talk of sucking on a soda straw and pulling the soda up, in reality we are removing the air pressure on the top of the soda column in the straw, and the atmospheric pressure is pushing the soda up.

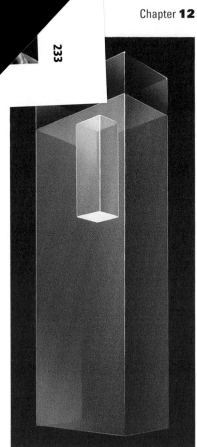

Figure 12-7 The force on the bottom of the box of fluid is equal to the weight of the fluid in the box plus the force of the atmosphere on the top of the box.

Figure 12-9 In a mercury barometer, the atmospheric pressure is balanced by the pressure due to the weight of the mercury column.

Figure 12-8 Two teams of eight horses could not separate von Guericke's evacuated half spheres.

Solid Liquids and Liquid Solids

Many substances exist between the ordinary boundaries of solids and liquids. When materials such as glass or wax cool, the molecules are frozen in space without arranging themselves into an orderly crystalline structure. These solids are amorphous, meaning that they retain some of the properties of liquids. A common example of an amorphous solid is the clear lollipop made by rapidly cooling liquid sugar. The average intermolecular forces in an amorphous material are weaker than those in a crystalline structure.

Despite the solid rigidity of an amorphous material, this form is more like a liquid than a solid because of its lack of order. In addition, the melting points of amorphous materials are not clearly defined. An amorphous material simply gets softer and softer, passing into the fluid state. Another characteristic of these solid liquids is that they actually do flow like a liquid, although many flow on time scales that make it difficult or impossible to detect. While it is widely believed that old church windows in Europe are thicker at the bottom than at the top due to centuries of flow, they were most likely assembled with the thicker edges at the bottom.

Other substances are liquids that retain some degree of orderliness, characteristic of solids. Liquid crystals can be poured like regular liquids. They lack positional order, but they possess an orientational order. Small electric voltages can align the rodlike molecules along a particular direction.

Liquid crystals have some interesting applications because polarized light behaves differently depending on whether it is traveling parallel or perpendicular to the alignment direction. For example, the orientation can be manipulated electrically to produce the numbers in a digital watch or an electronic calculator. You can verify that the light emerging from a liquid crystal display is polarized by looking at the display through polarized sunglasses. Changing the orientation of the sunglasses will vary the intensity of the image.

The display on this LCD TV is an application of liquid crystal technology.

Question How high a straw could you use to suck soda?

Answer Because soda is mostly water, we assume that it has the same density as water. Therefore, the straw could be 10 meters high—but only if you have very strong lungs. A typical height is more like 5 meters.

Underwater explorers must use vessels such as this bathysphere at the great depths of the ocean floor.

As you dive deeper in water, the pressure increases for the same reasons as in air. Because atmospheric pressure can support a column of water 10 meters high, we have a way of equating the two pressures. The pressure in water must increase by the equivalent of 1 atmosphere (atm) for each 10 meters of depth. Therefore, at a depth of 10 meters, you would experience a pressure of 2 atmospheres, 1 from the air and 1 from the water. The pressures are so large at great depths that very strong vessels must be used to prevent the occupants from being crushed.

Question   What is the pressure on a scuba diver at a depth of 30 meters (100 feet)?

Answer   The pressure would be (30 meters)/(10 meters per atmosphere) = 3 atmospheres due to the water plus 1 atmosphere due to the air above the water, for a total of 4 atmospheres.

Flawed Reasoning

Jeff designs a new scuba setup that is so profoundly simple he is surprised that no one has thought of this before. He has attached one end of a long garden hose to a large block of Styrofoam to keep the hose above the water level. He will breathe through the other end of the hose as he explores the depths. **What is wrong with Jeff's simple design?**

Answer   If Jeff dives 10 meters below the surface, the water will push inward on him with 2 atmospheres of pressure. Therefore, the air in his lungs will be at a pressure of 2 atmospheres. Because the air in the hose will be at a pressure of 1 atmosphere, he will expel air from his lungs and not be able to breathe!

Sink and Float

Floating is so commonplace to anyone who has gone swimming that it might not have occurred to ask, "Why do things sink or float?" "Why does a golf ball sink and an ocean liner float?" "And how is a hot-air balloon similar to an ocean liner?"

Anything that floats must have an upward force counteracting the force of gravity, because we know from Newton's first law of motion (Chapter 3) that an object at rest has no unbalanced forces acting on it. To understand why things float therefore requires that we find the upward **buoyant force** opposing the gravitational force.

The buoyant force exists because the pressure in the fluid varies with depth. To understand this, consider the cubic meter of fluid in Figure 12-10. The pres-

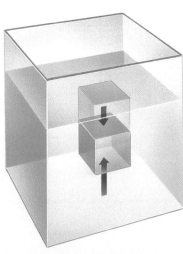

Figure 12-10 The buoyancy force on the cube is due to the larger pressure on the lower surface.

sure on the bottom surface is greater than on the top surface, resulting in a net upward force. The downward force on the top surface is due to the weight of the fluid above the cube. The upward force on the bottom surface is equal to the weight of the column of fluid above the bottom of the cube. The difference between these two forces is just the weight of the fluid in the cube. Therefore, the net upward force must be equal to the weight of the fluid in the cube.

These pressures do not change if a cube of some other material replaces the cube of fluid. Therefore, the net upward force is still equal to the weight of the fluid that was replaced. This result is known as **Archimedes' principle**, named for the Greek scientist who discovered it.

> The buoyant force is equal to the weight of the displaced fluid.

◂ Archimedes' principle

When you place an object in a fluid, it displaces more and more fluid as it sinks lower into the liquid, and the buoyant force therefore increases. If the buoyant force equals the object's weight before it is fully submerged, the object floats. This occurs whenever the density of the object is less than that of the fluid.

We can change a sinker into a floater by increasing the amount of fluid it displaces. A solid chunk of steel equal in weight to an ocean liner clearly sinks in water. We can make the steel float by reshaping it into a hollow box. We don't throw away any material; we only change its volume. If we make the volume big enough, it will displace enough water to float.

Ice floats because of a buoyant force. When water freezes, the atoms arrange themselves in a way that actually takes up more volume. As a result, ice has a lower density than liquid water and floats on the surface. This is fortunate; otherwise, ice would sink to the bottom of lakes and rivers, freezing the fish and plants.

PHYSICS | ON YOUR OWN

The next time you go swimming or use a hot tub, estimate your density by observing how much of your body floats above the surface of the water.

The buoyant force is present even when the object sinks! For example, any object weighs less in water than in air. You can verify this by hanging a small object by a rubber band. As you lower it into a glass of water, the rubber band is stretched less because the buoyant force helps support the object.

Question A piece of iron with a mass of 790 grams displaces 100 grams of water when it sinks. What does the iron weigh in air and under water?

Answer In air the weight is given by mg = (0.79 kilogram) × (10 [meters per second] per second) = 7.9 newtons. In water this is reduced by the weight of the displaced water. Therefore, we have 7.9 newtons − 1 newton = 6.9 newtons.

Changing the shape of the steel makes it a floater. A solid piece of steel with the mass of the ocean liner sinks.

Flawed Reasoning

Two wooden blocks with the same size and shape are floating in a bucket of water. Block A floats low in the water, and block B floats high, as shown in the figure:

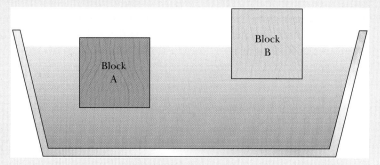

Three students have just come from an interesting lecture on Archimedes' principle and are discussing the buoyant forces on the blocks.

Erin: "Block B is floating higher in the water. It must have the greater buoyant force acting on it."

Diego: "You are forgetting Newton's first law. Neither block is moving, so the buoyant force must balance the gravitational force in both cases. The buoyant forces must be equal to each other."

Ashley: "Archimedes taught us that the buoyant force is *always* equal to the weight of the fluid displaced. Block A is displacing a lot more water than block B, so block A has the larger buoyant force."

Do you agree with any of these students?

Answer Ashley is correct. Archimedes' principle *always* applies, regardless of whether an object sinks or floats. Block A displaces the most water, so it experiences the larger buoyant force. Diego starts out with correct ideas but then draws a faulty conclusion. The buoyant force on either block must equal the gravitational force on that block (by Newton's first law), so block A must be heavier. Because both blocks have the same volume, block A must be made of a denser wood. Perhaps block A is made of oak, and block B is made of pine.

Bernoulli's Effect

The pressure in a stationary fluid changes with depth but is the same if you move horizontally. If the fluid is moving, however, the pressure can also change in the horizontal direction. Suppose we have a pipe that has a narrow section like the one shown in Figure 12-11. If we put pressure gauges along the pipe, the surprising finding is that the pressure is lower in the narrow region of the pipe. If

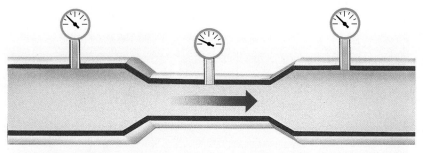

Figure 12-11 The pressure is smaller in the narrow region of the pipe where the velocity of the fluid is greater.

How Fatty Are You?

Exercising does not automatically reduce your weight. One outcome of exercising is the conversion of fatty tissue into muscle without changing your weight. Because healthy people have more muscle, it is important to be able to determine the percentage of body fat. That's a question that just stepping on the bathroom scale won't tell you. However, a 2000-year-old technique developed by Archimedes does work.

Around 250 BC, Archimedes was chief scientist for King Hieron of Syracuse (now modern Sicily). As the story goes, the king was concerned that his crown was not made of pure gold but had some silver hidden under its surface. Not wanting to destroy his crown to find out if he had been cheated, he challenged his scientist to find an alternative procedure. Everybody knows the legend of Archimedes leaping from his bathtub and shouting, "Eureka, I have found it!"

The key to Archimedes' solution is determining the average density of the crown or, in our case, your body. There is no problem getting your weight. A simple bathroom scale will do. The tricky part is determining your volume. You, like the king's crown, are oddly shaped (sorry!), not matching any of the geometric volumes you studied in school.

Archimedes discovered that an object immersed in water feels an upward buoyant force. If you were to stand on a scale while totally submerged, you would weigh less because the buoyant force supports part of your weight. This buoyant force is equal to the weight of the water your body displaces. From your weights in air and under water, your volume can be calculated.

Human performance scientists consider the body to be made of fat and "muscle." (Everything but fat—skin, bone, and organs—is grouped as muscle.) From the study of cadavers, the density of human fat is found to be about 90% the density of water, whereas the density of "muscle" is about 110% the density of water. The more fat you have, the lower your average density will be. The percentage of fat for healthy adults should be between 15% and 20% for men and between 22% and 28% for women. Champion distance runners and bicyclists have about 5–8% fat.

the fluid is not compressible, the fluid must be moving faster in the narrow region; that is, the same amount of fluid must pass by every point in the pipe, or it would pile up. Therefore, the fluid must flow faster in the narrow regions. This might lead one to conclude incorrectly that the pressure would be higher in this region. Swiss mathematician and physicist Daniel Bernoulli stated the correct result as a principle.

> The pressure in a fluid decreases as its velocity increases.

◄ Bernoulli's principle

We can understand **Bernoulli's principle** by "watching" a small cube of fluid flow through the pipe (Figure 12-12). The cube must gain kinetic energy as it speeds up entering the narrow region. Because there is no change in its gravitational potential energy, there must be a net force on the cube that does work on it. Therefore, the force on the front of the cube must be less than on the back. That is, the pressure must decrease as the cube moves into the narrow region. As

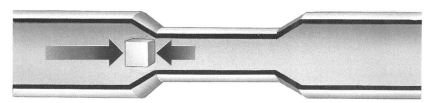

Figure 12-12 The cube of fluid entering the narrow region of the pipe must experience a net force to the right.

The Curve Ball

When a ball moves through air, strange things can happen. Perhaps the most common examples are the curve in baseball, the slice in golf, and the topspin serve in tennis. Isaac Newton wrote about the unusual behavior of spinning tennis balls, and baseball players and scientists have debated the behavior of the curve ball since the first baseball was thrown more than a hundred years ago. The balls naturally follow projectile paths due to gravity (Chapter 5), but it's the extra motion that is the bane of all batters.

Early on, the debate centered on whether the curve ball even existed. Scientists, believing that the only forces on the ball were gravity and air resistance (drag), argued that the curve ball must be just an optical illusion. "Not true!" retorted the baseball players. "It's like the ball rolled off a table just in front of the plate."

When there's a debate about the material world, the best procedure is to devise an experiment—that is, to ask the question of the material world itself. In the early 1940s *Life* magazine commissioned strobe photos of a curve ball and concluded that the scientists were right: The curve ball is an optical illusion. Not to be outdone, *Look* magazine commissioned its own photos and concluded that the scientists (and *Life* magazine!) were wrong.

More recently, three scientists reexamined the question. They found a dark warehouse, a bank of strobe lights, and—most important—a professional pitcher. After careful analysis the verdict was clear: The ball does indeed curve away from the projectile path, and the deviation is created by the ball's spin. If the ball travels at 75 mph, it takes the ball about $\frac{1}{2}$ second to travel the 60 feet to the batter. During this time, the ball rotates about 18 times and deviates from the projectile path by about 1 foot.

The direction of the deviation depends on the orientation of the spin. The deviation is always perpendicular to the axis of the spin; therefore, spin around a vertical axis moves the ball left or right. Because this would only change the point of contact with a horizontal bat, it is not very effective. Spin around a horizontal axis causes the ball to move up or down. Backspin (the bottom of the ball moving toward the batter) causes the ball to stay above the projectile path, which helps the batter. The best situation (for the pitcher!) is

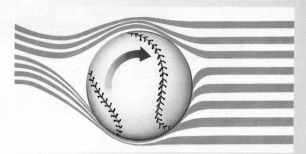

The spinning ball causes the airflow to deflect upward, imparting a downward force on the ball.

topspin. This increases the drop as the ball approaches the batter.

The question of when the ball deviates was also answered by this experiment. Batters have claimed for years that the ball travels along its normal path and then breaks at the last moment. Scientists claim that the forces—both gravity and the one caused by the spin—are constant; thus, the path is a continuous curve. Our study of projectile motion has shown that a ball falls farther during each succeeding second. This is compounded by the extra drop due to the spin. Thus, the vertical speed is much faster near the plate. But the drop is a continuous one; it does not abruptly change.

We are now left with the question of what causes the downward force. The baseball's cotton stitches—216 on a regulation ball—grab air, creating a layer of air that is carried around the spinning ball. (An insect sitting on the spinning ball would feel no wind—just like dust on a fan's blades is undisturbed by the fan's rotation.) To account for the force, we must look at the turbulence, or wake, behind the ball. The airflow over the top of the ball has a larger speed relative to the surrounding air and breaks up sooner than the airflow under the bottom as shown in the figure. This causes the wake behind the ball to be shifted upward. According to Newton's third law, the momentum imparted to the wake in the upward direction causes an equal momentum to be imparted downward on the ball and the curve ball drops.

Let the games begin!

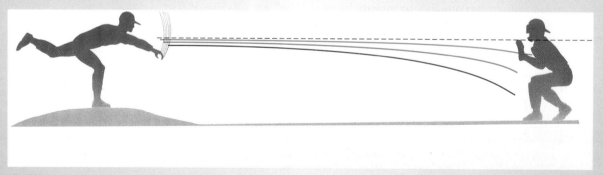

The blue curve represents the ball's path due to its spin without any gravity. The green curve is the ball's path due to gravity without spin. The red curve shows the combined effect of spin and gravity.

the cube of fluid exits the narrow region, it slows down. Therefore, the pressure must increase again.

There are many examples of Bernoulli's effect in our everyday activities. Smoke goes up a chimney partly because hot air rises but also because of the Bernoulli effect. The wind blowing across the top of the chimney reduces the pressure and allows the smoke to be pushed up. This effect is also responsible for houses losing roofs during tornadoes (or attacks by big bad wolves). When a tornado reduces the pressure on the top of the roof, the air *inside* the house lifts the roof off.

A fluid moving past an object is equivalent to the object moving in the fluid, so the Bernoulli effect should occur in these situations. A tarpaulin over the back of a truck lifts up as the truck travels down the road due to the reduced pressure on the outside surface of the tarpaulin produced by the truck moving through the air. This same effect causes your car to be sucked toward a truck as it passes you going in the opposite direction. The upper surfaces of airplane wings are curved so that the air has to travel a farther distance to get to the back edge of the wing. Therefore, the air on top of the wing must travel faster than that on the underside and the pressure on the top of the wing is less, providing lift to keep the airplane in the air.

A tornado caused the difference in the air pressures inside and outside this house that tore off the roof.

PHYSICS | **ON YOUR OWN**

Hold a piece of typing paper in front of your face, as shown in Figure 12-13. What happens as you blow over the top of the paper? Why?

PHYSICS | **ON YOUR OWN**

Some vacuum cleaners can be reversed so that they blow air. If you have access to such a vacuum cleaner, place a table-tennis ball in the stream of air when it is aimed vertically upward. Why does the ball stay in the stream? Can you tilt the airstream without dropping the ball? This project can also be done by blowing through a straw.

Figure 12-13 What happens when you blow over the top of the paper?

Summary

Density is an inherent property of a substance and is defined as the amount of mass in one unit of volume.

Elements combine into substances that can exist in four states of matter: solids, liquids, gases, and plasmas. The transitions between states occur when energy is supplied to or taken from substances. When a solid is heated above its melting point, interatomic bonds break to form a liquid in which atoms and molecules are free to move about. Upon further heating, the molecules totally separate to form a gas. In the plasma state, the atoms have been torn apart, producing charged ions and electrons. Although rare on Earth, plasma is the most common state in the Universe.

The electric forces between atoms bind all materials together. If the atoms are ordered, a crystalline structure results. Liquids take the shape of their container, and most lack an ordered arrangement of their molecules. The intermolecular forces in a liquid create a surface tension that holds the molecules to the liquid. A gas fills the container holding it, assuming its shape and volume. All gases are compressible and can be readily mixed with each other. The viscosity of a fluid determines how easily it pours and what resistance it offers to objects moving through it.

The pressure in a liquid or gas varies with depth because of the weight of the

Physics Now™ Assess your understanding of this chapter's topics with sample tests and other resources found by logging into PhysicsNow at **http://physics.brookscole.com/kf6e**.

fluid above that point. At sea level the average atmospheric pressure is about 101 kilopascals (14.7 pounds per square inch).

An object in a fluid experiences a buoyant force equal to the weight of the fluid displaced; therefore, all objects weigh less in water than in air. The buoyant force exists because the pressure in a fluid varies with depth. The pressure on the bottom surface of an object is greater than on its top surface. Objects less dense than the fluid float. The pressure in a moving fluid decreases with increasing speed.

Chapter 12 Revisited

The crystal's macroscopic shape results from a growth process that adds to the overall structure of the crystal, atom by atom. Study of these crystalline shapes gives scientists clues about the way atoms combine.

KEY TERMS

Archimedes' principle: The buoyant force is equal to the weight of the displaced fluid.

Bernoulli's principle: The pressure in a fluid increases as its velocity decreases.

buoyant force: The upward force exerted by a fluid on a submerged or floating object. (See Archimedes' principle.)

crystal: A material in which the atoms are arranged in a definite geometric pattern.

density: A property of a material equal to the mass of the material divided by its volume. Measured in kilograms per cubic meter.

gas: Matter with no definite shape or volume.

liquid: Matter with a definite volume that takes the shape of its container.

plasma: A highly ionized gas with equal numbers of positive and negative charges.

solid: Matter with a definite size and shape.

viscosity: A measure of the internal friction within a fluid.

CONCEPTUAL QUESTIONS

1. What are the four states of matter?
2. Is the average kinetic energy of the molecules in a liquid greater or smaller than in a solid of the same material? Why?
3. Does the aluminum in a soda can or in an automobile engine have the larger density? Why?
4. Which has a greater density, a tiny industrial diamond used in grinding powders or a 3-carat diamond in a wedding ring? Explain.
5. Aluminum and magnesium have densities of 2.70 and 1.75 grams per cubic centimeter, respectively. If you have equal masses of each, which one will occupy the larger volume? Explain.
6. Gold and silver have densities of 19.3 and 10.5 grams per cubic centimeter, respectively. If you have equal volumes of each, which one will have the larger mass? Explain.
7. Although the uranium atom is more massive than the gold atom, gold has the larger density. What does this tell you about the two solids?
8. Why do soda bottles break when the soda freezes?
9. Are crystal structures of ice and table salt the same? How do you know?

10. How does the crystal structure of mica differ from that of table salt?
11. How does the structure of diamond differ from that of graphite?
12. What does the observation that mica can be separated into thin sheets tell you about the crystal structure of mica?
13. What evidence do you have to indicate that the molecular bonding in solid oxygen is less than in solid nitrogen?
14. Is the bonding between molecules in liquid nitrogen stronger or weaker than that in liquid oxygen? Explain.
15. What shape would you expect a drop of water to take if it were suspended in the air in a space shuttle?
16. Why does water bead up when it is spilled on a waxed floor?

17. If you fill a glass with water level with the top of the glass, you can carefully drop several pennies into the glass without spilling any water. How do you explain this?

18. Why does soapy water bead up less than plain water on a countertop?
19. When you half-fill a glass with water, the water creeps up where it meets the glass. What can you conclude about the relative strengths of the intermolecular forces between the water molecules and the adhesive forces between the water molecules and the glass?
20. When you half-fill a glass with mercury, the mercury curls down where it meets the glass. What can you conclude about the relative strengths of the interatomic forces between the mercury atoms and the adhesive forces between the mercury atoms and the glass? (*Caution*: mercury is toxic and should *not* be handled.)

Questions 19 and 20

21. How does a gas differ from a plasma?
22. What state of matter forms the Van Allen belts?
23. Use the concept of pressure to explain why it is more comfortable to walk in bare feet across a paved driveway than across a gravel driveway.
24. What must happen to the area of a tire touching the ground if you reduce the pressure in the tire? Explain.
25. You place a small amount of water in a 1-gallon can and bring it to a rapid boil. You take the can off the stove and screw the cap on tightly (the order is important here!). As the steam inside cools, it condenses into water, causing the can to collapse. Why?
26. At sea level each square inch of surface experiences a force of 14.7 pounds due to air pressure. You are carrying a cookie sheet loaded with chocolate chip cookies. The surface area of the cookie sheet is 250 square inches, which means that the downward force exerted by the air column above the cookie sheet is 3675 pounds! Why doesn't the cookie sheet feel this heavy?
27. You repeat von Guericke's experiment (see Figure 12-8) using somewhat smaller half spheres and find that two teams of eight horses are just strong enough to pull the half spheres apart. You then transport the apparatus to Denver, which is at an elevation of 1 mile. Will you need more or fewer horses to pull the half spheres apart? Why?
28. A classmate explains that if your bathroom scale reads 150 pounds when you stand on it at sea level, it will read only 75 pounds on the top of an 18,000-foot mountain where atmospheric pressure is reduced by one-half. What is wrong with your classmate's reasoning? Would you expect the scale reading to be reduced at all as a result of the decrease in atmospheric pressure?
29. On a weather map, you see areas of low pressure marked with an L and areas of high pressure marked with an H. By convention, the pressures reported are always corrected to the value they would be at sea level. If this were

not the case, what letter would you see permanently above the mile-high city of Denver? Explain.

▲ 30. Mountaineers often carry altimeters that measure altitude by measuring atmospheric pressure. If a low-pressure weather system moves in, will the altimeter report an altitude that is higher or lower than the true altitude? Explain.

31. Are your ears going to hurt more due to water pressure if you are swimming 12 feet down in your swimming pool or 12 feet down in the middle of Lake Superior? Explain.

32. Why doesn't the water pressure crush a scuba diver at a depth of 30 meters?

33. Compare the pressures at the bottom of the two glasses shown below. Assume that both are filled to the same depth with the same fluid.

34. Two identical wooden barrels are fitted with long pipes extending out their tops. The pipe on the first barrel is 1 foot in diameter, and the pipe on the second barrel is only $\frac{1}{2}$ inch in diameter. When the larger pipe is filled with water to a height of 20 feet, the barrel bursts. To burst the second barrel, will water have to be added to a height less than, equal to, or greater than 20 feet? Explain.

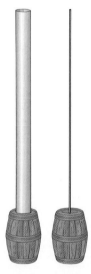

35. Fresh water has a density of 1000 kilograms per cubic meter at 4°C and 998 kilograms per cubic meter at 20°C. In which temperature water would you feel the greater pressure at a depth of 10 meters? Why?

▲ 36. Salt water is more dense than fresh water. This means that the mass of 1 cubic centimeter of salt water is larger than that of 1 cubic centimeter of fresh water. Would a scuba diver have to go deeper in salt water or in fresh water to reach the same pressure? Why?

37. Why can't water be sucked to a height greater than 10 meters even with a very good suction pump?

38. At sea level even a perfect vacuum can raise water only 10 meters up a straw. At an elevation of 5000 feet in Bozeman, Montana, can water be raised to a height greater than, equal to, or less than 10 meters? Explain your reasoning.

39. If you have a water well that is much deeper than about 5 meters, you put the pump at the bottom of the well and have it push the water up. Why is this better than placing the pump at the top?

40. You place a long straw in a glass of water and find that, no matter how hard you suck, you cannot drink the water. You place the same straw in an unknown liquid X and find that you can drink. If you combine liquid X and water in a glass, which one will float on the surface?

41. Some toys contain two different-colored liquids that do not mix. If the purple liquid always sinks in the clear liquid as shown in the toy, what can you say about the densities of the liquids?

42. Spilled gasoline can sometimes be seen as a colorful film on rain puddles. What does this tell you about the density of gasoline?

43. Salt water is slightly more dense than fresh water. Will a boat float higher in salt water or fresh water?

44. Use Archimedes' principle to explain why an empty freighter sits higher in the water than a loaded one.

45. Salt water is slightly more dense than fresh water. Will a 50-ton ship feel a greater buoyant force floating in a freshwater lake or in the ocean?

46. Salt water is slightly more dense than fresh water. Will a 12-pound bowling ball feel a greater buoyant force sitting on the bottom of a freshwater lake or on the bottom of the ocean?

47. When you blow air from your lungs, you are changing both your mass and your volume. Which of these effects explains why this causes you to sink to the bottom of the swimming pool?

48. What happens to the depth of a scuba diver who takes a particularly deep breath?

49. Use the data in Table 12-1 to determine whether a block of lead would float in a lake of liquid mercury. What

about a block of gold? (*Caution*: do *not* try this experiment; mercury is very toxic!)

50. Use the data in Table 12-1 to determine which would float higher in a lake of mercury, a block of copper or a block of silver.

51. A submarine could be made to surface by either increasing the buoyant force or decreasing the weight. When a submarine's ballast tanks are blown out, which is happening?

▲ 52. A scuba diver achieves neutral buoyancy by adjusting the volume of air in her air vest so that the buoyant force equals her weight. If she then kicks her feet and swims down an additional 20 feet, will the net force now be upward, zero, or downward? Explain.

53. You have two cubes of the same size, one made of aluminum and the other of lead. Both cubes are allowed to sink to the bottom of a water-filled aquarium. Which cube, if either, experiences the greater buoyant force? Why?

54. You have two cubes of the same size, one made of wood and the other of aluminum. Both cubes are placed in a water-filled aquarium. The wooden block floats, and the aluminum block sinks. Which cube, if either, experiences the greater buoyant force?

▲ 55. An ice cube is floating in a glass of water. Will the water level in the glass rise, go down, or stay the same as the ice cube melts? Why?

▲ 56. You are sitting in a boat in your swimming pool. There are six gold bricks in your boat. (You are rich!) If you throw the gold into the swimming pool, does the water level in the pool rise, lower, or stay the same? Explain.

57. You place a dime flat on a tabletop a couple of inches from the edge. With your mouth near the edge of the table, you blow sharply across the top of the dime. Why does the dime pop up in the air? Try this.

58. Why does your car get pulled sideways when a truck passes you going in the opposite direction on a two-lane highway?

59. Why do table-tennis players put a lot of topspin on their shots?

60. A partial vacuum can be created by installing a pipe at a right angle to a water faucet and turning on the water, as shown in the figure. What is the physics behind this?

61. The Green Building at the Massachusetts Institute of Technology (MIT) is a tall tower built on an inverted U-shape base that is open to the Charles River Basin. Why might the doors in the opening have opened "by themselves" on windy days before revolving doors were installed to correct the design flaw?

62. Why would an aneurysm (a widening of an artery) be especially subject to rupturing?

EXERCISES

1. What is the density of a substance that has a mass of 84 g and a volume of 8 cm^3? Use Table 12-1 to identify this substance.
2. A small ball has a mass of 6.75 g and a volume of 0.3 cm^3. Can you identify the material using Table 12-1?
3. A bowling trophy has a mass of 180 g. When placed in water, the trophy displaces 600 cm^3. What is the average density of the trophy?
4. If Archimedes' crown had a mass of 1 kg and a volume of 120 cm^3, was the crown made of pure gold? Explain.
5. A solid ball with a volume of 0.4 m^3 is made of a material with a density of 3000 kg/m^3. What is the mass of the ball?
6. What is the mass of a lead sinker with a volume of 3 cm^3?
7. Given that most people are just about neutrally buoyant, it is reasonable to estimate the density of the human body to be about that of water. Use this assumption to find the volume of a 90-kg person.
8. A cube with a mass of 48 g is made from a metal with a density of 6 g/cm^3. What is the volume of the cube and the length of each edge?
9. If 1000 cm^3 of a gas with a density of 0.0009 g/cm^3 condenses to a liquid with a density of 0.9 g/cm^3, what is the volume of the liquid?
10. A cube of ice, 10 cm on each side, is melted into a measuring cup. What is the volume of the liquid water?
11. Calculate the weight of a column of fresh water with cross-sectional area 1 m^2 and height 10 m. What pressure does this create at the bottom of the column of water? How does this compare to atmospheric pressure?
12. Calculate the height of a column of mercury with cross-sectional area 1 m^2 such that it has the same weight as the column of water in Exercise 11.
13. Given that atmospheric pressure drops by a factor of 2 for every gain in elevation of 18,000 ft, what is the height of a mercury column in a barometer located in an unpressurized compartment of an airliner flying at 36,000 ft?
▲ 14. Two barometers are made with water and mercury. If the mercury column is 30 in. tall, how tall is the water column?
▲ 15. Each cubic inch of mercury has a weight of 0.5 lb. What is the pressure at the bottom of a column of mercury 30 in. tall if there is a vacuum above the mercury?
▲ 16. If 1 m^3 of water has a mass of 1000 kg, what is the pressure at a depth of 150 m? Is the atmospheric pressure important?
17. An object has a mass of 150 kg and a volume of 0.2 m^3. What is its average density? Will this object sink or float in water?
18. Will an object with a mass of 1000 kg and a volume of 1.6 m^3 float?
19. A 500-g wooden block is lowered carefully into a completely full beaker of water and floats. What is the weight of the water, in newtons, that spills out of the beaker?
20. A 400-cm^3 block of aluminum ($D = 2.7$ g/cm^3) is lowered carefully into a completely full beaker of water. What is the weight of the water, in newtons, that spills out of the beaker?
21. A ball of wax is lowered carefully into a completely full beaker of water where it floats. This causes 18 cm^3 of water to spill out. The same ball of wax is then lowered carefully into a completely full beaker of ethyl alcohol ($D = 0.79$ g/cm^3) where it sinks, causing 20 cm^3 of alcohol to spill out. Which of these two experiments allows you to find the wax's mass, and which allows you to find its volume? Find the density of the wax.
22. A yellow object is lowered carefully into a completely full beaker of water, where it floats. This causes 28 cm^3 of water to spill out. The same object is then lowered carefully into a completely full beaker of gasoline ($D = 0.68$ g/cm^3) where it sinks, causing 40 cm^3 of gasoline to spill out. In which liquid does the yellow object experience the greater buoyant force?
23. A cubic meter of copper has a mass of 8930 kg. The block of copper is lowered into a lake by a strong cable until the block is completely submerged. Draw a free-body diagram for the block. Find the buoyant force on the block and the tension in the cable.
24. A ball fully submerged in a bathtub has a volume of 5 cm^3 and a mass of 30 g. Draw a free-body diagram for the ball. What is the normal force of the tub on the ball?

13 | Thermal Energy

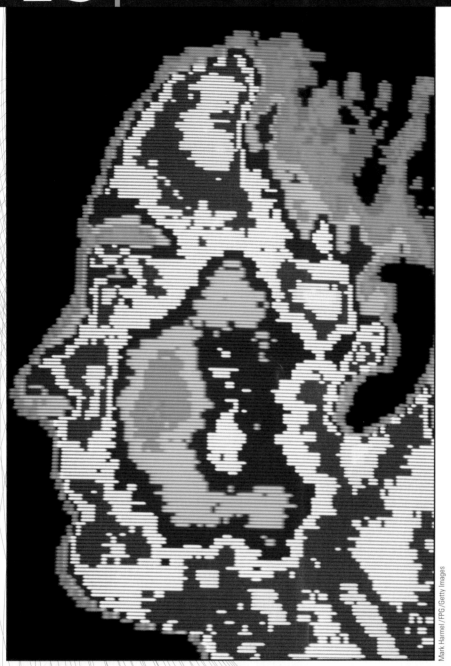

False-color thermograph of a human head.

This false-color thermograph of a human head shows the temperature variations of the surface. What factors control the rate at which radiation is emitted or absorbed and the resulting temperature changes?

(See page 264 for the answer to this question.)

PhysicsNow™ Test your understanding of this chapter by logging into PhysicsNow at **http://physics.brookscole.com/kf6e**, selecting the chapter, and clicking on the "Take a Pre-Test" link.

If we examine any system of moving objects very carefully or if we look at it for long enough, we find that mechanical energy is not conserved. A pendulum bob swinging back and forth does in fact come to rest. Its original mechanical energy disappears.

Other examples show the same thing. Rub your hands together. You are doing work—applying a force through a distance—but clearly your hands do not fly off with some newfound kinetic energy. Similarly, take a hammer and repeatedly strike a metal surface. The moving hammer has kinetic energy, but upon hitting the surface, its kinetic energy disappears. What happens to the energy? It is not converted to potential energy as happened in Chapter 7 because the energy doesn't reappear. So either the kinetic energy truly disappears and total energy is not conserved or it is transferred into some form of energy that is not a potential energy.

There are similarities in the examples given above. When you rub your hands together, they feel hot. The metal surface and the hammer also get hotter when they are banged together. The pendulum bob is not as obvious; the interactions are between the bob and the surrounding air molecules and between the string and the support. But closer examination shows that, once again, the system gets hotter.

At first glance it is tempting to suggest that temperature, or maybe the change in temperature, could be equated with the lost energy. However, neither of these ideas works. If the same amount of energy is expended on a collection of different objects, the resulting temperature increases are not equal. Suppose, for example, that we rub two copper blocks together. The temperature of the copper blocks increases. If we repeat the experiment by expending the same amount of mechanical energy with two aluminum blocks, the change in temperature will not be the same. The temperature change is an indication that something has happened, but it is not equal to the lost energy.

The Nature of Heat

Early ideas about the nature of heat centered on the existence of a fluid that was supposedly transferred between objects at different temperatures. Over centuries people had noted that a kettle of water could become very hot, boil, and turn to steam or that a snow bank could absorb the Sun's heat for an extended period and slowly melt. Fire transferred something to the hot water to make it boil; sunshine imparted something to the snow to liquefy it over time. This "fluid" was studied intensively in the era of early steam-engine technology. It became known as *caloric*, from the Latin *calor* meaning "heat." It was invisible and presumably massless because experimenters could not detect any changes in the mass of an object that was heated.

Count Rumford, an 18th-century British scientist, pioneered a study of work and heat. At that time he was in charge of boring cannons at a military arsenal in Munich and was struck by the enormous amount of heat produced during the boring process. Rumford decided to investigate this. He placed a dull boring tool and a brass cylinder in a barrel filled with cold water. The boring tool was forced against the bottom of the cylinder and rotated by two horses. These are the results described by Rumford:

> At the end of 2 hours and 30 minutes it [the water in the barrel] actually boiled! . . . It would be difficult to describe the surprise and astonishment expressed by the countenances of the by-standers, on seeing so large a quantity of cold water heated, and actually made to boil without any fire.

Rumford showed that large quantities of heat could be produced by mechanical means without fire, light, or chemical reaction. (This is a large-scale version of the simple hand-rubbing experiment.) The importance of his experiment was

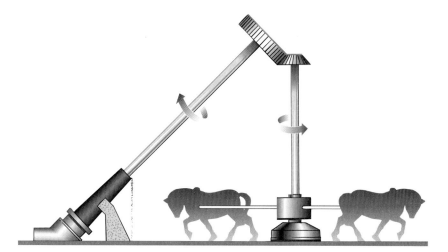

Rumford investigated the nature of heat while boring cannons.

the demonstration that the production of heat seemed inexhaustible. As long as the horses turned the boring tool, heat was generated without any limitation. He concluded that anything that could be produced without limit could not possibly be a material substance. Heat was not a fluid but something generated by motion.

In our modern physics world view, **heat** is energy *flowing* between two objects due to a difference in temperature. We measure the amount of energy gained or lost by an object by the resulting temperature change in the object. By convention, 1 **calorie** (cal) is defined as the amount of heat that raises the temperature of 1 gram of water by 1°C. In the U.S. customary system, the unit of heat, called a **British thermal unit** (Btu), is the amount of energy needed to change the temperature of 1 pound of water by 1°F. One British thermal unit is approximately equal to 252 calories.

✓ Extended presentation on Mechanical Work and Heat available in the *Problem Solving* supplement

Question How many calories are required to raise the temperature of 8 grams of water by 5°C?

Answer To raise the temperature of 1 gram by 5°C requires 5 calories. Therefore, 8 grams requires 5 calories/gram × 8 grams = 40 calories.

Mechanical Work and Heat ✓ MATH

The Rumford experiment used the *work* supplied by the horses to raise the temperature of the water, clearly demonstrating an equivalent way of heating the water. The water got hotter *as if* it were heated by a fire, but there was no fire.

There is a close connection between work and heat. Both are measured in energy units, but neither resides in an object. In Chapter 7 we saw that work was a measure of the energy "flowing" from one form to another. For example, the gravitational force does work on a free-falling ball, causing its kinetic energy to increase—thus, potential energy changes to kinetic energy. Similarly, heat does not reside in an object but flows into or out of an object, changing the internal energy of the object. This internal energy is sometimes known as **thermal energy**, and the area of physics that deals with the connections between heat and other forms of energy is called **thermodynamics**.

Although Rumford's experiment hinted at the equivalence between mechanical work and heat, James Joule uncovered the quantitative equivalence 50 years later. Joule's experiment used a container of water with a paddle-wheel arrangement like that shown in Figure 13-1. The paddles are connected via pulleys to a weight. As the weight falls, the paddle wheel turns, and the water's temperature goes up. The potential energy lost by the falling weight results in a rise

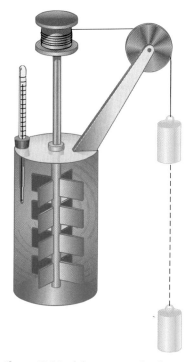

Figure 13-1 Joule's apparatus for determining the equivalence of work and heat. The decrease in the gravitational potential energy of the falling mass produces an increase in the energy of the water.

in the temperature of the water. Because Joule could raise the water temperature by heating it or by using the falling weights, he was able to establish the equivalence between the work done and the heat transferred. Joule's experiment showed that 4.2 joules of work are equivalent to 1 calorie of heat.

▶ 1 calorie = 4.2 joules

There are other units of energy. The Calorie used when referring to the energy content of food is not the same as the calorie defined here. The food Calorie (properly designated by the capital C to distinguish it from the one used in physics) is equal to 1000 of the physics calories. A piece of pie rated at 400 Calories is equivalent to 400,000 calories of thermal energy, or nearly 1.7 million joules of mechanical energy.

Question   Because joules and calories are both energy units, do we need to retain both of them?

Answer   No. However, both are currently used for historical reasons. Europeans are much further along than Americans in converting from Calories to kilojoules in the labeling of food.

Temperature Revisited

If we bring two objects at different temperatures into contact with each other, there is an energy flow between them, with energy flowing from the hotter object to the colder. We know from the structure of matter (Chapter 11) that the molecules of the hotter object have a higher average kinetic energy. Therefore, on the average, the more-energetic particles of the hotter object lose some of their kinetic energy when they collide with the less-energetic particles of the colder object. The average kinetic energy of the hotter object's particles decreases and that of the colder object's particles increases until they become equal. On a macroscopic scale, the temperature changes for each object: the hotter object's temperature drops, and the colder object's temperature rises. The flow of energy stops when the two objects reach the same temperature, a condition known as **thermal equilibrium**. Atomic collisions still take place, but on the average the particles do not gain or lose kinetic energy.

Let's assume that we have two objects, labeled A and B, that cannot be placed in thermal contact with each other. How can we determine whether they would be in thermal equilibrium if we could bring them together? Let's also assume that we have a third object, labeled C, that can be placed in thermal contact with A and that A and C are in thermal equilibrium. If C is now placed in thermal contact with B and if B and C are also in thermal equilibrium, then we can conclude that A and B are in thermal equilibrium. This is summarized by the statement of the **zeroth law of thermodynamics**.

▶ zeroth law of thermodynamics

> If objects A and B are in thermal equilibrium with object C, then A and B are in thermal equilibrium with each other.

Although this statement might seem to be so obvious that it is not worth elevating to the stature of a law, it plays a very fundamental role in thermodynamics because it is the basis for the definition of temperature. Two objects in thermal equilibrium have the same temperature. On the other hand, if two objects are not in thermal equilibrium, they must have different temperatures. The zeroth law was developed later in the history of thermodynamics but labeled with a zero because it is more basic than the other laws of thermodynamics.

JOULE | A New View of Energy

James Prescott Joule (1818–1889) was the second son of a family of wealthy brewers in the village of Salford near Manchester, England, an industrial region. Joule was tutored as a youngster by John Dalton, a noted, elderly chemist, and determined early in life to pursue physical science as a serious hobby.

Today Joule is remembered for his experiments and theories on work, energy, and heat. Michael Faraday was an early, major source of inspiration, and Joule conducted experiments on heating in conducting wires—mostly made of copper or platinum. His energy sources were Voltaic batteries and the dynamo. This early work earned him the reputation of a pioneer in battery design and efficiency. His views on the equivalence of heat and work derived from these early electrochemical experiments.

Manchester was a city of steam engines, so it was natural that Joule and his friends in the local scientific and engineering group would discuss theories of heat and means of improving engine performance. His own work led in a general way to the concept of thermal efficiency, but he contributed little to steam technology because he thought engineers were better suited to technical improvements. The physicists' job was to extend the power of theory.

Joule excelled in precision measurement. He was fortunate that a local Manchester firm could manufacture calorimeters and thermometers for him. By 1840 he had calibrated temperature differences to an accuracy of 0.01°F, and in later experiments he used calibrations one-half this size. Many of his peers at that time underestimated the importance of precision measure-

James Prescott Joule

ments. Joule's appointment to the first major British commission on scientific standards validated his efforts.

As his conceptual grasp of the issues evolved, Joule recognized and measured the equivalence between mechanical work and heat in several ways. One of the most striking was the accurate measurement of the increase in temperature of water as it fell over a waterfall into a pool. His first observations were in France, but he speculated about Niagara Falls, which he had not visited, and accurately predicted the increase.

At a meeting during which Joule delivered a paper that most in the audience did not understand completely, he met an important collaborator, William Thomson, known later as Lord Kelvin. Kelvin's mathematical skills combined with Joule's careful measurements solidified understanding of the first and second laws of thermodynamics. Among other contributions, their work presaged the rise of mechanical refrigeration later in the 19th century. Current concepts of energy and the use of precision measurements owe much to the quiet and unassuming work performed by James Prescott Joule.

—*Pierce C. Mullen, historian and author*

Sources: Mary B. Hesse, *Forces and Fields: The Concept of Action at a Distance in the History of Physics* (New York: Philosophical Library, 1962); Henry John Steffens, *James Prescott Joule and the Concept of Energy* (New York: Science History Publications, 1979).

Heat, Temperature, and Internal Energy

Heat and temperature are not the same thing. Heat is an energy, whereas temperature is a macroscopic property of an object. Two objects can be at the same temperature (the same average atomic kinetic energy) and yet transfer vastly different amounts of energy to a third object. For example, a swimming pool of water and a coffee cup of water at the same temperature can melt very different amounts of ice.

PHYSICS | ON YOUR OWN

Mix equal amounts of water at different temperatures to see if the equilibrium temperature is midway between the hot and cold temperatures. Styrofoam cups isolate the system from the surroundings pretty well. The losses to the surroundings can be further minimized if one temperature is above room temperature and the other is below. Next try unequal amounts of water and predict the final temperature.

When we consider the total microscopic energy of an object—such as translational and rotational kinetic energies, vibrational energies, and the energy stored in molecular bonds—we are talking about the **internal energy** of the object. There are two ways of increasing the internal energy of a system. One way is to heat the system; the other is to do work on the system. The law of conservation of energy tells us that the total change in the internal energy of the system is equal to the change due to the heat added to the system plus that due to the work done on the system. This is called the **first law of thermodynamics** and is really just a restatement of the law of conservation of energy.

first law of thermodynamics ▶ The increase in the internal energy of a system is equal to the heat added plus the work done on the system.

This law sheds more light on the nature of internal energy. Let's assume that if 10 calories of heat are added to a sample of gas, its temperature rises by 2°C. If we add the same 10 calories to a sample of the same gas that has twice the mass, we discover that the temperature rises by only 1°C (Figure 13-2). Adding the same amount of heat does not produce the same rise in temperature. This makes sense because the larger sample of gas has twice as many particles, and therefore each particle receives only half as much energy on the average. The average kinetic energy, and thus the temperature, should increase by half as much. An increase in the temperature is an indication that the internal energy of the gas has increased, but the mass must be known to say how much it increases.

Absolute Zero

The temperature of a system can be lowered by removing some of its internal energy. Because there is a limit to how much internal energy can be removed, it is reasonable to assume that there is a lowest possible temperature. This temperature is known as **absolute zero** and has a value of −273°C, the same temperature used to define the zero of the Kelvin scale.

The existence of an absolute zero raised the challenge of experimentally reaching it. The feasibility of doing so was argued extensively during the first three decades of the 20th century, and it was eventually concluded that it was impossible. This belief is formalized in the statement of the **third law of thermodynamics**.

Figure 13-2 Adding equal amounts of heat to different amounts of a material produces different temperature changes.

third law of thermodynamics ▶ Absolute zero may be approached experimentally but can never be reached.

There appears to be no restriction on how close experimentalists can get, only that it cannot be reached. Small systems in low-temperature laboratories have reached temperatures less than a billionth of a degree from absolute zero.

A substance at absolute zero has the lowest possible internal energy. Originally, it was thought that all atomic motions would cease at absolute zero. The development of quantum mechanics (Chapter 24) showed that all motion does not cease; the atoms sort of quiver with the minimum possible motion. In this state the atoms are packed closely together. Their mutual binding forces arrange them into a solid block.

Specific Heat

Suppose we have the same number of molecules of two different gases and each gas is initially at the same temperature. If we add the same amount of heat to each gas, we find that the temperatures do not rise by the same amount. Even though the gases undergo the same change in their internal energies, their molecules do not experience the same changes in their average translational kinetic energies. Some of the heat appears lost. Actually, the heat is transformed into other forms of energy. If the gas molecules have more than one atom, part of the internal energy is transformed into rotational kinetic energy of the molecules and part of it can go into the vibrational motion of the atoms (Figure 13-3). Only a small fraction of the increase in internal energy for most real gases goes into increasing the average kinetic energy that shows up as an increase in temperature.

The amount of heat it takes to increase the temperature of an object by 1°C is known as the **heat capacity** of the object. The heat capacity depends on the amount and type of material used to construct the object. An object with twice the mass will have twice the heat capacity, provided both objects are made of the same material.

We can obtain an intrinsic property of the material that does not depend on the size or shape of an object by dividing the heat capacity by the mass of the object. This property is known as the **specific heat** and is the amount of heat required to increase the temperature of 1 gram of the material by 1°C.

By definition, the specific heat of water is numerically 1; that is, 1 calorie raises the temperature of 1 gram of water by 1°C. The specific heat for a given material in a particular state depends slightly on the temperature but is usually assumed to be constant. The specific heats of some common materials are given in Table 13-1. Notice that the SI units for specific heat are joules per kilogram-kelvin. These are obtained by multiplying the values in calories per gram-degree Celsius by 4186. Note also that the value for water is quite high compared with most other materials.

Table 13-1

Specific Heats for Various Materials

Material	Specific Heat (cal/g · °C)	(J/kg · K)
Solids		
Aluminum	0.215	900
Copper	0.092	385
Diamond	0.124	519
Gold	0.031	130
Ice	0.50	2090
Silver	0.057	239
Liquids		
Ethanol	0.75	3140
Mercury	0.033	138
Water	1.00	4186
Gases		
Air	0.24	1000
Helium	1.24	5190
Nitrogen	0.25	1040
Oxygen	0.22	910

Question What is the rise in temperature when 20 calories are added to 10 grams of ice at −10°C?

Answer This is the same as adding 2 calories to each gram. Because $\frac{1}{2}$ calorie is required to raise the temperature of 1 gram of ice by 1°C, the 2 calories will raise its temperature by 4°C.

When we bring two different materials into thermal contact with each other, they reach thermal equilibrium but don't normally experience the same changes in temperature because they typically have different heat capacities. However, conservation of energy tells us that the heat lost by the hotter object is equal to the heat gained by the colder object. (We're assuming that no energy is lost to the environment.)

The specific heats of the materials on Earth's surface account for the temperature extremes lagging behind the season changes. The first day of summer in the Northern Hemisphere usually occurs on June 21. On this day the soil receives the largest amount of solar radiation because it is the longest day of the year and the sunlight arrives closest to the vertical. And yet the hottest days of summer typically occur several weeks later. It takes time for the ground to warm up because it requires a lot of energy to raise its temperature each degree.

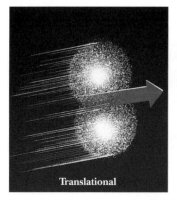
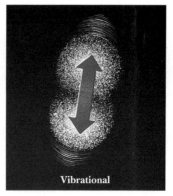

Figure 13-3 Three forms of internal energy for a diatomic molecule.

WORKING IT OUT | Specific Heat

The specific heat c is obtained by dividing the heat Q added to the material by the product of the mass m and the resulting change in temperature ΔT:

$$c = \frac{Q}{m\Delta T}$$

For example, if it requires 11 cal to raise the temperature of an 8-g copper coin 15°C, we can calculate the specific heat of copper:

$$c = \frac{Q}{m\Delta T} = \frac{11 \text{ cal}}{(8 \text{ g})(15°C)} = 0.092 \text{ cal/g} \cdot °C$$

Note that this agrees with the entry in Table 13-1.

We can rearrange our definition of specific heat to obtain an expression for the heat required to change the temperature of an object by a specific amount. For instance, suppose that you have a cup of water at room temperature that you want to boil. How much heat will this require? Let's assume that the cup contains $\frac{1}{4}$ L of water at 20°C and that we can ignore the heating of the cup itself. The mass of the water is 250 g, and the boiling point of water is 100°C at 1 atm. Therefore, the temperature change is 80°C, and we have

$$Q = cm\Delta T = \left(1 \frac{\text{cal}}{\text{g} \cdot °C}\right)(250 \text{ g})(80°C) = 20{,}000 \text{ cal} = 20 \text{ kcal}$$

The 20 kcal of energy must be supplied by a stove or microwave oven.

Flawed Reasoning

A lab manual asks students to mix 400 grams of warm ethanol at 60°C with 300 grams of room-temperature water at 20°C. Before performing this experiment, two students are making predictions for the final temperature of the mixture.

Dylan: "The final temperature will be *higher* than 40°C. If the masses were equal, the final temperature would be halfway between 20°C and 60°C, but there is more ethanol than water."

Jessica: "No, the final temperature of the mixture will be *lower* than 40°C. Water has a higher specific heat, so the water will have a smaller change in temperature."

Both of these students are wrong. **Find the flaw in their reasoning.**

Answer Dylan is focusing on the relative masses, whereas Jessica is focusing on the relative specific heats. Both factors play a role in determining the heat capacity of an object. The equilibrium temperature will be closer to the initial temperature of the fluid with the larger heat capacity. The heat capacities C of the ethanol and water are

$$C_{ethanol} = c_{ethanol} m_{ethanol} = \left(0.75 \frac{\text{cal}}{\text{g} \cdot °C}\right)(400 \text{ g}) = 300 \frac{\text{cal}}{°C}$$

$$C_{water} = c_{water} m_{water} = \left(1 \frac{\text{cal}}{\text{g} \cdot °C}\right)(300 \text{ g}) = 300 \frac{\text{cal}}{°C}$$

In this example both fluids have the same heat capacity, so the final equilibrium temperature will be 40°C, midway between the two initial temperatures.

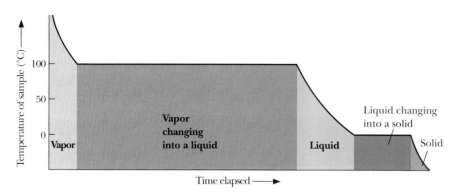

Figure 13-4 A graph of the temperature of water versus time as thermal energy is removed from the water. Notice that the temperature remains constant while the steam condenses to liquid water and while the liquid water freezes to form ice.

Change of State

We continue our investigation of internal energy by continually removing energy from a gas and watching its temperature. If we keep the pressure constant, the volume and temperature of the gas decrease rather smoothly until the gas reaches a certain temperature. At this temperature there is a rapid drop in volume and *no* change in temperature. Drops of liquid begin to form in the container. As we continue to remove energy from the gas, more and more liquid forms, but the temperature remains the same. When all the gas has condensed into liquid, the temperature drops again (Figure 13-4). The change from the gaseous state to the liquid state (or from the liquid to the solid), or vice versa, is known as a **change of state**.

While the gas was condensing into a liquid, energy was continually leaving the system, but the temperature remained the same. Most of this energy came from the decrease in the electric potential energy between the molecules as they got closer together to form the liquid. This situation is analogous to the release of gravitational potential energy as a ball falls toward Earth's surface. The energy that must be released or gained per unit mass of material is known as the **latent heat**. The values of the latent heat for melting and vaporization are given in Table 13-2.

PHYSICS | ON YOUR OWN

Folklore has it that hot water freezes faster than cold water. Investigate this by placing equal amounts of hot and cold water in identical containers in your freezer or outside on a cold night. Is there any truth to the folklore? Does your answer depend on the type of container or how hot and cold the water is?

The same processes occur when you heat a liquid. If you place a pan of water on the stove, the temperature rises until the water begins to boil. The

Table 13-2 | Melting Points, Boiling Points, and Latent Heats for Various Materials

Material	Melting Point (°C)	Latent Heat (Melting)		Boiling Point (°C)	Latent Heat (Vaporization)	
		(kJ/kg)	(cal/g)		(kJ/kg)	(cal/g)
Nitrogen	−210	25.7	6.14	−196	199	47.5
Oxygen	−218	13.8	3.3	−183	213	50.9
Water	0	334	79.8	100	2,257	539
Aluminum	660	396	94.6	2,467	10,900	2,600
Gold	1,064	63	15	2,807	1,710	409

The melting of snow and ice in Glacier National Park is a slow process because of the latent heat required to change the ice to liquid water.

temperature then remains constant as long as the water boils. It doesn't matter whether the water boils slowly or rapidly. (Because the rate at which foods cook depends only on the temperature of the water, you can conserve energy by turning the heat down as low as possible while still maintaining a boil.) During the change of state, the additional energy goes into breaking the bonds between the water molecules and not into increasing the average kinetic energy of the molecules. Each gram of water requires a certain amount of energy to change it from liquid to steam without changing its temperature. In fact, this is the same amount of energy that must be released to convert the steam back into liquid water. Furthermore, the temperature at which steam condenses to water is the same as the boiling point. The melting and boiling points for some common substances are given in Table 13-2.

Question   At the boiling temperature, what determines whether the liquid turns into gas or the gas turns into liquid?

Answer   If heat is being supplied, the liquid will boil to produce additional gas. However, if heat is being removed, some of the gas will condense to form additional liquid.

A similar change of state occurs when snow melts. The snow does not suddenly become water when the temperature rises to 0°C (32°F). Rather, at that temperature the snow continues to take in energy from the surroundings, slowly changing into water as it does. Incidentally, we are fortunate that it behaves this way; otherwise, we would have gigantic floods the moment the temperature rose above freezing! The latent heat required to melt ice explains why ice can keep a drink near freezing until the last of the ice melts.

On nights when the temperature is predicted to drop below freezing, owners of fruit orchards in California and Florida turn on sprinklers to keep the fruit from freezing. As the water freezes, heat is given off that maintains the temperature of the fruit at 0°C, a temperature above that where the fruit freezes. Once the water is completely frozen, the fruit is still protected because ice does not conduct heat very well. The ice serves as a "sweater" for the fruit. However, if the air temperature drops too low, the fruit will be ruined.

PHYSICS | ON YOUR OWN

> Determine the number of calories needed to melt 1 gram of ice. Place some ice in a Styrofoam cup of water after you have measured the mass of ice, the mass of water, and the temperature of the water. To minimize the loss of thermal energy to the surroundings, place a piece of Styrofoam (or another cup) over the top while the ice melts. Does the loss of thermal energy to the surroundings cause your value to be too high or too low?

Thermal energy is transported along the branding iron by conduction because the brand is hotter than the handle.

Conduction

Thermal energy is transported from one place to another via three mechanisms: conduction, convection, and radiation. Each of these is important in some circumstances and can be ignored in others.

If temperature differences exist within a single, isolated object such as a branding iron held in a campfire, thermal energy will flow until thermal equilibrium is achieved. We say that the thermal energy is conducted through the material. **Conduction** takes place via collisions between the particles of the material. The molecules and electrons at the hot end of the branding iron collide

Figure 13-5 The left end of this ceramic dish is ice cold while the right end is very hot. This occurs because ceramic is a very poor conductor of thermal energy.

with their neighbors, transferring some of their kinetic energy, on the average. This increased kinetic energy is passed along the rod via collisions until the end in your hand gets hot.

The rate at which energy is conducted varies from substance to substance. Solids, with their more tightly packed particles, tend to conduct thermal energy better than liquids and gases. The mobility of the electrons within materials also affects the thermal conductivity. Metals such as copper and silver are good thermal **conductors** as well as good electrical conductors. Conversely, electrical insulators such as glass and ceramic are also good thermal **insulators**. A glassblower can hold a glass rod in a flame for a very long time without getting burned. The ceramic bowl in Figure 13-5 has regions at drastically different temperatures.

The differences in the conductivity of materials explain why aluminum and wooden benches in a football stadium do not feel the same on a cold day. Before you sit on either bench, they are at the same temperature. When you sit down, some of the thermal energy in your bottom flows into the bench. Because the wooden bench does not conduct the heat very well, the spot you are sitting on warms up and feels more comfortable. On the other hand, the aluminum bench continually conducts heat away from your bottom, making your seat feel cold.

PHYSICS | ON YOUR OWN

Select a few kitchen utensils made of different materials and hold them next to your cheek. Because they all came from the kitchen, they must be at the same temperature. What do you feel? Can you categorize the materials according to their conductivities?

The rate at which thermal energy is conducted through a slab of material depends on many physical parameters besides the type of material. You might correctly guess that it depends on the area and thickness of the material. A larger area allows more thermal energy to pass through, and a greater thickness allows less. Experimentation tells us that the difference in the temperatures on the two sides of the slab also matters—the greater the difference, the greater the flow. Table 13-3 gives the thermal conductivity of a variety of common materials.

In our everyday lives, we are more concerned with reducing the transfer of thermal energy than increasing it. We wear clothing to retain our body heat, and we insulate our houses to reduce our heating and air-conditioning bills. Fig-

Table 13-3

Thermal Conductivities for Various Materials

Material	Conductivity (W/m · °C)
Solids	
Silver	428
Copper	401
Aluminum	235
Stainless steel	14
Building Materials	
Polyurethane foam	0.024
Fiberglass	0.048
Wood	0.08
Window glass	0.8
Concrete	1.1
Gases	
Air (stationary)	0.026
Helium	0.15

Figure 13-6 Melting snow patterns reveal differences in thermal conduction. For example, the old garage has been converted to living space, but it appears not to have been insulated.

Two climbers prepare their gear while sitting in a snow cave on Passo Superior.

ure 13-6 shows regions of a roof after a snowstorm. The unmelted patches exist where there is better insulation or, in the case of an unheated porch or garage, where there is little or no temperature difference. Table 13-3 allows us to compare the heat loss through slabs of different materials of the same size and thickness for the same difference in temperatures.

Examination of Table 13-3 reveals that static air is a pretty good insulator. This insulating property of air means that porous substances with many, small air spaces are good insulators, and it explains why the goose down used in sleeping bags keeps you so warm. It also explains how fishnet long underwear keeps you warm. The air trapped in the holes keeps your body heat from being conducted away.

Snow contains a lot of air space between the snowflakes, which makes snow a good thermal insulator. Mountaineers often dig snow caves to escape from severe weather. Likewise, snow-covered ground does not freeze as deep as bare ground.

Question Why do people who spend time outdoors in cold weather wear many layers of clothing?

Answer In addition to the flexibility of adding and removing layers to get the required insulation, the air spaces between the layers contribute to the overall insulation.

Convection

Thermal energy can also be transferred in fluids by **convection**. In convection the energy is transported by the movement of the fluid. This movement could be forced, as in heating systems or the cooling system in an automobile, or it could happen because of the changes that occur in the density of the fluid when it is heated or cooled. As the gas near the flame of a candle is heated, it becomes less dense and rises due to the buoyant force (Chapter 12).

Convection in Earth's atmosphere plays a fundamental role in our global climate as well as our daily weather. Convection currents arise from the uneven heating of Earth's surface. Glider pilots, hang-glider fliers, and birds of prey (such as hawks and eagles) use convection currents called thermals to provide them with the lift they need to keep aloft.

Local winds near a large body of water can be caused by temperature differences between the water and the land. The specific heat of water is much greater than that of rock and soil. (Convection currents in the water also moderate the changes in the water temperature.) During the morning, the land warms up faster than the water. The hotter land heats the air over it, causing the air to rise.

Glider pilots search for thermals to gain altitude.

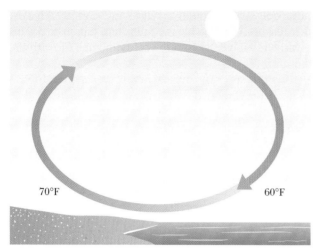 

Figure 13-7 The difference in temperature between the land and the water causes breezes to blow (a) onshore during the morning and (b) offshore during the evening.

The result is a pleasant breeze of cooler air coming from the water (Figure 13-7[a]). During the evening, the land cools faster, reversing the convection cycle (Figure 13-7[b]).

Question What role does convection play in bringing a pot of water to a boil?

Answer As the flame or heating element warms up the water near the bottom of the pan, it becomes less dense and rises. This circulation causes all of the water to warm up at the same time.

Radiation

The third mechanism for the transfer of thermal energy involves electromagnetic waves. As we will see in Chapter 22, these waves can travel through a vacuum, and thus **radiation** is still effective in situations where the conduction and convective processes fail. The electromagnetic radiation travels through space and is converted back to thermal energy when it hits other objects. Most of the heat that you feel from a cozy fire is transferred by radiation, especially if the fire is behind glass or in a stove.

All objects emit radiation. Although radiation from objects at room temperature is not visible to the human eye, it can be viewed with special night glasses or recorded on infrared-sensitive film. When colors are artificially added, we can distinguish the different temperatures of objects as shown in Figure 13-8.

As the temperature of the object rises, more and more of the radiation becomes visible. Objects such as the heating coils on kitchen stoves glow with a red-orange color. Betelgeuse, the red star marking the right shoulder of the constellation Orion, has a surface temperature of approximately 3000 K. As an object gets hotter, the color shifts to yellow and then white. Our Sun appears white (above Earth's atmosphere) with a temperature of 5800 K. The hottest stars appear blue and have temperatures exceeding 30,000 K.

Figure 13-8 This infrared photograph of a toaster shows the different temperatures of its parts.

The radiation that Earth receives from the Sun is typical of an object at 5800 K. If Earth absorbed all of this energy, it would continue to get hotter and hotter until life as we know it could not exist. However, Earth also radiates. Its temperature rises until it radiates as much energy into space as it receives; it reaches equilibrium.

A phenomenon known as the *greenhouse effect* can have a major effect on the equilibrium condition. Visible light easily passes through the windows of a greenhouse or a car, heating the interior. However, the infrared radiation given off by the interior does not readily pass through the glass and is trapped inside. Only when the temperature reaches a high value is equilibrium established. This is one reason why we are warned to never leave pets or children in a car with the windows rolled up on a hot day.

A similar thing happens with Earth. The atmosphere is transparent to visible light, but the water vapor and carbon dioxide in the atmosphere tend to block the infrared radiation from escaping, causing Earth's temperature to increase. The high surface temperatures on Venus are due to the greenhouse effect of its thick atmosphere. It is feared that increases in the carbon dioxide concentration in Earth's atmosphere (due to such things as the burning of fossil fuels) will cause global warming that in turn will cause unwanted changes in Earth's climate. Such alterations in the climate could change the types of crops that will grow and melt the polar ice caps, flooding coastal cities!

Wind Chill

The meteorologist on television announces the temperatures for the day and then adds that it's going to feel even colder because of the wind. If the air is a certain temperature, why does it matter whether the wind is blowing?

Your body is constantly producing heat that must be released to the environment to keep your body from overheating. The primary way that your body gets rid of excess heat is through evaporation. For each liter (about 1 quart) of water that evaporates, roughly 600 kilocalories of heat are absorbed from your body. While most of us correctly associate this mechanism with sweating, a surprising 25% of the heat lost by evaporation in a resting individual is due to the evaporation of water from the linings of our lungs into the air we exhale. Vigorous activity can produce sweating at a rate in excess of 2 liters per hour, removing 1200 kilocalories per hour, a rate tens of times larger than for a resting individual.

Another form of heat loss is due to convection of air away from the body. Even with no wind, the air leaves your skin on a cold day because its density changes as your body warms the air. A third form of heat loss is radiation loss. If your body is warmer than the surrounding objects such as the walls in a room, your body radiates energy to the walls. This is why you feel cold some mornings even though the air in the room has been heated to normal room temperature. The walls take some time to warm up, and you will continue to radiate to them until they warm up.

The wind near your body greatly alters the effectiveness of these heat transfers. In stationary air, the layer of air next to your skin becomes warm and moist, reducing the further loss of heat to this layer of air. However, if there is a wind, the wind brings new air to your skin that is colder and drier. Warming and adding moisture to this new air requires additional heat from your body.

In the mid-1940s a single index was created—the *windchill factor*—to express the cooling effects for various ambient temperatures and wind speeds in terms of an "equivalent" temperature with no wind. We can use the data shown in Table 13-4 to find the windchill temperature for a thermometer reading of 25°F on a day when the wind is blowing at 40 mph. Look along the top of the table until you locate the 25°F and then move down this column to the row labeled 40 mph along the left-hand side of the table. The entry at the intersection of this row and this column is the equivalent temperature. Therefore, the cooling effects are equivalent to a temperature of 6°F on a calm day.

Just as wind can increase the rate at which heat leaves the human body, relative humidity slows the evaporation of perspiration, decreasing the rate. The *heat index* in Table 13-5 combines the effects of temperature and relative humidity to yield an apparent temperature, similar to the windchill factor. Heat stroke is highly likely with continued exposure to heat indexes of 130°F or greater. Sun-

Table 13-4 | Windchill

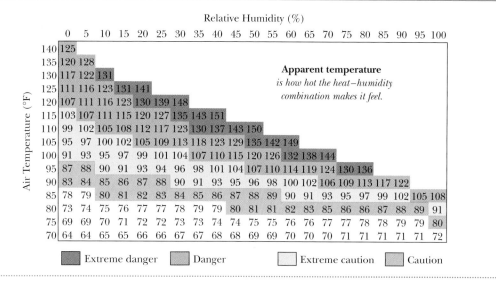

Table 13-5 | Heat Index

stroke, heat cramps, and heat exhaustion are likely for heat indexes between 105°F and 130°F and possible for values between 90°F and 105°F. It should be noted that these values are based on shady, light-wind conditions. Exposure to full sunshine can increase the heat-index values by up to 15°F.

Flawed Reasoning

Susan has just completed sculpting a figurine out of candle wax when she notices that the melting point for the wax is only 125°F. The weather report predicts tomorrow's high temperature to be 100°F with 65% relative humidity, giving an apparent temperature on the heat index of 138°F. **Should Susan store her masterpiece in the refrigerator?**

Answer The heat index applies only to humans and other animals that use the sweating process to cool themselves. Susan's statue will not melt.

Thermal Expansion

All objects change size as they change temperature. When the temperature increases, nearly all materials expand. But not all materials expand at the same rate. Solids, being most tightly bound, expand the least. All gases expand at the same rate, following the ideal gas equation developed in Chapter 11. Each material's characteristic **thermal expansion** is reflected in a number called its *coefficient of expansion*. The coefficient of expansion gives the fractional change in the size of the object per degree change in temperature.

Expansion slots allow bridges to change length with temperature changes without damage.

WORKING IT OUT | Thermal Expansion

Because the coefficient of thermal expansion tells us how much a unit length of material expands as the temperature is raised 1°C, the expansion for a particular object is given by

$$\Delta L = \alpha L \Delta T$$

where ΔL is the change in length, α is the coefficient of thermal expansion, L is the original length, and ΔT is the change in temperature. There is a similar expression for the volume expansion of liquids.

As an example, the coefficient of expansion for steel is 0.000 011 meter for each meter of length for each degree Celsius rise in temperature. This means that a bridge that is 50 m long expands by 0.000 55 m, or 0.55 mm, for each degree of temperature increase. If the temperature increases by 40°C from night to day, the bridge expands by 22 mm (almost 1 in.).

We can also obtain this answer using the relationship for thermal expansion:

$$\Delta L = \alpha L \Delta T = \left(\frac{0.000\ 011}{°C}\right)(50\ \text{m})(40°C) = 0.022\ \text{m}$$

Thermal expansion has many consequences. Civil engineers avoid the possibility of a bridge buckling by including expansion slots and by mounting one end of the bridge on rollers. The gaps between sections of concrete in highways and sidewalks allow the concrete to expand and contract without breaking or buckling. The romantic clickety clack of train rides is due to expansion joints between the rails.

Question   Why are telephone wires higher in winter than in summer?

Answer   The wires expand with the hotter temperatures in summer and therefore hang lower.

We use the differences in the thermal expansions of various materials to our advantage. Some thermostats are constructed of two different metal strips bonded face to face as shown in Figure 13-9. Because the metals have different coefficients of expansion, they expand by different amounts, causing the bimetallic strip to bend. Placing electrical contacts in appropriate places allows the thermostat to function as an electric switch to turn a furnace, heater, or air conditioner on and off at specified temperatures.

Question   How does running hot water on a jar lid loosen it?

Answer   The metal expands more than the glass, and the lid pulls away from the jar.

Figure 13-9 A bimetallic strip is used in thermostats to control furnaces.

Freezing Lakes

Life as we know it depends on the unique thermal expansion properties of water. All materials change size when their temperatures change. Because density is the ratio of mass to volume and because the mass of an object does not change when heated or cooled, a change in size means a change in the object's density.

As stated in the chapter, most objects expand when heated and contract when cooled. The behavior of water is not so simple. Over most of its liquid range, water behaves as expected, decreasing in volume as its temperature decreases. As the water is cooled below 4°C, however, it expands!

This unusual property affects the way lakes freeze. While cooling toward 4°C, the surface water becomes denser and therefore sinks (Chapter 12), cooling the entire lake. However, once the entire lake becomes 4°C, the surface water expands as it cools further and becomes less dense. Therefore, the cooler water floats on the top and continues to cool until it freezes. Lakes freeze from the top down. However, because ice is a good thermal insulator, most lakes do not freeze to the bottom. If water were like most other materials, the very cold water would sink, and lakes would freeze from the bottom up, creating a challenging evolutionary problem for all aquatic and marine life.

Because water is denser than ice, the ice floats and covers lakes and ponds in winter. The layer of ice insulates the water below from freezing air temperatures.

Summary

The law of conservation of mechanical energy does not apply whenever frictional effects are present. Often the transformation of mechanical energy to thermal energy is accompanied by temperature changes that produce observable changes in the object. Heat and temperature are not the same thing. Heat is an energy, and temperature is a macroscopic property of the object. The number of calories required to raise the temperature of 1 gram of a substance by 1°C is known as its specific heat.

The first law of thermodynamics tells us that the total change in the internal energy of a system is the sum of the heat added to the system and the work done on the system. This is just a restatement of the law of conservation of energy. Performing 4.2 joules of work on a system is equivalent to adding 1 calorie of thermal energy. Part of this energy increases the average kinetic energy of the atoms; the absolute temperature is directly proportional to this average kinetic energy. Other parts of this energy break the bonds between the molecules and cause substances to change from solids to liquids to gases. At higher temperatures, molecules, atoms, and even nuclei break apart.

There is a limit to how much internal energy can be removed from an object, and thus there is a lowest possible temperature—absolute zero, or −273°C—the same as the zero on the Kelvin scale. A substance at absolute zero has the lowest possible internal energy.

The temperature of a substance does not change while it undergoes a physical change of state. The energy that is released or gained per gram of material is known as the latent heat.

The natural flow of thermal energy is always from hotter objects to colder ones. In the process called conduction, thermal energy is transferred by collisions between particles; in convection the transfer occurs through the movement of the particles; and in radiation the energy is carried by electromagnetic waves.

PhysicsNow™ Assess your understanding of this chapter's topics with sample tests and other resources found by logging into PhysicsNow at http://physics.brookscole.com/kf6e.

Chapter 13 Revisited

The rate at which an object radiates energy is determined by the difference in temperature between the object and its surroundings, the surface area of the object, and characteristics of the surface. The change in temperature of the object depends primarily on the amount of energy radiated away, the mass of the object, and the specific heat of the material.

KEY TERMS

absolute zero: The lowest possible temperature: 0 K, −273°C, or −459°F.

British thermal unit: The amount of heat required to raise the temperature of 1 pound of water by 1°F.

calorie: The amount of heat required to raise the temperature of 1 gram of water by 1°C.

change of state: The change in a substance between solid and liquid or liquid and gas.

conduction: The transfer of thermal energy by the collisions of the atoms or molecules within a substance.

conductor: A material that allows the easy flow of thermal energy. Metals are good conductors.

convection: The transfer of thermal energy in fluids by means of currents such as the rising of hot air and the sinking of cold air.

heat: Energy flowing due to a difference in temperature.

heat capacity: The amount of heat required to raise the temperature of an object by 1°C.

insulator: A material that is a poor conductor of thermal energy. Wood and stationary air are good thermal insulators.

internal energy: The total microscopic energy of an object, which includes its atomic and molecular translational and rotational kinetic energies, vibrational energy, and the energy stored in the molecular bonds.

latent heat: The amount of heat required to melt (or vaporize) 1 gram of a substance. The same amount of heat is released when 1 gram of the same substance freezes (or condenses).

radiation: The transport of energy via electromagnetic waves.

specific heat: The amount of heat required to raise the temperature of 1 gram of a substance by 1°C.

thermal energy: Internal energy.

thermal equilibrium: A condition in which there is no net flow of thermal energy between two objects. This occurs when the two objects obtain the same temperature.

thermal expansion: The increase in size of a material when heated.

thermodynamics: The area of physics that deals with the connections between heat and other forms of energy.

thermodynamics, first law of: The increase in the internal energy of a system is equal to the heat added plus the work done on the system.

thermodynamics, third law of: Absolute zero may be approached experimentally but can never be reached.

thermodynamics, zeroth law of: If objects A and B are each in thermodynamic equilibrium with object C, then A and B are in thermodynamic equilibrium with each other. All three objects are at the same temperature.

CONCEPTUAL QUESTIONS

1. In an avalanche, the snow and ice begin at rest at the top of the mountain and end up at rest at the bottom. What happens to the gravitational potential energy that is lost in this process?
2. What happens to the sound energy from your stereo speakers?
3. What evidence did Rumford have that heat was not a fluid?
4. Suppose a student was careless in re-creating Joule's experiment and allowed the masses to speed up quickly as they dropped toward the floor. If he equated the change in gravitational potential energy with the change in thermal energy, would he have found 1 calorie to be greater than or less than 4.2 joules? Explain.
5. How are the concepts of work and heat the same? How are they different?
6. What would you expect to find if you measure the temperature of the water at the top and bottom of Niagara Falls? Explain your reasoning.

7. It might be argued that the only time you measure the undisturbed temperature of a system is when the reading on the thermometer does not change when it is placed in

thermal contact with the system. Use the zeroth law to explain why this is so.

8. Imagine a universe where the zeroth law of thermodynamics was not valid. Would the concept of temperature still make sense in this universe? Why or why not?
9. Could two objects be touching but not be in thermal equilibrium? Explain.
10. Is it possible for a bucket of water in Los Angeles and a bucket of water in New York City to be in thermal equilibrium? Explain.
11. Why is it incorrect to talk about the flow of temperature from a hot object to a colder object?
12. On the inside back cover of this textbook are conversion factors between different units. Why is there no conversion factor between joules and kelvin?
13. What is the difference between heat and temperature?
14. The same amount of heat flows into two different buckets of water, which are initially at the same temperature. Will both buckets necessarily end up at the same temperature? Explain.
15. Patrick claims, "Two buckets of water must have the same heat if they are at the same temperature." Victoria counters, "That's true only if both buckets contain the same amount of water." With which, if either, of these students do you agree? Explain.
16. How do the internal energies of a cup of water and a gallon of water at the same temperature compare?
17. Under what conditions is the first law of thermodynamics valid?
18. Work is done on a system without changing the internal energy of the system. Does heat enter or leave the system during this process? Use the first law of thermodynamics to justify your answer.
19. Does it take more thermal energy to raise the temperature of 5 grams of water or 5 grams of ice by 6°C? Explain.
20. Which of the following does not affect the amount of internal energy an object has: its temperature, the amount of material, its state, the type of material, or its shape?
21. Liquid X and gas Y have identical specific heats. Would 100 calories of heat raise the temperature of 1 liter of liquid X by the same amount as 1 liter of gas Y? Explain your reasoning.
22. One kilogram of material A at 80°C is brought into thermal contact with 1 kilogram of material B at 40°C. When the materials reach thermal equilibrium, the temperature is 52°C. Which material, if either, has the greater specific heat? Explain.
23. A hot block of aluminum is dropped into 1000 grams of water at room temperature in a thermally insulated container where it reaches thermal equilibrium. If 500 grams of water had been used instead, would the amount of heat transferred to the water be greater than, equal to, or less than it was before? Why?
24. A hot block of iron is dropped into room-temperature water in a thermally insulated container where it reaches thermal equilibrium. If twice as much water had been used, would the water's temperature change be greater than, equal to, or less than it was before? Why?
25. Why do climates near the coasts tend to be more moderate than those in the middle of the continent?
26. Why does the coldest part of winter occur during late January and February when the shortest day is near December 21?
27. Given that the melting and freezing temperatures of water are identical, what determines whether a mixture of ice and water will freeze or melt?
28. If you make the mistake of removing ice cubes from the freezer with wet hands, the ice cubes stick to your hands. Why does the water on your hands freeze rather than the ice cubes melt?

29. Why can an iceberg survive for several weeks floating in seawater that's above freezing?

30. The boiling point for liquid nitrogen at atmospheric pressure is 77 K. Is the temperature of an open container of liquid nitrogen higher than, lower than, or equal to 77 K? Explain.
31. One hundred grams of ice at 0°C is added to 100 grams of water at 80°C. The system is kept thermally insulated from its environment. Will the equilibrium temperature of the mixture be greater than, equal to, or less than 40°C? Explain your reasoning.
32. An ice cube at 0°C is placed in a Styrofoam cup containing 200 grams of water at 60°C. When the system reaches thermal equilibrium, its temperature is 30°C. Was the mass of the ice cube greater than, equal to, or less than 200 grams? Explain your reasoning.
33. Why is steam at 100°C more dangerous than water at 100°C?
34. A new liquid is discovered that has the same boiling point and specific heat as water but a latent heat of vaporization of 10 calories per gram. Assuming that this new liquid is safe to drink, would it be more or less convenient than water for boiling eggs? Why?

35. A system is thermally insulated from its surroundings. Is it possible to do work on the system without changing its internal energy? Is it possible to do work on the system without changing its temperature? Explain.
36. In Washington, D.C., the weather report sometimes states that the temperature is 95°F and the humidity is 95%. Why does the high humidity make it so uncomfortable?
37. Use a microscopic model to explain how a metal rod transports thermal energy from the hot end to the cold end.
38. Why would putting a rug on a tiled bathroom floor make it feel less chilly to bare feet?
39. Rank the following materials in terms of their insulating capabilities: static air, glass, polyurethane foam, and concrete.
40. Which of the following is the best thermal conductor: fiberglass, stainless steel, wood, or silver?
41. If the temperature is 35°F and the wind is blowing at 20 mph, the equivalent windchill temperature is 24°F. Will a glass of water freeze in this situation? Explain your reasoning.
42. You hear on the morning weather report that the outside temperature is −5°F with a windchill-equivalent temperature of −40°F. You know that your old car, which is parked outside, will not start if the temperature of the battery drops below −15°F. Will your car start this morning? Why or why not?
43. The respective thermal conductivities of iron and stainless steel are 79 W/m·°C and 14 W/m·°C. Use these data to explain why you need to use potholders for pots with iron handles but not for pots with stainless steel handles.
44. Why might a cook put large aluminum nails in potatoes before baking them?
45. In northern climates drivers often encounter signs that read, "BRIDGE FREEZES BEFORE ROADWAY." Why does this occur?
46. You have just made yourself a hot cup of coffee and are about to add the cream, which is at room temperature. Suddenly the phone rings and you have to leave the room for a while. Is it better to add the cream to the coffee before you leave or after you get back if you want your coffee as hot as possible? Why?
47. When pilots fly under clouds, they often experience a downdraft. Why is this?

48. It is midafternoon and you are canoeing down a river that empties into a large lake. You are having a hard time making progress because of a stiff wind in your face. Is this situation likely to get better or worse at sunset? Explain.
49. A black car and a white car are parked next to each other on a sunny day. The surface of the black car gets much hotter than the surface of the white car. Which mode of energy transport is responsible for this difference?
50. Suppose you make a videotape in a café using infrared-sensitive equipment. How could you tell which coffee mugs are full?
51. A Thermos bottle is usually constructed from two nested glass containers with a vacuum between them, as shown in the figure. The walls are usually silvered as well. How does this construction minimize the loss of thermal energy?

Questions 51 and 52

52. Will a Thermos bottle (shown in the figure) keep something cold as well as it keeps it hot? Explain.
53. The metal roof on a wooden shed makes noises when a cloud passes in front of the Sun. Why?
54. Why might a glass dish taken from the oven and put into cold water shatter?
▲ 55. Suppose the column in an alcohol-in-glass thermometer is not uniform. How would the spacing between the degrees on a wide portion of the thermometer compare with those on a narrow portion?
▲ 56. When a mercury thermometer is first put into hot water, the level of the mercury drops slightly before it begins to climb. Why?

EXERCISES

1. How much heat is required to raise the temperature of 400 g of water from 20°C to 30°C?
2. If the temperature of 500 g of water drops by 8°C, how much heat is released?
▲ 3. How much work is required to push a crate with a force of 200 N across a floor a distance of 6 m? How many calories of thermal energy does the friction produce?
▲ 4. How many joules of gravitational potential energy are converted to kinetic energy when 100 g of lead shot falls from a height of 50 cm? How many calories are released to the surroundings if none of this kinetic energy is converted to other forms when the shot hits the floor?
▲ 5. A physics student foolishly wants to lose weight by drinking cold water. If he drinks 1 L (1000 cm^3) of water at 10°C below body temperature, how many Calories will it take to warm the water?
6. A typical jogger burns up food energy at the rate of about 40 kJ per minute. How long would it take to run off a piece of cake if it contains 400 Calories?
7. During a process, 20 J of heat are transferred into a system, while the system itself does 12 J of work. What is the change in the internal energy of the system?
8. What is the change in the internal energy of a system if 15 J of work are done on the system and 6 J of heat are removed from the system?
9. When an ideal gas was compressed, its internal energy increased by 180 J and it gave off 150 J of heat. How much work was done on this gas?
10. If the internal energy of an ideal gas increases by 150 J when 240 J of work is done to compress it, how much heat is released?
11. It takes 250 cal to raise the temperature of a metallic ring from 20°C to 30°C. If the ring has a mass of 90 g, what is the specific heat of the metal?
12. If it takes 3400 cal to raise the temperature of a 500-g statue by 44°C, what is the specific heat of the material used to make the statue?
13. How many calories will it take to raise the temperature of a 50-g gold chain from 20°C to 100°C?
14. How many calories would it take to raise the temperature of a 400-g aluminum pan from 293 K to 373 K?
15. Six grams of liquid X at 35°C are added to 3 grams of liquid Y at 20°C. The specific heat of liquid X is 2 cal/g·°C, and the specific heat of liquid Y is 1 cal/g·°C. If each gram of liquid X gives up two calories to liquid Y, find the change in temperature of each liquid.
16. In Exercise 15, imagine that liquid X continues to transfer energy to the other liquid 12 calories at a time. How many transfers would be required to reach a common temperature? What is this equilibrium temperature?
17. Eighty grams of water at 70°C are mixed with an equal amount of water at 30°C in a completely insulated container. The final temperature of the water is 50°C.
    a. How much heat is lost by the hot water?
    b. How much heat is gained by the cold water?
    c. What happens to the total amount of internal energy of the system?
▲ 18. If 200 g of water at 100°C are mixed with 300 g of water at 50°C in a completely insulated container, what is the final equilibrium temperature?
19. A kettle containing 3 kg of water has just reached its boiling point. How much energy, in joules, is required to boil the kettle dry?
20. How much heat would it take to melt a 1-kg block of ice?
21. You wish to melt a 3-kg block of aluminum, which is initially at 20°C. How much energy, in joules, is required to heat the block to its melting point of 660°C? How much energy, in joules, is required to melt the aluminum without changing its temperature?
▲ 22. How much heat is required to convert 400 g of ice at −5°C to water at 5°C? (*Hint*: review Exercise 21.)
23. What is the change in length of a metal rod with an original length of 2 m, a coefficient of thermal expansion of 0.00002/°C, and a temperature change of 20°C?
24. A steel railroad rail is 24.4 m long. How much does it expand during a day when the low temperature is 50°F (18°C) and the high temperature is 91°F (33°C)?

14 Available Energy

The energy of the hot water and steam from a geyser can be used to run engines, yet the internal energy of a pail of room-temperature water, although quite high, isn't as useful as an energy source. Why is this true, and what would it take to extract this internal energy and make it available for performing useful work?

(See page 282 for the answer to this question.)

Old Faithful in Yellowstone National Park, Wyoming. Hot water from geysers can be used as an energy source.

MECHANICAL energy can be completely converted into the internal energy of an object. This is clearly demonstrated every time an object comes to rest due to frictional forces. In the building of a scientific world view, the belief in a symmetry often leads to interesting new insights. In this case it seems natural to ask if the process can be reversed. Is it possible to recover this internal energy and get some mechanical energy back? The answer is yes, but it is not an unqualified yes.

Imagine that we try to run Joule's paddle-wheel experiment (Figure 14-1) backward. Suppose we start with hot water and wait for the weight to rise up from the floor. Clearly, we don't expect this to happen. The water can be heated to increase its internal energy, but the hot water will not rotate the paddle wheel. Water can be very hot and thus contain a lot of internal energy, but mechanical work does not spontaneously appear.

The first law of thermodynamics doesn't exclude this; it places no restrictions on which energy transformations are possible. As long as the internal energy equals or exceeds that needed to raise the weight, there would be no violation of the first law if some of the internal energy were used to do this. And yet it doesn't happen. The energy is there, but it is not available. Apparently the availability of energy depends on the form that it takes.

This aspect of nature, unaccounted for so far in the development of our world view, is addressed in the second law of thermodynamics. It's a subtle law that has a rich history, resulting in three different statements of the law. The first form deals with heat engines.

Physics Now™ Test your understanding of this chapter by logging into PhysicsNow at **http://physics.brookscole.com/kf6e**, selecting the chapter, and clicking on the "Take a Pre-Test" link.

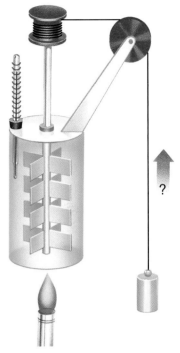

Figure 14-1 Why won't the weight rise when we heat the water?

Heat Engines

We can extract some of an object's internal energy under certain circumstances. Internal energy naturally flows from a higher-temperature region to one of lower temperature. A hot cup of coffee left on your desk cools off as some of its thermal energy flows to the surroundings. The coffee continues to cool until it reaches thermal equilibrium with the room. Energy flows out of the hot region because of the presence of the cold one. But no work—no mechanical energy—results from this flow.

Many schemes have been proposed for taking part of the heat and converting it to useful work. Any device that does this is called a **heat engine**. The simplest and earliest heat engine is traced back to Hero of Alexandria. About AD 50, Hero invented a device similar to that shown in Figure 14-2. Heating the water-filled container changes the water into steam. The steam, escaping through the two tubes, causes the container to rotate. From our modern perspective, we might judge this to be more of a toy than a machine. The importance of this device, however, was that mechanical energy was in fact obtained—the container rotated. Apparently, Hero did do something useful with his heat engine; there are stories that he used pulleys and ropes to open a temple door during a religious service. (Probably much to the shock of the worshipers!)

Figure 14-2 A modern version of Hero's heat engine rotates under the action of the escaping steam.

PHYSICS | ON YOUR OWN

Construct the Hero heat engine illustrated in Figure 14-3. Punch holes large enough for small straws on opposite sides of a 1-pound coffee can. Put a piece of straw in each hole and bend and tape them so that they both point in the clockwise direction. Fill the can about one-quarter full of water and seal the lid with tape. Suspend the can over a heating element as shown. The escaping steam will cause the can to rotate rapidly. *Be careful* of the hot water and steam.

The Industrial Revolution began with the move away from animal power toward machine power. The first heat machines were used to pump water out of

Figure 14-3 A coffee-can steam engine.

Figure 14-4 The essential features of Watt's steam engine. Opening valve 1 lets steam into the chamber, raising the piston. Closing valve 1 and opening valve 2 reduces the pressure, allowing the piston to fall.

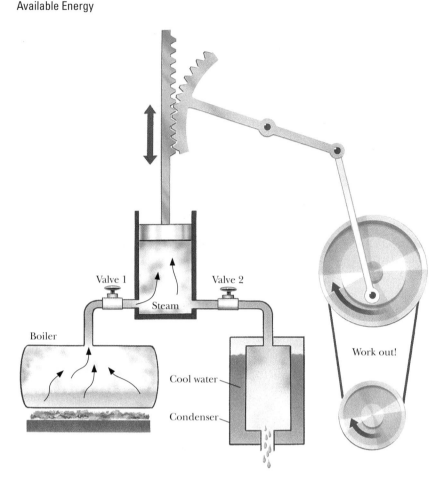

The steam locomotive is a heat engine.

The internal combustion engine in a Honda automobile is an example of a modern heat engine.

mines in England. Waste coal was cheap at the pithead, and the engines were inefficient. By the 1790s James Watt had developed steam engines that were more thermally efficient and were increasingly used in applications requiring powerful and reliable energy sources, such as water systems, mills, and forges.

Figure 14-4 shows the essential features of the most successful of the early steam engines. Invented by Watt in 1769, it had a movable piston in a cylinder. The flow of heat was not a result of direct thermal contact between objects but of the transfer of steam that was heated by the hot region and cooled by the cold region. By alternately heating and cooling the steam in the cylinder, the piston was driven up and down, producing mechanical work.

The opening of the American West was helped by another heat engine, the iron horse, or steam locomotive. A more modern heat engine is the internal combustion engine used in automobiles. Replacing the wood and steam, gasoline explodes to produce the high-temperature gas. The explosions move pistons, allowing the engine to extract some of this energy to run the automobile. The remaining hot gases are exhausted to the cooler environment.

Although there are many types of heat engine, all can be represented by the same schematic diagram. To envision this, recall that all heat engines involve the flow of energy from a hotter region to a cooler one. Figure 14-5(a) shows how we represent this flow. A heat engine is a device placed in the path of this flow to extract mechanical energy. Figure 14-5(b) shows heat flowing from the hotter region; part of this heat is converted to mechanical work, and the remainder is exhausted to the colder region. You can verify that heat is thrown away by an automobile's engine by putting your hand near (but not directly on) the exhaust pipe. The exhausted gases are hotter than the surrounding air. Without the cool region, the flow would stop, and no energy would be extracted.

Question How much work does a heat engine perform if it extracts 100 joules of energy from a hot region and exhausts 60 joules to a cold region?

Answer Conservation of energy requires that the work be equal to 40 joules, the difference between the input and the output.

Ideal Heat Engines

The first law of thermodynamics requires that the sum of the mechanical work and the exhausted heat be equal to the heat extracted from the hot region. But exhausted heat means wasted energy. When engines were developed at the beginning of the Industrial Revolution, engineers asked, "What kind of engine would get the maximum amount of work from a given amount of heat?"

In 1824 French army engineer Sadi Carnot published the answer to this question. Carnot found the best possible engine by imagining the whole process as a thought experiment. He assumed that his engine would use idealized gases and there would be no frictional losses due to parts rubbing against each other. His results were surprising. Even under these ideal conditions, the heat engine must exhaust some heat.

Carnot's work led to one version of the **second law of thermodynamics** and an understanding of why internal energy cannot be completely converted to mechanical energy. That is, even the best theoretical heat engine cannot convert 100% of the incoming heat into work.

> It is impossible to build a heat engine to perform mechanical work that does not exhaust heat to the surroundings.

◂ heat-engine form of the second law of thermodynamics

Stated slightly differently, the fact that the engine must throw away heat means that no heat engine can run between regions at the same temperature. This is unfortunate, because it would be a boon to civilization if we could "reach in" and extract some internal energy from a single region. For example, think of all the energy that would be available in the oceans. If we could build a single-temperature heat engine, it could be used to run an ocean liner. Its engine could

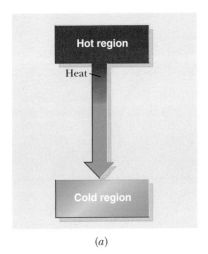
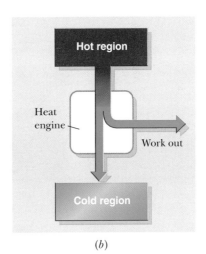

(a)   (b)

Figure 14-5 (a) A schematic showing the natural flow of thermal energy from a higher-temperature region to a lower-temperature region. (b) The white square represents the many types of heat engine. The heat engine converts part of the heat from the hot region to mechanical work and exhausts the remaining heat to the cold region.

Figure 14-6 An impossible engine in an ocean liner extracts energy from ocean water, makes ice cubes, and propels the ship.

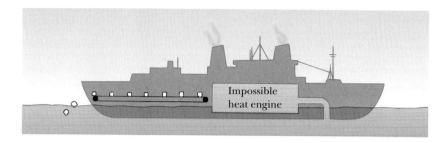

take in ocean water at the front of the ship, extract some of the water's internal energy, and drop ice cubes off the back, as shown in Figure 14-6. Notice that this hypothetical engine isn't intended to get something for nothing; it only tries to get out what is there. But the second law says that this is impossible.

Perpetual-Motion Machines

Since the beginning of the Industrial Revolution, inventors and tinkerers have tried to devise a machine that would run forever. This search was fueled by the desire to get something for nothing. A machine that did some task without requiring energy would have countless applications in society, not to mention the benefits that would accrue to its inventor. No such machine has ever been invented; in fact, the only "successful" perpetual-motion machines have been devices that were later shown to be hoaxes.

In the 17th century, English physician Robert Fludd proposed the device shown in Figure 14-7. Fludd wanted to turn the waterwheel to move the millstone and at the same time return the water to the upper level. It didn't work. Machines like Fludd's violate the first law of thermodynamics by trying to get more energy out of a device than is put into it. Every machine is cyclic by definition. After doing something the machine returns to its original state to start the whole process over again. Fludd's scheme was to have the machine start with a certain amount of energy (the potential energy of the water on the upper level), do some work, and then return to its original state. But if it did some work, some energy was spent. Returning the machine to its original state would require getting something for nothing, a violation of the first law.

Other perpetual-motion machines have been less ambitious. They were sim-

Figure 14-7 Fludd's proposed perpetual-motion machine cannot work because it would violate the first law of thermodynamics.

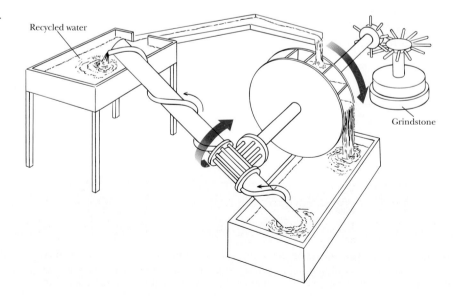

ply supposed to run forever without extracting any useful output. They also didn't work. Even if we ask nothing of such a machine except to run—and thus stay within the confines of the first law of thermodynamics—it still will not run forever. These machines failed because they were attempting to violate the second law of thermodynamics.

To illustrate this, suppose we invent such a machine as a thought experiment. Figure 14-8 is a diagram of one possibility. A certain amount of energy is initially put into the system, and somehow, the system is started. The paddle wheel turns in the water and, as the Joule experiment showed us, produces an increase in the temperature of the water. Mechanical energy has been converted to internal energy as indicated by the water's temperature rise. Thermal energy from the hot water flows to the pad below the water. The pad is a heat engine designed to capture this energy and convert it to mechanical energy. The mechanical energy is in turn used to drive gears that rotate the paddles.

Does the machine continue to run? Notice that this is a perfect machine, and we can assume that there are no frictional losses; so if it fails, it does so because of some fundamental reason. Such engines have been built—and they have all failed. The problem lies with the heat engine in the pad. Any heat engine, no matter how good, returns less mechanical energy than the thermal energy it receives. Therefore, our machine runs down, and we have failed. We do not fail because we try to create energy but because we cannot use the energy that we have. Once the energy is converted to internal energy, it is no longer fully available.

Even in a simpler machine without a heat engine, there would still be moving parts. Any rubbing of moving parts increases the machine's internal energy. All this internal energy would have to be converted back into mechanical energy if the machine were to run forever. The second law tells us that this cannot happen.

All machines and heat engines, no matter how complicated, cannot avoid the constraints imposed by the first and second laws of thermodynamics.

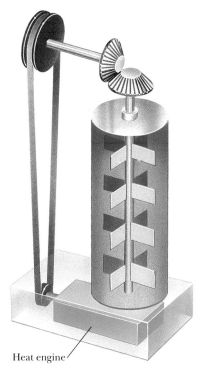

Figure 14-8 This proposed perpetual-motion machine would violate the second law of thermodynamics.

Flawed Reasoning

A classmate claims to suddenly understand the second law of thermodynamics. He explains, "All machines necessarily have moving parts that experience friction. This is why a heat engine can never transform its entire heat input into work." **Do you think your classmate understands the second law?**

Answer Your classmate seems to understand the first law of thermodynamics (energy conservation) but is missing the key point of the second law. The second law of thermodynamics states that even in the absence of any friction it is still impossible to transform thermal energy entirely into mechanical energy. Some of the input energy must be exhausted.

Real Engines

All engines can be rated according to their efficiencies. In general, efficiency can be defined as the ratio of the output to the input. In the case of heat engines, the **efficiency** η of an engine is equal to the ratio of the work W produced divided by the heat Q extracted from the hot region:

$$\eta = \frac{W}{Q}$$

◀ efficiency = $\dfrac{\text{work out}}{\text{heat in}}$

✓ Extended presentation available in the *Problem Solving* supplement

Carnot's ideal heat engine has the maximum theoretical efficiency. This efficiency can be expressed as a simple relationship using Kelvin temperatures:

> efficiency of an ideal heat engine ▶

$$\eta = 1 - \frac{T_c}{T_h}$$

The Carnot efficiency depends only on the temperatures of the two regions. It is larger (that is, closer to 1) if the temperature of the exhaust region T_c is low or that of the hot region T_h is high. The largest efficiency occurs when the temperatures are as far apart as possible. Usually, this efficiency is multiplied by 100 so that it can be stated as a percentage. Real engines produce more waste heat than Carnot's ideal engine, but his relationship is still important because it sets an upper limit for the efficiencies of real engines.

Today, steam engines are used primarily to drive electric generators in power plants. Although these engines are certainly not Carnot engines, their efficiencies can be increased by making their input temperatures as high as possible and their exhaust temperatures as low as possible. Due to constraints imposed by the properties of the materials used in their construction, these engines cannot be run much hotter than 550°C. Exhaust temperatures cannot be much lower than about 50°C. Given these constraints, we can calculate the maximum theoretical efficiency of the steam engines used in electricity-generating plants. Remembering to convert the temperatures from Celsius to Kelvin, we have

$$\eta = 1 - \frac{T_c}{T_h} = 1 - \frac{323 \text{ K}}{823 \text{ K}} = 0.61 = 61\%$$

A coal-fired electricity-generating plant.

Actual steam engines have efficiencies closer to 50%. Besides the efficiency of the engine, we must consider the efficiency of the boiler for converting the chemical energy of the fuel into heat. Typical oil- or coal-fired power plants have overall efficiencies of about 40% or less. Nuclear power plants use uranium as a fuel to make steam. Safety regulations require that they run at lower temperatures, so they are 5–8% less efficient than the oil- or coal-fired plants. In other words, nuclear plants exhaust more waste heat for each unit of electricity generated.

Question What is the maximum theoretical efficiency for a heat engine running between 127°C and 27°C?

Answer Being careful to convert these temperatures to the Kelvin scale, we have

$$\eta = 1 - \frac{T_c}{T_h} = 1 - \frac{300 \text{ K}}{400 \text{ K}} = 1 - 0.75 = 0.25 = 25\%$$

Cooling towers at the nuclear power generating facility on Three Mile Island near Harrisburg, Pennsylvania.

Not all engines are heat engines. Those that don't go through a thermodynamic process are not heat engines and can have efficiencies closer to 100%. Electric motors are close to being 100% efficient if you look only at the electric motor. However, the overall efficiency of the electric motor must include the energy costs of generating the electricity and the energy losses that occur in the electric transmission lines. Because most electricity is produced by steam engines, the overall efficiency is quite low.

Refrigerators   ✓ MATH

There are devices that extract heat from a cooler region and deposit it into a hotter one. These are essentially heat engines running backward. **Refrigerators** and air conditioners are common examples of such devices. By reversing the directions of the arrows in Figure 14-5(b), we produce the schematic of the process shown in Figure 14-9.

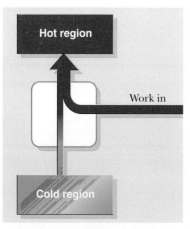

Figure 14-9 A refrigerator uses mechanical work to transfer heat from a colder region to a hotter one.

Refrigerators move heat in a direction opposite to its natural flow. The inside of your refrigerator is the cold region, and your room is the hot region. The refrigerator removes heat from the inside and exhausts it to the warm region outside. If you put your hand by the base of the refrigerator, you can feel the heat being transferred to the kitchen. Similarly, if you walk by the external part of an air conditioner, you feel heat being expelled outside the house. In both cases the warm region gets hotter, and the cool gets colder.

The icebox of olden days was a very simple device; place a block of ice on the top shelf and your produce on the bottom shelf. The modern refrigerator is almost as simple. It is constructed from the four basic parts illustrated in Figure 14-10. The *evaporator* (1) is a long metal tube that allows the low-pressure, cold liquid refrigerant to have thermal contact with the air inside the refrigerator. When a liquid evaporates, it absorbs heat from its surroundings. A refrigerant, such as Freon-12, is often used because it evaporates at a low temperature (−29°C). The *compressor* (2) pressurizes the gas, and the resulting high-pressure gas passes through the *condenser* (3), a second long metal tube outside the refrigerator. Thermal energy is released into the kitchen as the gas condenses back into a liquid. The liquid cools as it passes through the *expansion valve* (4), and the process repeats.

Refrigerators require work to move energy from a lower-temperature region to a higher-temperature region. Therefore, the heat delivered to the higher-temperature region is larger than that extracted from the lower-temperature region. The extra heat is the amount of work done on the system. In fact, this process leads to an equivalent statement of the second law of thermodynamics.

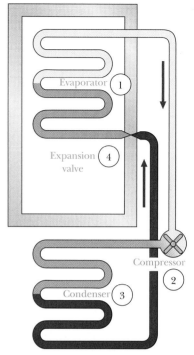

Figure 14-10 A schematic for a refrigerator.

> It is impossible to build a refrigerator that can transfer heat from a lower-temperature region to a higher-temperature region without using mechanical work.

◄ refrigerator form of the second law of thermodynamics

A **heat pump** is used in some homes to both cool during the summer and heat in winter. It is simply a reversible heat engine. In the summer the heat pump functions as an air conditioner by extracting heat from inside your house. In the winter it runs in reverse (like a refrigerator), extracting energy from the outside air and putting it into your house.

It might seem strange to be able to extract heat from the cold air. Remember that heat and temperature are not the same. Even on the coldest days, the outside air still has enormous amounts of internal energy. The heat pump transfers some of this internal energy from outside to inside. Actually, there is an economical limit. If the outside temperature is too low, it takes more energy to run the heat pump than it does to use electric baseboard heaters. This situation can be corrected if a warmer region such as underground water is available as a source of the internal energy.

Heat pumps can be used to cool houses in summer and heat them in winter. They can extract energy from the outside air even in cold weather.

Question How much energy does an air conditioner exhaust if it requires 200 joules of mechanical work to extract 1000 joules of energy from a house?

Answer Conservation of energy requires that the exhaust equal 1200 joules, the sum of the inputs.

Order and Disorder

Whether it is a heat engine or a refrigerator, the second law of thermodynamics essentially says that we can't break even. If we are trying to get mechanical energy from a thermal source, we have to throw some heat away. If we want to cool something, we have to do work to counteract the natural flow of energy. But

A new deck of cards has a high degree of order.

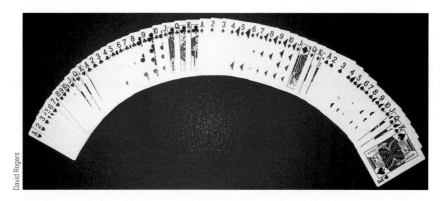

these two forms of the second law offer little insight into why this is so. The reason lies in the microscopic behavior of the many-particle systems that make up matter. To understand the second law of thermodynamics, we need to look at the details of such systems.

We begin by looking at systems in general. Any system consists of a collection of parts. Assuming that these parts can be shifted around, the system has a number of possible arrangements. We will examine systems and determine their organization because, as we will see, it is their organizational properties that lie at the heart of the second law. An organizational property can be such a thing as the position of objects in a box or the height of people in a community.

Suppose your system is a new deck of playing cards. The cards are arranged according to suits and within suits according to value. The organizational property in this case is the position of the card within the deck. If somebody handed you a card, you would have no trouble deciding where it came from. If the deck were arranged by suit but the values within the suits were mixed, you would not be able to pinpoint the location of the card, but you could say from which fourth of the deck it came. In a completely shuffled deck, you would have no way of knowing the origin of the card. The first arrangement is very organized, the second less so, and the last one very disorganized. We call a system that is highly organized an **ordered system** and one that shows no organization a **disordered system**.

Another way of looking at the amount of organization is to ask how many equivalent arrangements are possible in each of these situations. When the deck is arranged by suit and value, there is only one arrangement. When the values are shuffled within suits, there are literally billions of ways the cards could be arranged within each suit. When we completely shuffle the deck, the number of possible arrangements becomes astronomically large (8×10^{67}). The more disordered system is the one with the larger number of equivalent arrangements.

A simple system of three coins on a tray can model microscopic order. An obvious organizational property of this system is the number of heads (or tails) facing up. How many different arrangements are there? If the coins are different, there are eight different arrangements for the three coins as shown in Figure 14-11.

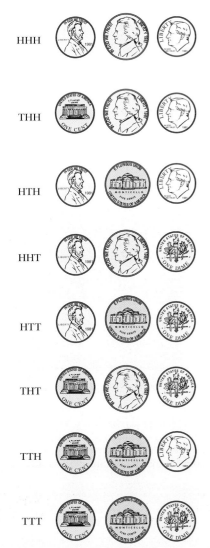

HHH

THH

HTH

HHT

HTT

THT

TTH

TTT

Figure 14-11 The eight possible arrangements of three different coins.

Question How many different arrangements can you make with three colored blocks: one *r*ed, one *b*lue, and one *g*reen?

Answer The arrangements are *rbg*, *rgb*, *bgr*, *brg*, *grb*, and *gbr*. Therefore, there are six possible arrangements.

In real situations the macroscopic property—for example, the total energy—doesn't depend on the actual properties of a particular atom. To apply our coin analogy to this situation, we would examine the different arrangements without identifying individual coins. In other words, we should simply count the number of heads and tails. How many different arrangements are there now? Four. Of the eight original arrangements, two groups of three are now indistinguishable as shown in Figure 14-12. Each of these groups has three *equivalent states*. The arrangements with all heads or all tails occur in only one way and thus have only one state each. Arrangements with only one state have the highest order. The arrangements that occur in three ways have the lowest order.

Understanding the microscopic form of the second law of thermodynamics depends on realizing that the order of real systems is constantly changing. This dynamic nature of systems occurs because macroscopic objects are composed of an immense number of atomic particles that are continually moving and therefore changing the order of the system. We can use our coins to understand this.

To model the passage of time, we shake the tray to flip the three coins. If we repeatedly start our system with a particular high-order arrangement, we observe that after one flip, the amount of order in the system usually decreases. Each of the eight states has an equal probability of occurring. However, the probability of a low-order arrangement occurring is not equal to that of a high-order arrangement, simply because there are more low-order possibilities. Of the eight possible states, only two yield high-order arrangements. The chance is only 2 in 8, or 25%, of obtaining a high-order arrangement. The other 75% of the time we expect to obtain two heads and a tail or two tails and a head.

PHYSICS | **ON YOUR OWN**

> Flip three coins 80 times. How many times do you get all heads? All tails? Two heads and one tail? Two tails and one head? Are your numbers consistent with the predicted odds? Would you expect the consistency to improve if you threw the coins 8000 times?

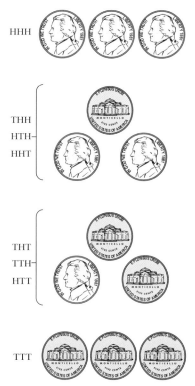

Figure 14-12 The four possible arrangements of three indistinguishable coins.

If we increase the number of coins, the probabilities change. With four coins, for example, the number of possible states grows to 16. Adding one additional coin doubles the total number of combinations because we are adding a head or a tail to each of the previous combinations. But again, there are only two arrangements with the highest order, those having all heads or all tails. The probability now of obtaining a high-order arrangement is 2 in 16, or about 12%. As the number of elements in the system increases, the probability of obtaining a high-order arrangement gets smaller. Conversely, the probability of obtaining a low-order arrangement increases with the size of the system.

Question   What is the probability of obtaining all heads or all tails with five coins?

Answer   Adding the additional coin doubles the total number of possible combinations to 32. Therefore, we have 2 chances out of 32, or 6%, of obtaining all tails or all heads.

For example, if we have 8 coins, the probability of obtaining the highest-order distribution (all heads or all tails) is less than 1%. If we increase the number of coins to 21, we have only one chance in a million of getting all heads or all tails. If the number of coins increases to that of the number of air molecules in a typical bedroom (10^{27}), the probability decreases to 1 in 10^{82}.

As another example, let's consider the air molecules in your room. Presumably, any air molecule can be anywhere in the room. A high-order arrangement might have all the molecules in one small location; a low-order arrangement would have them spread uniformly throughout the available space in one of many equivalent states. The number of low-order arrangements is astronomically higher than the number of high-order ones.

> **Flawed Reasoning**
>
> Sandra is explaining a get-rich-quick scheme to her friend Lynda.
>
> **Sandra:** "I have been flipping this nickel and the last 19 flips have all been heads! We should take this nickel to Las Vegas and bet someone that the next flip will be tails. We can't lose! The chance of a nickel landing heads 20 times in a row is less than one in a million."
>
> **Lynda:** "Let's stay home. If that nickel is not rigged, then any toss is as likely to give tails as heads. You would have only a 50% chance of winning your bet, even if the last 19 tosses turned up heads."
>
> **Should they go to Vegas?**
>
> **Answer** Lynda is right. The results of past tosses of the nickel do not affect the next toss. Unless the nickel is weighted, every toss has an equal probability of being heads or tails. It's amazing that Sandra was able to get heads 19 times in a row. The odds against that are nearly 500,000 to 1.

Entropy

We now introduce a new concept, called **entropy**, as a measure of a system's organization. A system that has some recognizable order has low entropy—the more disorganized the system, the higher its entropy. As with gravitational potential energy, we are concerned only with changes in entropy; the actual numerical value of the entropy does not matter.

We argued in the last section that the order of a system of many particles tends to decrease with time. Therefore, the entropy of the system tends to increase. This tendency is expressed by another version of the second law of thermodynamics.

▶ entropy form of the second law of thermodynamics

> The entropy of an isolated system tends to increase.

The word *tends* needs to be stressed. There is nothing that says the entropy *must* increase. It happens because of the overwhelming odds in its favor. It is still *possible* for all the air molecules to be in one corner of your room. (That would leave you gasping for air if you weren't in that corner!) Nothing prohibits this from happening. Fortunately, the likelihood of this actually happening is vanishingly small.

This entropy version looks so different from the forms of the second law developed for heat engines and refrigerators that you might think that they are different laws. The connection between them is that the motion associated with internal energy has a low-order arrangement, whereas macroscopic motion requires a high degree of organization. Atomic motions are random; at any instant, particles are moving in many directions with a wide range of speeds. If we could take a series of snapshots of these particles and scramble the snapshots, we could not distinguish them; one snapshot is just like another. There appears to be no order to the motions or positions of the particles.

The macroscopic motion of an object, on the other hand, gives organization to the motions of the object's atoms. Although the particles are randomly moving in all directions on the atomic level, they also have the macroscopic motion of the object. All the atoms in a ball falling with a certain kinetic energy are essentially moving in the same direction (Figure 14-13). When the ball strikes the

Arrow of Time

The concept of entropy gives insight into the character of time. Imagine a motion picture of a soccer game played backward. It looks silly because the order of the events doesn't match our experiences. There is a direction to time. But why do things happen in a particular sequence? According to most of the laws of nature we have established, other sequences of events *could* happen: The parabolic curve of the thrown football is the same regardless of which direction the film is played. Yet the game—played backward—is unreal. Even a repeating, cyclic event like a child on a swing has clues telling you whether the movie is running forward or backward. If the child ceases pumping, the swinging dies down because of frictional effects. Left without an input of energy, everyday motions stop when the macroscopic energy gets transformed into internal energy.

In principle, a stationary pendulum bob could transfer some of its internal energy into swinging the bob. This would (simply) require that the bob's atoms all move in the same direction at the same time. They are all moving but in random directions. It could happen that at some time they could all be moving in the same direction. It could happen, but it doesn't. The entropy form of the second law of thermodynamics tells us why: all systems tend toward disorder. This fact of nature gives a direction to time.

The direction of time is implicit in many events. Bright new paint has a radiant color because of the particular molecules and their particular arrangement. As time passes, there is a mixing of the molecules with the air and a rearranging of the molecules in the paint—and the paint fades. Structures crumble; their original shapes are very ordered, but as time passes, they deteriorate. Order is lost. The culprit is chance, and the consequence is increasing entropy. Time has a direction because the Universe has a natural tendency toward disorder.

A desk work area is an example of a system that tends toward disorder.

ground, the energy is not lost; the collective motions are randomized. The atoms move in all directions with equal likelihood.

Although it is certainly physically possible for all the particles to once again move in a single direction at some future time (without somebody picking it up), it is so unlikely that we never expect to see it happen. If it did happen, however, some of the internal energy of the ball's particles would then be available to do some mechanical work. This would be very bizarre; at some moment, a ball initially at rest would suddenly fly off with some kinetic energy.

Decreasing Entropy

Although the overall entropy of a system tends to increase, isolated pockets of activity can show a decrease in entropy—that is, an increase in order. But these increases in order are paid for with an overall decrease in the order of the Universe. Life, for example, is a glorious example of increasing order. It begins as a simple fertilized egg and evolves into a complicated human being. Clearly, order is increasing. But this increase is accomplished through the use of energy. To bring about the increase in human order, a great deal of energy becomes less available for performing useful work. The order of the environment decreases accordingly.

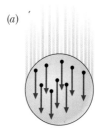
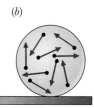

Figure 14-13 (a) The motions of atoms in a falling ball have a high degree of order. (b) The motions after the ball hits the ground are random with a low degree of order.

Quality of Energy

With the exceptions of a geothermal source of hot water heating a home or a windmill pumping water out of a well, most energy that we use in our lives has been converted from its original form to the form we use. An obvious example is the electric energy that lights our homes. This energy often comes from a fossil fuel that is burned to heat water to form steam, which turns a generator to produce electricity.

It is not possible to convert, or even transfer, energy without losing some of the energy in the process. (By *losing energy* we mean that some of the energy is transformed to some form other than the one desired.) Even charging a battery—storing energy for later use—requires more energy than gets stored in the battery. A more obvious loss is that due to frictional forces. If we use a waterwheel to turn an electric generator, some of the input energy is transformed to thermal energy by the frictional forces in the rubbing of the mechanical parts. Losses also occur when electric energy is transported from one place to another. A final loss comes from the thermal bottleneck described by the second law of thermodynamics. Whenever we convert thermal energy to *any* other form of energy, the second law of thermodynamics tells us that we always lose some energy to other forms: It is impossible to make the conversion with 100% efficiency.

All conversions of energy are not equivalent. Some energy conversions are less efficient and thus more costly to our world's energy budget. To obtain an accurate measure of the efficiency of the energy you use, you need to go back to the original source and calculate the overall efficiency of delivering it to your home and its final use. Suppose, for example, that the initial conversion is 50% efficient. For every unit of energy you start with, you have to give up one-half unit. If the process of transporting the energy to your home is also 50% efficient, the amount that you get to use is one-half of one-half, or one-fourth, of the original. To calculate the total efficiency, you multiply the fractional efficiencies of each stage together.

The table shows the relative efficiencies of two methods of heating water in your home. The efficiency of each step is shown as well as the overall efficiency through that step. Notice that the lowest efficiency occurs during the generation of electricity in the coal-fired plant. The extra steps of making and transporting the electricity reduce the overall efficiency of electric heat by a factor of $2\frac{1}{2}$ compared with heating with natural gas. Therefore, the quality of natural gas for heating is much higher than that of electricity produced from coal.

Gas water heaters are more efficient than electric ones.

Efficiency for Heating Water

	Step Efficiency (%)	Accumulative Efficiency (%)
Electric		
Mining of coal	96	96
Transportation of coal	97	93
Generation of electricity	33	31
Electric transmission	85	26
Heating	92	24
Gas		
Production of natural gas	96	96
Transportation	97	93
Heating	64	60

As a simpler example, imagine that you have two buckets of water, one hot and the other cold. If you place a heat engine between the buckets, you can do some work. After the heat engine operates for a while, the buckets reach the same temperature. The system now has a higher entropy. You can decrease the entropy by heating one bucket and cooling the other. But you can only do this by increasing the entropy of the Universe. We saw this with refrigerators. You can make the hot region hotter and the cold region colder but only after mechanical work is put in. That use of work lowers the availability of the energy in the Universe to do mechanical work.

Entropy and Our Energy Crisis

Save energy! That is the battle cry these days. But why do we have to *save* energy? According to the first law of thermodynamics, energy is neither created nor destroyed. The total amount of energy is constant. There is no need to conserve energy; it happens naturally. The crux of the matter is really the second law. There are pockets of energy in our environment that are more valuable than others. Given the proper conditions, we can use this energy to do some useful work for society. But if we later add up all the energy, we still have the same amount. The energy used to drive our cars around town is transformed via frictional interactions and exhaust, the consumption of food results in our body temperatures being maintained as well as moving us around, and so on. All the energy is present and accounted for.

Water naturally flows downhill. The water is still there at the bottom of the hill, but it is less useful. If all the water is at the same level, there is no further flow. Similarly, if all the energy in the Universe is spread out uniformly, we can get no more "flow" from one pocket to another.

The real energy issue is the preservation of the valuable pockets of energy. We can burn a barrel of oil only once. When we burn it, the energy becomes less useful for doing work, and the entropy of the Universe goes up. The second law doesn't tell us how fast entropy should increase. It does not tell us how fast we must use up our pockets of energy; it only says that it *will* occur. The decision of finding an acceptable rate is left to us. The battle cry for the future should be, "Slow down the increase in entropy!"

Summary

Mechanical energy can be completely converted into internal energy, but it is not possible to recover all this internal energy and get the same amount of mechanical energy back. The energy is there, but it is not available.

We can extract some of an object's internal energy using a heat engine. Part of the heat is converted to mechanical work, and the remainder is emitted as exhaust. The first law of thermodynamics requires that the sum of the mechanical work and the exhaust heat be equal to the heat extracted. The second law of thermodynamics says that we cannot build a heat engine without an exhaust. As a consequence, it is impossible to completely convert heat to mechanical energy.

People have tried to construct two kinds of perpetual-motion machines. The first kind violates the first law of thermodynamics by trying to get more energy out of a device than is put into it. The perpetual-motion machines of the second kind do not try to do any useful work but just try to run forever. They fail because they violate the second law of thermodynamics.

The efficiency of an ideal heat engine—a Carnot engine—depends on the temperatures of the input heat source and the exhaust region. The maximum theoretical efficiency increases as the difference in these two temperatures increases and is given by $\eta = 1 - T_c/T_h$, where T_h is the input temperature and T_c is the exhaust temperature in kelvin. The efficiencies of real engines are always less than this.

Refrigerators and air conditioners are essentially heat engines running backward. These devices require work to move energy from a lower-temperature region to a higher-temperature region. An equivalent form of the second law says that we cannot build a refrigerator that moves energy from a colder region to a hotter region without work being done.

The third form of the second law says that the entropy of an isolated system tends to increase. Entropy is a measure of the order of a system. Entropy increases as the order within a system decreases. The entropy of a system can decrease, but the decrease is paid for with an overall increase in the entropy of the Universe. Life is an example of decreasing entropy (increasing order).

PhysicsNow™ Assess your understanding of this chapter's topics with sample tests and other resources found by logging into PhysicsNow at **http://physics.brookscole.com/kf6e**.

Our energy crisis is really related to the using up of the pockets of energy in our environment that are more valuable than others. We still have the same amount of energy as we've always had.

Chapter 14 Revisited

Performing useful work with the internal energy in an object requires that the energy be able to flow from the object to a region at a lower temperature. Regardless of the amount of internal energy in an object, if there is no colder region nearby, the energy is useless for running an engine. On the other hand, the colder the exhaust region, the larger is the fraction of the internal energy that can be used.

KEY TERMS

disordered system: A system with an arrangement equivalent to many other possible arrangements.

efficiency: The ratio of the work produced to the energy input. For an ideal heat engine, the Carnot efficiency is given by $1 - T_c/T_h$.

entropy: A measure of the order of a system. The second law of thermodynamics states that the entropy of an isolated system tends to increase.

heat engine: A device for converting heat into mechanical work.

heat pump: A reversible heat engine that acts as a furnace in winter and an air conditioner in summer.

ordered system: A system with an arrangement belonging to a group with the smallest number (possibly one) of equivalent arrangements.

refrigerator: A heat engine running backward.

thermodynamics, second law of: There are three equivalent forms: (1) It is impossible to build a heat engine to perform mechanical work that does not exhaust heat to the surroundings. (2) It is impossible to build a refrigerator that can transfer heat from a lower-temperature region to a higher-temperature region without using mechanical work. (3) The entropy of an isolated system tends to increase.

CONCEPTUAL QUESTIONS

1. What does a heat engine do?
2. In a heat engine, 140 joules of energy are extracted from a hot region. According to the first law of thermodynamics, what is the maximum amount of work that can be done by this engine? Is this result consistent with the second law of thermodynamics? Explain.
3. Why is it not possible to run an ocean liner by taking in seawater at the bow of the ship, extracting internal energy from the water, and dropping ice cubes off the stern?
4. One possible end to the Universe is for it to reach thermal equilibrium; that is, it would have a uniform temperature. Would this temperature be absolute zero? Explain.
5. State the heat engine form of the second law of thermodynamics in your own words.
6. Give an example that clearly illustrates the meaning of the heat engine form of the second law of thermodynamics.
7. Does the following statement agree with the second law of thermodynamics? No engine can transform its entire heat input into work.
8. Would it be possible to design a heat engine that produces no thermal pollution? Explain.
9. It is possible to float heat engines on the ocean and extract some of the internal energy of the water by extending a tube well beneath the ocean's surface. Why is it necessary for the heat engine to have this tube to satisfy the second law of thermodynamics?
10. A hurricane can be thought of as a heat engine that converts thermal energy from the ocean to the mechanical motion of its winds. Use this idea to explain why the wind speeds decrease as the hurricane moves away from the equator.

11. Many people have tried to build perpetual-motion machines. What restrictions does the first law of thermodynamics place on the possibility of building a perpetual-motion machine?

12. What restrictions does the second law of thermodynamics place on the possibility of building a perpetual-motion machine?
13. Explain how the following simplified statements of the first and second laws of thermodynamics are consistent with the versions given in this chapter.
    First: You cannot get ahead.
    Second: You cannot even break even.
14. A student proposes to run an automobile without using any fuel by building a windmill on top of the car. The car's motion will cause the windmill to rotate and generate electricity. The electricity will run a motor, maintaining the car's motion, which in turn causes the windmill to rotate. What, if anything, is wrong with this proposal?

15. What does it mean to say that the human body is a heat engine with an efficiency of 20%?
16. Why are nuclear power plants less efficient than coal-fired plants?
17. What happens to the efficiency of an ideal heat engine as its input temperature is decreased while its exhaust temperature is held fixed?
18. If the input temperature of an ideal heat engine is fixed, what happens to its efficiency as its exhaust temperature is decreased?
19. Heat engine A operates between 300°C and 20°C, whereas heat engine B operates between 300°C and 80°C. Which engine has the greater possible theoretical efficiency? Explain.
20. You are building a heat engine in which the temperature difference between the hot and cold regions is 100 K. Will it be more efficient to have your cold region as cold as possible or as hot as possible? Why?
21. An engineer claims that she could build a more efficient automobile engine if the materials science division could develop piston materials that could withstand higher-operating temperatures. Why would this help?
22. A car company has just designed an ultra-fuel-efficient car, and they wish to advertise the best possible miles per gallon. If the engine can be thought of as a heat engine with a constant operating temperature, would it be better to run the trial on a hot day or a cold day? Why?
23. How is the following statement equivalent to the heat-engine form of the second law of thermodynamics? The efficiency of a heat engine must be less than 1.
▲24. With his paddle-wheel apparatus, Joule determined that 4.2 joules of mechanical work are equivalent to 1 calorie of heat. Imagine that he had mistakenly used a heat engine instead and had measured the heat flowing into the engine and the work done by the engine to determine the conversion factor. Would this have produced a conversion factor for 1 calorie that was greater than, equal to, or less than 4.2 joules? Why?
25. You are installing a central air-conditioning system in which the main unit sits outside your home. For maximum cooling, should you locate the unit on the sunny or the shady side of the house? Why?
26. Bob moves into a new home that is heated with an electric heat pump. He decides that because no heat pump can be perfectly efficient, he will disconnect the heat pump and use the electricity to run a baseboard heater instead. Will Bob's energy bill increase, remain the same, or decrease? Why?
27. Would it be possible to keep a room cool by leaving the door of the refrigerator open? Why or why not?
28. An air-conditioner mechanic is testing a unit by running it on the workbench in an isolated room. What happens to the temperature of the room?
29. A salesperson tries to sell you a "new and improved" air conditioner that does not need a window opening. The unit just sits in the corner of the room and keeps it cool. Use the second law of thermodynamics to convince the salesperson that this will not work.
30. Imagine you are heating your home with a heat pump that uses a small amount of work to transfer heat from the cold outside air to the warm inside air. Your friend suggests that you set up a second heat engine using the air inside the house as the hot region and the outside air as the cold region to provide the necessary work to drive the heat pump. Which law or laws of thermodynamics does this money-saving scheme violate?
31. State the refrigerator form of the second law of thermodynamics in your own words.
32. Give an example that clearly illustrates the meaning of the refrigerator form of the second law of thermodynamics.
33. In what way is the following statement equivalent to the refrigerator form of the second law of thermodynamics? The natural direction for the flow of heat is from hotter objects to colder objects.
34. Is the following statement equivalent to the refrigerator form of the second law of thermodynamics? In moving energy from a cold region to a hot region, the energy delivered to the hot region must be the sum of the work performed plus the energy extracted from the cold region. Explain your reasoning.
35. The coefficient of performance for a refrigerator is defined as the ratio of the heat extracted from the colder system to the work required. Will this number be greater than, equal to, or less than 1 for a good refrigerator? Explain.
36. Why is the efficiency of a heat engine always less than 1, whereas the coefficient of performance for a heat pump is not so constrained?
37. You have two friends who always play the state lottery. Janice's strategy is to always select last week's winning number for this week's draw. In contrast, Jeremy has researched the winning number combinations for the last 10 years and always selects a combination that has not yet won. Which strategy, if either, is more likely to be a winner? Explain.
38. You watch a friend flipping coins and notice that heads has come up four times in a row. Does this mean that it

is more likely that tails will come up on the next throw? Explain.

▲ 39. What sums of two dice have the highest and lowest order?

▲ 40. What sums of three dice have the highest order?

41. Why does the sum of two dice equal 7 more often than any other number?

42. Your friend challenges that, given 25 chances to roll two dice, you cannot roll a sum of 7 at least three times. Should you accept the challenge? Why or why not?

43. Why do "boxcars" (a pair of sixes) occur so rarely when throwing two dice?

44. On average, how many times would you expect to roll "boxcars" (a pair of sixes) with two dice if you rolled the dice a total of 180 times?

45. State the entropy form of the second law of thermodynamics in your own words.

46. Give an example that clearly illustrates the meaning of the entropy form of the second law of thermodynamics.

47. One end of a steel bar is held over a flame until it is red hot. We know from Chapter 13 that when the bar is removed from the flame, the thermal energy will diffuse along the bar until the entire bar has the same equilibrium temperature. Use a microscopic model to explain why the bar's entropy (that is, its disorder) is increasing during this equilibration process.

48. What happens to the entropy of the Universe as the orange liquid diffuses into the clear liquid?

49. What happens to the entropy of the Universe as an ice cube melts in water? Explain.

50. A cold piece of metal is dropped into an insulated container of hot water. After the system has reached an equilibrium temperature, has the entropy of the Universe increased or decreased? Explain.

▲ 51. Describe a system in which the entropy is decreasing. Is this system isolated from its surroundings?

52. What happens to the entropy of a human growing from childhood to adulthood? Is this consistent with the second law of thermodynamics? Explain.

53. When water freezes to ice, does the order of the water molecules increase or decrease? What does this imply about the change in entropy in the rest of the Universe?

54. A ringing bell is inserted into a large glass of water. The bell and the water are initially at the same temperature and are insulated from their surroundings. Eventually, the bell stops vibrating, and the water comes to rest.
    a. What happens to the mechanical energy of the bell?
    b. What happens to the temperature of the system?
    c. What happens to the entropy of the system?

55. If you slide a crate across the floor, kinetic energy is converted to thermal energy as it comes to rest. Why will adding thermal energy to a stationary crate not cause it to move?

▲ 56. Are Mexican jumping beans a violation of the second law of thermodynamics? Explain.

57. Imagine that you could film the motion of the gas molecules in the room. Would you be able to tell whether the film was running forward or backward? Would it make a difference if air were being released from a balloon? Explain.

58. You have an aquarium with a divider down the middle. One side is filled with hot water, and the other is filled with cold water. Imagine that as the divider is removed you can film the individual collisions between water molecules. When watching the film, how could you tell whether it was running forward or backward?

59. Which of the following statements explains why we are currently experiencing a worldwide energy crisis?
    a. The amount of energy in the world is decreasing rapidly.
    b. The entropy of the world is increasing rapidly.
    c. The entropy of the world is decreasing rapidly.

60. How does slowing the increase in entropy help solve the world's energy crisis?

61. Which results in the larger increase in the entropy of the Universe: heating a liter of room-temperature water to boiling using natural gas or using electricity? Why?

62. Why is heating water on a gas stove more efficient than heating it on an electric stove?

63. It is the middle of winter, and you live in a house with electric baseboard heating. Your friend chides you for being wasteful for turning on the oven to 400°F for 45 minutes just to cook a single baked potato. How do you respond?

64. Since childhood we've been told to turn out the lights when we leave a room. Does this really reduce the electricity bill during the winter for a house with electric heating? Why?

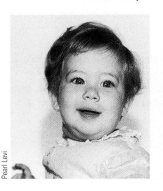

EXERCISES

1. What input energy is required if an engine performs 200 kJ of work and exhausts 400 kJ of heat?
2. How much work is performed by a heat engine that takes in 2000 J of heat and exhausts 600 J?
3. An engine takes in 7000 cal of heat and exhausts 4000 cal of heat each minute it is running. How much work does the engine perform each minute?
4. A heat engine requires an input of 9 kJ per minute to produce 3 kJ of work per minute. How much heat must the engine exhaust per minute?
5. What is the efficiency of a heat engine that does 50 J of work for every 200 J of heat it takes in?
6. An engine exhausts 1200 J of energy for every 3600 J of energy it takes in. What is its efficiency?
7. An engine has an efficiency of 40%. How much energy must be extracted to do 900 J of work?
▲ 8. An engine operates with an efficiency of 25%. If the engine does 600 J of work every minute, how many joules per minute are exhausted to the cold region?
9. An engine takes in 600 cal and exhausts 450 cal each second it is running. How much work does the engine perform each minute? What is the engine's efficiency?
10. How much work does an engine produce each second if it takes in 8000 cal and exhausts 5000 cal each second? What is the efficiency of the engine?
11. An engineer has designed a machine to produce electricity by using the difference in the temperature of ocean water at different depths. If the surface temperature is 20°C and the temperature at 50 m below the surface is 12°C, what is the maximum efficiency of this machine?
12. A heat engine takes in 1000 J of energy at 1000 K and exhausts 600 J at 500 K. What are the actual and maximum theoretical efficiencies of this heat engine?
13. An ideal heat engine has a theoretical efficiency of 60% and an exhaust temperature of 27°C. What is its input temperature?
14. What is the exhaust temperature of an ideal heat engine that has an efficiency of 50% and an input temperature of 400°C?
15. How much work is required by a refrigerator that takes in 1000 J from the cold region and exhausts 1500 J to the hot region?
16. A refrigerator uses 600 J of work to remove 2400 J of heat from a room. How much heat does it exhaust?
17. How much work per second (power) is required by a refrigerator that takes 800 J of thermal energy from a cold region and exhausts 1500 J to a hot region each second?
18. A heat pump requires 500 W of electric power to deliver heat to your house at a rate of 2400 J per second. How many joules of energy are extracted from the cold air outside each second?
19. The coefficient of performance for a heat pump is defined as the ratio of the heat extracted from the colder system to the work required. If a heat pump requires an input of 400 W of electric energy and has a coefficient of performance of 3, how much energy is delivered to the inside of the house each second?
20. If a refrigerator requires an input of 200 J of electric energy each second and has a coefficient of performance of 5, how much heat energy is extracted from the refrigerator each second?
21. Show that four coins can be arranged in 16 different ways.
22. Show that the combination of four coins with the lowest order (two heads and two tails) is the one with the largest number of arrangements.
23. What is the probability of rolling a total of 7 with two dice?
24. What is the probability of rolling a sum of 10 with two dice?
▲ 25. The total number of possible states for three dice is $6 \times 6 \times 6 = 216$. What is the probability of throwing a sum equal to 5?
▲ 26. The total number of possible states for three dice is $6 \times 6 \times 6 = 216$. What is the probability of throwing a sum equal to 15?

Interlude

Sculling on Lake Powell creates interesting wave patterns.

Waves—Something Else That Moves

Imagine standing near a busy highway trying to get the attention of a friend on the other side. How could you signal your friend? You might try shouting first. You could throw something across the highway, you could make a loud noise by banging two rocks together, you could shine a flashlight at your friend, and so on.

Signals can be sent by one of two methods. One method includes ways material moves from you to your friend—such as throwing a pebble. The other method includes ways energy moves across the highway without any accompanying material. This second method represents phenomena that we usually call **waves**.

The study of waves has greatly expanded the physics world view. Surprisingly, however, waves do not have a strong position in our commonsense world view. It's not that wave phenomena are uncommon, but rather that many times the wave nature of the phenomena is not recognized. Plucking a guitar string, for example, doesn't usually invoke images of waves traveling up and down the string. But that is what happens. The buzzing of a bee probably does not generate thoughts of waves either. We will discover interesting examples of waves in unexpected situations.

Waves are certainly common enough—we grow up playing with water waves and listening to sound waves—but most of us do not have a good intuitive understanding of the behavior of waves. Ask yourself a few questions about waves: Do they bounce off materials? When two waves meet, do they crash like billiard balls? Is it meaningful to speak of the speed of a wave? When speaking of material objects, the answers to such questions seem obvious, but when speaking of waves, the answers require closer examination.

We study waves for two reasons. First, because they are there, studying waves adds to our understanding of how the world works. The second reason is less obvious. As we delve deeper and deeper into the workings of the world, we reach limits beyond which we cannot directly observe phenomena. Even the best imaginable magnifying instrument is too weak to allow direct observation of the subatomic worlds. Our search to understand these worlds yields evidence only by indirect methods. We must use our common experiences to model a world we cannot see. In many cases the modeling process can be reduced to asking whether the phenomenon acts like a wave or acts like a particle.

To answer the question of whether something acts like a wave or a particle, we must expand our commonsense world view to include waves. After you study such common waves as sound waves, we hope you will be ready to "hear" the harmony of the subatomic world.

> It's not that wave phenomena are uncommon, but rather that many times the wave nature of the phenomena is not recognized.

15 Vibrations and Waves

Water drops falling onto the surface of water produce waves that move outward as expanding rings. But what is moving outward? Does the wave disturbance carry energy or matter? What happens when two waves meet? How does wave motion differ from particle motion?

(See page 308 for the answer to this question.)

Circular waves are formed by falling water drops.

If you stretch or compress a spring and let go, it vibrates. If you pull a pendulum off to one side and let it go, it oscillates back and forth. Such vibrations and oscillations are very common motions in our everyday world. If these vibrations and oscillations affect surrounding objects or matter, a wave is often generated. Ripples on a pond, musical sounds, laser light, exploding stars, and even electrons all display some aspects of wave behavior.

Waves are responsible for many of our everyday experiences. Fortunately, nature has been kind; all waves have many of the same characteristics. Once you understand one type, you will know a great deal about the others.

We begin our study with simple vibrations and oscillations. We then examine common waves, such as waves on a rope, water waves, and sound waves, and later progress to more exotic examples, such as radio, television, light, and even "matter" waves.

PhysicsNow™ Test your understanding of this chapter by logging into PhysicsNow at **http://physics.brookscole.com/kf6e**, selecting the chapter, and clicking on the "Take a Pre-Test" link.

Simple Vibrations

✓ **MATH**

✓ Extended presentation available in the *Problem Solving* supplement

If you distort an object and release it, elastic forces restore the object to its original shape. In returning to its original shape, however, the inertia of the displaced portion of the object causes it to overshoot, creating a distortion in the opposite direction. Again, restoring forces attempt to return the object to its original shape and, again, the object overshoots. This back-and-forth motion is what we commonly call a **vibration**, or an **oscillation**. For all practical purposes, the labels are interchangeable.

A mass hanging on the end of a vertical spring exhibits a very simple vibrational motion. Initially, the mass stretches the spring so that it hangs at the position where its weight is just balanced by the upward force of the spring, as shown in Figure 15-1. This position—called the **equilibrium position**—is analogous to the undistorted shape of an object. If you pull downward (or push upward) on the mass, you feel a force in the opposite direction. The size of this restoring force increases with the amount of stretch or compression you apply. If the applied force is not too big, the restoring force is proportional to the distance the mass is moved from its equilibrium position. If the force is too large, the spring will be permanently stretched and not return to its original length. In the discussion that follows, we assume that the stretch of the system is not too large. Many natural phenomena obey this condition, so little is lost by imposing this constraint.

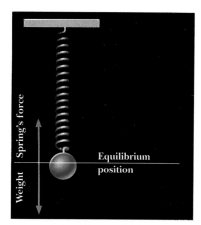

Figure 15-1 At the equilibrium position, the upward force due to the spring is equal to the weight of the mass.

Imagine pulling the mass down a short distance and releasing it as shown in Figure 15-2(a). Initially, a net upward force accelerates the mass upward. As the mass moves upward, the net force decreases in size (b), becoming zero when the mass reaches the equilibrium position (c). Because the mass has inertia, it overshoots the equilibrium position. The net force now acts downward (d) and slows the mass to zero speed (e). Then the mass gains speed in the downward direction (f). Again, the mass passes the equilibrium position (g). Now the net force is once again upward (h) and slows the mass until it reaches its lowest point (a). This sequence (Figure 15-2[a through a]) completes one **cycle**.

Actually, a cycle can begin at any position. It lasts until the mass returns to the original position *and* is moving in the same direction. For example, a cycle might begin when the mass passes through the equilibrium point on its way up (c) and end when it next passes through this point on the way up. Note that the cycle does not end when the mass passes through the equilibrium point on the way down (g). This motion is known as periodic motion, and the length of time required for one cycle is known as the **period** *T*.

◀ period is the time to complete one cycle

If we ignore frictional effects, energy conservation (Chapter 7) tells us that the mass travels the same distance above and below the equilibrium position. This distance is marked in Figure 15-2 and is known as the **amplitude** of the vibration. In real situations the amplitude decreases and eventually the motion

Figure 15-2 A time sequence showing one complete cycle for the vibration of a mass on a spring. The clocks show that equal time intervals separate the images.

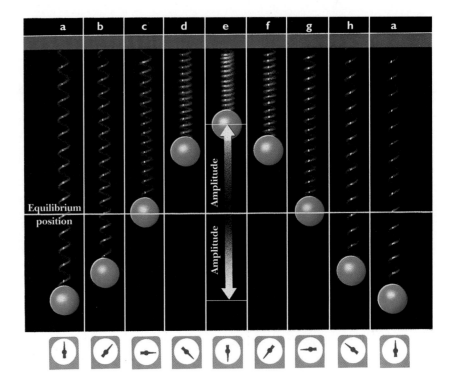

dies out because of the frictional effects that convert mechanical energy into thermal energy.

We can describe the time dependence of the vibration equally well by giving its **frequency** f, the number of cycles that occur during a unit of time. Frequency is often measured in cycles per second, or hertz (Hz). For example, concert A (the note that orchestras use for tuning) has a frequency of about 440 hertz, household electricity oscillates at 60 hertz, and your favorite FM station broadcasts radio waves near 100 million hertz.

There is a simple relationship between the frequency f and the period T: one is the reciprocal of the other:

$$f = \frac{1}{T}$$

$$T = \frac{1}{f}$$

▶ frequency = $\frac{1}{\text{period}}$

▶ period = $\frac{1}{\text{frequency}}$

To illustrate this relationship, let's calculate the period of a spring vibrating at a frequency of 4 hertz:

$$T = \frac{1}{f} = \frac{1}{4 \text{ Hz}} = \frac{1}{4 \text{ cycles/s}} = \frac{1}{4} \text{ s}$$

This calculation shows that a frequency of 4 cycles per second corresponds to a period of $\frac{1}{4}$ second. This makes sense because a spring vibrating four times per second should take $\frac{1}{4}$ of a second for each cycle. (When we state the period, we know it refers to one cycle and don't write "second per cycle.")

Question   What is the period of a mass that vibrates with a frequency of 10 times per second?

Answer   Because the period is the reciprocal of the frequency, we have

$$T = \frac{1}{f} = \frac{1}{10 \text{ Hz}} = 0.1 \text{ s}$$

We might guess that the time it takes to complete one cycle would change as the amplitude changes, but experiments show that the period remains essentially constant. It is fascinating that the amplitude of the motion does *not* affect the period and frequency. (Again, we have to be careful not to stretch the system too much.) This means that a vibrating guitar string always plays the same frequency regardless of how hard the string is plucked.

Although the period for a mass vibrating on the end of a spring does not depend on the amplitude of the vibration, we might expect the period to change if we switch springs or masses. The stiffness of the spring and the size of the mass do change the rate of vibration.

The stiffness of a spring is characterized by how much force it takes to stretch it by a unit length. For moderate amounts of stretch or compression, this value is a constant known as the **spring constant** k. In SI units this constant is measured in newtons per meter. Larger values correspond to stiffer springs.

In trying to guess the relationship between the spring constant, mass, and period, we would expect the period to decrease as the spring constant increases because a stiffer spring means more force and therefore a quicker return to the equilibrium position. Furthermore, we would expect the period to increase as the mass increases because the inertia of a larger mass will slow the motion.

WORKING IT OUT | Period of a Mass on a Spring ✓ MATH

The mathematical relationship for the period of a mass on a spring can be obtained theoretically and is verified by experiment:

$$T = 2\pi\sqrt{\frac{m}{k}}$$

◀ period of a mass on a spring

where π is approximately 3.14.

As an example, consider a 0.2-kg mass hanging from a spring with a spring constant of 5 N/m:

$$T = 2\pi\sqrt{\frac{m}{k}} = 2\pi\sqrt{\frac{0.2 \text{ kg}}{5 \text{ N/m}}} = 6.28\sqrt{\frac{1}{25} \text{ s}^2} = 1.26 \text{ s}$$

Therefore, this mass–spring combination vibrates with a period of 1.26 s, or a frequency of 0.793 Hz.

Question What is the period of a 0.1-kg mass hanging from a spring with a spring constant of 0.9 N/m?

Answer 2.09 s.

The Pendulum

The pendulum is another simple system that oscillates. Students are often surprised to learn (or to discover by experimenting) that the period of oscillation does not depend on the amplitude of the swing. To a very good approximation, large- and small-amplitude oscillations have the same period if we keep their amplitudes less than 30 degrees. This amazing property of pendula was first discovered by Galileo when he was a teenager sitting in church watching a swinging chandelier. (Clearly, he was not paying attention to the service.) Galileo tested his hypothesis by constructing two pendula of the same length and swinging them with different amplitudes. They swung together, verifying his hypothesis.

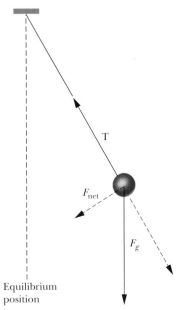

Figure 15-3 The net force on the pendulum bob accelerates it toward the equilibrium position.

A strobe photograph of a pendulum taken at 20 flashes per second. Note that the pendulum bob moves the fastest at the bottom of the swing.

period of a pendulum ▶

Let's consider the forces on a pendulum when it has been pulled to the right as shown in Figure 15-3. The component of gravity acting along the string is balanced by the tension in the string. Therefore, the net force is the component of gravity at right angles to the string and directed toward the lower left. This restoring force causes the pendulum bob to accelerate toward the left. Although the restoring force on the bob is zero at the lowest point of the swing, the bob passes through this point (the equilibrium position) because of its inertia. The restoring force now points toward the right and slows the bob.

We found in free fall that objects with different masses fall with the same acceleration because the gravitational force is proportional to the mass. Therefore, we might expect that the motion of a pendulum would not depend on the mass of the bob. This prediction is true and can be verified easily by making two pendula of the same length with bobs of the same size made out of different materials so that they have different masses. The two pendula will swing side by side.

Question   Why do we suggest using different materials?

Answer   If we use the same type of material, the size has to be different to get different masses. Different sizes might also affect the period. When doing an experiment, it is important to keep all but one factor constant.

We also know from our experiences with pendula that the period depends on the length of the pendulum; longer pendula have longer periods. Therefore, the length of the pendulum can be changed to adjust the period.

PHYSICS | ON YOUR OWN

Construct a simple pendulum and investigate the validity of some of the statements made in this section. Does the period depend on the amplitude of the swing, the length of the string, or the mass of the bob?

Because the restoring force for a pendulum is a component of the gravitational force, you might expect that the period depends on the strength of gravity, much as the period of the mass on the spring depends on the spring constant. This hunch is correct and can be verified by taking a pendulum to the Moon, where the acceleration due to gravity is only one-sixth as large as that on Earth.

WORKING IT OUT | Period of a Pendulum   ✓ MATH

The period of a pendulum is given by

$$T = 2\pi\sqrt{\frac{L}{g}}$$

As an example, consider a pendulum with a length of 10 m:

$$T = 2\pi\sqrt{\frac{L}{g}} = 2\pi\sqrt{\frac{10 \text{ m}}{10 \text{ m/s}^2}} = 6.28\sqrt{1 \text{ s}^2} = 6.28 \text{ s}$$

Therefore, this pendulum would oscillate with a period of 6.28 s.

Question   What would you expect for the period of a 1.7-m pendulum on the Moon?

Answer   6.28 s.

Clocks

Keeping time is a process of counting time intervals, so it is reasonable that periodic motions have been important to timekeepers. Devising accurate methods for keeping time has kept many scientists, engineers, and inventors busy throughout history. The earliest methods for keeping time depended on the motions in the heavens. The day was determined by the length of time it took the Sun to make successive crossings of a north–south line and was monitored with a sundial. The month was determined by the length of time it took the Moon to go through its phases. The year was the length of time it took to cycle through the seasons and was monitored with a calendar, a method of counting days.

As science and commerce advanced, the need grew for increasingly accurate methods of determining time. An early method for determining medium intervals of time was to monitor the flow of a substance such as sand in an hourglass or water in a water clock. Neither of these, however, was very accurate, and because they were not periodic, they had to be restarted for each time interval. It is interesting to note that Galileo kept time with a homemade water clock in many of his early studies of falling objects.

The next generation of clocks took on a different character, employing oscillations as their basic timekeeping mechanism. Galileo's determination that the period of a pendulum does not depend on the amplitude of its swing led to Christian Huygens's development in 1656 of the pendulum clock, 14 years after Galileo's death. One of the difficulties Huygens encountered was to develop a mechanism for supplying energy to the pendulum to maintain its swing.

Seafarers spurred the development of clocks that would keep accurate time over long periods. To determine longitude requires measuring the positions of prominent stars and comparing these positions with their positions as seen from Greenwich, England, *at the same time*. Because pendulum clocks did not work on swaying ships, several cash prizes were offered for the design and construction of suitable clocks. Beginning in 1728, John Harrison, an English instrument maker, developed a series of clocks that met the criteria, but he was not able to collect his money until 1765. One of Harrison's clocks was accurate to a few seconds after 5 months at sea.

Any periodic vibration can be used to run clocks. Grandfather clocks use pendula to regulate the hands and are powered by hanging weights. Mechanical watches have a balance wheel fastened to a spring. Electric clocks use 60-hertz alternating electric current. Digital clocks use the vibrations of quartz crystals or resonating electric circuits.

Modern time is kept with atomic clocks, which use the frequencies of atomic transitions (see Chapter 23), and are extremely insensitive to such changes in the clocks' environment as pressure and temperature. Atomic clocks are accurate to better than a second in 60 million years.

PHYSICS | ON YOUR OWN

Make a list of the different types of mechanisms used to run clocks. How would you adjust the natural frequency of each clock to make it run at the correct rate?

A replica of an early mechanical clock.

Modern time is kept by extremely accurate atomic clocks such as this F1 operated by the National Institute of Standards and Technology. It is accurate to 1 second in 60 million years.

Resonance

We discovered with the mass on a spring and the pendulum that each system had a distinctive, natural frequency. The natural frequency of the pendulum is determined by its length and the acceleration due to gravity. Pulling the bob back and releasing it produces an oscillation at this particular frequency.

A child on a swing is an example of a life-size pendulum. If the child does not pump her legs and if no one pushes her, the amplitude of the swing continually decreases, and the child comes to rest. As every child knows, however, pumping or pushing greatly increases the amplitude. A less obvious fact is that the size of the effort—be it from pumping or pushing—is not important, but its timing is crucial. The inputs must be given at the natural frequency of the swing. If the child pumps at random times, the swinging dies out. This phenomenon of a large increase in the amplitude when a periodic force is applied to a system at its natural frequency is called **resonance**.

Resonance can also be achieved by using impulses at other special frequencies, but each of these has a definite relationship to the natural frequency. For example, if you push the child on the swing every other time, you are providing inputs at one-half of the natural frequency; every third time gives inputs at one-third of the natural frequency; and so on. Each of these frequencies causes resonance.

If children pump at the right frequencies, they can increase the amplitudes of their motions.

Question   What happens to the amplitude of the swing if you push at twice the natural frequency?

Answer   In this case you would be pushing twice for each cycle. One of the pushes would negate the other, and the swing would stop.

PHYSICS | **ON YOUR OWN**

The next time you go to a playground, experiment with the phenomenon of resonance. Try pushing or pumping a swing at a frequency other than its natural one. What happens? If you're the one on the swing, what happens to the resonant frequency when you stand up?

More complex systems also have natural frequencies. If someone strikes a spoon on a table, you are not likely to mistake its sound for that of a tuning fork. The vibrations of the spoon produce sounds that are characteristic of the spoon. All objects have natural frequencies. The factors that determine these frequencies are rather complex. In general, the dominant factors are the stiffness of the material, the mass of the material, and the size of the object.

Resonance can have either good or bad effects. Although your radio receives signals from many stations simultaneously, it plays only one station at a time. The radio can be tuned so that its resonant frequency matches the broadcast frequency of your favorite station. Tuning puts the radio in resonance with one particular broadcast frequency and out of resonance with the frequencies of the competing stations. On the other hand, if the radio has an inferior speaker with one or two strong resonant frequencies, it will distort the sounds from the radio station by not giving all frequencies equal amplification.

Suppose you have a collection of pendula of different lengths, as shown in Figure 15-4. Notice that two of these pendula have the same length and thus the same natural frequency. The pendula are not independent because they are all tied to a common string. The motion of one of them is felt by all the others through pulls by the string. If you start the left-hand pendulum swinging, its back-and-forth motion creates a tug on the common string with a frequency equal to its natural, or resonant, frequency. Pendula with different frequencies jiggle a little bit but are not affected very much. However, the pendulum with the same frequency resonates with the input frequency, drastically increasing its amplitude. In exactly the same way, objects resonate when input frequencies are the same as any of their natural frequencies.

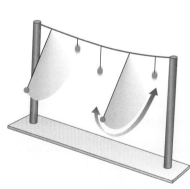

Figure 15-4 Pendula with the same natural frequency resonate with each other.

Figure 15-5 A wave pulse travels along a line of dominoes.

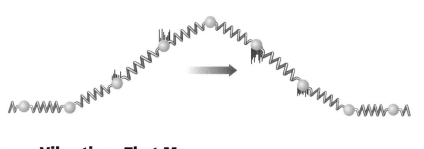

Figure 15-6 A wave disturbance can move along a chain of balls and springs.

Waves: Vibrations That Move

Most **waves** begin with a disturbance of some material. Some disturbances, such as the clapping of hands, are one-time, very abrupt events, whereas others, such as the back-and-forth vibration of a guitar string, are periodic events. The simplest wave is a single pulse that moves outward as a result of a single disturbance.

Imagine a long row of dominoes lined up as shown in Figure 15-5. Once a domino is pushed over, it hits its neighbor, and its neighbor hits its neighbor, and so on, sending the disturbance along the line of dominoes. The key point is that "something" moves along the line of dominoes—from the beginning to the end—but it is not any individual domino.

Actually, the domino example is not completely analogous to what happens in most situations involving waves because there is no mechanism for restoring the dominoes in preparation for the next pulse. We can correct this omission by imagining a long chain of balls connected by identical springs as shown in Figure 15-6. As the disturbance moves from left to right, individual balls are lifted up from their equilibrium positions and then returned to these equilibrium positions. The springs allow each ball to pull its neighbor away from equilibrium, just as the dominoes passed the disturbance from neighbor to neighbor by striking each other. After the pulse passes, the springs provide the restoring force that returns each ball to equilibrium. Notice that the pulse travels along the chain of balls without any of the balls moving in the direction of the pulse. Figure 15-6 shows the shape of the chain of balls and springs at the time the center ball reaches its maximum displacement.

In a similar manner, a pebble dropped into a pond depresses a small portion of the surface. Each vibrating portion of the surface generates disturbances in the surrounding water. As the process continues, the disturbance moves outward in circular patterns, such as those shown in Figure 15-7.

This type of disturbance, or pulse, occurs in a number of common, everyday events. A crowd transmits single pulses when a small group begins pushing. This push spreads outward through the crowd much like a ripple moves over a pond's surface. Similarly, the disturbance produced by a clap sends a single sound pulse through the air. Other examples of nonrecurrent waves include tidal bores, tidal waves, explosions, and light pulses emitted by supernovas (exploding stars).

Although a wave moves outward from the original disturbance, there is no overall motion of the material. As the wave travels through the medium, the particles of the material vibrate about their equilibrium positions. Although the wave travels down the chain, the individual balls of the chain return to their original positions. The wave transports energy rather than matter from one place to another. The energy of an undisturbed particle in front of the wave is increased as the wave passes by and then returns to its original value. In a real medium,

Figure 15-7 Water drops produce disturbances that move outward in circular patterns.

Tacoma Narrows Bridge

Resonant effects can sometimes have disastrous consequences. In 1940 a new bridge across one of the arms of Puget Sound in the state of Washington was opened to traffic. It was a suspension bridge with a central span of 850 meters (2800 feet). Because the bridge was designed for two lanes, it had a width of only 12 meters (40 feet). Within a few months after it opened, early morning winds in the Sound caused the bridge to oscillate in standing wave patterns that were so large in amplitude that the bridge failed structurally and fell into the water below, as shown in the figure.

But why did this bridge fail when other suspension bridges are still standing (including the bridge that now spans the Sound at the location of the original)? The bridge was long, narrow, and particularly flexible. Motorists often complained about the vertical oscillations and nicknamed the bridge "Galloping Gertie." However, the amplitudes of the vertical oscillations were relatively small until that fateful morning. The wind was blowing along the arm of the Sound (perpendicular to the length of the bridge) at moderate to high velocities but was not near gale force. One might speculate that fluctuations in the wind speed matched the natural frequency of the bridge, causing it to resonate. However, the wind was reasonably steady, and wind fluctuations are normally quite random. Furthermore, the forces would be horizontal, and the oscillations were vertical.

The best explanation involves the formation and shedding of vortices in the wind blowing past the bridge. *Vortices* are the eddies that you get near the ends of the oars when you row a boat. Vortices rotate in opposite directions in the wind blowing over and under the bridge. As each vortex is shed, it exerts a vertical impulse on the bridge. Therefore, if the frequency of vortex formation and shedding is near the natural frequency of the bridge for vertical oscillations, a standing wave will form just like those on a guitar string. (The frequency does not have to match exactly; it only needs to be close. How close depends on the details of the bridge construction.)

The bridge would have been fine except for another unfortunate circumstance. Besides the vertical standing wave, there were also torsional, or twisting, standing waves on the bridge. Normally, the frequencies of the two standing waves are quite different. But for the Tacoma Narrows Bridge, the two frequencies were fairly close (8 per minute for the vertical motion compared with 10 per minute for the twisting motion). This allowed some of the energy from the vertical motion to be transferred to the twisting motion that eventually led to the mechanical failure of the bridge.

The Tacoma Narrows Bridge collapsed when winds set up resonant vibrations.

however, some of the energy of the wave is left behind as thermal energy in the medium.

There are two basic wave types. A wave in which the vibration of the medium is perpendicular to the motion of the wave is called a **transverse wave**. Waves on a rope are transverse waves. A wave in which the vibration of the medium is along the same direction as the motion of the wave is called a **longitudinal wave**. Both types can exist in the chain of balls. If a ball is moved vertically,

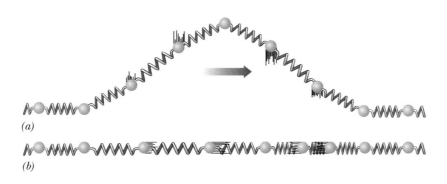

Figure 15-8 In a transverse wave (a) the medium moves perpendicular to the direction of propagation of the wave, whereas in a longitudinal wave (b) the medium's motion is parallel to the direction of propagation.

a transverse wave is generated (Figure 15-8[a]). If the ball is moved horizontally, the wave is longitudinal (Figure 15-8[b]).

Transverse waves can move only through a material that has some rigidity; transverse waves cannot exist within a fluid because the molecules simply slip by each other. Longitudinal waves, on the other hand, can move through most materials because the materials can be compressed and have restoring forces.

Question   Is it possible to have transverse waves on the surface of water?

Answer   Transverse surface waves are possible because the force of gravity tends to restore the surface to its flat equilibrium shape. Actually, the motion of the individual water molecules is a combination of transverse motion and longitudinal motion; the water molecules follow elliptical paths.

One-Dimensional Waves

Because all waves have similar properties, we can look at waves that are easy to study and then make generalizations about other waves. Imagine a clothesline tied to a post as in Figure 15-9. A flick of the wrist generates a single wave pulse that travels away from you. On an idealized rope, the wave pulse would maintain its shape and size. On a real rope, the wave pulse slowly spreads out. We will ignore this spreading in our discussion. The wave's speed can be calculated by dividing the distance the pulse travels by the time it takes.

The speed of the wave can be changed. If you pull harder on the rope, the pulse moves faster; the speed increases as the tension in the rope increases. The speed also depends on the mass of the rope; a rope with more mass per unit length has a slower wave speed. Surprisingly, the amplitude of the pulse does not have much effect on the speed.

These observations make sense if we consider the vibrations of a small portion of the rope. The piece of rope is initially at rest and moves as the leading edge of the pulse arrives. How fast the rope returns to its equilibrium position determines how the pulse passes through the region (and hence, the speed of the

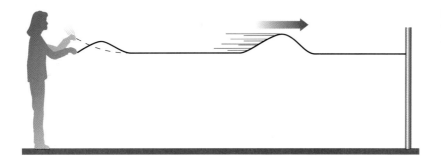

Figure 15-9 While pieces of the rope vibrate up and down, the wave moves along the rope.

pulse). The more massive the rope, the more sluggishly it moves. Also, if the rope is under a larger tension, the restoring forces on the piece of rope are larger and cause it to return to its equilibrium position more quickly.

Flawed Reasoning

A physics teacher has offered his class a prize if they can send a transverse pulse down a long spring and then send a second pulse down the same spring in such a manner as to catch up with the first pulse. Three students have taken up the challenge.

Carey: "I will make the first pulse with a slow movement of my hand and then make a second pulse with a very quick jerk on the spring. That should send the second pulse down the spring at a quicker speed."

Christopher: "I think the amplitude of the pulse is what matters, not how fast you move your hand. Send the first pulse down with a big amplitude and then send the second pulse down with a small amplitude."

Cody: "The textbook claims that pulse speeds don't depend on how the pulse was created but only on the tension and the mass density of the spring. We can't change the mass density after we send the first pulse, but we could tighten the spring. Send the first pulse and then pull the spring tighter before we send the second pulse."

All three students are wrong. **Find the flaws in their claims.**

Answer Cody correctly points out the flaws in his classmates' reasoning. The speed of a pulse down a stretched spring does not depend on the size or shape of the pulse or the manner in which it was created. Cody's suggestion of stretching the spring will indeed increase the speed of transverse pulses on the spring, but this will speed up *both* pulses. The physics teacher tricked them with an impossible challenge.

Figure 15-10 A wave pulse is inverted when it reflects from a fixed end. Note that the steep edge leads on the way in and on the way out.

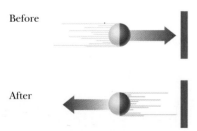

Figure 15-11 In contrast to a wave, the blue half of the ball leads on the way in and trails after reflection from the wall.

When a pulse hits the end that is attached to the post, it bounces off and heads back. This reflected pulse has the same shape as the incident pulse but is inverted as shown in Figure 15-10. If the incident pulse is an "up" pulse (a **crest**), the reflected pulse is a "down" pulse (a **trough**). If the end of the rope is free to move up and down, the pulse still reflects but no inversion takes place.

What about the front and back of the wave pulse? To observe this you would generate a pulse that is not symmetric. The pulses shown in Figure 15-10 are steeper in front than in back. The steeper edge is away from you when the pulse moves down the rope and toward you when the reflected pulse returns. The leading edge continues to lead.

These inversions contrast with the behavior of a ball when it "reflects" from a wall. If the ball is not spinning, the top of the ball remains on top, but the leading edge is interchanged. Figure 15-11 shows that the blue half leads before the collision, whereas the green half leads afterward.

PHYSICS | ON YOUR OWN

You can investigate the behavior of waves on a long clothesline by tying one end to a rigid post. (As an alternative, you can use a long Slinky on a floor.) Do small pulses and big pulses travel away from you at the same speed? Do the shapes of the pulses remain the same? Do reflected waves pass through waves you send down the line? Is a crest inverted when it reflects from the post?

Probing the Earth

Imagine drawing a circle to represent Earth. Further, imagine drawing a dot to show how far Earth's interior has been explored by direct drilling and sampling techniques. Where would you place the dot? The dot should be placed on the original circle. Earth is about 6400 kilometers (4000 miles) in radius, and we have drilled into its interior only about 12.2 kilometers, less than 0.2% of the distance to the center. Therefore, we must learn about Earth's interior using indirect means such as looking at signals from explosions and earthquakes.

Three kinds of waves are produced in an earthquake. One type travels along the surface, and the other two travel through Earth's interior; one of the interior waves is a longitudinal wave, and the other is a transverse wave. These waves move outward in all directions from the earthquake site and are received at numerous earthquake-monitoring sites around the world. The detection of these waves and their arrival times provide clues about Earth's interior.

Two major things happen to the waves: First, partial reflections occur at boundaries between distinctly different regions. Second, the waves change speed as the physical conditions—such as the elasticity and density—change. Changing a wave's speed usually results in the wave changing direction, a phenomenon known as refraction, which we will discuss in Chapter 18. As the waves go deeper into the interior, they speed up, causing them to change direction.

The longitudinal waves are called the primary waves, or P waves, and are created by the alternating expansion and compression of the rocks near the source of the shock. This push–pull vibration can be transmitted through solids, liquids, and gases. P waves move with the highest speeds and therefore are the first to arrive at a seismograph station. P waves move at about 5 kilometers per second (11,000 mph!) near the surface and speed up to about 7 kilometers per second toward the base of the upper crust.

The secondary waves, or S waves, are transverse waves. In this case the rock movement is perpendicular to the direction the wave is traveling. S waves travel through solids but cannot propagate through liquids and gases because fluids lack rigidity.

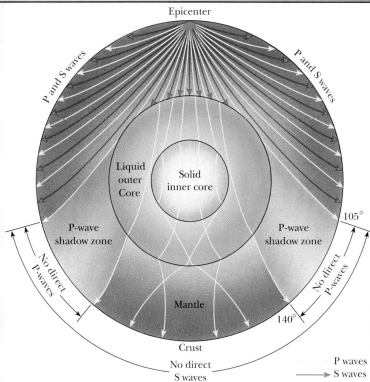

Cross section of Earth showing the paths of some waves produced by an earthquake.

We wouldn't be able to infer much about Earth's interior if only one signal arrived at each site. There would be a number of paths that could account for the characteristics and timing of the signal. Fortunately, many sites receive multiple signals that allow large computers to piece the information together to form a model of Earth's interior. Information that is not received is also important. After an earthquake there are many sites that do not receive any transverse signals. This tells us that they are located in a shadow region behind a liquid core as shown in the figure.

Source: H. Levin, *The Earth through Time* (Philadelphia: Saunders, 1992).

Superposition

Suppose you send a crest down the rope and, when it reflects as a trough, you send a second crest to meet it. An amazing thing happens when they meet. The waves pass through each other as if the other were not there. This is shown in Figure 15-12. Each pulse retains its own shape, clearly demonstrating that the pulses are not affected by the "collision." A similar thing happens when you throw two pebbles into a pond. Even though the wave patterns overlap, you can still see a set of circular patterns move outward from *each* splash.

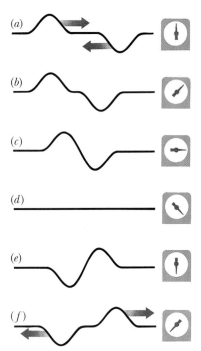

Figure 15-12 The two wave pulses on the rope pass through each other as if the other were not present.

In contrast, imagine what would happen if two particles—say, two Volkswagens—were to meet. Particles don't exhibit this special property of waves. It would certainly be a strange world if waves did not pass through each other. Two singers singing at the same time would garble each other's music, or the sounds from one might bounce off those from the other.

During the time the wave pulses pass through each other, the resulting disturbance is a combination of the individual ones; it is a **superposition** of the pulses. As shown in Figure 15-13, the distance of the medium from the equilibrium position, the **displacement**, is the algebraic sum of the displacements of the individual wave pulses. If we consider displacements above the equilibrium position as positive and those below as negative, we can obtain the shape of the resultant disturbance by adding these numbers at each location along the rope.

If two crests overlap, the disturbance is bigger than either one alone. A crest and a trough produce a smaller disturbance. If the crest and the trough are the same size and have symmetric shapes, they completely cancel at the instant of total overlap. A high-speed photograph taken at this instant yields a picture of a straight rope. This phenomenon is illustrated in Figure 15-12(d). This is not as strange as it may seem. If we take a high-speed photograph of a pendulum just as it swings through the equilibrium position, it would appear that the pendulum was not moving but simply hanging straight down. In either case, longer exposures would blur, showing the motion.

Periodic Waves

A rope moved up and down with a steady frequency and amplitude generates a train of wave pulses. All the pulses have the same size and shape as they travel down the rope. The drawing in Figure 15-14 shows a **periodic wave** moving to the right. New effects emerge when we examine periodic waves. For one thing, unlike the single pulse, periodic waves have a frequency. The frequency of the wave is the oscillation frequency of any piece of the medium.

An important property of a periodic wave is the distance between identical positions on adjacent wave pulses, called the **wavelength** of the periodic wave. This is the smallest distance for which the wave pattern repeats. It might be measured between two adjacent crests, or two adjacent troughs, or any two identical spots on adjacent pulses as shown in Figure 15-15. The symbol used for wavelength is the Greek letter lambda λ.

The speed of the wave can be determined by measuring how far a particular crest travels in a certain time. In many situations, however, the speed is too fast, the wavelength too short, or the amplitude too small to allow us to follow the motion of a single crest. We then use an alternative procedure.

Suppose you take a number of photographs at the same frequency as the vertical vibration of any portion of the rope. You would find that all the pictures look the same. During the time the shutter of the camera was closed, each portion of the rope went through a complete cycle, ending in the position it had during the previous photograph. But this means that each crest moved from its original position to the position of the crest in front of it. That is, the crest moved a distance equal to the wavelength λ. Because the time between exposures is equal to the period T, the wave's speed v is

$$v = \frac{\lambda}{T}$$

Because the frequency is just the reciprocal of the period, we can change the equation to read

$$v = \lambda f$$

▶ speed = wavelength × frequency

Although we developed this relationship for waves on a rope, there is nothing special about these waves. This relationship holds for all periodic waves, such as radio waves, sound waves, and water waves.

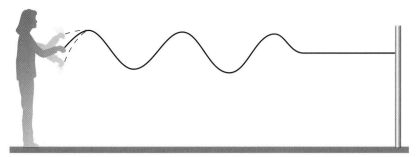

Figure 15-14 A periodic wave on a rope can be generated by moving the end up and down with a constant frequency.

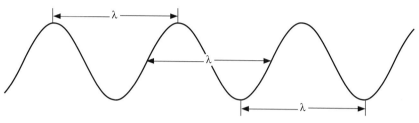

Figure 15-15 The wavelength of a periodic wave is the distance between any two identical spots on the wave.

WORKING IT OUT | Speed of a Wave ✓ MATH

If you know any two of the three quantities in the wave equation, you can use this relationship to calculate the third. As an example, let's calculate the speed of a wave that has a frequency of 40 Hz and a wavelength of $\frac{3}{4}$ m. Multiplying the wavelength and the frequency gives us the speed:

$$v = \lambda f = (\tfrac{3}{4}\text{ m})(40\text{ Hz}) = 30 \text{ m/s}$$

Question If water waves have a frequency of 5 Hz and a wavelength of 8 cm, what is the wave speed?

Answer $v = \lambda f = (8\text{ cm})(5\text{ Hz}) = 40 \text{ cm/s}.$

Flawed Reasoning

The following question appears on the midterm exam: "A periodic wave is traveling to the right on a long, stretched rope. Two small pieces of yarn are tied to the rope, one at point A and the other at point B, as shown in the figure:

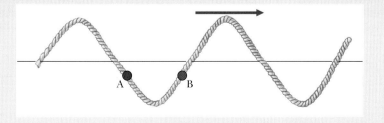

continues on next page

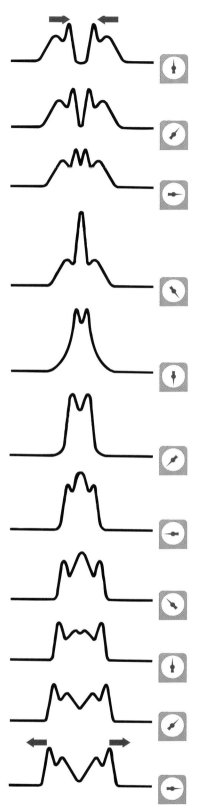

Figure 15-13 This time sequence shows that the superposition of two wave pulses yields shapes that are the sum of the individual shapes.

"Draw an arrow for each piece of yarn, indicating the direction of its velocity when the picture was taken."

Brianna gives the following answer to this question: "Because the wave is moving to the right, the pieces of yarn must also be moving to the right. The wave is carried by the rope."

What is wrong with Brianna's reasoning, and what is the correct answer to the exam question?

Answer The wave is a transverse wave, meaning that the medium moves perpendicular to the direction of the wave. Therefore, the pieces of yarn can only move up or down. If we look at the wave at a slightly later time (when it has moved a little to the right), we see that the yarn at point A has moved upward and the yarn at point B has moved downward:

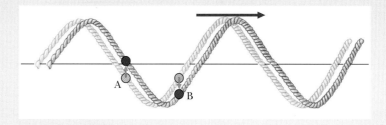

Standing Waves

When a periodic wave is confined, new effects emerge due to the superposition of the reflected waves with the original ones. Let's return to the example of a periodic wave moving down a rope toward a rigid post. When the periodic wave reflects from the post, it superimposes with the wave heading toward the post. The complete pattern results from the superposition of the original wave and reflections from both ends. In general, we get a complicated pattern with a small amplitude, but certain frequencies cause the rope to vibrate with a large amplitude. Figure 15-16 shows multiple images of a resonating rope. Although the superimposing waves move along the rope, they produce a resonant pattern that does not move along the rope. Because the pattern appears to stand still (in the horizontal direction), it is known as a **standing wave**.

It might seem strange that two identical waves traveling in opposite directions combine to produce a vibrational pattern that doesn't travel along the rope. We can see how this happens by using the superposition principle to find the results of combining the two traveling waves. Let's start at a time when the crests of the traveling wave moving to the right (the blue line in Figure 15-17) line up with the crests of the wave moving to the left (yellow line). (The blue and yellow lines lie on top of each other and are shown as a single green line.) Adding the displacements of the two traveling waves yields a wave that has the same basic shape but twice the amplitude. This is shown by the black line in Figure 15-17(a).

Figure 15-17(b) shows the situation a short time later. The blue wave has moved to the right, and the yellow wave has moved the same distance to the left. The superposition at this time produces a shape that still looks like one of the traveling waves but does not have as large an amplitude as before. A short time

Figure 15-16 A strobe drawing of a standing wave on a rope shows how the shape of the rope changes with time. The shape does not move to the left or right.

later, the crests of one wave line up with the troughs of the other. At this time the two waves cancel each other, and the rope is straight. Although the rope is straight at this instant, some parts of the rope are moving up while others are moving down. The remaining drawings in Figure 15-17 show how this pattern changes through the rest of the cycle as time progresses.

Notice that some portions of the rope do not move. Even though each traveling wave by itself would cause all pieces of the rope to move, the waves interfere to produce no motion at these points. Such locations are known as **nodes** and are located on the vertical lines indicated by N in Figure 15-17. The positions on the rope that have the largest amplitude are known as **antinodes** and are on the lines marked by A. Notice that the nodes and antinodes alternate and are equally spaced.

There is a relationship between the shape of the resonant pattern, or standing wave, and the moving periodic waves that superimpose to create it. The "wavelength" of the standing wave is equal to the wavelength of the underlying periodic wave. This can be seen in Figure 15-17.

Unlike the pendulum or the mass on a spring where there was only one resonant frequency, periodic waves that are confined have many different resonant frequencies. Strobe photographs of the standing wave with the lowest, or **fundamental**, **frequency** show that the rope has shapes like those drawn in Figure 15-18. The images show how the shape of the rope changes during one half cycle. At position 1 the rope has the largest possible crest. The displacement continually decreases until it becomes zero and the rope is straight (between positions 3 and 4). The rope's inertia causes it to overshoot and form a trough that grows in size. At the end of the half cycle, the rope is in position 6 and beginning its upward journey. The process then repeats.

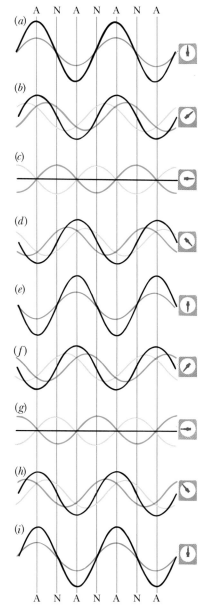

Question How does the distance between adjacent nodes or antinodes compare with the wavelength?

Answer Because one antinode is up when the adjacent ones are down, each antinodal region corresponds to a crest or a trough. Therefore, the distance between adjacent antinodes or adjacent nodes is one half wavelength. The distance between adjacent nodes and antinodes is one quarter wavelength.

This pattern has the lowest frequency and thus the longest wavelength of the resonant modes. Notice that one half wavelength is equal to the length of the rope, or the wavelength is twice the length of the rope. Because this is also the wavelength of the traveling waves, the longest resonant wavelength on a rope with nodes at each end is twice the length of the rope.

If we slowly increase the frequency of the traveling wave, the amplitude quickly decreases. The vibrational patterns are rather indistinct until the next resonant frequency is reached. This new resonant frequency has twice the frequency of the fundamental and is known as the second **harmonic**. Six shapes

Figure 15-17 This set of strobe drawings shows how two traveling waves (the blue and yellow lines) combine to form a standing wave (the black line). Only the black line would be visible in an actual photograph.

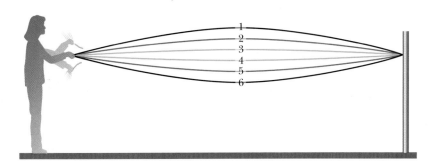

Figure 15-18 The shapes of a rope oscillating as a standing wave of the lowest frequency.

Figure 15-19 The shapes of a rope for a standing wave with the second resonant frequency.

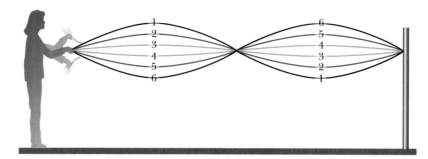

of the rope for this standing-wave pattern are shown in Figure 15-19. When the rope is low on the left side, it is high on the right, and vice versa. Because the rope has the shape of a full wavelength, the wavelength of the traveling waves is equal to the length of the rope. Note that this pattern has one more node and one more antinode than the fundamental standing wave.

We can continue this line of reasoning. A third resonant frequency can be reached by again raising the frequency of the traveling wave. This third harmonic frequency is equal to three times the fundamental frequency. This standing wave has three antinodes and four nodes, including the two at the ends of the rope. Other resonant frequencies occur at whole-number multiples of the fundamental frequency.

Remember that the product of the frequency and the wavelength is a constant. This is a consequence of the fact that the speeds of all waves on this rope are the same. Therefore, if we increase the frequency by some multiple while keeping the speed the same, the wavelength must decrease by the same multiple. The fundamental wavelength is the largest, and its associated frequency is the smallest. As we march through higher and higher frequencies, we get shorter and shorter wavelengths. The wavelengths of the higher harmonics are obtained by dividing the fundamental wavelength by successive whole numbers; the wavelength of the second harmonic is one-half the wavelength of the fundamental wavelength.

Question How does the wavelength of the third harmonic compare with the length of the rope?

Answer The wavelength of the third harmonic is one-third the length of the fundamental wavelength. Because the fundamental wavelength is twice the length of the rope, the wavelength of the third harmonic would be two-thirds the length of the rope.

Interference   ✓ MATH

Standing waves on a rope are an example of the superposition, or **interference**, of waves in one dimension. If we use a two-dimensional medium—say, the surface of water—we can generate some new effects.

Suppose we use two wave generators to create periodic waves on the surface of water in a ripple tank like the one in Figure 15-20(a). Because the two waves travel in the same medium, they have the same speed. We also assume that the two sources have the same frequency and that they are **in phase**; that is, both sources produce crests at the same time, troughs at the same time, and so on. The superposition of these waves creates the interference pattern shown in the photograph and drawing of Figure 15-20. The bright regions are produced by the crests, whereas the dark regions are produced by the troughs.

In some places, crest meets crest to form a supercrest, and one half period

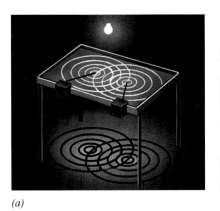 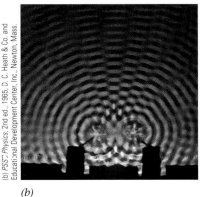 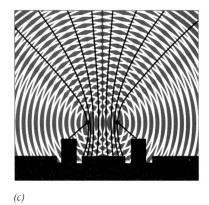

(a)   *(b)*   *(c)*

Figure 15-20 (a) A lightbulb above a ripple tank produces light and dark lines on the floor due to the water waves. (b) The interference pattern produced by two point sources of the same wavelength and phase. (c) The locations of the nodal lines in this pattern are shown in black; the central antinodal line is shown in red.

later, trough meets trough to form a supertrough. This meeting point is a region of large amplitude; the two waves form antinodal regions. In other places, crest and trough meet. Here, if the two waves have about the same amplitude, they cancel each other, resulting in little or no amplitude; the two waves form nodal regions.

Because of the periodic nature of the waves, the nodal and antinodal regions have fixed locations. These stationary interference patterns can be observed only if the two sources emit waves of the same frequency; otherwise, one wave continually falls behind the other, and the relationships between the two waves change. The two wave sources do not have to be in phase; there can be a time delay between the generation of crests by one source and the other as long as the time delay is constant. For simplicity we usually assume that this time delay is zero; that is, the two sources are in phase.

The regions of crests and troughs lie along lines. One such antinodal line lies along the perpendicular to the midpoint of the line joining the two sources. This is the vertical red line in Figure 15-20(c). This central antinodal line is the same distance from the two sources. Therefore, crests generated at the same time at the two sources arrive at the same time at this midpoint to form supercrests. Similarly, two troughs arrive together, creating a supertrough.

Consider a point P off to the right side of the central line, as shown in Figure 15-21. Although nothing changes at the sources, we get a different result. Crests from the two sources no longer arrive at the same time. Crests from the left-hand source must travel a greater distance and therefore take longer to get to P. The amount of delay depends on the difference in the two path lengths.

If the point P is chosen such that the distances to the sources differ by an amount equal to one half wavelength, crests overlap with troughs and troughs overlap with crests at point P. The waves cancel. There are many points that have this path difference. They form nodal lines that lie along each side of the central line as shown by the black lines in Figure 15-20(c). An antinodal line occurs when the path lengths differ by one wavelength; the next nodal line when the paths differ by $1\frac{1}{2}$ wavelengths, and so on.

Question Would the central line still be an antinodal line if the two sources were completely out of phase—that is, if one source generates a crest at the same time as the other generates a trough?

Answer The central line would now be a nodal line because crests and troughs would arrive at the same time.

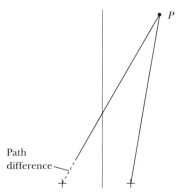

Figure 15-21 Whether the region at P is a nodal or antinodal region depends on the difference in the path lengths from the two sources.

Figure 15-22 The nodal lines (black) are more widely spaced for longer wavelengths. The central antinodal line (red) remains in the same location. Compare this with Figure 15-20.

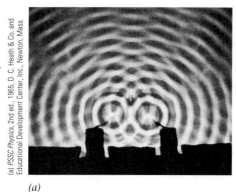
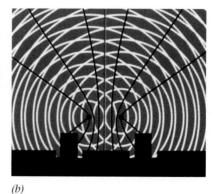

(a) (b)

The photograph in Figure 15-22 shows the interference pattern for water waves with a longer wavelength than those in Figure 15-20. The nodal lines are now more widely spaced for the same source separation. Therefore, longer wavelengths produce wider patterns. Actually, the width of the pattern depends on the relative size of the wavelength and the source separation. As the ratio of the wavelength to the separation gets bigger, the nodal lines spread out. If the wavelength is much larger than the separation, the pattern is essentially that of a single source, whereas if it is much smaller, the nodal lines are so close together they cannot be seen.

Diffraction

In the photographs in Figure 15-23, periodic water waves move toward a barrier. We see that the waves do not go straight through the opening in the barrier but spread out behind the barrier. This bending of a wave is called **diffraction** and is definitely not a property of particles. If a BB gun is fired through an opening in a barrier, the pattern it produces is a precise "shadow" of the opening if we assume that no BBs bounce off the opening's edges.

The amount of diffraction depends on the relative sizes of the wavelength and the opening. If the wavelength is much smaller than the opening, very little diffraction is evident. As the wavelength gets closer to the size of the opening, the amount of diffraction gets bigger. In Figure 15-23(c) the opening and the wavelength are approximately the same size, and the diffraction is evident.

Notice that diffraction produces nodal and antinodal lines similar to those observed in the interference patterns from two point sources. In this case there is a broad central antinodal region with nodal lines on each side, and the dif-

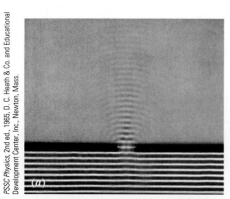
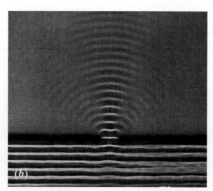
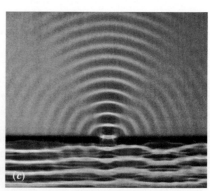

Figure 15-23 Ripple tank patterns of water moving upward and passing through a narrow barrier. Note that the amount of diffraction increases as the wavelength gets longer.

fraction pattern is created by different portions of the wave interfering with themselves. The spacing of these lines is determined by the ratio of the wavelength and the width of the opening.

PHYSICS | ON YOUR OWN

Watch for diffraction effects when ocean waves strike offshore rocks. Are the "shadows" very sharp?

Summary

Vibrations and oscillations are described by the length of time required for one cycle, the period T (or its reciprocal, the frequency f), and the amplitude of the vibration, the maximum distance the object travels from the equilibrium point. When vibrations are small, the period is independent of the amplitude. The pendulum and a mass hanging on a spring are examples of systems that vibrate.

All systems have a distinctive set of natural frequencies. A simple system such as a pendulum has only one natural frequency, whereas more complex systems have many natural frequencies. When a system is excited at a natural frequency, it resonates with a large amplitude.

Waves are vibrations moving through a medium; it is the wave (energy) that moves through the medium, not the medium itself. Transverse waves vibrate perpendicular to the direction of the wave, whereas longitudinal waves vibrate parallel to the direction of the wave. The speed of a periodic wave is equal to the product of its wavelength and its frequency, $v = \lambda f$.

Waves pass through each other as if the other were not there. When they overlap, the shape is the algebraic sum of the displacements of the individual waves. When a periodic wave is confined, resonant patterns known as standing waves can be produced. Portions of the medium that do not move are called nodes, whereas portions with the largest amplitudes are known as antinodes. The fundamental standing wave has the lowest frequency and the longest wavelength.

Two identical periodic-wave sources with a constant phase difference produce an interference pattern consisting of large-amplitude antinodal regions and zero-amplitude nodal regions. The spacing of the interference pattern depends on the relative size of the wavelength and the source separation.

Waves do not go straight through openings or around barriers but spread out. This diffraction pattern contains nodal and antinodal regions and depends on the relative sizes of the wavelength and the opening.

The waves are diffracted as they pass through the openings in the seawall, producing an interesting shape at the beach.

Physics Now™ Assess your understanding of this chapter's topics with sample tests and other resources found by logging into PhysicsNow at **http://physics.brookscole.com/kf6e**.

Chapter 15 Revisited

When waves move in a medium, the medium oscillates in place. No material is transported from one location to another; it is the disturbance that moves. Unlike with particles, when two waves pass through the same region at the same time, the individual disturbances are added together. Afterward, each wave retains its own identity.

KEY TERMS

amplitude: The maximum distance from the equilibrium position that occurs in periodic motion.

antinode: One of the positions in a standing-wave or interference pattern where there is maximum movement; that is, the amplitude is a maximum.

crest: The peak of a wave disturbance.

cycle: One complete repetition of a periodic motion. It may start any place in the motion.

diffraction: The spreading of waves passing through an opening or around a barrier.

displacement: In wave (or oscillatory) motion, the distance of the disturbance (or object) from its equilibrium position.

equilibrium position: A position where the net force is zero.

frequency: The number of times a periodic motion repeats in a unit of time. It is equal to the inverse of the period.

fundamental frequency: The lowest resonant frequency for an oscillating system.

harmonic: A frequency that is a whole-number multiple of the fundamental frequency.

in phase: Two or more waves with the same wavelength and frequency that have their crests lined up.

interference: The superposition of waves.

longitudinal wave: A wave in which the vibrations of the medium are parallel to the direction the wave is moving.

node: One of the positions in a standing-wave or interference pattern where there is no movement; that is, the amplitude is zero.

oscillation: A vibration about an equilibrium position or shape.

period: The shortest length of time it takes a periodic motion to repeat. It is equal to the inverse of the frequency.

periodic wave: A wave in which all the pulses have the same size and shape. The wave pattern repeats itself over a distance of one wavelength and over a time of one period.

resonance: A large increase in the amplitude of a vibration when a force is applied at a natural frequency of the medium or object.

spring constant: The amount of force required to stretch a spring by one unit of length. Measured in newtons per meter.

standing wave: The interference pattern produced by two waves of equal amplitude and frequency traveling in opposite directions. The pattern is characterized by alternating nodal and antinodal regions.

superposition: The combining of two or more waves at a location in space.

transverse wave: A wave in which the vibrations of the medium are perpendicular to the direction the wave is moving.

trough: A valley of a wave disturbance.

vibration: An oscillation about an equilibrium position or shape.

wave: The movement of energy from one place to another without any accompanying matter.

wavelength: The shortest repetition length for a periodic wave. For example, it is the distance from crest to crest or trough to trough.

CONCEPTUAL QUESTIONS

1. If the net force on a mass oscillating at the end of a vertical spring is zero at the equilibrium point, why doesn't the mass stop there?

2. If the restoring force on a pendulum is zero when it is vertical, why doesn't it quit swinging at this point?

3. A mass is oscillating up and down on a vertical spring. When the mass is above the equilibrium point and moving downward, in what direction does the net force on the mass act? When the mass is above the equilibrium point and moving upward, what is the direction of the net force on the mass? Explain.

4. A mass is oscillating up and down on a vertical spring. When the mass is below the equilibrium point and moving downward, what is the direction of its acceleration? Is the mass speeding up or slowing down? Explain.

5. A mass is oscillating up and down on a vertical spring. If the mass is increased, will the period of oscillation increase, decrease, or stay the same? Will the frequency increase, decrease, or stay the same? Explain.

6. A grandfather clock (with a pendulum) keeps perfect time on Earth. If you were to transport this clock to the Moon, would its period of oscillation increase, decrease, or stay the same? Would its frequency increase, decrease, or stay the same? Explain.

7. You hang a 1-kilogram block from a spring and find that the spring stretches 15 centimeters. What mass would you need to stretch the spring 45 centimeters?

8. Which spring would you expect to have the greater spring constant, the one in the suspension of your Chevy or the one in your watch? Why?

9. Assume that you pull the mass on a spring 1 centimeter from the equilibrium position, let go, and measure the period of the oscillation. Would you expect the period to be larger, the same, or smaller if you pulled the mass 2 centimeters from the equilibrium position? Why?
10. The amplitude of a real pendulum decreases because of frictional forces. How does the period of this real pendulum change?
11. What is the period of the hand on a clock that measures the seconds? What is its frequency?
12. What is the period of the hand on a clock that measures the minutes? What is its frequency?
13. Suppose your grandfather clock runs too fast. If the mass on the pendulum can be moved up or down, which way would you move it to adjust the clock? Explain your reasoning.
▲ 14. How does the natural frequency of a swing change when you move from sitting down to standing up?

15. You find that the exhaust system on your 1979 Chrysler Cordoba tends to rattle loudly when the tachometer, which measures the engine's frequency, reads 2000 rpm. It is relatively quiet at frequencies above or below 2000 rpm. Use the concept of resonance to explain this.
16. Why do soldiers "break step" before crossing a suspension bridge?
17. You hold one end of a spring in your hand and hang a block from the other end. After lifting the block up slightly and releasing it, you find that it oscillates up and down at a frequency of 4 hertz. At which of the following frequencies could you jiggle your hand up and down and produce resonance: 10 hertz, 8 hertz, 3 hertz, 2 hertz, or 1 hertz?
18. You stand to the side of the low point of a child's swing and always push the child in the same direction. Which of the following multiples of the fundamental frequency will not produce resonance: $\frac{1}{3}$, $\frac{1}{2}$, 1, or 2?
19. When you yell at your friend, are the air molecules that strike his ear the same ones that were in your lungs? Explain.
20. What is being transported along the surface of a lake when a wave moves across the surface?
21. Sonar devices use underwater sound to explore the ocean floor. Would you expect sonar to be a longitudinal or a transverse wave? Explain.

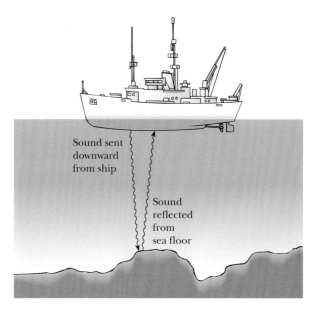

22. You fasten one end of a long spring to the base of a wall and stretch it out along the floor, holding the other end in your hand. Describe how you would generate a transverse pulse on the spring. Describe how you would generate a longitudinal pulse on the spring.
23. Is it possible for a shout to overtake a whisper? Explain.
24. You generate a small transverse pulse on a long spring stretched between a doorknob and your hand. How could you generate a second pulse that would overtake the first pulse?
25. Which one or more of the following properties affect the speed of waves along a rope: amplitude of the pulse, shape of the pulse, tension in the rope, or the mass per unit length of the rope? Why?
26. You move your hand up and down to send a pulse along a long spring stretched between a doorknob and your hand. Which of the following would generate a slower-traveling pulse: Move your hand the same distance as before but do it more slowly; move your hand a smaller distance at the same speed as before; or move closer to the doorknob to decrease the tension in the spring.
27. You send a pulse of amplitude 5 centimeters down the right side of a spring. A moment later you send an identical pulse on the same side. The first pulse reflects from the fixed end and returns along the spring. When the reflected pulse meets the second pulse, will the resulting amplitude be less than, equal to, or greater than 5 centimeters? Explain your reasoning.
28. Imagine that the string in Figure 15-14 is tied to a pole with a loose loop such that the end is free to move up and down. Two identical pulses of amplitude 10 centimeters are sent down the string. The first pulse reflects from the free end and meets the second pulse. Will the resulting amplitude be less than, equal to, or greater than 10 centimeters? Explain your reasoning.
29. The pulse in the figure is traveling on a string to the right

toward a fixed end. Draw the shape of the pulse after it reflects from the boundary.

30. A pulse in the shape of a crest is sent from left to right along a stretched rope. A trough travels in the opposite direction so that the pulses meet in the middle of the rope. Would you expect to observe a crest or a trough arrive at the right-hand end of the rope? Explain.
31. If shapes (a) and (b) in the figure correspond to idealized wave pulses on a rope, what shape is produced when they completely overlap?

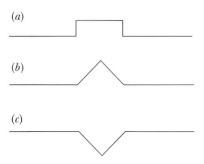

32. Repeat Question 31 for shapes (a) and (c).
33. Which of the following properties are meaningful for periodic waves but not for single pulses: frequency, wavelength, speed, amplitude?
34. In the following list of properties of periodic waves, which one is independent of the others: frequency, wavelength, speed, amplitude?
35. Two waves have the same speed but one has twice the frequency. Which wave has the longer wavelength? Explain.
36. If the frequency of a periodic wave is cut in half while the speed remains the same, what happens to the wavelength?
37. If the speed of a periodic wave doubles while the period remains the same, what happens to the wavelength?
38. What happens to the wavelength of a periodic wave if both the speed of the wave and the frequency are cut in half?
▲ 39. Travelers, spaced 10 feet apart, are all walking at 3 mph relative to a moving sidewalk. When the moving sidewalk ends, they continue to walk at 3 mph. An observer standing next to the moving sidewalk notes that the travelers are passing by at a frequency of 1 hertz. A second observer stands just beyond the end of the moving sidewalk and notes the frequency at which the travelers pass. Would this frequency be greater than, equal to, or less than 1 hertz? Is the spacing between the travelers after leaving the moving sidewalk greater than, equal to, or less than 10 feet? Explain.

▲ 40. A waterproof buzzer has a membrane that vibrates at a constant frequency of 440 hertz. The buzzer is placed in a bucket of water. (The speed of sound is much greater in water than in air.) Will the frequency of the sound heard in the air be greater than, equal to, or less than 440 hertz? Will the wavelength of the sound in air be greater than, equal to, or less than the wavelength in the water? Explain.
41. Draw a diagram to represent the standing-wave pattern for the third harmonic of a rope fixed at both ends. How many antinodes are there?
42. Draw a diagram to represent the standing-wave pattern for the fourth harmonic of a rope fixed at both ends. How many nodes are there?
43. How much higher is the frequency of the fifth harmonic on a rope than the fundamental frequency?
44. How much higher is the frequency of the sixth harmonic on a rope than that of the second?
45. Standing waves can be established on a rope that is fixed on one end but free to slide up and down a pole on the other. The fixed end remains a node, while the free end must be an antinode. Draw diagrams to represent the standing-wave patterns for the two lowest frequencies.
46. How does the fundamental wavelength of standing waves on a string with one end fixed and the other free compare with the fundamental wavelength if the same string is held with both ends fixed?
47. How does the wavelength of the fourth harmonic on a rope with both ends fixed compare with the length of the rope?
48. How does the wavelength of the fourth harmonic on a

rope with both ends fixed compare with that of the second harmonic?

49. A longitudinal standing wave can be established in a long aluminum rod by stroking it with rosin on your fingers. If the rod is held tightly at its midpoint, what is the wavelength of the fundamental standing wave? Assume that there are antinodes at each end of the rod and a node where the rod is held.

50. What is the wavelength of the fundamental standing wave for the rod in Question 49 if it is held midway between the center and one end? Will the resulting pitch be higher or lower than when the rod was held at its midpoint? Explain.

51. Two point sources produce waves of the same wavelength and are in phase. At a point midway between the sources, would you expect to find a node or an antinode? Explain.

52. Two point sources produce waves of the same wavelength and are completely out of phase (that is, one produces a crest at the same time as the other produces a trough). At a point midway between the sources, would you expect to find a node or an antinode? Why?

▲ 53. What happens to the spacing of the antinodal lines in an interference pattern when the two sources are moved farther apart? Explain.

▲ 54. As you increase the frequency, what happens to the spacing of the nodal lines in an interference pattern produced by two sources? Explain.

55. An interference pattern is produced in a ripple tank. As the two sources are brought closer together, does the separation of the locations of maximum amplitude along the far edge of the tank decrease, increase, or remain the same? Why?

56. As the frequency of the two sources forming an interference pattern in a ripple tank increases, does the separation of the locations of minimum amplitude along the far edge of the tank increase, decrease, or remain the same? Why?

▲ 57. What happens to the spacing of the antinodal lines in a diffraction pattern when the two slits are moved farther apart? Explain.

▲ 58. As you increase the frequency, what happens to the spacing of the nodal lines in a diffraction pattern? Explain.

EXERCISES

1. If a mass on a spring takes 3 s to complete two cycles, what is its period?
2. If a mass on a spring has a frequency of 5 Hz, what is its period?
3. A Foucault pendulum with a length of 9 m has a period of 6 s. What is its frequency?
4. A mass on a spring bobs up and down over a distance of 20 cm from the top to the bottom of its path twice each second. What are its period and amplitude?
5. A spring hanging from the ceiling has an unstretched length of 80 cm. A mass is then suspended at rest from the spring, causing its length to increase to 89 cm. The mass is pulled down an additional 3 cm and released. What is the amplitude of the resulting oscillation?
6. A mass oscillates up and down on a vertical spring with an amplitude of 4 cm and a period of 2 s. What total distance does the mass travel in 10 s?
7. What is the period of a 0.4-kg mass suspended from a spring with a spring constant of 40 N/m?
8. A boy with a mass of 50 kg is hanging from a spring with a spring constant of 200 N/m. With what frequency does the boy bounce up and down?
9. By what factor would you have to change the spring constant to double the period for a mass on a spring?
10. By what factor would you have to change the mass to triple the frequency for a mass on a spring?
11. A pendulum has a length of 5 m. What is its period?
12. A girl with a mass of 40 kg is swinging from a rope with a length of 2.5 m. What is the frequency of her swinging?
▲ 13. The highly idealized wave pulses shown in the figure at a time equal to zero have the same amplitudes and travel at 1 cm/s. Draw the shape of the rope at 2, 4, 5, and 8 s.

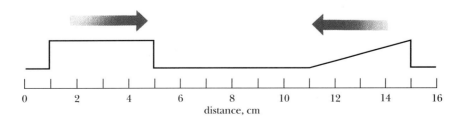

▲ 14. Work Exercise 13 but change the rectangular pulse from a crest to a trough.

15. A train, consisting of identical 10-m boxcars, passes you such that 25 boxcars pass you each minute. Find the speed of the train.

16. You observe that 25 crests of a water wave pass you each minute. If the wavelength is 10 m, what is the speed of the wave?

17. A periodic wave on a string has a wavelength of 50 cm and a frequency of 2 Hz. What is the speed of the wave?

18. If the breakers at a beach are separated by 5 m and hit shore with a frequency of 0.3 Hz, at what speed are they traveling?

19. What is the distance between adjacent crests of ocean waves that have a frequency of 0.2 Hz if the waves have a speed of 3 m/s?

20. Sound waves in iron have a speed of about 5100 m/s. If the waves have a frequency of 300 Hz, what is their wavelength?

21. For sound waves, which travel at 343 m/s in air at room temperature, what frequency corresponds to a wavelength of 1 m?

22. What is the period of waves on a rope if their wavelength is 0.8 m and their speed is 2 m/s?

▲ 23. A rope is tied between two posts separated by 2 m. What possible wavelengths will produce standing waves on the rope?

▲ 24. A 2-m-long rope is tied to a very thin string so that one end is essentially free. What possible wavelengths will produce standing waves on this rope?

▲ 25. What is the fundamental frequency on a 4-m rope that is tied at both ends if the speed of the waves is 20 m/s?

▲ 26. Tweety Bird hops up and down at a frequency of 0.5 Hz on a power line at the midpoint between the poles, which are separated by 20 m. Assuming Tweety is exciting the fundamental standing wave, find the speed of transverse waves on the power line.

16 Sound and Music

A military marching band at a ticker-tape parade in New York City.

Sounds are all around us. Some are pleasant and some irritate and distract us. How do the sounds of music differ from those sounds we call noise, and why do musicians use different-sized instruments?

(See page 329 for the answer to this question.)

PhysicsNow™ Test your understanding of this chapter by logging into PhysicsNow at **http://physics.brookscole.com/kf6e**, selecting the chapter, and clicking on the "Take a Pre-Test" link.

WHEN we think of sound, we generally think of signals traveling through the air to our ears. But sound is more than this. Sound also travels through other media. For instance, old-time Westerns showed cowboys and Indians listening for the iron horse by putting their ears to the rails or the ground; two rocks clapped together underwater are easily heard by swimmers below the surface; a fetus inside a mother's womb can be examined with ultrasound; and the voices of people talking in the next room are often heard through the walls. Sounds can be soothing and musical, but they can also be irritating or even painful.

Sound is a wave phenomenon. For example, we talk of the pitch, or frequency, of sounds, which is definitely a wave characteristic. But what other evidence do we have? The conclusive evidence is that sound exhibits superposition, something that we know distinguishes waves from particles.

Sound

A vibrating object produces disturbances in the surrounding air. When the surface moves outward, the air molecules are pushed away, creating a *compression*. When its surface moves in the opposite direction, a partial void, or *rarefaction*, is created near the object. Pressure differences cause the air molecules to rush back into the region only to get pushed out again. Thus, the air molecules vibrate back and forth near the object's surface as illustrated by the tuning fork in Figure 16-1. The compressions and rarefactions travel away from the vibrating surface as a sound wave. Because the vibrations of the air molecules are along the same direction as the motion of the wave, sound is a longitudinal wave.

▶ sound is a longitudinal wave

As with all waves, it is energy—not mass—that is transported. The individual molecules of the air are not moving from one place to another; they simply vibrate back and forth. It is the disturbance that moves across the room when you talk. This disturbance, or wave, moves with a certain speed.

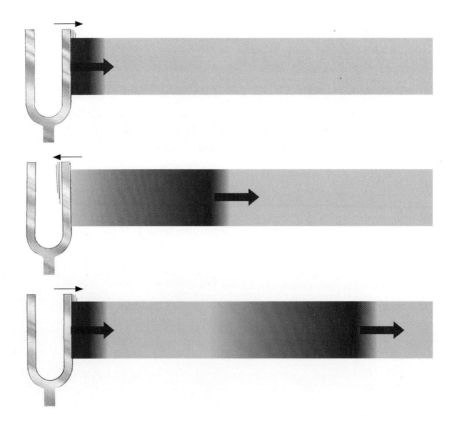

Figure 16-1 Sound is a longitudinal wave in which the air molecules vibrate along the direction the wave is traveling, producing compressions and rarefactions that travel through the air.

PHYSICS | ON YOUR OWN

Investigate the behavior of longitudinal waves with a Slinky. For example, fix one end and send a compression wave pulse down the Slinky, bouncing it off the end. If you send two wave pulses down the Slinky, the first pulse will reflect and meet with the trailing one. Do these pulses superimpose the same way that transverse pulses do? Can you generate standing waves on the Slinky similar to those discussed in Chapter 15?

Speed of Sound   ✓ MATH

✓ Extended presentation available in the *Problem Solving* supplement

Echoes demonstrate that sound waves reflect off surfaces and that they move with a finite speed. You can use this phenomenon to measure the speed of sound in air. If you know the distance to the reflecting surface and the time it takes for the echo to return, you can calculate the speed of sound. The speed of sound is 343 meters per second (1125 feet per second), or 1235 kilometers per hour (767 mph), at room temperature.

◄ speed of sound = 343 m/s

Experiments have shown that the speed of sound does not depend on the pressure, but it does depend on the temperature and the type of gas. Sound is slower at lower temperatures. At the altitude of jet airplanes, where the temperature is typically −40°C (−40°F), the speed of sound drops to 310 meters per second, or 1020 kilometers per hour (690 mph). The speed is higher for gases with molecules that have smaller masses. The speed of sound in pure helium at room temperature is three times that in air.

Flawed Reasoning

Heidi and Russell are discussing the speed of sound in air:

Heidi: "A sound wave travels through air via collisions of air molecules. If the air is compressed, the molecules are closer together. The sound wave should speed up because each molecule does not need to travel as far to collide with its neighbor."

Russell: "The textbook claims that the speed of sound in air depends only on the temperature of the air, not on its pressure."

Find the error in Heidi's reasoning.

Answer Imagine a relay race in which every runner runs *exactly* 5 mph. If one of the teams has twice as many runners (spaced half as far apart), the race still ends in a tie. Each runner on the smaller team runs twice as far before passing the baton, but there are half as many baton transfers. Similarly, each molecule in the compressed gas reaches its neighbor in less time, but more neighbors are involved. The air temperature determines the average molecular speed, and this determines how quickly the wave travels.

Knowing the speed of sound in air allows you to calculate your distance from a lightning bolt. Because the speed of light is very, very fast, the time light takes to travel from the lightning to you is negligible. Therefore, the time delay between the arrival of the light flash and the sound of the thunder is essentially all due to the time it takes the sound to travel the distance. Given the speed of sound, we can use the definition of speed from Chapter 2 to calculate that it takes sound approximately 3 seconds to travel 1 kilometer (5 seconds for a mile).

Sound waves also travel in other media. The speed of sound in water is about 1500 meters per second, much faster than in air. The speed of sound in solids is usually quite a bit higher—as high as 5000 meters per second—and the sound is quite a bit louder. Hearing the sounds of an approaching train through the rails works well because the sound moving through the rail does not spread out like sound waves in air and experiences less scattering along the path. You can easily experience this phenomenon in the classroom. Have a friend scratch one end of a meter stick while you hold the other end next to your ear. The effect is striking.

Sonar (sound navigation ranging) uses the echoes of sound waves in water to determine the distances to underwater objects. These sonar devices emit sound pulses and measure the time required for the echo to return. Sonar is used to determine the depth of the water, search for schools of fish, and locate submarines.

Question   If it takes the thunder 9 seconds to reach you, how far away is the lightning?

Answer   Because it takes the thunder 3 seconds to travel a kilometer, the lightning bolt must have been 3 kilometers away (a little less than 2 miles).

PHYSICS | **ON YOUR OWN**

Hook an elastic band to your eyetooth and pluck it. The sound is much louder to you than to a nearby friend because you hear the sound that travels through your jawbone, but your friend hears only the sound traveling through air. What happens when you stretch the band?

Hearing Sounds

Sound perception is a complex phenomenon whose study involves a wide variety of sciences, including physiology, psychology, and the branch of physics called acoustics. Experiments with human hearing, for example, clearly show that our perception is often different from the simple interpretations of the measurements taken by instruments. Our perception of pitch depends mainly on frequency but is also affected by other properties, such as loudness. And loudness depends on the amplitude of the wave (as well as the response of our ears). When our ear–brain systems tell us that one sound is twice as loud as another, instruments show that the power output is nearly eight times as great.

Our ears intercept sound waves from the air and transmit their vibrations through internal bone structures to special hairlike sensors. The ear canal acts as a resonator, greatly amplifying frequencies near 3000 hertz. This amplified sound wave moves the eardrum, which is located at the end of the ear canal, as shown in Figure 16-2. The eardrum is connected to three small bones in the middle ear. When sound reaches the middle ear, it has been transformed from a wave in air to a mechanical wave in the bones. These bones then move a smaller oval window inside the ear. The leverage advantage of the bone structure and the concentration of the pressure vibration onto a smaller window further amplify the sound, increasing our ability to hear faint sounds. The final transformation of mechanical sound waves to nerve impulses takes place in the inner ear. The pressure vibrations in the fluid of the inner ear resonate with different hairlike sensors, depending on the frequency of the sound.

The range of frequencies that we can hear clearly depends on the resonant structures within our ears. When a frequency is too high or too low, the sound wave is not amplified like those within the audible range. The audible range is normally from 20 hertz to 20,000 hertz, although it varies with age and the individual. The sensitivity of our ears varies over this range, with low sensitivity occurring at both ends of the range. As we get older, our ability to hear higher

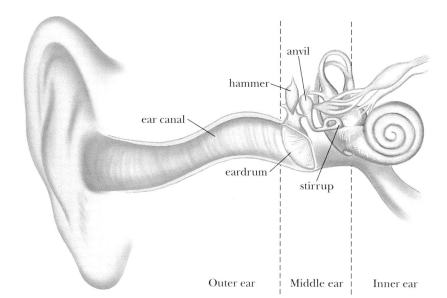

Figure 16-2 The structure of the human ear.

frequencies decreases. However, modern digital hearing aids can be individually programmed to compensate for a person's hearing loss at each frequency.

The Recipe of Sounds

Suppose you are in a windowless room but can hear sounds from the outside. You would have no trouble identifying most of the sounds you hear. A bird sounds different from a foghorn, a trumpet different from a baritone. Why is it that you can recognize these different sources of sound? They might be producing different notes; a bird sings much higher than a foghorn. They may make different melodies; some birds have a rather complex sequence of notes in their call, whereas the foghorn produces one continuous note.

What if all the sound makers outside your room are restricted to a single note? Do you suppose that you could still identify the different sources? Most likely. When each source produces the note, it is accompanied by other higher, resonant frequencies. You actually hear a superposition of these frequencies. The reason you can distinguish among the sources is that each sound has a unique combination of intensities of the various harmonics (Chapter 15). One sound might be composed of a strong fundamental frequency and weaker higher harmonics, and another may have a particularly strong second harmonic. Each sound has a particular recipe of resonant frequencies that combine to make the total sound.

The vibrating strings in a piano, violin, guitar, or banjo have their own combinations of the fundamental frequency and higher harmonics. The relative intensities of these harmonics depend primarily on the way the string was initially vibrated and on the vibrational characteristics of the body of the instrument. The initial part of the sound is called the *attack*, and its character is determined, in part, by how the sound is produced. Bowing, for example, produces a sound different from plucking or striking. The manner in which the various components of the sound decay also differs from one instrument to another. The wave patterns produced by various musical instruments are illustrated in Figure 16-3.

Our voices have the same individual character. We recognize different voices because of their particular recipe of harmonics. Research has shown that this recipe changes under emotional stress. Some people have suggested that these changes in the recipe of sound of a voice can be used in lie detection. Electronic devices can decompose the composite waveform—the superposition of all the harmonics—and evaluate the relative strength of each harmonic. This tech-

Digital hearing aid.

Animal Hearing

Animals have different ranges of sensitivity for hearing than humans. Dogs and bats, for example, have hearing ranges that extend to ultrasonic frequencies—frequencies above those that we can hear.

Unlike humans, most animals use their hearing as an aid in gathering food and escaping danger. Bats squeak at ultrasonic frequencies and detect the echoes from small flying insects (Figure A); robins cock their heads in early spring as they listen for the very faint sounds of worms in the ground, and owls have two different types of ears, providing them with "binocular" hearing for finding mice moving through grassy fields.

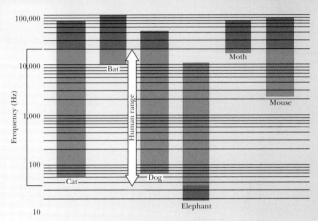

Figure B The frequency ranges of hearing for some animals.

Figure A The bat emits sound waves (shown in blue) that reflect off the insect and return (shown in red) to the bat, giving away the insect's position.

Animal researchers have found that homing pigeons and elephants hear sounds that are very, very low frequency—infrasonic frequencies. Low-frequency sounds do not attenuate as rapidly and therefore travel much farther than high-frequency sounds. In the case of pigeons, hearing the sounds of skyscrapers swaying in the wind in distant cities may provide them with navigational bearings. There is evidence that elephants communicate with each other over distances of miles using a low-frequency rumble. Figure B displays the frequency ranges that some animals can hear.

---

nique spots changes in higher harmonics that are virtually undetectable in the composite waveform.

How does music differ from other sounds? This question is difficult because what is music to one person might be noise to you. In general, music can be defined as that collection of periodic sounds that is pleasing to the ear.

Most cultures divide the totality of musical frequencies into groups known as octaves. A note in one octave has twice the frequency of the corresponding note in the next lower octave. For example, the pitches labeled C in ascending adjacent octaves have frequencies of 262, 524, and 1048 hertz, respectively. In Western cultures most music is based on a scale that divides the octave into 12 steps. It may seem strange that an octave spans 12 notes, because the word *octave* derives from the Latin word meaning "eight." An octave contains 7 notes and 5 sharps and flats. Indian and Chinese music have different divisions within their octaves.

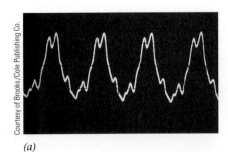 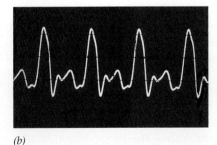 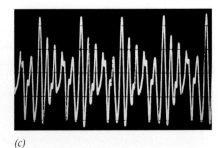

(a) (b) (c)

Figure 16-3 Wave patterns of (a) a violin, (b) a trumpet, and (c) a bassoon. These photographs show only general characteristics; the details depend on the placement of the microphone.

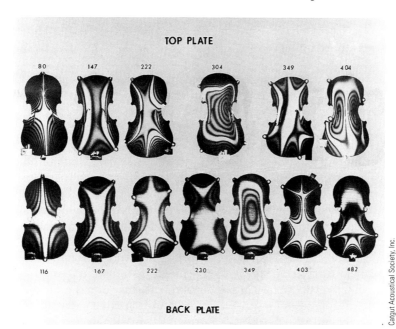

Figure 16-4 The vibrations in the body of the violin are made visible by holographic techniques.

Through the years, people have created instruments to produce sounds that were pleasing to them. Nearly all of these instruments involve the production of standing waves. Although there is an enormous variety of instruments, most of them can be classified as string, wind, or percussion instruments.

Stringed Instruments

When a string vibrates, it moves the air around it, producing sound waves. Because the string is quite thin, not much air is moved, and consequently the sound is weak. In acoustic stringed instruments, this lack of volume is solved by mounting the vibrating string on a larger body. The vibrations are transmitted to the larger body, which can move more air and produce a louder sound. Figure 16-4 shows a variety of vibrations produced in a violin's body by its vibrating strings. The design of the instrument produces variations in the instrument's vibrational patterns and thus changes the character of the sound produced. There was something special about the way Antonio Stradivari made his violins that made their sound more pleasing than others.

Modern electric guitars do not use the body of the instrument to transmit the vibrations of the strings to the air. The motions of the vibrating strings are converted to oscillating electric signals by pickup coils mounted under the strings. This signal is then amplified and sent to the speakers.

Whether acoustic or electric, plucking a guitar string creates vibrations. This initial distortion causes waves to travel in both directions along the string and reflect back and forth from the fixed ends. The initial shape of the string is equivalent to a unique superposition of many harmonic waves. Figure 16-5 shows the shape of a plucked string at the moment of release and the contributions of the first five harmonics. (For this particular shape, the fourth harmonic is zero everywhere and does not contribute.)

These harmonic waves travel back and forth on the string, creating standing waves with nodes at the two ends of the string. The first three standing waves are shown in Figure 16-6. Because the distance between nodes is one-half the wavelength, the wavelength of the fundamental, or first harmonic (a), is twice as long as the string, or $\lambda_1 = 2L$. The wavelength of the second harmonic (b) is $\lambda_2 = L$, the length of the string, and therefore half as long as that of the fundamental. The wavelength of the third harmonic (c) is $\lambda_3 = \frac{2}{3}L$, one-third of the fundamental wavelength.

◀ harmonic wavelengths $\lambda_1, \frac{\lambda_1}{2}, \frac{\lambda_1}{3}, \ldots$

Question What is the wavelength corresponding to the third harmonic on a 60-centimeter-long wire?

Answer It must be one-third the length of the fundamental wavelength. Because the fundamental wavelength is twice the length of the string, we obtain 2(60 centimeters)/3 = 40 centimeters.

The frequencies of the various harmonics can be obtained from the relationship for the wave's speed, $v = \lambda f$. As long as the vibrations are small, the speeds of the different waves on the plucked string are all the same. Therefore, $v = \lambda_1 f_1 = \lambda_2 f_2 = \lambda_3 f_3 = \cdots$. Because the second harmonic has half the wavelength, it must have twice the frequency. The third harmonic has one-third the wavelength and three times the frequency. The harmonic frequencies are whole-number multiples of the fundamental frequency.

▶ harmonic frequencies $f_1, 2f_1, 3f_1, \ldots$

There is a simple way to verify that more than one standing wave is present on a vibrating string. By lightly touching the string in certain places, particular standing waves can be damped out and thus make others more obvious. For example, suppose you touch the center of the vibrating guitar string. We can see from Figure 16-6(a) that the fundamental will be damped out because it has an antinode at the middle. The second harmonic, however, has a node at the middle (Figure 16-6[b]), so it is unaffected by your touch. The third harmonic is damped because it also has an antinode at the middle of the string. In fact, all odd-numbered harmonics are damped out, and all even-numbered ones remain.

When you do this experiment, you hear a shift in the lowest frequency. With the initial pluck, the fundamental is the most prominent frequency. After touching the middle of the string, the fundamental is gone, so the lowest frequency is now that of the second harmonic, a frequency twice the original frequency. In musical terms we say that the note shifts upward by one octave.

PHYSICS | ON YOUR OWN

Investigate the harmonics produced on a guitar string. Pluck the string and listen for the fundamental (first) harmonic. While the string is still vibrating, gently touch the string at its midpoint. You should be able to hear

(a) String plucked ¼ from an end

(b) First harmonic

(c) Second harmonic

(d) Third harmonic

(e) Fourth harmonic

(f) Fifth harmonic

Figure 16-5 The shape of the plucked string (a) is the superposition of the first five harmonics (b–f). Note that the fourth harmonic does not contribute to this particular shape.

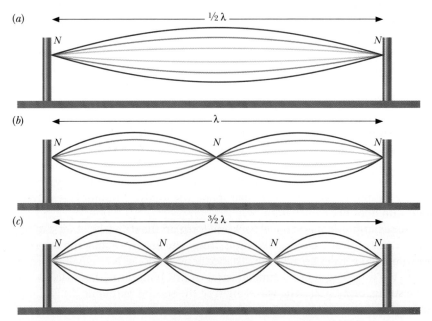

Figure 16-6 The first three standing waves on a guitar string. The wavelengths of these harmonics are $2L$, L, and $\frac{2}{3}L$, respectively.

the second harmonic. Find the higher harmonics by plucking the string again and gently touching it $\frac{1}{3}, \frac{1}{4}, \frac{1}{5}, \ldots$ of the way from one end.

Music, of course, consists of many notes. The string we have been discussing can play only one note at a time because the fundamental frequency determines the musical note. Instruments must be able to play many different frequencies to be useful. To play other notes, we must have more strings or a simple way of changing the vibrational conditions on the string. Most stringed instruments use similar methods for achieving different notes. Pianos, harps, and harpsichords have many strings. Striking or plucking different strings produces different notes.

A guitar usually has six strings, some more massive than others to extend the range of frequencies. Still, only six strings aren't enough to produce many interesting songs. A guitarist must be able to change the note produced by each string. By fingering the string the guitarist shortens the vibrating portion of the string, creating new conditions for standing waves. The new fundamental wavelength is now twice this *shortened* length and therefore smaller than before. This smaller wavelength produces a higher frequency.

The guitarist can change the speed of the waves by changing the tension of the string. An increase in the tension increases the speed and therefore increases the frequency. It takes too long to change the tension in the middle of a song, so changes in tension are only used to tune the instruments. An exception is the washtub bass shown in Figure 16-7.

Question   What are the different ways to increase the fundamental frequency of a note played on a guitar string?

Answer   To increase the fundamental frequency of a note on a guitar string, you can increase the tension, finger the string, or use a string with less mass per unit length.

Wind Instruments

✓ **MATH**

Wind instruments—such as clarinets, trumpets, and organ pipes—are essentially containers for vibrating columns of air. Each has an open end that transmits the sound and a method for exciting the air column. With the exception of the organ pipe, wind instruments also have a method for altering the fundamental frequency. In many ways wind instruments are analogous to the stringed instruments. A spectrum of initial waves is created by a disturbance. The instrument governs the standing waves that are generated; all other frequencies are quickly damped out. The sound we hear is a combination of the frequencies of these standing waves.

But there are three main differences between wind and string instruments. First, unlike vibrating strings, the vibrational characteristics of air cannot be altered to change the speed of the waves. Only the length of the vibrating air column can be changed. Second, a string has a node at each end, but there is an antinode near the open end of the wind instrument. At an antinode the vibration of the air molecules is a maximum. A node occurs at the closed end, where there is no vibration of the air molecules. Finally, the disturbance in wind instruments produces longitudinal waves in the air column instead of the transverse waves of the stringed instruments. A longitudinal wave at a single moment is represented in Figure 16-8.

Drawing longitudinal standing waves is difficult because the movement of the air molecules is along the length of the pipe. If we draw them in this manner for even one period, crests and troughs overlap and produce a very confusing drawing. When illustrating longitudinal waves in a windpipe, we normally

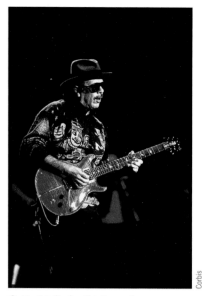

Guitarist Carlos Santana changes notes by pushing the strings against the frets to shorten the lengths that vibrate.

Figure 16-7 The note played by this one-string bass is changed by varying the tension (and thus the wave speed) in the string.

Figure 16-8 A drawing of a longitudinal sound wave. The dark regions represent compressions (high density); the lighter regions represent rarefactions (low density).

Loudest and Softest Sounds

The intensity of sound is measured in terms of the sound energy that crosses 1 square meter of area in 1 second and is measured by a sound level meter. At a reference value of 1000 hertz, the faintest sounds that can be heard by the human ear have intensities a little less than one-trillionth (10^{-12}) of a watt per square meter. This results from a variation in pressure of about 0.3 billionth of an atmosphere and corresponds to a displacement of the air molecules of approximately one-billionth of a centimeter, which is less than the diameter of a molecule! The ear is a very sensitive instrument.

On the loud side, the ear cannot tolerate sound intensities much greater than 1 watt per square meter without experiencing pain. This corresponds to a variation in pressure of about one-thousandth of an atmosphere, 1 million times as big as for sounds that can be barely heard. The displacements of the molecules are also 1 million times as large.

A source of sound that is transmitting at 100 watts in a spherically symmetric pattern is painful to our ears at a distance of 1 meter and theoretically audible at about 3000 kilometers (2000 miles!) if we assume no losses in moving through the air. Psychoacoustic scientists report that when the intensity of sound is increased about eight times people report a doubling in the loudness of the sound.

Because of the very large range of sensitivity of the ear as well as our perceptual scale, the scale scientists use to distinguish different sound levels is based on multiples of 10. A sound that has 10 times as much intensity as another sound has a level of 10 decibels, or 10 dB (pronounced "dee bee"), higher than the first sound. Because every increase of 10 decibels corresponds to a factor-of-10 increase in intensity, an increase of 20 decibels means that the intensity increases by $10 \times 10 = 100$ times. Sound-level meters use this same nonlinear scale. The table gives some representative values of sound intensity.

A sound level of 85 decibels is safe for unlimited exposure. However, a sound level of 100 decibels is safe for only 2 hours, and an increase to 110 decibels reduces the safe period to 30 minutes. Note that a typical rock concert has an intensity 10 times as large as this. It is also interesting to note that 75% of hearing losses are due to exposure to loud noises and not due to aging. Earplugs typically reduce sound levels by 20–30 decibels and should be used whenever sound levels exceed safe limits.

Decibel Levels for Some Common Sounds

Source	Decibels	Energy Relative to Threshold	Sensation
Nearby jet taking off	150	1 quadrillion	
Jackhammer	130	10 trillion	
Rock concert, automobile horn	120	1 trillion	Pain
Police siren, video arcade	110	100 billion	
Power lawn mower, loud music	100	10 billion	
Screaming baby	90	1 billion	Endangers hearing
Traffic on a busy street, alarm clock	80	100 million	Noisy
Vacuum cleaner	70	10 million	
Conversation	60	1 million	
Library	40	10,000	Quiet
Whisper	30	1000	Very quiet
Breathing	10	10	Just audible
	0	1	Hearing threshold

draw them *as if* the air movements were transverse. This should not be overly confusing as long as you remember that these drawings are basically graphs of the displacements of the air molecules versus position. In Figure 16-9 two curves are drawn to represent the range of displacement of the standing waves. Note that the curves meet at the closed end of the pipe. There is a node at that spot, indicating that the air molecules near the wall do not move.

Consider a closed organ pipe—one that is closed at one end and open at the other. The largest wavelength that produces a node at the closed end and an antinode at the open end is four times the length of the pipe, or $4L$. This is illustrated in Figure 16-9(a). The next-smaller wavelength (Figure 16-9[b]) is four-thirds the length of the pipe, or $\frac{4}{3}L$, and the wavelength in Figure 16-9(c) is four-fifths the length of the pipe, or $\frac{4}{5}L$.

As with the stringed instruments, we can generate a relationship among the various frequencies by examining the relationship between the wave's speed, its wavelength, and its frequency: $v = \lambda f$. The possible wavelengths are $1, \frac{1}{3}, \frac{1}{5}, \frac{1}{7}, \ldots$ times the fundamental wavelength. Because the wave speeds are constant, the corresponding frequencies are 1, 3, 5, 7, . . . times the fundamental frequency.

Question What is the wavelength corresponding to the fifth harmonic in a 50-centimeter-long closed organ pipe?

Answer It must be one-fifth the length of the fundamental wavelength. Because the fundamental wavelength is four times the length of the pipe, we obtain 4(50 centimeters)/5 = 40 centimeters.

The closed organ pipe does not have resonant frequencies that are 2, 4, 6, . . . times the fundamental frequency; the even-numbered harmonics are missing because these frequencies would not produce a node at the closed end and an antinode at the open end. On the other hand, the open organ pipe (one that is open at both ends) has all harmonics.

PHYSICS | **ON YOUR OWN**

Create a soft-drink bottle instrument. The bottles produce musical notes when you blow across their tops. You can adjust the frequencies of the notes by filling the bottles with different amounts of water, thus changing the length of the vibrating column of air above the water.

Percussion Instruments

Percussion instruments are characterized by their lack of the harmonic structure of the string and wind instruments. Percussionists employ a wide range of instruments, including drums, cymbals, bells, triangles, gongs, and xylophones. Although all of these resonate at a variety of frequencies, the higher frequencies are not whole-number multiples of the lowest frequency.

Each of the percussion instruments behaves in a different way. The restoring

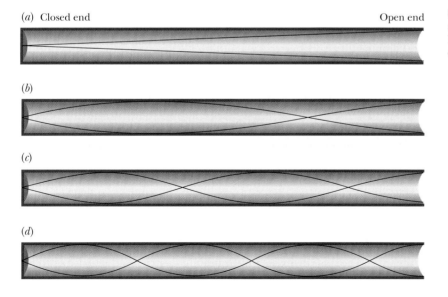

Figure 16-9 Graphs of the first four standing waves in a closed organ pipe. Note that there is a node at the closed end and an antinode at the open end.

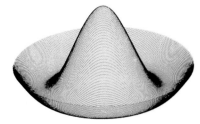

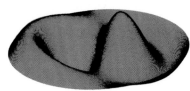

Figure 16-10 A circular drumhead can vibrate in many different modes with nonharmonic frequencies.

force in a drum is provided by the tension in the drumhead. The two-dimensional standing-wave patterns are produced by the reflection of transverse waves from the edges of the drum and are characterized by nodal lines. These nodal lines are the two-dimensional analogs of the nodal points in vibrating strings.

The nodal lines for a circular membrane are either along a diameter or circles about the center; there is always a nodal line around the edge of the drumhead. Figure 16-10(a) shows the fundamental mode in which the center of the drumhead moves up and down symmetrically. Figure 16-10(c) shows a mode with an antinodal line along a diameter. When one-half of the drumhead is up, the other half is down. This mode has a frequency 1.593 times that of the fundamental frequency. The second symmetric mode is shown in Figure 16-10(b) and has a frequency 2.295 times the fundamental frequency.

Beats ✓ MATH

When we listen to two steady sounds with nearly equal frequencies, we hear a periodic variation in the volume. This effect is known as **beats** and should not be confused with the rhythm of the music that you might dance to; beats are the result of the superposition of the two waves. Because the two waves have different frequencies, there are times when they are in step and add together and times when they are out of step and cancel. The result is a periodic cancellation and reinforcement of the waves that is heard as a periodic variation in the loudness of the sound.

This is illustrated by the drawings in Figure 16-11. It is important to realize that these drawings do not represent strobe pictures of the waves. The horizontal line represents time, not position. The drawings represent the variation in the amplitude of the sound at a single location—for instance, as it reaches one of your ears. Figures 16-11(a) and (b) show each wave by itself, and Figure 16-11(c) shows the superposition of these two waves. Your ears hear a frequency that is the average of the two frequencies and that varies in amplitude with a beat frequency.

We can obtain the beat frequency by examining Figure 16-11. If the time between vertical lines is 1 second, wave (a) has a frequency of 10 hertz and wave (b) 11 hertz. At the beginning of a second, a crest from wave (a) cancels a trough from wave (b), and the sound level is zero. Wave (a) has a lower frequency and continually falls behind. At the end of 1 second, it has fallen an entire cycle behind, and the waves once again cancel. The difference of 1 hertz in their frequencies shows up as a variation in the sound level that has a frequency of 1 hertz. This same process is valid for any frequencies that differ by 1 hertz. For example, the beat frequency produced when two sound waves of 407 hertz and 408 hertz are combined is also 1 hertz.

If the frequencies differ by 2 hertz, it takes only $\frac{1}{2}$ second for the lower frequency wave to fall one cycle behind. Therefore, this pattern happens two times per second, or with a beat frequency of 2 hertz. This reasoning can be generalized to show that the beat frequency is equal to the difference in the two frequencies.

beat frequency = Δf ▶

Piano tuners employ this beat phenomenon when tuning pianos. The tuner produces the desired frequency by striking a tuning fork and then adjusts the piano wire's tension until the beats disappear. Modern electronics have now made it possible for a tone-deaf person to make a living tuning pianos.

Question Why don't we hear beats when adjacent keys on a piano are hit at the same time?

Answer The beats do exist, but the beat frequency is too high for us to notice.

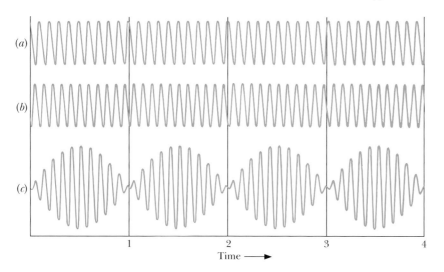

Figure 16-11 The superposition of two sound waves of different frequencies produces a sound wave (c) that varies in amplitude.

PHYSICS | ON YOUR OWN

Play the same note on two strings of a guitar. As you adjust the tension of one of them, can you hear beats?

Doppler Effect

What we hear is not necessarily the sound that was originally produced, even when only a single source is involved. Recall the sound of a car as it passes you. This variation in sound is especially noticeable with race cars and some motorcycles because they emit distinctive roars. Imagine standing by the side of a road as a car passes you with its horn blasting constantly. Two things about the horn's sound change. First, as the car approaches, its sound gets progressively

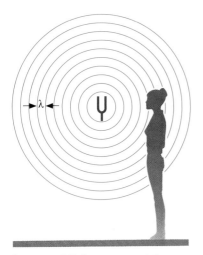

Figure 16-12 If the source and the ear are stationary relative to each other, the ear hears the same frequency as emitted by the source.

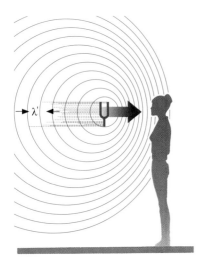

Figure 16-13 If the source of sound moves toward the right, the waves bunch up on the right-hand side and spread out on the left-hand side. The person hears a frequency that is higher than that emitted by the source.

louder; as it leaves, the sound gets quieter. The volume changes simply because the wave spreads out as it moves away from the source. If you are far from the source, the energy of the sound waves intercepted by your ears each second is smaller.

The second change in the sound of the horn might not be as obvious. The frequency *that you hear* is not the same as the frequency that is actually emitted by the horn. The frequency you hear is higher as the car approaches you and lower as the car recedes from you. This shift in frequency due to the motion is called the **Doppler effect**, after the Austrian physicist and mathematician Christian Doppler.

The pitch of the sound we hear is determined by the frequency with which crests (or troughs) hit our ears. Our ears are sensitive to the frequency of a wave, not to the wavelength. Figure 16-12 shows a two-dimensional drawing of sound waves leaving a tuning fork. A small portion of the sound is intercepted by the ear. If the tuning fork and the ear are stationary relative to each other, the frequency heard is the same as that emitted.

When the tuning fork is moving, the listener hears a different frequency. Because the tuning fork moves during the time between the generation of one crest and the next, the sound waves crowd together in the forward direction and spread out in the backward direction, as shown in Figure 16-13. If the tuning fork moves toward the listener, the ear detects the sound waves that are crowded together. The frequency with which the waves hit the ear is therefore higher than was actually emitted by the source. Similarly, if the tuning fork moves away from the listener, the observed frequency is lower.

Notice, however, that in both Figures 16-12 and 16-13 the spacings do not change with the separation of the source and receiver. Because the shift does not depend on the distance, the Doppler-shifted frequency does not change as the source gets closer or farther away.

Question If you were flying a model airplane on a wire so that it traveled in circles about you, would you hear a Doppler shift?

Answer Because the airplane is not moving toward or away from you, you would not hear a Doppler shift. However, someone standing off to the side watching you would hear a Doppler shift.

We will see in later chapters that the Doppler effect also occurs for light. The observation that the light from distant galaxies is shifted toward lower frequencies (their colors are red-shifted) tells us that they are moving away from the Galaxy and that the Universe is expanding.

The Doppler effect also occurs when the receiver moves and the source is stationary. If the listener moves toward a stationary tuning fork, her ear intercepts wave crests at a higher rate, and she hears a higher frequency. If she recedes from the tuning fork, the crests must continually catch up with her ear, which therefore intercepts the crests at a reduced rate, and she consequently hears a lower frequency. Notice that the shift in the frequency is still constant as long as the velocities are constant. Of course, the loudness of the signal decreases as the distance to the tuning fork increases.

When a wave bounces off a moving object, it experiences a similar Doppler shift in frequency due to successive crests (or troughs) having longer or shorter distances to travel before reflecting. By monitoring these frequency shifts, the speed of objects toward or away from the original source can be determined. Because the Doppler effect occurs for all kinds of waves, this technique is used in many situations, from catching speeding motorists using radar (an electromagnetic wave) to monitoring the movement of dolphins using sound waves in water.

Breaking the Sound Barrier

A common misunderstanding is that the sonic boom occurs when the aircraft "breaks" the sound barrier. The boom actually continues as long as the plane flies at supersonic speeds. However, at a given location, the loud noise is heard only when the edge of the pressure cone reaches the ear, as illustrated in the figure. This short, loud noise results from the accumulated superposition of the wave crests generated by the plane.

The first person to break the sound barrier was Chuck Yeager on October 14, 1947. He flew a Bell XS-1 rocket plane at 12,800 meters over Edwards Air Force Base in California at a speed of 1080 kilometers per hour (670 mph). This speed was also known as Mach 1.015 because it was 1.015 times the speed of sound at this altitude on that day. The supersonic Concorde cruised at speeds up to Mach 2.2 (1450 mph), and the Russian MIG-25B has been tracked on radar at speeds up to Mach 3.2 (2110 mph).

This F/A 18 Hornet is transitioning from subsonic to supersonic flight. The atmospheric conditions were just right for the shock waves produced by the cockpit and the wings/fuselage to produce condensation.

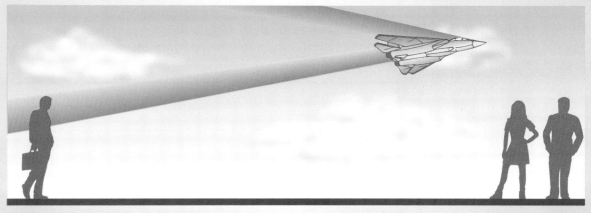

A plane traveling faster than the speed of sound produces a shock wave that is heard after the plane passes overhead. The couple on the right has not yet heard the sonic boom or any other sound from the plane.

Flawed Reasoning

The following question appears on the final exam: "You and a friend are standing a city block apart. An ambulance with its siren blaring drives down the street toward you and away from your friend. The ambulance passes you and continues down the street at a constant speed. Is the frequency that you hear higher than, lower than, or the same as the frequency your friend hears?"

Matt gives the following answer to this question: "According to the Doppler shift, as the siren gets farther away, the wavelengths get farther apart. The longer the wavelength, the lower the frequency. I will therefore hear a higher frequency than my friend."

What is wrong with Matt's reasoning, and what is the correct answer to the exam question?

continues on next page

Answer The Doppler shift does not depend on the distance to the source. The shift in frequency depends only on the velocity of the source relative to the observer. The ambulance is moving away from both of you with the same speed. Therefore, both of you will hear the same lower frequency, but the siren will be louder to you.

Shock Waves

When a source of waves moves faster than the speed of the waves in the medium, the next crest is generated in front of the leading edge of the previous crest. This causes the expanding waves to superimpose and form the conical pattern shown in Figure 16-14(a). The amplitude along the cone's edge can become very large because the waves add together with their crests lined up. The edge of the cone is known as a **shock wave** because it arrives suddenly and with a large amplitude.

Shock waves are common in many media. When speedboats go much faster than the speed of the water waves, they create shock waves commonly known as wakes. The Concorde, a supersonic plane no longer flying, traveled much faster than the speed of sound in air and therefore produced shock waves. Some people are concerned about the effects of the sonic boom when the edge of the pressure cone reaches the ground.

The return of a space shuttle to Earth produces a double sonic boom. The nose produces one boom, and the engine housings near the rear of the spacecraft produce the other. Listen for this the next time you watch a television broadcast of a shuttle returning.

Question Would you expect a spacecraft traveling to the Moon to produce a shock wave during its entire trip?

Answer No. Because there is no air in most of the space between Earth and the Moon, the spacecraft would not produce any sound.

A stroboscopic photograph of a bullet traveling through the hot air above a candle. The formation of the shock wave tells us that the bullet was traveling faster than the speed of sound.

PhysicsNow™ Assess your understanding of this chapter's topics with sample tests and other resources found by logging into PhysicsNow at **http://physics.brookscole.com/kf6e**.

Summary

Sound is a longitudinal wave that travels through a variety of media. The speed of sound is 343 meters per second in air at room temperature, 4 times faster in water, and more than 10 times faster in solids. In a gas the speed of sound depends on the temperature (slower at lower temperatures) and the type of gas (faster in molecules with less mass).

Ears detect sound waves and send electric signals to the brain. The range of

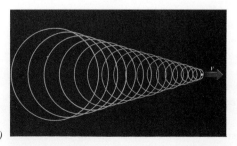
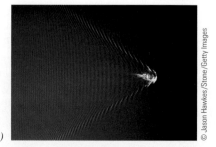

Figure 16-14 (a) When the source moves faster than the speed of sound, the waves form a cone known as the shock wave. (b) The presence of the bow wave (analogous to a shock wave) indicates that the boat is traveling faster than the speed of the water waves.

frequencies that can be detected depends on the resonant structures in the animals' ears. The audible range in humans is roughly 20–20,000 hertz, although it varies with age and the individual.

You can recognize different sources of sound because each source produces a unique combination of intensities of the various harmonics—its own recipe of harmonics.

Music generally differs from other sounds in its periodicity. Nearly all musical instruments—string, wind, or percussion—involve the production of standing waves. In many of these instruments, the standing-wave frequencies, or harmonics, are whole-number multiples of the fundamental frequency. The frequency of the second harmonic is twice the fundamental frequency. The string's actual shape is a superposition of all the standing waves and looks quite different from those of the individual harmonics.

Changing the standing-wave conditions generally produces different notes. The relationship $v = \lambda f$ tells us that if the speed or the wavelength is changed, the frequency must change. A smaller wavelength gives a higher frequency when the speed is held constant.

Two steady sounds with nearly equal frequencies superimpose to form beats, a variation in the loudness of the sound that has a frequency equal to the difference in the frequencies.

The frequency of the sound from a moving object is shifted in frequency according to the Doppler effect. The frequency is shifted upward as the emitting object approaches and downward as it recedes from the receiver. The shift in the frequency is constant as long as the velocity is constant. The Doppler effect occurs for all kinds of waves.

When a source of waves moves faster than the speed of the waves in the medium, a conical shock wave is formed. When the edge of the cone created by supersonic aircraft reaches us, we hear a sonic boom.

Chapter 16 Revisited

By studying the sounds produced by instruments such as guitars, pianos, and organs, we learn that musical sounds often have combinations of frequencies that are whole-number multiples of the lowest frequency. The size of the body of the instrument determines the range of frequencies that it amplifies.

KEY TERMS

beats: A variation in the amplitude resulting from the superposition of two waves that have nearly the same frequencies. The frequency of the variation is equal to the difference in the two frequencies.

Doppler effect: A change in the frequency of a periodic wave due to the motion of the observer, the source, or both.

shock wave: The characteristic cone-shaped wave front that is produced whenever an object travels faster than the speed of the waves in the surrounding medium.

CONCEPTUAL QUESTIONS

1. What is the evidence that sound is a wave phenomenon?
2. Your lab partner argues that sound is a transverse wave. How would you convince him otherwise?
3. Which of the following has the most effect on the speed of sound in air: amplitude, frequency, wavelength, or temperature?
4. Does a 220-hertz sound wave move faster than, slower than, or at the same speed as a 440-hertz sound wave? Explain.
5. Is the speed of sound faster in air or water?
6. Why does the voice of a clown who has inhaled helium sound so high pitched?

7. Why do you see the lightning before you hear the thunder?
8. When the time interval between seeing the lightning and hearing the thunder is short, the thunder is loud. Why?
9. Even though you may be far away from an orchestra, the trumpet and the trombone do not sound "out of step" with each other. What does this tell us about sound waves?
10. You visit the Grand Canyon on a particularly hot day and, like all good tourists, you yell, "Hello!" Would you expect to hear the echo sooner or later than on a cooler day? Why?
11. If earplugs are advertised to reduce sound levels by 20 decibels, by how much do they reduce the intensity of the sound?
12. A friend tells you that a real estate developer has reassured her that the increased traffic due to the new mall on her street will increase the average traffic noise a mere 20%, from 50 decibels to 60 decibels. How would you advise her?
13. Is it possible for a 440-hertz sound wave to be louder than an 880-hertz sound wave?
14. Which of the following properties of a sound wave determines its loudness (or intensity): wavelength, speed, amplitude, or frequency?
15. Which of the following properties of a sound wave determines its pitch: wavelength, speed, amplitude, or frequency?
16. Which is longer—the wavelength of infrasound or ultrasound?
17. The speed of sound is much greater in water than in air. If you were to jump into a pool holding a 500-hertz buzzer, would the frequency of the sound you hear while underwater be greater than, less than, or equal to 500 hertz? Explain your reasoning.
18. For sound waves in room-temperature air, is it possible to change the wavelength of a sound without changing the frequency? Explain your reasoning.
19. What frequency is two octaves higher on the musical scale than 100 hertz?
20. Which harmonic is one octave higher in frequency than the fundamental?
21. How can you increase the speed of the waves on a guitar string?

22. What can you do to raise the frequency of a note on a guitar string?
23. How does the wavelength of the fundamental standing wave on a violin string compare with the length of the string?
24. Two standing waves are created on the same guitar string. Will the frequency of the one with the shorter wavelength be higher than, lower than, or the same as the frequency of the one with the longer wavelength? Explain.
▲ 25. On what two places on a guitar string of length L could you place your finger on an antinode to damp out the second harmonic? Would the fourth harmonic be damped out if you placed your finger at either of these locations?
▲ 26. Where on a guitar string of length L would you place your finger to damp out the third harmonic? Do these places differ in their effect on other harmonics?
27. Does plucking the string harder affect the fundamental frequency of a guitar string? Explain.
28. Does increasing the tension of the string affect the wavelength of the fundamental standing wave on a guitar string?
29. Does the second or the third harmonic on a guitar string have traveling waves with the higher speed, or are the speeds the same?
▲ 30. You can tune the frequency of a guitar string without changing its length. To tune a flute, however, you have to slightly adjust the length of the tube. How do you account for this difference?
31. Would you expect to find nodes or antinodes at the ends of a guitar string? Explain.
32. Would you expect to find a node or an antinode at each end of an organ pipe that is closed at one end?
33. How does the wavelength of the fundamental standing wave for an organ pipe that is open at both ends compare to the length of the pipe?

34. How does the wavelength of the fundamental standing wave for an organ pipe that is closed at one end and open at the other compare to the length of the pipe?
35. How does the fundamental wavelength of an organ pipe, which is open at both ends, change as you close one of the ends?
36. How does the fundamental frequency of an organ pipe, which is open at both ends, change as you close one of the ends?
37. You have an organ pipe that resonates at frequencies of 300, 450, and 600 hertz but nothing in-between. It may resonate at lower and higher frequencies as well. Is the pipe open at both ends or open at one end and closed at the other? How can you tell?
38. You have an organ pipe that resonates at frequencies of 500, 700, and 900 hertz but nothing in-between. It may resonate at lower and higher frequencies as well. Is the pipe open at both ends or open at one end and closed at the other? How can you tell?

39. What happens to the fundamental frequency of an organ pipe as the temperature in the room decreases? Explain.
40. What would happen to the fundamental wavelength of an organ pipe if you fill it with helium instead of air? What would happen to the fundamental frequency? Explain.
41. The fundamental wavelength for a bar on a xylophone that is suspended at points midway between the ends and the middle of the bar is _____ the length of the bar.
42. What is the fundamental wavelength for standing waves on a rod held at its center?
43. Electronic signal generators can produce pure frequencies with great precision. Why then is it so difficult to make an electronic piano that sounds like an acoustic piano?
44. Why does middle C played on an oboe sound different from middle C played on a piano?
45. As the frequencies of two waves get farther apart, what happens to the beat frequency?
46. As the frequencies of two waves get closer together, what happens to the beat frequency?
47. You are singing along with Celine Dion on the car radio as she sings a long steady note. You know that you are a little off-key because you are hearing beats at 2 hertz. Are you singing a little too high, a little too low, or can't you tell? Explain.
48. You have two pairs of tuning forks, one pair at 345 and 348 hertz and the other pair at 489 and 491 hertz. Which pair of tuning forks, when sounded together, produces the higher beat frequency?
▲ 49. The two wires corresponding to one key on a piano are out of tune. If we decrease the tension of the wire producing the lower frequency, will the beat frequency increase, decrease, or stay the same? Why?
▲ 50. You have two tuning forks that, when played together, produce beats at a frequency of 4 hertz. You place a small piece of putty on one of the tuning forks, which lowers its frequency, and find that the beat frequency increases. Did you place the putty on the tuning fork with the lower or the higher initial frequency? Explain.
51. Describe the sounds you would hear if a train passed you with its whistle blowing.
52. A tuning fork rings with a frequency of 400 hertz. As the tuning fork moves away from you, would the frequency you hear be lower than, equal to, or higher than 400 hertz? Explain your reasoning.
53. Which of the following properties of the wave does not change in the Doppler effect: wavelength, speed, or frequency? Explain.
54. An automobile sounding its horn is moving toward you at a constant speed. How does the frequency you hear compare with that heard by the driver?
55. A police car with a fixed frequency siren drives straight toward you. It slows down gradually and stops right next to you. As the car is slowing down, what do you hear happening to the frequency and the loudness of the siren?
56. You tie a battery-powered, fixed-frequency buzzer to the end of a string and swing it around in a large horizontal circle as shown. Your friend stands well off to the right and listens to the tone. At which of the labeled locations will the buzzer be when your friend hears the highest frequency? (Assume that the time it takes the sound to travel from the buzzer to your friend's ear can be ignored.)

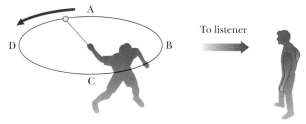

▲ 57. Billy and Elaine are holding electric buzzers that sound at slightly different frequencies. When they are both standing still, the two buzzers produce a beat frequency of 4 hertz. Elaine begins to run away, and Billy hears the beat frequency gradually increase. Whose buzzer has the higher frequency? Explain.
58. If a wind is blowing from a siren toward your ear, does the frequency of the siren change? Why?
59. Explain why a sonic boom sounds much like an explosion.
60. What does the observation of a wake tell you about the speed of a boat compared with the speed of water waves?
61. A jet that has been traveling faster than the speed of sound for several miles passes overhead. Describe what you would hear from the moment the plane is directly overhead until it is several miles away.
62. A jet is flying along close to the ground at just below the speed of sound. Describe how the pilot could "break the sound barrier" without flying faster.

EXERCISES

1. The musical note middle C has a frequency of 262 Hz. What is its period of vibration?
2. What is the period of a tuning fork that is vibrating at 1200 Hz?
3. What is the wavelength of a musical note with a frequency of 1048 Hz?
4. How long is a wavelength of infrasound with a frequency of 3 Hz?
5. What frequency would you need to produce a sound wave in room-temperature air with a wavelength of 1 m?
6. Radio waves travel at the speed of light, which is 3×10^8 m/s. What is the wavelength for FM station WFIZ broadcasting at 97.3 MHz? (The prefix M is *mega* and means 10^6.)
7. What is the longest wavelength that can be heard by a normal ear?
8. What is the shortest wavelength that the average human can hear?
9. You observe that the delay between a lightning flash and the thunder is 12 s. How far away is the lightning?

10. If the "shot heard round the world" could actually travel around the world, how long would it take? (Assume that the circumference of Earth is 40,000 km.)
11. If the sonar signal sent straight down from a boat takes 0.8 s to return, how deep is the lake?
▲ 12. Your sonic range finder measures the distance to a nearby building at 20 m. The range finder is calibrated for sound traveling 343 m/s, but on this very cold day (−37°C) the speed of sound is only 309 m/s. How far away is the building?

13. If ear protectors can reduce the sound intensity by a factor of 1000, by how many decibels is the sound level reduced?
14. You go out and spend a bucket of money on a new stereo amplifier that advertises that it can produce maximum volumes 20 dB greater than your current system. How many times more powerful must this amplifier be?
15. If the fundamental frequency of a guitar string is 125 Hz, what harmonic frequencies are possible?

▲ 16. You measure all of the possible resonant frequencies for a guitar string in the range 500–1000 Hz and find that it will resonate at only 600, 750, and 900 Hz. What is the fundamental frequency for this string?
17. What is the fundamental frequency for a 40-cm banjo string if the speed of waves on the string is 470 m/s?
18. If the fundamental frequency of a 60-cm-long guitar string is 500 Hz, what is the speed of the traveling waves?
19. What harmonic frequencies are possible in a closed organ pipe that is 1 m long?
20. What harmonic frequencies are possible in an open organ pipe that is 1 m long?
21. What length of closed organ pipe is required to produce the note B_4 with a frequency of 493.88 Hz?
22. How long is an open organ pipe with a fundamental frequency of 493.88 Hz?
▲ 23. You have an organ pipe that resonates at frequencies of 375, 450, and 525 Hz but nothing in-between. It may resonate at lower and higher frequencies as well. What is the fundamental frequency for this pipe?
▲ 24. You have an organ pipe that resonates at frequencies of 500, 700, and 900 Hz but nothing in-between. It may resonate at lower and higher frequencies as well. What is the fundamental frequency for this pipe?
25. A tuning fork has been damaged and its frequency slightly changed. What could its altered frequency be if it produces two beats per second with a tuning fork that is known to vibrate at 262 Hz?
26. You have a tuning fork of unknown frequency. When you ring it alongside a tuning fork with known frequency of 360 Hz, you hear beats at a frequency of 2 Hz. When you ring it alongside a tuning fork with known frequency of 355 Hz, you hear beats at a frequency of 3 Hz. What is the unknown frequency?

Interlude

Rainbow over a farm in Pennsylvania.

The Mystery of Light

Throughout history our knowledge, attitudes, and values have been reflected in our sciences and in our arts. The most obvious example of this parallelism is light. Light and one of its characteristics, color, have followed parallel trends in science and art. In the Middle Ages, for example, light was sacred and mysterious; artists placed it in some heavenly plane. When St. John the Evangelist said, "God is light," his words reflected the belief that the presence of light in a place of worship was a sign of the presence of the Holy Spirit. Beautiful, brightly colored leaded-glass windows became the norm in the great churches being built at the time.

Light, as seen in these early works, had a surrealistic behavior. It wasn't until the Renaissance that artists studied the effects of light on objects, how it illuminates some sides of an object but not others. During the late 1600s, when light was shown to take a finite time to travel through space, paintings were just beginning to show objects casting shadows on the ground.

The next time you look at a star, think about the starlight that reaches your eyes. It probably began its journey hundreds or thousands of years ago, in many cases long before civilization began on Earth.

Still the scientific questions remained: What is light? Does light behave as a wave or as a stream of particles? It was known that light travels from the Sun to Earth and that space is a very, very good vacuum. If light is a wave, how can it travel through a vacuum where there is nothing to wave? The history of this debate about the wave or particle nature of light goes back at least to Newton's time. Newton, perhaps influenced by his successful work on apples and moons, believed that light was a stream of very tiny particles. His contemporary, Dutch scientist Christian Huygens, believed that light was a wave. The controversy was not resolved until 100 years later. (However, the question of the nature of light reemerged during the 20th century with the development of quantum mechanics.)

And what is the origin of the colors? Before Newton's

A prism spreads light out into its component colors.

work with clear glass prisms, color was thought to be something emanating from objects. A ruby's redness was only due to the characteristics of the ruby. As we will see, this is partially correct; the character of the light shining on the ruby also plays a role.

The next time you look at a star, think about the starlight that reaches your eyes. It probably began its journey hundreds or thousands of years ago, in many cases long before civilization began on Earth. When it arrives at your eye, its journey is over. The light is absorbed in your retina, producing an electric signal that travels to your brain. However, this light is carrying a much more complex message than simply the location of the star in the heavens. With the exception of a few artificial satellites within the Solar System and a variety of cosmic particles striking Earth, our entire knowledge of the cosmos comes from the information carried by light.

We continue to build our physics world view by studying this mysterious phenomenon in the next three chapters; then we will use our newly gained knowledge of light to probe the structure of matter at the atomic level . . . and beyond.

17 Light

The reflection of light at the surface of the water acts like a mirror to produce a virtual image.

When a magician thrusts a sword through the lovely assistant or spooks sit next to you as you traverse the haunted mansion's dark innards, the explanation is, "It's done with mirrors." But how does light produce these illusions that are so convincing?

(See page 354 for the answer to this question.)

Shadows   **337**

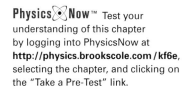

PhysicsNow™ Test your understanding of this chapter by logging into PhysicsNow at **http://physics.brookscole.com/kf6e**, selecting the chapter, and clicking on the "Take a Pre-Test" link.

ALTHOUGH the phenomenon of light is so common to our everyday experience and has played such a central role in the histories of religion, art, and science, it is actually quite elusive. Even the act of seeing has confused people. We talk of looking *at* things—of looking *into* a microscope, of sweeping our glance *around* the room—as if seeing were an active process, much like beaming something in the direction of interest. This notion goes back to some early ideas about light in which rays were supposedly emitted by the eyes and found the objects seen. The idea is still common, perhaps perpetuated in part by the Superman stories. In these stories Superman supposedly has the ability to emit powerful X rays from his eyes, enabling him to see through brick walls or, in the modern movie version, through clothes.

Seeing is actually a rather passive activity. What you see depends on the light that enters your eyes and not on some mysterious rays that leave them. The light is emitted whether or not your eyes are there to receive it. You simply point your eyes toward the object and intercept some of the light. In fact, light passing through clean air is invisible. If a flashlight is shined across a room, you don't see the light passing through the air. You see only the light that strikes the wall and bounces back into your eyes. (Sometimes you can see the beam's path because part of the light scatters from dust, fog, or smoke particles in the air and is sent toward your eyes.)

Shadows

One of the earliest studies of light was how it moved through space. By observing shadows and the positions of the light sources and the objects causing the shadows, it is easy to deduce that light travels in straight lines. In drawings illustrating the paths of light, it is convenient to use the idea of **light rays**. Because there are an infinite number of paths, we draw only enough to illustrate the general behavior.

An opaque object illuminated by a point source of light blocks some of the rays from reaching a screen behind it, producing a shadow like the one shown in Figure 17-1. The shadow has the same shape as the cross section of the object but is larger.

Most sources of light are not points but extend over some space. However, we can think of each small portion of the source as a point source casting its own sharp shadow. All these point-source shadows are superimposed on the screen behind the object. The darkest region is where all of the shadows overlap. This is known as the **umbra** (Figure 17-2). Surrounding the umbra is the **penumbra**, where only some of the individual shadows overlap.

Rather than looking at the shadow, imagine standing in the shadow look-

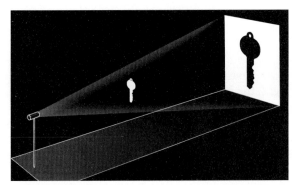

Figure 17-1 The shadow produced by a point source of light is very sharp.

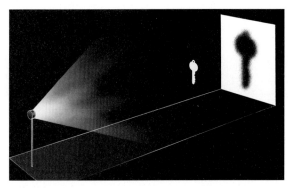

Figure 17-2 The shadow produced by an extended source of light has a dark central umbra surrounded by a lighter penumbra.

Eclipses

The most spectacular shadows are eclipses, especially total solar eclipses. During a solar eclipse, the Moon's shadow sweeps a path across a portion of Earth, as shown in Figure A. If you are in the path of the umbra, the Sun is totally obscured (Figure B). Observers to the side of the umbra's path but in the path of the penumbra see a partial eclipse.

One of Aristotle's arguments that Earth is a sphere involved the shadow during lunar eclipses (Figure C). Here, Earth's shadow falls onto the face of the Moon. Because (1) the shape of the shadow is always circular and (2) the only solid that always casts a circular shadow is a sphere, Aristotle correctly concluded that Earth must be a sphere.

Few people experience a solar eclipse, although the population on half of Earth can see a lunar eclipse. To see a solar eclipse, observers must be directly in the shadow of the Moon. This covers only a small portion of Earth's surface. During a lunar eclipse, we observe Earth's shadow on the Moon (Figure D). Anyone who can see the Moon can therefore see this eclipse.

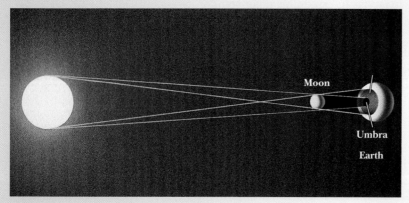

Figure A A total solar eclipse occurs when the umbra of the Moon's shadow falls on Earth.

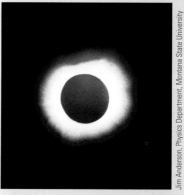

Figure B A total eclipse of the Sun.

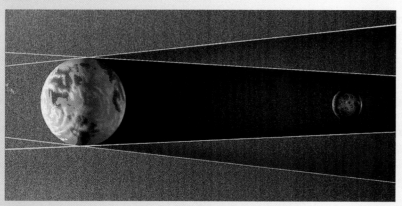

Figure C During a lunar eclipse, the Moon is in Earth's shadow.

Figure D During a lunar eclipse, we can see part of Earth's circular shadow.

ing back toward the source. If your eye is located in the umbra, you will not see any portion of the source of light; the object between you and the source blocks out all of the light. If your eye is in the penumbra, you will see part of the light source.

Shadows that we observe on Earth are not totally black and devoid of light. We can, for example, see things in shadows. This is due to the light that scatters into the shadow from the atmosphere or from other objects. On the Moon, however, the shadows are much darker because there is no lunar atmosphere. Astronauts exploring the Moon's surface have to be careful. Stepping into their own shadows can be dangerous because the extreme blackness of the shadow would hide everything within it—sharp rocks, uneven terrain, even a deep hole.

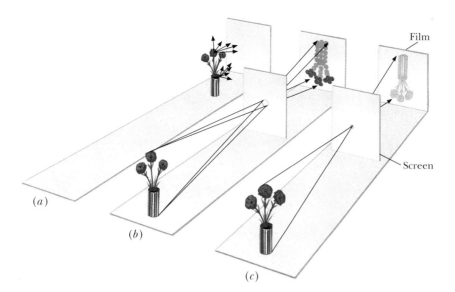

Figure 17-3 (a) Each part of the film receives light from many parts of the vase and flowers, and no image is recorded. (b) The screen restricts the light so that each part of the film receives light from a small portion of the scene. (c) Reducing the size of the hole produces a sharper but dimmer image.

Pinhole Cameras

✓ Extended presentation available in the *Problem Solving* supplement

Light that strikes most objects leaves in a great many directions. (This must be true because we can see the object from many different viewing directions.) If we place a piece of photographic film in front of a vase of flowers, as in Figure 17-3(a), the film will be completely exposed, leaving no record of the scene. Light from one part of the vase hits the film at the same place as light from many other parts of the vase and flowers. Every spot on the film receives light from virtually every spot that faces the film.

We can get an image by controlling which light rays hit the film. A screen with a small hole in it is placed between the vase and the film, as shown in Figure 17-3(b). Now only the light from a small portion of the vase reaches a given region of the film. Making the hole smaller, as in Figure 17-3(c), further reduces the portion of the vase exposed to a given spot on the film. If the hole is made small enough, a recognizable image of the vase is formed on the film. This technique can be used to make a pinhole camera by enclosing the film in a light-tight box with a hole that can be opened and shut. The photograph in Figure 17-4 was taken with a pinhole camera made from a shoe box.

The problem with pinhole cameras is the very small amount of light that reaches the film. Exposure times generally must be long, requiring that the scene be relatively static. We will see in Chapter 18 that this problem was eventually overcome by using a lens.

Pinhole cameras were used before the invention of film. If you add another hole to the box, as illustrated in Figure 17-5, you can see the image on the back wall. During solar eclipses, some observers make use of this process to watch the partial phases of the eclipse safely. On a grander scale, an entire room can be converted into a pinhole camera known as a *camera obscura*. Renaissance artists used smaller versions to help them draw landscapes and portraits; they traced the images formed on the back wall.

Figure 17-4 A photograph made with a pinhole camera. The exposure time was 2 seconds.

PHYSICS | ON YOUR OWN

Convert your room into a camera obscura by covering the windows with black plastic garbage bags. Make a hole about the size of a quarter in one of the bags. On a sunny day you should get a good image of the outdoors on the wall opposite the window. Experiment with the size of the hole to obtain the best picture.

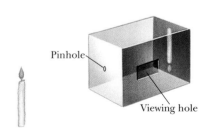

Figure 17-5 A pinhole in one end of the box produces an inverted image on the opposite end.

Figure 17-6 A beam of light striking a rough surface is scattered in many directions.

Reflections

Some things emit their own light—a candle, a lightbulb, and the Sun, to mention a few. But we see most objects because they reflect some of the light that hits them. The incident light, or the light striking the object, is scattered in many directions by the relatively rough surface of the object, a process known as **diffuse reflection**. If the surface of the object is very smooth, a light beam reflects off it much like a ball rebounds from a wall. Presumably then, when light hits rougher surfaces, the same thing happens. But with rough surfaces, different portions of the incident light reflect in many directions, as shown in Figure 17-6.

By looking at the reflection of a thin beam of light (a good approximation to a "ray" of light) on a smooth surface, we can discover a rule of nature. Figure 17-7(a) shows the reflections of light beams hitting a mirror at three different angles. Clearly, the angle between the incident and the reflected ray is different in each situation. However, if we examine only one case at a time, we notice that the angles the incident and reflected rays make with the surface of the mirror are equal. If we measure the angles of the incident ray and the reflected ray for many such situations, we discover that the two angles are always equal.

In our illustration of this phenomenon, we used the angles made with the reflecting surface. It is more convenient to consider the angles between the rays and the **normal**, an invisible line perpendicular to the surface that touches the surface where the rays hit (Figure 17-7[b]). The reflected ray lies in the same plane as the normal and the incident ray. Using these angles we state the **law of reflection**.

law of reflection ➤ | The angle of reflection is equal to the angle of incidence.

Question   Assume that you are stranded on an island. Where would you aim a mirror to signal a searching aircraft with sunlight?

Answer   The normal to the mirror would have to be directed at the point midway between the Sun and the aircraft. Luckily, this is easier to do than it might seem.

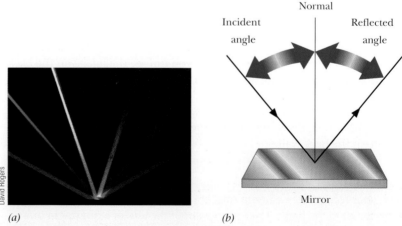

Figure 17-7 (a) Reflections of three thin beams of light hitting a mirror at different angles from the left. (b) For each one the angle of reflection is equal to the angle of incidence.

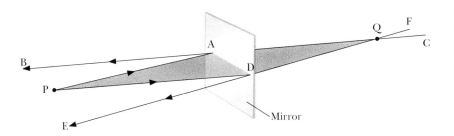

Figure 17-8 A ray diagram showing that the location of the image Q is determined by the crossing of the two lines of sight. The image Q is as far behind the mirror as the object P is in front.

Flat Mirrors

When we look at smooth reflecting surfaces, we don't see light rays; we see images. We can see how images are produced by looking at the paths taken by light rays. In Figure 17-8 we locate the image Q of a single point P in front of a flat mirror. Light leaves the point P in all directions; some of the light strikes the mirror at point A, then reflects and travels in the direction of B. An eye at B would receive the light coming along the direction AB.

Remember that seeing is a passive activity. Our eye–brain system records only the direction from which the light arrives. We do not know from how far away the light originated, but we do know that it came from someplace along the line from B to C. However, we can say the same thing about the light that arrives from another direction, say, at E. The eye perceives this light as coming from someplace along the line EF. Because the only place that lies on both lines is Q, our brain says that the light originated at point Q. This is the location of the image. We see an image of P located behind the mirror at point Q. After the light reflects from the mirror, it has all of the properties it would have had if the object had actually been at Q.

The image has a definite location in space. Figure 17-8 shows that the point Q is located the same distance behind the mirror as the point P is in front of the mirror. A straight line drawn between P and Q is normal to the mirror. These observations allow us to locate the image quickly. For example, in Figure 17-9 we can locate the image of the pencil by first locating the image of its tip and then the image of its eraser. The size of the image is the same as that of the object (although it is farther away from you and looks smaller). Note that the pencil doesn't have to be directly in front of the mirror. However, the mirror must be between the entire image and the observer's eyes, as illustrated in Figure 17-10.

PHYSICS | ON YOUR OWN

Use tape to outline the minimum region on a full-length mirror that is required for you to be able to see from the top of your head to your toes. Does the size or location of this region depend on how far from the mirror you stand? How does the height of this region compare with your height?

Figure 17-9 The image of a pencil formed by a flat mirror is located behind the mirror.

PHYSICS | ON YOUR OWN

Stand a plane mirror vertically, as shown in Figure 17-9. Find two identical objects that are taller than the mirror. Stand the first object in front of the mirror and then place the second object behind the mirror so that it appears to be in the same location as the image of the first object. If you have placed the second object correctly, you should be able to look at this object from many different angles and see that the object and image appear to be in the same location. It should be hard to tell where the image stops and the object begins. Where is this special location relative to the first object and the mirror?

Figure 17-10 A flat mirror forms an image of an object even if the object is not located directly in front of the mirror.

Flawed Reasoning

Cassandra complains to her mother: "Physics is hard. I am supposed to light a match in front of a mirror and find the location of the flame's image. When I move my head to the left, the image of the flame is located on the left-hand side of the mirror. When I move my head to the right, it is located on the right-hand side of the mirror. How can I find the location of something that keeps moving?"

Help Cassandra's mother convince her that physics is easy.

Answer Suppose Cassandra looks at a tree through her kitchen window. By moving her head back and forth, she can align the tree with the left-hand side or the right-hand side of the window. The tree, however, is not moving. The mirror is like the kitchen window, and the flame is like the tree. The light that reflects from the mirror behaves as if it came directly from the image. The image is not on the mirror's surface; it is located behind the mirror and does not move. Cassandra should think of the mirror as a window through which she can view this "image world."

A magician's trick illustrates the realism of an image by presenting the audience with a live, talking head on a table, as in Figure 17-11(a). Like many other illusions, the trick lives up to the cliché that "it's all smoke and mirrors." But where is the person's body? Your eye–brain system is tricked by the images. You think that the table is an ordinary one with legs and a wall behind it. You "know" this because you can see the wall between the table's legs. Figure 17-11(b) shows the set-up with the mirrors removed. The table has mirrors between its legs so that the walls you see under the table are really images of the side walls.

Multiple Reflections

When a light beam reflects from two or more mirrors, we get interesting new optical effects from the multiple reflections. If the two mirrors are directly opposite each other, such as in some barbershops and hair salons, we get an infinite number of images, with successive images being farther and farther away from the object. To see how this works, remember that the light appearing to come from an image has the same properties as if it actually came from an object *at the location of the image*. Each mirror forms images of everything in front of it, *including* the images formed by the other mirror. Each of the mirrors forms an image of the object. The light from these images forms an additional set of images behind the opposing mirrors, and on and on, as illustrated in Figure 17-12.

Question Would the opposing mirrors in Figure 17-12 allow you to see the back of your head?

Answer Yes, provided the mirrors are tilted a bit so that your head doesn't get in the way of your view. The second image in the mirror in front of you will be of the back of your head.

Figure 17-11 (a) The woman's head appears to sit on the table. How is this done? (b) The same scene with both mirrors removed.

Figure 17-12 Two opposing parallel mirrors form an infinite number of images. Notice the reversals in the orientations of the triangles.

Curved Mirrors

(a)

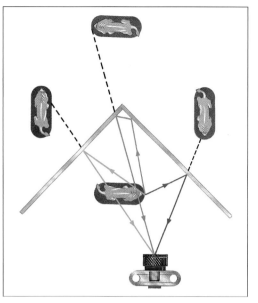

(b)

Figure 17-13 (a) Two mirrors at right angles to each other produce three images of the lion. (b) The red, green, and blue lines show sample paths taken by rays that enter the camera lens.

In Figure 17-13(a) we have placed a figurine of a lion in front of two mirrors that form a right angle. Figure 17-13(b) shows the paths taken by the rays that entered the camera. Notice that there are three images; each mirror forms one image, and then each mirror forms an image of these images. (Remember that a mirror does not have to actually extend between the image and the object. However, it is often useful to imagine that each mirror is extended.) If the angle between the mirrors is precisely 90 degrees, these latter two images overlap to form a single image beyond the corner.

If the angle between the mirrors is made smaller, the overlapping images beyond the corner separate. When the angle between the mirrors reaches 60 degrees, we once again get overlapping images beyond the corner and a total of five images, as shown in Figure 17-14.

Figure 17-14 When the mirrors form an angle of 60 degrees, five images of the lion are produced.

PHYSICS | ON YOUR OWN

Examine the image of your face formed by two mirrors at right angles to each other. What do you see when you touch one of your cheeks? How does the image differ from the image you see in a flat mirror?

Curved Mirrors

Fun-house mirrors are interesting because of the distortions they produce. The distortions are not caused by a failure of the law of reflection but result from the curvature of the mirrors. Some distortions are desirable. If the distortion is a magnification of the object, we can see more detail by looking at the image.

Cosmetics mirrors and some rearview mirrors on cars are simple curved mirrors that don't produce bizarre distortions but do change the image size. A cosmetics mirror uses the concave side—the reflecting surface is on the inside of the sphere—to generate a magnified image of your face. The convex reflecting surface—the outside of the sphere—always produces a smaller image but has a bigger field of view. Convex mirrors are quite often used on cars and trucks and on "blind" street corners because they provide a wide-angle view.

Question Sometimes mirrors are installed in stores to inhibit shoplifting. Are these concave or convex mirrors?

Answer The mirrors must be convex to provide a wide view of the store.

The curved surfaces of fun-house mirrors produce interesting images.

Retroreflectors

An interesting consequence of having two mirrors at right angles is that an incoming ray (in a plane perpendicular to both mirrors) is reflected parallel to itself, as shown by the three rays in the figure. This works for all rays if we add a third mirror to form a "corner" reflector, like putting mirrors on the ceiling and the two walls in the corner of a room. These *retroreflectors* are used in the construction of bicycle reflectors so that the light from a car's headlights is reflected back to the car driver and not off to the side as it would be with a single mirror. Examination of many reflectors reveals a surface covered with holes in the shape of the corners of cubes. Reflectors used on clothing and the surfaces of stop signs are often covered with a layer of reflective beads. The surfaces in the regions between the beads work like corner reflectors.

An outer-space application of retroreflectors involves an experiment to accurately measure the distance to the Moon. The Apollo astronauts placed panels of retroreflectors on the Moon to allow scientists on Earth to bounce a laser beam off the Moon and receive the reflected signal back on Earth. Ordinary mirrors would not have worked because the astronauts could not have aimed them well enough to send the beam back to Earth. Additionally, the Moon's wobbly rotation would send the reflected beam in many directions. Because retroreflectors work for all incident angles, they could be simply laid on the Moon's surface. We now know that the Earth–Moon distance increases by 3 to 4 centimeters per year.

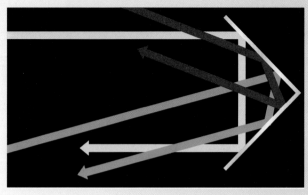

Each of the three rays is reflected parallel to itself.

Convex mirrors allow clerks to watch for potential shoplifters.

Figure 17-15 shows the essential geometry for a concave spherical mirror; the reflecting surface is the inside of a portion of a sphere. The line passing through the center of the sphere C and the center of the mirror M is known as the **optic axis**. Light rays parallel to the optic axis are reflected through a common point F called the **focal point**. The focal point is located halfway between the mirror and the center of the sphere. The distance from the mirror to the focal point is known as the **focal length** and is equal to one-half the radius R of the sphere.

Concave mirrors can be used to focus light. Some solar collectors use them to concentrate sunlight from a large area onto a smaller heating element. Because the Sun is very far away, its rays are essentially parallel, and a concave spherical mirror can focus the sunlight at the focal point. A cylindrical mirror focuses sunlight to a line instead of a point. A pipe containing a fluid can be placed along this line to carry away the thermal energy. If the heat energy is used to generate electricity, the higher temperature makes the process more efficient.

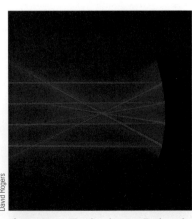 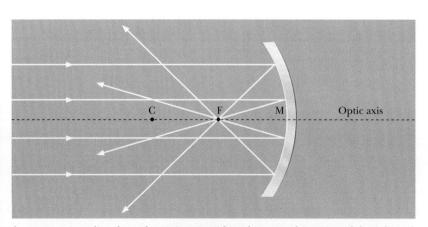

Figure 17-15 The focal point F of a spherical concave mirror lies along the optic axis midway between the center of the sphere C and the center of the mirror M. Rays parallel to the optic axis are focused at the focal point F.

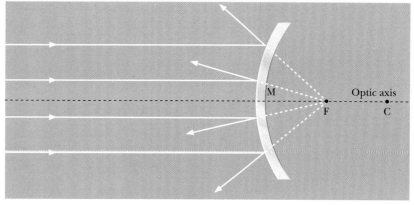

Figure 17-16 The focal point F of a spherical convex mirror lies along the optic axis midway between the center of the sphere C and the center of the mirror M. Rays parallel to the optic axis are reflected as if they come from the focal point F.

Light rays are reversible. The law of reflection is still valid when the incident and reflected rays are reversed. So the shape that focuses parallel rays to a point will take rays from that point and send them out as parallel rays. This idea is used in automobile headlights. The bulb is placed near the focal point of the mirror, producing a nearly parallel beam.

The optic axis of a convex mirror also passes through the center of the sphere C, the focal point F, and the center of the mirror M, but in this case, the focal point and the center of the sphere are on the back side of the mirror. (Again, the focal point is halfway between the center of the sphere and the center of the mirror.) Rays parallel to the optic axis are now reflected *as if they came from* the focal point behind the mirror, as shown in Figure 17-16.

Images Produced by Mirrors

A concave mirror can form two different types of image, depending on how close the object is to the mirror. Imagine walking toward a very large concave mirror. At a large distance from the mirror, you will see an image of your face that is inverted and reduced in size (Figure 17-17). This image is formed by light from your face reflecting from the mirror and converging to form an image in front of

Figure 17-17 An inverted image is formed when the head is located outside the focal point of a concave mirror. Note that both the face and the floor tiles are inverted.

Cylindrical mirrors are used to concentrate sunlight at this solar farm.

Figure 17-18 A magnified, erect image is formed when the head is located inside the focal point of a concave mirror.

the mirror. This image is known as a **real image**, because the rays reflect from the mirror and converge to form the image. The rays then diverge, behaving as if your face were actually at the image location.

As you walk toward the mirror, the image of your face moves toward you and gets bigger. As you pass the center point of the mirror, the image moves behind you. When your face is closer to the mirror than the focal point, the image is similar to that in a flat mirror except that it is magnified (Figure 17-18). As with the flat mirror, the rays diverge after reflecting, and the image is located behind the mirror's surface. In this case there can be no light at the location of the image because it is formed behind the mirror. The light only appears to come from the location of the image. This second type of image is called a **virtual image**. As you continue approaching the mirror, the image moves closer and gets smaller.

The essential difference between a real and a virtual image is whether the light actually comes from the image location or only appears to come from there. If the rays diverge upon reflecting, they never come together to form a real image. They will, however, appear to originate from a common location behind the mirror. Reflected rays that converge to form a real image can be seen on a piece of paper placed at the image location because the light actually converges at that location. However, if you put a piece of paper at the location of a virtual image, you get nothing because there is no light there.

Locating the Images

A simple way of locating an image without measuring any angles is by looking at a few special rays. Light leaves each point on the object in all directions; those rays that strike the mirror form an image. Although any of these rays can be used to locate the image of a point, three are easy to draw and are therefore useful in locating the image. Because all rays from a given point on the object are focused at the same place for a real image (or appear to come from the same point for a virtual image), we need only find the intersection of any two of them. The third one can be drawn as a check. (In actual drawings these three rays do not always meet at a point. However, they give a pretty good location for the image if the object is small enough that the special rays strike the mirror near the optic axis.)

The three rays that are useful in these ray diagrams are shown in Figure 17-19. The easiest ray to draw is the red one lying along a radius of the sphere. It strikes the mirror normal to the surface and reflects back on itself. Another easy ray is the blue one that approaches the mirror parallel to the optic axis. It is reflected toward the focal point. The third ray (shown in green) is a reverse version of the second one; a ray passing through the focal point reflects

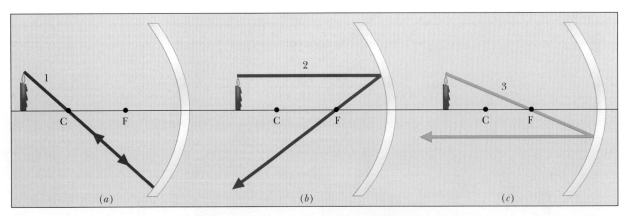

Figure 17-19 The three light rays used in drawing ray diagrams for concave mirrors.

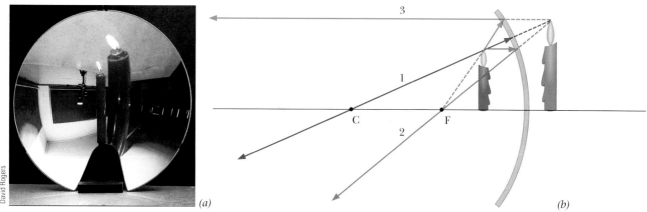

Figure 17-20 The image of a candle inside the focal point of a concave spherical mirror is virtual, erect, and magnified.

parallel to the optic axis. (This ray does not actually need to pass through the focal point. If the object is closer than the focal point, the ray still lies along the line from the focal point to the mirror.)

If the mirror is small, some of these special rays may not strike the actual surface of the mirror. For the purposes of the ray diagram, we extend the mirror because a larger mirror with the same focal length would produce the same image. In fact, the mirror could be so small that none of the three easily drawn rays hits the mirror.

The descriptions of these rays can be abbreviated as follows:

1. Along radius—back on itself
2. Parallel to optic axis—through focal point
3. Through focal point—parallel to optic axis

◀ rays for curved mirrors

To illustrate the use of these ray diagrams, consider an object located inside the focal point of a concave mirror (Figure 17-20[a]). Figure 17-20(b) shows the three special rays that we use to locate the image of the tip of the candle. The rays are color-coded to correspond to the descriptions given above. Because the base of the candle is on the optic axis, we know that the image of the base is also on the optic axis. So finding the location of the tip of the candle gives the image location, orientation, and magnification. For the case illustrated in Figure 17-20, we can see that the rays intersect behind the mirror, forming a virtual image that is erect and magnified.

As the candle is moved away from the mirror, the image size and the distance of the image behind the mirror increase. As the candle approaches the focal point, the image becomes infinitely large and infinitely far away. You can verify this by drawing a ray diagram.

When the candle is beyond the focal point, a real image is formed (Figure 17-21[a]). The ray diagram in Figure 17-21(b) shows that the reflected rays do not diverge as in the previous case but come together, or converge. These rays actually cross at some point in front of the mirror to form a real, inverted image.

Question Where would you place an object to get an image at the same location?

Answer At the center of the sphere. This can be checked with a ray diagram.

Fortunately, locating images formed by convex mirrors is the same process used for images formed by concave ones. A ray diagram showing how to locate the image is given in Figure 17-22(b). The same three rays are used. You must remember only that the focal point is now behind the mirror. With ray diagrams

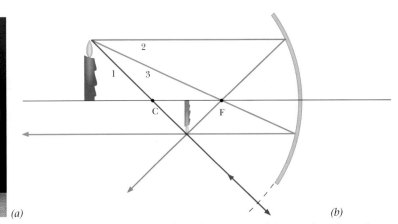

Figure 17-21 The image of a candle outside the focal point of a concave spherical mirror is real, inverted, and may be larger or smaller than the object.

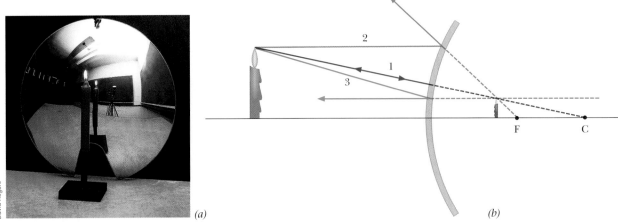

Figure 17-22 The image of a candle in front of a convex spherical mirror is always virtual, erect, and reduced in size. Ray 2 reflects as if it came from the focal point, and ray 3 starts toward the focal point.

you can verify that images formed by convex mirrors are always erect, virtual, and reduced in size.

Speed of Light ✓ MATH

Besides traveling in a straight line in a vacuum, light moves at a very, very high speed. In fact, people originally thought that the speed of light was infinite—that it took no time to travel from one place to another. It was clear to these observers that light travels very much faster than sound; lightning striking a distant mountain is seen much before the thunder is heard.

Because of its very great speed, early attempts to measure the speed of light failed. Galileo made one interesting attempt when he tried to measure the speed of light by sending a light signal to an assistant on a nearby mountain. The assistant was instructed to uncover a lantern upon receiving Galileo's signal. Galileo measured the time that elapsed between sending his signal and receiving that of his assistant. Knowing the distance to the mountain, he was able to calculate the speed of light. Upon repeating the experiment with a more distant mountain, however, he found the same elapsed time! Had the speed of light increased? No. Galileo correctly concluded that the elapsed time was due to his assistant's reaction time. Therefore, the time it took light to travel the distance was either zero or much smaller than he was able to measure.

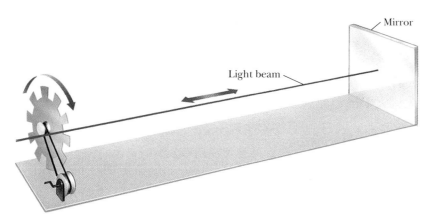

Figure 17-23 A schematic drawing of Fizeau's apparatus for measuring the speed of light.

About 40 years later, in 1675, Danish astronomer Ole Roemer made observations of the moons of Jupiter that showed that light had a finite speed. Roemer found that the period of revolution of a moon around Jupiter was shorter during the part of the year when Earth approached Jupiter and longer when Earth receded from Jupiter. He expected the period to be constant like that of our Moon about Earth. He correctly concluded that the period of the Jovian moon was constant and that the variations observed on Earth were due to the varying distance between Earth and Jupiter. When Earth approaches Jupiter, the light emitted at the beginning of the period must travel farther than that emitted at the end of the period. Therefore, the light emitted at the end of the period arrives sooner than it would if Earth and Jupiter remained the same distance apart. This difference in distance makes the period appear shorter. Once the radius of Earth's orbit was determined, the speed of light could be calculated.

French physicist Hippolyte Fizeau performed the first nonastronomical measurement of the speed of light in 1849. He sent light through the gaps in the teeth of a rotating gear to a distant mirror. The mirror was oriented to send the light directly back (Figure 17-23). At moderate speeds of rotation, the returning light would strike a tooth. But at a certain rotational speed, the light would pass through the next gap. Knowing the speed of the gear and the distance to the mirror, Fizeau was able to calculate the speed of light to a reasonable accuracy.

The speed of light has been measured many, many times. As the methods improved, the uncertainty in its value became less than 1 meter per second. In 1983 an international commission set the speed of light in a vacuum to exactly 299,792,458 meters per second and used it with atomic clocks to define the length of the meter. The speed of light is usually rounded off to 3×10^8 meters per second (186,000 miles per second). If light traveled in circles, it could go around Earth's equator 7.5 times in 1 second. It's no wonder that early thinkers thought light traveled at an infinite speed. Although the speed of light is finite, it can be considered infinite for many everyday experiences.

◄ speed of light

Question Given that radio waves travel at the speed of light, would you expect there to be longer than normal pauses in a conversation between two people talking via radio from the two U.S. coasts? How about in a conversation with astronauts on the Moon?

Answer Because light can travel around the world 7.5 times in 1 second, we would expect it to travel across the United States and back in a small part (about $\frac{1}{30}$) of a second. Therefore, the pauses would seem normal. The round-trip time for the signal to the Moon is a little less than 3 seconds; this would produce noticeable pauses.

(a) (b)

Figure 17-24 (a) A brightly colored box and dice illuminated with yellow light. (b) The same box and dice illuminated with white light.

Color

Color is all around us and adds beauty to our lives. Color and the way we perceive it is an area of research involving many disciplines, including physics, chemistry, physiology, and psychology. One example of our body's role in detecting color is the fact that there is no such thing as white light. What we perceive as white is really the summation of many different colors reaching our eyes. The color of an object is determined by the color, or colors, of the light that enters our eyes and the way that this is interpreted by our brains.

Besides the additive effect within our brain, there is also the issue of what reaches our eyes. If you have ever tried to match the color of two pieces of clothing under different kinds of lighting, you know it can be difficult. They may match very well in the store but be quite different in sunlight. The colors we perceive are determined by two factors: the color present in the illuminating light and the colors reflected by the object. A red sweater is red because the pigments in the dye absorb all colors except red. When viewed under white light, the sweater looks red. If the illuminating light does not contain red, the sweater will appear black because all the light is absorbed. A brightly colored box and dice look different under different illuminating lights (Figure 17-24).

Although most lights give off all colors, the colors do not have the same relative intensities found in sunlight. Fluorescent lights are often brighter in the blue region and therefore highlight blues. At the risk of taking some of the romance out of candlelit dinners, we note that your date's warm glow is due to the yellow-red light from the candles and may have nothing to do with your date's feelings toward you.

Often an object that appears to be a single color reflects several different colors. Although different colors may enter our eyes, we do not see each of these colors. Our eye–brain systems process the information, and we perceive a single color sensation at each point. The color perceived may appear to have nothing in common with the component colors. For instance, if an object reflects red and green, it will appear yellow. This behavior is in sharp contrast with the sense of hearing. Our ear–brain combination can hear and distinguish many different sounds coming from the same place at the same time.

Placing colored filters in front of spotlights or slide projectors and allowing the colored beams from each one to overlap demonstrates the additive effects of color. One combination of light beams that produces most of the colors that we perceive is red, green, and blue. Figure 17-25 illustrates the colors that are seen on the screen. For instance, red and green yield yellow, blue and green yield cyan (a bluish green), and red and blue yield magenta (a reddish purple). All three colors together produce white!

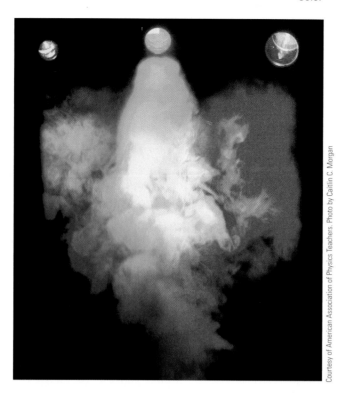

Figure 17-25 The overlap of colored lights produces new colors.

Question What color would you expect to see if you remove some of the blue light from white light?

Answer This leaves red and green behind, creating a yellow color.

Two colors that produce white light when added together are called **complementary colors**. This process is illustrated by positive and negative color film of the same scene; the colors on one are the complements of those on the other, as illustrated in Figure 17-26.

Figure 17-26 The colors in the positive image on the film are the complements of those in the negative image.

Figure 17-27 The array of colors on your television comes from the addition of three basic colors—red, green, and blue.

By using dimmer switches to vary the brightness of each beam, we can generate a very wide range of colors. This process is the basis of color television. Examine a color TV screen with a magnifying glass and you will see that it is covered with arrays of red, green, and blue dots or lines like those in Figure 17-27.

It may seem strange to find that mixing certain colors yields white; from childhood we have learned that mixing many different colored paints together does not yield white, but rather a dark brownish color. Mixing paints is different from mixing colored lights. Mixing paints is a *subtractive* process, whereas with light beams you are *adding* colors. When white light strikes a red object, the pigment in the object subtracts all colors except red and reflects the red back to the viewer. Likewise, a red filter subtracts all colors except the red that passes through it. Each additional color pigment or filter subtracts out more colors from the incident light. This suggests that the primary colors of light—red, green, and blue—are not the best colors to use as primary colors of paint. Red paint absorbs both green and blue. Magenta paint, on the other hand, reflects both blue and red, absorbing only one primary color of light, green. Likewise, cyan paint absorbs only red and reflects both blue and green, and yellow paint absorbs only blue light and reflects both red and green. Indeed, most color printing is done with four colors of ink—cyan, magenta, yellow, and black. (Although black could be created by combining the other three colors, using a separate black improves image quality.)

Flawed Reasoning

What is wrong with the following statement? "If I take a green banana into a dark room and shine red light on it, it should appear yellow—that is, ripe."

Answer It is true that overlapping green and red lights produce yellow. However, green bananas are not a source of green light. Green bananas absorb all colors of light except green, so they appear green under white light. However, if we shine only red light on the bananas, they will absorb this red light and appear black (overripe).

PHYSICS | ON YOUR OWN

Use a magnifying glass to look carefully at the color pictures in any magazine. What colors do you see when you look at a part of the picture that appears red? Green? Blue? Brown?

The Sun appears redder as it sets because the sunlight passes through more of the atmosphere.

This brief coverage of color perception allows us to answer the questions "Why is the sky blue?" and "Why is the Sun yellow?" The Sun radiates light that is essentially white. Because it appears yellow, we can assume that some of the complementary color has somehow been removed. The complement of yellow is blue—the color of the sky. The molecules in the atmosphere are more effective in scattering blue light than red light. As the sunlight passes through the atmosphere, more and more of the blue end of the spectrum is removed, leaving the transmitted light with a yellowish color (Figure 17-28). When we look away from the Sun, the sky has a bluish cast because more of the blue light is scattered into our eyes. This effect is enhanced when the rays have a longer path through the atmosphere: the Sun turns redder near sunrise and sunset. The redness also increases with increased numbers of particles in the air (such as dust). The additional dust in the air during harvest time produces the spectacular harvest moons. Although much less romantic, the same effect results from the air pollution near urban industrial sites.

These ideas also account for the color of water. Because water absorbs red light more than the other colors, the water takes on the color that is complementary to red—that is, cyan. Consequently, underwater photographs taken without artificial lighting look bluish green.

Question If red light were scattered more than blue light, what color would the Sun and sky appear?

Answer The sky would appear red due to the scattered light, and the Sun would appear cyan because of the removal of more of the red end of the spectrum.

Figure 17-28 The sky is blue and the Sun yellow because air scatters more blue light than red light.

Summary

Light is seen only when it enters our eyes. Because we know that light travels in straight lines, we can use rays to understand the formation of shadows and images.

When light reflects from smooth surfaces, it obeys the law of reflection, which states that the angles the incident and reflected rays make with the normal to the surface are equal. The reflected ray lies in the same plane as the normal and the incident ray.

Mirrors produce real and virtual images. Light converges to form real images that can be projected, whereas light only appears to come from virtual images. The virtual image formed by a flat mirror is located on the normal to the mirror that passes through the object. The image is located the same distance behind the mirror as the object is in front of the mirror. The sizes of the image and the object are the same.

Images formed by spherical mirrors can be located by drawing three special rays: (1) along the radius—back on itself; (2) parallel to the optic axis—through the focal point; and (3) through the focal point—parallel to the optic axis. The focal point is located halfway between the surface and the center of the sphere.

Light travels through a vacuum at 299,792,458 meters per second.

The additive effects of color mean that adding red and green lights yields yellow, blue and green lights yield cyan, and red and blue lights yield magenta. All three colors produce white. Mixing paints is a subtractive process, whereas mixing light beams is an additive process. The sky is blue and the Sun is yellow because the molecules in the atmosphere preferentially scatter blue light, leaving its complement.

Physics Now™ Assess your understanding of this chapter's topics with sample tests and other resources found by logging into PhysicsNow at **http://physics.brookscole.com/kf6e**.

Chapter 17 Revisited

When light reflects from a smooth mirror or clean piece of glass, its direction is changed, giving the viewer false information about the location of the source of the light. For instance, reflections of the walls of a box make the box appear to be empty when, in fact, a rabbit is hiding behind the mirrors. This is the basis of many visual illusions.

KEY TERMS

complementary color: For lights, two colors that combine to form white.

diffuse reflection: The reflection of rays from a rough surface. The reflected rays do not leave at fixed angles.

focal length: The distance from a mirror to its focal point.

focal point: The location at which a mirror focuses rays parallel to the optic axis or from which such rays appear to diverge.

law of reflection: The angle of reflection (measured relative to the normal to the surface) is equal to the angle of incidence. The incident ray, the reflected ray, and the normal all lie in the same plane.

light ray: A line that represents the path of light in a given direction.

normal: A line perpendicular to a surface or curve.

optic axis: A line passing through the center of a curved mirror and the center of the sphere from which the mirror is made.

penumbra: The transition region between the darkest shadow and full brightness. Only part of the light from the source reaches this region.

real image: An image formed by the convergence of light.

umbra: The darkest part of a shadow where no light from the source reaches.

virtual image: The image formed when light only appears to come from the location of the image.

CONCEPTUAL QUESTIONS

1. A professor shines a light beam across the front of a lecture hall. Why can you see the light on the wall but not in the air?
2. If you shine a laser pointer across the room, you see only a red spot on the far wall. However, you can see the path of the beam if you create a dust cloud with a pair of chalk erasers. Explain how the cloud allows you to see the beam.
3. You place a plane mirror flat on the floor ahead of you and shine a laser beam toward its center. Why do you not see a red dot on the face of the mirror?
4. If you spread a fine layer of dust on the mirror in Question 3, you suddenly see a red dot on the face of the mirror. Why does this happen?
5. Which of the following will cast a shadow that has an umbra but no penumbra: the Sun, a lightbulb, a campfire, or a point source of light? Explain.
6. You hold your hand 3 feet above the ground and look at the shadow cast by the Sun. You repeat this inside using the light from a 2-by-4-foot fluorescent light box in the ceiling. In which case will the penumbra be more pronounced? Which of these two sources is acting more like a point source? Explain.
7. You are in a dark room with a single incandescent 60-watt bulb in the center of the ceiling. You hold a book directly beneath the bulb and begin lowering it toward the floor. As the book is lowered, what happens to the size of the umbra?
8. Repeat Question 7 using a 2-by-4-foot fluorescent light box in place of the incandescent bulb.
9. Under what conditions will the shadow of a ball on a screen not have an umbra? What does this have to do with the observation that some solar eclipses are not total for any observer on Earth?
10. During some solar eclipses, the angular size of the Moon is smaller than that of the Sun. What would observers on Earth see if they stood directly in line with the Sun and Moon?
11. What effect does enlarging the hole in a pinhole camera have on the image?
12. What happens to the image produced by a pinhole camera when you move the back wall of the camera closer to the pinhole?
13. You form an L using two fluorescent light tubes in front of the opening of a pinhole camera as shown. Sketch the image formed on the film.

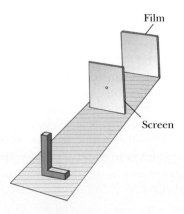

14. You are sitting under a large maple tree on a sunny day. You notice that the light filtering through the leaves does not have the sharp lines you would expect from maple

leaves. Instead, the pattern consists of many small round circles of light. Use the concept of a pinhole camera to explain this.

15. When the incident ray IO reflects from the mirror in the figure, the reflected ray lies along the line O_____.

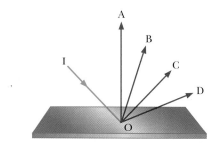

16. Which letter corresponds to the location of the image of the object O in the figure?

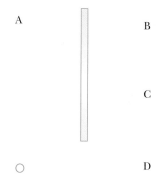

17. In the figure, observers at locations A and B are attempting to see the image of the star in the plane mirror. An obstruction is placed in front of the mirror as shown. Which observers, if either, can see the star's image?

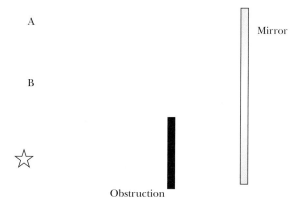

18. In talking about Question 17, Miguel claims, "The star still has an image directly behind the mirror because the light doesn't have to travel from the object to the image." Antonio counters, "The obstruction prevents light from going directly to the mirror, so no image can be formed." Jocelyn argues, "There has to be an image, it just can't be directly behind the obstruction." Which students, if any, do you agree with?

19. If a 0.7-meter-tall child stands 0.6 meter in front of a vertical plane mirror, how tall will the image of the child be?

20. How do the size and location of your image change as you walk toward a flat mirror?

21. At which of the lettered locations in the figure would an observer be able to see the image of the star in the mirror?

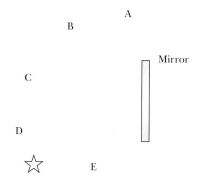

22. Where is the image of the arrow shown in the figure located? Shade in the region where an observer could see the entire image.

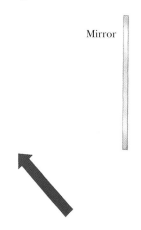

23. What is the magnification of a flat mirror? What is its focal length?

▲ 24. If you walk toward a flat mirror at a speed of 1.2 meters per second, at what speed do you see your image moving toward you?

25. How does the height of the shortest mirror in which a woman can see her entire body compare with her height? Does your answer depend on how far she stands from the mirror?

▲ 26. The word AMBULANCE is often written backward on the front of the vehicle so that it can be read correctly in a rearview mirror. Why do we have to switch the left and right but not the up and down?

27. You are standing in a room that has large plane mirrors on opposite walls. Why do the images produced appear to get progressively smaller? Are these images real or virtual?

28. How many images would be formed by two mirrors that form an angle of 45 degrees?

29. Why are the back surfaces of automobile headlights curved?

30. If rays of light parallel to the optic axis converge to a point after leaving the mirror, what kind of mirror is it?

31. Most of us find that we have to really strain our eyes to focus on objects located close to our noses. You hold two

mirrors 1 foot in front of your face. One is a plane mirror, and the other is a concave mirror with a 3-inch focal length. In which case are you more likely to have to strain your eyes to see the image of your nose?

32. Can the image produced by a convex mirror ever be larger than the object? Explain.
33. What type of mirror would you use to produce a magnified image of your face?
▲ 34. The image produced by a convex mirror is always closer to the mirror than the object. Then why is it that the convex mirrors used on cars and trucks often have the warning "Caution: Objects Are Closer Than They Appear" printed on them?
35. What are the size and location of the image of your face when your face is very close to a concave mirror? How do the size and location change as you move away from the mirror?
36. If you hold your face very close to a convex mirror, what are the size and location of the image of your face? How do the size and location change as you move away from the mirror?
37. What is the fundamental difference between a real image and a virtual one?
38. Can both real and virtual images be photographed? Explain.
39. A searchlight uses a concave mirror to produce a parallel beam. Where is the bulb located?
40. What happens to the location of the real image produced by a concave mirror if you move the object to the original location of this image?
41. Why does the arrival of the sound from a bass drum in a distant band not correspond to the blow of the drummer?
42. Astronomers claim that looking at distant objects is the same as looking back in time. In what sense is this true?
43. The *Sojourner* rover that explored the surface of Mars as part of NASA's Pathfinder mission had to make decisions on its own rather than be driven by remote control from Earth. Why?

44. Without asking, how could you tell whether you were talking to astronauts on the Moon or on Mars?
45. What color is produced by the overlap of a blue spotlight and a red spotlight?
46. A substance is known to reflect red and blue light. What color would it have when it is illuminated by white light? By red light?
47. A surface appears yellow under white light. How will it appear under red light? Under green light? Under blue light?
48. An actress wears a blue dress. How could you use spotlights to make the dress appear to be black?
49. A Crest toothpaste tube viewed under white light has a red *C* on a white background. What would you see if you used red light?

50. When you place a blue filter in the light from a projector, it produces a blue spot on the wall. If you use a red filter, you get a red spot. What would you see on the wall if you passed the light through both filters?
51. When you mix red and green light from separate projectors, you get a yellow spot on the wall. However, if you mix red and green paint, you get a muddy brown color. How do you account for this difference?
52. What color do you expect to get if you mix magenta and cyan paints?
53. What is the complementary color to red?
54. What is the complementary color to magenta?
55. If you remove all of the green light from white light, what color would you see?
56. A lens for a spotlight is coated so that it does not transmit yellow light. If the light source is white, what color is the spot?
57. If the atmosphere primarily scattered green light instead of blue light, what color would the sky and Sun appear?
▲ 58. How would the color of sunlight change if the atmosphere were much more dense?

EXERCISES

1. A 5-cm-diameter ball is located 30 cm from a point source and 90 cm from a wall. What is the size of the shadow on the wall?
2. A 5-cm-diameter ball is located 50 cm in front of a pinhole camera. If the film is located 10 cm from the pinhole, what is the size of the image on the film?
3. Use a ruler and a protractor to verify that the image produced by a flat mirror is as far behind the mirror as the object is in front.
4. George's eyes are located 60 inches from the floor. His belt buckle is 36 inches from the floor. Determine the maximum distance from the floor that the bottom of a plane mirror can be placed such that George can see the belt buckle's image in the mirror. (*Hint*: you can verify that it does not matter how far George stands from the mirror.)
5. An object and an observer are located 2 m in front of a plane mirror, as shown in the figure. If the observer is 3 m from the object, find the distance between the observer and the location of the object's image.

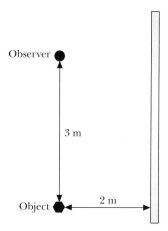

6. In Exercise 5 find the distance that the light travels from the object to the observer.
7. In the figure associated with Exercise 5, light leaves the object, reflects from the mirror, and reaches the observer. Use a protractor to find the angle of reflection.
8. Use a compass and a protractor to verify that the three rules for drawing ray diagrams for spherical mirrors satisfy the law of reflection.
9. What is the radius of the spherical surface that would produce a mirror with a focal length of 4 m?
10. A telescope mirror is part of a sphere with a radius of 6 m. What is the focal length of the mirror?
11. An object is located three times the focal length from a concave spherical mirror. Draw a ray diagram to locate its image. Is the image real or virtual, erect or inverted, magnified or reduced in size? Explain.
12. An object is located midway between the focal point and the center of a concave spherical mirror. Draw a ray diagram to locate its image. Is the image real or virtual, erect or inverted, magnified or reduced in size? Explain.
13. A 4-cm-tall object is placed 75 cm from a concave mirror with a focal length of 25 cm. Draw a ray diagram to find the location and size of the image.
14. How would your answers to Exercise 13 change if the same object were 120 cm from a concave mirror with a focal length of 40 cm?
15. Draw a ray diagram to locate the image of a 5-cm-tall object located 60 cm from a convex mirror with a focal length of 30 cm.
16. How would your answers to Exercise 15 change if the same object were 90 cm from a concave mirror with a focal length of 45 cm?
17. A convex mirror has a focal length of 60 cm. Draw a ray diagram to find the location and magnification of the image of an object located 30 cm from the mirror.
18. Repeat Exercise 17 for a concave mirror.
19. If you place an object 40 cm in front of a concave spherical mirror with a focal length of 20 cm, where will the image be located?
20. You have a concave spherical mirror with a focal length of 30 cm. Where could you place a candle to make it appear to burn at both ends?
21. If Galileo and his assistant were 15 km apart, how long would it take light to make the round-trip? How does this time compare with reaction times of about 0.2 s?
22. Suppose Galileo, in the experiment described in Exercise 21, had assumed that the entire 0.2-s delay was due to the travel time of light rather than to his assistant's reaction time. What value would he have calculated for the speed of light?
23. Approximately how long would it take a telegraph signal to cross the United States from the East Coast to the West Coast? (Telegraph signals travel at about the speed of light.)
24. Mars and Earth orbit the Sun at radii of 228 million km and 150 million km, respectively. When, in the future, your friend from Mars calls you on the phone and you answer, "Hello," what are the minimum and maximum times you will have to wait for your friend to reply?
25. How far does light travel in 1 year? This distance is known as a light-year and is a commonly used length in astronomy.
26. How far does light travel in 1 nanosecond—that is, in one-billionth of a second?

18 | Refraction of Light

When light travels from one transparent material to another—say, from air to water—some interesting visual effects are created. Fish in an aquarium look bigger, and a tree that has fallen into a lake looks bent. Is it ever possible that the light cannot travel from one transparent material to another?

(See page 378 for the answer to this question.)

Refraction of light in the atmosphere causes the oval shape of the setting Sun.

W HEN light strikes a transparent material, it usually changes direction. This change accounts for many interesting effects ranging from the apparent distortion of objects to the beauty of an afternoon rainbow. This bending of light that occurs at the surface of a transparent object is called **refraction**.

Refraction can be studied by looking at the paths the light takes as the incident angle is varied, as shown in Figure 18-1. As in reflection the angles are measured with respect to the normal to the surface. In this case the normal is extended into the material, and the angle of refraction is measured with respect to the extended normal. The amount of bending is zero when the angle of incidence is zero; that is, light incident along the normal to the surface is not bent. As the angle of incidence increases relative to the normal, the amount of bending increases; the angle of refraction differs more and more from the angle of incidence.

Physics⚛Now™ Test your understanding of this chapter by logging into PhysicsNow at **http://physics.brookscole.com/kf6e**, selecting the chapter, and clicking on the "Take a Pre-Test" link.

Index of Refraction ✓ MATH

✓ Extended presentation available in the *Problem Solving* supplement

The amount of bending that occurs when light enters the material depends on the incident angle and an optical property of the material called the **index of refraction**. (We will refine the definition of the index of refraction in the next chapter.) A mathematical relationship can be written that predicts the refracted angle given the incident angle and the type of material. This rule, called *Snell's law*, is not as simple as the rule for reflection because it involves trigonometry. A simpler way to express the relationship is to construct a graph of the experimental data. Of course, although graphs are easier to use, they often have the disadvantage of being less general. In this case a graph has to be made for each sub-

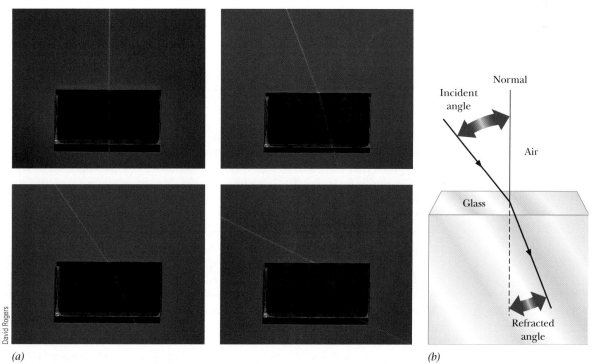

(a)   *(b)*

Figure 18-1 The amount of refraction depends on the angle of incidence. Notice that some of the incident light is reflected and some is refracted.

Figure 18-2 This graph shows the relationship between the angle of refraction and the angle of incidence for light entering air, water, and glass from a vacuum.

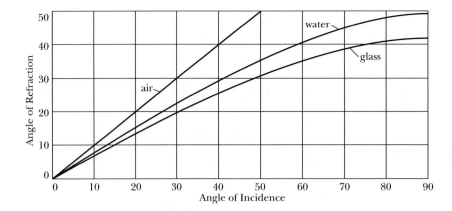

stance. The graph in Figure 18-2 gives the angle of refraction in air, water, and glass for each angle of incidence in a vacuum. Although the curves for water and glass have similar shapes, light is refracted more on entering glass than water.

If no refraction takes place, the index of refraction is equal to 1. You can see from the graph that very little bending occurs when light goes from a vacuum into air; the index of refraction of air is only slightly greater than 1. Because the index of refraction of air is very close to 1, air and a vacuum are nearly equivalent. Therefore, we will use the graph in Figure 18-2 for light entering water or glass from either air or a vacuum. The index of refraction for water is 1.33; for different kinds of glass, it varies from 1.5 to 1.9. The curve for glass on the graph is drawn for an index of 1.5. The index of refraction for diamond is 2.42. A larger index of refraction means more bending for a given angle of incidence. For example, the graph indicates that light incident at 50 degrees (50°) has an angle of refraction of 31 degrees in glass and 35 degrees in water. Thus, the light is bent 19 degrees going into glass (index = 1.5) and only 15 degrees going into water (index = 1.33).

Question   What is the angle of refraction for light incident on glass at 30 degrees? How much does the ray bend?

Answer   The graph in Figure 18-2 gives an angle of refraction of approximately 20 degrees. Therefore, the ray bends 30 degrees − 20 degrees = 10 degrees from its original direction.

Light entering a transparent material from air bends *toward* the normal. What happens if light originates in the material and exits into the air? Experiments show that the paths of light rays are reversible. The photographs in Figure 18-1 can be interpreted as light inside the glass passing upward into the air. (If this were really the case, however, there would also be a faint reflected beam in the glass.) This example shows that when light moves from a material with a higher index of refraction to one with a lower index, the light leaving the material is bent *away from* the normal. Because of the reversibility of the rays, you can still use the graph in Figure 18-2 to find the angle of refraction; simply reverse the labels on the two axes.

Question   If a ray of light in water strikes the surface at an angle of incidence of 40 degrees, at what angle does it enter the air?

Answer   Locate the 40-degree angle on the *vertical* axis of the graph in Figure 18-2 and move sideways until you encounter the curve for water. Then, moving straight down to the horizontal axis, we obtain an angle of 58 degrees.

Another consequence of this reversibility is that light passing through a pane of glass that has parallel surfaces continues in its original direction after emerging. The glass has the effect of shifting the light sideways as shown in Figure 18-1.

The refraction of light produces interesting optical effects. A straight object partially in water appears bent at the surface. The photograph of a pencil in Figure 18-3 illustrates this effect. Looking from the top, we see that the portion of the pencil in the water appears to be higher than it actually is.

Figure 18-3 A straight pencil appears to be bent at the surface of the water.

This phenomenon can also be seen in the photographs of identical coins, one underwater (Figure 18-4[a]) and the other in air (Figure 18-4[b]). Even though the coins are the same distance from the camera, the one underwater appears closer and larger. The drawing in Figure 18-4(c) shows some of the rays that produce this illusion. This effect also makes fish appear larger—although never as large as the unlucky fisherman would like you to believe.

Let's examine the reason for the coin's appearing larger when it is in the water. Is it due to the image being closer, or is the image itself bigger? It is fairly straightforward to see that the increase in size is due to the image being closer. To see that the image hasn't increased in size, we need to remind ourselves that rays normal to the surface are not refracted. Therefore, if we use vertical rays to locate the images of all points on the rim of the coin, each image will be directly above the corresponding point on the rim. This means that the image has the same size as the coin.

Question If you keep your stamp collection under thick pieces of glass for protection, will the stamps appear to have their normal sizes?

Answer No. Just like the coin in water, the stamps appear to be closer and are therefore apparently larger in size.

Total Internal Reflection

There are situations where light can't pass between two substances even if they are both transparent. This occurs at large incident angles when the light strikes a material with a lower index of refraction, such as from glass into air as shown at the lower surface in Figure 18-5. At small angles of incidence, both reflection and refraction take place. The refracted angle is larger than the incident angle as shown in Figure 18-5(a). As the incident angle increases, the refracted angle increases even faster. At a particular incident angle, the refracted angle reaches

(a)

(b)

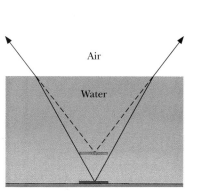

(c)

Figure 18-4 A coin underwater (a) appears closer than an identical coin in air (b). (c) Some of the rays that produce the virtual image.

Figure 18-5 (a) Light traveling from glass into air at the lower surface bends away from the normal. (b) When the incident angle is larger than some critical angle, the light is totally reflected. None of the light passes through the surface.

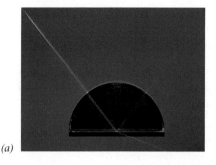
(a)

(b)

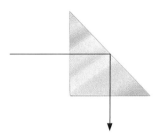

Figure 18-6 A prism acts like a flat mirror when the light is totally internally reflected.

90 degrees. Beyond this incident angle—called the **critical angle**—the light no longer leaves the material; the light is totally reflected as shown in Figure 18-5(b). This is called **total internal reflection**.

The critical angle can be found experimentally by increasing the incident angle and watching for the disappearance of the emerging ray. Because the graph in Figure 18-2 works for both directions, we can find the critical angle by looking for the angle of refraction for an incident angle of 90 degrees. The intersection of the curve with the right-hand edge of Figure 18-2 indicates that the critical angle for our glass is about 42 degrees. The critical angle for diamond is only 24 degrees.

Question   What is the critical angle for water?

Answer   The graph in Figure 18-2 shows that the angle of refraction in water never exceeds 49 degrees, so this is the critical angle.

This total internal reflection has many applications. For example, a 45-degree right prism can act as a mirror, as shown in Figure 18-6. If the incident angle of 45 degrees is greater than the critical angle, when the light beam hits the back surface, the beam is totally reflected. This reflecting surface has many advantages over ordinary mirrors. It doesn't have to be silvered, it is easier to protect than an external surface, and it is also more efficient for reflecting light.

Another application of this principle is to "pipe" light through long narrow fibers of solid plastic or glass, as shown in Figure 18-7. Light enters the fiber from one end. Once inside, the light doesn't escape out the side because the angle of incidence is greater than the critical angle. The rays finally exit at the end of the fiber because there the incident angles are smaller than the critical angle. Fiber-optic applications are found in photography, medicine, telephone transmissions, and even decorative room lighting.

Atmospheric Refraction

We live at the bottom of an ocean of air. Light that reaches us travels through this air and is modified by it. Earth's atmosphere is not uniform. Under most conditions the atmosphere's density decreases with increasing altitude. As you

Strands of glass optical fibers are used to carry voice, video, and data signals in telecommunication networks. Typical fibers have diameters of 60 micrometers.

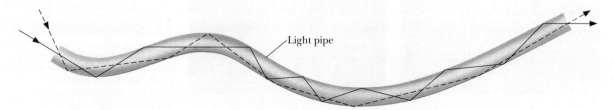

Figure 18-7 Light may be "piped" through solid plastic or glass rods using total internal reflection.

might guess, the index of refraction depends on the density of a gas because the less dense the gas, the more like a vacuum it becomes. We therefore conclude that the index of refraction of the atmosphere gradually decreases the higher we go.

Refraction occurs whenever there is any change in the index of refraction. When there is an abrupt change, as at the surface of glass, the change in the direction of the light is abrupt. But when the change is gradual, the path of a light ray is a gentle curve. The gradual increase in the index of refraction as light travels into the lower atmosphere means that light from celestial objects such as the Sun, Moon, and stars bends toward the vertical. Figure 18-8 shows that this phenomenon makes the object appear higher in the sky than its actual position. Astronomers must correct for atmospheric refraction to get accurate positions of celestial objects.

This shift in position is zero when the object is directly overhead and increases as it moves toward the horizon. Atmospheric refraction is large enough that you can see the Sun and Moon before they rise and after they set. Of course, without knowing where the Sun and Moon should be, you are not able to detect this shift in position. You can see, however, distortions in their shapes when they are near the horizon, as shown in the photographs in Figure 18-9. Because the amount of refraction is larger closer to the horizon, the apparent change in position of the bottom of the Moon is larger than the change at the top. This results in a shortening of the diameter of the Moon in the vertical direction and gives the Moon an elliptical appearance.

There are other changes in the atmosphere's index of refraction. Because of the atmosphere's continual motion, there are momentary changes in the density of local regions. Stars get their twinkle from this variation. As the air moves, the index of refraction along the path of the star's light changes, and the star appears to change position slightly and to vary in brightness and color—that is, to twinkle. Planets do not twinkle as much because they are close enough to Earth to appear as tiny disks. Light from different parts of the disk averages out to produce a steadier image.

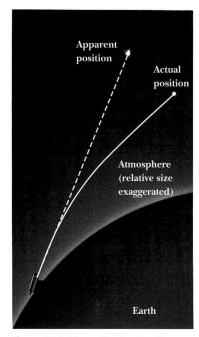

Figure 18-8 Atmospheric refraction changes the apparent positions of celestial objects, making them appear higher in the sky.

Dispersion

Although the ancients knew that jewels produced brilliant colors when sunlight shone on them, they were wrong about the origin of the colors. They thought the colors were part of the jewel. Newton used a prism to show that the colors don't come from jewels but rather from light itself—that the colors are already present in sunlight. When sunlight passes through a prism, the light refracts and

Figure 18-9 The atmospheric distortion of the setting Moon increases as the Moon gets closer to the horizon, giving the Moon an elliptical shape.

Mirages

Normal atmospheric conditions are altered when Earth's surface is very warm or very cold, producing some interesting optical effects. When the surface is cold, such as over cold water, the density of the atmosphere decreases more rapidly than usual. This bends light rays traveling near the surface so that even terrestrial objects shift in position. For example, an object floating on the water may appear in midair, an effect known as *looming*, shown in Figure A.

A layer of very warm air near the ground can cause the opposite effect. If the air near the ground is warmed, the index of refraction is low near the ground. This can cause the light near the ground to be bent upward. The combination of the direct and refracted light produces *mirages* like the one shown in Figure B. This is why motorists see "wet" spots on the road in the distance when the weather is particularly hot; the mirage of the sky looks like water on the road. Sometimes light from the part of an object near the ground, such as the base of the tree in Figure B, cannot reach your eyes. In this case the object appears to be floating in the air.

Mirages can often be seen along dark walls facing the Sun. Have a friend hold a bright object near the wall. Stand next to the wall about 10 meters from your friend and look with one eye. You may need to move your eye toward or away from the wall to see the mirage.

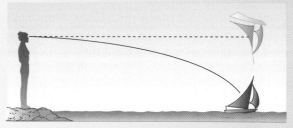

Figure A Looming occurs when the colder air near the surface causes the light rays to bend downward.

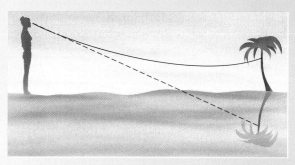

Figure B Mirages occur when the hot air near the surface causes the light rays to bend upward.

is split up into a spectrum of colors ranging from red to violet, a phenomenon known as **dispersion** (Figure 18-10). To eliminate the idea that the prism somehow produced the colors, Newton did two experiments. He took one of the colors from a prism and passed it through a second prism, demonstrating that no new colors were produced. He also recombined the colors and obtained white light. His experiments showed conclusively that white light is a combination of all colors. The prism just spreads them out so that the individual colors can be seen.

The name ROY G. BIV is a handy mnemonic for remembering the order of the colors produced by a prism or those in the rainbow: red, orange, yellow, green, blue, indigo, and violet. (Indigo is included mostly for the mnemonic; people can seldom distinguish it from blue or violet.)

The light changes direction as it passes through the prism because of refraction at the faces of the prism. Dispersion tells us that the colors have slightly different indexes of refraction in glass. Violet light is refracted more than red and therefore has a larger index. ("Blue bends better" is an easy way of remembering this.) The brilliance of a diamond is due to the small critical angle for internal reflection and the separation of the colors due to the high amount of dispersion.

Rainbows

Figure 18-10 A prism separates white light into the colors of the rainbow.

Sometimes after a rain shower, you get to see one of nature's most beautiful demonstrations of dispersion, a rainbow. Part of its appeal must be that it appears to come from thin air. There seems to be nothing there but empty sky.

In fact, rainbows result from the dispersion of sunlight by water droplets in the atmosphere. The dispersion that occurs as the light enters and leaves the droplet separates the colors that compose sunlight. You can verify this by making your own rainbow. Turn your back to the Sun and spray a fine mist of water

Figure 18-11 A rainbow's magic is that it seems to appear out of thin air. Notice the secondary rainbow on the right.

A rainbow formed in the spray from a sprinkler hose.

from your garden hose in the direction opposite the Sun. Each color forms part of a circle about the point directly opposite the Sun (Figure 18-11). The angle to each of the droplets along the circle of a given color is the same. Red light forms the outer circle and violet light the inner one. Figure 18-12 shows the paths of the red and violet light. The other colors are spread out between these two according to the mnemonic ROY G. BIV. Each droplet disperses all colors. Your eyes, however, are only in position to see one color coming from a particular droplet. For instance, if the droplet is located such that a line from the Sun to the droplet and a line from your eyes to the droplet form an angle of 42 degrees, the droplet appears red (Figure 18-13). If this angle is 40 degrees, the droplet appears violet. Intermediate angles yield other colors.

Whether or not you believe there is a pot of gold at the end of the rainbow, you will never be able to get there to find out. As you move, the rainbow "moves." In your new position, you see different droplets producing your rainbow.

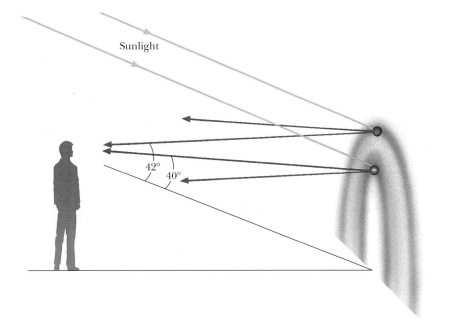

Figure 18-13 The color of each water droplet forming the rainbow depends on the viewing angle.

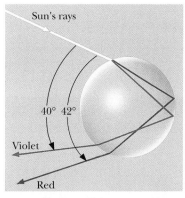

Figure 18-12 Dispersion of sunlight in a water droplet separates the sunlight into a spectrum of colors.

Flawed Reasoning

A friend calls you at 8:00 a.m. and tells you to go outside and observe a very beautiful rainbow in the east. **Would you hire this friend as a hiking guide?**

Answer Your friend has serious compass issues. The Sun comes up in the east. You see rainbows by looking away from the Sun. Indeed, the center of the rainbow will lie along a line passing through the Sun and your head. Therefore, at 8:00 a.m. you will see the rainbow in the west.

PHYSICS | **ON YOUR OWN**

Produce a "rainbow" by shining a flashlight at a quart jar filled with water. Covering the flashlight with an index card with a slit cut in it produces a narrow vertical beam that makes the colors more prominent.

If you are willing to get wet, it is possible to see a complete circular rainbow. Near noon on a sunny day, spray the space around you with a fine mist. Looking down, you will find yourself in the center of a rainbow. A circular rainbow can sometimes be seen from an airplane.

If viewing conditions are good, you can see a secondary rainbow that is fainter and larger than the first (Figure 18-11). It is centered about the same point, but the colors appear in the reverse order. This rainbow is produced by light that reflects twice inside the droplets.

Question If you see a rainbow from an airplane, where do you expect to see the shadow of the airplane?

Answer Because the center of the rainbow is always directly opposite the Sun, the shadow of the airplane will be at the center of the rainbow.

Halos

Sometimes a large halo can be seen surrounding the Sun or Moon. These halos and other effects, such as *sundogs* and various arcs, are caused by the refraction of light by ice crystals in the atmosphere.

Atmospheric ice crystals have the shape of hexagonal prisms. Each one looks like a slice from a wooden pencil that has a hexagonal cross section. Light hitting the crystal is scattered in many different directions, depending on the angle of incidence and which face it enters and exits. Light entering and exiting alternate faces, as shown in Figure 18-14, has a minimum angle of scatter of 22 degrees. Although light is scattered at other angles, most of the light concentrates near this angle.

To see a ray of light that has been scattered by 22 degrees, you must look in a direction 22 degrees away from the Sun. Light scattering this way from crystals randomly oriented in the atmosphere forms a 22-degree halo around the Sun, as shown in Figure 18-15. The random nature of the orientations ensures that at any place along the halo there will be crystals that scatter light into your eyes. Dispersion in the ice crystals produces the colors in the halo.

Occasionally one also sees "ghost" suns located on each side of the Sun at the same height as the Sun, as seen in Figure 18-15. Ice crystals that have vertically oriented axes produce these *sundogs*. These crystals can refract light into your eyes only when they are located along or just outside the halo's circle at the same altitude as the Sun.

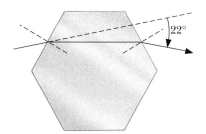

Figure 18-14 Light passing through alternate surfaces of a hexagonal ice crystal changes direction by at least 22 degrees.

Figure 18-15 A photograph of the 22-degree halo and its associated sundogs taken with a fisheye lens. A building was used to block out the Sun's direct rays.

An even larger but dimmer halo at 46 degrees exists but is less frequently seen. It is formed by light passing through one end and one side of the crystals. Other effects are produced by light scattering through other combinations of faces in crystals with particular orientations.

Lenses

When light enters a material with entrance and exit surfaces that are not parallel, unlike a pane of glass, the direction of the light beam changes. Two prisms and a rectangular block can be used to focus light, as shown in Figure 18-16. However, most other rays passing through this combination would not be focused at the same point. The focusing can be improved by using a larger number of blocks or by shaping a piece of glass to form a lens.

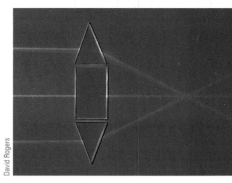

Figure 18-16 Two prisms and a rectangular block form a primitive lens.

We see the world through lenses. This is true even for those of us who don't wear glasses, because the lenses in our eyes focus images on our retinas. Other lenses extend our view of the Universe—microscopes for the very small and telescopes for the very distant.

Although many lens shapes exist, they can all be put into one of two groups: those that converge light and those that diverge light. If the lens is thicker at its center than at its edge, as in Figure 18-17(a), it is a converging lens. If it is thinner at the center, as in Figure 18-17(b), it is a diverging lens.

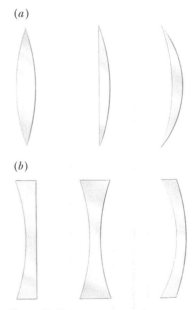

Figure 18-17 (a) Converging lenses are thicker at the middle. (b) Diverging lenses are thinner at the middle.

Question Lenses in eyeglasses are made with one convex surface and one concave surface. How can you tell if the lenses are converging or diverging?

Answer Check to see if they are thicker at the center than at the edges. If they are thicker at the center, they are converging.

Lenses have two focal points—one on each side. A converging lens focuses incoming light that is parallel to its **optic axis** at a point on the other side of the lens known as the *principal* **focal point** (Figure 18-18). The distance from the center of the lens to the focal point is called the **focal length**. We can find the other focal point by reversing the direction of the light and bringing it in from the right-hand side of the lens. The light then focuses at a point on the left-hand side of the lens that we refer to as the "other" focal point in drawing ray diagrams.

For a diverging lens, incoming light that is parallel to the optic axis appears to diverge from a point on the same side of the lens (Figure 18-19). This point is known as the principal focal point, and the focal point on the other side is known as the "other" one. You can show by experiment that the two focal points are the same distance from the center of the lens if the lens is thin. A lens is considered to be thin if its thickness is very much less than its focal length.

The shorter the focal length, the "stronger" the lens; that is, the lens focuses light parallel to the optic axis at a point closer to the lens.

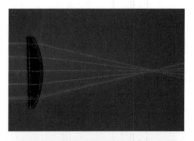

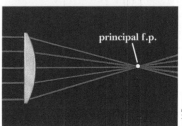

Figure 18-18 Light parallel to the optic axis of a converging lens is focused at the principal focal point.

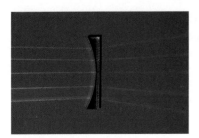
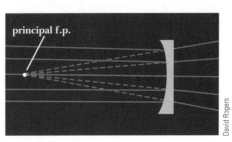

Figure 18-19 Light parallel to the optic axis of a diverging lens appears to come from the principal focal point.

PHYSICS | **ON YOUR OWN**

Determine the focal length of a magnifying glass or other converging lens by focusing sunlight to the smallest possible spot. Turn the lens around to see if you get the same distance on the other side.

Images Produced by Lenses ✓ MATH

The same ray-diagramming techniques used for curved mirrors in the previous chapter will help us locate the images formed by lenses. Again, three of the rays are easily drawn without measuring angles. The intersection of any two determines the location of the image. Figure 18-20 shows the three rays.

First, a ray passing through the center of the lens continues without deflection. Next, for a converging lens, a ray parallel to the optic axis passes through the principal focal point. And third, a ray coming from the direction of the other focal point leaves the lens parallel to the optic axis (Figure 18-20[a]). (The optic axis passes through the center of the lens and both focal points.) Notice that the second and third rays are opposites of each other. For a diverging lens, the second ray comes in parallel to the optic axis and leaves as if it came from the principal focal point, and the third ray heads toward the other focal point and leaves parallel to the optic axis (Figure 18-20[b]).

These rays are very similar to the ones used for mirrors. There are two main differences: the first ray passes through the center of the lens and not the center of the sphere as it did for mirrors, and there are now two focal points instead of one. We can still give abbreviated versions of these rules (the words in parentheses refer to diverging lenses).

rays for lenses ➤

1. Through center—continues
2. Parallel to optic axis—through (from) principal focal point
3. Through (toward) other focal point—parallel to optic axis

These rules assume that the lens is thin. The first rule neglects the offset that takes place when a light ray passes through parallel surfaces of glass at other than normal incidence (see Figure 18-1). For the purposes of drawing these rays, the bending of the light is assumed to take place at a plane perpendicular to the optic axis and through the center of the lens. A vertical dashed line indicates this plane.

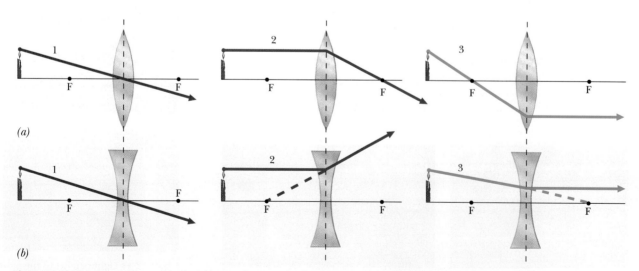

Figure 18-20 The three rays used in drawing ray diagrams for (a) converging and (b) diverging lenses.

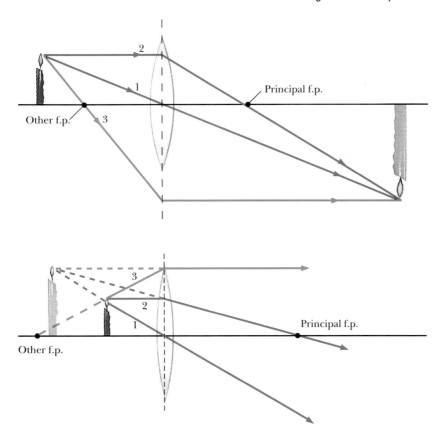

Figure 18-21 The image formed by a candle located outside the focal point of a converging lens is real, inverted, and may be magnified or reduced in size.

Figure 18-22 The image formed by a candle located inside the focal length of a converging lens is virtual, erect, and magnified.

We can apply these rays to locate the image of a candle that is located on the optic axis outside the focal point of a converging lens. The ray diagram in Figure 18-21 shows that the image is located on the other side of the lens and is real and inverted. (See the previous chapter for a discussion of the types of image.) Whether the image is magnified depends on how far it is from the focal point. As the candle is moved away from the lens, the image moves closer to the principal focal point and gets smaller.

If the candle is moved inside the focal point as illustrated in Figure 18-22, the image appears on the same side of the lens. This is the arrangement that is used when a converging lens is used as a magnifying glass. The lens is positioned such that the object is inside the focal point, producing an image that is virtual, erect, and magnified.

A diverging lens always produces a virtual image as shown in Figure 18-23. The image changes location and size as the object is moved, but the image remains erect and virtual.

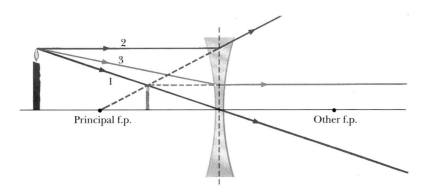

Figure 18-23 The image formed by a diverging lens is virtual, erect, and reduced in size.

Question Is the lens used in a slide projector converging or diverging?

Answer It must be converging because it forms a real image on the screen.

Notice that one of the rays in Figure 18-21 does not pass through the lens. This isn't a problem because there are many other rays that do pass through the lens to form the image. Ray diagramming is just a geometric construction that allows you to locate images, a process that can be illustrated with an illuminated arrow and a large-diameter lens as drawn in Figure 18-24. A piece of paper at the image's location allows the image to be easily seen. If the lens is then covered with a piece of cardboard with a hole in it, the image is still in the same location, is the same size, and is in focus. The light rays from the arrow that form the image are those that pass through the hole. The image is not as bright because less light now forms the image. The orange lines illustrate the paths of some of the other rays.

Flawed Reasoning

The following question appears on the final exam: "Three long light filaments are used to make a Y that is placed in front of a large converging lens such that it creates a real image on the other side of the lens. The meeting point of the three filaments lies on the optic axis of the lens. A piece of cardboard is then used to cover up the bottom half of the lens. Describe what happens to the image of the Y."

Three students give their answers:

Jacob: "The cardboard will block the light from the lower filament, so the image will appear as a V."

Emily: "The real image formed by a converging lens is inverted. The image would now appear to be an upside-down V."

Michael: "The image is inverted, so the light from the lower filament must pass through the top half of the lens and the light from the upper two filaments will be blocked by the cardboard. The image will appear as an I."

All three students have answered incorrectly. **Find the flaws in their reasoning.**

Answer A point source of light sends light to all parts of the lens's surface. This light converges at a single point on the other side of the lens (the image location). Covering half the lens blocks half the light, but the other half still forms an image at the same location. The three long filaments can be thought of as a collection of many point sources. They still form the same image (an upside-down Y). The image will be dimmer because half the light is blocked.

Figure 18-24 The orange rays form the image. The black ones are used because they are easy to draw.

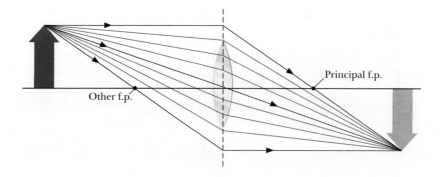

Cameras

We saw in the last chapter that pinhole cameras produce sharp images if the pinhole is very small. The amount of light striking the film, however, is quite small. Very long exposure times are needed, which means that the objects in the scene must be stationary. The amount of light reaching the film can be substantially increased (and the exposure time substantially reduced) by using a converging lens instead of a pinhole.

The essential features of a simple camera are shown in Figure 18-25. This camera has a single lens at a fixed distance from the film. The distance is chosen so that the real images of faraway objects are formed at, or at least near, the film. These cameras are usually not very good for taking close-up shots, such as portraits, because the images are formed beyond the film and are therefore out of focus at the film. More expensive cameras have an adjustment that moves the lens relative to the film to position (focus) the image on the film.

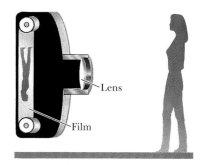

Figure 18-25 The essential features of a simple camera.

Question If the focal length of the lens in a simple camera is 50 millimeters, how far is it from the lens to the film for a subject that is very far from the camera?

Answer If the objects are effectively at infinity, the light from each point will be focused at a distance equal to the focal length. Thus, the film should be about 50 millimeters from the center of the lens.

Ideally, all light striking the lens from a given point on the object should be focused to a given point on the film. However, real lenses have a number of defects, or **aberrations**, so that light is not focused to a point but is spread out over some region of space.

A lens cannot focus light from a white object to a sharp point because of dispersion. A converging lens focuses violet light at a point closer to the lens than it does red light. This chromatic aberration produces images with colored fringes. Because the effect is reversed for diverging lenses and the amount of dispersion varies with material, lens designers minimize chromatic aberration by combining converging and diverging lenses made of different types of glass.

A spherical lens (or a spherical mirror for that matter) does not focus all light parallel to the optic axis to a sharp point. Light farther from the optic axis is focused at a point closer to the lens than light near the optic axis. Using a combination of lenses usually corrects this spherical aberration; using a diaphragm to decrease the effective diameter of the lens also reduces it. Although this sharpens the image, it also reduces the amount of light striking the film. New techniques for reducing spherical aberration by grinding lenses with nonspherical surfaces and by making lenses in which the index of refraction of the glass changes with the distance from the optic axis have been developed.

Our Eyes

Leonardo da Vinci stated in the 15th century that the lens of an eye forms an image inside the eye that is transmitted to the brain. He believed that this image must be upright. It was a century before it was shown that he was half right: The lens forms an image inside the eye, but the image is upside down. The inverted nature of the image was demonstrated by removing the back of an excised animal's eye and viewing the image. The inverted world received by our retinas is interpreted as right-side up by our eye–brain system. The essential features of this remarkable optical instrument include the cornea, the lens, and some fluids,

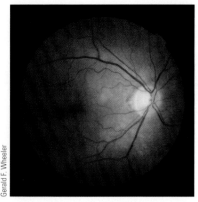

The retina of a human eye.

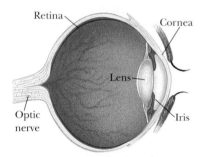

Figure 18-26 Schematic drawing of the human eye.

	Sphere	Cylinder	Axis
R	−6.50	+3.25	089
L	−5.75	+2.75	074
R	+2.00		
L	+2.00	Bifocals	

The prescription for your author's eyeglasses. The spherical and cylindrical corrections are given in diopters. Axis is the number of degrees the axis of the cylinder is rotated from the vertical. The bifocal correction is added to the others.

Figure 18-27 A test pattern for astigmatism. If you see some lines blurred while other lines are sharp and dark, you have some astigmatism.

which act collectively as a converging lens to form real, inverted images on the retina (Figure 18-26).

PHYSICS | ON YOUR OWN

You can observe that the real image formed on the retina of your eye is inverted. Punch a pinhole in an index card and hold it about 15 centimeters (6 inches) in front of one eye while looking at a bright light source. Hold the head of the pin in line with the hole and midway between the hole and your eye. You should be able to see the inverted shadow of the pinhead on your retina.

When you look at a distant object, nearby objects are out of focus. Only distant objects form sharp images on the surface of the retina. The nearby objects form images that would be behind the retina, and the images on the retina are therefore fuzzy. This phenomenon occurs because the locations of images of objects at various distances depend on the distances between the lens and the objects and on the focal length of the lens. The lens in the eye changes its shape and thus its focal length to accommodate the different distances.

Opticians measure the strength of lenses in *diopters*. The lens strength in diopters is equal to the reciprocal of the focal length measured in meters. For example, a lens with a focal length of 0.2 meter is a 5-diopter lens. In this case a larger diopter value means that the lens is stronger. Converging lenses have positive diopters, and diverging lenses have negative diopters. Diopters have the advantage that two lenses placed together have a diopter value equal to the sum of the two individual ones.

In the relaxed eye of a young adult who does not wear corrective lenses, all the transparent materials have a total "power" of +60 diopters. Most of the refraction (+40 diopters) is due to the outer element of the eye, the cornea, but the relaxed lens contributes +20 diopters.

The eye can vary the strength of the lens from a relaxed value of +20 diopters to a maximum of +24 diopters. When the relaxed eye views a distant object, the +60 diopters produce an image at 1.7 centimeters (0.7 inch), which is the distance to the retina in a normal eye. The additional +4 diopters allow the eye to view objects as near as 25 centimeters (10 inches) and still produce sharp images on the retina.

The ability of the eye to vary the focal length of the lens decreases with age as the elasticity of the lens decreases. A 10-year-old eye may be able to focus as close as 7 centimeters (+74 diopters), but a 60-year-old eye may not be able to focus any closer than 200 centimeters ($6\frac{1}{2}$ feet). An older person often wears bifocals when the eyes lose their ability to vary the focal length.

The amount of light entering the eye is regulated by the size of the pupil. As with the ear, the range of intensities that can be viewed by the eye is very large. From the faintest star that can be seen on a dark, clear night to bright sunlight is a range of intensity of approximately 10^{10}.

PHYSICS | ON YOUR OWN

Another common visual defect is *astigmatism*. When some of the refracting surfaces are not spherical, the image of a point is spread out into a line. Use the pattern in Figure 18-27 to check for astigmatism in your eyes. Lines along the direction in which images of point sources are spread remain sharp and dark, but the others become blurred. Are your two eyes the same?

Eyeglasses

Our optical system is quite amazing. The lens in our eye can change its shape, altering its focal length to place the image on the retina. Sometimes, however, the eye is too long or too short, and the images are formed in front of or behind the retina. When the eye is too long, the images of distant objects are formed in front of the retina, as shown in Figure A. Such a person has *myopia* (nearsightedness) and can see things that are close but has trouble seeing distant objects. When the eye is too short, the person has *hyperopia* (farsightedness) and has trouble seeing close objects. Distant objects can be imaged on the retina, but close objects form images behind the retina (Figure B).

Our knowledge of light and refraction allows us to devise instruments—eyeglasses—that correct these deficiencies. The nearsighted person wears glasses with diverging lenses to see distant objects (Figure A). The farsighted person's sight is corrected with converging lenses (Figure B).

Even those people with perfect vision early in life lose some of the lens's range, particularly the ability to shorten the focal length. These people have trouble creating a focused image on their retina for objects that are close. Many older people wear bifocals when their eyes lose the ability to shorten the focal length. The upper portion of the lens is used for distant viewing, while the lower portion is used for close work or reading. When people work at intermediate distances such as looking at computer screens, they sometimes wear trifocals.

The difficulty of making glasses to correct vision is increased when a person has astigmatism. This occurs when some of the refracting surfaces are not spherical and the image of a point is spread out into a line. This visual defect is corrected by adding a cylindrical curvature to the spherical curvature of the lens. You can check your (or a friend's) glasses for correction for astigmatism by looking through them in the normal way but holding the glasses at a distance from your head. When you rotate the lens about a horizontal axis, you will see background distortion if the lens has an astigmatic correction.

Many people wear contact lenses to correct their vision. The use of contact lenses has created some interesting challenges in correcting for astigmatism and the need to wear bifocals. When correcting for astigmatism, it is very important that the cylindrical correction has the correct orientation. Contact lenses can be weighted at one place on the edge to keep the lens oriented correctly.

People who need bifocals can sometimes be fitted with different corrections in each eye; one eye is used for close vision and the other is used for distant vision. The eye–brain system switches from one eye to the other as the situation demands.

Finally, a medical procedure uses laser technology to correct nearsightedness. In this procedure, the laser is used to make radial cuts in the cornea of the eye to reduce its curvature.

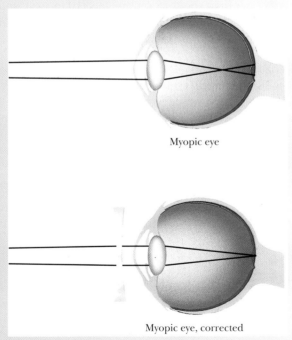

Figure A The myopic eye forms the images of distant objects in front of the retina. This is corrected with diverging lenses.

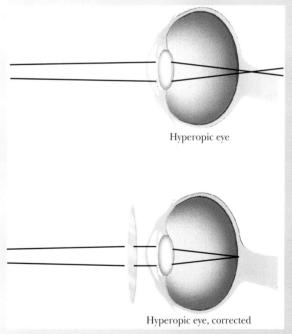

Figure B The hyperopic eye forms images of close objects in back of the retina. This is corrected with converging lenses.

Figure 18-28 A candle viewed (a) at a distance of 25 centimeters and (b) through a magnifying glass.

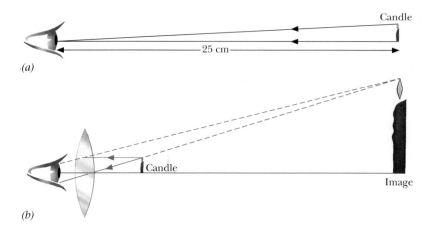

Magnifiers

It has been known since the early 17th century that refraction could bend light to magnify objects. The invention of the telescope and microscope produced images of regions of the Universe that until then had been unexplored. Galileo used the newly discovered telescope to see Jupiter's moons and the details of our Moon's surface. English scientist Robert Hooke spent hours peering into another unexplored world with the aid of the new microscope.

The size of the image on the retina depends on the object's physical size and on its distance away. The image of a dime held at arm's length is much larger than that of the Moon. What really matters is the angular size of the object—that is, the angle formed by lines from your eye to opposite sides of the object. The angular size of an object can be greatly increased by bringing it closer to your eye. However, if you bring it closer than about 25 centimeters (10 inches), your eye can no longer focus on it, and its image is blurred. You can get both an increased angular size and a sharp image by using a converging lens as a magnifying glass. When the object is located just inside the focal point of the lens, the image is virtual, erect, and has nearly the same angular size as the object. Moreover, as shown in Figure 18-28, the image is now far enough away that the eye can focus on it and see it clearly.

An even higher magnification can be achieved by using two converging lenses to form a compound microscope, as shown in Figure 18-29. The object is located just outside the focal point of the objective lens. This lens forms a real image that is magnified in size. The eyepiece then works like a magnifying glass to further increase the angular size of this image.

Figure 18-29 Schematic of a compound microscope.

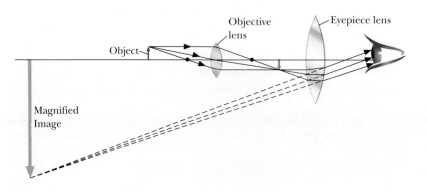

PHYSICS | **ON YOUR OWN**

Place a drop of water on a piece of transparent plastic food wrap and use it to magnify the print in this text. Do drops of other clear liquids yield higher magnifications?

Telescopes

There are many varieties of telescope. A simple one using two converging lenses is known as a **refracting telescope**, or refractor. Figure 18-30 shows that this type of telescope has the same construction as a compound microscope except that now the object is far beyond the focal point of the objective lens. Like the microscope the refractor's objective lens produces a real, inverted image. Although the image is much smaller than the object, it is much closer to the eye. The eyepiece acts like a magnifying glass to greatly increase the angular size of the image. The magnification of a telescope is equal to the ratio of the focal lengths of the objective lens and the eyepiece. To get high magnification, the focal length of the objective lens needs to be quite long.

Binoculars were designed to provide a long path length in a relatively short instrument. The diagram in Figure 18-31 shows that this is accomplished by using the internal reflections in two prisms to fold the path.

Large-diameter telescopes are desirable because they gather a lot of light, al-

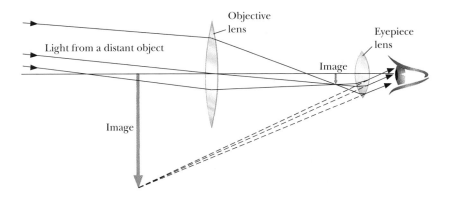

Figure 18-30 Schematic of an astronomical telescope.

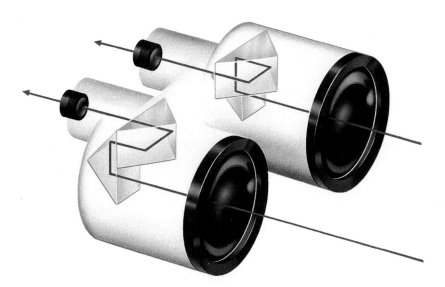

Figure 18-31 Schematic of prism binoculars.

The Hubble Space Telescope

At a cost of $1.6 billion, the Hubble Space Telescope was placed in an orbit 575 kilometers (357 miles) above Earth by the space shuttle *Discovery* in April 1990. Its 15-year-long mission is to provide astronomers with observations of the Universe without the disturbances caused by Earth's atmosphere. The atmosphere absorbs most of the radiation reaching us from space, except for two broad bands in the radio region and around the visible region. A variety of experiments were planned that ranged from viewing distant, faint objects to accurately measuring the positions of stars. The design specifications indicated that the Hubble Space Telescope would be able to see objects seven times as far as could be observed from Earth's surface.

However, these experiments were seriously hampered by a defect in the telescope's primary mirror. Although the error in its shape was only 0.002 millimeter (about one-fortieth the thickness of a human hair) at the edge of the 2.4-meter diameter mirror, it caused light from the edge of the mirror to focus 38 millimeters beyond light from the center. This left the telescope with an optical defect—spherical aberration—that created fuzzy halos around images of stars, and blurred images of extended objects like galaxies and giant clouds of gas and dust. Only 15% of the light from a star was focused into the central spot compared with the design value of 70%. However, because this type of aberration is well known, computer enhancement was used to sharpen images of brighter objects, producing some rather remarkable views and some very good science, but the technique did not work for faint objects.

Fortunately, the Hubble Space Telescope was designed to be serviced by the space shuttles, and new optics were designed to compensate for the error. In December 1993 the space shuttle *Endeavour* docked with the Hubble Space Telescope to repair the optics and replace a number of mechanical and electric components that had failed or required scheduled replacement. The repairs required five space walks involving two astronauts each. The total repair mission had a cost of $700 million. The repaired Hubble Space Telescope is able to see much farther into space.

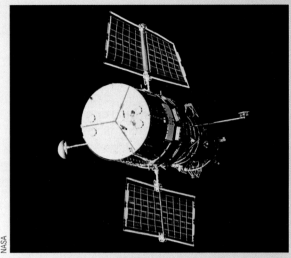

The Hubble Space Telescope.

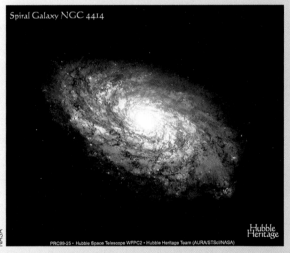

Careful study of images of this spiral galaxy (NGC 4414) taken by the Hubble Space Telescope allowed astronomers to determine that it is 60 million light-years from Earth.

lowing us to see very faint objects or to shorten the exposure time for taking pictures. The problem, however, is making a large-diameter glass lens. It is very difficult, if not impossible, to make a piece of glass of good enough quality. Also, a lens of this diameter is so thick that it sags under its own weight. Therefore, most large telescopes are constructed with concave mirrors as objectives and are known as **reflecting telescopes**, or reflectors. The use of a concave mirror to focus the incoming light has several advantages: the construction of a mirror requires grinding and polishing only one surface rather than two; a mirror can be supported from behind; and finally, mirrors do not have the problem of chromatic aberration. Figure 18-32 illustrates several designs for reflecting telescopes.

The world's largest refractor has a diameter of 1 meter (40 inches), whereas the largest reflector has a diameter of 6 meters (236 inches). This is just about the limit for a telescope with a single objective mirror; the costs and manufacturing difficulties are not worth the gains. Telescope makers have recently built tele-

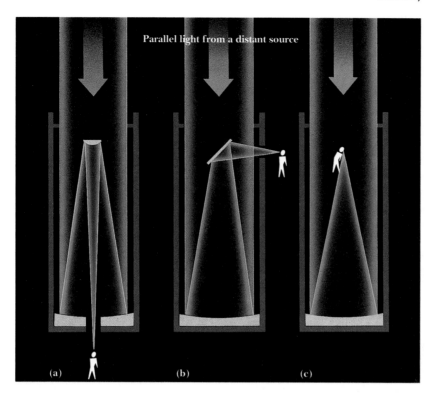

Figure 18-32 Schematics of a (a) Cassegrain reflector, (b) Newtonian reflector, and (c) prime-focus telescope.

scopes in which the images from many smaller mirrors are combined to increase the light-gathering capabilities.

Summary

When light strikes a transparent material, part of it reflects and part refracts. The amount of refraction depends on the incident angle and the index of refraction of the material. Light entering a material of higher index of refraction bends toward the normal. Because the refraction of light is a reversible process, light entering a material with a smaller index of refraction bends away from the normal. For light in a material with the larger index of refraction, total internal reflection occurs whenever the angle of incidence exceeds the critical angle.

The refraction of light at flat surfaces causes objects in or behind materials of higher indexes of refraction to appear closer, and therefore larger. The apparent locations of celestial objects are changed by refraction in the atmosphere.

White light is separated into a spectrum of colors because the colors have different indexes of refraction, a phenomenon known as dispersion. Rainbows are formed by dispersion in water droplets. Each color forms part of a circle about the point directly opposite the Sun. Halos are caused by the refraction of sunlight in ice crystals.

Ray diagrams can be used to locate the images formed by lenses. The rays are summarized by the following rules: (1) through center—continues; (2) parallel to optic axis—through (from) principal focal point; and (3) through (toward) other focal point—parallel to optic axis.

Cameras and our eyes contain converging lenses that produce real, inverted images. Converging lenses can be used as magnifiers of objects located inside the focal points. Lenses can be combined to make microscopes and telescopes.

Physics⚛Now™ Assess your understanding of this chapter's topics with sample tests and other resources found by logging into PhysicsNow at **http://physics.brookscole.com/kf6e**.

Chapter 18 Revisited

The most obvious consequence of the passage of light between two transparent materials is that the direction of the light changes at the interface. This may produce virtual images that are closer, making the fish look bigger and the tree look bent. It is possible for light to get "trapped" if it is in the material with the larger index of refraction and if the angle of incidence is larger than the critical angle.

KEY TERMS

aberration: A defect in a mirror or lens causing light rays from a single point to fail to focus at a single point in space.

critical angle: The minimum angle of incidence for total internal reflection to occur.

dispersion: The spreading of light into a spectrum of colors.

focal length: The distance from the center of a lens to its focal point.

focal point: The location at which a lens focuses rays parallel to the optic axis or from which such rays appear to diverge.

index of refraction: An optical property of a substance that determines how much light bends upon entering or leaving it.

optic axis: A line passing through the center of a lens and both focal points.

reflecting telescope: A type of telescope using a mirror as the objective.

refracting telescope: A type of telescope using a lens as the objective.

refraction: The bending of light that occurs at the interface between transparent media.

total internal reflection: A phenomenon that occurs when the angle of incidence of light traveling from a material with a higher index of refraction into one with a lower index of refraction exceeds the critical angle.

CONCEPTUAL QUESTIONS

1. A narrow beam of light emerges from a block of glass in the direction shown in the figure. Which arrow best represents the path of the beam within the glass?

   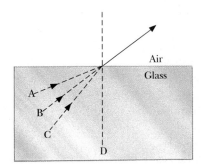

2. A mirror is lying on the bottom of a fish tank that is filled with water. If IN represents a light ray incident on the top of the water, which possibility in the figure best represents the outgoing ray?

   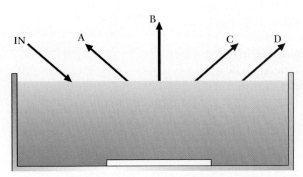

3. You place a waterproof laser and a glass prism flat on the bottom of an empty aquarium, as shown in the figure. The light leaving the prism follows path B. If you filled the aquarium with water, which path would the light leaving the prism now follow?

   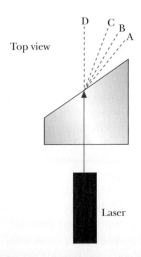

4. Figure 18-2 shows the refraction curves for air, water, and glass. If we were to draw the curve for diamond, would it appear above or below the curve for glass? Why?

5. Why do clear streams look so shallow?

6. Suppose you are lying on the bottom of a swimming pool looking up at a ball that is suspended 1 meter above the surface of the water. Does the ball appear to be closer, farther away, or still 1 meter above the surface? Explain.

7. Is the critical angle greater at a water–air surface or a glass–air surface?
8. For what range of incident angles is light totally reflected at a water–air surface?
9. Telephone companies are using "light pipes" to carry telephone signals between various locations. Why does the light stay in the pipe?
10. There is a limit to how much a fiber-optic cable can be bent before light "leaks" out because bending the pipe allows light to strike the surface at angles less than the critical angle. If you were laying fiber-optic cable under water instead of in air, would this be a greater or lesser problem? Why?
11. The distance from Earth to Mars varies from 48 million miles to 141 million miles as the two planets orbit the Sun. At which distance would Mars appear to twinkle more? Why?
12. We observe stars twinkle when viewed from Earth, but astronauts in a space shuttle do not observe this. Why?
▲ 13. As you look toward the west, you see two stars one above the other with a 5-degree separation. As the two stars move closer to the western horizon, will their apparent separation increase, decrease, or stay the same? Why?
▲ 14. In the absence of an atmosphere, a star moves across the sky from horizon to horizon at a constant speed. How does the star appear to move in the presence of an atmosphere?
15. How is the time of sunrise affected by atmospheric refraction?
16. How does the presence of an atmosphere affect the length of day and night?
17. How does the refraction of light in the atmosphere affect the appearance of the Sun or Moon as it approaches the horizon?
18. During a total lunar eclipse, the Moon lies entirely within Earth's umbra and yet is still faintly visible. If Earth lacked an atmosphere, the Moon would not be visible at all. Explain how Earth's atmosphere allows the Moon to be seen during an eclipse.
19. You are trying to spear fish from a boat, and you spy a fish about 2 meters from the boat. Should you aim high, low, or directly at the fish? Why?
▲ 20. If you were going to send a beam of light to the Moon when it is just above the horizon, would you aim high, low, or directly at the Moon? Explain.
21. Does a beam of white light experience dispersion as a result of reflecting from a mirror? Explain.
22. Why is a diamond more brilliant than a clear piece of glass having the same shape?

▲ 23. A fiber-optic cable is used to transmit yellow light. At one sharp bend, the incident angle is exactly at the critical angle for yellow light. If the pipe were bent any more, the yellow light would "leak" out. If the cable were used with blue light, would this bend be a problem? What if it were used with red light? Explain.
24. While transmitting white light down a fiber-optic cable, you bend the cable too much in one place, and some of the light "leaks" out. Which is the first color of light to "leak" out?
25. On a hot day, travelers in the desert may "see" a pond even though there is nothing but sand. Draw a ray diagram showing how this can happen.

26. Sometimes mariners report seeing an island floating in the air. Draw a ray diagram showing how this can happen.
27. If your line of sight to a water droplet makes an angle of 41 degrees with the direction of the sunlight, what color would the raindrop appear to be?
▲ 28. You are looking at a rainbow from the ground floor of an apartment building and notice that a kite is hovering right in the green portion. If you were to go up to the second floor, would you be likely to see the kite hovering in the red portion or the blue portion of the rainbow? Explain.
29. Why is a shadow of your head always in the center of a rainbow?
30. If you were flying in an airplane and saw a rainbow at noon, what shape would it be and where would you be looking?
▲ 31. At what time of day might you expect to see the top of a rainbow set below the horizon? In what direction would you look to see it?
▲ 32. At what time of day and in what direction would you look to see the top of a rainbow rise above the horizon?
33. To produce a hologram, a narrow beam of laser light must be spread out enough to expose the surface of a piece of film. What type of lens would accomplish this? Explain your reasoning.
34. What type of lens would be helpful in starting a campfire? Why?
35. You find a converging lens in the storeroom and wish to determine its focal length. Describe how you could use two lasers to accomplish this.
36. You place a laser at the principal focus of a converging lens and aim the beam toward any part of the lens. Describe the beam's path after passing through the lens.

37. You find a diverging lens in the storeroom and wish to determine its focal length. Describe how you could use two lasers to accomplish this.
38. You wish to use a diverging lens to redirect the light from six lasers to produce beams that are parallel to the optic axis of the lens. How should you aim the lasers to accomplish this?
39. What type of lens would you use to construct an overhead projector? Explain your reasoning.
40. Can a prism be used to form an image? Explain.
41. Where does a ray arriving parallel to the optic axis of a converging lens go after passing through the lens?
42. What path does a ray take after passing through a converging lens if it passes through a focal point before it enters the lens?
43. A converging lens is used to form a sharp image of a candle. What effect does covering the upper half of the lens with paper have on the image?
44. How does covering all but the center of a lens affect the image of an object?
45. Two converging lenses with identical shapes are made from glasses with different indexes of refraction. Which one has the shorter focal length? Why?
46. You are building a device in which the dimensions of a diverging lens cannot be changed. Upon testing the device, however, you discover that the focal length of your lens is too short. You decide to fix this problem by grinding a lens of identical shape from different glass. Should you use a glass with a larger or a smaller index of refraction? Why?
▲ 47. Consider the image of a candle as shown in Figure 18-21. Explain why you would not be able to see the image if your eye were located to the left of the image location.
▲ 48. Consider the image of a candle as shown in Figure 18-21. To clearly see the image, should you locate your eye at the image location or to the right of the image location? Explain.
49. What kind of image is formed by the retina of an eye?
50. How might you convince a friend that the image formed by a camera is a real image?
51. What is the purpose of the pupil in the eye?
52. What is the purpose of a diaphragm in a camera?
53. Why does a telescope that uses a mirror to focus the light not exhibit chromatic aberration whereas a telescope that uses a lens to focus the light does?
54. You measure the focal length of a converging lens by finding the crossing point of parallel beams of red laser light. If you measure the focal length using lasers that emit green light, would the measured value be greater than, less than, or the same as before? Why?
55. Are reading glasses used by older people converging or diverging? Explain.
56. Without glasses, the light entering our eyes comes directly from the object. With glasses, the light comes from the image formed by the glasses. Is this image real or virtual? Explain.
57. When a person has cataract surgery, the lens of the eye is removed and replaced with a plastic one. Would you expect this lens to be converging or diverging? Explain.
58. Stamp and coin collectors often wear special glasses that allow them to see the details of the stamps and coins. Are the lenses in these glasses converging or diverging? Why?
59. The ray diagram for a magnifying glass is shown in Figure 18-22. As the object is moved toward the focal point, the direction of emerging ray 2 does not change whereas the direction of emerging ray 1 does. By looking at where these rays now intersect, determine whether the image gets larger or smaller as the object is moved closer to the focal point.
60. If the magnifying lens in Figure 18-22 were replaced with a lens of shorter focal length, the direction of emerging ray 1 would not change whereas the direction of emerging ray 2 would. By looking at where these rays would now intersect, determine whether the magnification would be increased or decreased.
▲ 61. The figure shows the words MAGNESIUM DIOXIDE viewed through a solid plastic rod. Why does MAGNESIUM appear upside down, while DIOXIDE appears right-side up?

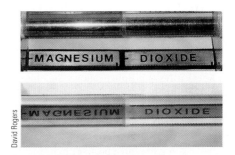

▲ 62. When a single converging lens is used to focus white light, the image has a colored fringe due to chromatic aberration. Describe the changes in the color of the fringe as a screen is moved through the focal point.

EXERCISES

1. If light in air is incident at 60°, at what angle is it refracted in water? In glass?
2. Light in air is incident on a surface at an angle of 30°. What is its angle of refraction in glass? In water?
▲ 3. You are spear fishing in waist-deep water when you spot a fish that appears to be down at a 45° angle. You recognize that the light coming from the fish to your eye has been refracted and you must therefore aim at some angle below the apparent direction to the fish. What is this angle?
4. Light from the bottom of a swimming pool is incident on the surface at an angle of 30°. What is the angle of refraction?
5. Use Figure 18-2 to estimate the critical angle for glass with an index of refraction of 1.6.
6. Use Figure 18-2 to estimate the angle of refraction for light in air incident at 50° to the surface of glass with an index of refraction of 1.6.
7. A prism made of glass with an index of refraction of 1.5 has the shape of an equilateral triangle. A light ray is in-

cident on one face at an angle of 48°. Use a protractor and Figure 18-2 to find the path through the prism and out an adjacent side. What is the exit angle?

8. A prism made of plastic with an index of refraction of 1.33 has the shape of a cube. A light ray is incident on one face at an angle of 70°. Use a protractor and Figure 18-2 to find the path through the prism and out an adjacent side. What is the exit angle?

9. You are scuba diving below a fishing pier. You look up and see a fishing pole that appears to be 2 m above the surface of the water. Use a ray diagram to show that the pole is actually closer to the surface of the water.

▲ 10. Using data from Figure 18-2 to determine the exact angles, redraw Figure 18-4(c) to locate the image of something sitting on the bottom of a pond. Use your scale diagram and a ruler to show that an object in water appears to be about $\frac{3}{4}$ as deep as it actually is. (Note that the index of refraction for water is $\frac{4}{3}$.)

11. An object is located midway between the other focal point and the center of a converging lens. Draw a ray diagram showing how you locate the image. Estimate the magnification of the image from your diagram.

12. Use a ray diagram to find the location and magnification of an object located three focal lengths from a converging lens.

13. Use a ray diagram to locate the image of an arrow placed 80 cm from a diverging lens with a focal length of 40 cm.

14. The focal length of a converging lens is 40 cm. Use a ray diagram to locate the image of an object placed 80 cm from the center of this lens.

▲ 15. Over what range of positions can an object be located so that the image produced by a converging lens is real and smaller than the object?

▲ 16. Over what range of positions can an object be located so that the image produced by a converging lens is real and magnified?

17. In Figure 18-21 a weakly diverging lens is inserted at the principal focal point of the converging lens. Use a ray diagram to show that this results in the real image being shifted to the right.

▲ 18. Use ray diagrams to show that a diverging lens cannot produce a magnified image of a real object.

19. Draw a ray diagram to find the image of an object located inside the focal point of a diverging lens. Is the image real or virtual? Erect or inverted? Magnified or reduced in size? Explain.

20. Where is the image of an object located at the focal point of a diverging lens? Is the image real or virtual? Erect or inverted? Magnified or reduced in size? Explain.

21. How many diopters are there for a converging lens with a focal length of 0.2 m?

22. If a lens has a focal length of 12.5 cm, how many diopters does it have?

23. A converging lens of focal length 20 cm is placed next to a converging lens of focal length 50 cm. What is the effective focal length for this combination?

▲ 24. What focal-length lens would you need to place next to a converging lens of focal length 25 cm to create an effective focal length of 20 cm for the combination?

25. A converging lens of focal length 50 cm is placed next to a diverging lens of length 25 cm. What is the effective focal length for this combination? Is it diverging or converging?

26. You ordered a converging lens of focal length 2 m, but the company delivered a lens whose focal length was only 1 m. You don't have time to wait for a replacement, so you decide to correct the problem by placing a diverging lens next to the lens they sent you. What must the focal length be for this diverging lens?

19 A Model for Light

Viewing a plastic model of a structure through crossed pieces of Polaroid filter shows the stress patterns in the structure. This photograph shows a model of the flying buttresses of the Notre Dame cathedral.

Light is one of our most common phenomena, and our knowledge of light is responsible for our understanding of the beautiful effects in soap bubbles and for technological advances into extending human perception. But, what is light?

(See page 397 for the answer to this question.)

IN the previous two chapters, we learned a great deal about light by simply observing how it behaves, but we did not ask, "What is light?" This question is easy to ask, but the answer is more difficult to give. For example, we can't just say, "Let's look!" What we "see" is the stimulation caused by light entering our eyes, not light itself. To understand what light is, we need to look for analogies, to ask ourselves what things behave like light. In effect, we are building a model of a phenomenon that we can't observe directly. This same problem occurs quite often in physics because many components of nature are not directly observable. In the case of light, we need a model that accounts for the properties that we studied in the previous two chapters. When we look at the world around us, we see that there are two candidates: *particles* and *waves*.

So the question becomes, does light behave as if it were a stream of particles or a series of waves? Newton thought that light was a stream of particles, but other prominent scientists of his time thought that it behaved like waves. Because of Newton's great reputation, his particle model of light was the accepted theory during the 18th century. However, many of the early observations could be explained by a particle *or* by a wave model. Scientists continually looked for new observations that could distinguish between the two theories. We will examine the experimental evidence to see whether light behaves like particles or waves. The process of deciding is as important as the answer.

PhysicsNow™ Test your understanding of this chapter by logging into PhysicsNow at **http://physics.brookscole.com/kf6e**, selecting the chapter, and clicking on the "Take a Pre-Test" link.

Reflection

We know from such common experiences as echoes and billiards that both waves and particles can bounce off barriers. But this fact is not enough. Our model for the behavior of light must agree with our conclusion that the angle of reflection is equal to the angle of incidence. A particle model for light can account for this if the reflecting surface is frictionless and perfectly elastic. A wave model of light also has no problem accounting for the law of reflection. The photograph in Figure 19-1(a) shows a straight wave pulse on the surface of water striking a smooth, straight barrier. The wave was initially moving toward the top of the photograph and is being reflected toward the right. We see that the angle between the incident wave and the barrier is equal to that between the reflected wave and the barrier. In the corresponding ray diagram (Figure 19-1[b]), we draw the rays perpendicular to the straight wave fronts—that is, in the direction the wave is moving.

Observing that light reflects from surfaces gives us no clues as to its true nature; both particles and waves obey the law of reflection.

(a)

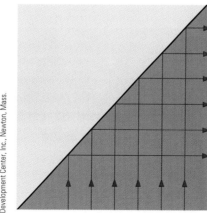

(b)

Figure 19-1 (a) The horizontal white line is a straight water-wave pulse moving toward the top of the picture. The part that has already hit the black barrier is reflected toward the right. (b) The corresponding ray diagram has rays perpendicular to the wave fronts and shows that the angle of reflection is equal to the angle of incidence.

✓ Extended presentation available in the *Problem Solving* supplement

Refraction

✓ MATH

We learned a great deal in the previous chapter about the behavior of light when it passes through boundaries between different transparent materials. Newton thought that the particles of light experienced a force as they passed from air into a transparent material. This inferred force would occur only at the surface, act perpendicular to the surface, and be directed into the material. This force would cause the particles to bend toward the normal, as shown in Figure 19-2. In this scheme the light particles would also experience this force upon leaving the material. Because the force acts into the material, it now has the effect of bending the particles away from the normal. Furthermore, when Newton calculated the dependence of the amount of **refraction** on the angle of incidence, his answer agreed with the curves in Figure 18-2. So the model gives not only the correct qualitative results but also the correct quantitative results.

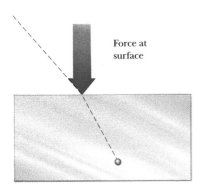

Figure 19-2 A force perpendicular to the surface would cause light particles to be refracted.

Water waves also refract. The photograph in Figure 19-3 shows the refraction of water waves. The boundary that runs diagonally across the photograph separates the shallow region on the left from the deeper region on the right. Once again, the numerical relationship between the angle of refraction and the angle of incidence is in agreement with the law of refraction.

There is an important difference between the wave and particle predictions; the speeds of the particles and waves do not change in the same way. Figures 19-2 and 19-3 both correspond to light entering a substance with a higher index of refraction. In the particle model, particles bend toward the normal because they speed up. Thus, the particle model predicts that the speed of light should be faster in substances with higher indexes of refraction.

In waves the opposite is true. Because the crests in Figure 19-3 are continuous across the boundary, we know the frequency of the waves doesn't change. However, the wavelength does change; it is shorter in the shallow region to the left of the boundary. Because $v = \lambda f$ (Chapter 15), a decrease in the wavelength means a decrease in the speed of the wave. This means that the speed of the waves in the shallow region is smaller, and therefore the wave model predicts that the speed of light should be slower in substances with higher indexes of refraction.

▶ light travels slower in materials

Because the two models predict opposite results, we have a way of testing them; the speed of light can be measured in various materials to see which model agrees with the results. The speed of light in a material substance was not measured until 1862, almost two centuries after the development of the two theories. French physicist Jean Foucault measured the speed of light in air and water and found the speed in water to be less. This dealt a severe blow to the particle model of light and consequently caused a modification of the physics world view.

Figure 19-3 (a) Water waves refract when passing from deep to shallow water. (b) The corresponding ray diagram shows that the waves refract toward the normal.

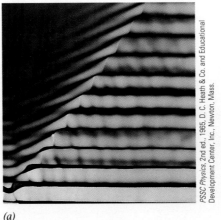

(a)

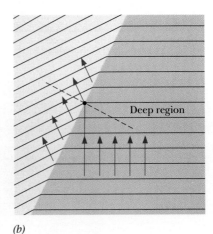

(b)

American physicist Albert Michelson improved on Foucault's measurements and found a ratio of 1.33 for the speed of light c in vacuum to the speed of light v in water. This value is equal to the **index of refraction** n of water as predicted by the wave model, and thus $n = c/v$. Because the speed of light in a vacuum is the maximum speed, the indexes of refraction of substances must be greater than 1. This gives us a way of determining the speed of light in any material once we know its index of refraction. The speed of light in a substance is equal to its speed in a vacuum divided by the index of refraction, $v = c/n$.

$n = c/v$

In the previous chapter, we discovered that different colors have slightly different indexes of refraction within a material, which results in **dispersion**. Because the index of refraction is related to the speed, we can now conclude that different colors must have different speeds in a material.

Question   Does red or blue light have the slower speed in glass?

Answer   Because blue light is refracted more than red, it has the higher index of refraction. Therefore, blue light has a slower speed.

Interference ✓ MATH

Although the speed of light in materials provided definitive support early on for the fact that light behaved like a wave, there was other supporting evidence in the debate. Newton's main adversary in this debate was largely unsuccessful at convincing others of the importance of the evidence. We now examine some of the other properties of waves that differ from those of particles and show that light exhibits these properties.

We begin with interference. The interference of two sources of water waves is shown in the photograph and drawing in Figure 15-20. Attempts to duplicate these results with two lightbulbs, however, fail. But if light is a wave phenomenon, it should exhibit such interference effects. Light from the two sources does superimpose to form an interference pattern, but the pattern isn't stationary. Stationary patterns are produced only when the two sources have the same wavelength and a constant phase difference; that is, the time interval between the emission of a crest from one source and the emission of a crest from the second source does not change. Because the phase difference between the lightbulbs varies rapidly, the pattern blurs out, and the region looks uniformly illuminated.

The interference of two light sources was first successfully demonstrated by Thomas Young in 1801 (more than a half century before Foucault's work on the speed of light in materials) when he let light from *one* pinhole impinge on two other pinholes (Figure 19-4[a]). Passing the light through the first pinhole produced light at the second pinholes that was reasonably in phase. The pattern shown in Figure 19-4(b) consists of colored bars formed by light from the two slits interfering to form antinodal regions. These antinodal regions have large amplitudes and appear bright. Each color forms its own set of colored bars. Antinodal regions for the different rainbow colors superimpose to produce the colors we perceive. Modern versions of this experiment are in complete agreement with the wave model.

If the slits are illuminated with a single color, the pattern looks much like the one you would expect to find along the far edge of the ripple tank in Figure 15-20. As the color of the light changes, the pattern on the screen changes size (Figure 19-5). Red light produces the widest pattern, and violet light produces the narrowest one. The sizes produced by other colors vary in the same order as the colors of the rainbow.

Figure 19-4 (a) A schematic of Young's experiment that demonstrated the interference of light. (b) The interference pattern produced by white light incident on two slits.

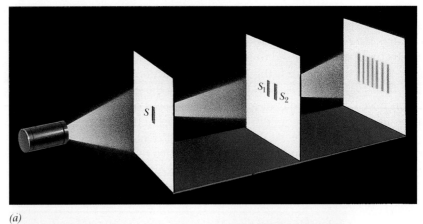

(a)

(b)

Question What happens to the width of one of these interference patterns if the distance between the two slits is increased?

Answer We learned in Chapter 15 that the spacing of the nodal lines depends on the ratio of the wavelength to the slit separation. Therefore, the wider spacing will produce a narrower pattern.

Recall that in the two-source interference patterns in Chapter 15 the nodal lines on each side of the central line were created by a difference in the path lengths from the two sources that was equal to one-half wavelength (Figure 15-21). Increasing the wavelength causes these nodal lines to move farther away from the central line; the pattern widens. Therefore, the shifting of the dark regions on our screen with color signals a change in wavelength. We conclude that the color of light depends on its wavelength, with red being the longest and violet the shortest.

Measurements of the interference pattern and the separation of the slits can be used to calculate the wavelength of the light. Experiments show that visible light ranges in wavelength from 400 to 750 nanometers (nm), where a nanome-

Figure 19-5 The two-slit interference pattern produced by red light (a) is wider than that produced by blue light (b).

ter is 10^{-9} meter. It takes more than 1 million wavelengths of visible light to equal 1 meter. Knowing the speed of light to be 3.0×10^8 meters per second, we can use the relationship $c = \lambda f$ to calculate the corresponding frequency range to be roughly $(4.0 \text{ to } 7.5) \times 10^{14}$ hertz.

PHYSICS | **ON YOUR OWN**

Perform Young's two-slit experiment by taping the slits to one end of the cardboard tube from a roll of paper towels. To make the two slits, blacken a microscope slide (or other piece of glass) with a candle flame. Scratch two viewing slits on the blackened side by holding two razor blades tightly together and drawing them across the slide. Cut a slit about 2 millimeters wide in a piece of paper and tape it to the other end of the tube. Aim this end of the tube at a light source and look through the end with the two slits to see the interference pattern (Figure 19-6).

Figure 19-6 A device for viewing two-slit interference.

You can vary the wavelength of the light by using pieces of red and blue cellophane as filters. Putting a thin spacer between the razor blades can vary the spacing.

Diffraction

✓ MATH

Young's experiment points out another aspect of waves. His interference pattern was only possible because light from one pinhole overlapped that from the other. Light spreads out as it passes through the pinholes. In other words, light exhibits diffraction just like the water waves of Figure 15-23.

The drawings in Figure 19-7 show the diffraction patterns produced by red light passing through a narrow slit. The slit was wider in Figure 19-7(b)! But this difference in the patterns makes sense if you examine the photographs of water waves in Figure 15-23. Contrary to what common sense might tell us,

Figure 19-7 Diffraction patterns produced by red light incident on (a) a narrower and (b) a wider slit. The narrower slit produced the wider pattern.

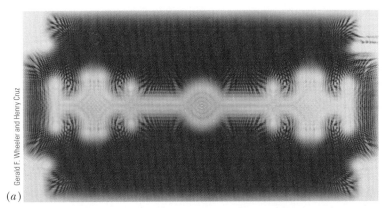
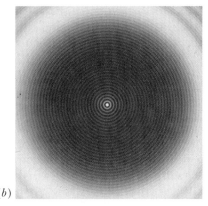

Figure 19-8 Photographs of diffraction patterns in the shadows of (a) a razor blade and (b) a penny.

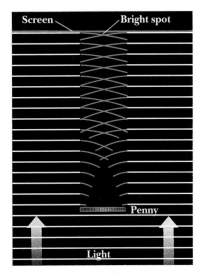

Figure 19-9 Diagram illustrating how wave interference and diffraction produce a bright spot at the center of the penny's shadow.

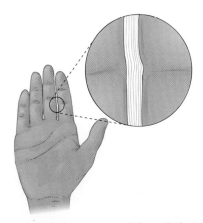

Figure 19-10 Light viewed through the thin slit between two fingers exhibits diffraction effects.

the narrower slit produces the wider pattern. These patterns are in agreement with the wave model and contrary to what we would expect if light traveled only in straight lines. In the case of particles, the wider slit would produce the wider pattern, not the narrower one.

Question Does red light or blue light produce the wider diffraction pattern?

Answer The width of the diffraction pattern depends on the ratio of the wavelength to the width of the slit. Because red light has the longer wavelength, it would produce the wider pattern.

Water waves also show that diffraction takes place around the edges of barriers. To observe this effect with light, we should look along the edge of the shadow of an object. This effect is usually not observed with light because most shadows are produced by broad sources of light. The resulting shadows have smooth changes from umbra through penumbra to full brightness. Therefore, we should look at shadows produced by point sources to eliminate the penumbra.

The photographs in Figure 19-8 were created by putting photographic paper in the shadows of a razor blade and a penny. Red light from a laser was used because the light waves are in phase and can easily be made to approximate a point source. You can see that the edges of the shadows show the effects of the interference of diffracted light. Only a wave model of light can explain these effects.

Notice the center of the shadow of the penny. Even though the penny is solid, the shadow has a bright spot in the center. Can the wave model explain this bright spot? The diagram in Figure 19-9 shows how the light diffracts around the edge of the penny. The light coming from each point on the edge travels the same distance to the center of the shadow. These waves arrive in phase and superimpose to form the bright spot. This is added support for the wave model of light.

PHYSICS | ON YOUR OWN

Look at a lightbulb through the slit between two fingers held in front of your eyes, as in Figure 19-10. How does the image of the lightbulb change as you slowly change the spacing between your fingers? Notice the dark, vertical lines in the image. What happens to those lines as your fingers move closer together?

To observe two-dimensional diffraction patterns, look at a distant light through a fine mesh screen or a piece of cloth held at arm's length.

Diffraction Limits

The wave nature of light and the size of the viewing instrument limit how small an object we can see, even with the best instruments. Because of the wave nature of light, there is always some diffraction. The image is spread out over a region in space; the region is larger for longer wavelengths and smaller openings.

Consider two objects with small *angular* sizes (they could be very big but so far away that they look small) separated by a small angular distance. Each of these will produce a diffraction pattern when its light passes through a small opening such as in our eye or a telescope. The diagram and photograph in (a) correspond to the case in which the diffraction patterns can still be clearly distinguished. In (c) the overlap is so extensive that you cannot resolve the individual images. The limiting case is shown in (b); the separation of the centers of the two patterns is less than the width of the central maximum of either pattern. At this limit the first minimum of each pattern lies on top of the maximum of the other pattern.

This minimum angular separation can be calculated mathematically. The light coming from two point sources can be resolved if the angle between the objects is greater than 1.22 times the ratio of the wavelength of the light to the diameter of the aperture. Imagine viewing an object from a distance of 25 centimeters, the nominal distance of closest vision. If we use an average pupil size of 5 millimeters and visible light with a wavelength of 500 nanometers, we calculate that you can distinguish a separation of 0.03 millimeter, about the radius of a human hair. A similar calculation tells us that we should be able to distinguish the headlights of an oncoming car at a distance of 10 kilometers.

From this relationship you can see why astronomers want bigger telescopes. With a larger mirror, the resolving angle is smaller and they can distinguish more detail in distant star clusters. For instance, the resolving angle of a 5-meter telescope in visible light is about 0.02 arcsecond, where 1 arcsecond is $\frac{1}{3600}$ degree. In practice, this resolution is never obtained because turbulence in Earth's atmosphere limits the resolution to about 1 arcsecond. This is one of the major reasons for placing the Hubble Space Telescope in orbit.

Diffraction effects also place lower limits on the sizes of objects that can be examined under an optical microscope because the details to be observed must be separated by more than the diffraction limits set by the microscope.

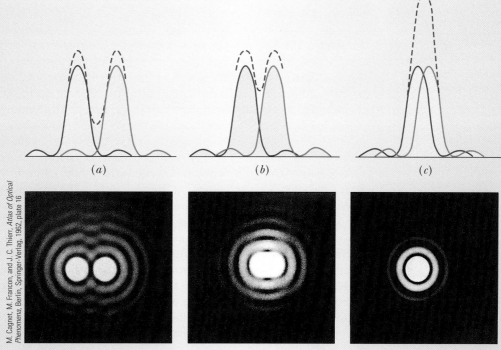

Two overlapping diffraction patterns (a) can just barely be resolved (b) if the central maximum of each pattern lies on the first minimum of the other. If the patterns are closer, they appear to be a single object (c).

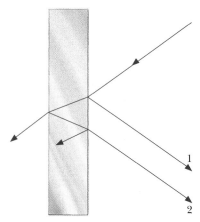

Figure 19-11 Light incident on a thin film is reflected and transmitted at each surface. Rays 1 and 2 interfere to produce more or less light, depending on the thickness of the film and the wavelength of the light in the film.

Figure 19-12 Different colors are reflected from different thicknesses of the thin film. The very thin region at the top of the soap film appears black because the reflected light is inverted at the front surface but not at the back surface.

Thin Films

Soap bubbles, oil slicks, and in fact, thin films of any transparent material exhibit beautiful arrays of color under certain conditions—an effect not of dispersion but of interference.

Consider a narrow beam of red light incident on a thin film as shown in Figure 19-11. At the first surface, part of the light is reflected and part is transmitted. The same thing happens to the transmitted part when it attempts to exit the film; part is transmitted and part is reflected, and so on. Considering just the first two rays that leave the film on the incident side, we can see that they interfere with each other. Because of the thickness of the film, ray 2 lags behind ray 1 when they overlap on the incident side of the film. Ray 2 had to travel the extra distance within the film. At a certain thickness, this path difference is such that the crests of one ray line up with the troughs from the other and the two rays cancel. In this case little or no light is reflected. At other thicknesses, the crests of one ray line up with the crests from the other and the two rays reinforce. In this case light is reflected from the surface.

Because different colors have different wavelengths, a thickness that cancels one color may not be the thickness that cancels another. If the film is held vertically (Figure 19-12), the pull of gravity causes the film to vary in thickness, being very thin at the top and increasing toward the bottom. In fact, when the film breaks, it begins at the top where it has been stretched too thin. If the film varies in thickness and the incident light is white, different colors will be reflected for different thicknesses, producing the many colors observed in soap films.

Question Why does part of the thin film in Figure 19-12 appear white when white is not one of the rainbow colors?

Answer We observe white when most of the colors overlap.

Notice that the very thin region at the top of the soap film in Figure 19-12 has no light reflecting from the film. At first this seems confusing. Because there is essentially no path difference between the front and back surfaces, we might expect light to be reflected. However, the first ray is inverted when it reflects; that is, crests are turned into troughs and vice versa. This process is analogous to the inversion that takes place when a wave pulse is reflected from the fixed end of a rope (Chapter 15). Light waves are inverted when they reflect from a material with a higher index of refraction. At the back surface, the rays reflect from air (a lower index of refraction than the soap), and no inversion takes place.

Because the ray reflected from the front surface is inverted while the ray reflected from the back surface is not inverted, a crest from the front surface will overlap a trough from the back surface if the thickness of the film is much smaller than a wavelength. The two reflected rays will cancel, and no light will be reflected. If the light is normal to the surface and the thickness of the film is increased until it is one-quarter of a wavelength thick, the light will now be strongly reflected. The ray that reflects from the back surface must travel an extra one-half wavelength, and these crests and troughs will be delayed so that they now line up with those reflected from the front surface. It is important to note that we are referring to the wavelength of the light *in the film*. Because the frequency of the light is the same in the air and in the film while the speed is reduced, the wavelength in the film is equal to that in the air divided by the index of refraction, $\lambda_f = \lambda/n$.

This colorful phenomenon also occurs after a rain has wetted the highways. Oil dropped by cars and trucks floats on top of the puddles. Sunlight reflecting from the top surface of the oil and from the oil–water interface interferes to produce the colors. Again, variations in the thickness of the oil slick produce the array of colors.

Interference also occurs for the transmitted light. Light passing directly through the thin film (Figure 19-13) can interfere constructively or destructively with light that is reflected twice within the film. The effects are complementary to those of the reflected light. When the thickness of the film is chosen to minimize the reflection of a certain color of light, the transmission of that color is maximized. This process is a consequence of the conservation of energy. The light must go someplace.

The colors on the puddle are caused by the interference of light reflected from the top and bottom of a thin oil film floating on the surface.

> **Flawed Reasoning**
>
> A factory worker is coating camera lenses with a special film to filter out red light by reflecting the red light from the surface. The boss has instructed him to make the film 120 nanometers thick. The worker decides that "more is always better" and that the boss should not be so stingy with the coating material. He therefore makes the coating twice as thick. **Will the customers get a better red filter than they bargained for?**
>
> **Answer** The boss is not being stingy. The coating filters red light if its thickness is one-quarter the wavelength of the red light in the material. This causes the light that reflects from the back of the coating to be in phase (crest lined up with crest) with the light that reflects from the front of the coating. If the thickness of the coating is doubled, these two reflections will be out of phase, and the red light will pass through the lens instead of being reflected.

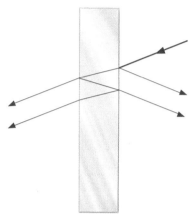

Figure 19-13 The light transmitted through a thin film also displays interference effects.

Some modern office buildings have windows with thin films to reduce the amount of light entering the offices. The visors in the helmets of space suits are coated with a thin film to protect the eyes of the astronauts. Thin films are also very important to lens makers because the proper choice of material and thickness allows them to coat lenses so that they do not reflect certain colors or conversely so that they do not transmit certain colors. It is common to coat the lenses in eyeglasses with a thin film to stop the transmission of ultraviolet light. Lenses in high-quality telescopes and binoculars are coated to reduce the reflection (and therefore enhance the transmission) of visible light, increasing the brightness of images.

An interesting example of thin-film interference was observed by Newton (even though he did not realize that it supported the wave model of light). When a curved piece of glass such as a watch glass is placed in contact with a flat piece of glass, a thin film of air is formed between the two. An interference pattern is produced when the light reflects from the top and bottom of the air gap. The interference pattern that is formed by a wedge-shaped air gap is shown in Figure 19-14. Such patterns are commonly used to test the surface quality of lenses and mirrors.

The thin film on the visor protects the astronaut's eyes.

Figure 19-14 (a) Interference lines are produced by light reflecting from the top and bottom surfaces of an air wedge between two pieces of glass. (b) A drawing of the side view of the air wedge showing the interference of one ray.

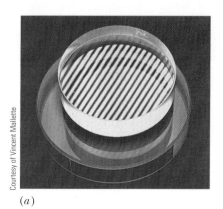

(a)

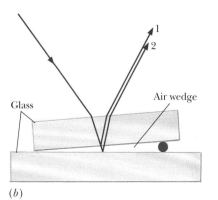

(b)

Question Would you expect to find a dark spot or a bright spot when you look at the reflected light from an air gap of almost zero thickness?

Answer The light waves will be inverted when they reflect off the second interface, but not the first one. Therefore, you would see a dark spot in the reflected light and a bright spot in the transmitted light.

Polarization

We have established that light is a wave phenomenon, but we have not discussed whether it is transverse or longitudinal—that is, whether the vibrations take place perpendicular to the direction of travel or along it. We will determine this by examining a property of transverse waves that does not exist for longitudinal waves and then see if light exhibits this behavior.

Transverse waves traveling along a horizontal rope can be generated so that the rope vibrates in the vertical direction, the horizontal direction, or in fact, in any direction in-between. If the vibrations are in only one direction, the wave is said to be *plane polarized*, or often just **polarized**. This property becomes important when a wave enters a medium in which various directions of polarization are not treated the same. For instance, if our rope passes through a board with a vertical slit cut in it (Figure 19-15), the wave passes through the slit if the vibration is vertical but not if it is horizontal.

What happens if the slit is vertical but the polarization of the wave is someplace between vertical and horizontal? Imagine that you are looking along the length of the rope, and we represent the polarization of the wave by an arrow (a vector) along the direction of vibration. This polarization can be imagined as a superposition of a vertical vibration and a horizontal vibration, as illustrated in Figure 19-16. The vertical slit allows the vertical vibration to pass through while blocking the horizontal vibration. Therefore, the wave has a vertical polarization after it passes through the vertical slit. The amplitude of the transmitted wave is equal to the amplitude of the vertical vibration of the incident wave.

Determining whether light can be polarized is a little difficult because our eyes are unable to tell if light is polarized or not. Nature, however, provides us with materials that polarize light. The mere existence of polarized light demonstrates that light is a transverse wave. If it were longitudinal, it could not be polarized. Commercially available light-polarizing material, such as Polaroid filters, consist of long, complex molecules whose long axes are parallel. These molecules pass light waves with polarizations perpendicular to their long axes but absorb those parallel to it.

▶ light is a transverse wave

We can use a piece of Polaroid filter (or the lens from Polaroid sunglasses) to

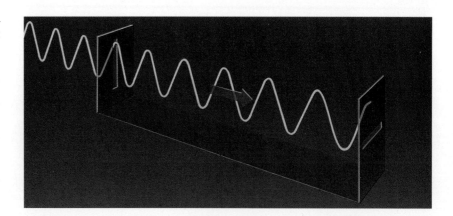

Figure 19-15 Waves with vertical polarization pass through a vertical slit, but not through a horizontal one.

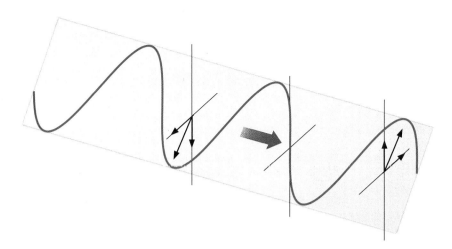

Figure 19-16 The plane-polarized wave can be broken up into two perpendicular component waves: one in the vertical plane and one in the horizontal plane. The displacements of these component waves are shown at two sample locations.

analyze various light sources to see if they are polarized. If the light is polarized, the intensity of the transmitted light will vary as the Polaroid filter is rotated. This simple procedure shows that common light sources such as ordinary incandescent lamps, fluorescent lights, candles, and campfires are unpolarized. However, if we examine the light reflected from the surface of a lake, we find that it is partially polarized in the horizontal direction. Light reflected from nonmetallic surfaces is often partially polarized in the direction parallel to the surface. This is the reason why the axes of polarization of Polaroid sunglasses are in the vertical direction. Boaters know that Polaroid sunglasses remove the glare from the surface of water and allow them to see below the surface.

PHYSICS | ON YOUR OWN

Examine a variety of light sources through the lens of Polaroid sunglasses to see if the sources are polarized. Looking away from the direction of the Sun, check to see if skylight is polarized. If you have two sets of glasses, you can verify that the lenses are polarizing by rotating one with respect to the other to see if they block out light.

(a) (b)

Photographs taken through a plate-glass window (a) with and (b) without a polarizing filter on the camera lens. Notice that the polarizing filter eliminates the glare.

Holography

Wouldn't it be fantastic to have a method for catching waves, preserving the information they carry, and then at some later time playing them back? In fact, high-fidelity equipment routinely records a concert and plays it back so well that the listeners can hardly tell the difference. Similarly, high-fidelity photography should record the light waves coming from a scene and then play them back so well that the image is nearly indistinguishable from the original scene.

Conventional photography, however, does not do this. Although modern chemicals and papers have created images with extremely fine resolution, nobody is likely to mistake a photograph for the real object because a photographic image is two dimensional, whereas the scene is three dimensional. This loss of depth means that you can view the scene in the photograph from only one angle, or perspective. You cannot look around objects in the foreground to see objects in the background. The objects in the scene do not move relative to one another as you move your point of view.

Holography is a photographic method that produces a three-dimensional image that has virtually all the optical properties of the scene. This process was conceived by Dennis Gabor in 1947; he received a Nobel Prize in 1971 for this work. Gabor chose the word *hologram* to describe the three dimensionality of the image by combining the Greek roots *holo* (complete) and *gram* (message).

Although Gabor could make a hologram out of a flat transparency, the lack of a proper light source prevented him from making one of an object with depth. With the invention of the laser in 1960, interest in holography was renewed. We will postpone the discussion of laser operation until Chapter 24. At the moment we need only know that it is a device that produces light with a single color and a constant phase relationship.

Figure A shows the essential features of a set-up for making holograms. Notice that there is no lens between the object and the film. Light from the object reflects onto all portions of the film. Although this *object beam* carries the information about the object, if this were all that happened, the film would be completely exposed, and virtually no information about the object would be recorded.

Another portion of the beam from the laser (the *reference beam*) illuminates the film directly. Once again, if this were all that happened, the film would not record the scene. However, the light from the laser has a single well-defined wavelength (color) with the crests lined up. Therefore, the light reflected from the object and the light in the reference beam produce an interference pattern that is recorded

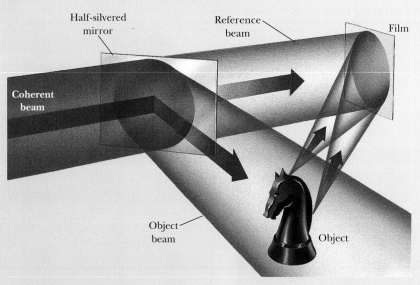

Figure A Making a hologram.

Until now we have discussed materials in which one direction passes the light and the other absorbs it. Some materials allow both polarizations to pass through, but the orientations have different speeds. This can have the effect of rotating the plane of polarization. The cellophane on cigarette packages and some types of transparent adhesive tape have this property.

You can make an interesting display by crumpling a cigarette wrapper and looking at it between two Polaroid filters, as in Figure 19-17(a). The Polaroid filter on the side near the light polarizes the light striking the display. When this light is viewed with the second Polaroid filter, the different thicknesses of cellophane

fore, viewing a hologram is just like viewing the original scene through a window. The photographs in Figure C were taken of a single hologram. Note that the relative locations of the chessmen change as the camera position changes. This is what we would see if we were looking at the actual chessboard.

Because information from each point in the scene is recorded in all points of the hologram, a hologram can be broken, and each piece will produce an image of the entire scene. Of course, something must be lost. Each piece will allow the scene to be viewed only from the perspective of that piece. This is analogous to looking at a scene through a window that has been covered except for a small hole.

Advances in holography have made it possible to display holograms with ordinary sources of white light instead of the much more expensive lasers. It is now possible to hang holograms in your home as you would paintings. In fact, they have become so inexpensive that they have appeared on magazine covers. Holograms can also be made in the shape of cylinders so that you can walk around the hologram and see all sides of the scene. In some of these, individuals in the scene move as you walk around the hologram. Holographic movies are possible, and holographic television should be possible in the future.

In addition to being a fantastic art medium, holography has found many practical applications. It can be used to measure the three-dimensional wear patterns on the cylinder walls of an engine to an accuracy of a few ten-thousandths of a millimeter (a hundred-thousandth of an inch), perform nondestructive tests of the integrity of machined parts or automobile tires, study the shape and size of snowflakes while they are still in the air, store vast quantities of information for later retrieval, and produce three-dimensional topographical maps.

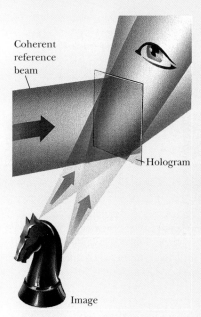

Figure B A hologram is viewed by placing it in the reference beam. The light passing through the film produces a virtual image at the location of the original scene.

in the film. This interference pattern contains the three-dimensional information about the scene.

The hologram is viewed by placing it back in a reference beam. The pattern in the film causes the light passing through it to be deflected so that it appears to come from the original scene (Figure B).

Because information from each point in the scene is recorded in all points of the hologram, a hologram contains information from all perspectives covered by the film. There-

Figure C The relative positions of the chessmen in the holographic image change as the point of view changes.

have different colors, as shown in Figure 19-17(b). The amount of rotation depends on the thickness of the cellophane and the wavelength of the light. Certain thicknesses rotate certain colors by just the right amount to pass through the second Polaroid filter. Others are partially or completely blocked. The colors of each section change as either Polaroid filter is rotated.

Glasses and plastics under stress rotate the plane of polarization, and the greater the stress, the more the rotation. Plastic models of structures, such as cathedrals, bones, or machined parts, can be analyzed to discover where the stress is greatest.

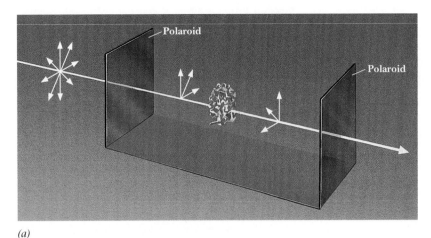

Figure 19-17 (a) Cellophane or transparent tape rotates the plane of polarization. (b) A pattern formed by different thicknesses of cellophane viewed between two pieces of Polaroid filter.

Our perceptions of this audiocassette are enhanced by viewing it between crossed Polaroid filters to observe the stress patterns in the plastic.

Physics☒Now™ Assess your understanding of this chapter's topics with sample tests and other resources found by logging into PhysicsNow at http://physics.brookscole.com/kf6e.

PHYSICS | **ON YOUR OWN**

Place a piece of Polaroid filter (or the lens from Polaroid sunglasses) over the lens of an empty slide projector. Place a jar of clear Karo syrup in the light beam and view the transmitted light through another piece of Polaroid filter. Different path lengths through the syrup produce different amounts of rotation of the plane of polarization.

Looking Ahead

We have not finished our search to understand light. Notice, for example, that although we established that the wave model of light explains all the phenomena we have observed so far, we still have not said what it is that is waving! We know that light travels from the distant stars to our eyes through incredible distances in a vacuum that is much better than any we can produce on Earth. How is this possible?

We will see in Chapter 23 that understanding the nature of light was the key to our modern understanding of atoms. This reexamination of the model for light occurred at the beginning of the last century and began with Einstein asking many of the same questions that we addressed here.

Summary

Developing a theory of light involves building a model from known experiences that can be compared with the behavior of light. Both particle and wave models can account for the law of reflection and the law of refraction. However, only a wave model can correctly account for the speed of light in transparent materials. The index of refraction n is the ratio of the speed of light c in a vacuum (300 million meters per second) to its speed v in the material; that is, $n = c/v$. Dispersion of light indicates that different colors have different speeds in a material.

Other properties of waves that differ from those of particles further support the fact that light is a wave phenomenon. Interference of two coherent light sources produces a stationary pattern. If different colors are used, the pattern changes size, demonstrating that colors have different wavelengths. Red light has the largest wavelength and produces the widest pattern, and violet light produces the narrowest one. Visible light ranges in wavelength from 400 to 750 nanometers.

Light exhibits diffraction, which is the spreading out of a wave as it passes through narrow openings and around the edges of objects. The width of the diffraction pattern depends on the ratio of the wavelength to the size of the opening—the narrower the opening, the wider the pattern. Because red light has the longest visible wavelength, it would produce the widest pattern for a given opening.

Thin films of transparent materials exhibit beautiful arrays of colors due to the interference of light rays reflecting from the two surfaces. A ray that reflects from a material with a higher index of refraction is inverted. Different wavelengths are strongly reflected or transmitted at different thicknesses of film.

Light exhibits polarization, demonstrating that it is a transverse wave. Polarizing materials pass light waves with polarizations perpendicular to one axis, but absorb those parallel to it. Common light sources are usually unpolarized. However, light reflected from the surface of a lake or glass is partially polarized parallel to the surface.

Chapter 19 Revisited

Light is probably the most fascinating and elusive phenomenon in nature. The answer to the question "What is light?" actually changes with the techniques used to examine the question. In this chapter we found very good evidence to support the conclusion that light is a wave phenomenon. In future chapters we'll return to this question and delve further into the mystery of light.

KEY TERMS

dispersion: The spreading of light into a spectrum of colors. The variation in the speed of a periodic wave due to its wavelength or frequency.

index of refraction: An optical property of a substance that determines how much light bends upon entering or leaving it. The index is equal to the ratio of the speed of light in a vacuum to its speed in the substance.

polarized: A property of a transverse wave when its vibrations are all in a single plane.

refraction: The bending of light that occurs at the interface between two transparent media. It occurs when the speed of light changes.

CONCEPTUAL QUESTIONS

1. Newton believed that light beams consist of tiny particles. If these beams travel in straight lines, what does that imply about their speed?
2. How does the particle theory of light account for the diffuse reflection of light?
3. Argue that the law of reflection would not hold for particles rebounding from a surface that is not frictionless or not perfectly elastic.
4. When a particle reflects elastically from a smooth surface, the component of the particle's momentum parallel to the surface is conserved while the component of the particle's momentum perpendicular to the surface is reversed. Use this information to argue that Newton's particle theory of light is consistent with the observation that the angle of incidence equals the angle of reflection.
5. If particles incident at 45 degrees from the normal strike a completely elastic surface that has friction, will the angle of reflection (with respect to the normal) be greater than, equal to, or less than 45 degrees? Explain.
6. If particles incident at 45 degrees from the normal strike a frictionless surface that is not completely elastic, will the angle of reflection (with respect to the normal) be greater than, equal to, or less than 45 degrees? Explain.
7. How does Newton's idea of light particles explain the law of refraction?
8. Explain how Newton's idea of light particles predicts that the speed of light in a transparent material will be faster than in a vacuum.
9. Does the wave's frequency or its wavelength remain the same when the wave crosses from one medium into another? Explain.
10. In which region of Figure 19-3(a) (top left or bottom right) are the waves traveling at the higher speed? Explain.
11. Which color of light, red or blue, travels faster in a diamond? Explain your reasoning.
12. Do you expect the speed of light in glass to be slower than, faster than, or the same as that in diamond? Why?
13. What property of a light wave determines its brightness?

▲ 14. Does the amplitude of a light wave increase, decrease, or stay the same upon reflection from a transparent material? Explain.

15. Starting with the observation that waves that have been bent toward the normal have a shorter wavelength than the incident waves, explain how the wave model for light predicts that the speed of light in glass will be slower than the speed in a vacuum.

16. Imagine that Newton knew that light travels slower in glass than in air but was unaware of the law of refraction. In what direction would he have predicted light to bend when passing from air into glass?

17. Does total internal reflection result from light trying to pass from a slow medium to a fast medium or from a fast medium to a slow medium? Explain.

▲ 18. Different colors of light have different critical angles for total internal reflection. Is the critical angle greater for colors of light that travel faster or slower in the medium? Explain.

19. What is the physical difference between red and blue light?

20. How does the slow speed of light in diamonds affect their brilliance?

21. Why do we not notice any dispersion when white light passes through a windowpane?

22. What does the dispersion of light tell us about the speeds of various colors of light in a material?

23. Will the converging lens in the figure focus blue light or red light at a closer distance to the lens? Explain.

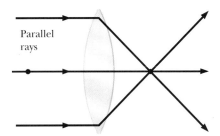

Questions 23 and 24

24. Using blue light, you determine the focal point for the lens in the figure. If you were to shine a green laser beam from this focal point to a point near the top of the lens, would the emerging beam be bent toward or away from the optic axis? Explain.

25. Red light is used to form a two-slit interference pattern on a screen. As the two slits are moved closer together, does the separation of the bright bands on the screen decrease, increase, or remain the same? Why?

26. What happens to the separation of the bright bands in a two-slit interference pattern if the slits are made narrower but their separation remains the same?

27. Would yellow light or green light produce the wider two-slit interference pattern? Why?

28. We observe that the two-slit interference pattern produced by blue light is narrower than that produced by red light. What does this tell us about red and blue light?

29. What determines whether two light beams with the same wavelength tend to cancel or reinforce each other?

30. Why don't we notice interference patterns when we turn on two lights in a room?

31. If light and sound are both wave phenomena, why can we hear sounds around a corner but cannot see around a corner?

32. Approximately how narrow should a slit be for the diffraction of visible light to be observable?

33. Blue light is used to form a single-slit diffraction pattern on a screen. As the slit is made narrower, does the separation of the bright bands on the screen decrease, increase, or remain the same? Explain.

34. Would orange light or blue light produce the wider diffraction pattern? Why?

35. Would a slit with a width of 300 nanometers or of 400 nanometers produce a wider diffraction pattern when illuminated by light of the same wavelength? Why?

▲ 36. Which of the following single-slit diffraction experiments would produce the wider diffraction pattern: 800-nanometer light passing through a 500-nanometer-wide slit, or 450-nanometer light passing through a 400-nanometer-wide slit? Why?

37. Why can't an ordinary microscope using visible light be used to observe individual molecules?

38. Is it better to use red light or blue light to minimize diffraction effects while photographing tiny objects through a microscope? Why?

39. A common technique used by astronomers for overcoming diffraction limits is to electronically combine the light from more than one telescope. This effectively increases the diameter of the aperture to the distance between the telescopes. If the signals from two 5-meter-diameter telescopes located 100 meters apart were being combined when one of the telescopes stopped functioning, by what factor would the minimum resolvable angle be increased?

40. Why are the diffraction effects of your eyes more important during the day than at night?
41. Will you observe multicolored patterns if you illuminate a thin soap film with monochromatic light? Why?
42. A thin film of oil on top of a bucket of water will produce multicolored patterns. However, a bucket full of oil will produce no such effect. Explain the difference.
43. Assume that you have the thinnest film that strongly reflects blue light. Would you need to make the film thinner or thicker to completely reflect red light? Why?
▲ 44. You are coating glass with a film of higher index of refraction. You make the thinnest film that will produce a strong reflection for a particular monochromatic light source. You then gradually increase the film's thickness until you find another strong reflection. How many times thicker is this film than the original?
▲ 45. A glass pane with index of refraction 1.5 is coated with a thin film of a material with index of refraction 1.6. The coating is as thin as possible to produce maximum reflection for red light. If this same material is used to coat a different kind of glass with index of refraction 1.9, the light reflected from the back surface of the film now experiences an inversion. Does the coating have to be thicker or thinner in this case to produce strong reflection? Explain.
▲ 46. The office workers in a skyscraper complain that the morning sun shines too brightly into their work areas. The problem is resolved by applying a thin film to each windowpane. The film has an index of refraction smaller than the glass and is designed to reflect yellow light when applied to the glass. If a sheet of this film is held in front of a yellow spotlight, would any of the light pass through the film? Explain.
47. A thin film in air strongly reflects orange light. Will it still reflect orange light when it is placed in water?
▲ 48. A thin, transparent film strongly reflects yellow light in air. What does the film do when it is applied to a glass lens that has a higher index of refraction than the film?
49. If all the labels had come off the sunglasses in the drug store, how could you tell which ones were polarized?
50. Can sound waves be polarized?
51. The digital displays at fuel pumps often use liquid crystal displays (LCDs) to show the price. Because the light from LCDs is polarized, they can often be impossible to read while wearing Polaroid sunglasses. What could you do to read the display without removing your glasses?
52. How could you use Polaroid sunglasses to tell if light from the sky is polarized?
53. How would you distinguish a hologram from a flat transparency?
54. If each point on a holographic film contains the entire image, what is gained by making the hologram larger?
55. What kind of light is required to make a hologram of a three-dimensional object?
56. What kind of light is required to display a hologram?
57. To gather enough light to expose the film, long time exposures are often necessary to make holograms of inanimate objects. Why is a very powerful laser required to make a hologram of a person's face?
58. Which of the following phenomena does not show a difference between the wave theory and particle theory of light: reflection, refraction, interference, diffraction, or polarization?

EXERCISES

1. What is the speed of light in glass with an index of refraction of 1.7?
2. What is the speed of light in water?
3. The speed of light in diamond is 1.24×10^8 m/s. What is the index of refraction for diamond?
4. Zircon is sometimes used to make fake diamonds. What is its index of refraction if the speed of light in zircon is 1.6×10^8 m/s?
5. If it takes light 5 ns (1 nanosecond = 10^{-9} s) to travel 1 m in an optical cable, what is the index of refraction of the cable?
6. If an optical cable has an index of refraction of 1.5, how long will it take a signal to travel between two points on opposite coasts of the United States separated by a distance of 5000 km?
▲ 7. The index of refraction for red light in material X is measured at 1.80. It is determined that blue light travels 5×10^6 m/s slower than red light in this material. What is the index of refraction for blue light in material X?
8. For crown glass, the index of refraction for violet light is 1.532 and the index of refraction for red light is 1.515. How much faster is red light than violet light in this medium?
9. What is the wavelength of the radio signal emitted by an AM station broadcasting at 1420 kHz? Radio waves travel at the speed of light.
10. What is the wavelength of light that has a frequency of 5×10^{14} Hz?
11. The red light from a helium–neon laser has a wavelength of 633 nm. What is its frequency?
12. What is the frequency of the yellow light with a wavelength of 590 nm that is emitted by sodium lamps?
13. What is the wavelength of the red light from a helium–neon laser when it is in glass with an index of refraction of 1.7? The wavelength in a vacuum is 633 nm.
14. A transparent material is known to have an index of refraction equal to 1.9. What is the wavelength of light in this material if it has a wavelength of 650 nm in a vacuum?
15. Light from a sodium lamp with a wavelength in a vacuum of 590 nm enters diamond in which the speed of light is 1.24×10^8 m/s. What is the wavelength of this light in diamond?
16. What is the wavelength of light in water if it has a frequency of 5×10^{14} Hz?
17. For distant objects, the angular size in degrees can be approximated as $57° \times w/d$, where w is the width of the object and d is its distance. What is the angular size of the headlights on a car 10 km away if the headlights are 1.2 m apart?
18. The minimum angular separation in arcseconds ($\frac{1}{3600}$ degree) is found by first finding the ratio of the wavelength of light to the diameter of the aperture and then multiplying by 2.5×10^5. Using visible light with a wavelength of 550 nm, calculate the minimum angular separation for an eye with a pupil size of 5 mm.
19. Using the information in Exercise 18, find the theoretical resolution of a telescope with a 10-m-diameter mirror for visible light at 550 nm.
20. What is the theoretical resolution of a radio telescope with a 10-m-diameter collecting dish for radio waves with a wavelength of 21 cm? (See Exercise 18.)

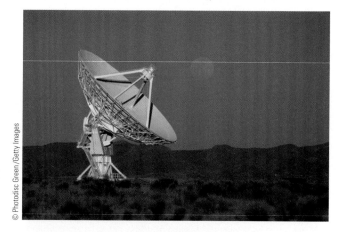

21. What is the thinnest soap film that will strongly reflect light with a wavelength of 400 nm in the film?
22. What is the thinnest soap film that will strongly reflect red light from a helium–neon laser? The wavelength of this light is 633 nm in air and 470 nm in soapy water.
▲ 23. You are coating a glass lens of index of refraction 1.6 with a film of material of index of refraction 1.7. You start with the thinnest film possible that creates a strong reflection for 500-nm light. You gradually increase the film thickness until you again get strong reflection. What is the thickness of the film now?
▲ 24. Repeat Exercise 23 for a glass lens of index of refraction 1.8.

Interlude

The hairs on this girl's head repel each other because they have the same charge.

An Electrical and Magnetic World

If our common experiences are any guide, electrical and magnetic effects appear insignificant. They don't seem to be a common property of all matter, nor do they usually seem very strong. Magnets are used to lift a couple of paper clips, and static electricity lifts bits of paper.

If we ignore the electricity supplied by both power companies and lightning, electricity seems only to give us occasional annoying shocks when we scuff our feet on rugs or slide across vinyl seat covers. Furthermore, the effects are intermittent; shocks usually occur only in winter or when it is especially dry. Even then, they occur only with a few materials: we feel a shock when we touch metal doorknobs but not when we touch wooden doors.

Magnetic properties seem more permanent: a piece of metal that is magnetic in the winter is still magnetic in the summer. But again, these properties don't seem to be universal. Only a few materials are magnetic, and the magnetic forces don't seem very strong.

Most people held these attitudes in Newton's time. The investigation of electrical and magnetic properties of matter was not considered important. Gravity reigned supreme as the governing force in the Universe. All objects attract each other via the gravitational force. It holds objects to Earth's surface, it holds our life-supporting atmosphere around Earth, and it even controls the motions of the planets. But, in fact, gravity is the weakest force. It is the electric and magnetic forces that control the structure of matter on the atomic level and therefore the structure of matter in the macroscopic world.

> It is the electric and magnetic forces that control the structure of matter on the atomic level and therefore the structure of matter in the macroscopic world.

We are electrical creatures. What tastes good or bad, what's medicine and what's poison, and the contractions of our muscles or our vocal cords are all controlled by electrical interactions. Even our thoughts are due to electric signals. Magnets and magnetism seem more removed from our daily lives, but images of our brains can be taken using magnetic techniques. As we will learn in Chapter 22, the seemingly separate phenomena of electricity and magnetism are in fact intimately connected.

As we have seen, the age of Newton, Galileo, and Huygens ushered in a new way of thinking about and seeing everyday phenomena. The one belief they all shared was that the Universe is composed of atoms. However, atoms and the void between them implied some mysterious action at a distance, and no one could adequately account for many optical, electrical, and magnetic phenomena. Waves and atoms did not seem to mix well with the concept of a void, and by 1800 it was clear that some new theory was required.

The next three chapters follow a path similar to the one followed by the early pioneers in this area. We first look at electricity and magnetism separately. Then we look for connections between the two. As we develop the connections, you will find yourself returning to components of the physics world view that we developed in earlier chapters. All of this discussion will set the stage for the study of the basic building block of nature, the atom.

20 | Electricity

A lightning storm produces very high voltages, creating very large and dangerous electric currents.

We see electrical effects when sparks occur. On very dry days, we may feel a shock when we touch a metal doorknob after walking across a carpet. If we look closely, we can see a spark jump between our hand and the doorknob. Although they may surprise us, the sparks do not hurt us. On the other hand, nature produces very long and dangerous sparks during lightning storms. What determines the length of the sparks?

(See page 421 for the answer to this question.)

THE early Greeks knew that amber—a fossilized tree sap currently used in jewelry—had the interesting ability to attract bits of fiber and hair after it was rubbed with fur. This was one way of recognizing an object that was electrified. In modern terminology we say the object is **charged**. This doesn't explain what **charge** is, but is a handy way of referring to this condition.

In 1600 English scientist William Gilbert published a pioneering work, *De Magnete*, in which he pointed out that this electrical effect was not an isolated property of amber but a much more general property of matter. Materials such as gems, glass, and sealing wax could also be charged. Rubbing together two objects made from different materials was the most common way of charging an object. In fact, both objects become charged.

After Gilbert, many experimenters joined in the activity of investigating electricity. However, the question of what happens when an object gains or loses this electrical property remained unanswered. A modern response might include words like *electron* or *proton*. But we have to be careful that we are not simply substituting new words or phrases for old ones. To answer this question fully and thus expand our world view, we need to carefully examine our electrical world.

Electrical Properties

In an effort to explain electricity, Gilbert proposed the existence of an electric fluid in certain types of objects. He suggested that rubbing an object removed some of this fluid, leaving it in the region surrounding the object. Bits of fiber were attracted to the object by the draft caused by the fluid returning to the object. Although many other electrical phenomena could not be explained with this idea, it was the beginning of attempts to model the underlying, invisible processes that caused the electrical effects.

Little progress was made for more than a century. In the 1730s it was shown that charge from one object would be transferred to a distant object if metal wires connected them but not if silk threads connected them. Materials that are able to transfer charge are known as **conductors**; those that cannot are called nonconductors, or **insulators**. It was discovered that metals, human bodies, moisture, and a few other substances are conductors.

The discovery that moisture is a conductor explains why electrical effects vary from day to day. You normally experience bigger shocks in the winter when the humidity is naturally low. On more humid days, any charge that you get by scuffing your feet is quickly dissipated by the moisture in the air surrounding your body.

Fuel trucks are grounded to remove the charge built up by the fuel flowing through the hoses.

PhysicsNow™ Test your understanding of this chapter by logging into PhysicsNow at **http://physics.brookscole.com/kf6e**, selecting the chapter, and clicking on the "Take a Pre-Test" link.

Many charged objects have to be suspended from insulators such as silk threads or plastic bases or they quickly lose their charge to the earth, or ground. In fact, we speak of **grounding** an object to ensure that it is not charged. Other objects hold their charges without being insulated. When we charge a plastic rod by rubbing it with a cloth, the charge stays on the rod even if we hold it. But our bodies are conductors. Why doesn't the charge flow to ground through our bodies? It stays on the rod because the rod is an insulator; charge generated at one end remains there. The charge can be removed by moving our hands along the charged end. As we touch the regions that are charged, the charges flow through our bodies to ground.

A metal rod cannot be charged by holding it in our hands and rubbing it with a cloth because metal conducts the charge to our hands. A metal rod can be charged if it is mounted on an insulating stand or if we hold it with an insulating glove; that is, the rod must be insulated from its surroundings.

Even the flow of liquid through pipes is enough of a rubbing action to charge the liquid. Fuel trucks tend to build up charge as they dispense fuel. If this charge becomes large enough, sparking may occur and cause a fire or explosion. To avoid this danger, a conducting wire is connected from the truck to the ground to allow the charge to flow into the ground.

PHYSICS | **ON YOUR OWN**

Notice the sparks when you do one or more of the following in a dark room: (1) Pull apart some clothes that have just come from a clothes dryer. (2) Rapidly pull some transparent tape from its roll. (3). Comb your hair and bring the comb near a metal object.

Two Kinds of Charge

It was also the early 1730s before there was any mention that charged objects could repel one another. Electricity, like gravity, was believed to be only attractive. It may seem strange to us now because both the attractive and repulsive aspects of electricity are easy to demonstrate. If you comb your hair, the comb becomes charged and can be used to attract small bits of paper. After contacting the comb, some of these bits are then repelled by the comb.

This phenomenon can be investigated more carefully using balloons and pieces of wool. If we rub a balloon with a piece of wool, the balloon becomes charged; it attracts small bits of paper and sticks to walls or ceilings. If we suspend this balloon by a thread and bring uncharged objects near it, the balloon is attracted to the objects (Figure 20-1[a]). Everything seems to be an attractive effect.

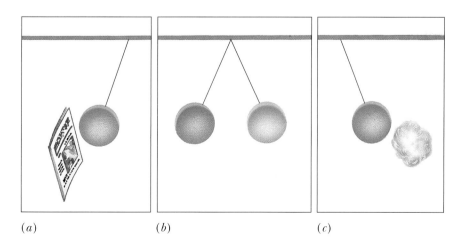

Figure 20-1 A charged balloon is (a) attracted by uncharged objects, (b) repelled by an identically charged balloon, and (c) attracted by the wool used in charging it.

(a)   (b)   (c)

Charging another balloon in the same way demonstrates the new effect: the two balloons repel one another (Figure 20-1[b]). Because we believe that any two objects prepared in the same way are charged in a like manner, we are led to the idea that like-charged objects repel one another.

Whenever we charge an object by rubbing it with another, both objects become charged. If we examine the pieces of wool, we find that they are also charged: they each attract bits of paper.

Question Will the two pieces of wool attract or repel each other?

Answer Because they have been charged in a like manner, they will repel each other.

The piece of wool and the balloon, however, attract each other after being rubbed together (Figure 20-1[c]). If they had the same kind of charge, they should repel. We are therefore led to the idea that there must be two different kinds of charge and that the two kinds attract each other. These experiments can be summarized by stating

> Like charges repel; unlike charges attract.

PHYSICS | **ON YOUR OWN**

Perform the experiments with the wool and balloons described in the text.

Conservation of Charge

Like Gilbert, Benjamin Franklin believed that electricity was a single fluid and that an excess of this fluid caused one kind of charged state, whereas a deficiency caused the other. Because he could not tell which was which, he arbitrarily named one kind of charge positive and the other kind negative. By convention the charge on a glass rod rubbed with silk or plastic film is positive (Figure 20-2), whereas that on an amber or rubber rod rubbed with wool or fur is negative.

At first glance these names seem to have no advantage over other possible choices, such as black and white or yin and yang. They were, however, more significant. Franklin's use of the fluid model led him to predict that charge should be conserved. The amount of electric fluid should remain the same; it is just transferred from one object to another. If you start with two uncharged objects and rub them together, the amount of excess fluid on one is equal to the deficiency on the other. In other words, the positive and negative charges are equal. Using his system of positive and negative numbers, we can add them and see that the total charge remains zero.

We no longer believe in Franklin's fluid model; the fluid model was abandoned because it was not able to account for later experimental observations. However, Franklin's idea about the **conservation of charge** has been verified to a very high precision. It is one of the fundamental laws of physics. In its generalized form, it can be stated

Figure 20-2 The charge on the glass rod is called positive. This is indicated by a few plus signs.

conservation of charge ▶

> In an isolated system the total charge is conserved.

In our modern physics world view, all objects are composed of negatively charged electrons, positively charged protons, and uncharged neutrons. The elec-

FRANKLIN | The American Newton

When Benjamin Franklin (1706–1790) arrived in Paris as Revolutionary America's ambassador, he was known to the sympathetic French as a famous scientist who "stole the scepter from tyrants and lightning from the gods." Precisely because he was the "American Newton," he was the most effective representative the young nation in the making could have sent.

Born in Boston, Franklin made his fortune and reputation in Philadelphia—then the largest English-speaking city outside of London. He was a one-man chamber of commerce who greeted newcomers, brought promising young people into his circle of friends, and helped organize everything from the first philosophical society in the country to a street-lighting district, a fire department, and a mental hospital. His publications on electricity brought him membership in the Royal Philosophical Society of London. (Newton had been president just 50 years earlier.) He invented lightning rods and used them on his own house to great effect.

Franklin believed that electricity was a fluid that pervaded all bodies in varying quantities. A body would seek to maintain

Benjamin Franklin

electrical equilibrium. If there were a deficit, the body would be in a negative state, and electricity would flow into the body. If there were a surplus, electricity would flow out. He also believed that unlike gravitation, which is only attractive, electrified matter could repel as well as attract other electrified matter. It could also attract nonelectrified matter.

As were others of his era, he was also interested in heat, illumination, weather, and other aspects of science. Like Michael Faraday a bit later, Franklin was not a mathematician and so could not give definitive form to his arguments. Even so, this well-known patriot was the best known of early American scientists and a fascinating human being.

—Pierce C. Mullen, historian and author

Sources: Carl Van Doren, *Benjamin Franklin* (New York: Viking Press, 1938) and *The Autobiography of Benjamin Franklin* (New York: Dover Publishing, Thrift Editions, 1966).

---

tron's charge and the proton's charge have the same size. An object is uncharged (or neutral) because it has equal amounts of positive and negative charges, not because it contains no charges. For example, atoms are electrically neutral because they have equal numbers of electrons and protons.

Positively charged objects may have an excess of positive charges or a deficiency of negative charges; that is, an excess of protons or a deficiency of electrons. We simply call them positively charged because the electrical effects are the same in both situations.

The modern view easily accounts for the conservation of charge when charging objects. The rubbing simply results in the transfer of electrons from one object to the other; whatever one object loses, the other gains.

Question If you remove one electron from a neutral quarter and one electron from a neutral Buick, which (if either) has the greater net charge?

Answer In either case, each proton is paired with an electron except for one. The net charge on either object would just be the charge of a single proton.

Induced Attractions

Attraction is more common than repulsion because charged objects can attract *uncharged* objects. How do we explain the observation that charged objects attract uncharged objects? Consider a positively charged rod and an uncharged metal ball. As the rod is brought near the ball, the rod's positive charges attract the negative charges and repel the positive charges in the ball. Because the charges in the ball are free to move, this results in an excess of negative charges on the near side and an excess of positive charges on the far side of the ball (Figure 20-3). Because charge is conserved, the excess negative charge on one side is equal to the excess positive on the other. Experiments with balloons show that the electric force varies with distance; the force gets weaker as the balloons are

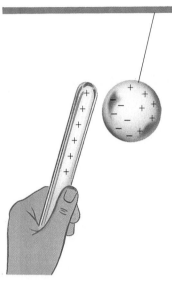

Figure 20-3 A charged rod attracts a neutral metal ball because of the induced separation of charges.

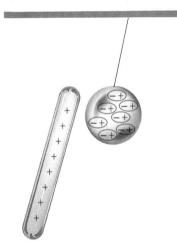

Figure 20-4 An insulating object is attracted because of the separation of charges within the molecules.

moved farther apart. Thus, the ball's negative charge is attracted to the rod more than its positive charge is repelled. These induced charges result in a net attraction of the uncharged ball toward the rod.

Experiments show that if the ball is made of an insulating material the attraction still occurs. In insulators the charges are not free to move across the object, but there can be motion on the molecular level. Although the molecules are uncharged, the presence of the charged rod might induce a separation of charge within the molecule. Other molecules are naturally more positive on one end and more negative on the other. The positive rod rotates these polar molecules so that their negative ends are closer to the rod, as shown in Figure 20-4. This once again results in a net attractive force.

PHYSICS | ON YOUR OWN

Run a comb through your hair and then hold it close to a thin stream of water from a faucet. How do you explain what happens?

A particularly graphic example of the attraction of an uncharged insulator is illustrated in Figure 20-5. A long two-by-four is balanced on a watch glass so that the board is free to rotate. When a charged rod is brought near one end, the resulting attractive force produces a torque that rotates the board.

Question Assuming that the two-by-four was attracted by a positively charged rod in the previous example, what direction would it rotate if a negatively charged rod was used?

Answer It would rotate in the same direction. The interaction between any charged object and an uncharged object is always attractive.

We can now return to the charged comb that attracted and then repelled bits of paper. Initially, the bits of paper are uncharged. They are attracted to the comb by the induced charge. When the bits of paper touch the comb, they acquire some of the charge on the comb and are repelled because the comb and the paper have the same sign of charge and like charges repel one another.

Flawed Reasoning

Your teacher demonstrates that the south pole of a bar magnet is attracted to a charged glass rod. She then asks your class to predict what will happen when the charged glass rod is brought near the north pole of the magnet. The best student in the class answers, "The glass rod is positively charged, so the south pole of the magnet must be negatively charged. I know that a north pole will attract a south pole, so north poles of magnets must be positively charged. I predict that the glass rod will repel the north pole of the magnet."

Your teacher then performs the experiment and shows that the glass rod *attracts* the north pole. She then shows that a negatively charged rubber rod also attracts both poles of the magnet, demonstrating that the magnet must be electrically neutral. Clearly the best student in the class made a mistake in his reasoning. **What is it?**

Answer An electrical *attraction* between two objects only indicates that at least one of the objects is charged. The class already knew that the glass rod was charged, so the first demonstration teaches nothing about the charge (if any) on the magnet. If the charged rod had *repelled* one of the poles of the magnet, it would indicate that the magnet must be charged. However, bar magnets are normally not charged and are attracted by any charged rod.

Figure 20-5 A standard two-by-four balanced on a curved dish can be rotated by the attractive force of a charged rod.

The Electroscope

In most experiments we transfer so few charges that the total attraction or repulsion is small compared with the pull of gravity. Detecting that an object is charged becomes difficult unless it is very light—like bits of paper or a balloon. We get around this difficulty with a device called an *electroscope*, which gives easily observable results when it is charged. By bringing the object in question near the electroscope, we can deduce whether it is charged or not by the effect it has on the electroscope.

The essential features of an electroscope are a metal rod with a metal ball on top and two very light metal foils attached to the bottom. Figure 20-6 shows a homemade electroscope constructed from a chemical flask. The glass enclosure

Figure 20-6 An electroscope made from a flask, a metal rod, and two pieces of thin metal foil.

Figure 20-7 When an electroscope is charged by direct contact, it has the same charge as the rod.

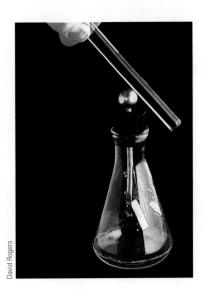
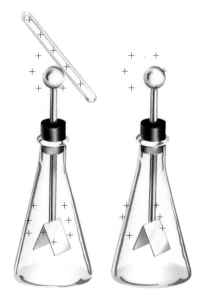

protects the very light foils from air currents and electrically insulates the foils and rod from the surroundings.

If the electroscope isn't charged, the foils hang straight down under the influence of gravity. When we touch the ball of the electroscope with a charged rod, some of the rod's excess charge is shared with the electroscope. The charges flow from the rod onto the electroscope because the charges on the rod repel each other and therefore distribute themselves over the largest region possible, which includes the ball, the metal rod, and the foils. Because the foils have the same kind of charge, they repel each other and swing apart (Figure 20-7). The separation of the foils indicates that the electroscope is charged; the amount of separation is a rough indication of the amount of excess charge. This happens when the electroscope is touched with a positive rod as well as a negative one.

When a charged object is touched to an electroscope, the electroscope takes on the same kind of charge as the object. However, the electroscope shows only the presence of charge; it does not indicate the sign. We can determine the sign by slowly bringing a rod with a known positive or negative charge toward the ball of the charged electroscope and watching the motion of the foils.

For example, as a positive rod nears the electroscope, it induces a redistribution of charge in the electroscope. It repels the positive charge and attracts the negative charge, causing an increase in positive charge (or a decrease in negative charge) on the foils. If the foils were originally positive, the extra positive charge would cause them to move farther apart. If the foils were originally negative, the extra positive charge would cancel part of the negative charge and the foils would move closer together.

Question   When a negative rod is slowly brought near an electroscope that is initially charged, the foils move closer together. What is the charge on the electroscope?

Answer   The negative rod repels negative charges on the electroscope's ball, increasing their concentration on the foils. Because this negative charge reduces the charge on the foils, they must have been positive initially.

A charged rod will also separate the foils on an uncharged electroscope. As a positive rod is brought near an uncharged electroscope, some of the negative charges in the electroscope are attracted to the ball on top; the ball has a net negative charge, and the foils have a net positive charge, as shown in Figure 20-8(a).

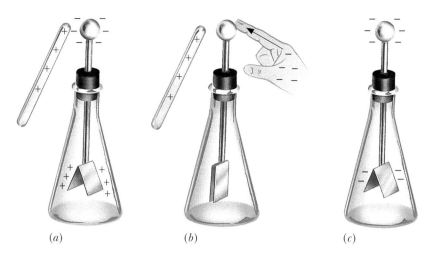

Figure 20-8 (a) The positively charged rod brought near the electroscope attracts negative charges to the ball, leaving the foils with an excess positive charge. (b) Touching the ball with a finger allows negative charges to flow into the electroscope, (c) leaving it with a net negative charge.

Note that the net charge on the electroscope is still zero because we did not touch it with the rod. We have only moved the charges around in the electroscope by bringing the charged rod near the ball. When the charged rod is removed, the charges redistribute themselves and the foils fall. An analogous phenomenon happens with a negative rod.

Imagine that while the positive rod is held near the electroscope you touch the ball with your finger, as shown in Figure 20-8(b). Negative charges are attracted by the large positive charge on the rod and travel from the ground through your body to the electroscope. If you first remove your finger and then the rod, the electroscope is now charged and the foils repel each other. This is known as charging the electroscope by induction. The electroscope acquires a charge *opposite* to that on the rod.

The Electric Force ✓MATH

✓Extended presentation available in the *Problem Solving* supplement

Simple observations of the attraction or repulsion of two charged objects indicate that the size of the electric force depends on distance. For instance, a charged object has more effect on an electroscope as it gets nearer. But we need to be more precise. How does the size of this force vary as the separation between two charged objects changes? And how does it vary as the amount of charge on the objects varies?

In 1785 French physicist Charles Coulomb measured the changes in the electric force as he varied the distance between two objects and the charges on them. He verified that if the distance between two charged objects is doubled (without changing the charges), the electric force on each object is reduced to one-fourth the initial value. If the distance is tripled, the force is reduced to one-ninth, and so on. This type of behavior is known as an **inverse-square relationship**; *inverse* because the force gets smaller as the distance gets larger, *square* because the force changes by the square of the factor by which the distance changes.

Coulomb also showed that reducing the charge on one of the objects by one-half reduced the electric force to one-half its original value. Reducing the charge on each by one-half reduced the force to one-fourth the original value. This means that the force is proportional to the product of the two charges.

Question Coulomb was not able to measure charge directly. What technique could he have used to reduce the charge to one-half its original value?

Answer He could have used identical conducting spheres and put the charged sphere into contact with a neutral one. By symmetry each sphere would have one-half the original charge.

These two effects are combined into a single relationship known as Coulomb's law:

Coulomb's law ▶

$$F = k\frac{q_1 q_2}{r^2}$$

In this equation, q_1 and q_2 represent the amount of charge on objects 1 and 2, r is the distance between their centers, and k is a constant (known as Coulomb's constant) whose value depends on the units chosen for force, charge, and distance.

Each object feels the force due to the other. The forces are vectors and act along the line between the centers of the two objects. The force on each object is directed toward the other if the charges have opposite signs and away from each other if the charges have the same sign (Figure 20-9). Because the two forces are due to the interaction between the two objects, the forces are an action–reaction pair. According to Newton's third law, the forces are equal in magnitude, point in opposite directions, and act on different objects.

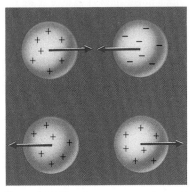

Figure 20-9 The forces on two charged objects are equal in size and opposite in direction in accordance with Newton's third law.

Because the existence of an elementary, fundamental charge was not known until the 20th century, the unit of electric charge, the **coulomb** (C), was chosen for convenience in use with electric circuits. (We will formally define the coulomb later.) Using the coulomb as the unit of charge, the value of Coulomb's constant is determined by experiment to be

Coulomb's constant ▶

$$k = 9 \times 10^9 \frac{\text{N} \cdot \text{m}^2}{\text{C}^2}$$

The coulomb is a tremendously large unit for the situations we have been discussing. For instance, the force between two spheres, each having 1 coulomb of charge and separated by 1 meter, is

$$F = k\frac{q_1 q_2}{r^2} = \left(9 \times 10^9 \frac{\text{N} \cdot \text{m}^2}{\text{C}^2}\right) \frac{(1\text{ C})(1\text{ C})}{(1\text{ m})^2} = 9 \times 10^9 \text{ N}$$

This is a force of 1 million tons!

The charges that we have been discussing are much less. When you scuff your shoes on a carpet, the charges transferred are about a millionth of a coulomb. But even this charge is very large compared with the charge on a single electron or proton. The size of the charge on an electron or a proton is now known to be 1.6×10^{-19} coulomb. Conversely, it would take the charge on 6.24×10^{18} protons (or electrons) to equal a charge of 1 coulomb.

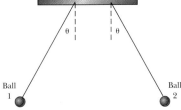

Figure 20-10

Flawed Reasoning

The following question appears on the final exam: "Two small, identical metal balls are hung next to each other from silk threads. Both balls are given the same net charge, such that they repel each other, as shown in Figure 20-10. If the net charge on ball 2 is reduced by half, will the angle that ball 1 makes with the vertical be smaller than, the same as, or larger than the angle that ball 2 makes with the vertical?"

Peter gives the following answer: "Ball 2 will still hang at the same angle as before. The charge on ball 1 has not changed, so it exerts the same force on ball 2. Ball 2 has half the charge, so it exerts half the force on ball 1. Therefore, ball 1 will hang at a smaller angle than ball 2." **Should Peter get full credit for this problem?**

Answer Coulomb's law automatically satisfies Newton's third law. The force exerted on ball 1 by ball 2 must be equal in magnitude (and opposite in direction) to the force exerted on ball 2 by ball 1. Reducing the charge on ball 2 by half will reduce the force felt by *each* ball by half. Both balls hang at the same, smaller angle.

Electricity and Gravity

The mathematical form of Coulomb's law is the same as that of the law of universal gravitation discussed in Chapter 5. Therefore, the drawing in Figure 5-2 illustrating the dependence of the inverse-square law on distance also holds for electricity if Earth is replaced by a charged object. There are, however, several important differences between gravity and electricity that govern the roles of these two forces in the Universe.

There are two kinds of electric charge; positive and negative. Opposite charges attract each other and like charges repel each other. There is, however, only one kind of mass. Negative mass does *not* exist, and the gravitational force is never repulsive. Because of the induced charge distribution possible with two kinds of charge, conducting materials can shield a region from all external electric forces. Suppose we were to build a large metal room. If this room were placed near a large electric charge, the charges in the metal walls would redistribute in such a way as to cancel the effect of the external charge for all locations in the room. The cancellation only works because there are two kinds of electric charge. This result is not possible for gravitational forces; a gravitation-free chamber cannot be created, because there is no negative mass. Antigravity spaceships, although they make good material for science fiction writers, are not possible.

Another difference between electricity and gravity lies in the behavior of objects in the respective fields. The gravitational force is proportional to an object's mass, and therefore objects of all sizes and compositions have the same acceleration in a gravitational field. For example, all objects have the same free-fall acceleration near Earth's surface. In contrast, the electric force is proportional to the charge on an object and *not* to the object's mass. Therefore, charged objects have different accelerations in an electric field. For example, if we were to release a proton and an electron near a charged object, the electron would have a much larger acceleration. The forces on the electron and the proton are the same size, but the electron's mass is much smaller, resulting in a much larger acceleration. In this case, the accelerations would be in opposite directions because the charges have opposite signs.

Finally, all charges occur as integral multiples of one fixed size. Experiments have shown that the charges on electrons and protons are the same size. The

Professor Sprott is protected from the high-voltage discharge by the metallic cage.

The electric force dominates on the atomic scale, such as in this large crystal.

The gravitational force dominates in the Solar System and beyond, such as in this galaxy.

charges on other ordinary subatomic particles are either this same size or integral multiples of this size; the charges can be 2 times or 3 times the charge on the proton but never 1.5 times this charge. Mass also occurs in lumps, but there are many different sizes; for instance, the masses of the electron, the proton, and the neutron. These can combine in many different ways and the total masses are not multiples of a single unit mass.

A consequence of these differences is the different roles these two forces play in our lives. The electric force is the dominant force in the atomic world; it determines the properties of atoms and molecules. On the other hand, the gravitational force dominates on the macroscopic scale of people, planets, and galaxies. The reason electricity and gravity switch roles in these two domains is that macroscopic objects are essentially uncharged; that is, there are approximately equal numbers of positive and negative charges on most large objects. Even though the total attractive force between all the positive charges in Earth and all the negative charges in the Moon is tremendous, it is equal and opposite to the repulsive force between the negative charges in Earth and the negative charges in the Moon. The net electric force is essentially zero because of this cancellation. However, because there is only one kind of mass, there is nothing to cancel the much weaker gravitational force between the atoms in Earth and atoms in the Moon. So as we move from people to planets and eventually to galaxies, the gravitational force becomes the dominant force.

WORKING IT OUT | Gravitational and Electric Forces

As a numerical example, let's calculate the sizes of the electric and gravitational forces between an electron and a proton in a hydrogen atom. We start by assuming that the electron and proton are separated by a distance of 5.29×10^{-11} m and use the known masses and charges of the electron and proton. (These are given on the back end pages of this text.)

$$F_g = G\frac{m_1 m_2}{r^2} = \left(6.67 \times 10^{-11}\, \frac{\text{N} \cdot \text{m}^2}{\text{kg}^2}\right) \frac{(9.11 \times 10^{-31}\, \text{kg})(1.673 \times 10^{-27}\, \text{kg})}{(5.29 \times 10^{-11}\, \text{m})^2}$$

$$= 3.63 \times 10^{-47}\, \text{N}$$

$$F_e = k\frac{q_1 q_2}{r^2} = \left(9 \times 10^9\, \frac{\text{N} \cdot \text{m}^2}{\text{C}^2}\right) \frac{(1.6 \times 10^{-19}\, \text{C})^2}{(5.29 \times 10^{-11}\, \text{m})^2} = 8.23 \times 10^{-8}\, \text{N}$$

Therefore, the electric force is more than 10^{39} times the strength of the gravitational force. (That's 1000 trillion trillion trillion times bigger!) Because the two forces change in the same way as the separation of the electron and proton changes, the separation does not matter and the electric force is always this much stronger.

The Electric Field

Implicitly, we have assumed the force between two charges to be the result of some kind of direct interaction—sort of an action-at-a-distance interaction. This type of interaction is a little unsettling because there is no direct pushing or pulling mechanism in the intervening space. Electrical effects are evident even in situations in which there is a vacuum between the charges.

In many cases, it proves to be both conceptually and computationally simpler to separate the electrical interaction that one object feels as the result of another into two distinct steps. First, one of the objects generates, by virtue of its charge, an **electric field** at every point in space. Second, another object interacts, by virtue of its charge, with the electric field to experience the force. The electric field is created by electric charges and exerts forces on other electric charges. This method is analogous to what we did with the gravitational force near the end of Chapter 5.

If this were the only purpose of the field idea, it would play a minor role in our physics world view. In fact, it probably seems like we are trading one unsettling idea for another. However, as we continue our studies, we will find that the electric field takes on an identity of its own. As we will learn in Chapter 22, electric and magnetic fields can travel through space as waves.

We define the electric field **E** at every point in space as the force exerted on a unit positive charge placed at the point. This is equivalent to the way that the gravitational field was defined, with the unit mass replaced by a unit positive charge.

◂ electric field = force on unit positive charge

Because force is a vector quantity, the electric field is a vector field; it has a size and a direction at each point in space. You could imagine the space around a positive charge as a "porcupine" of little arrows pointing outward, as shown in Figure 20-11. The arrows farther from the charge would be shorter to indicate that the force is weaker there.

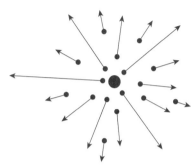

Question   What does the electric field surrounding a negative charge look like?

Answer   An electric field surrounding a negative charge looks just like the electric field surrounding a positive charge except that all of the arrows are reversed.

Figure 20-11 Sample electric field vectors around a positive source charge.

The values for an actual electric field can be measured with a test charge. The unit of charge that we have been using is 1 coulomb. This is a very large amount of charge, and if we actually used 1 coulomb as our test charge, it would most likely move the charges that generated the field, thus disturbing the field. Therefore, we use a much smaller charge, such as 1 microcoulomb, and obtain the size of the field by dividing the measured force **F** by the size q of the test charge:

$$\mathbf{E} = \frac{\mathbf{F}}{q}$$

◂ electric field = $\dfrac{\text{force}}{\text{charge}}$

Notice that the units of electric field are newtons per coulomb (N/C).

If we know the sizes and locations of the charges creating the electric field, we can also calculate the value of the field at any point of interest by assuming that we place a 1 coulomb charge at the location and calculating the force on this charge using Coulomb's law. In doing this, we can take advantage of the fact that each charge acts independently; the effects simply add. This means that we calculate the contribution of each charge to the field and then add these contributions vectorially.

Once we know the value of the electric field at any point, we can calculate the force that any charge q would experience if placed at that point:

$$\mathbf{F} = q\mathbf{E}$$

◂ electric force = charge × electric field

This is read as, "The force on an object is equal to the net charge q on the object times the electric field **E** at the location of the object."

As an example, let's assume that we have generated a uniform electric field that points downward and has a size of 1000 newtons per coulomb. If we place an object in this field that has a positive charge of 1 microcoulomb, the object will experience a downward force of

$$F = qE = (10^{-6}\,\text{C})\left(10^3\,\frac{\text{N}}{\text{C}}\right) = 10^{-3}\,\text{N}$$

If we change the charge on the object, it is very easy to calculate the new force; we do not have to deal with the charges that produced the electric field.

Figure 20-12 Electric field lines represent the total electric field in a region of space.

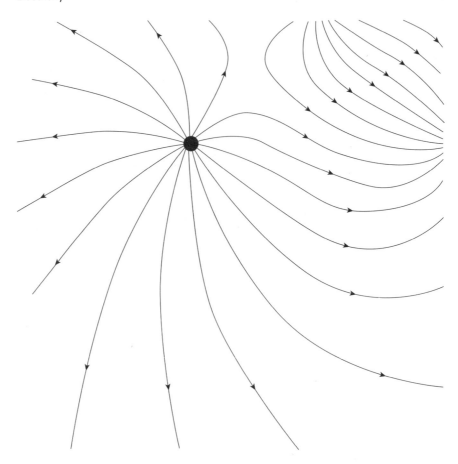

Electric Field Lines

If we are only interested in what is going on at a single point, the electric field representation is very helpful. However, it becomes cumbersome if we are interested in a region of space because each point in space may have a different electric field vector associated with it. To draw this, we would need to draw a different vector (possibly different in both magnitude and direction) at each point and these would tend to overlap! To deal with this, we introduce an alternative representation that uses **electric field lines**.

Imagine a region containing charged particles that are fixed in place and create an electric field at every point in the region. We now draw an electric field line. Find the direction of the electric field at a starting point and take a small step in the direction of this vector. At the new point, again find the direction of the electric field and take a small step in this direction. Continuing this process creates a series of points that we connect with a smooth line. This line is an electric field line. The final step is to put a small arrow on the line to indicate the direction of travel. Starting at a new point in the region leads to a new electric field line. If we do this from enough different starting points, we can get a sample of lines throughout the region, as shown in Figure 20-12.

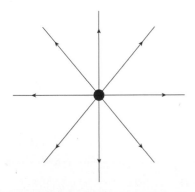

Figure 20-13 Electric field lines due to an isolated positive source charge.

This sounds like a long, involved process—and it is. The good news is that we can often learn a great deal through qualitative sketches of electric field lines, which are much easier than doing the detailed calculations. We can use our intuition to draw the electric field lines surrounding an isolated positive source charge. A positive test charge would be repelled directly away from the positive source charge, so the electric field lines should start on the source charge and continue radially outward to infinity, as shown in Figure 20-13. The electric field vector at any point in space is then tangent to the electric field line that would pass through that point. This is consistent with the drawing in Figure 20-11 of the electric field vectors for a positive source charge. (If we are interested in a point that does not lie on one of the electric field lines that have been drawn, we can find the approximate direction of a line that passes through the point by looking at the surrounding lines.)

Notice that the electric field lines in Figure 20-13 start at locations equally spaced around the source charge. When the electric field lines are drawn this way they give us visual information about the magnitude of the electric field vector at any point. Notice that the field lines are close together near the source charge, where the electric field is strong, and the field lines are far from each other at locations away from the source charge, where the electric field is weak. Indeed, any point in space that is 3 centimeters from the source charge should have the same size electric field. The spherical symmetry of our field lines ensures that the spacing between adjacent field lines is the same for all these points. In general, the strength of the electric field (that is, the length of the vector) is greater in regions where the electric field lines are closer together. Another way to say this is that the electric field is proportional to the density of electric field lines.

Question How would the diagram in Figure 20-13 change if the positive source charge were replaced by a negative source charge?

Answer The arrows in the diagram would point in opposite directions.

When more than one source charge is present in a region, the field lines represent the total electric field in the region due to all the source charges. At locations very close to one of the source charges, the electric field lines should still be radially symmetric about that source charge (as its contribution will dominate). The number of field lines originating on a positive source charge or ending on a negative source charge should be proportional to the magnitude of the charge. In other words, the electric field strength at a location 1 centimeter from a +2 coulomb source charge should be twice as big as the electric field strength 1 centimeter from a +1 coulomb source charge. Figure 20-14 shows the electric field lines around two source charges, one positive and one negative.

Question If the positive source charge in Figure 20-14 has a charge of +6 microcoulombs, what is the magnitude of the charge on the negative source charge?

Answer The number of electric field lines originating on a positive charge or ending on a negative charge is proportional to the size of the electric charge.

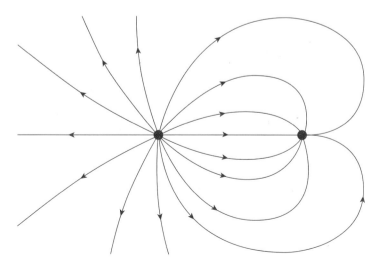

Figure 20-14 The electric field lines indicate that the magnitude of the positive charge is twice that of the negative charge.

Sixteen field lines leave the positive source charge, but only eight of those field lines end on the negative source charge. The negative charge must be half the size of the positive charge. It must have a charge of −3 microcoulombs.

Notice that electric field lines always begin on positive charges and end on negative charges. If the region you are considering contains more positive than negative charges, some lines will leave the region. If the region contains more negative charges, some lines will come in from outside the region.

Question Why do electric field lines never cross?

Answer The electric field lines represent the total electric field in a region due to all source charges. The electric field vector at any point in space is tangent to the electric field line through that point. If two electric field lines crossed at any point, the electric field at that point would have to be tangent to both electric field lines. It would have to point in two directions at a single location. This is not possible.

A very common situation where electric field lines provide physical insight is the case of charged parallel metal plates. If electrons are taken from one metal plate and placed on the other, both plates end up with the same amount of excess electric charge (positive on one plate and negative on the other). The charge will spread out on the facing surfaces of the plates until the charge density is uniform. Electric field lines originate on positive charge and end on negative charge, so the uniform charge distribution on the plates dictates that the electric field lines between the plates be parallel to each other, perpendicular to the plates, and uniformly spaced, as shown in Figure 20-15.

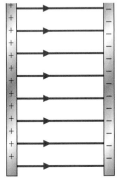

Figure 20-15 The electric field lines between charged metal plates indicate that the electric field is uniform between the plates.

These electric field lines are very simple, and they represent an electric field that is uniform in strength everywhere between the two plates. The simplicity of the electric field lines predicts a result that is contrary to common sense: if you place a small positive test charge at a location halfway between the plates, it experiences the same electric field (and hence the same electric force) as it would very close to the positive plate (or, in fact, anywhere else between the plates). It is possible to show that this is indeed the case mathematically, but it is much easier to use the concept of electric field lines.

Electric Potential ✓ MATH

We must do work to lift a 1-kilogram block in Earth's gravitational field, and this work increases the gravitational potential energy of the block. When the block is released, it falls and the gravitational potential energy is changed to kinetic energy. Lifting a 5-kilogram block the same distance requires five times the work and the block gains five times the gravitational potential energy. In other words, knowing the energy that is gained by the 1-kilogram block allows us to calculate the energy that would be gained by any other mass moved between the same two points.

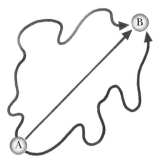

Figure 20-16 The work performed in taking a charge from A to B does not depend on the path.

In the same way, it requires work to move a charged particle in an electric field and this work changes the **electric potential energy** of the particle. When we release the particle, this electric potential energy can be converted to kinetic energy. Therefore, we define the electric potential energy the same way we did for gravity. The electric potential energy of a charged object is equal to the work done in bringing the object from some zero reference location to the object's location. As with gravitational potential energy, the value of the electric potential energy does not depend on the path, but it does depend on the reference location, the location of the object, and the charge on the object (Figure 20-16).

electric potential energy = ▶ work from reference point

As with gravitational potential energy, the actual value of the electric po-

Lightning

Everybody has seen lightning (Figure A) and has probably been frightened by this spectacular display of energy at some time or another. Lightning is unpredictable and seems to occur randomly and instantaneously. Everything but the thunder is usually over in less than half a second. And yet the effects are not inconsequential: each year lightning causes about $80 million of damage in the United States. On the average, lightning kills 85 people a year and injures another 250 in the United States alone.

We still have a lot to learn about lightning. For instance, we don't know what causes the initial buildup of charge, what determines the paths the lightning takes, or what triggers the lightning bolts.

We do know that, during a storm, clouds develop a separation of electric charge, with the tops of the clouds positively charged and the bottoms negatively charged, as shown in Figure B. The production of a lightning bolt begins when the negative charge on the bottom of the cloud gets large enough to overcome air's resistance to the flow of electricity (Figure C) and electrons begin flowing toward Earth along a zigzag, forked path at about 60 miles per second! As the electrons flow downward, they collide with air molecules and ionize them, producing more free electrons. Even though the current may be as large as 1000 amperes, this is not the lightning bolt we see.

Meanwhile, as the electrons approach the ground, the ground becomes more and more positively charged due to the repulsion of electrons in the ground. This positively charged region moves up through any conducting objects on the ground—houses, trees, and people—into the air (Figure D). When the downward-moving electrons meet the upward flowing positive regions at an altitude of 100 meters or so, they form a complete circuit (Figure E), and the lightning

Figure A Lightning bolts can generate currents up to 200,000 amperes.

begins. In less than 1 millisecond, up to 1 billion trillion electrons may reach the ground; the current may be as high as 200,000 amperes. Although the flow of charge is downward, the point of contact between the cloud charge and the ground charge is flowing *upward* at about 50,000 miles per second. (Observers will report that the lightning bolt is moving downward because the pathway that is lit is the initial forked pathway coming from the cloud.) The upward surge heats the air to 50,000°F. One meter of the lightning's path may shine as bright as 1 million 100-watt lightbulbs (Figure F). This rapid heating of the air along the lightning's path also produces a shock wave that we hear as the thunder. What we see as a single lightning bolt is usually several bolts in rapid succession along the same path.

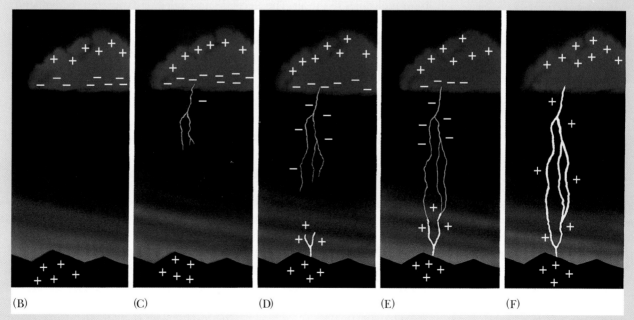

(B)　　　(C)　　　(D)　　　(E)　　　(F)

Figures B–F The development of a lightning bolt.

tential energy is not important in physical problems; it is only the difference in energy between points that matters. If it requires 10 joules of work to move a charged object from point A to point B, the electric potential energy of the object at point B is 10 joules higher than at point A. If point A is the zero reference point, the electric potential energy of the object at point B is 10 joules.

Because objects with different charges have different electric potential energies at a given point, it is often more convenient to talk about the energy available due to the electric field without reference to a specific charged object. The **electric potential** V at each point in an electric field is defined as the electric potential energy EPE divided by the object's charge q:

electric potential = $\dfrac{\text{electric PE}}{\text{charge}}$ ▸

$$V = \frac{EPE}{q}$$

Notice that it doesn't matter which charged object we use to define the electric potential. This quantity is numerically equal to the work required to bring a positive test charge of 1 coulomb from the zero reference point to the specified point. The units for electric potential are joules per coulomb (J/C), a combination known as a **volt** (V). Because of this, we often speak of the electric potential as a *voltage*.

Defining the electric potential allows us to obtain the electric potential energy for any charged object by multiplying the potential by the charge. Once again, it is only the *potential difference* that matters. For instance, a 12-volt battery has an electric potential difference of 12 volts between its two terminals. This means that 1 coulomb of charge moving from one terminal to the other would gain or lose (1 coulomb)(12 volts) = 12 joules of energy.

Question How much work is required to move 3 coulombs of positive charge from the negative terminal of a 12-volt battery to the positive terminal?

Answer Each coulomb requires 12 joules of work, so 3 coulombs require 36 joules: $W = qV = $ (3 coulombs)(12 volts) = 36 joules.

If the potential difference, or voltage, between two points is high and the points are close together, the electric field can be strong enough to tear electrons from molecules in the air. The electrons are pulled one way, and the remaining positive ions are pulled the other way. Because electrons have much less mass than ions, they quickly accelerate to high velocities. As they accelerate they collide with other molecules, knocking additional electrons loose. This forms a cascade of electrons that we call a *spark*. Dry air breaks down in this fashion when the electric field reaches about 30,000 volts per centimeter.

PHYSICS | **ON YOUR OWN**

The next time the weather is dry and you get sparks when touching doorknobs, try holding a key in your hand when you touch metal objects. Why does this hurt less? Estimate your electric potential by estimating the length of the sparks.

Summary

Objects become electrically charged or uncharged by transferring charges. An uncharged object has equal numbers of positive and negative charges. Positively charged objects may have an excess of positive charges or a deficiency of negative charges. In an isolated system, the total charge is conserved. Charges can flow through conductors such as metal wires but not through insulators such as silk threads.

In electricity, unlike gravity, there are two different types of charge, and the direction of the force depends on the relative types of the charges. Like charges repel; unlike charges attract. A charged object can attract an uncharged object due to an induced separation of charges.

The electric force has the same mathematical form as the gravitational force. The electric force, however, can be repulsive as well as attractive. Electric charge comes in whole-number multiples of one fixed size. The electric charges on electrons and protons have the same size but are opposite in sign.

Electric charges are surrounded by an electric field that is equal to the force experienced by a unit positive charge. This electric field is a vector field with a size and a direction at each point in space. The units for electric field are newtons per coulomb. Electric field lines represent the electric field in a region of space. The electric field at any location points in a direction tangent to the field line at that location, and the strength of the electric field is proportional to the local density of electric field lines.

The electric potential energy of a charged object in an electric field is equal to the work done in bringing the object from some zero reference location. Only the differences in electric potential energy matter in physical situations. The electric potential equals the electric potential energy divided by the object's charge. This quantity is numerically equal to the work required to bring a positive test charge of 1 coulomb from the zero reference point. The units for electric potential are joules per coulomb, or volts.

Chapter 20 Revisited

Sparks occur when the electric field becomes strong enough to pull electrons from atoms. These electrons are accelerated by the electric field, obtaining sufficient kinetic energy to knock electrons from other atoms. This avalanche of electrons produces the visible sparks that we see. Electric fields of 30,000 volts per centimeter are strong enough to break down atoms in dry air. Because the electric field is proportional to the electric potential difference and inversely proportional to the distance, it takes large voltages to produce long sparks.

KEY TERMS

charge: A property of elementary particles that determines the strength of its electric force with other particles possessing charge. Measured in coulombs, or in integral multiples of the charge on the proton.

charged: Possessing a net negative or positive charge.

conductor: A material that allows the passage of electric charge. Metals are good conductors.

conservation of charge: In an isolated system, the total charge is conserved.

coulomb: The SI unit of electric charge; the charge of 6.24×10^{18} protons.

electric field: The space surrounding a charged object, where each location is assigned a value equal to the force experienced by one unit of positive charge placed at that location; measured in newtons per coulomb.

electric field lines: A representation of the electric field in a region of space. The electric field is tangent to the field line at any point and its magnitude is proportional to the local density of field lines.

electric potential: The electric potential energy divided by the object's charge; the work done in bringing a positive test charge of 1 coulomb from the zero reference location to a particular point in space; measured in joules per coulomb, or volts.

electric potential energy: The work done in bringing a charged object from some zero reference location to a particular point in space; measured in joules.

grounding: Establishing an electrical connection to the earth to neutralize an object.

insulator: A material that does not allow the passage of electric charge. Ceramics are good insulators.

inverse-square relationship: A relationship in which a quantity is related to the reciprocal of the square of a second quantity; the electric force is proportional to the inverse square of the distance from the charge.

volt: The SI unit of electric potential, 1 joule per coulomb.

CONCEPTUAL QUESTIONS

1. A handheld glass rod can be charged by rubbing it with silk or a plastic bag while holding it in your hands. Would you conclude from this that glass is a conductor or an insulator? Why?
2. Inexperienced physics teachers will often demonstrate the use of the electroscope by touching it with a charged glass rod at a single point. More experienced teachers will typically drag the length of the rod across the top of the electroscope to increase the desired effect. Why does this help?
3. Why is it not possible to charge just one end of a metal rod?
4. Could you use fur to charge a metal rod that is held in your bare hand? Explain.
5. Why is it easier to charge a balloon on a dry day than on a humid day?
6. Why is it easier to demonstrate electrostatic phenomena in Fairbanks, Alaska, than in Honolulu, Hawaii?
7. Before an aircraft is fueled from a truck, a wire from the truck is attached to the aircraft. Why?
8. Why do clothes sometimes stick together when removed from a dryer?
9. If one rubber rod is charged with fur and another with plastic, they attract each other. How would you explain this?
10. You use a piece of silk to charge a glass rod. Does this leave the silk with a positive or a negative charge? Explain.
11. When you rub balloons with wool and place them near bits of paper, you find three categories of behavior. Balloons attract wool and bits of paper but repel other balloons. Pieces of wool attract balloons and bits of paper but repel other pieces of wool. Bits of paper are attracted to both the balloons and the pieces of wool but do not react with other bits of paper. When we have three distinct behaviors, why do we need only two types of charge in our model?
12. How would you know if you discovered a third kind of charge? What other objects would you expect the third kind of charge to attract and repel?
13. Your classmate claims that magnets are really just charged rods with opposite charges at the two ends. What experiment could you perform to convince your classmate otherwise?
14. You find that the north pole of a bar magnet is attracted to a charged glass rod. You then discover that the south pole of the magnet is also attracted to the same rod. Will the north pole be attracted or repelled or experience no force if brought near a charged rubber rod? Explain.
15. You have three small balls, each hanging from an insulating thread. You find that balls 1 and 2 repel one another and that balls 2 and 3 repel one another. Will balls 1 and 3 attract or repel one another? Explain.
▲ 16. You have three small balls, each hanging from an insulating thread. You find that balls 1 and 2 attract one another and that balls 1 and 3 attract one another. If balls 2 and 3 are brought together, what are the possible outcomes?
17. The nucleus of a helium atom contains two protons and two uncharged neutrons. If two electrons surround the nucleus, what is the total charge of the atom?

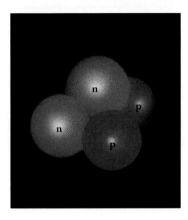

18. In the modern view of electricity, does an object with a negative charge have an excess of positive charge, an excess of negative charge, a deficiency of negative charge, or a deficiency of positive charge? Explain.
19. Describe how a charged balloon sticks to a wall.
20. When a charged comb is brought near bits of paper, the bits are first attracted to the comb and then repelled after the bits come in contact with the comb. Describe how the charges on the comb and on the bits of paper change during this process.
21. Why are neutral objects attracted to both negatively and positively charged objects?
22. Describe how a negatively charged rod attracts (a) an uncharged conducting object and (b) an uncharged insulating object.
23. How could you use an electroscope of known positive charge to determine the sign of a charged rod?
24. How could you use a negatively charged rod to determine whether a charged electroscope has a negative or a positive charge without changing the charge on the electroscope?
25. Describe how you would use a negatively charged rod to put a negative charge on an electroscope.
26. Describe how you would use a negatively charged rod to give an electroscope a positive charge.
27. An electroscope initially has a net negative charge. Why do the foils come together and stay together when a human hand touches the electroscope?
28. An electroscope is initially given a net positive charge. You find that, as you bring your hand near the top of the electroscope without touching it, the foils come slightly closer together. When you remove your hand, they move back apart. How do you account for this?
▲ 29. You have two metal spheres on insulating stands. You bring them each in turn near the ball of a positively charged electroscope. In each case the deflection of the leaves is reduced slightly. You then bring them each in turn near the ball of a negatively charged electroscope and find that sphere A causes the deflection to increase whereas sphere B causes it to decrease. What can you say about the net charge on each of the two metal spheres?

▲ 30. You hold a charged rubber rod near the ball of an electroscope and then briefly touch the ball with your finger. After removing the rubber rod, you bring a charged glass rod near the ball. Will the deflection of the leaves increase, decrease, or stay the same? Why?

▲ 31. When Coulomb was developing his law, he did not have an instrument for measuring charge. And yet he was able to obtain spheres with $\frac{1}{2}, \frac{1}{3}, \frac{1}{4}, \ldots$ of some original charge. How might he have used a set of identical spheres to do this?

▲ 32. You have three identical metal spheres on insulating stands. The spheres hold charges $Q_A = -2q$, $Q_B = -q$, and $Q_C = 4q$. First, sphere A is brought into contact with sphere C and separated. Second, sphere C is brought into contact with sphere B and separated. What is the resulting charge on sphere B?

33. Two uniformly charged insulating spheres are held atop fixed posts. The charge on one of the spheres is three times the charge on the other. Which diagram correctly represents the magnitude and direction of the electric forces on the two spheres? Explain your answer.

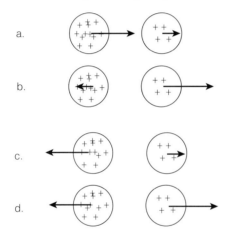

34. You have positively charged objects. Object B has twice the charge of object A. When they are brought close together, object A experiences an electric force of 10 newtons. Is the electric force experienced by object B greater than, equal to, or less than 10 newtons? Explain.

35. Assume that you have two identically charged objects separated by a certain distance. How would the force change if the objects were three times as far away from each other?

36. Two charged objects are very, very far from any other charges. If the distance between them is cut to one-half its original value, what happens to the electric force between the charges?

37. Assume that you have two identically charged objects separated by a certain distance. How would the force change if one object had three times the charge and the other object kept the same charge?

38. Two charged objects are very, very far from any other charges. If the charge on each object is doubled, what happens to the electric force between the objects?

39. Three identical charges are arranged as shown in the figure. The distance from A to B is twice the distance from A to C. Which of the vectors best represents the force on charge A due to B and C? Justify your answer.

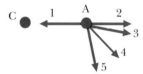

40. Two identical charges are fixed at the locations indicated in the figure. A third charge, identical to the other two, is placed first at position A and then position B. For each case, determine the direction of the net electric force on the third charge. If the force is zero, state that explicitly.

41. Although the formulas for electric and gravitational forces have the same form, there are differences in the two. What are some of these differences?

42. What are the similarities between electric and gravitational forces?

43. When we approach another person, we are not aware of the gravitational and electric forces between us. What are the reasons in each case?

44. Even though electric forces between electrons and protons are very much stronger than gravitational forces, gravitational forces determine the motions in the Solar System. Why?

45. Why are the accelerations of all charged objects near a charged sphere not the same?

▲ 46. In this chapter we compared the electric and gravitational forces between an electron and a proton. Why is the result valid for all separations?

47. A metal sphere with a charge of +3 coulombs experiences an electric force of 15 newtons directed to the left. If the charge on the sphere is increased to 6 coulombs, what force will it experience?

48. A metal sphere with a charge of +2 coulombs experiences an electric force of 25 newtons directed to the left. If the charge on the sphere is changed to −8 coulombs, what force will the sphere experience?

49. The electric field lines between two parallel charged plates are everywhere perpendicular to the plates and parallel to each other as shown in the figure. A small positively charged particle placed at the midpoint between the plates experiences an electric force of magnitude F. If the charged particle were instead placed at location A, close

to the positive plate, would the force on the particle be greater than, equal to, or less than *F*? Use the concept of field lines to justify your answer.

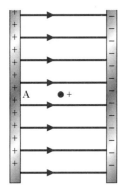

50. For the situation described in Question 49, the charged particle is released from rest at the midpoint and strikes the negative plate at speed *v*. If the particle had instead been released from rest at point A, would it have struck the negative plate at a speed greater than, equal to, or less than *v*? Explain your reasoning.
51. An electron and a proton are released in a region of space where the electric field is vertically upward. How do the electric forces on the electron and proton compare?
52. How do the accelerations of the electron and proton in Question 51 compare?
53. A portion of an electric field is shown below. Four locations within the field are marked with x's and labeled a, b, c, and d. Draw the vectors representing the force on a positive test charge when placed at each of these locations. Explain your reasoning.

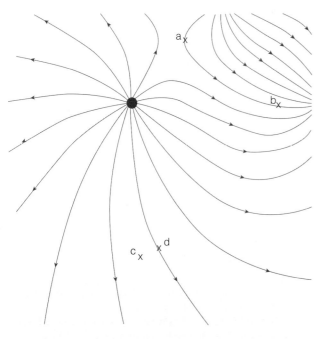

54. The field lines (solid lines) shown below represent the electric field in a certain region of space. A charged particle is released with an initial velocity and follows the trajectory from points 1 to 4 as shown by the dashed line. Draw vectors to represent the force on the charged particle at each location. What is the sign of the charged particle? Explain how you can tell.

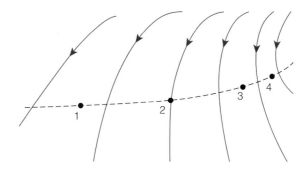

55. The figure shows the electric field lines in a region of empty space. Is the electric field in the region between A and B greater than, equal to, or less than the electric field in the region between C and D? Explain.

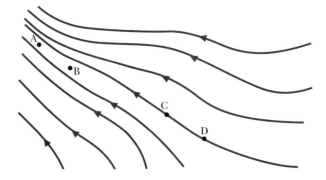

56. In the figure described in Question 55, points A and B and points C and D are the same distance apart. Is the potential difference between A and B greater than, equal to, or less than the potential difference between C and D? Explain.
▲ 57. A proton is observed moving at speed v_o at point A in space where the electric potential is 750 volts. It moves to point B where the electric potential is 550 volts. Is its speed at point B greater than, equal to, or less than v_o? Explain.
▲ 58. An electron is observed moving at speed v_o at point A in space where the electric potential is 750 volts. It moves to point B where the electric potential is 550 volts. Is its speed at point B greater than, equal to, or less than v_o? Explain.
59. How is the value of the electric potential at each point in space defined?
60. What is the definition of the value of the electric potential energy of a charged object at each point in space?
▲ 61. A proton is released from rest in a uniform electric field. Does the proton's electric potential energy increase or decrease? Does the proton move toward a location with a higher or lower electric potential? Explain.
▲ 62. An electron is released from rest in a uniform electric field. Does the electron's electric potential energy increase or decrease? Does the electron move toward a location with a higher or lower electric potential? Explain.

EXERCISES

1. A sodium ion contains 11 protons, 12 neutrons, and 10 electrons. What is the net charge of the ion?
2. What is the net charge of an iodine ion containing 53 protons, 74 neutrons, and 54 electrons?
3. The nucleus of a certain type of plutonium atom contains 94 protons and 150 neutrons. What is the total charge of the nucleus?
4. How many electrons does it take of have a total charge of −1 coulomb?
5. What is the electric force of attraction between charges of +3 C and −6 C separated by a distance of 2 m?
6. By what distance must two charges of +4 C be separated so that the repulsive force between them is 3.6×10^{10} N?
7. The nucleus of lithium contains three protons and three neutrons. When two electrons are removed from the neutral lithium atom, the remaining electron has an average distance from the nucleus of 0.018 nm. What is the force between the electron and the nucleus at this separation?
8. In an ionic solid like ordinary table salt (NaCl), an electron is transferred from one atom to the other. If a distance of 0.1 nm separates the atoms, how strong is the electric force between them?
9. How much stronger is the electric force between 2 protons than the gravitational force between them?
10. Calculate the ratio of the electric force to the gravitational force between 2 electrons.
11. A 5-mC charge experiences a force of 4 N directed north. What is the electric field (magnitude and direction) at the location of the charge? Find the force on a −20-mC charge that replaces the 5-mC charge.
12. A −20-mC charge experiences a force of 25 N directed west. What is the electric field (magnitude and direction) at the location of the charge? If this charge is removed and not replaced, what is the electric field at this location?
13. What is the electric field at a distance of 4 cm from 2 mC of negative charge?
14. What is the electric field 2 m from 4 C of positive charge?
15. What is the electric field at a distance of 0.2 nm from a carbon nucleus containing six protons and six neutrons?
16. What is the electric field 5.3×10^{-11} m from a proton?
17. What is the electric field midway between charges of 2 C and 6 C separated by 2 m?

18. What is the electric field midway between an electron and a proton separated by 0.2 nm?
▲ 19. What is the force on a proton located in an electric field of 5000 N/C? What is the proton's acceleration?
▲ 20. How would the values obtained in Exercise 19 change if the proton were replaced by an electron?
21. The electric potential energy of an object at point A is known to be 40 J. If it is released from rest at A, it gains 50 J of kinetic energy as it moves to point B. What is its potential energy at B?
22. If the object in Exercise 21 had a charge of +2 C, what would be the potential difference between points A and B?
23. How much work does a 12-V battery do in pushing 2 mC of charge through a circuit containing one lightbulb?
24. Points A and B each have an electric potential of +9 V. How much work is required to take 3 mC of charge from A to B?
25. What was your electric potential relative to a metal pipe if a spark jumped 1.3 cm (0.5 in.) through dry air from your finger to the pipe?
26. How far can a spark jump in dry air if the electric potential difference is 6×10^5 V? (This is the reason why high-voltage sources are surrounded by a vacuum or an insulating fluid.)

21 | Electric Current

This night scene of Boathouse Row in Philadelphia is an example of the artistic use of lighting.

We live in an electrical world. It is hard to imagine a world without electricity. How does the movement of electrons produce the light that allows us to work and play indoors and at night?

(See page 442 for the answer to this question.)

NEW phenomena often enter our world view accidentally, a result of someone with an open mind making an observation or conducting an experiment. Initially, these phenomena are often just curiosities, new challenges to our understanding of the world. Many of them give us new ways to do things and eventually find widespread use as our understanding of them increases. In some cases, however, a whole new technology is born. The electrical properties we studied in Chapter 20 were such curiosities. In the early days of investigation into these phenomena, little use was made of electricity other than building devices to amaze and shock friends. In the past 200 years, electricity has developed into a major technological presence in our society. Today, the United States, with 5% of the world's population, uses about one-fourth of the electricity generated in the world. This means that the average U.S. citizen uses five times the electricity the average world citizen does, including those in developed countries. When electricity usage of U.S. citizens is compared with that of citizens in third-world countries, the per capita usage is very much larger.

An important advance in understanding electricity came with the development of batteries. Batteries could make electric charges flow continuously as a current rather than in short bursts. This development eventually led to modern electric circuits that have had an obvious impact on our lives. Electric currents can produce heat and light, run motors and radios, operate sound systems and computers, and so on. The list is nearly endless.

PhysicsNow™ Test your understanding of this chapter by logging into PhysicsNow at **http://physics.brookscole.com/kf6e**, selecting the chapter, and clicking on the "Take a Pre-Test" link.

An Accidental Discovery

Near the end of the 18th century, Italian anatomist Luigi Galvani announced that an excised frog's leg twitched when he touched it. It was well known from firsthand experiences with static electricity that twitches in humans could result from electric charges. But this was just a leg! Galvani was convinced that he had discovered the secret life force that he believed existed within all animals. Galvani found that a twitch occurred when he touched the frog's leg with a metal object but not when he touched it with an insulator. Furthermore, the metal had to be different from the metal hook holding the other end of the leg. It took two *different* metals to produce the twitch.

Another Italian scientist, Alessandro Volta, heard of these results and performed many experiments with very sensitive electroscopes, searching for evidence that electric charge resides in animal tissue. He eventually convinced himself that the electricity was not in the frog's leg but was the result of touching it with the two metals.

Volta tried many different materials and discovered that an electric potential difference was produced whenever two different metals were joined. Some combinations of metals produced larger potential differences than others. One of his demonstrations was to place his tongue between pieces of silver and zinc that were held together at one end. The flow of electricity caused his tongue to tingle. You may have felt a similar sensation when a piece of metal or metal foil touched a metal filling in your tooth. A more detailed explanation of this process would take us beyond the scope of this book, but it is enough to realize that some chemical process moves charges from one metal to the other, creating a potential difference.

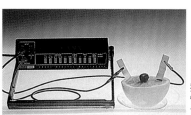

A modern version of Volta's demonstration that zinc and copper rods inserted into a grapefruit produce a potential difference. (The cherry is for decorative purposes only.)

PHYSICS | ON YOUR OWN

Make a battery by sticking a paper clip and a piece of copper wire into a lemon. You can feel the voltage by placing the tips of the wires to your tongue.

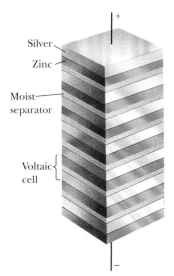

Figure 21-1 A schematic of Volta's battery composed of zinc and silver plates with moist paper between them.

A car battery can be recharged with a battery charger.

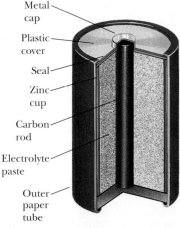

Figure 21-2 Schematic of a standard flashlight-type dry cell.

Batteries

Volta used his discovery that two dissimilar metals produce an electric potential difference to make the first battery. He made a single cell by putting a piece of paper that had been soaked in a salt solution between pieces of silver and zinc. He then stacked these cells in a pile like the one shown in Figure 21-1. By doing this he was able to produce a larger potential difference. The potential difference of all the cells together was equal to the sum of the potential differences due to the individual cells.

A cell's potential difference depends on the choice of metals. Those commonly used in batteries have potential differences of $1\frac{1}{2}$ or 2 volts. Putting lead and lead oxide plates (known as electrodes) into dilute sulfuric acid (the electrolyte) makes a simple 2-volt cell. Connecting six of these cells makes a 12-volt car battery. The positive electrode of one cell is connected to the negative electrode of the next. The connectors sticking out of the first and last cells (the terminals) are usually marked with a + or − sign to indicate their excess charges.

Question How many $1\frac{1}{2}$-volt cells does it take to make a 9-volt battery?

Answer Because the voltages of the cells add, it takes six cells.

Batteries that can be recharged are called *storage batteries*. This recharging is usually accomplished with an electric charger with the negative terminals of the charger and battery connected to each other and the positive terminals connected to each other. The charger forces the current to run backward through the battery, reversing the chemical reactions and refreshing the battery. This electric energy is stored as chemical energy in the battery for later use.

Nonrechargeable flashlight batteries (sometimes called *dry cells*) are constructed with a carbon rod down the center, as shown in Figure 21-2. Carbon is a conductor and replaces one of the metals. A moist paste containing the electrolyte surrounds the rod. The other electrode is the zinc can surrounding the paste. The cell is then covered with an insulating material. These $1\frac{1}{2}$-volt cells are often used end to end (both pointing in the same direction) to provide the 3 volts used in many flashlights.

The voltage produced by an individual cell depends on the materials used and not on its size. The size determines the total amount of chemicals used and therefore the total amount of charge that can be transferred. We have seen that the voltage can be increased by placing cells end to end in a row, an arrangement called **series**. Cells (or batteries) can also be placed side by side, or **parallel**, as shown in Figure 21-3. The cells must all point in the same direction. This parallel arrangement does not increase the voltage but does increase the effective size of the battery. The larger amount of chemicals means that the batteries will last longer before they run down.

Question When might you wish to connect batteries in parallel?

Answer This is especially useful when the batteries are difficult to change, as in remote radio transmitters on mountaintops.

Household electricity has a number of differences from the electricity produced by batteries. The electricity in homes is supplied at a much higher voltage, typically 110 volts. This electricity is supplied by huge electric generators in dams or power plants burning coal, oil, gas, or nuclear fuel. (We will discuss generators in the next chapter.) In household circuits the flow of charge (current) reverses direction 120 times a second. This is called *60-cycle alternating current* (ac) as opposed to the *direct current* (dc) supplied by batteries that flows in a single direction.

Because most of the uses of electricity we will discuss in this chapter depend on only the flow of charge and not on its direction, we will simplify our discussions by referring to flashlight batteries. Another advantage of using batteries is that they have low voltages, so you can experiment with some of the ideas mentioned here without fear of being shocked. **Household electricity**, on the other hand, is dangerous and **should not be used in experiments**.

Pathways

The invention of batteries provided experimenters with a way of producing a continuous flow of charge: a **current**. Although the properties of interacting electric charges are unchanged from the last chapter, new effects are observed when charges move through conducting pathways in a continuous fashion. The simplest way to learn about these is to experiment with a simple flashlight battery, some wire, and a few flashlight bulbs. The basic properties of household electricity are then simple extensions of these ideas.

Figure 21-3 Two cells in parallel have the same voltage as a single cell but last twice as long.

Imagine that you have a battery, a wire, and a bulb. From your previous experiences can you predict an arrangement that will light the bulb? A common response is to connect the bulb to the battery, as shown in Figure 21-4. The bulb doesn't light. It doesn't matter which end of the battery is used or which part of the bulb is touched. The charges don't flow from one end of the battery to the bulb and cause it to light. Suppose you hold the wire to one end of the battery and the metal tip of the bulb to the other end. Touching the free end of the wire to various parts of the bulb will eventually yield success. Two of the four possible arrangements for lighting the bulb are shown in Figure 21-5.

Comparing the methods for lighting the bulb with the ways that don't light it indicates that we must use two parts of each object: the two ends of the wire, the two ends of the battery, and the two metal parts of the bulb. The two parts of the bulb are the metal tip and the metal around the base (this is threaded on many bulbs). Whenever all six ends are connected in pairs, forming a continuous loop (no matter how you do it), the bulb lights.

The continuous conducting path (known as a **complete circuit**) allows the electric charges to flow from one end of the battery to the other end. If the bulb is to light, the conducting path must go through the bulb. You can verify this by examining a bulb with unfrosted glass, as illustrated in Figure 21-6. The part of the bulb that glows is a thin wire (called the *filament*) that is sup-

Figure 21-4 An unsuccessful attempt to light a bulb.

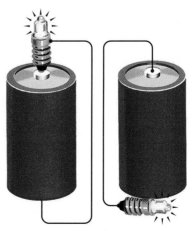

Figure 21-5 Successful ways to light a bulb.

Figure 21-6 Schematic drawing showing the continuous conducting path through a bulb.

ported by two thicker wires coming out of the base. The bulb has to be taken apart to see that one of these support wires is connected to the metal tip and the other to the metal side of the bulb. The rest of the bulb is made from insulating materials so that there is a single conducting path through the filament within the bulb.

Question Would the bulb still light if you turned the battery in the opposite direction in the arrangements shown in Figure 21-5?

Answer Either end of the battery works because reversing the battery does not change the conducting path.

PHYSICS | ON YOUR OWN

Connect a flashlight bulb to a flashlight battery, leaving a gap in the circuit. Test a variety of materials to see if they are conductors by inserting them in the gap as shown in Figure 21-7.

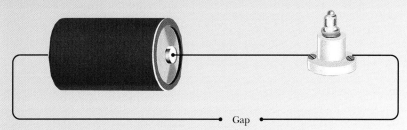

Figure 21-7 When a conducting material is placed between wires at the gap, the bulb lights.

Combining the concept of a complete circuit with the idea of the conservation of charge leads to the conclusion that electricity flows out one end of the battery and back into the other end. The charge that leaves one end of the battery returns to the other end. Charge does not get lost or used up along the way.

Question How does the information about pathways explain why the bulb in Figure 21-4 doesn't light?

Answer Because there is no pathway for the charges to flow back to the other end of the battery, there is no complete circuit.

A Water Model

The flow of charge in a complete circuit is analogous to the flow of water in a closed system of pipes. The battery is analogous to a pump, the wires to the pipes, and the bulb (something that transforms energy in the flow to another form) to a paddle wheel that turns when the water is flowing (Figure 21-8). In a closed system of pipes, the water that leaves the pump returns to the intake side of the pump.

The water model is useful for clarifying the difference between current, charge, and voltage. The current is a measure of the amount of charge flowing past a given section of the circuit in a unit time. If we measure a charge ΔQ passing the point in an interval of time Δt, the current I in the circuit is

> current = charge / time

$$I = \frac{\Delta Q}{\Delta t}$$

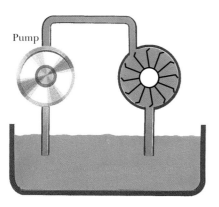

Figure 21-8 The flow of charge is analogous to the flow of water through a closed system.

This expression is analogous to the rate at which water flows through a pipe. The flow of water is measured in units such as liters per second. Electric current is measured in **coulombs** per second, a unit known as the **ampere** (A). Flashlight batteries usually provide less than 1 ampere of current, typical household circuits are usually limited to a maximum of 20 amperes, and a car battery provides more than 100 amperes while starting the car. (The precise definition of the ampere will be described in the next chapter.)

The voltage between two points in a circuit is a measure of the change in electric potential between these two points. That is, voltage is a measure of the work done in moving a unit electric charge between the two points. The work is equal to the distance the charge moves along the circuit multiplied by the force on the unit electric charge due to the electric field that exists in the circuit. You can also think of voltage as being analogous to pressure in our water model.

Although the water model is useful, it is important to realize its limitations. The electric circuit is always "filled" with charges; it does not need to be filled as you would a garden hose. A break in an electric circuit causes the electricity to stop flowing; it doesn't spill out of the end of the wire like water would from a broken pipe. An electric circuit is not analogous to a sprinkler system used to water lawns. A valve blocks the flow of water, whereas a switch puts a gap in the conducting path to stop the flow of electricity.

Another difference is that one or both types of charge may be free to move in a conductor. In plasmas and ionized liquids and gases, both charges can move. It wasn't until 1879 that experiments determined that only some of the negatively charged electrons are mobile in metals. For most macroscopic effects, positive charges moving in one direction are equivalent to negative charges moving in the opposite direction. For example, consider two neutral metal spheres. If you move 1 coulomb of positive charge from sphere A to sphere B, B has an excess of 1 coulomb of positive charge. Sphere A has a deficiency of 1 coulomb of positive charge or, equivalently, an excess of 1 coulomb of negative charge. The net charges on the two spheres are identical to moving 1 coulomb of negative charge from B to A. Because of this equivalence, as well as the historic origins of the subject, when discussing the macroscopic effect of current, we adopt the convention of assuming that the direction of the electric current is the same as the flow of the positive charges.

Resistance

✓ Extended presentation available in the *Problem Solving* supplement

If a bulb is left connected to a battery, the battery runs down in less than a day. (The actual time depends on the particular bulb and battery used.) The bulb has the same brightness for most of this time, but near the end it dims and goes out. However, if you connect just a wire across the ends of the battery, the battery runs down in less than 1 hour, and the wire often gets too hot to touch. Because the battery is drained much faster and something is obviously happening in the wire, we infer that the current is larger through the wire than through the path-

Table 21-1 | The Resistance of Various 100-Meter-Long Wires with a Diameter of 1 Millimeter

Metal	Resistance (Ω)
Silver	2.02
Copper	2.16
Gold	3.11
Aluminum	3.59
Tungsten	7.13
Iron	12.7
Nichrome	191

Resistors are used in electronic circuits.

way with the bulb. The bulb offers more **resistance** to the flow of electricity. It is important to notice that the amount of current delivered by a battery depends on what is connected to the battery. Increasing the resistance in the circuit decreases the current through the battery, and decreasing the resistance in the circuit increases the current through the battery.

This notion of resistance also makes sense according to the water model. More water flows through wide pipes each minute than through narrow pipes. The bulb's filament is a very thin section of wire and should offer more resistance. Our analogy also tells us that long pipes offer more resistance than short pipes. Therefore, we would expect the resistance of wires to increase with length and decrease with diameter. The other factor—not obvious from our water model—is that the resistance depends on the type of material used for the wires. These ideas are verified by experiments with wires of different sizes and materials. Table 21-1 compares the resistances of wires with the same length and diameter but made of different metals.

Resistance is a result of the interaction of the pathway with the flow of charge. The electrons in a wire feel a net force due to the repulsion of the negative terminal and the attraction of the positive terminal and are accelerated; the electrons experience forces due to the electric field that exists in the wire. However, the electrons don't go very far before they bump into atoms, which causes them to lose speed and to be deflected in random directions. Although the average speed of the electrons due to their thermal motion is quite high, all of these collisions keep them from moving very fast along the wire; a typical average speed along the wire is on the order of millimeters per second. This impedance to the flow of charge determines the resistance of the wire.

As the electrons move through a wire, they undergo collisions that transfer kinetic energy to the atoms and cause the wire's temperature to increase. If there is enough energy transferred, the wire gets hot enough to glow. Heating coils and elements in stoves, ovens, toasters, baseboard heaters, and lights do this. Resistors are used to control the voltages and currents in circuits used in such devices as radios and curling irons.

Resistance R is defined to be the voltage V across an object divided by the current I through the object:

resistance = voltage / current ▶

$$R = \frac{V}{I}$$

Resistance is the number of volts across an object required to drive 1 ampere of current through the object, and is therefore measured in volts per ampere, a unit known as the **ohm** (Ω, uppercase Greek omega). This definition is always valid but is most useful when the resistance is constant or relatively constant. In this case this relationship is known as **Ohm's law**. The resistances of pieces of metal, carbon, and some other substances are approximately constant if they are maintained at a constant temperature. The resistance of the filament in a lightbulb increases as the filament heats up.

WORKING IT OUT | Ohm's Law ✓ MATH

Ohm's law provides us with a relationship between the resistance of, the potential difference (voltage) across, and the current through a circuit element. For example, suppose we have a 12-V battery and want to produce a current of 1 A in a particular circuit. What resistance should the circuit have?

$$R = \frac{V}{I} = \frac{12\text{ V}}{1\text{ A}} = 12\ \Omega$$

continues on next page

Question If a lightbulb draws a current of 0.5 A when connected to a 110-V circuit, what is the resistance of its filament?

Answer $R = V/I = (110\text{ V})/(0.5\text{ A}) = 220\ \Omega$.

Ohm's law can be rearranged to find any of the three quantities when the other two are given. Suppose a heating element has a resistance of 8 Ω when hot. What current will it draw when connected to 110 V?

$$I = \frac{V}{R} = \frac{110\text{ V}}{8\ \Omega} = 13.8\text{ A}$$

Question If a 3-V flashlight bulb has a resistance of 9 Ω, how much current will the bulb draw?

Answer $I = V/R = (3\text{ V})/(9\ \Omega) = \frac{1}{3}$ A.

The Danger of Electricity

Electricity is dangerous because our bodies are electric machines. Our muscles constrict when neurons "fire." This reaction is normally triggered by a complex chain of chemical reactions but can also occur when electric charges flow through the muscle. This process is the origin of twitches! This situation can become dangerous when the muscle is part of the heart or breathing system. In these cases the current can also affect the pacing signals, interrupting the normal processes. Currents as small as 50 milliamperes can interrupt breathing in some individuals.

Often people feel that the danger with electricity is the voltage. Although high voltages are dangerous, it is the current through a body that is lethal. Ohm's law tells us that the current is equal to the voltage divided by the resistance of the path. In this case the body is part of the path. If the conditions are such that the total resistance is low, even a low voltage can cause a dangerously high current. The resistance of dry skin is high enough that you can place a finger across the terminals of a 9-volt battery without feeling anything. However, if you touch the terminals of the 9-volt battery to your tongue, you will feel it. The moisture of the tongue lowers the resistance.

One way of preventing high currents from passing through your body, especially through your heart, is to keep one hand in your pocket. Wearing rubber-soled shoes is also good.

A Model for Electric Current

We can use flashlight bulbs to develop a simple model of more complicated electric circuits. Each bulb serves as a visual indicator of the current through the bulb. Bulbs don't glow at all until the current exceeds a certain value; after that the brightness increases with increased current. Although the relationship between the current and the bulb's brightness is complicated, it is reasonable to assume that if one bulb is glowing more brightly than an identical bulb, it must have more current. We say that "more flow means more glow."

We begin by creating a standard to which we can refer; the brightness of a single bulb connected to a single battery will represent a standard current. We will also assume that all batteries and bulbs are identical.

Two bulbs can be connected to a battery so that there is a single path from the battery through one bulb, through the second bulb, and back to the other

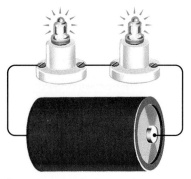

Figure 21-9 Two bulbs wired in series are equally bright but dimmer than the standard.

end of the battery. In this arrangement (Figure 21-9), the two bulbs are said to be in series with each other. We observe that the two bulbs have the same brightness and that they are dimmer than the single bulb in the standard circuit. This decrease in brightness indicates that there is less current in the series circuit than in our standard circuit. This conclusion is supported by the observation that the battery's lifetime in this series circuit is longer than the lifetime of the battery in the standard circuit. Therefore, the resistance of two bulbs in series is greater than that of a single bulb. (It is tempting to say that there is twice as much resistance in the series circuit because there are two bulbs. However, the resistance of the lightbulbs changes with brightness, consequently the resistance does not double.)

Of special note is the observation that the two bulbs have the same brightness. This tells us that the current through each bulb is the same; whatever charge flows through the first bulb flows through the second bulb. Electric charge is not being used up or lost along the pathway in agreement with the conservation of charge. The same amount of charge flows back into the battery as left it. It is the flow of charge through a bulb that causes the filament to heat up and glow. If the same amount of charge flows through each bulb each second, the two identical filaments will reach the same temperature and glow with the same brightness.

Flawed Reasoning

Alan connects a single bulb to a battery and notices that it is fairly bright. He then adds a second bulb in series with the first and finds that both bulbs are now dimmer and equally so. He makes the following conclusion: "The single bulb got all the current from the battery. The two bulbs must now *share* the current, each getting half. It makes sense that they should be dimmer."

Alan thinks he understands what is going on, but he has made a serious error in his reasoning. **Can you find it?**

Answer The word *share* has two meanings. We can share a book or we can share a pizza. If we share a book, you read it first, and then I read it. If we share a pizza, we first divide the pizza in two pieces and each take a piece. Alan believes that the current through the battery is always the same size and that the two bulbs are "sharing" this current as we would share a pizza. In fact, the bulbs share the current as we would share a book; first the current passes through one bulb (heating its filament) and then through the other (heating its filament). The two bulbs are dimmer than the single bulb because the increased resistance in the circuit reduces the current through the battery.

Figure 21-10 Two bulbs wired in parallel are equally bright and have the same brightness as the standard.

Another consequence of the equal brightness is that bulbs cannot be used to determine a direction for the flow of electricity. Everything can be explained equally well assuming a flow of negative charges in one direction (the actual situation in wires), a flow of positive charges in the other direction, or both charges flowing simultaneously in opposite directions (as happens in fluids). In fact, because it is only the motion of charges that matters, one charge moving back and forth would also work. In household electricity the negative charges move back and forth with a frequency of 60 hertz.

Two bulbs can also be connected so that each bulb has its own path from one end of the battery to the other, as shown in Figure 21-10. In this arrangement the two bulbs are said to be wired in parallel. In contrast to the bulbs in series, the current in one bulb does not pass through the other, which can be seen by disconnecting either bulb and observing that the other is not affected. The two bulbs in parallel are equally bright, and each is as bright as our standard. Because each bulb has its own path, the battery in this circuit supplies twice as much current as the battery in the standard circuit. This can be verified experimentally; in this arrangement the battery runs down in one-half its normal lifetime.

Adding the extra bulb in parallel has increased the current through the battery, indicating that the resistance of the circuit must have decreased. Although a bulb can be correctly thought of as a resistance to the flow of charge, the addition of one more bulb in a circuit can either increase or decrease the total resistance of the circuit, depending on how we add the bulb. When the new bulb is added in series (adding a new resistance on an existing line) the resistance of the circuit increases and the current through the battery decreases. When the new bulb is added in parallel (on a new path that did not exist before) the resistance of the circuit goes down and the current through the battery increases. Even though the new path contains resistance (the new bulb), it represents a new opportunity for flow that did not exist before.

Consider the following analogy. When a popular movie ends, people jostle to get out of the theater through the front door. The finite width of the door represents resistance to the flow of people. If the back door is also opened, the flow of people out of the building increases, even if the back door is very narrow. It is a new opportunity for flow. Any path that is added in parallel will lower the total resistance of a circuit, regardless of how much resistance is contained on the new path.

Three or more bulbs can be connected in series or parallel or in combinations of series and parallel. The relative brightnesses of these bulbs can be predicted using the ideas that we have discussed. As an example, consider the combination of bulbs in Figure 21-11. Bulb A will be the brightest because all of the current must pass through it, but it will not be as bright as our standard. Bulbs B and C will be dimmer because the current splits—part going through bulb B and part through bulb C. The current flowing into the junction J must be equal to that flowing out of the junction. This is just a consequence of the conservation of charge. Because the bulbs are identical, the resistances of the two paths are equal, and the current will split equally; bulbs B and C will be equally bright.

In general, when charge reaches the junction between two parallel branches of unequal resistance, more charge will flow through the easier branch. We say that "current favors the path of least resistance." This does not mean that all of the current takes the easier path; some current takes the more difficult path. If parallel path 1 has twice the resistance as path 2, path 1 will have one-half the current of path 2.

The German physicist Gustav Kirchhoff formalized two rules for analyzing the current in circuits. **Kirchhoff's junction rule** states

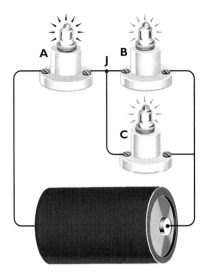

Figure 21-11 Bulbs B and C are in parallel, and the pair is in series with bulb A.

Figure 21-12 When the wire on the left is connected from one end of the battery to the other, producing a short circuit, the light goes out.

> The sum of the currents entering any junction in a circuit must equal the sum of the currents leaving that junction.

◀ Kirchhoff's junction rule

As we have seen, this is a consequence of the conservation of charge. We will discuss Kirchhoff's other rule in the next section.

If one of the paths in a circuit is a conducting wire without a lightbulb as shown in Figure 21-12, the path has very little resistance and virtually all of the current takes this path. This is known as a **short circuit**. When there is a short circuit, so little current flows through any bulbs in parallel to the short that the bulbs go out.

In general, parallel paths are not independent of each other. Making a change in one parallel path usually affects the current in the other path. The only exception to this rule is the case we started with in Figure 21-10, where each path had direct connections to both sides of an ideal battery. These paths are said to be "parallel across the battery," and each path acts as if it has its own battery (for reasons that we will discuss later).

> **Flawed Reasoning**
>
> Three students are considering what will happen to the brightness of bulb B in Figure 21-11 when bulb C is removed from its socket:
>
> **Terry:** "The current splits after passing through A. When C is removed, all the current must pass through B, so B will get brighter."
>
> **Judy:** "Removing bulb C removes a pathway from the circuit, which will increase the resistance of the circuit. The current through the battery will go down, and B will get dimmer."
>
> **Kay:** "You're both half right. The current through A will go down, but B will now get all the current instead of sharing it with C. It's like choosing between some of a large pizza or all of a small one. We need more information to make a decision."
>
> **Which student (if any) do you agree with?**
>
> **Answer** Kay correctly recognizes the conundrum. We know that the current through the battery (and hence through bulb A) must decrease when bulb C is removed, but we do not know by how much it will decrease. The model that we have developed so far does not allow us to make a prediction; we will need to add something to our model.

A Model for Voltage

If you leave your house and climb a nearby mountain, you gain a certain amount of gravitational potential energy. As you return to your house, you lose this same amount of gravitational potential energy, regardless of which path you choose to descend the mountain. It is the same with electric circuits. A 12-volt battery delivers 12 joules of electric potential energy to every coulomb of charge that passes through it. As that coulomb of charge travels through the circuit and returns to the battery, it must lose 12 joules of electric potential energy, regardless of which path it takes. This electric potential energy is delivered to the resistive elements in the circuit (like bulbs) and converted to heat and light. This basic application of the conservation of energy is called **Kirchhoff's loop rule**:

Kirchhoff's loop rule ▶ | Along any path from the positive terminal to the negative terminal of a battery, the voltage drops across the resistive elements encountered must add up to the battery voltage.

This means that a single bulb connected to a battery will have a voltage drop equal to the voltage of the battery. We also know from experience that a bulb will glow more brightly when connected to a 12-volt battery than when it is connected to a 6-volt battery. This suggests that the bulb's brightness can be used as an indication of the voltage drop across the bulb. If one bulb is brighter than another, it must have a larger voltage drop across it (assuming again that both bulbs are identical). This important idea, coupled with Kirchhoff's loop theorem, allows us to answer many questions that are left unanswered by our present model for electric current.

Let's return to the conundrum presented in the last Flawed Reasoning. Our model for electric current allows us to predict that bulb A in Figure 21-11 will get dimmer when bulb C is removed from its socket. As Judy explains, removing bulb C removes a pathway from the circuit, increasing the resistance of the circuit. The current through the battery must decrease, and because all of the current passes through bulb A, bulb A must get dimmer. The current model is not, however, able to predict what happens to bulb B. The total current in the circuit has decreased, but bulb B's share of that current has increased (from 50% to 100%). Which effect wins out? Our model for electric voltage answers the question without ambiguity. The sum of the voltage across bulb A and the voltage across bulb B must be equal to the battery voltage by Kirchhoff's loop rule. If bulb A gets dimmer when bulb C is removed, the voltage across bulb A must decrease. Because the voltage drop across bulbs A and B must be equal to the battery voltage, the voltage across bulb B must increase by the same amount and bulb B will therefore get brighter.

We are now in a position to understand the independence of branches that are parallel across an ideal battery—why each parallel branch acts as if it has its own battery. The voltage drops across each branch must equal the battery voltage by Kirchhoff's loop rule, regardless of what changes are made to the other branches. An ideal 6-volt battery is one that is able to maintain a potential difference of 6 volts across its terminals regardless of how much current is drawn by the circuit. In practice, real batteries have trouble delivering large currents due to limitations of the chemical processes inside the battery. This limitation manifests itself in the form of a drop in the battery voltage as the current increases. The addition of a second bulb in parallel to our standard bulb will actually dim the first bulb slightly if you are using an alkaline battery. The effect will not be noticeable with a rechargeable NiCad battery. The NiCad battery is said to be more ideal, and many branches can be connected across a NiCad battery before the effect appears. This assumes that each of those branches contains resistance and draws a small amount of current. If a parallel branch contains nothing but a conducting wire, it will demand more current than the battery can supply and the voltage of the battery drops dramatically, causing the bulbs on the other branches to go out. This is called "shorting out the battery" and is similar to what happens on a hot summer day when Los Angeles has a "brown out" caused by the huge demand for current to run air conditioners.

The circuits in houses are wired in parallel so that electric devices can be turned on and off without affecting one another. As each new appliance is turned on, the electric company supplies more current. Each parallel circuit, however, is wired in series with a circuit breaker, as shown in Figure 21-13, to deliberately put a "weak link" in the circuit. If too many devices are plugged into one circuit, they will draw more current than the wires can safely carry. For example, the wires might heat up at a weak spot and start a fire. The circuit breaker interrupts the circuit, shutting everything off.

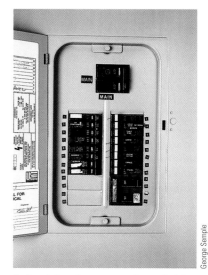

A circuit breaker protects household circuits from drawing too much current and starting a fire.

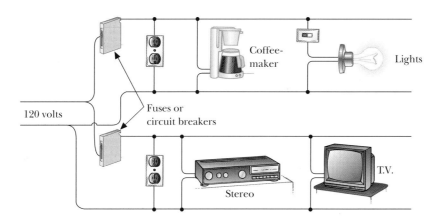

Figure 21-13 Appliances are connected to household circuits in parallel. The circuit is protected from drawing too much current by a circuit breaker wired in series.

The Real Cost of Electricity

The real cost of electricity varies a great deal, depending on the source. We are most familiar with purchasing electric energy from the local energy company for use in our homes. But what is the cost of this electric energy? Examining your electric bill shows that you are charged something between 8¢ and 25¢ per kilowatt-hour. Let's use a representative value of 10¢ per kilowatt-hour. Knowing that 1 kilowatt-hour is equal to 3.6 million joules, we can divide these two numbers to learn that we can use about 36 million joules for each dollar we spend.

We can also buy electric energy packaged in a variety of batteries. In this case we pay for properties of the batteries that appeal to the manufacturer of the electric appliances. This might include the batteries' size, voltage, lifetime, or their ability to maintain current as they wear out. To discover what we are paying for the electric energy, we need to know the prices and the total electric energy that the batteries provide during their lifetimes. As shown in the text, the energy can be obtained by multiplying the total charge by the voltage rating of the battery. The total charge is obtained by multiplying the current rating by the battery's lifetime.

A standard flashlight battery—usually a carbon–zinc D cell—costs about 75¢ and has a rating of 1.5 volts. Its lifetime depends on the way it is used. Flashlight D cells are normally used for short intervals over a much longer time. Under these conditions the battery could supply 375 milliamperes for about 400 minutes, yielding a total charge of 9000 coulombs and a total energy of 13,500 joules. Therefore, you are getting 18,000 joules per dollar, 2000 times the cost of household electricity. Alkaline batteries were a major improvement in performance over their carbon–zinc cousins, primarily in applications requiring fairly large, continuous currents, such as portable stereo radios. Alkaline batteries typically last about 50–75% longer than a standard battery but under some conditions can last much longer. However, their costs are also higher. Are alkaline batteries a good buy? Only if their increase in cost is less than their corresponding increase in lifetime.

The development of mercury, silver, and lithium batteries allowed manufacturers to pack more energy in smaller volumes. These devices have a much more stable current over their lifetimes, so they are especially useful in devices such as wristwatches and photocells in cameras, which require a particular value for the current. A watch battery delivers 1.3 volts and 10 millionths of an ampere for a year, giving a total energy of 410 joules. If these batteries sell for about $2, they have an energy-cost rating of 205 joules per dollar, or 175,000 times the cost of household electricity!

PHYSICS ON YOUR OWN

Obtain two three-way switches used in wiring hallway lights. (The switches act like the one in Figure 21-14.) Wire them to a flashlight battery and bulb so that either switch can turn the light on or off independently of the other. Be sure that there are no short circuits when the light is off.

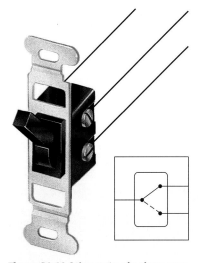

Figure 21-14 Schematic of a three-way switch. The wire on the left can be connected to either wire on the right by throwing the switch.

Some modern Christmas tree lights are wired in parallel. If one bulb burns out, it is the only bulb that goes out. The older style was wired in series; when one bulb went out, they all went out. That was rather inconvenient, because when one bulb burned out, all of them had to be tested until the defective one was found. The latest style of Christmas tree lights has the bulbs wired in series, but when a bulb burns out, a short circuit is formed across the burned-out bulb that allows the rest of the bulbs to stay lit.

Over the years, a convention for drawing circuit elements has been developed. Like all symbols, the electrical symbols capture the functional essence of the element and omit nonessential characteristics. A lightbulb, for example, has no directional characteristic, but a battery does. Their symbols reflect this difference. Figure 21-15 gives the common symbols paired with a diagram like those we have been using.

Circuits are also drawn with sharper corners than exist in the actual circuits. This is simply a technique that has helped communication among experimenters.

Electric Power

When you buy electricity, it is energy that you buy. This electric energy is converted to heat, light, or motion. Why is it, then, that we talk about power? As we discussed in Chapter 7, **power** is the amount of energy used per unit time. But we need to be careful here; conservation of energy tells us that energy can neither be created nor destroyed. Formally, power is the amount of energy transformed from one form to another form divided by the elapsed time. In electricity the energy the electrons gain from the electric field is converted to such forms as kinetic energy, heat, sound, and light. Most electric devices in our houses are rated by their power usage. Power is measured in **watts** (W). Many household lightbulbs are rated 60, 75, or 100 watts. Electric heaters and hair dryers might use 1500 watts.

The electric meter connected between your house and the energy, or power, company records the energy you use much like the odometer in your car records the miles you drive. Electric energy is usually billed at 8 to 25 cents per kilowatt-hour. A typical household with an electric range and an electric clothes dryer (but no electric heat) uses about 900 kilowatt-hours per month.

Figure 21-15 Common symbols used in diagrams of electric circuits. (*Top to bottom*: resistor, battery, switch.)

This electric meter measures the energy used in a house or business. This meter reads 65,401 kilowatt-hours.

PHYSICS | ON YOUR OWN

Monitor the use of electric energy in your home by taking periodic readings of the electric meter. How much electric energy does your household use in a typical day?

You can also try to turn off all electric devices in your home. You can monitor your progress by looking at how fast the wheel in your electric meter turns. If you have an electric clothes dryer, turn it on and see how rapidly the wheel turns.

WORKING IT OUT | Electric Power ✓ MATH

We can obtain an expression for electric power by using the definition for electric energy we developed in Chapter 20 and the definition for current. Because energy is the charge multiplied by the voltage, we have

$$P = \frac{\Delta E}{\Delta t} = \frac{\Delta QV}{\Delta t} = \frac{\Delta Q}{\Delta t} V$$

This equation tells us that the power is equal to the current times the voltage:

$$P = IV$$

Therefore, an electric appliance that draws a current of 10 A at a voltage of 110 V uses energy at a rate of

$$P = IV = (10 \text{ A})(110 \text{ V}) = 1100 \text{ W}$$

◀ power = current × voltage

Question What power is required to operate a clock radio if it draws 0.05 A from the household circuit?

Answer $P = IV = (0.05 \text{ A})(110 \text{ V}) = 5.5 \text{ W}$.

continues on next page

By rearranging the expression for power, you can determine the current through a 100-W bulb:

$$I = \frac{P}{V} = \frac{100 \text{ W}}{110 \text{ V}} = 0.91 \text{ A}$$

We can obtain an alternative expression for electric power by using the relationship that the voltage is equal to the current multiplied by the resistance

$$P = IV = I(IR) = I^2R$$

◀ power = current squared × resistance

In this form it is more obvious why a bulb in a circuit glows yet the connecting wires do not. The current through the circuit is the same everywhere, but the section that has the highest resistance receives the most thermal energy per unit time. Because the filament is made of a very fine wire of high-resistance material, it has the highest resistance; its temperature goes up, and it glows. The connecting wires also heat up, but they just don't get hot enough to glow or even feel warm.

There are times when the energy losses in connecting wires become important. Sending electric energy long distances through wires from a power plant could result in significant energy losses. Two things can be done to minimize these losses. First, the wires should have as little resistance as possible, which means using large-diameter wires and low-resistance materials. Second, transformers (discussed in Chapter 22) allow the utility company to send the same energy through the wire by raising the voltage and simultaneously lowering the current. Lower current means less thermal loss in the transmission lines. Economic and engineering considerations dictate the extent each of these methods is used to minimize energy loss. Electricity is carried between energy companies and towns at very high voltages (up to 765,000 volts). The voltage is lowered by transformers for distribution around towns and then lowered once more to the safer 110 volts used in homes.

WORKING IT OUT | Cost of Electric Energy ✓ MATH

A device running at constant power uses an amount of energy equal to the power multiplied by the time; the longer we use it, the more energy it consumes. For example, a 60-W bulb burning for 10 h uses

$$\Delta E = P \Delta t = (60 \text{ W})(10 \text{ h}) = 600 \text{ Wh} = 0.6 \text{ kWh}$$

of electric energy. It is important to recognize that a kilowatt-hour (kWh) is a unit of energy, not one of power. Because 1 kW is equal to 1000 J/s and 1 h is the same as 3600 s, 1 kWh is 3.6 million J.

We can now calculate the monthly cost of electric energy for this typical household using 900 kWh per month. Let's assume that electric energy costs 10¢ per kWh:

$$\text{cost} = (900 \text{ kWh})(\$0.10/\text{kWh}) = \$90$$

Question At this price what would it cost to run a 1500-W heater continuously during an 8-h night?

Answer $E = Pt = (1500 \text{ W})(8 \text{ h}) = 12{,}000 \text{ Wh} = 12 \text{ kWh}$. Cost = $(12 \text{ kWh})(10¢/\text{kWh}) = \$1.20$.

Summary

Volta used the electric potential difference between two different metals to build the first battery. Batteries make charges flow continuously as a current, producing heat and light and running motors and other appliances.

The voltage produced by an individual cell depends on the materials used but not on its size. The size determines the amount of chemicals contained in the cell and therefore the total amount of charge that can be transferred. The voltages of cells in series are additive. Cells placed in parallel do not increase the voltage but increase the effective size of the battery.

The electricity in homes is usually supplied at 110 volts and as 60-hertz alternating current. The electricity from a battery is direct current and usually at a much lower voltage.

The resistance of wires increases with increasing length, decreases with increasing cross-sectional area, and depends on the type of material. Resistance R is the ratio of the voltage V across an object to the current I through the object, $R = V/I$. Resistance is measured in volts per ampere, or ohms (Ω).

Electric charges can flow continuously only when there is a complete pathway, or circuit, made of conducting materials. The current is determined by the voltage and the total resistance of the circuit, $V = IR$. The voltage between two points in a circuit is equal to the change in electric potential between these two points, which is the work done in moving a unit of positive charge through the circuit. The most charge flows through the path with the least resistance. Conservation of charge requires that electricity flow out one end of the battery and back into the other. Charge does not get lost or used up along the way.

Conservation of energy requires that any electric potential energy gained by a charge as it passes through the battery must be lost by the charge as it passes through the circuit. The voltage drops across the resistive elements along any path between the terminals of a battery must sum to the voltage of the battery.

One or both types of charge may be free to move in a conductor. In plasmas and ionized liquids and gases, both charges can move, whereas only some of the negatively charged electrons are mobile in metals. For most macroscopic effects, positive charges moving in one direction are equivalent to negative charges moving in the opposite direction.

A power company sells energy, not power. Power is the rate of using energy and is used to rate most electric devices in our houses. Power is measured in watts (W) or kilowatts (kW), whereas household electric energy usage is measured in kilowatt-hours (kWh).

PhysicsNow™ Assess your understanding of this chapter's topics with sample tests and other resources found by logging into PhysicsNow at **http://physics.brookscole.com/kf6e**.

Chapter 21 Revisited

When electrons move through a conductor, they collide with the atoms in the conductor, raising the average kinetic energy of the atoms and therefore the temperature of the conductor. The rise in temperature depends on the thermal properties of the material and its electrical resistance. Filaments in lightbulbs are made of very thin wires of high-resistance materials to enhance the heating. If the rise in temperature is great enough, the wire glows, emitting light.

KEY TERMS

ampere: The SI unit of electric current, 1 coulomb per second.

complete circuit: A continuous conducting path from one end of a battery (or other source of electric potential) to the other end of the battery.

coulomb: The SI unit of electric charge. The amount of charge passing a given point in a conductor carrying a current of 1 ampere.

current: A flow of electric charge. Measured in amperes.

Kirchhoff's junction rule: The sum of the currents entering any junction in a circuit must equal the sum of the currents leaving that junction.

Kirchhoff's loop rule: Along any path from the positive terminal to the negative terminal of a battery, the voltage drops across the resistive elements encountered must add up to the battery voltage.

ohm: The SI unit of electric resistance. A current of 1 ampere will flow through a resistance of 1 ohm under 1 volt of potential difference.

Ohm's law: The resistance of an object is equal to the voltage across it divided by the current through it.

parallel: Two circuit elements are wired in parallel when the current can flow through one or the other but not both. Elements that are wired parallel to each other are directly connected to each other at both terminals.

power: The rate at which energy is converted from one form to another. In electric circuits the power is equal to the current times the voltage. Measured in joules per second, or watts.

resistance: The impedance to the flow of electric current. The resistance is equal to the voltage across the object divided by the current through it. Measured in volts per ampere, or ohms.

series: An arrangement of resistances (or batteries) on a single pathway so that the current flows through each element.

short circuit: A path in an electric circuit that has very little resistance.

watt: The SI unit of power, 1 joule per second.

CONCEPTUAL QUESTIONS

1. Does the size of a cell affect its voltage, its lifetime, or the current it delivers?
2. When you stack three flashlight batteries in the same direction, you get a voltage of $3 \times 1\frac{1}{2}$ volts = $4\frac{1}{2}$ volts. What voltage do you get if one of the batteries is turned to face in the opposite direction?
3. You have two 9-volt cells. What voltage would you get if you wired them in series?
4. If you have four 1.5-volt cells, what voltage would you get if you wired them in parallel?
5. What are the differences and similarities of the electricity supplied by a flashlight battery and a car battery?
6. What are the differences and similarities of the electricity supplied by a car battery and your local energy company?
7. Which bulbs in the figure are the brightest? Which are the dimmest but still glow? Why?
8. Which bulbs in the figure do not light? Explain.
9. Which arrangements of batteries in the figure will burn out most quickly? Which will last the longest? Why?
10. Which arrangements of batteries in the figure will light the bulb for the longest time? Explain.

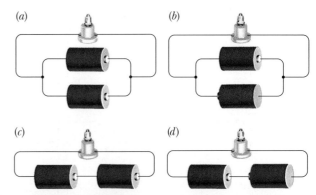

Questions 7–10 Identical bulbs are connected to various combinations of identical batteries.

11. There are four different ways of connecting a battery, a bulb, and a wire so that the bulb lights. Two of these are shown in Figure 21-5. What are the other two?
12. Write a short statement describing how you would connect a battery, a bulb, and a single wire so that the bulb

lights. Your statement should include all four ways without favoring any one way.

13. What is the difference, if any, in a circuit between a bulb burning out and removing the bulb from its socket?
14. Suppose that you ran speaker wires from the stereo in the family room to your bedroom, but you forgot to mark the wires. What could you do to tell which wires you should connect to which speaker?
▲ 15. You are given a mystery box with six bolt heads on top as shown in the figure. You connect a battery and a bulb across each pair of bolt heads and find that the bulb lights for the following pairs: AB, CF, CE, and EF, but not for any of the other combinations. Draw diagrams showing all of the ways the box could be wired.

▲ 16. A friend examines a box similar to the one in Question 15 and finds that the bulb lights only for the following combinations: AB, CD, and DE. Can your friend's observations be correct? Explain.
17. What is the difference between a volt and an ampere?
18. Car batteries are often rated in ampere-hours. What does this mean?
19. In the water model for electricity, what are the analogs of a battery, a switch, a wire, and a lightbulb?
20. In the water model for electricity, what are the analogs of charge, current, and voltage?
21. Which of the following affects the resistance of a wire: diameter, type of metal, length, or temperature? Explain.
22. Some power tools run poorly when connected to very long extension cords. Why? What could you do to improve the situation?
23. If the resistance connected to a battery is cut in half, what happens to the current through the battery?
24. A resistor has a current of 2 amperes when connected to a single battery. If a second identical battery is connected in series with the first, what is the current through the resistor? What is the current if the batteries are connected in parallel?
25. How does the atomic theory of matter in Chapter 11 account for the observation that the resistance of metallic wires increases with temperature?
26. Why is it dangerous to use a blow dryer while taking a bath?
27. If the only voltage you have is 110 volts, how could you light some 5-volt lightbulbs without burning them out?
28. In England the voltage supplied to homes is 240 volts. What would happen if you used an American lightbulb in England?
29. In a circuit with a single lightbulb, the current through the bulb is found to be 2 amperes. Is the current that flows through the battery greater than, equal to, or less than 2 amperes? Explain.
30. Two bulbs in series are connected to a battery. The current in the first bulb is found to be 1 ampere. What is the current through the battery?
31. If a second bulb is added in series to a circuit with a single bulb, does the resistance of the circuit increase, decrease, or remain the same? Does the current through the battery increase, decrease, or remain the same? Why?
32. When a bulb is added in parallel to a circuit with a single bulb, does the resistance of the circuit increase, decrease, or remain the same? Does current increase, decrease, or remain the same? Why?
33. If a string of five bulbs in series is added in parallel to a single bulb connected to a battery, you find that the brightness of the original bulb does not change. Has the current through the battery increased, decreased, or stayed the same? Has the resistance of the circuit increased, decreased, or stayed the same? Explain.
34. Which circuit offers the greater resistance to the battery, two bulbs in series or two bulbs in parallel? Why?
35. For the circuits in the figure, is the current through the battery in circuit A greater than, equal to, or less than the current through the battery in circuit E? Explain your reasoning.

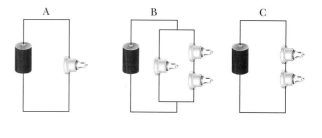

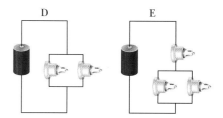

Questions 35–38 Identical batteries are connected to various combinations of identical bulbs.

36. For the circuits in the figure, is the current through the battery in circuit A greater than, equal to, or less than the current through the battery in circuit B? Explain your reasoning.
37. For the circuits in the figure, is the current through the battery in circuit C greater than, equal to, or less than the current through the battery in circuit E? Explain your reasoning.
38. For the circuits in the figure, is the current through the battery in circuit D greater than, equal to, or less than the current through the battery in circuit E? Explain your reasoning.
39. How would you connect two batteries and two bulbs to get the most light?
40. How would you wire two batteries and two bulbs to produce light for the longest time?

41. Which bulbs are the brightest and dimmest in the figure? Explain.

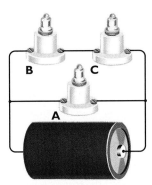

Questions 41–46

42. Would the resistance of the circuit in the figure increase, decrease, or stay the same if bulb A were removed from its socket? What if bulb C were removed instead? Explain.
43. What happens to the brightness of bulbs B and C in the figure when bulb A burns out? Explain.
44. What happens to the brightness of bulbs A and C in the figure when bulb B is removed from its socket? Explain.
45. If you put a wire across the two terminals of bulb A, what happens to the brightness of each of the bulbs in the figure? Why?
46. What happens to the brightness of each of the bulbs in the figure when you connect a wire across the terminals of bulb B?
47. How does the brightness of bulb C compare with the brightness of bulb D in the figure? Explain your reasoning.

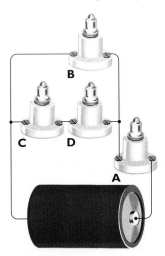

Questions 47–52

48. How does the brightness of bulb D compare with the brightness of bulb B in the figure? Explain your reasoning.
49. What happens to the brightness of bulbs B, C, and D in the figure when bulb A burns out? Explain.
50. What happens to the brightness of bulbs A, B, and C in the figure when bulb D is removed from its socket? Explain.
51. If you put a wire across the two terminals of bulb A, what happens to the brightness of each of the bulbs in the figure? Why?
52. What happens to the brightness of each of the bulbs in the figure when you connect a wire across the terminals of bulb B?
▲ 53. A box has three identical bulbs mounted on its top with the wires hidden inside the box. Initially, bulb A is the brightest, and bulbs B and C are equally bright. If you unscrew A, bulbs B and C go out. If you unscrew B, A gets dimmer and C gets brighter so that A and C are equally bright. If you unscrew C, A gets dimmer and B gets brighter so that A and B are equally bright. If you unscrew B and C, A goes out. How are the bulbs wired?
▲ 54. A box has three identical bulbs mounted on its top with the wires hidden inside the box. Initially, bulb A is the brightest, and bulbs B and C are equally bright. If you unscrew A, bulbs B and C remain the same. If you unscrew B, A remains the same and C goes out. If you unscrew C, A remains the same and B goes out. If you unscrew B and C, A remains the same. How are the bulbs wired?
55. Are the headlights in cars and trucks wired in series or parallel? Explain.
56. Why should you not replace a 5-ampere fuse in your car with a 10-ampere fuse?
57. What happens to the power supplied by a battery if the resistance connected to the battery is doubled? Explain.
58. A single resistor dissipates 4 watts of power when connected directly across a battery. If a second identical resistor is added in series with the first, what power will the original resistor dissipate?
59. A resistor dissipates 4 watts of power when connected to a single battery. If a second identical battery were connected in parallel with the first, what power would the resistor dissipate?
60. A resistor dissipates 4 watts of power when connected to a single battery. If a second identical battery is connected in series with the first, what power will the resistor dissipate?
61. Two bulbs have ratings of 60 watts and 120 watts. Which bulb carries the higher current? Explain.
▲ 62. Two bulbs have ratings of 60 watts and 120 watts. Which bulb has the higher resistance? Explain.

EXERCISES

1. What is the resistance of a lightbulb that draws 0.8 A when it is plugged into a 110-V outlet?
2. You plug a vacuum cleaner into a 110-V outlet with nothing else on the circuit. If the 15-A circuit breaker is tripped, what is the maximum possible resistance of the vacuum's motor?
3. The coils of a heater have a resistance of 12 Ω when hot. What current does the heater draw when plugged into a 110-V outlet?
4. You plug in your new 8-Ω hair dryer to a 110-V outlet and trip the 10-A circuit breaker. If you plug the hair dryer into a line with a 15-A breaker, will it trip the breaker? Assume nothing else is plugged into either circuit.
5. If a lightbulb has a resistance of 8 Ω and a current of 0.5 A, at what voltage is it operating?
6. A lightbulb has a resistance of 250 Ω. What voltage is required for the bulb to draw a current of 0.5 A?
7. Two $1\frac{1}{2}$-V batteries are connected in series to a 6-Ω resistor. How much current flows through each battery?
8. Two $1\frac{1}{2}$-V batteries are connected in parallel to a 6-Ω resistor. How much current flows through each battery?
9. Two 3-Ω resistors are connected in series to a 12-V battery. What single resistor, if connected to the battery alone (called the *equivalent resistance*), would draw this same current? What is the current through the battery?
10. A 2-Ω and a 4-Ω resistor are connected in series to a 12-V battery. What is the current through each resistor? Use Ohm's law to show that the voltage drops across the individual resistors add up to 12 V.
11. Two 4-Ω resistors are connected in parallel to a 12-V battery. Use the fact that the voltage across each of the resistors is 12 V to find the total current through the battery. What single resistor, if connected to the battery alone (called the *equivalent resistance*), would draw this same current?
12. A 3-Ω resistor is connected in parallel with a 12-Ω resistor and the combination is connected to a 12-V battery. How much current does the battery supply?
13. What is the power used by a toaster that draws a current of 7 A when connected to a 110-V line?
14. A heating element is rated at 1500 W. How much current does it draw when it is connected to a 110-V line?
15. If a clock draws a maximum current of 5 mA from a 110-V line, what is its maximum power consumption?
16. A VCR has a maximum power rating of 24 W when plugged into a 110-V outlet. What is the maximum current that it requires?
17. A 4-Ω resistor is connected to a 12-V battery. What is the current through the battery? What is the power dissipated by the resistor?
18. A 3-Ω resistor draws a current of 4 A when hooked up to a battery of unknown voltage. What is the battery's voltage? What is the power dissipated by the resistor?
19. What is the power rating of a heating coil with a resistance of 11 Ω that draws a current of 20 A?
20. A coffeemaker has a resistance of 10 Ω and draws a current of 11 A. What power does it use?
▲ 21. What is the resistance of a 60-W lightbulb in a 110-V circuit?
▲ 22. What is the resistance of the coil in a 1400-W heater?
23. If a hair dryer is rated at 1200 W, how much energy is used to operate the hair dryer for 8 min?
24. If a 60-W bulb is left on continuously in a secluded hallway, how much energy is used each month?
25. If electricity costs 15¢/kWh, how much does it cost to burn a 100-W bulb for one day?
26. A 1400-W heater for a sauna requires 40 min to heat the sauna to 190°F. What does this cost if electricity sells for 15¢/kWh?

22 Electromagnetism

During the 19th century, scientists discovered that electricity and magnetism were not separate phenomena but two different aspects of something called electromagnetism. It is electromagnetism that enables us to pop corn in our microwave ovens and brings television to us while we enjoy our popcorn. What is this connection between electricity and magnetism?

(See page 466 for the answer to this question.)

A radio telescope scanning the sky over Hawaii.

MOST of us have played with magnets. Through play we learn that these little pieces of material attract and repel each other and that they attract some objects but have no effect on others. For instance, magnets stick to refrigerator doors but not to the walls of a room; to paper clips but not to paper; and so on.

Although magnetic properties are interesting, they, like electricity, appear to be rather isolated phenomena. Unlike electricity, however, magnetism does seem more permanent. And although there are similarities between these two phenomena, they seem to be separate properties of matter: a magnet with no net electric charge does not deflect the foils of an electroscope. So magnetism appears to be a different phenomenon from electricity. Yet some experimenters in the 17th century were more fascinated by the similarities than the differences between magnetism and electricity, and they searched for a connection.

PhysicsNow™ Test your understanding of this chapter by logging into PhysicsNow at http://physics.brookscole.com/kf6e, selecting the chapter, and clicking on the "Take a Pre-Test" link.

A collection of permanent magnets.

Magnets

It has been known since at least the 6th century BC that lodestone, a naturally occurring mineral, attracts iron. It was also known that iron can be made magnetic by rubbing it with lodestone. The magnetized piece of iron then attracts other pieces of iron. The only elements that occur naturally in the magnetized state are iron, nickel, and cobalt. Many permanent magnets are made from alloys of these metals.

PHYSICS | ON YOUR OWN

Examine a variety of materials and compile a list of the kinds that are attracted by magnets. Is your list the same as the list of conductors found in Chapter 20?

Imagine having three unmarked, magnetized rods. If we arbitrarily mark one end of one of the rods with an X, we find that it attracts one end of each of the other two magnets (Figure 22-1) and it repels their other ends. Let's mark these ends A (for attract) and R (for repel), respectively. We then find that the two A ends repel each other, the two R ends repel each other, and the A and R ends attract each other. Because the two A ends were determined in the same way using the first magnet, they should be *alike*. Therefore, the ends of the magnets (called **magnetic poles**) behave like the electric charges we studied in Chapter 20. We can summarize the behavior of the magnetic poles with the simple statement that

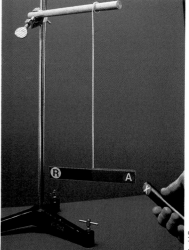

Figure 22-1 Magnetic poles can be identified by their attraction to and repulsion by known poles.

◄ magnets

Like poles repel; unlike poles attract.

Both electricity and magnetism seem to have two kinds of "charge." In electricity, however, it is possible to separate the two charges or at least to put more of one kind of charge on one object than on the other. But in magnetism the poles always come in pairs that have the same strength.

Question How might you search for a third kind of magnetic pole?

Answer You could search for a magnetized material that attracts or repels both north poles *and* south poles. No such material has been discovered.

Figure 22-2 Dividing a magnet results in two smaller magnets, each with both poles.

Suppose you try to separate the poles by breaking a magnet in half. You get two magnets, and each of these new magnets has two poles (Figure 22-2). If you break each of these in half, you get four magnets, and so on. Even if you continue dividing down to the atomic level, you always obtain magnets with two poles. It does not seem to be possible to isolate a single magnetic pole, a **magnetic monopole**. Many extensive searches for magnetic monopoles have been conducted. Although the existence of a magnetic monopole is not ruled out by the present, well-established theories, all searches have ended in failure and it does not seem likely that one will be discovered. The discovery of a magnetic monopole would most likely result in a Nobel Prize.

In naming the ends, or poles, of the magnets, we make use of the observation that a freely swinging magnet aligns itself along the north–south direction. The end that points north is called *north* and the other *south*. The compasses used for navigation are simply tiny magnets that are free to rotate. Magnets have been used as compasses since the 11th century, but it wasn't until 1600 that William Gilbert, an English physician, hypothesized that they work because Earth itself is a giant lodestone. Gilbert even made a spherically shaped piece of lodestone to show that a compass placed near it behaved as it did on Earth.

PHYSICS | ON YOUR OWN

You can make a simple compass by magnetizing a sewing needle and placing it on a cork floating in a glass of water. The needle can be magnetized by stroking it with a magnet.

Magnets are surrounded by **magnetic fields** in the same way electric charges are surrounded by electric fields. These magnetic fields can be detected using a small compass. The direction of a magnetic field at a particular point is the direction that the north pole of the compass points when placed at this point, and the torque that aligns the compass is a measure of the strength of the magnetic field. The photograph in Figure 22-3(a) was obtained by sprinkling

Figure 22-3 (a) Iron filings line up along the magnetic field lines surrounding a bar magnet. (b) The drawing indicates the direction of the magnetic field outside the magnet.

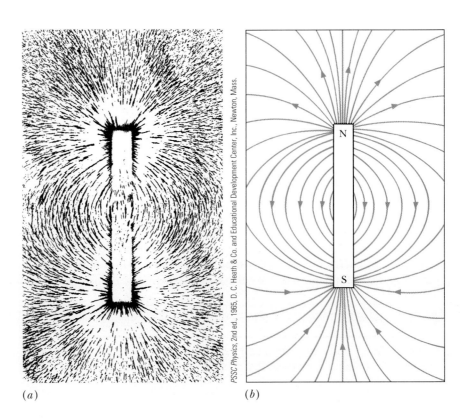

(a) (b)

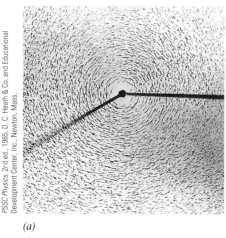
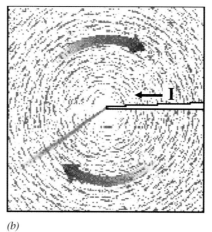

(a) *(b)*

Figure 22-4 (a) Iron filings show that the magnetic field lines surrounding a long, straight, current-carrying wire are circles around the wire. The dark line at the left is the shadow of the wire. (b) The diagram shows the direction of the magnetic field for a current flowing away from you (into the page).

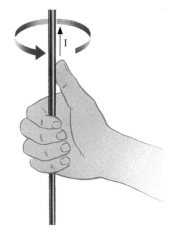

Figure 22-5 The direction of the magnetic field is given by the right-hand rule. If your thumb points in the direction of the current, your fingers encircle the wire in the direction of the field.

iron filings on a piece of glass placed over a bar magnet. The iron filings line up along the direction of the magnetic field.

Question How would the photograph in Figure 22-3 change if the poles of the magnet were reversed?

Answer The photograph would not change because the iron filings do not tell us which way the field points.

Electric Currents and Magnetism

Although electricity and magnetism were well known for centuries, any connection between the two phenomena eluded experimenters until the 19th century. In 1820 Hans Christian Oersted, a Danish scientist, discovered a connection while performing a lecture demonstration: a compass needle experiences a force when it is brought near a current-carrying wire. This means that the current-carrying wire produces a magnetic field in the surrounding space. This discovery was fascinating and yet confusing. It was known that stationary charges did not produce magnetic fields, so the motion of the charges must produce the field. The photograph in Figure 22-4(a) shows that the magnetic field lines form circles about the wire.

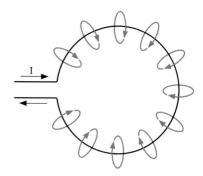

Figure 22-6 The magnetic field of this single loop of wire points into the page inside the loop and out of the page outside the loop.

The direction of the magnetic field is given by the *right-hand rule*: if you grasp the wire with the thumb of your right hand pointing in the direction of the positive current, your fingers encircle the wire in the direction of the magnetic field, as shown in Figure 22-5.

Bending the wire produces different field patterns. For instance, the magnetic field of a circular loop of wire is shown in Figure 22-6. The contributions to the magnetic field from each segment of the wire add together inside the loop to produce a field pointing into the page. Adding more loops to form the cylindrical structure, called a *solenoid* (Figure 22-7), produces a magnetic field like that of the bar magnet in Figure 22-3. The magnetic field inside the solenoid is relatively strong and rather uniform except near the ends.

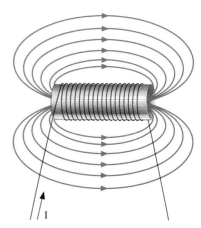

Figure 22-7 The magnetic field of a solenoid is like that of a bar magnet.

Observations like these led French physicist André Ampère to suggest that *all* magnetic fields originate from current loops. We now believe that magnetism originates in current loops at the atomic level. In a simplified model of the atom, we visualize these currents arising from electrons orbiting atomic nuclei or spinning about their axes. The macroscopic magnetic properties of an object are determined by the superposition of these atomic magnetic fields. If their orientations are random, the object has no net magnetization; if they are aligned, the object is magnetized.

Making Magnets

Like electric charge, our Universe is filled with magnetism. We do not normally detect the magnetism because of the random orientation of the atomic current loops. The atoms can be aligned, however, by the presence of a magnetic field. In most materials this alignment disappears when the magnetic field is removed. However, in some materials the atoms remain aligned and therefore retain their macroscopic magnetism.

A piece of iron can be magnetized by placing it in a strong magnetic field, by stroking it with a magnet, or by hitting it while it is in a magnetic field. Tapping an iron rod on the floor while holding it vertically will magnetize it because Earth's magnetic field has a vertical component. You can check for the resulting magnetization by holding a compass near the top and bottom of a metal object, such as a filing cabinet. The opening and closing of the file drawers provides the tapping; Earth's magnetic field does the rest. Figure 22-8 shows the orientation of compasses near the top and bottom of a file cabinet. On the other hand, an iron object that has been magnetized can lose its magnetism if it is dropped because the impact tends to randomize the alignment of the atoms. Heating the object can also randomize the atomic magnetic fields.

PHYSICS | ON YOUR OWN

Magnetize an iron rod by aligning it with Earth's magnetic field and hitting it on the end with a hammer. In the United States, Earth's magnetic field points north and down. Because Earth's field has a vertical component, the rod could also be magnetized by bouncing its end on a hard surface.

A versatile magnet can be constructed by wrapping wire around an iron core and connecting the ends of the wire to a battery, as illustrated in Figure 22-9. When current passes through the coil of wire, it generates a magnetic field along

Figure 22-8 The magnetic fields near the top and bottom of a metal file cabinet point in opposite directions.

A large electromagnet is used to load cargo containers onto a ship.

Figure 22-9 A very simple electromagnet can be constructed from a nail and some wire.

the axis of the coil. The magnetic field that is induced in the iron adds to that of the solenoid, increasing the strength of the magnetic field. These **electromagnets** are very useful because they can be turned on and off, the strength of the field can be varied by varying the current, and they can produce very large magnetic fields. Electromagnets can be used for such tasks as moving cars or sorting scrap iron from nonmagnetic metals.

The Ampere

A current-carrying wire exerts forces on compass needles (Figure 22-10). Therefore, by Newton's third law, a magnet should exert a force on the wire. This effect was very quickly verified by experiment. A wire between the jaws of a large horseshoe magnet jumps out of the gap when the current is turned on (Figure 22-11). This force is biggest when the wire is perpendicular to the magnetic field. The force on the wire is always perpendicular to the wire and to the magnetic field.

These experiences indicate that two current-carrying wires should exert forces on each other. They do. The magnetic field produced by each current exerts a force on the other. If the currents are in the same direction, they attract each other (Figure 22-12); if the currents are in opposite directions, they repel.

This effect is used to define the unit of current—the basic electrical unit in the metric system. Consider two long parallel wires separated by 1 meter and carrying the same current. If the force between these wires is 2×10^{-7} newton on each meter of wire, the current is defined as 1 **ampere** (A). The **coulomb** (C) is then defined as the amount of charge passing a given point in one of these wires during 1 second. This is the same as the charge on 6.24×10^{18} protons.

The source of the force on each wire is the magnetic field produced by the other wire. We can also use the interacting wires to define a field strength for magnetism. The strength of the magnetic field at a distance of 1 meter from a long straight wire carrying a current of 1 ampere is 2×10^{-7} **tesla** (T). Another unit for the magnetic field strength, the **gauss** (G), is often used instead of tesla, where 1 tesla = 10,000 gauss.

The magnets that are used to hold notes or pictures on refrigerator doors produce fields on the order of 0.3 tesla, whereas large laboratory magnets produce fields of 2.5 teslas. The theoretical limit for a permanent magnetic field is 5 teslas. Electromagnets made with ordinary wires have produced steady fields of 34 teslas, while those made with superconducting wires have not yet exceeded 22 teslas. Combinations of these two types of magnets have reached 45 teslas. At

Figure 22-10 A current-carrying wire produces a magnetic field surrounding the wire. Notice that the compasses above and below the wire point in opposite directions.

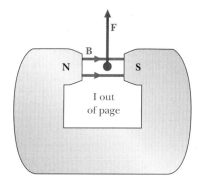

Figure 22-11 The magnetic field of the horseshoe magnet exerts a force on the current-carrying wire, causing it to jump out of the jaws of the magnet.

Superconductivity

Under normal conditions all conductors exhibit resistance to the passage of electric current. This makes sense if you visualize the conduction electrons bumping their way through a metal consisting of an array of vibrating atoms. The electrons are accelerated along the wire by the electric potential difference provided by a battery and then lose their acquired speeds in collisions. The surprise is that the electrical resistance of some materials goes to zero at very low temperatures; they become *superconductors*.

This knowledge is not a new breakthrough: superconductivity was first discovered by Dutch physicist Heike Kamerlingh Onnes in 1911. Three years earlier Onnes had developed a process for liquefying helium and was therefore able to study the properties of materials at very low temperatures. The first superconductor was solid mercury, which lost its electrical resistance at a critical temperature of 4.15 K. Experimenters searched for metals or alloys that exhibited superconductivity, and new superconductors were found with critical temperatures as high as 23 K.

In 1972 three American physicists, John Bardeen, Leon Cooper, and J. Robert Schrieffer, received the Nobel Prize for explaining superconductivity in 1957. Their theory—called the BCS theory—showed how electrons pair up to create a resonance effect that allows the electron pairs to travel effortlessly through the material. The electrical resistance does not simply drop to a very low value; it drops to zero! Once a current has been established in a superconducting material, it will persist without any applied voltage. Such supercurrents have been observed to last for years, which can have great practical applications in the generation of magnetic fields. Large magnetic fields require very large currents. In ordinary materials this means generating a lot of thermal energy at high cost and diverting the heat to avoid melting the magnets. In contrast, once the large currents have been established in superconducting magnets, they can be disconnected from the power supply; no further electric energy is needed.

Theoretical models for superconductivity predicted an upper limit for the critical temperature of about 30 K. Had this upper limit been true, superconductivity would have remained in the domain of very low temperatures. The only gases with boiling points this low are helium (which is expensive) and hydrogen (which is explosive). In 1986, nearly 80 years after the discovery of the first superconductor, a major new class of superconductors was discovered with much higher critical temperatures. These new superconductors are *ceramics* made from copper oxides mixed with such rare earth elements as lanthanum and yttrium and have critical temperatures as high as 125 K. This new limit is very significant because this temperature is higher than the boiling point of nitrogen, an abundant gas that is relatively inexpensive to liquefy and very safe to use, allowing liquid nitrogen to be used to keep a material superconducting. A tremendous increase in the use of superconductors will occur if a superconducting material can be made with a critical temperature above room temperature (or at least above those obtained by ordinary refrigeration).

One very important potential application of superconductivity is in the transportation of electric energy. With conventional transmission lines, much of the electricity produced at a distant power plant is lost to resistive heating effects in the wires that carry this energy to our homes. A superconducting transmission line would eliminate these losses. The problem, of course, is that we need to cool these conduits to make them superconducting. However, if we can use ordinary refrigeration, the cooling costs will be much less than the costs of resistive losses.

Superconductors have a second property that is very important. They expel magnetic fields when they become superconducting. This also means that magnetic fields cannot penetrate the superconductors and they will therefore repel magnets. This is dramatically shown by floating a permanent magnet above a superconductor, as shown in the figure. This effect may find applications in improving the magnetic levitation of trains for high-speed transportation. A prototype train has been constructed in Japan using helium-cooled superconducting magnets to levitate the train as well as to propel it at speeds up to 300 mph.

A permanent magnet levitates above a superconductor because its magnetic field cannot penetrate the superconducting material.

Many other applications have been proposed. The switching properties of superconducting materials could have a large impact on the field of computer electronics. It might also be possible to construct superconducting generators and motors. These new superconducting materials might also find applications in the field of medical imaging. However, many of these applications will require major technological breakthroughs in fabrication (for instance, ceramics are brittle and therefore difficult to form into wires) and in finding materials that will carry larger currents.

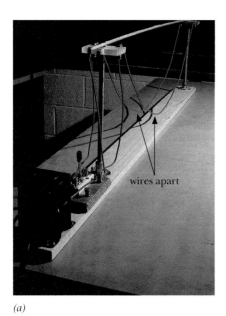
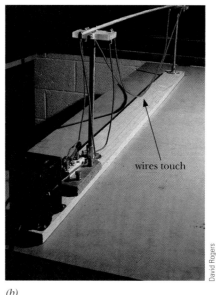

(a) (b)

Figure 22-12 (a) When the switch is open, there is no attractive force between the wires. (b) When the switch is closed, the currents are in the same direction, and the wires are attracted to each other.

the upper extreme, pulsed magnets have produced fields of 70 teslas. Even higher fields approaching 1000 teslas have been generated for extremely short times, but the forces are so large that the magnets destroy themselves.

The Magnetic Earth

As mentioned earlier, a simple compass shows that Earth acts as if it had a huge magnet in its core, as illustrated by Figure 22-13. The strength of Earth's magnetic field at Earth's surface is typically 5×10^{-5} tesla (0.5 gauss). In the United States its horizontal component points generally northward and an even larger component points downward.

Measurements of the magnetic field show that one of Earth's magnetic poles is located in northeastern Canada just north of Hudson Bay, about 1300 kilometers from the geographic North Pole. Earth's other magnetic pole is almost directly on the other side of Earth. A line through Earth connecting these two poles is tilted about 12 degrees from Earth's rotational axis, which passes through Earth's geographic North and South Poles.

Figure 22-13 Earth is a giant magnet with its south magnetic pole in the Northern Hemisphere. This early view of Earth's magnetism has been replaced by one in which the magnetic field is generated by electric currents in Earth's core.

Question   Is the magnetic pole located in northern Canada a north or a south magnetic pole?

Answer   Because this pole attracts the north pole of a compass and because opposite poles attract, this magnetic pole must actually be a south magnetic pole. However, the geographical location of the pole is still known as the magnetic North Pole.

The tilt of Earth's magnetic axis from its rotational axis makes finding true north with a compass a bit complicated. The only time your compass will point to true north is when the magnetic pole is between you and the geographic pole. Because of local variations in the magnetic field, this situation occurs along an irregular line running from Florida to the Great Lakes, as shown in Figure 22-14. If you are east of this line, your compass will point to the west of true north. If you are west of this line, your compass points to the east of true north. The difference in the directions to the magnetic and geographic poles is known as the *magnetic variation*. As an example, the magnetic variation in Bozeman, Montana, is currently 15.5 degrees east.

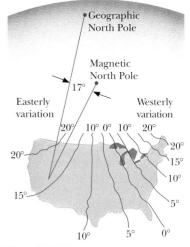

Figure 22-14 The magnetic variation between the directions to the geographic and magnetic North Poles changes across the United States.

Airplane pilots use magnetic headings and refer to their aeronautical maps to find the true headings. The large numbers painted on the ends of runways correspond to the magnetic directions of the runways divided by 10 and rounded.

At one time scientists imagined that Earth's magnetic field was caused by magnetized solid iron in Earth's interior, but they now believe this cannot be true. Earth's interior is known to be hot enough that the iron and nickel are in a liquid state. In a liquid state, the atomic magnetic fields do not remain aligned but rather take on random orientations, thus eliminating a macroscopic magnetic effect.

The best model is that large electric currents circulating in the molten interior cause Earth's magnetic field. Such currents could easily produce the field that we observe on Earth, as well as the general features of the magnetic fields of the other planets. However, there are some difficulties with this theory. No one understands the details of the mechanisms for producing and maintaining the currents.

The most puzzling aspect of Earth's magnetic field is its reversals, when the North and South Poles switch locations. There is strong evidence that Earth's magnetic field has reversed directions 171 times in the last 17 million years. This evidence comes from the rocks on both sides of the mid-Atlantic rift. As molten rock emerges from the rift, it cools and solidifies, and at that moment the direction of Earth's magnetization is locked into the rocks and preserved. Samples from the ocean's floor show that the directions of the magnetic fields in the rocks alternate as we approach the rift. Although we know the reversals have occurred, no one has been able to propose a satisfactory mechanism for these reversals or account for the source of the vast amount of energy required to reverse the field.

Charged Particles in Magnetic Fields ✓ MATH

Recall that a magnet has no effect on a charged object other than the normal electrostatic attraction between neutral and charged objects. On the other hand, a current-carrying wire in a magnetic field does experience a force. Because an electric current is a stream of charged particles, the motion of the charges must be important. As bizarre as it seems, the magnetic force on a charged particle is zero unless the charge is moving!

In addition, the direction of the magnetic force is probably not the direction you would predict. Recall that the current (the motion of charged particles) is along the wire in Figure 22-11 but that the wire experiences a force in a direction perpendicular to the wire and perpendicular to the magnetic field. Thus, we expect a charged particle moving in a magnetic field to experience a force that is at right angles to its velocity and to the magnetic field.

The strength of the force depends on the charge q, the strength of the magnetic field B, the speed v, and the angle between the field and the velocity. The magnetic force is maximum when the field and velocity are perpendicular:

$$F_{max} = qvB$$

▶ maximum magnetic force = charge × speed × magnetic field

When the velocity and the magnetic field are parallel, the force is zero.

The magnetic force produces some interesting phenomena. First, because the force is always perpendicular to the particle's velocity, it never changes the particle's speed, only its direction. If the particle's velocity is parallel to the magnetic field, the particle does not experience a force, and it moves in a straight line. On the other hand, if the velocity is perpendicular to the field, the particle experiences a centripetal force that causes it to move in a circular path (Figure 22-15[a]). If the velocity has components parallel and perpendicular to the magnetic field, these two motions will superimpose and the particle will follow a helical path (Figure 22-15[b]) along the magnetic field.

> ✓ Extended presentation available in the *Problem Solving* supplement

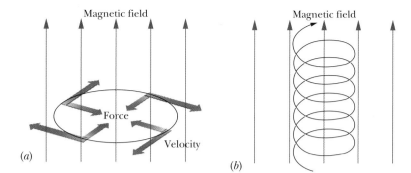

Figure 22-15 (a) The charged particle follows a circular path if its velocity is perpendicular to the magnetic field. (b) The path is helical if its velocity has components both perpendicular and parallel to the field.

The magnetic force on charged particles is the cause of dramatic effects known as the *aurora borealis* (northern lights) and the *aurora australis* (southern lights). As charged particles, mostly from the Sun but also from outer space, approach Earth, they interact with Earth's magnetic field. This interaction causes the charged particles to follow helical paths along Earth's magnetic field so that they strike Earth's atmosphere in the regions of the magnetic poles. The collisions of these cosmic rays with the oxygen and nitrogen molecules in the atmosphere produce the spectacular evening light shows high over the magnetic North and South Poles.

The magnetic forces on charged particles are important in many scientific and technologic devices, ranging from containment vessels used in developing the future technology of fusion reactors to particle accelerators used in research and to television sets in our homes. We will discuss many of these in the remaining chapters.

Magnetism and Electric Currents

In the evolution of the physics world view, a small number of basic themes have emerged that reflect our biases but also fuel our searches. One such theme is symmetry. A number of times, progress has been made while searching for symmetric effects. Early in the development of our understanding of electricity and magnetism, the questions of symmetry haunted some experimenters. In particular, they asked, "Given that an electric current produces a magnetic field, does a magnetic field produce an electric current?" To investigate this, you might try wrapping some wire around a magnet and connecting the wire to an *ammeter*, an instrument for measuring current. Although similar experiments were performed with large magnets and sensitive instruments, no effect was found.

British scientist Michael Faraday discovered the connection in 1831. He found that motion is the key to producing an electric current with a magnet. If a wire is moved through a magnetic field (but not parallel to the field), a current is produced in the wire. This current is due to the motion of the wire in the magnetic field; there are no batteries. The current is largest if the motion is perpendicular to the field and increases with the speed of the wire. Having the benefit of hindsight, we see that this is understandable because of the magnetic force on the charges in the wire.

Because the principle of relativity (Chapter 9) must be valid for electricity and magnetism, we know that it is the *relative* motion of the wire and the magnet that is important. It doesn't matter which is moving relative to the laboratory and which is stationary. Therefore, we should be able to generate a current in a stationary wire by moving the magnet. This principle can be demonstrated with a simple experiment. If a bar magnet is placed in a coil of wire and quickly withdrawn, a current is generated that is easily detected. Quickly inserting the magnet into the coil generates a current in the opposite direction. Reversing the

The aurora borealis is caused by cosmic rays following magnetic field lines toward the magnetic North Pole.

Figure 22-16 A magnet inserted into or removed from a coil of wire produces a current. (When the needle is centered, there is no current.)

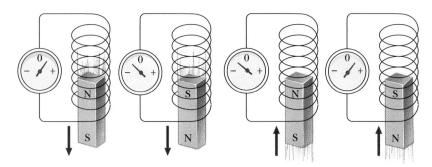

direction of the bar magnet also reverses the direction of the current. The four possibilities are shown in Figure 22-16.

Faraday also discovered that motion is not the only way of producing a current with a magnetic field. A current is produced if the strength of the magnetic field varies with time, even when there is no relative motion of the wire and the magnet. An increasing field produces a current in one direction; a decreasing field produces a current in the opposite direction.

After a long series of experiments conducted over several years, Faraday was able to generalize his results in terms of magnetic field lines. These lines are visualized as pointing along the direction of the magnetic field at all points in space. The number of lines in a given region of space represents the strength of the magnetic field. The lines are closer together where the magnetic field is stronger.

Faraday's law ▶

Faraday showed that if the number of magnetic field lines passing through a loop of wire changed *for any reason*, a current was produced in the loop (Figure 22-17). The voltage (and hence the current) generated in the loop depends on how fast the number of field lines passing through the loop changes—the faster the change, the larger the voltage. These phenomena are now known as Faraday's law.

Lenz's law ▶

The direction of the current induced in the coil by the changing number of magnetic field lines is given by Lenz's law, which states that the current always produces a magnetic field to oppose the change. For example, in Figure 22-17 the number of lines in the upward direction is increasing. Therefore, the induced current will produce a magnetic field that points downward to try to cancel out the increase. Notice that the current shown in Figure 22-17 does this. If the number of lines had been decreasing, the current would be induced in the opposite direction to produce a magnetic field in the upward direction to try to maintain the original field.

Figure 22-17 Changing the number of magnetic field lines passing through a loop of wire produces a current. The direction of the current is given by Lenz's law.

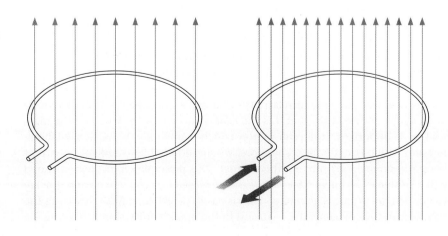

Question What would you see on the ammeter if you dropped a magnet through the coil in Figure 22-16?

Answer As the magnet enters the coil, the needle will swing to one side due to the increase in the number of field lines passing through the coil. As the magnet exits, the needle will swing to the other side because the number of lines decreases.

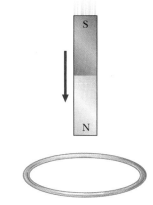

Figure 22-18

Flawed Reasoning

The following question appears on the midterm: "A copper loop lies flat on a table. If the north pole of a bar magnet is lowered quickly toward the loop, as shown in Figure 22-18, a current is induced in the loop in the direction given by Lenz's law. Find this direction and explain how you used Lenz's law."

Jose gives the following answer: "Magnetic field lines leave the north pole of a bar magnet, so the field lines through the loop are directed downward. Lenz's law states that an induced magnetic field will be generated in the opposite direction—that is, upward. By the right-hand rule, the induced current must be counterclockwise to produce this field."

The induced current is counterclockwise, as Jose claims, but not for the reason given. **Explain Lenz's law to Jose.**

Answer There are two magnetic fields in the loop: the *external* field of the bar magnet and the *induced* field due to the current. Jose believes that the induced field opposes the external field, but this is not what Lenz's law asserts. The induced field opposes any *change* in the external field, not the field itself. Jose gets the correct direction by luck. If the bar magnet is moving upward, Jose's reasoning yields the wrong answer. The external field in the loop is downward and growing weaker. Jose predicts that the induced field is upward to oppose the downward external field. Lenz tells us that the induced field is downward to oppose the weakening of the external field.

Transformers ✓ MATH

These discoveries about magnetism—especially its connections with electricity—have many practical uses. For example, they are fundamental to the operation of transformers used to change the voltage of alternating-current electricity. The same amount of electric power can be delivered through wires at low voltage and high current as at high voltage and low current (Chapter 21). For instance, 10 amperes at 12 volts supplies the same power as 1 ampere at 120 volts. The particular choice of what amperage and voltage are used depends on the circumstances. The large cylindrical containers on power poles are transformers for reducing the voltage to 120 volts before it enters our homes and businesses.

The schematic diagram in Figure 22-19 shows the essential features of a transformer. One coil, called the *primary coil*, is connected to a source of alternating current. The alternating current in the primary coil produces an alternating magnetic field that is transmitted to the *secondary coil* by the iron core. This alternating magnetic field produces an alternating voltage in the secondary coil.

The size of the voltage produced in the secondary coil depends on the ratio of the number of loops in the two coils. A transformer designed to reduce the voltage by a factor of 2 would have one-half as many loops in the secondary as in the primary coil.

Neighborhood transformers reduce the voltage for household use.

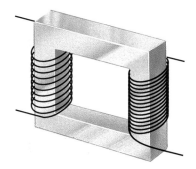

Figure 22-19 The essential features of a transformer.

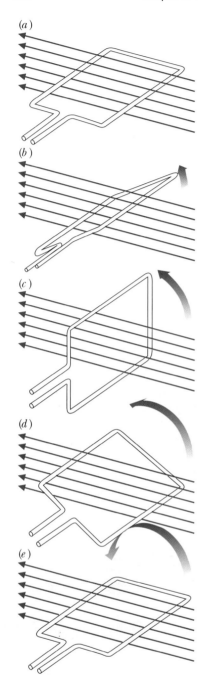

Figure 22-20 The number of magnetic field lines passing through a loop of wire changes as it rotates. This produces the alternating voltage shown in Figure 22-21.

Question Why won't a transformer work with direct-current electricity?

Answer Direct current will not produce the varying magnetic field necessary to induce a current in the secondary coil.

Generators and Motors

The electric generator is another application of Faraday's discovery. If we rotate the loop of wire in Figure 22-17 by 90 degrees, the number of magnetic field lines passing through the loop drops to zero. The change in the number of field lines passing through the loop produces a voltage around the loop, which produces a current. The loop is usually rotated by a steam turbine or falling water in a hydroelectric facility. The steam can be produced by many means: burning wood, coal, petroleum, or natural gas or by using heat from the Sun or nuclear reactors.

Figure 22-20 contains a series of drawings of the magnetic field and a wire loop in an electric generator that illustrates how the rotating loop produces electric voltages (and currents). Figure 22-20(a) shows the plane of the loop parallel to the magnetic field lines. The number of lines passing through the loop in this orientation is zero. As the loop rotates at a constant speed, the number of lines initially increases rapidly, producing a large voltage. As it continues to rotate (b), the number of lines passing through the loop continues to increase but at a slower rate. The voltage drops to zero when the plane of the loop is perpendicular to the field lines (c) and the number of lines through the loop is a maximum. The number of lines through the loop now decreases (d), and the voltage increases in the opposite direction. It increases to a maximum (e) and then returns to zero when the plane of the loop is perpendicular to the field lines once again. The voltage increases to a maximum in the opposite direction, and the entire cycle repeats. This generator produces the alternating voltage shown in Figure 22-21(a).

A simple change in the way the voltage is carried to the external circuit converts the generator to produce pulsating direct current (Figure 22-21[b]). A connector (called a *commutator*), shown in Figure 22-22, reverses the connections from the loop to the outside circuit each half turn. This pulsating current can then be electronically smoothed to produce a constant direct current like that from a battery.

A direct-current motor is basically a direct-current generator run backward. In fact, the discovery of the first such motor occurred during an exhibition in 1873 when a technician setting up a demonstration of generators hooked up one

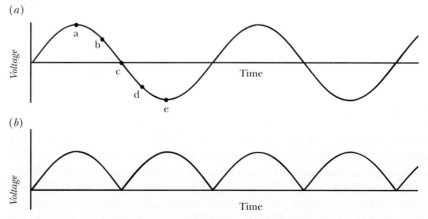

Figure 22-21 (a) A loop rotating in a magnetic field produces an alternating voltage. The letters show the voltage produced when the loop has the positions shown in Figure 22-20. (b) The pulsating direct current produced by using a commutator.

"Wireless" Battery Charger

An electric toothbrush has a base designed to hold the toothbrush handle when not in use. As shown in the figure, the handle has a plastic cylindrical hole that fits loosely over a matching plastic cylinder on the base. How does the toothbrush get charged when there are no metal contacts? When the handle is placed on the base, a changing current in a solenoid inside the base cylinder (the primary coil) induces a current in a coil inside the handle (the secondary coil). This induced current charges the battery in the handle. The size of the induced current is amplified by inserting a core of ferromagnetic material, like iron, into the primary coil. The primary coil then magnetizes the iron core such that its poles reverse 120 times each second (for 60-cycle current). This causes a greater change in the magnetic field inside the secondary coil.

Wireless charging is used in a number of other "cordless" devices. One significant example is the inductive charging used by some electric car manufacturers that avoids direct metal-to-metal contact between the car and the charger.

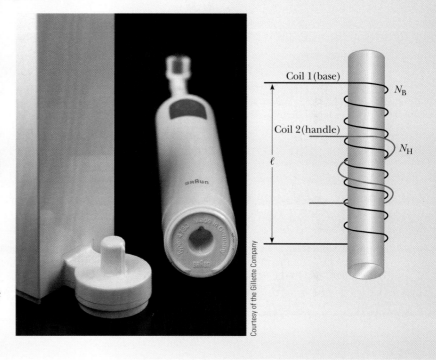

This electric toothbrush uses magnetic induction to charge its battery. The primary coil (Coil 1) of N_H turns is in the base unit and the secondary coil (Coil 2) of N_B turns is in the handle.

---

the wrong way and "discovered" a motor! In a generator we rotate the loop in a magnetic field, which produces a voltage that moves electric charges. In a motor the sequence is reversed; we apply a voltage, causing the charges to move. This current in a magnetic field produces a force that rotates the loop.

Figure 22-20 can be used to illustrate the operation of a motor. When a voltage is applied to the loop when it is in the position shown in Figure 22-20(a), the magnetic field exerts a torque on the current-carrying loop. The forces on the long sides of the loop are in opposite directions because the currents are in opposite directions. The torque decreases as the loop rotates and becomes zero when the plane of the loop is vertical. If nothing changes as the loop coasts through this position, it will oscillate and eventually come to a stop. However, the commutator reverses the direction of the current so that the torque continues to act in the same direction.

The similarity of a motor and a generator is important in some electric-powered vehicles. While the vehicle is speeding up or cruising, the engine acts like an electric motor. However, when the brakes are applied, the connections to the engine are changed so that it acts like a generator to recharge its batteries. The current exerts a torque on the engine, which in turn exerts a torque on the wheels to stop the vehicle. This is not a perpetual motion machine because only part of

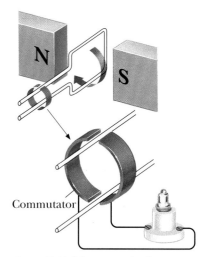

Figure 22-22 Schematic of a direct-current generator.

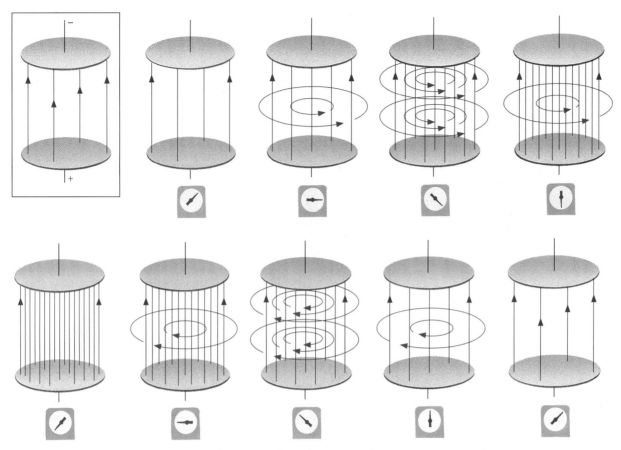

Figure 22-23 A steady electric field between the parallel plates shown in the boxed diagram at the left does not produce a magnetic field. An increasing electric field produces a magnetic field in one direction, whereas a decreasing field produces a magnetic field in the opposite direction.

the energy is recovered. However, regenerative braking can extend the range of the vehicle.

A Question of Symmetry

The connection between electricity and magnetism appears to be complete—a *changing* magnetic field produces an electric current, and an electric current produces a magnetic field. Notice, however, that the situation is not quite symmetric. It requires a changing magnetic field to produce the electric current, but the electric current produces a steady magnetic field.

Actually, there is more to this connection that becomes apparent only when everything is expressed in terms of the electric and magnetic fields. Consider two parallel plates connected to a battery, as shown in Figure 22-23. As current from the battery builds up charges on the plates, an electric field is generated in the region between the plates. Even though no charges are flowing between the plates, a magnetic field is produced in the region surrounding the plates that matches the magnetic field surrounding the wires. Therefore, the origin of the new field cannot be the charges. Furthermore, the magnetic field disappears when the current stops. The electric field between the plates continually increases as long as charge is flowing to the plates and remains constant when the charge stops flowing. Therefore, the magnetic field is produced by a *changing* electric field. As the charges leave the plates, the electric field decreases and a magnetic field in the opposite direction is produced.

We can make the analogy closer by looking at the region between the poles

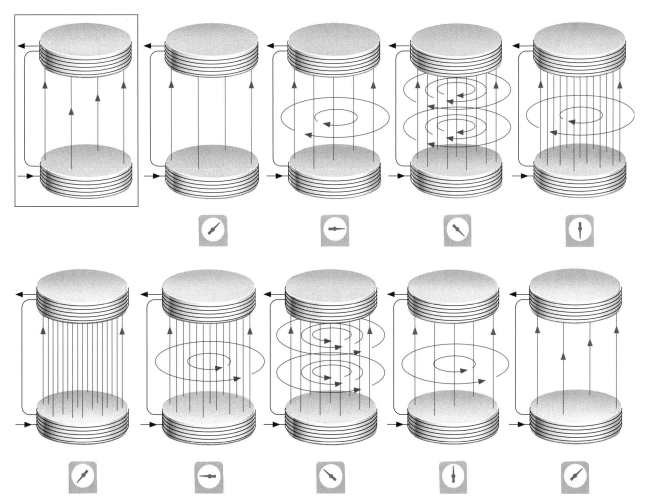

Figure 22-24 A steady magnetic field between the poles of the electromagnet shown in the boxed diagram at the left does not produce an electric field. An increasing magnetic field produces an electric field in one direction, whereas a decreasing magnetic field produces an electric field in the opposite direction.

of an electromagnet. As the magnetic field between the poles increases, an electric field is produced in the surrounding region (Figure 22-24). When the magnetic field reaches its maximum strength and is no longer changing, the electric field disappears. As the magnetic field decreases, an electric field in the opposite direction is produced.

It is important to realize that a changing magnetic field produces an electric field in *empty* space. There is no need for a wire. If a wire is present, the electric field exerts forces on the charges within the wire and produces a current. But the important point is that, even in the absence of the wire, the electric field is present.

If we use these results and focus on the fields rather than the currents, the situation is completely symmetric: a *changing* magnetic field generates an electric field, and a *changing* electric field generates a magnetic field. There is an intimate relationship between electricity and magnetism.

Electromagnetic Waves

If the magnetic field changes at a constant rate (that is, if it changes by the same amount each second), the electric field that is produced is constant. A rapidly changing magnetic field produces a large electric field, and a slowly changing magnetic field produces a smaller one. However, if the magnetic field starts out changing slowly and then increases its rate of change, the electric field starts out small and grows larger. Thus, it is possible for a changing magnetic field to produce a changing electric field.

The parallel plates in Figure 22-23 show the symmetric effect; a changing electric field produces a magnetic field. The rate of change of the electric field determines the size of the magnetic field. Therefore, a changing electric field can produce a changing magnetic field.

Question   How would you use an electric field to produce a magnetic field that increases in size?

Answer   The electric field must change slowly at the beginning and continually increase its rate of change.

We have discussed a lot of "rates of change," and this can be quite confusing, but the payoff for understanding the process is worth the effort. We have discovered a sequential chain of field productions: one changing field produces another changing field, and then this new changing field produces the first kind of field again, and so on. The two fields generate each other in empty space.

We have only argued that this is possible, but the process was rigorously deduced in the 1860s by James Clerk Maxwell. Maxwell was able to show that this was a consequence of a set of four equations that he and others had developed to describe electricity and magnetism and the many connections between them. The equations, called Maxwell's equations in honor of his contributions, summarize all of electricity and magnetism.

Maxwell combined these equations into a single equation that had the same form as the equations that describe periodic waves, whether they are waves on a rope, water waves, or sound waves. Oscillating electric and magnetic fields can combine to produce waves that travel through space (Figure 22-25). As calculated by Maxwell, these **electromagnetic waves** take the form of oscillating electric and magnetic fields that travel with a speed equal to that of light. In 1887 German physicist Heinrich Hertz was able to produce electromagnetic waves on one side of a room and detect them on the other. The existence of electromagnetic waves affirmed Faraday's belief that these fields had their own identities.

Electromagnetic waves are produced whenever electric charges are accelerated. If the charges have a periodic oscillatory motion, the wave will have a fixed

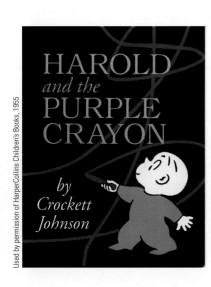

Flawed Reasoning

All of us struggle with misconceptions as we build our world views. Even Isaac Newton made mistakes in his reasoning. He believed that "luminiferous ether" must fill all space to provide a medium for the propagation of light waves. If we jiggle gelatin on one side of a bowl, the jiggle moves across the bowl. Newton felt it was impossible for the jiggle to move across an empty bowl. **Explain how an electromagnetic wave can propagate through empty space.**

Answer   The electromagnetic wave propagates through empty space by recreating itself. A changing electric field creates a changing magnetic field— it waves the magnetic field. This changing magnetic field in turn creates a changing electric field—it waves the electric field. And then the process starts over again. It is somewhat like the popular preschool story of *Harold and the Purple Crayon* by Crockett Johnson. In this story, Harold gets to where he wants to go by drawing himself a set of stairs with his purple crayon and his imagination.

MAXWELL | Unifying the Electromagnetic Spectrum

James Clerk Maxwell was born on June 13, 1831, into a prominent Edinburgh family. Clerk (pronounced "Clark") was his original family name before the Clerks and the Maxwells intermarried. His "hyphenated" name was taken as a result of legal manipulations that prevented the family from holding extensive property in one unit.

Maxwell was brought up in a thoroughly Scottish environment, with kilts, Gaelic, and a burr. He was a curious and gifted child. His father, who had serious scientific interests, encouraged this bent. He attended Edinburgh Academy and in 1847 entered Edinburgh University, where a distinguished faculty welcomed him. One of his professors was famous for his textbooks in physics—some of which were used in the United States until after World War II. In 1850 Maxwell went to Cambridge University, first to Peterhouse College and then to Newton's old haunt, Trinity College. He was a brilliant student of mathematics and began a lifelong study of mathematical models of Saturn's rings. He won the most prestigious mathematical award in British school mathematics and began what would be a wonderfully productive review of the pioneering electromagnetic experiments of Michael Faraday.

Maxwell returned as a professor to Aberdeen, Scotland, and then to King's College in London. He married but had no chil-

James Clerk Maxwell

dren. He retired to his Galloway estate to write the monumental *Treatise on Electricity and Magnetism*. His work in this field generated important advances in the electrical industry, in technology, and in every branch of physical science.

Although retired and in ill health, Maxwell designed and founded the Cavendish Laboratory and was its first director. This was the first major experimental laboratory explicitly designed with shielding from outside radiation and vibration.

He also developed an equally powerful kinetic theory of gases, which led in turn to further breakthrough work in thermodynamics. He continued the Saturn work and contributed to astrophysics, color vision, optics, photoelasticity, servomechanics, and in the long run, to color photography. He left a remarkable legacy. In tribute, Albert Einstein always displayed Maxwell's portrait alongside that of Isaac Newton. He endures as one of the greatest scientists in history.

—*Pierce C. Mullen, historian and author*

Source: C. W. Francis Everitt, *James Clerk Maxwell: Physicist and Natural Philosopher* (New York: Scribner, 1975).

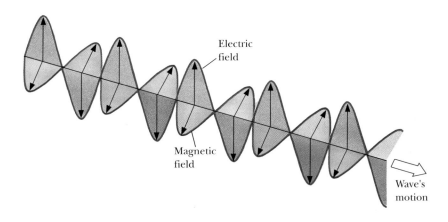

Figure 22-25 An electromagnetic wave propagating through space. The electric and magnetic fields are perpendicular to each other and to the direction the wave is moving.

frequency and therefore a fixed wavelength according to the relationship we developed in Chapter 15 between speed, wavelength, and frequency:

$$v = \lambda f$$

◀ speed = wavelength × frequency

Maxwell's equations require that the speed be that of light but place no restrictions on the frequency, as shown in the diagram of the electromagnetic spectrum in Figure 22-26. Although these waves all have the same basic nature, the different ranges in frequency are produced (and detected) by different means. The boundaries between the various named regions are not distinct and, in fact, they overlap quite a bit.

The lowest frequencies and the longest wavelengths belong to the radio waves. These are produced by large devices like broadcast radio antennas. Microwaves are also produced electronically, but the devices are smaller, ranging in size from a few millimeters to a few meters. These devices are used in microwave ovens, radar, and the long-distance transmission of telephone calls.

The frequencies of visible light extend from 4.0 to 7.5×10^{14} hertz. Although

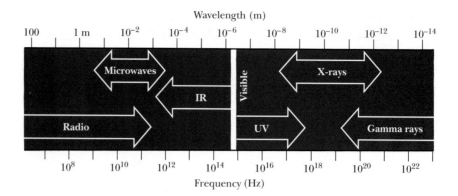

Figure 22-26 The electromagnetic spectrum.

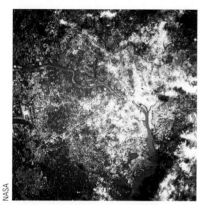

Infrared photo of the Washington, D.C., area looking north, showing the Potomac River in blue and vegetation in red. This view from a space shuttle covers 68 square miles.

visible light occupies only a very small region of the complete spectrum, it is obviously very important to us. This region is bounded on the low side by infrared radiation (IR), waves whose wavelengths are too long (beyond the red) to be seen by the human eye. Infrared radiation is most noticeable when given off by hot objects, especially those that are red hot. This is the radiation you feel across the room from a fire or a heating element. The radiation that lies beyond violet is known as ultraviolet (UV) waves. This is the component of sunlight that causes suntans (and sunburns if it is excessive). Visible light and its neighbors are produced at the atomic level, and their properties are valuable clues about the structure of matter at the atomic level.

We are all familiar with X rays from visits to the doctor or dentist. X rays have high frequencies and are very penetrating. They are produced by the rapid acceleration of electrons in X-ray machines and are emitted by atoms. Gamma rays are an even higher-frequency radiation that originate in the nuclei of atoms. We will study these in more detail in later sections of this textbook.

Radio and TV

Radio is a means of coding electromagnetic waves with the information in sound waves so that they can be transmitted through space, intercepted, and converted back into sound. Television is the same sort of process with the addition of the video information. Sound waves are changed to an electric signal by a microphone. In one version the sound waves cause a coil to vibrate in a magnetic field. This produces a current in the coil that can then be amplified. Audio frequencies are in the range of 20 hertz to 20 kilohertz.

If audio frequencies were broadcast directly, there could be only one radio station in any geographic region. Instead, the audio signal is combined with a broadcast signal. A station broadcasting at "1450 on your dial" sends out waves with a frequency of 1450 kilohertz (often called kilocycles). This is the *carrier frequency*. The audio signal is used to vary, or *modulate*, the carrier signal. Two modulation methods are *amplitude modulation* (AM) and *frequency modulation* (FM). The frequency of the sound determines the frequency of the modulation, and the loudness of the sound determines the amplitude of the modulation. In AM radio the audio signal causes the amplitude of the carrier signal to vary; in FM radio it causes the frequency of the carrier signal to vary (Figure 22-27).

A receiver dish for satellite television signals.

Question   Although modern police and fire sirens produce sound waves rather than radio waves, they are modulated. Is this modulation AM or FM?

Answer   Because these sirens have oscillating frequencies, they must be FM.

Stereo Broadcasts

The carrier frequencies of AM radio stations are separated by only 10 kilohertz. This means that the full 20-kilohertz range in audio frequencies cannot be broadcast without the stations overlapping. In practice, each station is limited to audio frequencies up to 5 kilohertz. The development of FM radio allowed the broadcasting of more of the audio frequency range (up to 15 kilohertz) as well as stereophonic signals.

But how do FM radio stations broadcast signals in stereo without affecting the performance of the older radios that were designed to receive monophonic signals? This is done by transmitting the two audio signals on the same carrier but separated by a fixed frequency. A 38-kilohertz frequency is added to one signal before broadcasting and then subtracted in the receiver. Similar techniques are used with the audio and video signals for TV broadcasts.

To create signals that can be played on either stereophonic or monophonic receivers, the signals that are broadcast are not the signals for the left and right channels but the sum and difference of the two channels. A monophonic receiver plays the sum of the two signals and ignores the difference signal, whereas a stereophonic receiver electronically adds and subtracts these signals to recover the left and right channels.

In either case, the final signal is then amplified and sent to the antenna, where it causes electrons to move up and down the antenna wire. The accelerations of these electrons produce electromagnetic waves that are broadcast. As these waves hit the antenna in your radio, they cause the electrons in the antenna to move back and forth, producing oscillating currents. Although all nearby stations are received simultaneously, the radio is tuned so that it resonates with only one of them at a time. Your radio amplifies this station's signal.

The radio contains electric circuits to filter out the carrier frequency and retain only the electrical version of the sound information. These signals are then sent to a speaker for reconversion to sound waves. In one version of a speaker, the electric signal generates a magnetic field that interacts with a magnet to move a diaphragm. This diaphragm then moves air to generate sound.

The allocation of the possible ranges of carrier frequencies to various types of broadcast is a governmental responsibility that is complicated by the history of the development of such advances in broadcasting techniques as FM stereo and TV. AM radio stations broadcast between 550 and 1500 kilohertz, FM radio between 88 and 108 megahertz, and TV in three regions between 54 and 890 megahertz. Other regions are assigned to citizens-band receivers, ships, airplanes, police, amateur radio operators, and satellite communication.

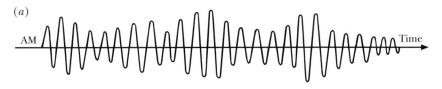

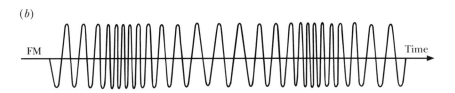

Figure 22-27 (a) AM and (b) FM radio waves.

Summary

Magnetic poles behave similarly to electric charges: like poles repel and unlike poles attract. However, magnetic poles always occur in pairs. Magnets attract some objects but have no effect on others.

Magnets have no effect on stationary charges and do not deflect the foils of an electroscope. On the other hand, a current-carrying wire produces a magnetic field and is attracted or repelled by other magnets or current-carrying wires. The force on the wire is always perpendicular to the wire and to the magnetic field. Two current-carrying wires attract each other if the currents are in the same direction; they repel if the currents are in opposite directions. The field strength for magnetism is defined as 2×10^{-7} tesla at a distance of 1 meter from a wire carrying a current of 1 ampere.

All magnetic fields originate from current loops. Naturally occurring magnetism originates in current loops at the atomic level. Earth's magnetism has a strength at the surface of about 5×10^{-5} tesla (0.5 gauss) and is caused by large electric currents circulating in Earth's molten interior.

A charged particle moving in a magnetic field experiences a force at right angles to its velocity and to the magnetic field. The strength of the force depends on the angle between the field and the particle's motion. It is maximum when they are perpendicular and zero when they are parallel.

If a wire and magnetic field move relative to each other, a current is produced in the wire, provided the motion is not parallel to either the wire or the field. This current is largest if the motion is perpendicular to the field and increases with the relative speed. A current also occurs in a loop of wire if the magnetic field inside the loop varies with time. An increasing field produces a current in one direction; a decreasing field produces a current in the opposite direction.

Field lines can be used to represent magnetic fields. The magnetic field is strongest in regions where the field lines are close together. If the number of magnetic field lines passing through a loop of wire changes *for any reason*, a current is produced in the loop. The voltage (and hence the current) generated in the loop depends on the rate of change—the faster the change, the larger the voltage.

There is a close connection between changing electric and magnetic fields. A changing magnetic field can generate a changing electric field, and a changing electric field can generate a changing magnetic field, creating electromagnetic waves that travel through empty space. These waves are produced whenever electric charges are accelerated. When the charges have a periodic oscillatory motion, the wave has a fixed frequency and wavelength. The spectrum of these waves ranges from very low-frequency radio waves, to visible light, to high-frequency X rays and gamma rays.

Chapter 22 Revisited

The connection between your favorite TV show and your microwave oven lies in the electromagnetic waves that are created by combining electric and magnetic effects. These waves can resonate with molecules in a microwave oven to raise the temperature of the popcorn and can be beamed through space to bring us television (and radio) programming.

KEY TERMS

ampere: The SI unit of electric current. The current in each of two parallel wires when the magnetic force per unit length between them is 2×10^{-7} newton per meter.

electromagnet: A magnet constructed by wrapping wire around an iron core. An electromagnet can be turned on and off by turning the current in the wire on and off.

electromagnetic wave: A wave consisting of oscillating electric and magnetic fields. In a vacuum, electromagnetic waves travel at the speed of light.

gauss: A unit of magnetic field strength; 10^{-4} tesla.

magnetic field: The space surrounding a magnetic object, where each location is assigned a value determined by the torque on a compass placed at that location. The direction of the field is in the direction of the north pole of the compass.

magnetic monopole: A hypothetical, isolated magnetic pole.

magnetic pole: One end of a magnet; analogous to an electric charge.

tesla: The SI unit of magnetic field.

CONCEPTUAL QUESTIONS

1. If you are given three iron rods, how could you use them to find the one that is not magnetized?
2. If the labels on a magnet are missing, how would you determine which pole is the north pole?
3. You have three iron bars, each of which may or may not be a permanent magnet. Each rod is painted green on one end and yellow on the other. You perform three experiments and find that the green end of bar A attracts the green end of bar B, the yellow end of bar A repels the green end of bar B, and the green end of bar B repels the yellow end of bar C. Which of these three results indicates that bar A must be a permanent magnet? Explain.
4. Consider the experiments described in Question 3. Would the green end of bar A attract, repel, or have no interaction with the green end of bar C? Explain.
5. You have three iron bars, each of which may or may not be a permanent magnet. Each rod is painted black on one end and white on the other. You perform three experiments and find that the white end of bar A repels the black end of bar B, the black end of bar A attracts the white end of bar C, and the black end of bar B attracts the white end of bar C. Is bar C a permanent magnet? Explain.
6. Consider the experiments described in Question 5. Would the white end of bar A attract, repel, or have no interaction with the white end of bar C? Explain.
7. If a bar magnet is broken into two pieces, how many magnetic poles are there?
8. We label magnetic poles such that like poles repel and unlike poles attract. Would it have been possible to use a labeling convention where like poles attract and unlike poles repel? Why or why not?
9. How is the direction of the magnetic field defined at each point in space?
10. Why is it not possible for two magnetic field lines to cross?
11. Magnetic field lines in a region are shown in the figure. A freely rotating compass needle is released in the orientation shown. Will the needle rotate clockwise, counterclockwise, or not at all? Explain.

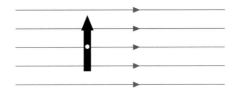

12. Oersted found that the magnetic field lines surrounding a current-carrying wire circle around and close on themselves. Because the field lines for a bar magnetic also close on themselves, what does this imply about the direction of the field lines inside the magnet? Do the field lines point from north to south or from south to north?
13. When you look down the axis of a solenoid so that the current circulates in the clockwise direction, are you looking in the direction of the magnetic field or opposite of it? Explain.
14. How would the photograph in Figure 22-4 change if the current were in the opposite direction?
15. In the figure, you are looking at the end of a long straight wire. The current in the wire is directed out of the page. At the point labeled A, is the direction of the magnetic field left, right, up, down, into the page, or out of the page? What is the direction at point B? Explain.

I (out of page)

16. A long straight wire carries a current in the direction shown in the figure. At the point labeled A, is the direction of the magnetic field left, right, up, down, into the page, or out of the page? What is the direction of the magnetic field at the point labeled B? Explain.

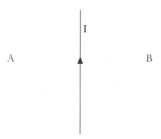

17. The figure shows the top view of a compass next to a long straight wire. An arrow indicates the north pole of the compass. When a switch is closed, a current is established in the wire, and the compass needle initially ro-

tates as indicated. Is the direction of the current in the wire into the page or out of the page? Explain.

Current-carrying wire

18. The figure shows the top view of a compass next to a long straight wire. The north pole of the compass is indicated by an arrow. When a switch is closed, the current is directed out of the page as indicated. Will the compass needle initially rotate clockwise or counterclockwise? Explain.

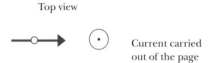
Current carried out of the page

19. In electric circuits and telephone lines, two wires carrying currents in opposite directions are twisted together. How does this reduce the magnetic fields surrounding the wires?
20. Can you use a compass to locate the electric wires located inside the walls of your house? Explain.
21. Two long straight wires carry identical currents in opposite directions as shown in the figure. At the points labeled A and B, is the direction of the magnetic field left, right, up, down, into the page, or out of the page? Explain.

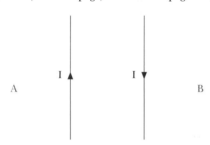

22. Two wires carry current into the page as shown. One wire carries a current I, and the other carries a current 2I. Which of the arrows best represents the direction of the magnetic field at point A, which is the same distance from both wires? Explain.

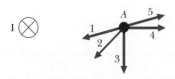

23. In Melville's *Moby Dick*, Captain Ahab regained the confidence of his crew when he fixed the compass that an electrical storm had damaged so that it pointed south. How might he have done this?

24. Lodestone (magnetite) is an igneous rock, one that forms from molten material. How do you suppose it became magnetized?
25. What do you expect will happen to the strength of the magnetism of a bar magnet that is dropped on a hard floor?
26. Would you expect the head of a steel hammer to be magnetized? Explain.
27. Two long straight wires carry currents in opposite directions. The current in wire 1 is twice the current in wire 2. If the net magnetic force per unit length on wire 1 is 4 newtons per meter, what is the net magnetic force per unit length on wire 2?
28. Three long straight wires each carry identical current in the directions shown in the figure. What is the direction of the net force on the wire at the right?

29. How is the unit of current defined?
30. How is the unit of charge defined?
31. One of Earth's magnetic poles is located in Antarctica. Is it a magnetic north pole or a magnetic south pole?
32. Why might it be more proper to call a north magnetic pole a "north-seeking" magnetic pole?
33. If you want to walk toward the geographic North Pole while in Portland, Oregon, what compass heading would you follow?
34. How far does a compass in New York City point away from the geographic North Pole?
35. Why are there more cosmic rays in Antarctica than in Hawaii?
36. We can model Earth's magnetic field as due to a single large current-carrying loop that runs around the equator just below the surface. In this model, is the direction of the current east to west or west to east? Why?
37. If a charged particle travels in a straight line, can you say that there is no magnetic field in that region of space? Explain.
38. Can you accelerate a stationary charged particle with a magnetic field? An electric field? Explain.
39. A proton and an electron with the same velocity enter a bending magnet with a magnetic field perpendicular to their velocity. Compare the motions of the proton and electron.
40. A magnet produces a magnetic field that points vertically upward. In what direction does the force on a proton act if it enters this region with a horizontal velocity toward the east?
41. A conducting loop is lying flat on the ground. The north pole of a bar magnet is brought down toward the loop. As the magnet approaches the loop, will the magnetic field created by the induced current point up or down? What is the direction of the current in the loop? Explain.

42. The magnet in Question 41 is now lifted upward. Which, if either, of your answers will change? Explain.
43. Consider the case where the south pole of a bar magnet is being moved toward a conducting copper ring. Do the field lines created by the induced current point toward the bar magnet or away? Will this induced field pull on the magnet or push against it? Explain.
44. Consider the case where the south pole of a bar magnet is being moved away from a conducting copper ring. Do the field lines created by the induced current point toward the bar magnet or away? Will this induced field pull back on the magnet or push it away? Explain.
45. A copper ring is oriented perpendicular to a uniform magnetic field. If the ring is suddenly moved in the direction of the field lines, will the magnitude of the net magnetic field in the center of the loop (the uniform field plus the induced field) be greater than, equal to, or less than the magnitude of the uniform field? Explain.

Questions 45 and 46

46. A copper ring is oriented perpendicular to a uniform magnetic field. The ring is quickly stretched such that its radius doubles over a short time. As the ring is being stretched, is the magnitude of the net magnetic field in the center of the loop (the uniform field plus the induced field) greater than, equal to, or less than the magnitude of the uniform field? Explain.
47. Quickly inserting the north pole of a bar magnet into a coil of wire causes the needle of a meter to deflect to the right. Describe two actions that will cause the needle to deflect to the left.
48. How could you produce a current in a solenoid by rotating a small magnet inside the coil?
49. When a transformer is plugged into a wall socket, it produces 9-volt alternating-current electricity for a portable tape player. Is the coil with the larger or smaller number of turns connected to the wall socket? Why?
50. If you had an old car that needed a 6-volt battery to drive the starter motor, could you use a 12-volt battery with a transformer to get the necessary output voltage? How?
51. What is the purpose of using a commutator in a motor?
52. What effect does a commutator have on the electrical output of a generator?
53. Describe an electromagnetic wave propagating through empty space.
54. How are electromagnetic waves generated?
55. Which of the following is *not* an electromagnetic wave: radio, television, blue light, infrared light, or sound?
56. Which of the following electromagnetic waves has the lowest frequency: radio, microwaves, visible light, ultraviolet light, or X rays? Which has the highest frequency?
57. What is the difference between X rays and gamma rays?
58. How fast do X rays travel through a vacuum?
59. How is sound encoded by an AM radio station?
60. How is sound transmitted by an FM radio station?
61. At what frequency does radio station FM 102.1 broadcast?
62. What does it mean when your local disc jockey says that you are "listening to radio 1380"?

EXERCISES

1. The record for a steady magnetic field is 45 T. What is this field expressed in gauss?
2. What is the magnetic field (expressed in gauss) of a refrigerator magnet with a magnetic field of 0.3 T?
3. The magnetic field at the equator of Jupiter has been measured to be 4.3 G. What is this field expressed in teslas?
4. The magnetic fields associated with sunspots are on the order of 1500 G. How many teslas is this?
5. An electron has a velocity of 3×10^6 m/s perpendicular to a magnetic field of 2 T. What force and acceleration does the electron experience?
6. What force and acceleration would a proton experience under the conditions of Exercise 5?
7. A metal ball with a mass of 2 g, a charge of 1 μC, and a speed of 40 m/s enters a magnetic field of 30 T. What are the maximum force and acceleration of the ball?
8. A very small ball with a mass of 0.1 g has a charge of 10 μC. If it enters a magnetic field of 10 T with a speed of 30 m/s, what are the maximum force and acceleration the ball experiences?
▲ 9. If you wanted the maximum magnetic force on the ball in Exercise 7 to equal the gravitational force on the ball, what charge would you need to give it?
▲ 10. What speed would be needed in Exercise 8 for the maximum magnetic force to be equal to the gravitational force?
▲ 11. An ink drop with charge $q = 3 \times 10^{-9}$ C is moving in a region containing both an electric field and a magnetic field. The strength of the electric field is 3×10^5 N/C, and the strength of the magnetic field is 0.2 T. At what speed must the particle be moving perpendicular to the magnetic field so that the magnitudes of the electric and magnetic forces are equal?
▲ 12. How would your answer to Exercise 11 change if the charge on the ink drop were doubled?
▲ 13. A particle with charge $q = 5$ μC and mass $m = 6 \times 10^{-5}$ kg is moving parallel to Earth's surface at a speed of 1000 m/s. What minimum strength of magnetic field would be required to balance the gravitational force on the particle?

▲ 14. What minimum strength of magnetic field is required to balance the gravitational force on a proton moving at speed 3×10^6 m/s?

15. A transformer is used to convert 120-V household electricity to 9 V for use in a portable CD player. If the primary coil connected to the outlet has 400 loops, how many loops does the secondary coil have?

16. The voltage in the lines that carry electric power to homes is typically 2000 V. What is the required ratio of the loops in the primary and secondary coils of the transformer to drop the voltage to 120 V?

17. A transformer is used to step down the voltage from 120 V to 6 V for use with an electric razor. If the razor draws a current of 0.5 A, what current is drawn from the 120 V lines? What is the ratio of the loops in the primary and secondary coils of the transformer?

18. Your 120-V outlet is protected by a 20-A circuit breaker. What is the maximum current that you can provide to an appliance using a transformer with one-tenth as many loops in the secondary as in the primary?

19. You are using a transformer with 800 loops in the primary and 80 loops in the secondary. If the 120-V input is supplying 2 A, what is the current in the appliance connected to the transformer?

20. You are using a transformer with 800 loops in the primary and 80 loops in the secondary. If the transformer is plugged into a 120-V outlet protected by a 20-A circuit breaker, what is the maximum power rating of an appliance that could be used with this transformer?

21. How long does it take a radio signal from Earth to reach the Moon when it is 384,000 km away?

22. The communications satellites that carry telephone messages between New York City and London have orbits above Earth's equator at an altitude of 13,500 km. Approximately how long does it take for each message to travel between these two cities via the satellite?

23. Many microwave ovens use microwaves with a frequency of 2.45×10^9 Hz. What is the wavelength of this radiation, and how does it compare with the size of a typical oven?

24. An X-ray machine used for radiation therapy produces X rays with a maximum frequency of 2.4×10^{20} Hz. What is the wavelength of these X rays?

25. The ultraviolet light that causes suntans (and sunburns!) has a typical wavelength of 300 nm. What is the frequency of these rays?

26. The radioactive isotope most commonly used in radiation therapy is cobalt-60. It gives off two gamma rays with wavelengths of 1.06×10^{-12} m and 9.33×10^{-13} m. What are the frequencies of these gamma rays?

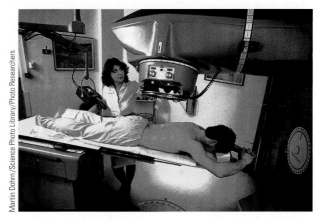

27. What is the range of wavelengths for AM radio?
28. What is the range of wavelengths for FM radio?
29. What is the wavelength of the carrier wave for an AM radio station located at 1090 on the dial?
30. What is the shortest wavelength used for TV broadcasts?

Interlude

The shape of a snowflake is determined by the quantum-mechanical interactions of water molecules with each other.

The Story of the Quantum

In 1887 German physicist Heinrich Hertz generated sparks in an electrostatic machine and successfully caused a spark to jump across the gap of an isolated wire loop on the other side of his laboratory. Hertz had transmitted the first radio signal a distance of a few meters. Hertz was not interested in the societal implications of sending signals; he was testing James Clerk Maxwell's prediction of the existence of electromagnetic waves. Hertz also mentioned the curious fact that when ultraviolet light illuminated his loop, more sparks were produced.

The 19th century had been the age of steam engines and heat. It is no wonder that tremendous strides were made in the study of thermodynamics—the behavior of gases under different conditions of pressure and temperature. By the 1890s entirely new areas of confusing phenomena provided huge intellectual challenges to scientists. In 1895 Wilhelm Roentgen discovered a new form of radiation emanating from an electromagnetic device known as a cathode-ray tube. The following year Henri Becquerel observed that a lump of uranium emitted some sort of energy that radiated through the wood of his desk and fogged his photographic plates. In 1900 Max Planck puzzled over unusual results when he calculated the spectrum of radiation from heated objects.

Some of the greatest minds of the 20th century grappled with these issues of radiation and matter, experimental fact and theory. They continually received information that was unbelievable. It was a crisis in the process of discovery itself, born in turmoil and confusion. What were they to believe? A philosopher of that era put it this way: "The senses do not lie. They just do not tell us the truth."

The only possible resolution was a revolution in the physics world view. Ideas that had been taken for granted were discarded, and the unbelievable became believable. The core of the revolution seemed innocent enough: Energy exists in "chunks," or quantum units. At first glance quantization may not seem that radical. The paintings of the French Impressionists look normal when viewed from afar, and our weight seems continuous even though it is really the sum of the weights of discrete atoms. But it is not just a matter of adding the many tiny components of this new discreteness to get the whole; it is the tiny components that control the character of the whole.

The quantum nature of energy was the first of three changes leading to this new physics. New insights into the nature of light and, finally, new insights into the nature of matter followed. Together these changes revealed the character of the Universe, from the shape of snowflakes to the existence of neutron stars.

In the remaining chapters, we will follow the development of these ideas. Although it might seem easier to simply learn "the answers," they are unbelievable without knowing the experiments, the conflicts, the debates, and the eventual resolutions that led to them. As you track these developments, keep in mind Sir Hermann Bondi's comment: "We should be surprised that the gas molecules behave so much like billiard balls and not surprised that electrons don't."

> We should be surprised that the gas molecules behave so much like billiard balls and not surprised that electrons don't.
> —Hermann Bondi

Heinrich Hertz

23 | The Early Atom

Diffraction gratings separate the different colors of light. These gratings can be used to probe the structure of atoms.

We get much of our information about atoms from the light they emit. In fact, the pretty arrays of color emitted by atoms—known as atomic fingerprints—distinguish one type of atom from the others and hold many clues about atomic structure. What does the light tell us about atoms?

(See page 494 for the answer to this question.)

PhysicsNow™ Test your understanding of this chapter by logging into PhysicsNow at http://physics.brookscole.com/kf6e, selecting the chapter, and clicking on the "Take a Pre-Test" link.

NEW world views don't emerge in full bloom—they start as seeds. Although a change may begin because of a single individual, it usually evolves from the collective efforts of many. Ideas are proposed. As they undergo the scrutiny of the scientific community, some are discarded and others survive, although almost always in a modified form. These survivors then become the beginnings of new ways of looking at the world.

During the early years of the last century, a number of phenomena were known or discovered that eventually led to several models for the atom. After exploring the phenomena, we will examine several atomic models that emerged in response to the experiments.

Periodic Properties

Once the values of the atomic weights of the chemical elements were determined (Chapter 11), it was tempting to arrange them in numerical order. When this was done, a pattern emerged; the properties of one element corresponded very closely to those of an element farther down the list and to another one even farther down the list. A periodicity of characteristics became apparent.

Russian chemist Dmitri Mendeleev wrote the symbols for the elements on cards and spent a great deal of time arranging and rearranging them. For some reason he omitted hydrogen, the lightest element, and began with the next lightest one known at that time, lithium (Li). He laid the cards in a row. When he reached the 8th element, sodium (Na), he noted that it had properties similar to lithium and therefore started a new row. As he progressed along this new row, each element had properties similar to the one directly above it. The 15th element, potassium (K), is similar to lithium and sodium, and so Mendeleev began a third row. His arrangement of the first 16 elements looked like Figure 23-1.

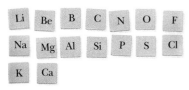

Figure 23-1 Mendeleev's arrangement of the elements.

The next known element was titanium, but it did not have the same properties as boron (B) and aluminum (Al); it was similar to carbon (C) and silicon (Si). Mendeleev boldly claimed that there was an element that had not yet been discovered. He went on to predict two other elements, even predicting their properties from those of the elements in the same column. Later, these elements were discovered and had the properties that Mendeleev had predicted. Mendeleev's work was the beginning of the modern periodic table of the elements. At that time no theory could explain why the elements showed these periodic patterns. However, the existence of the periodicity was a clue to scientists that there might be an underlying structure to atoms.

Atomic Spectra

Another thing fascinated and confused scientists around the beginning of the 20th century. When an electric current passes through a gas, the gas gives off a characteristic color. A modern application of this phenomenon is the neon sign. Although each gas emits light of a particular color, it isn't necessarily unique.

The various colors of light emitted by neon-type signs are due to different gases.

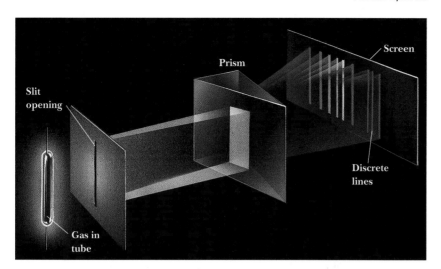

Figure 23-2 The emission spectrum of a gas consists of a small number of distinctly colored lines.

However, if you pass the light through a prism or a grating to spread out the colors, the light splits up into a pattern of discrete colored lines (Figure 23-2), which is always the same for a given gas but different from the patterns produced by other gases.

Question If a particular gas glows with a yellow color when a current is passed through it, will its spectrum necessarily contain yellow light?

Answer The spectrum will not necessarily contain yellow light because the yellow sensation could result from a combination of red and green spectral lines. In fact, red and green light from two different helium–neon lasers produces a beautiful yellow even though each laser has a very narrow range of frequencies.

Each gas has its own particular set of spectral lines, collectively known as its **emission spectrum**. The emission spectra for a few gases are shown in the schematic drawings in Figure 23-3. Because these lines are unique—sort of an atomic fingerprint—they identify the gases even if they are in a mixture. Vaporized ele-

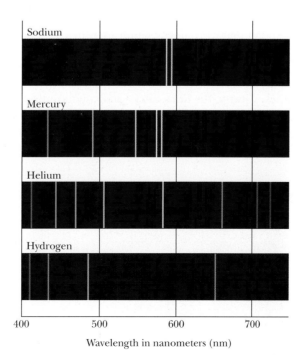

Figure 23-3 Schematic drawings of the emission spectra for sodium, mercury, helium, and hydrogen near the visible region.

Figure 23-4 The absorption spectrum consists of a rainbow spectrum with a number of dark lines.

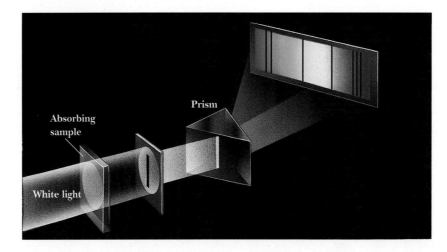

ments that are not normally gases also emit spectral lines, and these can be used to identify them.

PHYSICS | ON YOUR OWN

Observe the spectra of a variety of light sources through a diffraction grating. Inexpensive diffraction gratings mounted in 2-by-2 inch slide frames can be obtained from hobby stores. Don't look directly at the Sun!

Spectral lines also appear when white light passes through a cool gas (Figure 23-4). In this case, however, they do not appear as bright lines but as dark lines in the continuous rainbow spectrum of the white light; that is, light is missing in certain narrow lines. These particular wavelengths are absorbed or scattered by the gas to form this **absorption spectrum**.

All the lines in the absorption spectrum correspond to the lines in the emission spectrum for the same element. However, the emission spectrum has many lines that do not appear in the absorption spectrum, as illustrated in Figure 23-5.

Because the lines that appear in both spectra are the same, the absorption spectrum can also be used to identify elements. In fact, this is how helium was discovered. Spectral lines were seen in the absorption spectrum of sunlight that did not correspond to any known elements. These lines were attributed to a new element that was named helium after the Greek name for the Sun, *helios*. Helium was later discovered on Earth.

Spectral lines are a valuable tool for identifying elements. Although this is of great practical use, the scientific challenge was to understand the origin of the lines. Clearly, the lines carry fundamental information about the structure of atoms. The lines pose a number of questions. The most important question concerns their origin: what do spectral lines have to do with atoms? Also, why are certain lines in the emission spectra missing in the absorption spectra?

Figure 23-5 The absorption spectrum for an element has fewer lines than its emission spectrum.

Cathode Rays

During the latter half of the 19th century, two new technologies contributed to the next advance in our understanding of atoms: vacuum pumps and high-voltage devices. It was soon discovered that a high voltage applied across two electrodes in a partially evacuated glass tube produces an electric current and light in the tube. Furthermore, the character of the phenomenon changes radically as more and more of the air is pumped from the tube. Initially, after a certain amount of the air is removed, sparks jump between the electrodes, followed by a glow throughout the tubes like that seen in modern neon signs. As more air is pumped out, the glow begins to break up and finally disappears. When almost all of the air has been removed, only a yellowish green glow comes from one end of the glass tube.

These discoveries led to a flurry of activity. Experimenters made tubes with a variety of shapes and electrodes from a variety of materials. The glow always appeared on the end of the glass tube opposite the electrode connected to the negative terminal of the high-voltage supply. When the wires were exchanged, the glow appeared at the other end, showing that it depended on which electrode was negative. A metal plate placed in a tube cast a shadow (Figure 23-6). These effects indicated that rays were coming from the negative electrode. Because this electrode was known as the *cathode* (the other electrode is the *anode*), the rays were given the name **cathode rays**.

The path of the cathode rays could be traced by placing a fluorescent screen along the length of the tube. The screen glowed wherever it was struck by the cathode rays. A magnet was used to see if the rays were particles or waves. We saw in the previous chapter that a magnetic field exerts a force on a moving charged particle, but this does not occur for a beam of light. A magnet placed near the tube deflected the cathode rays, indicating that they were indeed charged particles.

Question Based on the information given, would you expect the cathode rays to be positively or negatively charged?

Answer Because they come from the cathode and the cathode is negatively charged, we expect the cathode rays to also have a negative charge. The directions of the magnetic deflections were also consistent with this expectation.

This being the case, cathode rays should also be deflected by an electric field. Early experiments showed no observable effect until British scientist J. J. Thom-

J. J. Thomson

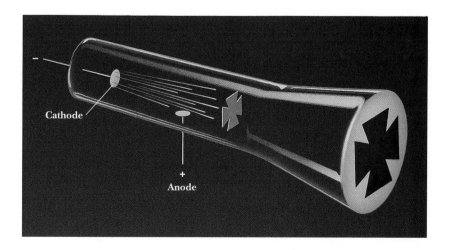

Figure 23-6 Cathode rays cast a shadow of the cross onto the end of the tube.

Figure 23-7 The cathode rays are deflected by an electric field.

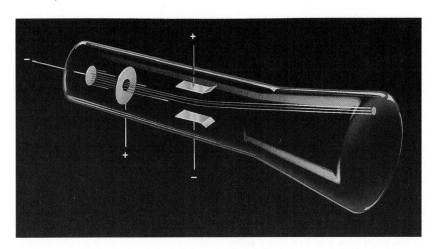

son used the tube sketched in Figure 23-7. By placing the electric field plates inside the tube and by using a much better vacuum, he could observe electric deflection of the cathode rays. The direction of the deflection was that expected for negatively charged particles.

The Discovery of the Electron

✓ Extended presentation available in the *Problem Solving* supplement

Thomson went beyond these qualitative observations. He compared the amount of deflection using a known electric field with that of a known magnetic field. Using these values, he obtained a value for the ratio of the charge to the mass for the cathode particles. Furthermore, Thomson showed that the *charge-to-mass ratio* was always the same regardless of the gas in the tube or the cathode material. This was the first strong clue that these particles were universal to all matter and therefore an important part of the structure of matter.

This ratio could be compared with known elements. The charge-to-mass ratio for positively charged hydrogen atoms (called hydrogen **ions**) had previously been determined; Thomson's number was about 1800 times as large. If Thomson assumed that the charges on the particles were all the same, the mass of the hydrogen ion was 1800 times that of the cathode particle. Conversely, if he assumed the masses to be the same, the charge on the cathode ray was 1800 times as large. (Of course, any combination in-between was also possible.)

Thomson believed that the first option—assuming that all charges are equal—was the correct one. (There were other data that supported this assumption.) This assumption meant that the cathode particles had a much smaller mass than a hydrogen atom, the lightest atom. At this point, he realized that the atom had probably been split!

This was a special moment in the building of the physics world view. The discussion that had originated at least as far back as the early Greeks was centered on the question of the existence of atoms, never on their structure. It had always been assumed that these particles were indivisible. Thomson's work showed that the indivisible building blocks of matter are divisible; they have internal structures.

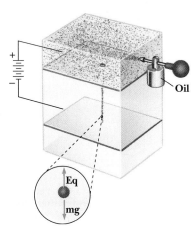

Figure 23-8 The Millikan oil-drop experiment showed that electric charges are quantized.

Around 1910 American experimentalist Robert Millikan devised a way of measuring the size of the electric charge, thus testing Thomson's assumption. Millikan produced varying charges on droplets of oil between a pair of electrically charged plates (Figure 23-8). By measuring the motions of individual droplets in the electric field, he determined the charge on each droplet and confirmed the widely held belief that electric charge comes in identical chunks, or **quanta**. A particle might have one of these elemental units of charge, or two, or three, but never two and a half. This smallest unit of charge is 1.6×10^{-19} coulomb, an

elementary charge ➤

extremely small charge; a charged rod would have more than billions upon billions of these charges.

The mass of the cathode particles could now be calculated from the value of the charge and the charge-to-mass ratio. The modern value for the mass of the cathode particle is 9.11×10^{-31} kilogram. This particle—a component part of all atoms with a mass 1800 times smaller than the hydrogen atom and carrying the smallest unit of electric charge—is known as an **electron**. The electron is so small that it would take a trillion trillion of them in the palm of your hand before you would notice their mass.

◀ mass of the electron

PHYSICS | ON YOUR OWN

Hold a small magnet near the face of a black-and-white television set while it's on. (*Caution*: do not do this with color television sets, because you can cause permanent damage.) Is the picture generated by light or by cathode rays traveling inside the television tube?

Thomson's Model

Thomson's work showing that atoms have structure was an exciting breakthrough. A natural question emerged immediately: what does an atom look like? One of its parts is the electron, but what about the rest? Thomson proposed a model for this structure, picturesquely known as the plum-pudding model, with the atom consisting of some sort of positively charged material with electrons stuck in it much like plums in a pudding (Figure 23-9). Because atoms were normally electrically neutral, he hypothesized that the amount of positive charge was equal to the negative charge and that atoms become charged ions by gaining or losing electrons.

Recall that gases were known to emit distinct spectral lines. Thomson suggested that forces disturbing the atom caused the electrons to jiggle in the positive material. These oscillating charges would then generate the electromagnetic waves observed in the emission spectrum. However, the model could not explain why the light was emitted only as definite spectral lines or why each element had its own unique set of lines.

Thomson also attempted to explain the repetitive characteristics in the periodic table. He suggested that each element in the periodic table had a different number of electrons. He hoped that with different numbers of electrons, different arrangements would occur that might show a periodicity and thus a clue to the periodicity shown by the elements. For example, three electrons might arrange themselves in a triangle, four in a pyramid, and so on. Perhaps there was a periodicity to the shapes. Patterns did occur in his model, but the periodicity did not match that of the elements.

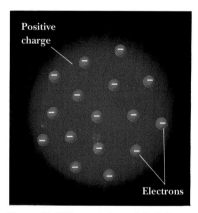

Figure 23-9 Thomson's model of the atom.

Rutherford's Model

Ernest Rutherford, a student of Thomson, did an experiment that radically changed his teacher's model of the atom. A key tool in Rutherford's work was the newly discovered radioactivity, the spontaneous emission of fast-moving atomic particles from certain elements. Rutherford had previously shown that one of these radioactive products is a helium atom without its electrons. These particles have two units of positive charge and are known as **alpha particles**.

Rutherford bombarded materials with alpha particles and observed how the alpha particles recoiled from collisions with the atoms in the material. This procedure was much like determining the shape of an object hidden under a table by rolling rubber balls at it (Figure 23-10). Rutherford used the alpha particles as

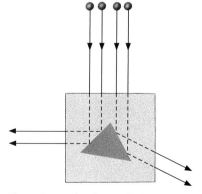

Figure 23-10 The shape of a hidden object can be determined by rolling rubber balls and seeing how they are scattered.

his probe because they were known to be about the same size as atoms. (One doesn't probe the structure of snowflakes with a sledgehammer.) The target needed to be as thin as possible to keep the number of collisions by a single alpha particle as small as possible, ideally only one. Rutherford chose gold because it could be made into a very thin foil.

The alpha particles are about 4 times the mass of hydrogen atoms and 7300 times the mass of electrons. Thomson's model predicted that the deflection of alpha particles by each atom (and hence by the foil) would be extremely small. The positive material of the atom was too spread out to be a very effective scatterer. Consider the following analogy. Use 100 kilograms of newspapers to make a wall one sheet thick. Glue table-tennis balls on the wall to represent electrons. If you were to throw bowling balls at the wall, you would expect them to go straight through with no noticeable deflection.

Initially, Rutherford's results matched Thomson's predictions. However, when Rutherford suggested that one of his students look for alpha particles in the backward direction, he was shocked with the results. A small number of the alpha particles were actually scattered in the backward direction! Rutherford later described his feelings by saying, "It was almost as incredible as if you fired a 19-inch shell at a piece of tissue paper and it came back and hit you!"

On the basis of his experiment, Rutherford proposed a new model in which the positive charge was not spread out but was concentrated in a very, very tiny spot at the center of the atom, which he called the **nucleus**. The scattering of alpha particles is shown in Figure 23-11. Returning to our analogy, imagine what would happen if all the newspapers were crushed into a small region of space: most of the bowling balls would miss the paper and pass straight through, but occasionally a ball would score a direct hit on the 100-kilogram ball of newspaper and recoil.

Rutherford determined the speed of the alpha particles by measuring their deflections in a known magnetic field. Knowing their speeds, he could calculate their kinetic energies. Because he knew the charges on the alpha particle and the nucleus of the gold atom, he could also calculate the electric potential energy between the two. These calculations allowed him to calculate the closest distance an alpha particle could approach the nucleus of the gold atom before its kinetic energy had been entirely converted to potential energy. These experiments gave him an upper limit on the size of the gold nucleus. The nucleus has a diameter approximately 100,000 times smaller than its atom. This means that atoms are almost entirely empty space.

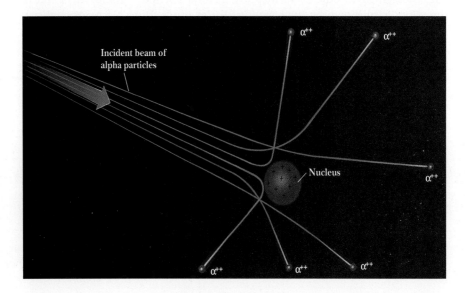

Figure 23-11 The positively charged alpha particles are repelled by the positive charges in the nucleus.

RUTHERFORD | At the Crest of the Wave

Well, I made the wave, didn't I?

—*Ernest Rutherford, first Baron Rutherford of Nelson*

Ernest Rutherford (1871–1937) grew up on a small farm near Nelson, New Zealand, at a time when that island nation was new and very much part of the expansive British Empire. He received a sound education in Nelson and won a scholarship to the University of New Zealand. His diligence and brilliance there earned him an imperial scholarship to Cambridge University, where J. J. Thomson, director of the Cavendish Laboratory, put him to work measuring electromagnetic waves. Teacher and student then moved on to analyze X-ray ionization in gases and aspects of the photoelectric effect. Henri Becquerel's discovery of radioactivity created a vast new arena for physicists to explore.

Rutherford had just begun to examine radium ionization because he had discovered that two types of radiation in this strange gas, alpha and beta, had different penetrating properties. The former was stopped by a thin layer of common air; the latter passed through a thin aluminum foil. Rutherford made a series of great discoveries involving alpha rays. He left England in 1898 and continued his work as a new professor at McGill University in Montreal, Quebec, Canada.

Rutherford's students and colleagues at McGill, the most notable of whom was Frederick Soddy, continued to work on explanatory mechanisms for radioactivity and published the first substantial study of that complex subject. In 1908 Rutherford received the Nobel Prize in chemistry. His address was a model of scientific reporting. He explained the importance of alpha radiations, their identity as helium atoms, and their use in analyzing atomic numbers. In 1909 Rutherford accepted a chair of physics at Manchester University, where he continued to work on the cutting edge of atomic energy questions and in 1912 attracted perhaps his most outstanding student, Niels Bohr. Another brilliant young physicist, H. G. Moseley, arrived in 1913. Moseley's death on the Gallipoli battlefront in World War I led to Rutherford's futile request that scientists be exempted from frontline service. That piece of advice was taken by the United States in World War II. Rutherford also contributed to wartime research in a number of areas, most notably submarine detection.

Lord Ernest Rutherford

The New Zealander was knighted in 1914 and returned to Cambridge in 1919 when Thomson retired. He was made a baron in 1931, which gave him a seat in the House of Lords. Rutherford, always gifted with a splendid clarity of vision and expression, lived out his life full of honors and continued scientific challenges. The most complex problem for him was a matter of resolving the key components into a simpler form so that they could be tackled. The great revolutions in physics—relativity, atomic physics, and quantum theory—all saw him at the center of ongoing work. His comment above was true: he had indeed been at the crest of the wave that swept over this heroic age of physics.

—*Pierce C. Mullen, historian and author*

Sources: David Wilson, *Rutherford: Simple Genius* (Cambridge, Mass.: MIT Press, 1983); Arthur S. Eve, *Rutherford* (New York: Macmillan, 1939).

Question If the nucleus were the size of a basketball, what would be the diameter of the atom?

Answer Assuming that a basketball has a diameter of 24 centimeters, the diameter of the atom would be (100,000)(0.24 meter) = 24 kilometers (15 miles) across.

Our work in electricity tells us that the negative electrons are attracted to the positive nucleus. Rutherford proposed that they don't crash together at the center of the atom because the electrons orbit the nucleus much like the planets in our Solar System orbit the Sun (Figure 23-12). The force that holds the orbiting electrons to the nucleus is the electric force. (The gravitational force is negligible for most considerations at the atomic level.)

Question If atoms are mostly empty space, why can't we simply pass through each other?

Answer The atom's electrons in the outer regions keep us apart. When two atoms come near each other, the electrons repel each other strongly.

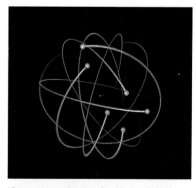

Figure 23-12 Rutherford's model of the atom. (The size of the nucleus is greatly exaggerated, and the electrons are known to be even smaller than the nucleus.)

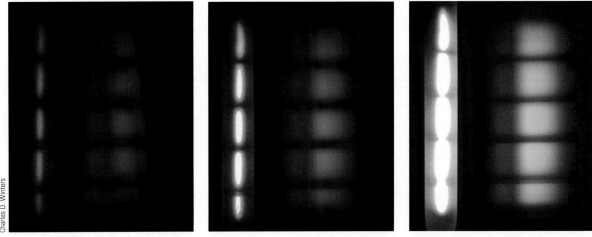

Figure 23-13 The color of the hot filament and its continuous spectrum change as the temperature is increased.

Rutherford's work led to a new explanation for the emission of light from atoms. The orbiting electrons are accelerating (Chapter 4), and accelerating charges radiate electromagnetic waves (Chapter 22). A calculation showed that the orbital frequencies of the electron correspond to the region of visible light. But there is a problem. If light is radiated, the energy of the electron must be reduced. This loss would cause the electron to spiral into the nucleus. In other words, this model predicts that the atom is not stable and should collapse. Calculations showed that this should occur in about a billionth of a second! Because we know that most atoms are stable for billions of years, there is an obvious defect in the model.

Rutherford's solar-system model of the atom also failed to predict spectral lines and to account for the periodic properties of the elements. The resolution of these problems was accomplished by atomic models proposed during the two decades following Rutherford's work.

Radiating Objects ✓ MATH

The Thomson and Rutherford models of the atom could explain some of the experimental observations, but they had serious deficiencies. Clearly, a new model was needed. Interestingly, the germination of the next model began in yet another area of physics, one having to do with the properties of the radiation emitted by a hot object.

All objects glow. At normal temperatures the glow is invisible; our eyes are not sensitive to it. As an object gets warmer, we feel this glow—we call it heat. At even higher temperatures, we can see the glow. The element of an electric stove or a space heater, for example, has a red orange glow. In all these cases, the object is emitting electromagnetic radiation. The frequencies, or colors, of the electromagnetic radiation are not equally bright, as can be seen in the photographs of Figure 23-13. Furthermore, the color spectrum shifts toward the blue as the temperature of the object increases.

All solid objects give off continuous spectra that are similar but not identical. However, the radiation is the same for all materials at the same temperature if the radiation comes from a small hole in the side of a hollow block. Typical distributions of the brightness of the electromagnetic radiation versus wavelength are shown in the graph of Figure 23-14 for three temperatures. When the temperature changes, the position of the hump in the curve changes; the wavelength with the maximum intensity varies inversely with the temperature—that is, as the temperature goes up, the highest point of the curve moves to smaller wavelengths.

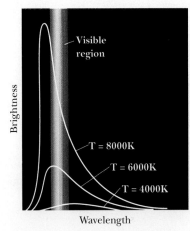

Figure 23-14 This graph shows the intensity of the light emitted by a hot object at various wavelengths for three different temperatures.

Question What happens to the frequency corresponding to the maximum intensity as the temperature increases?

Answer Because the product of the frequency and the wavelength must equal the speed of light, the frequency must increase as the wavelength decreases. Therefore, the frequency increases as the temperature increases.

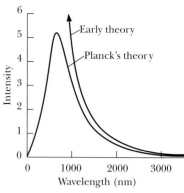

Figure 23-15 The theoretical ideas and the experimental data for the intensity of the light emitted by a hot body differed greatly in the low-wavelength (high-frequency) region.

The fact that the same distribution is obtained for cavities made of any material—gold, copper, salt, or whatever—puzzled and excited the experimenters. However, all attempts to develop a theory that would explain the shape of this curve failed. Most attempts assumed that the material contained some sort of charged atomic oscillators that generated light of the same frequency as their vibrations. Although the theories were able to account for the shape of the large-wavelength end of the curve, none could account for the small contribution at the shorter wavelengths (Figure 23-15). Nobody knew why the atomic oscillators did not vibrate as well at the higher frequencies.

A solution was discovered in 1900 by Max Planck, a German physicist. He succeeded by borrowing a new technique from another area of physics. The technique called for temporarily pretending that a continuous property was discrete in order to make particular calculations. After finishing the calculation, the technique called for returning to the real situation—the continuous case—by letting the size of the discrete packets become infinitesimally small.

Planck was attempting to add up the contributions from the various atomic oscillators. Unable to get the right answer, he temporarily assumed that energy existed in tiny chunks. Planck discovered that if he didn't take the last step of letting the chunk size become very small, he got the right results! Planck announced his success in very cautious tones—almost as if he were apologizing for using a trick on nature.

Planck's "incomplete calculation" was in effect an assumption that the atomic oscillators could vibrate with only certain discrete energies. Because the energy could have only certain discrete values, it would be radiated in little bundles, or quanta (singular, **quantum**). The energy E of each bundle was given by

$$E = hf$$

where h is a constant now known as Planck's constant and f is the frequency of the atomic oscillator. An oscillator could have an energy equal to 1, 2, 3, or more of these bundles, but not $1\frac{5}{8}$ or $2\frac{1}{2}$. The value of Planck's constant, 6.63×10^{-34} joule-second, was obtained by matching the theory to the radiation spectrum.

◂ energy quantum = Planck's constant × frequency

◂ Planck's constant

WORKING IT OUT | Energy of a Quantum ✓ MATH

We can get a feeling for the size of the energies of the atomic oscillators from our knowledge that red light has a frequency around 4.6×10^{14} Hz. Substituting these numbers into our relationship for the energy yields

$$E = hf = (6.63 \times 10^{-34} \text{ J} \cdot \text{s})(4.6 \times 10^{14} \text{ Hz}) = 3.05 \times 10^{-19} \text{ J}$$

Question What is the size of the energy quantum for violet light with a frequency of 7×10^{14} Hz?

Answer 4.64×10^{-19} J.

PLANCK | Founder of Quantum Mechanics

Max Planck personified science in the early 20th century. He fought a difficult battle against a technical problem in physics, and his victory ushered in a revolution that he neither wanted nor welcomed. He originated quantum mechanics; that is, he laid the foundation for modern physics and destroyed commonsense physical philosophy. He was awarded the Nobel Prize in physics, and his renown in the field was second only to that of Einstein.

Although he did not support Hitler, Planck stayed in Germany under the Hitler regime to provide an institutional basis for the rebirth of German science after the Third Reich. He lost a son in World War I, and a second was executed for plotting to assassinate Hitler. Late in the war, a bombing raid destroyed his priceless laboratory, correspondence, and manuscript collections. His was a noble and tragic life.

Planck's family background was filled with culture and learning. His father was a law professor at Kiel when Max, his sixth child, was born in 1858. He was gifted in many areas: mathematics, music, languages, history, and science. In his youth most friends would have predicted that he would become a humanist or a musician. He became a physicist because he wanted to solve problems that were intractable.

In his work with the radiation from heated bodies, he discovered that Maxwell's laws predicted a host of phenomena but not the equilibrium distribution Planck needed. His work in thermo-

Max Planck

dynamics resulted in two constants (and constants are rare indeed in science), with h the most commonly encountered. As he expressed it in 1910, "The introduction of the quantum of action h into the theory should be done as conservatively as possible; that is, alterations should only be made that have shown themselves to be absolutely necessary."

As the major figure in German physical science, Planck presided over great institutions and expenditures. He was a professor who led two dozen students through their doctorates. He served on an astounding 650 doctoral examination committees, and he championed, as only a powerful figure could, the careers of some great women and Jewish colleagues.

Max Planck lived long enough to see new science born from the ashes of the Third Reich. In 1946 the first institute to bear his name appeared in the British zone of occupation. At 89 he was the sole German to be invited to the 300th (belated) anniversary of Isaac Newton's birth. Planck died full of years, toil, pain, and honor in October 1947.

—*Pierce C. Mullen, historian and author*

Source: John L. Heilbron, *The Dilemmas of an Upright Man: Max Planck as Spokesman for German Science* (Berkeley: University of California Press, 1986).

Planck's idea explained why the contribution of the shorter wavelengths was smaller than predicted by earlier theories. Shorter wavelengths mean higher frequencies and therefore more energetic quanta. Because there is only a certain amount of energy in the macroscopic object and this energy must be shared among the atomic oscillators, the high-energy quanta are less likely to occur.

This discreteness of energy was in stark disagreement with the universally held view that energy was a continuous quantity and any value was possible. At first thought you may want to challenge Planck's result. If this were true, it would seem that our everyday experiences would have demonstrated it. For example, the kinetic energy of a ball rolling down a hill appears to be continuous. As the ball gains kinetic energy, its speed appears to increase continuously rather than jerk from one allowed energy state to the next. There doesn't seem to be any restriction on the values of the kinetic energy.

We are, however, talking about very, very small packets of energy. A baseball falling off a table has energies in the neighborhood of 10^{19} times as large as those of the quanta. These energy packets are so small that we don't notice their size in our everyday experiences. On our normal scale of events energy seems continuous.

The Photoelectric Effect ✓ MATH

Figure 23-16 The photoelectric effect. Light shining on a clean metal surface causes electrons to leave the surface.

Shortly before 1900 another phenomenon was discovered that connected electricity, light, and the structure of atoms. When light is shined on certain clean metallic objects, electrons are ejected from their surfaces (Figure 23-16). This effect is known as the **photoelectric effect**. The electrons come off with a range of energies up to a maximum energy, depending on the color of the light. If the light's intensity is changed, the rate of ejected electrons changes, but the maximum energy stays fixed.

Question   Does the name *photoelectric effect* make sense for this phenomenon?

Answer   Yes. The root *photo* is often associated with light; the most obvious example is photography. The *electric* root refers to the emitted electrons.

The ideas about atoms and light prevalent at the end of the 19th century were able to account for the possibility of such an effect. Because Planck's oscillators produced light, presumably light could jiggle the atoms. Light is energy, and when it is shined on a surface, some of this energy can be transferred to the electrons within the metal. If the electrons accumulate enough kinetic energy, some should be able to escape from the metal. But no theory could account for the details of the photoelectric process. For instance, if the light source is very weak, the old theories predict that it should take an average of several hours before the first electron acquires enough energy to escape. On the other hand, experiments show that the average time is much less than 1 second.

A range of kinetic energies made sense in the old theories—electrons bounced about within the metal accumulating energy until they approached the surface and escaped. There were bound to be differences in their energies. But a maximum value for the kinetic energy of the electrons was puzzling. The wave theory of light said that increasing the intensity of the light shining on the metal increases the kinetic energy of the electrons leaving the metal. However, experiments show that the intensity has no effect on the maximum kinetic energy. When the intensity is increased, the number of electrons leaving the metal increases, but their distribution of energy is unchanged.

One thing does change the maximum kinetic energy. If the frequency of the incident light increases, the maximum kinetic energy of the emitted electrons increases. This result was totally unexpected. If light is a wave, the amplitude of the wave should govern the amount of energy transferred to the electron. Frequency (color) should have no effect.

The established ideas about light were inadequate to explain the photoelectric effect. The conflict was resolved by Albert Einstein. In yet another outstanding paper written in 1905, he claimed that the photoelectric effect could be explained by extending the work of Planck. Einstein said that Planck didn't go far enough: not only are the atomic oscillators quantized but the experimental results could be explained only if light itself is also quantized! Light exists as bundles of energy called **photons**. The energy of a photon is given by Planck's relationship:

$$E = hf$$

◂ energy of a photon

Although it was universally agreed that light was a wave phenomenon, Einstein said that photons acted like particles, little bundles of energy.

Einstein was able to explain all the observations of the photoelectric effect. The ejection of an electron occurs when a photon hits an electron. A photon gives up its *entire* energy to a single electron. The electrons that escape have different amounts of energy, depending on how much they lose in the material before reaching the surface. Also, the fact that the first photon to strike the surface can eject an electron easily accounts for the short time between turning on the light and the appearance of the first electron. A maximum kinetic energy for the ejected electrons is now easy to explain. A photon cannot give an electron any more energy than it has, and the chances of an electron getting hit twice before it leaves the metal are extremely small.

Question How does this model explain the change in the maximum electron energy when the color of the light changes?

Answer This change occurs because the photon's color and energy depend on the frequency: the higher the frequency, the more energy the photon can give to the electron.

Einstein's model suggests a different interpretation of intensity. Because all photons of a given frequency have the same energy, an increase in the intensity of the light is due to an increase in the number of photons, not the energy of each one. Therefore, the model correctly predicts that more electrons will be ejected by the increased number of photons but their maximum energy will be the same.

Einstein's success introduced a severe conflict into the physics world view. Diffraction and interference effects (Chapter 19) required that light behave as a wave; the photoelectric effect required that light behave as particles. How can light be both? These two ideas are mutually contradictory. This is not like saying that someone is fat and tall, but rather like saying that someone is short and tall. The two descriptions are incompatible. This conflict took some time to resolve. We will examine this important development in the next chapter.

Meanwhile, Einstein's work inspired a new, improved model of the atom.

Bohr's Model   ✓ MATH

The model of the atom proposed by Rutherford had several severe problems. It didn't account for the periodicity of the elements, it was unable to give any clue about the origins of the spectral lines, and it implied that atoms were extremely unstable. In 1913 Danish theoretician Niels Bohr proposed a new model incorporating Planck's discrete energies and Einstein's photon into Rutherford's model.

Scientists are always using models when attempting to understand nature. The trick is to understand the limitations of the models. Bohr challenged the Rutherford model, stating that it was a mistake to assume that the atom is just a scaled-down solar system with the same rules; electrons do not behave like miniature planets. Bohr proposed a new model for the hydrogen atom based on three assumptions, or postulates.

First, he proposed that the angular momentum of the electron is quantized. Only angular momenta L equal to whole-number multiples of a smallest angular momentum are allowed. This smallest angular momentum is equal to Planck's constant h divided by 2π. Bohr's first postulate can be written

▶ Bohr's first postulate

▶ allowed angular momenta

$$L_n = n\frac{h}{2\pi}$$

where n is a positive integer (whole number) known as a **quantum number**.

In classical physics the angular momentum L of a particle of mass m traveling at a speed v in a circular orbit of radius r is given by $L = mvr$ (Chapter 8). Therefore, the restriction on the possible values of the angular momentum puts restrictions on the possible radii and speeds. Bohr showed that this restriction on the electron's angular momentum means that the only possible orbits are those for which the radii obey the relationship

▶ allowed radii

$$r_n = n^2 r_1$$

where r_1 is the smallest radius and n is the same integer that appears in the first postulate. This means that the electrons cannot occupy orbits of arbitrary size but only a certain discrete set of allowable orbits, as illustrated in Figure 23-17. The numerical value of r_1 is 5.29×10^{-11} meter.

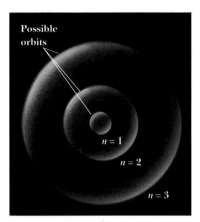
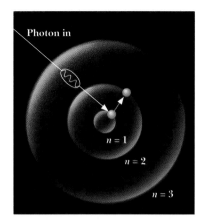
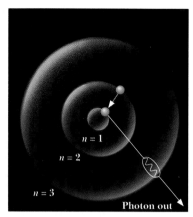

Figure 23-17 Electrons can only occupy certain orbits. These are designated by the quantum number n.

Figure 23-18 An electron can absorb a photon and jump to a higher-energy orbit.

Figure 23-19 When an electron jumps to a lower-energy orbit, it emits a photon.

There is also a definite speed associated with each possible orbit. This means that the kinetic energies are also quantized. Because the value of the electric potential energy depends on distance, the quantized radii mean that the potential energy also has discrete values. Therefore, there is a discrete set of allowable energies for the electron.

Question Would you expect the energy of the electron to increase or decrease for larger orbits?

Answer Because the electric force is analogous to the gravitational force, we can think about putting satellites into orbit around Earth. Higher orbits require bigger rockets and therefore more energy. Therefore, larger orbits should have more energy.

Bohr's second postulate states that an electron does not radiate when it is in one of the allowed orbits. This statement is contrary to the observation that accelerated charges radiate energy at a frequency equal to their frequency of vibration or revolution (Chapter 22). Bohr challenged the assumption that this property is also true in the atomic domain. Radically breaking away from what was accepted, Bohr said that an electron has a constant energy when it is in an allowed orbit.

◄ Bohr's second postulate

But atoms produce light. If orbiting electrons don't radiate light, how does an atom emit light? Bohr's third postulate answered this question: a single photon is emitted whenever an electron jumps down from one orbit to another. Energy conservation demands that the photon have an energy equal to the difference in the energies of the two levels. It further demands that jumps up to higher levels can occur only when photons are absorbed. The electron is normally in the smallest orbit, the one with the lowest energy. If the atom absorbs a photon, the energy of the photon goes into raising the electron into a higher orbit (Figure 23-18). Furthermore, it is not possible for the atom to absorb part of the energy of the photon. It is all or nothing.

◄ Bohr's third postulate

The electron in the higher orbit is unstable and eventually returns to the innermost orbit, or ground state. If the electron returns to the ground state in a single jump, it emits a new photon with an energy equal to that of the original photon (Figure 23-19). Again, this is different from our common experience. When a ball loses its mechanical energy, its temperature rises—the energy is transferred to its internal structure. As far as we know, an electron has no internal structure. Therefore, it loses its energy by creating a photon.

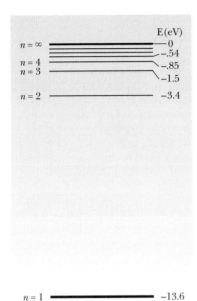

Figure 23-20 The energy-level diagram for the hydrogen atom.

Occasionally, the phrase "a quantum leap" is used to imply a big jump. But the jump needn't be big. There is nothing big about the quantum jumps we are discussing. A quantum leap simply refers to a change from one discrete value to another.

Ordinarily, an amount of energy is stated in joules, the standard energy unit. However, this unit is extremely large for work on the atomic level, and a smaller unit is customarily used. The **electron volt** (eV) is equal to the kinetic energy acquired by an electron falling through a potential difference of 1 volt. One electron volt is equal to 1.6×10^{-19} joule. It requires 13.6 electron volts to remove an electron from the ground state of the hydrogen atom.

Atomic Spectra Explained

Bohr's explanation accounted for the existence of spectral lines, and his numbers even agreed with the wavelengths observed in the hydrogen spectrum. To illustrate Bohr's model, we draw energy-level diagrams like the one in Figure 23-20. The ground state has the lowest energy and appears in the lowest position. Higher-energy states appear above the ground state and are spaced to indicate the relative energy differences between the states.

To see how Bohr's model gives the spectral lines, consider the energy-level diagram for the hypothetical atom shown in Figure 23-21(a). Suppose the electron has been excited into the $n = 4$ energy level. There are several ways that it can return to the ground state. If it jumps directly to the ground state, it emits the largest-energy photon. We will assume it appears in the blue part of the emission spectrum in Figure 23-21(b). This line is marked $4 \rightarrow 1$ on the spectrum, indicating that the jump was from the $n = 4$ level to the $n = 1$ level.

The electron could also return to the ground state by making a set of smaller jumps. It could go from the $n = 4$ level to the $n = 2$ level and then from the $n = 2$ level to the ground level. These spectral lines are marked $4 \rightarrow 2$ and $2 \rightarrow 1$. The energies of these photons are smaller, so their lines occur at the red end of the spectrum. The lines corresponding to the other possible jumps are also shown in Figure 23-21(b).

Question How does the total energy of the photons that produce the $4 \rightarrow 2$ and $2 \rightarrow 1$ lines compare with the photon that produces the $4 \rightarrow 1$ line?

Answer Conservation of energy requires that the two photons have the same total energy as the single photon.

A single photon is not energetic enough for us to see. In an actual situation, a gas contains a huge number of excited atoms. Each electron takes only one of the paths we have described in returning to the ground state. The total effect of all the individual jumps is the atomic emission spectrum we observe.

Figure 23-21 (a) Four possible ways for the electron in a hypothetical atom to drop from the $n = 4$ level to the $n = 1$ level. (b) The emission spectrum produced by these jumps.

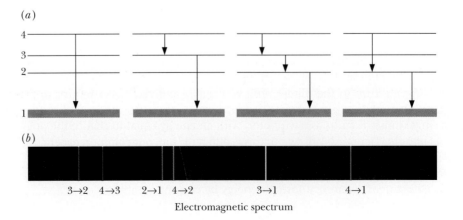

Bohr's scheme also explains the absorption spectrum. As we discussed earlier, the absorption spectrum is obtained when white light passes through a cool gas before a prism or a diffraction grating disperses it. In this case, selected photons from the beam of white light are absorbed and later reemitted. Because the reemitted photons are distributed in all directions, the intensities of these photons are reduced in the original direction of the beam. This removal of the photons from the beam leaves dark lines in the continuous spectrum.

The solution to the mystery of the missing lines in the absorption spectrum is easy using Bohr's model. For an absorption line to occur, there must be electrons in the lower level. Then they can be kicked up one or more levels by absorbing photons of the correct energy. Because almost all electrons in the atoms of the gas occupy the ground state, only the lines that correspond to jumps from this state show up in the spectrum. Jumps up from higher levels are extremely unlikely because electrons excited to a higher level typically remain there for less than a millionth of a second. Thus, there are fewer lines in the absorption spectrum (Figure 23-22) than in the emission spectrum, in agreement with the observations.

Flawed Reasoning

A large glass jar filled with helium gas is placed on the table and a very hot, glowing chunk of steel is placed behind it. (*Caution*: This should only be done by professionals.) Joshua and Darcy are discussing what they would see if they look through a diffraction grating at the light passing through the glass jar as the steel cools:

Joshua: "We should see the absorption spectrum for helium. The helium in the jar isn't changing, so the spectrum should remain constant as the steel cools."

Darcy: "But we learned that the spectrum for hot, glowing objects depends on temperature. As the steel cools, the entire spectrum, including the absorption lines, should move to longer wavelengths; a shift from blue toward red."

What will they see?

Answer There are two pieces to this puzzle. If they look directly at the piece of hot steel through the grating, they see a colorful rainbow spectrum. As the steel cools, the light will get dimmer, with the blue end of the spectrum fading the fastest. When the light passes through the jar of helium, certain wavelengths are removed by absorption, leaving dark lines in the rainbow pattern. The wavelengths that are absorbed by the helium do not change, so the absorption spectrum remains constant as the steel cools. However, the lines disappear as the colors dim because the photons of the required energy are no longer present in the light.

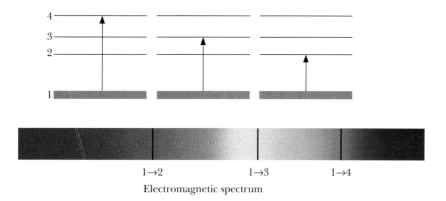

Figure 23-22 The absorption spectrum for the hypothetical atom in Figure 23-21.

BOHR | Creating the Atomic World

On January 16, 1939, the *Drottingholm* carried Niels Bohr into New York harbor. His arrival was eagerly awaited by a small group of physicists, among them several who had fled Hitler's spreading power. Bohr's hosts knew that he would bring the latest news from a Europe rapidly drifting toward war. Most important, he would relay the latest information on German science. He would tell them about work on the atom.

Niels Bohr is one of the best-known physicists of the 20th century. His pioneering work on atomic structure earned the Nobel Prize in 1922—the year after his friend Einstein received his. Bohr was comfortable with the commonsense violations of quantum physics and uncertainty. He believed in an underlying order, but it was a good deal messier than what Einstein could accept. For Bohr, God could play dice because probability ruled the Universe. Atomic physics, however, was no gamble. It was the future of humankind.

Niels Hendrik Bohr was born in the heart of Copenhagen in 1885. His father was a physiologist, and Bohr was reared in the embrace of a close professional family. He received his doctorate from the University of Copenhagen and then traveled to Cambridge to study with J. J. Thomson—the first Cavendish Professor. Bohr then moved on to the industrial city of Manchester to study with Ernest Rutherford, a pioneer in radiation studies. Armed with considerable laboratory and theoretical experience, Bohr returned to the University of Copenhagen as professor of physics. His work received a high honor when the Carlsberg Breweries' foundation constructed a new building for theoretical physics (1921) and a House of Honor (1932) for the most distinguished Dane—Niels Bohr.

Bohr was naturally gregarious, and upon his accession to the university professorship in Copenhagen, he recognized that his small country would be a natural gathering place for scientists who had been so alienated from one another in World War I. Young American men and women of science also joined this cosmopolitan and vibrant society as American science came of age. But it was an old friend, a Jewish physicist of considerable talent named Lise Meitner, who visited Bohr on her way to exile in Sweden and who gave him the information on the Berlin experiments of Christmastide 1938: The uranium atom had released fission by-products. Great energy had been released. This was the portentous news Bohr brought to New York that cold January day in 1939.

Bohr remained in Copenhagen until it became too dangerous, and then he and his wife fled in September 1943 to Sweden and a month later to England. British and American scientists of the highest rank wanted access to him and his knowledge of the German nuclear effort. Bohr worked assiduously counseling British and American officials on postwar nuclear policy. He was convinced that there was no secret to nuclear energy and that developed industrial societies could acquire that capability if they desired. His efforts were recognized in the Ford Motor Company's Atoms for Peace Award—a part of President Eisenhower's policy on atomic energy.

Niels Bohr

Niels Bohr was first and foremost a scientist, but he was a committed and engaged human being. More clearly than most, he saw that science has consequences for society and that scientists should accept the task of educating their publics about those consequences. Neither President Franklin Roosevelt nor Prime Minister Winston Churchill could fathom Bohr's concern; later world leaders did recognize his prescience and his humanity.

Bohr continues to fascinate us. In 1998 British playwright and author Michael Frayn presented a popular and controversial drama, *Copenhagen*. In 2000 it was a hit on the New York stage. The title encapsulates a hypothetical meeting in Copenhagen between Bohr and his former student Werner Heisenberg in 1941. Bohr knew that an Allied program to develop atomic energy for military purposes was under way in Britain and the United States. Many leading scientists in that program were former associates and students. Bohr also recognized Heisenberg's leadership in the nascent Nazi bomb project. The two had been close, almost like father and son—certainly like mentor and pupil. Germany's destruction of European Jewry was also known. It was a meeting fraught with tension, memories, and cutting-edge science. The drama in Copenhagen takes place with these two protagonists and a third, more anti-German participant, Bohr's wife, Margrethe.

Did Heisenberg come to renew this old friendship with Bohr simply to spy on the Allied program? Did he come to assure his old friend that he would do his best as head of the atomic research program in Germany to ensure that the program would fail? Did he come simply to secure a blessing from the most civilized man in the world—the man whose word could confer moral comfort? In a series of flashbacks and references to scientific work in physics over the previous two decades, the actors build their characters and enlarge the audience's understanding of the colossal stakes in this nuclear game. "Bohr, I have to know! I'm the one who has to decide! If the Allies are building a bomb, what am I choosing for my country?" Does a physicist have a moral right to create a weapon of such vast destructive power? Neither man can answer sufficiently to satisfy the other. Margrethe passes the final judgment: "He is one of them." Scientists have become the creator and destroyer of worlds.

—*Pierce C. Mullen, historian and author*

Sources: Ruth Moore, *Niels Bohr* (New York: Knopf, 1966; MIT paperback edition 1985); *Physics Today* (October 1985); R. Harre, *Scientific Thought 1900–1960: A Selective Survey* (Oxford: Oxford University Press, 1969); Michael Frayn, *Copenhagen* (New York: Random House–Anchor Books, 2000).

PHYSICS | ON YOUR OWN

Go to Professor Dean Zollman's "Visual Quantum Mechanics" website at www.phys.ksu.edu/perg/vqm/sls.html. Select the simulation titled "Gas Spectra—Emission" and find a set of energy levels that produce the emission spectrum of each of the gases available. Then return to the index and select the simulation titled "Gas Spectra—Absorption" and complete the equivalent activity.

The Periodic Table

Bohr's model helped unravel the mystery of the periodic properties of the chemical elements, a major unresolved problem for several decades. The table displaying the repeating chemical properties had grown considerably since Mendeleev's discovery and looked much like the modern one shown on the inside front cover of this text. About the time Bohr proposed his model, it became widely accepted that if the elements were numbered in the order of increasing atomic masses, their numbers would correspond to the number of electrons in the neutral atoms.

Bohr's theory gave a fairly complete picture of hydrogen (H): one electron orbiting a nucleus with one unit of positive charge (Figure 23-23). However, it was not able to explain the details of the spectrum of the next element, helium (He), because of the effects of the mutual repulsion of the two electrons. But the theory was successful for the helium atom with one electron removed (the He$^+$ ion). Bohr correctly assumed that the helium atom has two electrons in the lowest energy level. Because the electrons presumably have separate orbits, the collection of orbits of the same size was called a **shell**.

The next element, lithium (Li), has spectral lines similar to those of hydrogen. This similarity could be explained if lithium had one electron orbiting a nucleus with three units of positive charge and an inner shell of two electrons. The two inner electrons partially shield the nucleus from the outer electron, so the outer electron "sees" a net charge of approximately +1.

How many electrons can be in the second shell? The clue is found with the next element that has properties like lithium—sodium (Na), element 11. Sodium must have 1 electron outside two shells. Because the first shell holds 2, the second shell holds 8. The next lithium-like element is potassium (K), with 19 electrons. This time we have 1 electron orbiting 18 inner electrons: one shell of 2 and two shells of 8. And so on for succeeding elements.

Question How many electrons are there in the fourth shell?

Answer The next lithium-like element is rubidium (Rb), with 37 electrons. Therefore, the 36 inner electrons must be arranged in shells with 2, 8, 8, and 18 electrons.

In this scheme the chemical properties of atoms are determined by the electrons in their outermost shells. This idea can be extended to other vertical

Hydrogen   Helium   Lithium   Sodium

Figure 23-23 The electron shell structure for hydrogen, helium, lithium, and sodium.

columns in the periodic table. The outer shells of elements in the right-hand column (group 8) are completely filled. These elements do not react chemically with other elements and rarely form molecules with themselves. They are inert. They are also called the *noble elements* because they don't mix with the common elements. The properties of group-8 elements led to the idea that the most stable atoms are those with filled outer shells.

The elements in group 7 are missing one electron in the outer shell and, like the elements in group 1, are chemically very active. They become much more stable by gaining an electron to fill this shell. Therefore, they often appear as negative ions. On the other hand, the elements in group 1 are more stable if they give up the single electron in the outer shell to become positive ions. Therefore, group-1 atoms readily combine with group-7 atoms. The atom from group 1 gives up an electron to that from group 7, and these two charged ions attract each other. Common table salt—sodium chloride—is such a combination. As soon as the two atoms combine, their individual properties—reactive metal and poisonous gas—disappear because the outer structure has changed.

Question   What other salts besides sodium chloride might you expect to find in nature?

Answer   Most of the combinations of group-1 and group-7 elements should occur in nature. Therefore, we expect to find such salts as sodium fluoride, sodium bromide, potassium fluoride, potassium chloride, and rubidium chloride.

Atoms can also share electrons. Oxygen (O) is missing two electrons in the outer shell, and hydrogen (H) is either missing one or has an extra. They can both have filled outer shells if they share electrons. Two hydrogens combine with one oxygen to form the very stable water molecule H_2O.

These ideas were a great accomplishment for the Bohr model of the atom. Finally, there was a simple scheme for explaining the large collection of properties of the elements. The model provided a *physical* theory for understanding the *chemical* properties of the elements.

X Rays

The cathode-ray experiments that we discussed earlier in this chapter led to another important discovery that had an impact on the new atomic theory. In 1895 German scientist Wilhelm Roentgen saw a glow coming from a piece of paper on a bench near a working cathode-ray tube. The paper had been coated with a fluorescent substance. Fluorescence was known to result from exposure to ultraviolet light, but there was no ultraviolet light present. The glow persisted even when the paper was several meters from the tube, and the tube was covered with black paper!

Roentgen conducted numerous experiments demonstrating that although these new rays were associated with cathode rays, they were different. Cathode rays could penetrate no more than a few centimeters of air, but the new rays could penetrate several meters of air or even thin samples of wood, rubber, or metal. Using electric and magnetic fields, Roentgen could not deflect these rays, but he did find that if he deflected the cathode rays, the rays came from the new spot where the cathode rays struck the glass walls. The new rays were also shown to be emitted by many different materials. Roentgen named them **X rays** because of their unknown nature.

The discovery of X rays quickly created public excitement because people could see their bones and internal organs. This discovery was quickly put to use by the medical profession. In 1901 Roentgen received the first Nobel Prize in

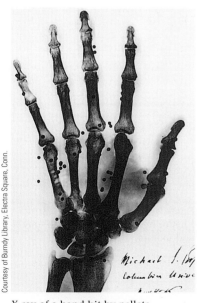

X ray of a hand hit by pellets.

physics. We now know that X rays are high-frequency electromagnetic radiation created when the electrons in the cathode-ray tube are slowed down as they crash into the metal target inside the tube.

A typical X-ray spectrum (Figure 23-24) consists of two parts: a continuous spectrum produced by the rapid slowing of the bombarding electrons and a set of discrete spikes. These spikes are from X-ray photons emitted by the atoms and therefore carry information about the element. They occur when a bombarding electron knocks an atomic electron from a lower energy level completely out of the atom. An electron in a higher energy level fills the vacancy. When this electron falls into the vacated level, an X-ray photon is emitted. The energies of these X rays can be used to calculate the energy differences of the atom's inner electron orbits, which are characteristic of the element.

Shortly after Bohr proposed his model of the atom, Henry Moseley, a British physicist, was able to use Bohr's model and X rays to resolve some discrepancies in the periodic table. In several cases the order of the elements determined by their atomic masses disagreed with their order determined by their chemical properties. For instance, according to atomic masses, cobalt should come after nickel in the periodic table, yet its properties suggested that it should precede nickel.

Bohr's model predicted that the energies of these characteristic X rays would increase as the number of electrons in the atom increased. Moseley bombarded various elements with electrons and studied the X rays that were given off. His experimental results agreed with Bohr's theoretical prediction. In almost all cases, the results matched the order obtained from the atomic masses; however, they showed that cobalt should come before nickel. As a result of this work, we now know that the correct ordering of the elements is by **atomic number** (the number of electrons in neutral atoms) and not by atomic mass.

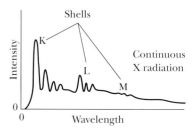

Figure 23-24 A typical X-ray spectrum contains a continuous background and characteristic lines from the inner atomic levels.

Summary

When an electric current passes through a gas, it gives off a characteristic emission spectrum. The absorption spectrum is obtained by passing white light through a cool gas, producing dark lines in the continuous, rainbow spectrum. The emission spectrum has all the lines in the absorption spectrum plus many additional ones.

Cathode rays are electrons with a charge-to-mass ratio about 1800 times that for hydrogen ions. Electrons have an electric charge of 1.6×10^{-19} coulomb and a mass of 9.11×10^{-31} kilogram. The cathode-ray experiments also led to the discovery of X rays, a form of high-frequency electromagnetic radiation.

Rutherford used alpha particles to show that the positive charge was concentrated in a nucleus with a diameter 100,000 times smaller than the atom. Thus, Rutherford's solar-system model of the atom is almost entirely empty space. His model failed to explain the chemical periodicity and details of the spectral lines.

All objects emit electromagnetic radiation. The spectrum emitted through a small hole in a heated hollow block is the same for all materials at the same temperature. The frequencies are not equally bright, and the maximum of the spectrum shifts toward the blue as the temperature of the cavity increases. The spectrum drops to zero on both the high-frequency (small-wavelength) and low-frequency (long-wavelength) sides.

Planck's analysis of the electromagnetic spectrum emitted by hot cavities provided the first clues that phenomena on the atomic level are quantized. The energy E of the quanta depends on Planck's constant h and the frequency f: $E = hf$.

Planck's idea was expanded to include the electromagnetic radiation by Einstein's efforts to explain the photoelectric effect. When light is shined on certain clean metallic surfaces, electrons are ejected with a range of kinetic energies up to a maximum energy, depending on the color of the light. If the light's inten-

Physics Now™ Assess your understanding of this chapter's topics with sample tests and other resources found by logging into PhysicsNow at http://physics.brookscole.com/kf6e.

sity is changed, the rate of ejecting electrons changes, but the maximum kinetic energy stays fixed. This is explained by assuming that light is also quantized, existing as bundles of energy called photons. The energy of a photon is given by Planck's relationship, $E = hf$. A photon gives its entire energy to one electron in the metal.

Bohr's model for the hydrogen atom incorporated Planck's discrete energies and Einstein's photons into Rutherford's model. It was based on three postulates: (1) Allowed electron orbits are those in which the angular momentum is an integral multiple of a smallest one. (2) An electron in an allowed orbit does not radiate. (3) When an electron jumps from one orbit to another, it emits or absorbs a photon whose energy is equal to the difference in the energies of the two orbits.

The Bohr model was successful in accounting for many atomic observations, especially the emission and absorption spectra for hydrogen and the ordering of the chemical elements.

Chapter 23 Revisited

The light emitted by an atom is created when an atomic electron moves from one energy level to a lower one. The size of the jump determines the energy of the photon and—via Planck's relationship—the color of the emitted light. Therefore, the colors provide us with information about the energy levels of the atom's electronic structure.

KEY TERMS

absorption spectrum: The collection of wavelengths missing from a continuous distribution of wavelength due to the absorption of certain wavelengths by the atoms or molecules in a gas.

alpha particle: The nucleus of the helium atom.

atomic number: The number of protons or the number of electrons in the neutral atom of an element. This number also gives the order of the elements in the periodic table.

cathode ray: An electron emitted from the negative electrode in an evacuated tube.

electron: The basic constituent of atoms that has a negative charge.

electron volt: A unit of energy equal to the kinetic energy acquired by an electron or proton falling through an electric potential difference of 1 volt. Equal to 1.6×10^{-19} joule.

emission spectrum: The collection of discrete wavelengths emitted by atoms that have been excited by heat or electric currents.

ion: An atom with missing or extra electrons.

nucleus: The central part of an atom containing the positive charges.

photoelectric effect: The ejection of electrons from metallic surfaces by illuminating light.

photon: A particle of light. The energy of a photon is given by the relationship $E = hf$, where f is the frequency of the light and h is Planck's constant.

quantum (pl., *quanta*): The smallest unit of a discrete property. For instance, the quantum of charge is the charge on the proton.

quantum number: A number giving the value of a quantized quantity. For instance, a quantum number specifies the angular momentum of an electron in an atom.

shell: A collection of electrons in an atom that have approximately the same energy.

X ray: A high-energy photon with a range of frequencies in the electromagnetic spectrum lying between the ultraviolet and the gamma rays.

CONCEPTUAL QUESTIONS

1. Today, we think of the periodic table as being arranged in order of increasing proton (or electron) number. Why did Mendeleev not use this approach?
2. If Mendeleev was ordering the elements according to their masses, why did he not produce a table with only one row?
3. What element in Mendeleev's periodic table is most similar to silicon (Si)?
4. What elements would you expect to have chemical properties similar to chlorine (Cl)?
5. Why are the spectral lines for elements sometimes called "atomic fingerprints"?

6. How could you determine whether there is oxygen in the Sun?
7. Would you obtain an emission or an absorption spectrum at location A in the setup shown in the figure?

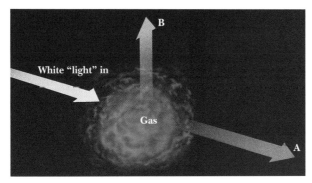

Questions 7 and 8

8. Would you obtain an emission or an absorption spectrum at location B in the setup shown in the figure?
9. The emission spectra shown in the figure were all obtained with the same apparatus. What elements can you identify in sample (a)? Are there any that you cannot identify?

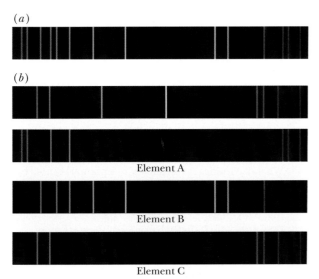

Questions 9 and 10

10. What elements can you identify in sample (b) of the figure? Are there any that you cannot identify?
11. How does the number of lines in the absorption spectrum for an element compare with the number in the emission spectrum?
12. What are the differences between an emission spectrum and an absorption spectrum?
13. When your authors were students, their textbooks did not have color. Could the spectra in Figure 23-3 still be used to identify atoms if they were in black and white? Explain.
14. Two graphs of brightness versus wavelength are shown in the figure. Identify which is an absorption spectrum and which is an emission spectrum.

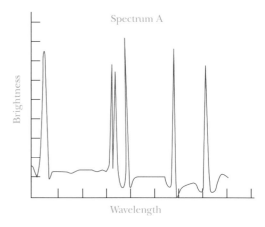

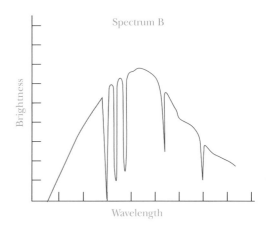

15. To which of the brightness versus wavelength graphs in Question 14 does the line spectrum in the figure correspond?

16. Sketch the brightness versus wavelength graph for the hydrogen spectrum shown in Figure 23-3.
17. Suppose you were a 19th-century scientist who had just discovered a new phenomenon known as Zeta rays (yes, we're making this up). What experiment could you perform to determine whether Zeta rays were charged particles or an electromagnetic wave? Could this experiment distinguish between neutral particles and an electromagnetic wave?
18. Imagine that you determined that the Zeta rays from Question 17 were charged particles. How would you determine the sign of the charge?
19. In the Millikan oil-drop apparatus shown in Figure 23-8, an electric field provides a force that balances the gravitational force on charged oil drops. Millikan found that he needed the electric field to point down toward the floor. Was the net charge on the oil drops a result of an excess or a deficit of electrons?
20. Why was it not necessary for Millikan to use oil drops with an excess of only one electron to determine the charge of a single electron?

21. Millikan's oil-drop experiment was used to determine the charge on a single electron. Why should Millikan also get credit for determining the electron's mass?
22. What do we mean when we say that a physical quantity is quantized?
23. Why did Thomson feel that electrons were pieces of atoms rather than atoms of a new element?
24. How does Rutherford's model of an atom explain why most alpha particles pass right through a thin gold foil? How does it account for some alpha particles being scattered backward?
25. Rutherford's model predicted that atoms should be unstable; the electrons should spiral into the nucleus in extremely short times. What caused this instability in Rutherford's model?
26. Rutherford's model provided an explanation for the emission of light from atoms. What was this mechanism and why was it unsatisfactory?
27. Why can the curves for the intensities of the colors emitted by hot solid objects not serve as "atomic fingerprints" of the materials?
28. If all objects emit radiation, why don't we see most of them in the dark?
29. As you move to the right along the horizontal axis of Figure 23-14, is the frequency increasing or decreasing? Explain your reasoning.
30. For an object at a temperature of 8000 K, use Figure 23-14 to determine whether the light intensity is greater for light in the ultraviolet or in the infrared.
31. You measure the brightness of two different hot objects; first with a blue filter and then with a red filter. You find that object A has a brightness of 25 in the blue and 20 in the red. Object B has a brightness of 12 in the blue and 3 in the red. The brightness units are arbitrary but the same for all measurements. Which is the hotter of the two objects?
32. The curves in Figure 23-14 show the intensities of the various wavelengths emitted by an object at three different temperatures. The region corresponding to visible light is indicated. How would the color of the object at 8000 K compare with the color of the object at 4000 K?
33. Why do astronomers often use the terms *color* and *temperature* interchangeably when referring to stars?

Questions 33 and 34

34. Why are blue stars thought to be hotter than red stars?
35. What assumptions did Planck make that enabled him to obtain the correct curve for the spectrum of light emitted by a hot object?
36. What assumptions did Einstein make that enabled him to account for the experimental observations of the photoelectric effect?
37. What property of the emitted photoelectrons depends on the intensity of the incident light?
38. What property of the emitted photoelectrons depends on the frequency of the incident light?
39. If a metal surface is illuminated by light at a single frequency, why don't all the photoelectrons have the same kinetic energy when they leave the metal's surface?
40. How is it possible that ultraviolet light can cause sunburn but no amount of visible light will?
41. You find that if you shine ultraviolet light on a negatively charged electroscope, the electroscope discharges even if the intensity of the light is low. Red light, however, will not discharge the electroscope even at high intensities. How do you account for this?
42. You find that if you shine ultraviolet light on a negatively charged electroscope, the electroscope discharges. Can you discharge a positively charged electroscope the same way? Why or why not?
43. What are the three assumptions of Bohr's model of the atom?
44. Why did Bohr assume that the electrons do not radiate when they are in the allowed orbits?
45. An electron in the $n = 3$ energy level can drop to the ground state by emitting a single photon or a pair of photons. How does the total energy of the pair compare with the energy of the single photon?
46. If electrons in hydrogen atoms are excited to the fourth Bohr orbit, how many different frequencies of light may be emitted?
47. How can the spectrum of hydrogen contain so many spectral lines when the hydrogen atom only has one electron?
48. What determines the frequency of a photon emitted by an atom?
49. Why does the spectrum of lithium (element 3) resemble that of hydrogen?
50. How does Bohr's model explain that there are more lines in the emission spectrum than in the absorption spectrum?
51. How many electrons would you expect to find in each shell for sulfur (S)?
52. Radon (element 86) is a gas. Would you expect the molecules of radon to consist of a single atom or a pair of atoms? Why?
53. Sodium does not naturally occur as a free element. Why?
54. What effective charge do the outer electrons in magnesium "see"? (Magnesium is element 12.)
55. What type of electromagnetic wave has a wavelength about the size of an atom? (An electromagnetic spectrum is given in Figure 22-26.)
56. Are X rays deflected by electric or magnetic fields? Explain.
57. How does an X ray differ from a photon of visible light?
58. Why would you not expect an X-ray photon to be emitted every time an inner electron is removed from an atom?
59. How does Bohr's atomic model account for the observation that the energy of the most energetic X rays increases with atomic number?

EXERCISES

1. What is the charge-to-mass ratio for a cathode ray?
2. What is the charge-to-mass ratio for a hydrogen ion (an isolated proton)?
3. Given that the radius of a hydrogen atom is 5.29×10^{-11} m and that its mass is 1.682×10^{-27} kg, what is the average density of a hydrogen atom? How does it compare with the density of water?
4. What is the average density of the hydrogen ion (an isolated proton) given that its radius is 1.2×10^{-15} m and that its mass is 1.673×10^{-27} kg? It is interesting to note that such densities also occur in neutron stars.
5. If you were helping your younger brother build a scale model of an atom for a science fair and wanted it to fit in a box 1 m on each side, how big would the nucleus be?
6. A student decides to build a physical model of an atom. If the nucleus is a rubber ball with a diameter of 1 cm, how far away would the outer electrons be?
7. What is the energy of a photon of orange light with a frequency of 5×10^{14} Hz?
8. What is the energy of the least energetic photon of visible light?
9. A photon of red light has energy 3×10^{-19} J. What is its frequency?
10. An X-ray photon has energy 1.5×10^{-15} J. What is its frequency?
11. A microwave photon has an energy of 2×10^{-23} J. What is its wavelength?
12. A photon of yellow light has a wavelength of 6×10^{-7} m. What is its energy?
13. What is the angular momentum of an electron in the ground state of hydrogen?
14. What is the angular momentum of an electron in the $n = 3$ level of hydrogen?
15. What is the radius of the $n = 3$ level of hydrogen?
16. What is the quantum number of the orbit in the hydrogen atom that has 100 times the radius of the smallest orbit?
17. What is the frequency of a photon of energy 4 eV?
18. What is the energy, in electron volts, of a yellow photon of wavelength 6×10^{-7} m?
19. According to Figure 23-20, it requires a photon with an energy of 10.2 eV to excite an electron from the $n = 1$ energy level to the $n = 2$ energy level. What is the frequency of this photon? Does it lie in, above, or below the visible range?
20. When a proton captures an electron, a photon with an energy of 13.6 eV is emitted. What is the frequency of this photon? Does it lie in, above, or below the visible range?
21. What is the ratio of the volumes of the hydrogen atom in the $n = 1$ state compared with those in the $n = 2$ state?
▲ 22. The diameter of the hydrogen atom is 10^{-10} m. In Bohr's model this means that the electron travels a distance of about 3×10^{-10} m in orbiting the atom once. If the orbital frequency is 7×10^{15} Hz, what is the speed of the electron? How does this speed compare with that of light?
23. What difference in energy between two atomic levels is required to produce an X ray with a frequency of 2×10^{18} Hz?
24. What is the frequency of the X ray that is emitted when an electron drops down to the ground state from an excited state with 1000 eV more energy?

24 The Modern Atom

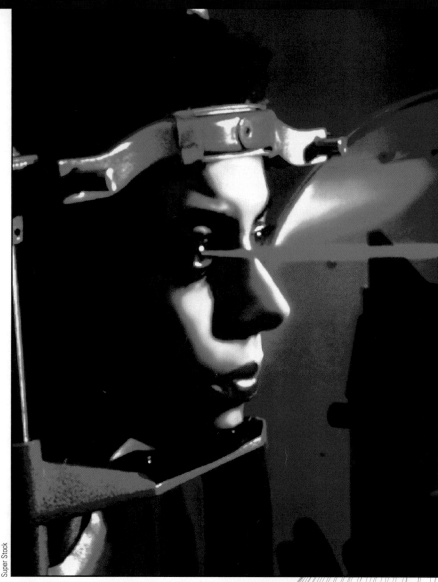

Modern lasers have allowed us to take truly three-dimensional pictures (holography), perform surgery inside our bodies, weld retinas in our eyes, cut steel plates for industry and fabric for blue jeans, and perform myriad other amazing tasks. What makes laser light so special?

(See page 518 for the answer to this question.)

A laser being used to perform eye surgery.

BOHR'S model of the atom lacked "beauty"; the way the theory blended classical and quantum ideas in a seemingly contrived way to account for the experimental results was troublesome. Bohr used the ideas of Newtonian mechanics and Maxwell's equations until he ran into trouble; then he abruptly switched to the quantum ideas of Planck and Einstein. For example, the electron orbits were like Newton's orbits for the planets. The force between the electron and the nucleus was the electrostatic force. According to Maxwell's equations, these electrons should have continuously radiated electromagnetic waves. Because this obviously didn't happen, Bohr postulated discrete orbits with photons being emitted only when an electron jumped from a higher orbit to a lower one. There was no satisfactory reason for the existence of the discrete orbits other than that they gave the correct spectral lines.

Bohr's model of the atom was remarkably successful in some areas, but it failed in others. Even though other models have replaced it, it was nevertheless important in advancing our understanding of matter at the atomic level.

PhysicsNow™ Test your understanding of this chapter by logging into PhysicsNow at http://physics.brookscole.com/kf6e, selecting the chapter, and clicking on the "Take a Pre-Test" link.

Successes and Failures

Accounting for the stability of atoms by postulating orbits in which the electron did not radiate was a mild success of the Bohr theory. The problem was that nobody could give a fundamental reason for why this was so. Why *don't* the electrons radiate? They are accelerating, and according to the classical theory of electromagnetism, they should radiate.

The Bohr model successfully provided the numerical values for the wavelengths in the spectra of hydrogen and hydrogen-like ions. It provided qualitative agreement for the elements with a single electron orbiting a filled shell, but it could not predict the spectral lines when there was more than one electron in the outer shell.

But all along new data were being collected. Close examination of the spectral lines, especially when the atoms were in a magnetic field, revealed that the lines were split into two or more closely spaced lines. The Bohr theory could not account for this. Also, the different spectral lines for a particular element were not of the same brightness. Apparently, some jumps were more likely than others. The Bohr theory provided no clue to why this was so.

Although Bohr's model successfully described the general features of the periodic table, it could not explain why the shells had a certain capacity for electrons. Why did the first shell contain two electrons? Why not just one, or maybe three? And what's so special about the next level that it accepts eight electrons?

There were other loose ends. Einstein's theory of relativity was established, and it was generally accepted that the speeds of the electrons were fast enough and the measurements precise enough that relativistic effects should be included. But Bohr's model was nonrelativistic. Clearly, this was a transitional model.

Louis de Broglie

De Broglie's Waves

✓ **MATH**

In 1923 a French graduate student proposed a revolutionary idea to explain the discrete orbits. Louis de Broglie's idea is easy to state but hard to accept: *electrons behave like waves*. De Broglie reasoned that if light could behave like particles, then particles should exhibit wave properties. He viewed all atomic objects as having this dualism; each exhibits wave properties *and* particle properties.

De Broglie's view of the hydrogen atom had electrons forming standing-wave patterns about the nucleus like the standing-wave patterns on a guitar string (Chapter 16). Not every wavelength will form a standing wave on the string; only those that have certain relationships with the length of the string will do so.

Imagine forming a wire into a circle. The waves travel around the wire in both directions and overlap. When a whole number of wavelengths fits along

✓ Extended presentation available in the *Problem Solving* supplement

◄ electrons behave like waves

Figure 24-1 The shaft holding the wire executes simple harmonic motion that creates waves traveling around the circular wire in both directions. Standing waves form when the circumference is equal to a whole number of wavelengths.

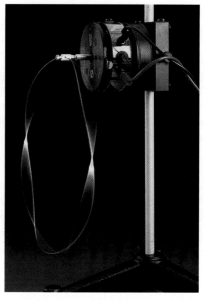

the circumference of the circle, a crest that travels around the circle arrives back at the starting place at the same time another crest is being generated. In this case the waves reinforce each other, creating a standing wave. This process is illustrated in Figure 24-1 for a wire and in Figure 24-2 for an atom.

De Broglie postulated that the wavelength λ of the electron is given by the expression

wavelength of an electron ▶

$$\lambda = \frac{h}{mv}$$

where h is Planck's constant, m is the electron's mass, and v is the electron's speed. With this expression de Broglie could reproduce the numerical results for the orbits of hydrogen obtained by Bohr. The requirement that the waves form standing-wave patterns automatically restricted the possible wavelengths and therefore the possible energy levels.

De Broglie's idea put the faculty at his university in a very uncomfortable position. If his idea proved to have no merit, they would look foolish awarding a Ph.D. for a crazy idea. The work was very speculative. Although he was able to account for the energy levels in hydrogen, de Broglie had no other experimental evidence to support his hypothesis. To make matters worse, de Broglie's educational background was in the humanities. He had only recently converted to

Figure 24-2 The electrons form standing waves about the nucleus.

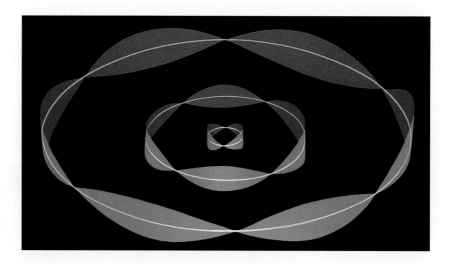

Seeing Atoms

Atoms are so small that they cannot be seen with the most powerful optical microscopes. This limitation is due to the wave properties of light that we discussed in Chapter 19. The relatively large wavelength of the light scatters from the atoms without yielding any information about the individual atoms. While electron microscopes have much higher resolution (electrons can be accelerated to high speeds by electric fields, giving them wavelengths more than 100 times shorter than wavelengths of visible light), even the most powerful electron microscope does not allow us to view individual atoms.

A new type of microscope, developed in the early 1980s, provides views of atomic surfaces with a resolution thousands of times greater than is possible with light. The developers of the *scanning tunneling microscope*, or STM, were awarded the Nobel Prize in 1986 (along with the inventor of the electron microscope).

The STM uses a very sharp probe (several atoms across at the tip) placed close to the surface (about 1 ten-billionth of a meter) to "view" the surface under investigation. As the probe is moved back and forth over the surface, electrons are pulled from the atoms to the probe, creating a current that is monitored by a computer. The current is greatest when the probe is directly over an atom and quite small when the probe is between atoms. By carefully mapping the surface, the structure of the surface can be plotted with a resolution smaller than the atomic dimensions. Computer displays like the one shown here clearly indicate the orderly arrangement of the atoms on the surface of the material. These measurements can also show the location of individual atoms that are deposited onto the surface. Such studies are important in understanding the physics of surfaces as well as of the catalytic processes used in manufacturing.

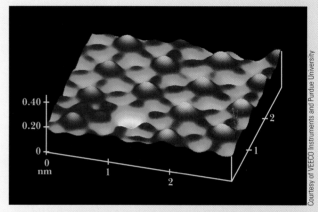

A scanning tunneling microscope image of iodine atoms absorbed on a platinum surface. Note the orderly arrangement of the atoms and that one of the atoms is missing.

---

the study of physics and was working in an area of physics that was not very well understood by the faculty. Further, de Broglie came from an influential noble family that had played a leading role in French history. He carried the title of prince. His thesis supervisor resolved the problem by showing the work to Einstein, who indicated that the idea was basically sound.

Although de Broglie's idea was appealing, it was still speculative. It was important to find experimental confirmation by observing some definitive wave behavior, such as interference effects, for electrons. An electron accelerated through a potential difference of 100 volts has a speed that yields a de Broglie wavelength comparable to the size of an atom. Therefore, to see interference effects produced by electrons, we need atom-sized slits. Because X rays have atom-sized wavelengths, they can be used to search for possible setups. Photographs of X rays scattered from randomly oriented crystals, such as the one shown in Figure 24-3, show that the regular alignment of the atoms in crystals can produce interference patterns, such as those produced by multiple slits. Therefore, we might look for the interference of electrons from crystals.

Experimental confirmation of de Broglie's idea came quickly. Two American scientists, C. J. Davisson and L. H. Germer, obtained confusing data from a seemingly unrelated experiment involving the scattering of low-energy electrons from nickel crystals. The scattered electrons had strange valleys and peaks in their distribution. While showing these data at a conference at Oxford, it was suggested to Davisson that this anomaly might be a diffraction effect. Further analysis of the data showed that electrons do exhibit wave behavior. The photograph in Figure 24-4 shows an interference pattern produced by electrons. De Broglie was one of a very few physicists to receive a Nobel Prize for his thesis work.

There is nothing in de Broglie's idea to suggest that it should apply only to

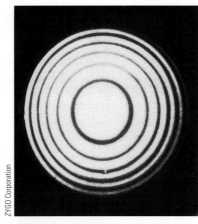

Figure 24-3 An interference pattern produced by light.

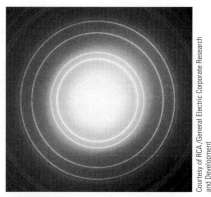

Figure 24-4 An interference pattern produced by electrons scattering from randomly oriented aluminum crystallites.

electrons. This relationship for the wavelength should also apply to all material particles. Although it may not apply to macroscopic objects, assume for the moment that it does. A baseball (mass = 0.14 kilogram) thrown at 45 meters per second (100 mph) would have a wavelength of 10^{-34} meter! This is an incredibly small distance. It is 1 trillion-trillionth the size of a single atom. This wavelength is smaller when compared with an atom than an atom is when compared with the Solar System. Even a mosquito flying at 1 meter per second would have a wavelength that is only 10^{-27} meter.

You shouldn't spend much time worrying that your automobile will diffract off the road the next time you drive through a tunnel. Ordinary-sized objects traveling at ordinary speeds have negligible wavelengths.

Question Why would you not expect baseballs thrown through a porthole to produce a measurable diffraction pattern?

Answer For diffraction to be appreciable, the wavelength must be about the same size as the opening.

Waves and Particles

Imagine the controversy de Broglie caused. Interference of light is believable because parts of the wave pass through each slit and interfere in the overlap region (Chapter 19). But electrons don't split in half. Every experiment designed to detect electrons has found complete electrons, not half an electron. So how can electrons produce interference patterns?

The following series of thought experiments was proposed by physicist Richard Feynman to summarize the many, many experiments that have been conducted to resolve the wave–particle dilemma. Although they are idealized, this sequence of experiments gets at crucial factors in the issue.

We will imagine passing various things through two slits and discuss the patterns that would be produced on a screen behind the slits. In each case there are three pieces of equipment: a source, two slits, and a detecting screen arranged as in Figure 24-5.

In the first situation, we shoot indestructible bullets at two narrow slits in a steel plate. Assume that the gun wobbles so that the bullets are fired randomly at the slits. Our detecting screen is a sand box. We simply count the bullets in certain regions in the sand box to determine the pattern.

Imagine that the experiment is conducted with the right-hand slit closed.

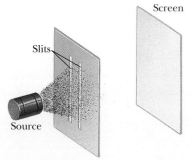

Figure 24-5 The experimental setup for the thought experiment concerning the wave–particle dilemma.

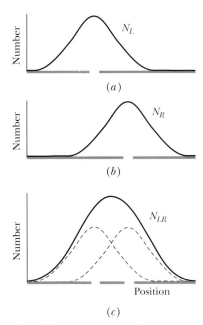

Figure 24-6 Distribution of bullets with (a) the left-hand slit open N_L, (b) the right-hand slit open N_R, and (c) both slits open N_{LR}.

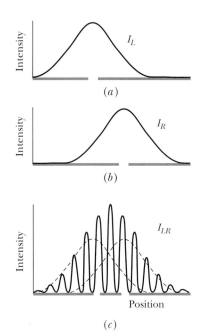

Figure 24-7 Intensity of water waves with (a) the left-hand slit open I_L, (b) the right-hand slit open I_R, and (c) both slits open I_{LR}.

After 1 hour we sift through the sand and make a graph of the distribution of bullets. This graph is shown in Figure 24-6(a). The curve is labeled N_L to indicate that the left-hand slit was open. Repeating the experiment with only the right-hand slit open yields the similar curve N_R shown in Figure 24-6(b).

Question What curve would you expect with both slits open?

Answer The curve with both slits open should be the sum of the first two curves.

When both slits are opened, the number of bullets hitting each region during 1 hour is just the sum of the numbers in the previous experiments—that is, $N_{LR} = N_L + N_R$. This is just what we expect for bullets or any other particles. This graph is shown in Figure 24-6(c).

◀ bullets add

Now imagine repeating the experiment using water waves. The source is now an oscillating bar that generates straight waves. The detecting screen is a collection of devices that measure the energy (that is, the intensity) of the wave arriving at each region. Because these are waves, the detector does not detect the energy arriving in chunks but rather in a smooth, continuous manner. The graphs in Figure 24-7 show the average intensity of the waves at each position across the screen for the same three trials.

Once again, the three trials yield no surprises. We expect an interference pattern when both slits are open. We see that the intensity with two slits open is not equal to the sum of the two cases with only one slit open, $I_{LR} \neq I_L + I_R$, water waves interfere but this is what we expect for water waves or for any other waves.

◀ water waves interfere

Both sets of experiments make sense, in part because we used materials from the macroscopic world—bullets and water waves. However, a new reality emerges when we repeat these experiments with particles from the atomic world.

This time we use electrons as our "bullets." An electron gun shoots electrons randomly toward two very narrow slits. Our screen consists of an array of devices

electrons interfere ▸

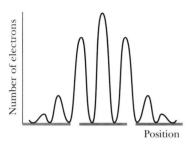

Figure 24-8 The distribution of electrons with two slits open.

photons interfere ▸

that can detect electrons. Initially, the results are similar to those obtained in the bullet experiment: electrons are detected as whole particles, not fractions of particles. The patterns produced with either slit open are the expected ones and are similar to those in Figure 24-6(a and b). The surprise comes when we look at the pattern produced with both slits open: we get an interference pattern like the one for waves (Figure 24-8)!

Note what this means: if we look at a spot on the screen that has a minimum number of electrons and close one slit, we get an *increase* in the number of counts at that spot. Closing one slit yields more electrons! This is not the behavior expected of particles.

One possibility that was suggested to explain these results is that the electrons are somehow affecting each other. We can test this by lowering the rate at which the source emits electrons so that only one electron passes through the setup at a time. In this case we might expect the interference pattern to disappear. How can an electron possibly interfere with itself? Each individual electron should pass through one slit or the other. How can it even know that the other slit is open? But the interference pattern doesn't disappear. Even though there is only one electron in the apparatus at a time, the same interference effects are observed after a large number of electrons are measured.

The set of experiments can be repeated with photons. The results are the same. The detectors at the screen see complete photons, not half photons. But the two-slit pattern is an interference pattern. Photons behave like electrons. Photons and electrons exhibit a duality of particle and wave behavior. Table 24-1 summarizes the results of these experiments.

Flawed Reasoning

Your friend argues: "The interference pattern for electrons passing through two slits isn't that mysterious. We know that two electrons interact via the electric force, and so an electron passing through one slit is going to affect the motion of an electron passing through the other slit."

What experimental evidence could you use to persuade your friend that the interference pattern is not caused by the interaction between electrons?

Answer Even if the beam is turned down so low that only one electron passes through the slits at a time, the same interference pattern is produced.

PHYSICS | ON YOUR OWN

Go to Professor Dean Zollman's "Visual Quantum Mechanics" website at www.phys.ksu.edu/perg/vqm/difsuite.html. Select the simulation entitled "Double Slit Diffraction" and explore the relationship between the energy of the electron beam and the pattern that results from the two-slit interference.

Table 24-1 | Summary of Experiments with Two Slits

Source	Detection	Pattern
Bullets	Chunks	No interference
Water waves	Continuous	Interference
Electrons or photons	Chunks	Interference

Probability Waves

Nature has an underlying rule governing this strange behavior. We need to unravel the mystery of why electrons (or photons, for that matter) are always detected as single particles and yet collectively they produce wavelike distributions. Something is adding together to give the interference patterns. To see this we once again look at water waves.

There is something about water waves that does add when the two slits are open—the displacements (or heights) of the individual waves at any instant of time. Their sum gives the displacement at all points of the interference pattern—that is, $h_{12} = h_1 + h_2$. The maximum displacement at each point is the amplitude of the resultant wave at that point.

With light and sound waves we observe the intensities of the waves, not their amplitudes. Intensity is proportional to the square of the amplitude. It is important to note that the resultant, or total, intensity is *not* the sum of the individual intensities. The total intensity is calculated by adding the individual displacements to obtain the resulting amplitude of the combined waves and then squaring this amplitude. Thus, the intensity depends on the amplitudes and the relative phase of the waves. As you can see in Figure 24-9, the intensity of the combined waves is larger than the sum of the individual intensities when the displacements are in the same direction and smaller when they are in opposite directions.

There is an analogous situation with electrons. The definition for a **matter-wave amplitude** (called a *wave function*) is analogous to that for water waves. (This matter-wave amplitude is usually represented by the Greek letter ψ, pronounced "sigh.") The square of the matter-wave amplitude is analogous to the intensity of a wave. In this case, however, the "intensity" represents the likelihood, or probability, of finding an electron at that location and time. There is one very important difference between the two cases. We can physically measure the amplitude of an ordinary mechanical wave, but there is *no* way to measure the amplitude of a matter wave. We can measure only the value of this amplitude squared.

These ideas led to the development of a new view of physics known as **quantum mechanics**, which are the rules for the behavior of particles at the atomic and subatomic levels. These rules replace Newton's and Maxwell's rules. A quantum-mechanical equation, called Schrödinger's equation after Austrian physicist Erwin Schrödinger, is a wave equation that provides all possible information about atomic particles.

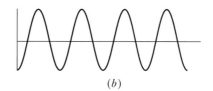
(a)

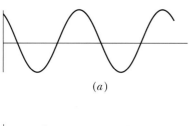
(b)

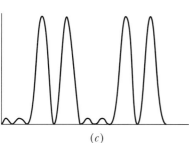
(c)

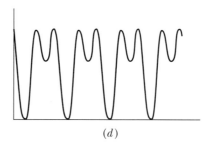
(d)

Figure 24-9 The two individual waves (a and b) are combined (c) by adding the displacements before squaring. Note the difference when the individual displacements are squared before adding (d).

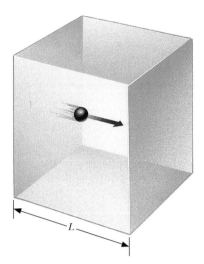

Figure 24-10 A particle in a box has quantized values for its de Broglie wavelength.

A Particle in a Box

It is instructive to look at another simple, though somewhat artificial, situation. Imagine you have an atomic particle that is confined to a box. Further imagine that the box has perfectly hard walls so that no energy is lost in collisions with the walls and that the particle moves in only one dimension, as shown in Figure 24-10.

From a Newtonian point of view, there are no restrictions on the motion of the particle; it could be at rest or bouncing back and forth with any speed. Because we assume that the walls are perfectly hard, the sizes of the particle's momentum and kinetic energy remain constant. In this classical situation, things happen much as we would expect from our commonsense world view. The particle is like an ideal Super Ball in zero gravity. It follows a definite path; we can predict when and where it will be at any time in the future.

Until the early 1900s, it was assumed that an electron would behave the same way. We now know that atomic particles have a matter-wave character and their properties—position, momentum, and kinetic energy—are governed by Schrödinger's equation. When these particles are confined, their wave nature guarantees that their properties are quantized. For example, the solutions of Schrödinger's equation for the particle in a box are a set of standing waves. In fact, the solutions are identical to those for the guitar string that we examined in Chapter 16. A discrete set of wavelengths fits the conditions of confinement.

Question How would you expect the wavelength of the fundamental standing wave to compare with the length of the box?

Answer As with the guitar string analogy, we would expect the wavelength to be twice the length of the box.

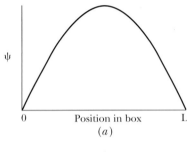

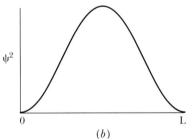

Figure 24-11 The (a) wave amplitude and (b) probability distribution for the lowest energy state for a particle in a box.

The fundamental standing wave corresponds to the lowest energy state (the ground state), and one-half of a wavelength fits into the box, as shown in Figure 24-11(a). The probability of finding the particle at some location depends on the square of this wave amplitude. These probability values (Figure 24-11[b]) help locate the particle. You can see that the most likely place for finding the atomic particle is near the middle of the box because ψ^2 is large there. The probability of finding it near either end is quite small.

A bizarre situation arises with the higher energy levels. The wave amplitude and its square for the next higher energy level are shown in Figure 24-12. Now the most likely places to find the particle are midway between the center and either end of the box. The least likely places are near an end or near the center. It is tempting to ask how the atomic particle gets from one region of high probability to the other. How does it cross the center where the probability of locating it is zero? This type of question, however, does not make sense in quantum mechanics. Particles do not have well-defined paths; they have only probabilities of existing throughout the space in question.

Figure 24-12 The (a) wave amplitude and (b) probability distribution for the first excited state for a particle in a box.

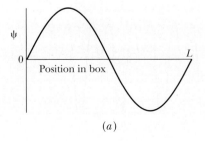
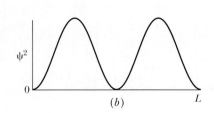

> **Question** Where are the most likely places to find the particle in the third energy level?
>
> **Answer** The square of the third standing wave will have maxima at the center and between the center and each side of the box (see Figure 16-6).

Suppose you try to find the particle. You would find it in one place, not spread out over the length of the box. However, you cannot predict where it will be. All you can do is predict the probability of finding it in a particular region. In addition, even if you know where it is at one time, you still cannot predict where it will be at a later time. Atomic particles do not follow well-defined paths as classical objects do.

The existence of quantized wavelengths means that other quantities are also quantized. Because de Broglie's relationship tells us the momentum is inversely proportional to the wavelength, momentum is quantized. The kinetic energy is proportional to the square of the momentum. Therefore, kinetic energy must also be quantized. Thus, the particle in a box has quantized energy levels that can be numbered with a quantum number to distinguish one level from another as we did for the Bohr atom.

The Quantum-Mechanical Atom

The properties of the atom are calculated from the Schrödinger equation just as those for the particle in a box. Quantum mechanics works; it not only accounts for all the properties of the Bohr model but also corrects most of the deficiencies of that earlier attempt.

But this success has come at a price. We no longer have an atom that is easily visualized; we can no longer imagine the electron as being a little billiard ball moving in a well-defined orbit. It is meaningless to ask "particle" questions such as how the electron gets from one place to another or how fast it will be going after 2 minutes. The electron orbits are replaced by standing waves that represent probability distributions. The best we can do is visualize the atom as an electron cloud surrounding the nucleus. This is no ordinary cloud; the density of the cloud gives the probability of locating an electron at a given point in space. The probability is highest where the cloud is the most dense and lowest where it is the least dense. An artist's version of these three-dimensional clouds is shown in Figure 24-13. (Although each drawing is confined to a finite space, the probability cloud extends to infinity, getting rapidly thinner the farther away you go from the center.)

The loss of orbiting electrons also means, however, the loss of the idea of an accelerating charge continuously radiating energy. The theory agrees with nature: atoms are stable.

As with the particle in a box, the simple act of confining the electron to the volume of the atom results in the quantization of its properties. If we allow the particle in the box to move in all three dimensions, we have standing waves in three dimensions and therefore three independent quantum numbers, one for each dimension. Similarly, the three-dimensionality of the atom yields three quantum numbers.

The particular form that these numbers take depends on the symmetry of the forces involved. In the case of atoms, the force is spherically symmetric and the three quantum numbers are associated with the energy n, the size ℓ of the angular momentum, and its direction m_ℓ.

However, these three quantum numbers were not adequate to explain all the features of the atomic spectra. The additional features could be explained by assuming that the electron spins on its axis. A fourth quantum number m_s

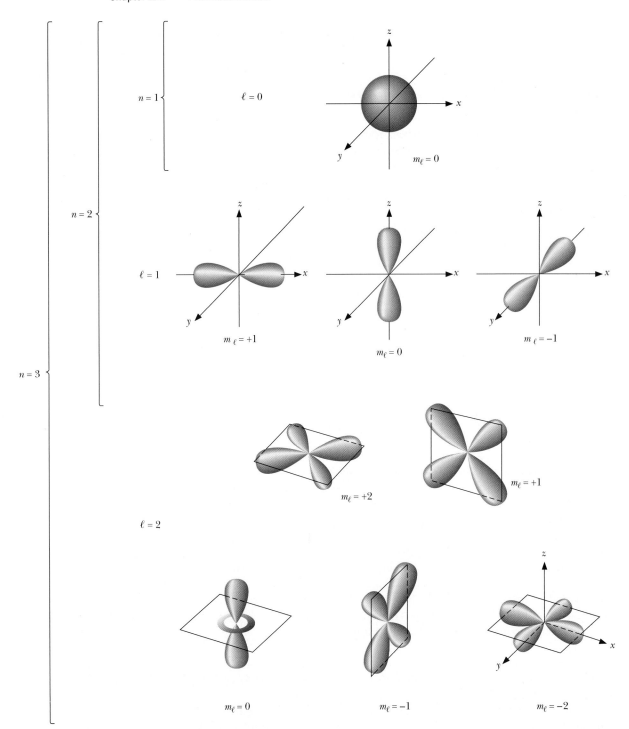

Figure 24-13 Probability clouds for a variety of electron states.

was added that gave the orientation of the electron spin. There are only two possible spin orientations, usually called "spin up" or "spin down." Although the classical idea of the electron spinning like a toy top does not carry over into quantum mechanics, the effects analogous to those of a spinning electron are accounted for with this additional quantum number. The quantum number is retained, and for convenience we use the Newtonian language of electron spin.

The most recent model of the atom combines relativity and the quantum mechanics of electrons and photons in a theory known as *quantum electrodynamics* (QED), which is even more abstract than the quantum-mechanical model. In

Psychedelic Colors

In Chapter 17 we discussed why objects have colors. The object's color depends on the colors reflected from the illuminating light, but there is an exception to this. When certain materials are illuminated with ultraviolet (UV) light, they glow brightly in a variety of colors. This process is called *fluorescence* and is responsible for the colors seen on many contemporary, psychedelic posters.

The "black lights" used to illuminate these posters radiate mostly in the UV region, which is invisible to our eyes. While the light gives off a little purple, there are certainly no reds, yellows, or greens present. And yet these bright colors appear in the posters. The atomic model explains their presence. UV photons are very energetic and can kick electrons to higher energy levels. If the electron returns by several jumps as shown in Figure A, it gives off several photons, each of which has a lower energy. When one or more of these photons lie in the visible region of the electromagnetic spectrum, the material fluoresces.

A laundry detergent manufacturer once claimed that its soap could make clothes "whiter than white" by pointing out that the soap contained fluorescent material. This makes sense scientifically because some of the UV light is converted to visible light, making the shirt give off more light in the visible region. You may wish to observe various types of clothing and laundry soap under UV light to see if they fluoresce. Do not look directly at the black light because UV light is dangerous to your eyes.

This process is also at work in fluorescent lights. The gas atoms inside the tube are excited by an electric discharge and emit UV photons. These photons cause atoms in the coating on the inside of the tube to fluoresce, giving off visible photons (Figure B).

A related phenomenon, *phosphorescence*, occurs with some materials. These materials continue to glow after the ultraviolet light is turned off. Some luminous watch dials are phosphorescent. In these materials the excited electrons remain much longer in upper energy states. In effect, phosphorescence is really a delayed fluorescence.

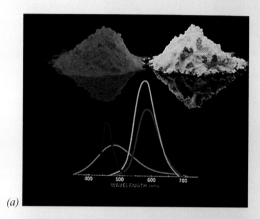
(a)

(b)

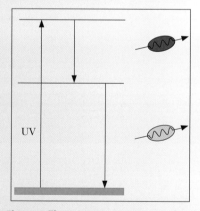

Figure A Fluorescence may occur when an electron makes several jumps in returning to its original state.

Figure B In a fluorescent lamp, a gas discharge produces ultraviolet light that is converted into visible light by a phosphor coating on the inside of the tube. (a) The two-component phosphor of praseodymium and yttrium fluorides produces the visible soft violet-pink light flooding the face of the researcher (b).

this theory, however, the concept of electron spin is no longer just an add-on but is a natural result of the combination of quantum mechanics and relativity.

The Exclusion Principle and the Periodic Table

Let's return to the periodicity of the chemical elements that we discussed in Chapter 23. The periodicity required that we think of electrons existing in shells, but the theory did not tell us how many electrons could occupy each shell. The introduction of the quantum numbers said that electrons could exist only in certain discrete states and that these states formed shells, but it did not say how many electrons could exist in each state.

In 1924 Wolfgang Pauli suggested that no two electrons can be in the same state; that is, no two electrons can have the same set of quantum numbers. This statement is now known as the Pauli **exclusion principle**. When this principle is applied to the quantum numbers obtained from Schrödinger's equation and the electron spin, the periodicity of the elements is explained, as we will demonstrate.

exclusion principle ▶

Table 24-2 gives the first two quantum numbers for the electrons in atoms of the first 30 elements. The values of the angular momentum quantum number ℓ are restricted by Schrödinger's equation to be integers in the range from 0 up to $n - 1$, and the values for the direction of the angular momentum m_ℓ are all integers from $-\ell$ to $+\ell$. Unlike the Bohr model, the angular momentum of the lowest energy state is zero, further supporting the notion that we cannot expect these atomic particles to act classically. For each value of n, ℓ, and m_ℓ, 2 spin states are available: spin up and spin down.

The good news is that the relationships between the quantum numbers explain why the orbital shells have different capacities, and this in turn explains the properties of the elements in the periodic table shown on the inside front cover of this text. The $n = 1$ state has only one angular momentum state and 2 spin states, so its maximum capacity is two electrons, both with $n = 1$ and $\ell = 0$ but with dif-

Wolfgang Pauli

Table 24-2 | Ground-State Quantum Numbers for the First 30 Elements

Atomic Number	Element	$n = 1$ $\ell = 0$	2 0	2 1	3 0	3 1	3 2	4 0
1	hydrogen (H)	1						
2	helium (He)	2						
3	lithium (Li)	2	1					
4	beryllium (Be)	2	2					
5	boron (B)	2	2	1				
6	carbon (C)	2	2	2				
7	nitrogen (N)	2	2	3				
8	oxygen (O)	2	2	4				
9	fluorine (F)	2	2	5				
10	neon (Ne)	2	2	6				
11	sodium (Na)	2	2	6	1			
12	magnesium (Mg)	2	2	6	2			
13	aluminum (Al)	2	2	6	2	1		
14	silicon (Si)	2	2	6	2	2		
15	phosphorus (P)	2	2	6	2	3		
16	sulfur (S)	2	2	6	2	4		
17	chlorine (Cl)	2	2	6	2	5		
18	argon (Ar)	2	2	6	2	6		
19	potassium (K)	2	2	6	2	6	0	1
20	calcium (Ca)	2	2	6	2	6	0	2
21	scandium (Sc)	2	2	6	2	6	1	2
22	titanium (Ti)	2	2	6	2	6	2	2
23	vanadium (V)	2	2	6	2	6	3	2
24	chromium (Cr)	2	2	6	2	6	5	1
25	manganese (Mn)	2	2	6	2	6	5	2
26	iron (Fe)	2	2	6	2	6	6	2
27	cobalt (Co)	2	2	6	2	6	7	2
28	nickel (Ni)	2	2	6	2	6	8	2
29	copper (Cu)	2	2	6	2	6	10	1
30	zinc (Zn)	2	2	6	2	6	10	2

ferent spins. The first element, hydrogen, has one electron in the lowest energy state, whereas helium, the second element, has both of these states occupied. Because there are no more $n = 1$ states available and the Pauli exclusion principle does not allow two electrons to fill any one state, this completes the first shell.

The next two electrons go into the states with $n = 2$ and $\ell = 0$ because these states are slightly lower in energy than the $n = 2$ and $\ell = 1$ states. This takes care of lithium (element 3) and beryllium (4). There are 6 states with $n = 2$ and $\ell = 1$ because m_ℓ can take on three values ($-1, 0, 1$), and each of these can be occupied by two electrons, one with spin up and the other with spin down. These 6 states correspond to the next six elements in the periodic table, ending with neon (element 10). This completes the second shell, and the completed shell accounts for neon being a noble gas.

The electrons in the next eight elements occupy the states with $n = 3$ and $\ell = 0$ or $\ell = 1$. This completes the third shell, ending with the noble gas argon (18). The 2 states with $n = 4$ and $\ell = 0$ are lower in energy than the rest of the $n = 3$ states and are filled next. These correspond to potassium (19) and calcium (20). Then the remaining 10 states corresponding to $n = 3$ and $\ell = 2$ are filled, yielding the transition elements scandium (21) through zinc (30). Thus, the quantum-mechanical picture of the atom accounts for the observed periodicity of the elements.

Question There are 18 states with $n = 3$. How many are there with $n = 4$?

Answer With $n = 4$ we can now have $\ell = 3$ in addition to the other values possible for $n = 3$. This gives us seven values of m_ℓ ranging from -3 to $+3$. Because each of these has two spin states, there are 14 additional states for a total of 32.

The Uncertainty Principle   ✓ MATH

The interpretation that atomic particles are governed by probability left many scientists dissatisfied and hoping for some ingenious thinker to rescue them from this foolish predicament. German physicist Werner Heisenberg showed that there was no rescue. He argued that there is a fundamental limit to our knowledge of the atomic world.

Heisenberg's idea—that there is an indeterminacy of knowledge—is often misinterpreted. The uncertainty is not due to a lack of familiarity with the topic nor is it due to an inability to collect the required data, such as the data needed to predict the outcome of a throw of dice or the Kentucky Derby. Heisenberg was proposing a more fundamental uncertainty—one that results from the wave–particle duality.

Imagine the following thought experiment. Suppose you try to locate an electron in a room void of other particles. To locate the electron, you need something to carry information from the electron to your eyes. Suppose you use photons from a dim lightbulb. Because this is a thought experiment, we can also assume that you have a microscope so sensitive that you will see a single photon that bounces off the electron and enters the microscope (Figure 24-14).

You begin by using a bulb that emits low-energy photons. These have low frequencies and long wavelengths. The low energy means that the photon will not disturb the electron very much when it bounces off it. However, the long wavelength means that there will be lots of diffraction when the photon scatters from the electron (Chapter 19). Therefore, you won't be able to determine the location of the electron very precisely.

To improve your ability to locate the electron, you now choose a bulb that emits more energetic photons. The shorter wavelength allows you to determine the electron's position relatively well. But the photon kicks the electron so hard

Werner Heisenberg

Figure 24-14 Heisenberg's thought experiment to determine the location of an electron.

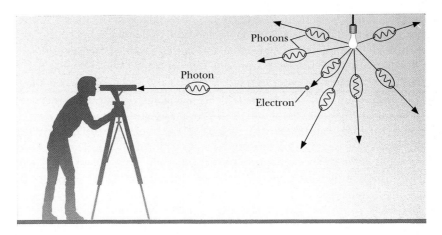

that you don't know where the electron is going next. The smaller the wavelength, the better you can locate the electron but the more the photon alters the electron's path.

Heisenberg argued that we cannot make any measurements on a system of atomic entities without affecting the system in this way. The more precise our measurements, the more we disturb the system. Furthermore, he argued, the measured and disturbed quantities come in pairs. The more precisely we determine one half of the pair, the more we disturb the other. In other words, the more *certain* we are about the value of one, the more *uncertain* we are about the value of the other. This is the essence of the uncertainty principle.

Two of these paired quantities are the position and momentum along a given direction. (Recall from Chapter 6 that the momentum for a particle is equal to its mass multiplied by its velocity.) As we saw in the thought experiment described above, the more certain your knowledge of the position, the more uncertain your knowledge of the momentum. The converse is also true.

This idea is now known as Heisenberg's **uncertainty principle**. Mathematically, it says that the product of the uncertainties of these pairs has a lower limit equal to Planck's constant. For example, the uncertainty of the position along the vertical direction Δy multiplied by the uncertainty of the component of the momentum along the vertical direction Δp_y must always be greater than Planck's constant h:

◀ uncertainty principle

$$\Delta p_y \, \Delta y > h$$

This principle holds for the position and component of momentum along the same direction. It does not place any restrictions on simultaneous knowledge of the vertical position and a horizontal component of momentum.

Another pair of variables that is connected by the uncertainty principle is energy and time, $\Delta E \, \Delta t > h$. This mathematical statement tells us that the longer the time we take to determine the energy of a given state, the better we can know its value. If we must make a quick measurement, we cannot determine the energy with arbitrarily small uncertainty. Stated in another way, the energy of a stable state that lasts for a very long time is very well determined. However, if the state is unstable and exists for only a very short time, its energy must have some range of possible values given by the uncertainty principle.

Question What does the uncertainty principle say about the energy of the photons emitted when electrons in the $n = 2$ state of hydrogen atoms drop down to the ground state?

Answer Because the electrons spend a finite time in the $n = 2$ state, the energy of that state must have a spread in energy. Therefore, the photons have a spread in energy that shows up as a nonzero width of the spectral line.

Flawed Reasoning

Two students are discussing the interpretation of Heisenberg's uncertainty principle:

Cassandra: "A particle cannot have a well-defined position and a well-defined momentum at the same time, which really goes against our commonsense."

Mary: "There's nothing really strange about the uncertainty principle—we learn in lab that all measurements have some degree of uncertainty."

Which of these students really understands the uncertainty principle?

Answer Mary is mistakenly thinking of the uncertainty principle as putting a limit on our ability to measure properties that, at least in principle, have well defined values. Cassandra understands that it is more fundamental than that and really does challenge our common sense.

The Complementarity Principle

Frustrating questions emerged as physicists built a world view of nature on the atomic scale. Is there *any* underlying order? How can one contemplate things that have mutually contradictory attributes? Can we understand the dual nature of electrons and light, which sometimes behave as particles and other times as waves?

There was no known way out. Physicists had to learn to live with this wave–particle duality. A complete description of an electron or a photon requires both aspects. This idea was first stated by Bohr and is known as the **complementarity principle**.

◄ complementarity principle

The complementarity principle is closely related to the uncertainty principle. As a consequence of the uncertainty principle, we discover that being completely certain about particle aspects means that we have no knowledge about the wave aspects. For example, if we are completely certain about the position and time for a particle, the wave aspects (wavelength and frequency) have infinite uncertainties. Therefore, wave and particle aspects do not occur at the same time.

Question If you could determine which slit the electron goes through, would this have any effect on the two-slit interference pattern?

Answer It would destroy the interference pattern. Knowledge of the particle properties (that is, the path of the electron) precludes the wave properties.

"BUT YOU CAN'T GO THROUGH LIFE APPLYING HEISENBERG'S UNCERTAINTY PRINCIPLE TO EVERYTHING."

The yin-yang symbol reflects the complementarity of many things.

The idea that opposites are components of a whole is not new. The ancient Eastern cultures incorporated this notion as part of their world view. The most common example is the yin-yang symbol of T'ai Chi Tu. Later in life, Bohr was so attracted to this idea that he wrote many essays on the existence of complementarity in many modes of life. In 1947, when he was knighted for his work in physics, he chose the yin-yang symbol for his coat of arms.

Determinism

Classical Newtonian mechanics and the newer quantum mechanics have been sources of much debate about the role of cause and effect in the natural world. With Newton's laws of motion came the idea that specifying the position and momentum of a particle and the forces acting on it allowed the calculation of its future motion. Everything was determined. It was as if the Universe was an enormous machine. This idea was known as the *mechanistic view*. In the 17th century, René Descartes stated, "I do not recognize any difference between the machines that artisans make and the different bodies that nature alone composes."

These ideas were so successful in explaining the motions in nature that they were extended into other areas. Because the Universe is made of particles whose futures are predetermined, it was suggested that the motion of the entire Universe must be predetermined. This notion was even extended to living organisms. Although the flight of a bumblebee seems random, its choices of which flowers to visit are determined by the motion of the particles that make up the bee. These generalizations caused severe problems with the idea of free will—that humans had something to say about the future course of events.

There were, however, some practical problems with actually predicting the future in classical physics. Because measurements could not be made with absolute precision, the position and momentum of an object could not be known exactly. The uncertainties in these measurements would lead to uncertainties about the calculations of future motions. However, at least in principle, certainty was possible. It was also impossible to measure the positions and momenta of all atomic particles in a small sample of gas, let alone all those in the Universe. However, the motion of each atom was predetermined, and therefore the properties of gases were predetermined. Even though humans could not determine the paths, nature knew them. The future was predetermined.

With the advent of quantum mechanics, the future became a statistical issue. The uncertainty principle stated that it was impossible *even in principle* to measure simultaneously the position and momentum of a particle. The mechanistic laws of motion were replaced by an equation for calculating the matter waves of a system that gave only *probabilities* about future events. Even if we know the state of a system at some time, the laws of quantum mechanics do not

Einstein felt that the probabilistic nature of modern physics did not reflect the reality of nature.

Is the flight of the bumblebee predetermined or a result of free will?

permit the calculation of a future, only the probabilities for each of many possible futures. The future is no longer considered to be predetermined but is left to chance.

One of the main opponents of this probabilistic interpretation was Albert Einstein. His objections did not arise out of a lack of understanding. He understood quantum mechanics very well and even contributed to its interpretation. He believed that the path of an electron was governed by some (hidden) deterministic set of rules (like an atomic version of Newton's laws), not by some unmeasurable probability wave. The new physics didn't fit into his philosophy of the natural world. Einstein's famous rebellious quote is "I, at any rate, am convinced that [God] is not playing at dice."

But nobody has ever found those hidden rules, and quantum mechanics continues to work better than anything else that has been proposed. The point is that hoping doesn't change the physics world view. Einstein spent a lot of time trying to disprove the very theory that he helped begin. He did not succeed. New work has shown that quantum mechanics is a complete theory, proving that there are no hidden variables.

Lasers

The understanding of the quantized energy levels in atoms and the realization that transitions between these levels involved the absorption and emission of photons led to the development of a new device that produced a special beam of light. Assume that we have a gas of excited atoms. Further assume that an electron drops from the $n = 3$ to the $n = 2$ level in the energy diagram in Figure 24-15(a). A photon is emitted in some random direction. It could escape the gas, or it could interact with another atom with an electron in one of two ways. It could be absorbed by an atom with an electron in the $n = 2$ level, exciting the electron to the $n = 3$ level. This excited atom would then emit another photon at a random time in a random direction. On the other hand, the original photon could stimulate an electron in the $n = 3$ level to drop to the $n = 2$ level, causing the emission of another photon (Figure 24-15[c]). This latter process is known as **stimulated emission**. Moreover, this new photon does not come out randomly. It has the same energy, the same direction, and the same phase as the incident photon; that is, the two photons are coherent. These photons can then stimulate the emission of further photons, producing a coherent beam of light.

The device that produces coherent beams of light is called a **laser**, which is the acronym derived from *l*ight *a*mplification by *s*timulated *e*mission of *r*adiation. A laser produces a beam of light that is very different from that emitted by an ordinary light source such as a flashlight. The laser beam has a very, very narrow range of wavelengths, is highly directional, and can be quite powerful. The laser beam is a single color because all the photons have the same energy. For instance, the helium–neon laser usually has a wavelength of 632.8 nanometers. The beam is highly directional because the stimulated photons move in the same direction as those doing the stimulating. The fact that all the photons have the same phase means that the amplitude of the resulting electromagnetic wave is very large. The intensity of a collection of coherent photons is obtained by adding the amplitudes and then squaring the sum. The intensity of a collection of incoherent photons is obtained by squaring the amplitudes and then adding these squares. The difference can be illustrated by considering a collection of five photons. In the laser beam, we have $(2 + 2 + 2 + 2 + 2)^2 = 100$, whereas in the ordinary light beam we have $(2^2 + 2^2 + 2^2 + 2^2 + 2^2) = 20$. The effect is even more drastic for the very large number of photons in a laser beam. These factors combine to allow us to clearly see a 1-milliwatt laser beam shining on the surface of a 100-watt lightbulb.

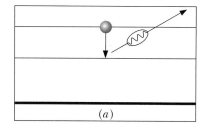
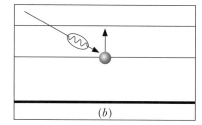
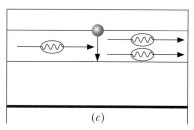

Figure 24-15 (a) Spontaneous emission. (b) Absorption. (c) Stimulated emission.

Light from an argon laser is carried by a fiber-optic cable to perform surgery deep within an ear.

PHYSICS | ON YOUR OWN

Shine the light from a small laser pointer (about 1 milliwatt of power) onto the surface of an incandescent lightbulb (60- or 100-watt). Can you see the laser beam clearly even though the power of the laser beam is much less than that of the lightbulb?

Making a working laser was more complicated than the above description implies. A method had to be found for "building" a beam of many, many photons. This was done by putting mirrors at each end of the laser tube to amplify the beam by passing it back and forth through the gas of excited atoms. One of the mirrors was only partially silvered, so that a small part of the beam was allowed to escape (Figure 24-16).

The construction of a laser was difficult because a photon is just as likely to be absorbed by an atom with an electron in the lower level as it is to cause stimulated emission of an electron in an excited level. Usually, most of the atoms are in the lower energy state, so most of the photons are absorbed and only a few cause stimulated emission. Therefore, building a laser depends on developing a *population inversion*, a situation in which there are many more electrons in the excited state than in the lower energy state. This is usually done by exciting the atom's electrons into a *metastable state* that decays into an unpopulated energy state. A metastable state is one in which the electrons remain for a long time. The electrons can be excited by using a flash of light, an electric discharge, or collisions with other atoms.

Lasers have a wide range of uses, from surveying to surgery. In surveying the light beam defines a straight line, and by pulsing the beam, the time for a round-trip can be used to measure distances. This same method is used on a gigantic scale to determine the distance to the Moon to an accuracy of a few centimeters. A very short pulse (less than 1 nanosecond long) of laser light is sent through a telescope toward the Moon's surface. The beam is so well collimated that it spreads out over an area only a few kilometers in diameter. The Apollo astronauts left a panel of retroreflectors (Chapter 17) on the surface that reflects the light back to the telescope, allowing the round-trip time to be measured. These measurements serve as a test of the validity of the general theory of relativity because the theory predicts the detailed orbit of the Moon.

Laser "knives" are used in optometry and surgery. Two leading causes of blindness are glaucoma and diabetes. In treating glaucoma a laser beam "drills" a small hole to relieve the high pressure that builds up in the eye. One of the complications of diabetes is the weakening of the walls of blood vessels in the eye to the point where they leak. The laser can be shined into the eye to cause coagulation to stop the bleeding. The energy in laser beams can also be used to

A laser-surveying instrument used to ensure proper placement of underground sewer pipes.

Figure 24-16 A schematic of a simple laser.

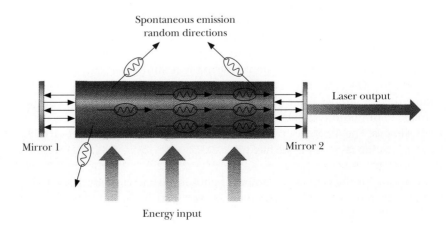

weld detached retinas to the back of the eye. In laser surgery the beam coagulates the blood as it slices through the tissue, greatly reducing bleeding. Laser beams can also be directed by fiber optics through tiny incisions to locations that would otherwise require major incisions and long healing times.

Lasers also have widespread use in the marketplace. Laser beams read the audio and video information stored in the pits on CDs and DVDs. Likewise, the lasers at store checkout counters read the bar codes on the product identification labels, allowing the computer to print out a short description of the object and its current price. This has greatly reduced billing errors as well as the time required to check out.

Lasers have made practical holography possible (Chapter 19) and opened up a whole new research tool in holographic measurements.

Summary

The Bohr model was successful in accounting for many atomic observations, especially the emission and absorption spectra for hydrogen and the ordering of the chemical elements. However, it failed to explain the details of some processes and could not give quantitative results for multielectron atoms. It described the general features of the periodic table but could not provide the details of the shells or explain how many electrons could occupy each shell. Primary among the Bohr model's failures was the lack of intuitive reasons for its postulates. Finally, it was nonrelativistic.

The replacement of Bohr's model of the atom began with de Broglie's revolutionary idea that electrons behave like waves. The de Broglie wavelength of the electron is inversely proportional to its momentum. The primary consequence of this wave behavior is that confined atomic particles form standing-wave patterns that quantize their properties.

Although successful, this new understanding led to a wave–particle dilemma for electrons and other atomic particles: they are always detected as single particles, and yet collectively they produce wavelike distributions. The behavior of these particles is governed by a quantum-mechanical wave equation that provides all possible information about the particles. For example, the probability of finding the particle at a given location is determined by the square of its matter-wave amplitude. As a consequence, atomic particles do not follow well-defined paths as classical particles do. This quantum-mechanical view of physics replaced the classical world view containing Newton's laws and Maxwell's equations.

There are four quantum numbers associated with an electron bound in an atom: the energy n, the size ℓ and direction m_ℓ of the angular momentum, and the orientation m_s of the spin. The periodicity of the chemical elements is explained by the existence of shells of varying capacity determined by the constraints of quantum mechanics and the Pauli exclusion principle. This principle states that no two electrons can have the same set of quantum numbers.

According to the complementarity principle, all atomic entities require both wave and particle aspects for a complete description, but these cannot appear at the same time. These ideas also led to Heisenberg's uncertainty principle, a statement that there is an indeterminacy of knowledge that results from the wave–particle duality. Mathematically, the product of the uncertainties of paired quantities has a lower limit equal to Planck's constant.

The understanding of the quantified energy levels in atoms led to the development of lasers, which produce coherent beams of light through the stimulated emission of radiation by electrons in excited, metastable states.

Physics Now™ Assess your understanding of this chapter's topics with sample tests and other resources found by logging into PhysicsNow at http://physics.brookscole.com/kf6e.

Chapter 24 Revisited

Laser light is special for a number of reasons. In many of the applications mentioned in the opening question, the laser light's key attribute is that it is (nearly) unidirectional, allowing a very concentrated beam. With proper optical focusing, this beam can be concentrated to a very small spot, drastically increasing the intensity—the amount of energy per unit area—striking a surface. In holography the most important attribute is its coherence, allowing a beam of light to be split and recombined while preserving the phase information.

KEY TERMS

complementarity principle: The idea that a complete description of an atomic entity such as an electron or a photon requires both a particle description and a wave description, but not at the same time.

exclusion principle: No two electrons can have the same set of quantum numbers.

laser: A device that uses stimulated emission to produce a coherent beam of electromagnetic radiation. Laser is the acronym from *l*ight *a*mplification by *s*timulated *e*mission of *r*adiation.

matter-wave amplitude: The wave solution to Schrödinger's equation for atomic and subatomic particles. The square of the matter-wave amplitude gives the probability of finding the particle at a particular location.

quantum mechanics: The rules for the behavior of particles at the atomic and subatomic levels.

stimulated emission: The emission of a photon from an atom due to the presence of an incident photon. The emitted photon has the same energy, direction, and phase as the incident photon.

uncertainty principle: The product of the uncertainty in the position of a particle along a certain direction and the uncertainty in the momentum along this same direction must be greater than Planck's constant; $\Delta p_x \Delta x > h$. A similar relationship applies to the uncertainties in energy and time.

CONCEPTUAL QUESTIONS

1. Make a list summarizing the successes and failures of the Bohr theory.
2. The theory of special relativity requires that time, distance, and energy be treated in a new way as a particle approaches the speed of light. Because these corrections were understood at the time that Bohr developed his model of the hydrogen atom, could he have included them and achieved better agreement with experimental results? Why or why not?
3. You find that the lowest frequency at which you can set up a standing wave in a wire loop, as shown in Figure 24-1, is 10 hertz. When you increase the driving frequency slightly above 10 hertz, the resonance goes away. What is the next frequency at which resonance will again appear? Explain.
4. A 48-centimeter-long wire loop is used to demonstrate standing waves, as shown in Figure 24-1. What are the three longest wavelengths that will produce standing waves?
5. One might be tempted to interpret the de Broglie wave of an electron as a modified orbital path around the nucleus in which the electron deviates up and down from a circular path as it orbits. Does this model overcome the difficulties of Bohr's model? Explain.
6. For standing waves on a guitar string, adjacent antinodes are always moving in opposite directions. Use this principle to explain why a standing-wave pattern with three antinodes cannot exist on a wire loop.
7. Your friend claims that light is a wave. What experimental evidence could you cite to demonstrate that light behaves like a particle?
8. Your friend claims that electrons are particles. What experimental evidence could you cite to demonstrate that electrons behave like waves?
9. In Chapter 10 we found that an infinite amount of energy is required to accelerate a massive particle to the speed of light. What does this imply about the mass of a photon?
10. De Broglie argued that his relationship between wavelength and momentum should apply to photons as well as to electrons. For massless particles, the relationship takes the form $p = h/\lambda$. Which has more momentum, a red photon or a blue photon? Explain.
11. When applied to photons, the de Broglie relationship $p = h/\lambda$ shows that mass is not required for a particle to have momentum, which disagrees with our classical definition for particles. What other quantity that classically depends on mass can also be attributed to a photon?
12. Which of the following technical terms can be used to describe both an electron and a photon: wavelength, velocity, mass, energy, or momentum? Explain.
13. Why is it not correct to say that an electron is a particle that sometimes behaves like a wave and light is a wave that sometimes acts like a particle?
14. Why do you think that the particle nature of the electron was discovered before its wave nature?

15. The wavelength of red light is 600 nanometers. An electron with a speed of 1.2 kilometers per second has the same wavelength. Will the electron look red? Explain.
16. An electron and a proton have the same speeds. Which has the longer wavelength? Why?
17. Bohr could never really explain why an electron was limited to certain orbits. How did de Broglie explain this?
18. What do standing waves have to do with atoms?
19. Why is the wave behavior of bowling balls not observed?
20. Why doesn't a sports car diffract off the road when it is driven through a tunnel?

21. When we perform the two-slit experiment with electrons, do the electrons behave like particles, waves, or both? What if we perform the experiment with photons?
22. Two students are discussing what happens when you turn down the rate at which electrons are fired at two slits. Student 1 claims, "Because you still get an interference pattern even with only one electron at a time, each electron must interfere with itself. As weird as it sounds, each electron must be going through both slits." Student 2 counters, "That's crazy. I can't be at class and on the ski slope at the same time. Each electron must pass through only one slit." Which student is correct? Explain.
23. What are the differences between an electron and a photon?
24. What are the similarities between an electron and a photon?
25. If the two-slit experiment is performed with a beam of electrons so weak that only one electron passes through the apparatus at a time, what kind of pattern would you expect to obtain on the detecting screen?
26. In the two-slit experiment with photons, what type of pattern do you expect to obtain if you turn the light source down so low that only one photon is in the apparatus at a time?
27. Two lightbulbs shine on a distant wall. To obtain the brightness of the light, do we add the displacements or the intensities? Explain.
28. Two coherent sources of light shine on a distant screen. If we want to calculate the intensity of the light at positions on the screen, do we add the displacements of the two waves and then square the result or do we square the displacements and then add them? Explain.
29. What meaning do we give to the square of the matter-wave amplitude?
30. The waves that we studied in Chapters 15 and 16 could be classified as either longitudinal or transverse. Why does this classification scheme have no meaning when applied to de Broglie's matter waves?
31. Where would you most likely find an electron in the lowest energy state for a one-dimensional box, as shown in Figure 24-11?
32. Where would you most likely find an electron in the first excited state for a one-dimensional box, as shown in Figure 24-12?
33. In discussing the first excited state of a particle in a one-dimensional box, your classmate claims, "The electron has to get from one region of high probability to the other without spending much time in-between. It must be accelerating, so it must radiate energy." What is wrong with this reasoning?
34. Why doesn't the quantum-mechanical model of the atom have the problem of accelerating charges emitting electromagnetic radiation?
35. Where would you most likely find the electron if it is in a quantum state with $n = 2$, $\ell = 1$, and $m_\ell = +1$, as shown in Figure 24-13?
36. Where would you most likely find the electron if it is in a quantum state with $n = 3$, $\ell = 2$, and $m_\ell = 0$, as shown in Figure 24-13?
37. How many electrons can have the quantum numbers $n = 4$ and $\ell = 2$?
38. How many electrons can have the quantum numbers $n = 4$ and $\ell = 3$?
39. Make a list showing the quantum numbers for each of the four electrons in the beryllium atom when it is in its lowest energy state.
40. Make a list showing the quantum numbers for each of the 10 electrons in the neon atom when it is in its lowest energy state.
41. Your friend argues: "The Heisenberg uncertainty principle shows that we can never be certain about anything. If this is true, we shouldn't believe anything that science teaches." How do you respond?
42. Like light, electrons exhibit diffraction when passed through a single slit. Use the Heisenberg uncertainty principle to explain why narrowing the slit (that is, improving the knowledge of the electron's position in a direction perpendicular to the beam) causes the diffraction pattern to get wider.
43. Explain why the Heisenberg uncertainty principle does not put any restrictions on our simultaneous knowledge of a particle's momentum and its energy.
44. According to the uncertainty principle, why does a tennis ball appear to have a definite position and velocity but an electron does not?
45. De Broglie's relationship gives the wavelength for an electron of given momentum. If we have some idea where the electron is, what does the uncertainty principle tell us about the electron's wavelength?

46. In principle you can balance a pencil with its center of mass directly above its point, even if this point is perfectly sharp. Explain why the uncertainty principle makes this feat impossible.
47. Explain why Bohr's model of the atom is not compatible with the uncertainty principle.
▲ 48. What does the uncertainty principle say about the energy of an excited state of hydrogen that exists only for a short time? How will this affect the emission spectrum?
49. Einstein once complained, "The quantum mechanics is very imposing. But an inner voice tells me that it is still not the final truth. The theory yields much, but it hardly brings us nearer to the secret of the Old One. In any case, I am convinced that He does not throw dice." What about quantum mechanics was troubling him?
50. If the Bohr model of the atom has been replaced by the newer quantum-mechanical models, why do we still teach the Bohr model?
51. What quantum-mechanical variable is complementary to energy?
52. What quantum-mechanical variable is complementary to momentum?
▲ 53. If we are certain about the energy of a particle, what does the Heisenberg uncertainty principle say about the uncertainty in its frequency?
▲ 54. If we are certain about the position of a particle, what does the Heisenberg uncertainty principle say about the uncertainty in its wavelength?

55. Why do some minerals glow when they are illuminated with ultraviolet light?
56. When ultraviolet light from a "black light" shines on crayons, visible light is emitted. How can you account for this?

57. Phosphorescent materials continue to glow after the lights are turned off. How can you use the model of the atom to explain this?
58. Can infrared rays cause fluorescence? Why or why not?
59. What is stimulated emission?
60. How does light from a laser differ from light emitted by an ordinary lightbulb?

EXERCISES

1. What is the de Broglie wavelength of a Volkswagen Beetle (mass = 1260 kg) traveling at 30 m/s (67 mph)?
2. A bullet for a 30-06 rifle has a mass of 10 g and a muzzle velocity of 900 m/s. What is its wavelength?
3. Nitrogen molecules (mass = 4.6×10^{-26} kg) in room-temperature air have an average speed of about 500 m/s. What is a typical wavelength for these nitrogen molecules?
4. What is the de Broglie wavelength for an electron traveling at 500 m/s?
5. What is the wavelength for a photon with energy 4 eV?
6. What is the wavelength for an electron with energy 4 eV?
7. What speed would an electron need to have a wavelength equal to the diameter of a hydrogen atom (10^{-10} m)?
8. What is the speed of a proton with a wavelength of 2 nm?
9. What is the size of the momentum for an electron that is in the lowest energy state for a one-dimensional box whose length is 2 nm?
10. What is the size of the momentum for an electron that is in the first excited energy state for a one-dimensional box whose length is 2 nm?
11. A child runs straight through a door with a width of 0.75 m. What is the uncertainty in the momentum of the child perpendicular to the child's path?
12. A Ford Escort passes through a tunnel with a width of 10 m. What uncertainty is introduced in the Escort's momentum perpendicular to the highway?
13. A proton passes through a slit that has a width of 10^{-10} m. What uncertainty does this introduce in the momentum of the proton at right angles to the slit?
14. Repeat Exercise 13 for an electron.
15. What is the minimum uncertainty in the position along the highway of a Ford Escort (mass = 1150 kg) traveling at 20 m/s (45 mph)? Assume that the uncertainty in the momentum is equal to 1% of the momentum.
16. What is the uncertainty in the momentum of an Acura NSX with a mass of 1200 kg parked by the curb? Assume that you know the location of the car with an uncertainty of 1 cm.

17. What is the uncertainty in the location of a proton along its path when it has a speed equal to 0.1% the speed of light? Assume that the uncertainty in the momentum is 1% of the momentum.
18. What is the uncertainty in each component of the mo-

mentum of an electron confined to a box approximately the size of a hydrogen atom, say, 0.1 nm on a side?

19. Consider an electron confined to a diameter of 0.1 nm, about the size of a hydrogen atom. If the electron's speed is on the order of the uncertainty in its speed, approximately how fast is it traveling? Assuming that the electron can still be treated without making relativistic corrections, find its kinetic energy (in electron volts).

20. Repeat Exercise 19 for a proton confined to a diameter of 10^{-14} m, about the size of a nucleus.

21. Electrons in an excited state decay to the ground state with the release of a photon. The uncertainty in the time that an electron spends in the excited state can be estimated by the average time electrons spend in that state. What is the spread in energy, in electron volts, of the photons from a state with a lifetime of 2×10^{-8} s?

22. The lifetime of an excited nuclear state is 5×10^{-12} s. What will be the spread in energy, in electron volts, of the photons from this state?

23. If the photons from an excited atomic state show a spread in energy of 2×10^{-4} eV, what is the lifetime (see Exercise 21) of the state?

24. The uncertainty principle allows for "violation" of conservation of energy for times less than the associated uncertainty. For how long could an electron increase its energy by 10 eV without violating conservation of energy?

Interlude

The giant Super Kamiokande neutrino detector is buried 1 kilometer under a mountain near Tokyo, Japan. Physicists are cleaning the faces of the photomultipliers before they are submerged.

The Subatomic World

The quantum-mechanical model of the atom has been very successful. From a large collection of seemingly unrelated phenomena has come a rather complete picture of the atom. This development has created a very profound change in our physics world view.

Our study of the physical world now goes one "layer" deeper, to that of the atomic nucleus. The search for the structure of the nucleus led to the discovery of new forces in nature, forces much more complicated than those of gravity and electromagnetism. It is not possible to write simple expressions for the nuclear forces like the one we wrote for gravity; the nuclear forces depend on much more than distance and mass. This discovery radically changed our basic concepts about forces. The models for the interactions between particles that began with Newton's action at a distance and progressed to the field concept in Maxwell's time now involve the exchange of fundamental particles.

Because the nuclear force is so strong, the energy associated with it is very large. The power of nuclear energy became evident with the explosion of nuclear bombs over Japan in World War II. The release of this energy in the form of bombs or nuclear power plants poses serious questions not only for scientists and engineers but also for our entire society.

In our search for the ultimate structure of all matter, we have discovered that atoms are not the most basic building blocks in nature. They are composed of electrons and very small, very dense nuclei. The nuclei are in turn composed of particles. These were originally believed to be electrons and protons. This notion was appealing because it required only three elemental building blocks: the electron, the proton, and the photon. Later, the neutron replaced the electron in the nucleus, but the overall picture for the fundamental structure was still appealing.

> The search for the structure of the nucleus led to the discovery of new forces in nature, forces much more complicated than those of gravity and electromagnetism.

But as scientists delved deeper and deeper into the subatomic world, new particles emerged. The number of "elementary" particles grew to more than 300! In current theories, most of these are composed of a small number of more basic particles called quarks.

The search for the elusive elementary particles is analogous to a child playing with a set of nested eggs. Each egg is opened only to reveal another egg. However, the child eventually reaches the end of the eggs. In one version the smallest egg contains a bunny. Is there an end to the search for the elementary particles? Do quarks represent the bunny?

Are the layers of matter like the nested dolls, or is there a final set of elementary particles?

25 The Nucleus

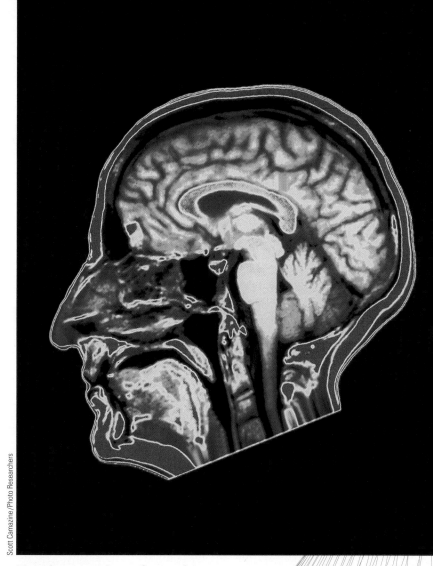

Magnetic resonance image of a normal human brain.

As our knowledge of nuclei has grown, whole new technologies have been developed to make use of this new knowledge. The new technologies range from obtaining medical images of our internal organs to determining the age of mummies or artifacts. How can nuclear techniques be used to determine the age of an artifact?

(See page 545 for the answer to this question.)

THE **nucleus** of an atom is unbelievably small. If you could line up a trillion (10^{12}) of them, they would only stretch a distance equal to the size of the period at the end of this sentence. Another way of visualizing these sizes is to imagine expanding things to more human sizes and then comparing relative sizes. If we imagine a baseball as big as Earth, the baseball's atoms would be approximately the size of grapes. Even at this scale, the nucleus would be invisible! To "see" the nucleus, we need to expand one of these grape-sized atoms until it is as big as the Superdome in New Orleans. The nucleus would be in the middle and would be about the size of a grape.

These analogies are a little risky because the quantum-mechanical view does not consider atomic entities to be classical particles. The images do, however, demonstrate the enormous amount of ingenuity required to study this submicroscopic realm of the Universe. You don't just pick up a nucleus and take it apart.

We learn about the nucleus by examining what pieces come out—either through naturally occurring radioactivity or artificially induced nuclear reactions. Studying the nucleus is more difficult than studying the atom because we already understood the electric force that held the atom together. There, the task was to find a set of rules that governs the behavior of the atom. These rules were provided by quantum mechanics. In nuclear physics the situation was turned around; quantum mechanics provided the rules, but at first there wasn't a good understanding of the forces.

The Discovery of Radioactivity

Discoveries in science are often not anticipated. Radioactivity is such a case. Radioactivity is a nuclear effect that was discovered 15 years before the discovery of the nucleus itself. Although the production of X rays is not a nuclear effect, their study led to the discovery of radioactivity. Four months after Roentgen's discovery of X rays (Chapter 23), a French scientist, Henri Becquerel, was looking for a possible symmetry in the X-ray phenomenon. Roentgen's experiment had shown that X rays striking certain salts produced visible light. Becquerel wondered if the phenomenon might be symmetric—if visible light shining on these fluorescing salts might produce X rays.

His experiment was simple. He completely covered a photographic plate with paper so that no visible light could expose it. Then he placed a fluorescing mineral on this package and took the arrangement outside into the sunlight (Figure 25-1). He reasoned that the sunlight might activate the atoms and cause them to emit X rays. The X rays would easily penetrate the paper and expose the photographic plate.

The initial results were encouraging; his plates were exposed as expected. However, during one trial, the sky was clouded over for several days. Convinced that his photographic plate would be slightly exposed, he decided to start over. But, being meticulous, he developed the plate anyway. Much to his surprise, he found that it was completely exposed. The fluorescing material—a uranium salt—apparently exposed the photographic plate without the aid of sunlight. Becquerel soon discovered that all uranium salts exposed the plates—even those that did not fluoresce with X rays.

Question What hypothesis does Becquerel's observation suggest about the origin of the radiation?

Answer Because the radiation came from all uranium salts, it is likely that the phenomenon is a property of uranium and not of the particular salt.

This new phenomenon appeared to be a special case of Roentgen's X rays. They penetrated materials, exposed photographic plates, and ionized air mole-

Figure 25-1 Becquerel's experimental setup to test whether X rays would penetrate the paper and expose the photographic plate.

cules; but unlike X rays, the new rays occurred naturally. There was no need for a cathode-ray tube, a high voltage, or even the Sun!

Becquerel did a series of experiments that showed that the strength of the radiation depended only on the amount of uranium present—either in pure form or combined with other elements to form uranium salts. Later it was realized that this was the first clue that radioactivity was a nuclear effect rather than an atomic one. Atomic properties change when elements undergo chemical reactions; this new phenomenon did not. The nuclear origin of this radiation was further supported by observations that the exposure of the photographic plates did not depend on outside physical conditions such as strong electric and magnetic fields or extreme pressures and temperatures.

Pierre Curie, a colleague of Becquerel, and Marie Sklodowska Curie developed a way of quantitatively measuring the amount of radioactivity in a sample of material. Marie Curie then discovered that the element thorium was also radioactive. As with uranium, the amount of radiation depended on the amount of thorium and not on the particular thorium compound. It was also not affected by external physical conditions.

The Curies found that the amount of radioactivity present in pitchblende, an ore containing a large percentage of uranium oxide, was much larger than expected from the amount of uranium present in the ore. They then attempted to isolate the source of this intense radiation, a task that turned out to be difficult and time-consuming. Although they were initially unable to isolate the new substance, they did obtain a sample that was highly concentrated. The new element was named polonium after Marie Curie's native Poland. Further work led to the discovery of another highly radioactive element, radium. A gram of radium emits more than a million times the radiation of a gram of uranium. A sample of radium generates energy at a rate that keeps it hotter than its surroundings. This mysterious energy source momentarily created doubt about the validity of the law of conservation of energy.

Flawed Reasoning

A question on the midterm exam asks: "If radium (which is radioactive) and chlorine (which is not radioactive) combine to form radium chloride, is the compound radioactive?"

Kay gives the following answer: "Not necessarily. We found that the reactive metal sodium combines with the poisonous gas chlorine to form common table salt. A compound can have very different properties than the elements that form it."

Kay's answer sounds very reasonable, but she is missing a very important concept. **What is it?**

Answer The *chemical* properties of elements are governed by the structure of their outer electron shells. These properties can change when compounds are formed because the outer shell structure is changed. Radioactivity, on the other hand, is independent of conditions outside the nucleus. Radium chloride is radioactive because it contains radium nuclei.

Types of Radiation

These radioactive elements were emitting two types of radiation. One type could not even penetrate a piece of paper; it was called **alpha (α) radiation** after the first letter in the Greek alphabet. The second type could travel through a meter of air or even thin metal foils; it was named **beta (β) radiation** after the second letter in the Greek alphabet.

CURIE | Eight Tons of Ore

Nothing in life is to be feared. It is only to be understood.

—Marie Sklodowska Curie

Maria Sklodowska (1867–1934) was born in Warsaw, the daughter of a rather rigid and demanding father and a brilliant, artistic mother. Well tutored at home, it was natural for her to want a university education. Warsaw was under Russian rule and higher education for women was impossible in Poland. Her sister Bronya shared her thirst for learning, and the two made a pact. Maria would work to support her older sister in the pursuit of a medical degree in Paris. With her new economic security, Bronya would then support Maria, who had long dreamed of studying at the Sorbonne.

Marie Curie

Maria taught briefly in a Warsaw school and served as a governess in eastern Poland. When she was 24, she eagerly accepted her sister and new brother-in-law's invitation to come and live with them in Paris. Because their apartment was two hours away from the Sorbonne and to save tram fare, Marie—the French version of her name—took an attic room in an ancient and unheated building. She endured bitter cold but rejoiced in her freedom to learn and to work: "It was like a new world to me, the world of science." In 1893 she took her first university degrees, receiving a first in physics and a runner-up in mathematics. Her intention was to obtain a teaching diploma and return to Poland. Fate intervened and she met a charming, shy, and handsome young professor, Pierre Curie; they were married in 1896. Pierre had no sooner finished his final degree when he convinced his new wife to pursue the same high goal.

She began research under the supervision of a distinguished committee—two of whom would be awarded Nobel Prizes. She sought to explore thoroughly the most critical aspects of the new phenomenon of radioactivity that was discovered by Henri Becquerel. After an exhaustive analysis of radiation intensity and the elements and compounds that produced it, Marie came to a brilliant conclusion: This strange energy came from the interior of the atoms.

With Pierre's assistance she began to assay possible sources of radioactive materials. They located a considerable amount of ore in an old and well-known mine at Joachimstal in Bohemia. The Curies had many tons of the ore shipped to Paris, where they literally dug, boiled, and refined this material to isolate one-tenth of a gram of the most intense radiation source—a radium salt. When she presented her thesis research in 1903, her committee noted that this was the most significant work ever submitted by a student at the Sorbonne. Later that year Henri Becquerel and Pierre and Marie Curie shared the Nobel Prize in Physics for their work with radioactivity—a word she was the first to use scientifically.

The Nobel Prize, the new and mysterious subject of radiation, and the love story of the two young scientists combined to promote the importance of the Nobel Prize and to focus an intense journalistic attention on the Curies. From that time onward, Marie—Pierre died tragically in 1906 when he was struck by a heavy, horse-drawn freight wagon—was the subject of popular interest to an extent unknown to scientists previously. Marie concentrated her incredible work ethic on the rearing of her two young daughters and discovering the ultimate element from which the intense radiation in pitchblende emanated. Her reclusive and reticent style later made Albert Einstein a better subject for story-hungry writers.

Despite taboos against women, Marie Curie's outstanding record secured her a position as the first female professor at the Sorbonne. She raised her two young daughters, Irene and Eve, in a circle of elite children of similar age. They were to be instructed only by close friends and university professors, and the standards of their scientific training were extraordinary. Irene Joliot-Curie won the Nobel Prize in Chemistry in 1935—and her husband was also a Nobel laureate. Marie was the first of three scientists to win a second Nobel Prize—this time in chemistry for her isolation of pure radium and polonium. Curium, element 96, was named in honor of Marie and Pierre Curie.

World War I shattered the European scientific community. Marie took her tremendous drive and dedication to the trenches. She set up mobile X-ray units and trained a generation of radiologists. Like Florence Nightingale a generation earlier, she took on any military authority who stood in the way of medical treatment.

She did not live quite long enough to see her daughter and son-in-law receive their Nobel Prizes. She died of leukemia on July 4, 1934. She was interred in the great Pantheon. Poland produced her, France nourished her, and the world inherits her great work.

—Pierce C. Mullen, historian and author

Sources: Rosalynd Pflaum, *Grand Obsession: Marie Curie and Her Work* (New York: Doubleday, 1989); Susan Quinn, *Marie Curie: A Life* (New York: Simon & Schuster, 1995); Eve Curie, *Madame Curie* (New York, Doubleday, 1938).

By 1900 several experimentalists had shown that beta radiation could be deflected by a magnetic field, demonstrating that it was a charged particle. These particles had negative charges and the same charge-to-mass ratio as the newly discovered electrons. Later it was concluded that these **beta particles** were in fact electrons that were emitted by nuclei.

In 1903 Rutherford showed that alpha radiation could also be deflected by magnetic fields and that these particles had a charge of +2. (The unit of charge is assumed to be the elementary charge—that is, the magnitude of the charge on the electron or proton.) Because they had a larger charge than electrons and yet were much more difficult to deflect, it was concluded that they were much more massive than electrons. Six years later Rutherford was able to show that **alpha particles** are the nuclei of helium atoms.

Figure 25-2 Alpha, beta, and gamma radiation behave differently in a magnetic field.

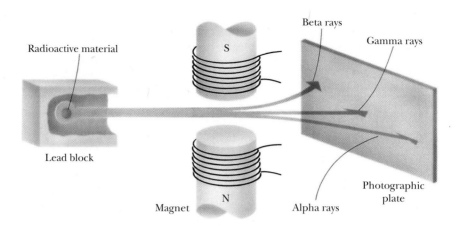

The third type of radiation was discovered in 1900. Naturally, it was named after the third letter of the Greek alphabet and is known as **gamma (γ) radiation**. This radiation has the highest penetrating power; it can travel through many meters of air or even through thick walls. Unlike the other two types of radiation, gamma radiation was unaffected by electric and magnetic fields (Figure 25-2). Gamma rays, like X rays, are now known to be very high-energy photons. Although the energy ranges associated with these labels overlap, gamma-ray photons usually have more energy than X-ray photons, which in turn have more energy than photons of visible light.

The Nucleus

The recognition that radioactivity was a nuclear phenomenon indicated that nuclei have an internal structure. Because nuclei emitted particles, it was natural to assume that nuclei were composed of particles. It was hoped there would be only a small number of different kinds of particles and that all nuclei would be combinations of these.

Rutherford's early work with alpha particles showed that he could change certain elements into others. While bombarding nitrogen with alpha particles, Rutherford discovered that his sample contained oxygen. Repeated experimentation convinced him that the oxygen was being *created* during the experiment. In addition to heralding the beginning of artificially induced transmutations of elements, this experiment led to the discovery of the proton.

Rutherford noticed that occasionally a fast-moving particle emerged from the nitrogen. These particles were much lighter than the bombarding alpha particles and were identified as **protons**. Rutherford and James Chadwick bombarded many of the light elements with alpha particles and found 10 cases in which protons were emitted. By 1919 they realized that the proton was a fundamental particle of nuclei.

The hydrogen atom has the simplest and lightest nucleus, consisting of a single proton. The next lightest element is helium. Its nucleus has approximately four times the mass of hydrogen. Because helium has two electrons, it must have two units of positive charge in its nucleus. If we assume that the helium nucleus contains two protons, we obtain the correct charge but the wrong mass—it would be only twice that of hydrogen. What accounts for the other two atomic mass units? One possibility is that the nucleus also contains two electron–proton pairs. Each pair would be neutral in charge and contribute 1 atomic mass unit. Thus, the helium nucleus might consist of four protons and two electrons.

This scheme had some appeal. It accounted for the electrons emitted from nuclei, and it could be used to "build" nuclei. However, it had several serious defects that caused it eventually to be discarded. For example, the spins of the electron and proton cause each of them to act like miniature magnets, and therefore the magnetic properties of a nucleus should be a combination of those of its protons and electrons. These magnetic values did not agree with the experimental results. The existence of electrons in the nucleus was also incompatible with the uncertainty principle (Chapter 24). If the location of the electron were known well enough to say that it was definitely in the nucleus, its momentum would be large enough to easily escape the nucleus. This left the "extra" mass unexplained.

The Discovery of Neutrons

By 1924 Rutherford and Chadwick began to suspect there was another nuclear particle. The **neutron** would have a mass about the same as the proton's but no electric charge. Chadwick received a Nobel Prize (1935) for the experimental verification of the neutron's existence in 1932. The "extra" mass in nuclei was now explained. Nuclei are combinations of protons and neutrons, as illustrated in Figure 25-3. Helium nuclei are made of 2 protons and 2 neutrons, oxygen nuclei have 8 protons and 8 neutrons, gold nuclei have 79 protons and 118 neutrons, and so on.

When it is not necessary to distinguish between neutrons and protons, they are often called **nucleons**. The number of nucleons in the nucleus essentially determines the atomic mass of the atom because the electrons' contribution to the mass is negligible. The scale of relative atomic masses has been chosen so that the atomic mass of a single atom in atomic mass units is nearly equal to the number of nucleons in the nucleus. This equivalence is attained by setting the mass of the neutral carbon atom with 12 nucleons equal to 12.0000 atomic mass units (Chapter 11). The masses of the electron, proton, and neutron are given in Table 25-1 for several different mass units. The masses of nuclei vary from 1 to about 260 atomic mass units and the corresponding radii vary from 1.2 to 7.7×10^{-15} meter.

Isotopes

While studying the electric and magnetic deflection of ionized atoms, J. J. Thomson discovered that the nuclei of neon atoms are not all the same. The neon atoms did not all bend by the same amount. Because each ion had the same charge, some neon atoms must be more massive than others. It was shown that the lighter

Table 25-1 | Masses of Electrons, Protons, and Neutrons in Kilograms, Atomic Mass Units, and Units in Which the Mass of the Electron Is 1

Particle	kg	amu	$m_e = 1$
Electron	9.109×10^{-31}	0.000 549	1
Proton	1.6726×10^{-27}	1.007 276	1836
Neutron	1.6750×10^{-27}	1.008 665	1839

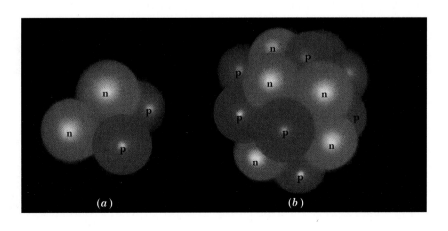

Figure 25-3 The helium nucleus (a) consists of two protons and two neutrons, and the oxygen nucleus (b) has eight protons and eight neutrons.

Figure 25-4 Hydrogen has three isotopes: (a) hydrogen, (b) deuterium, and (c) tritium.

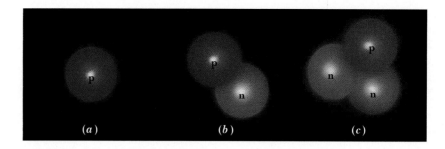

ones had masses of about 20 atomic mass units, whereas the heavier ones had masses of about 22 atomic mass units. This discovery destroyed the concept that all atoms of a given element were identical, an idea that had been accepted for centuries. We now know that all elements have nuclei that have different masses. These different nuclei are known as **isotopes** of the given element. Isotopes have the same number of protons but differ in the number of neutrons.

Question Given that neon is the 10th element, how many neutrons and protons would each of its isotopes have?

Answer Because each nucleon contributes about 1 atomic mass unit and neon must have 10 protons, the lighter isotope with 20 atomic mass units must have 10 neutrons and the heavier one with 22 atomic mass units must have 12 neutrons.

Although each isotope of an element has its own nuclear properties, the chemical characteristics of isotopes do not differ. Adding a neutron to a nucleus does not change the electric charge; therefore, the atom's electronic structure is essentially unchanged. Because the electrons govern the chemical behavior of elements, the chemistry of all isotopes of a particular element is virtually the same. Some differences, however, can be detected. There are slight changes in the electronic energy levels that show up in the atomic spectra. The extra mass of the heavier isotopes also slows the rates at which chemical and physical reactions occur.

Hydrogen has three isotopes (Figure 25-4). The most common one contains a single proton. The next most common has one proton and one neutron; it has about twice the mass of ordinary hydrogen. This "heavy" hydrogen, known as *deuterium*, is stable and occurs naturally as 1 atom out of approximately every 6000 hydrogen atoms. The third isotope has one proton and two neutrons. This "heavy, heavy" hydrogen, known as *tritium*, is radioactive.

Isotopes of other elements do not have separate names. The symbolic way of distinguishing isotopes is to write the chemical symbol for the element with two numbers added to the left of the symbol. A subscript gives the number of protons, and a superscript gives the number of nucleons—that is, the total number of neutrons and protons. The most common isotope of carbon has six protons and six neutrons and is written $^{12}_{6}C$. Because there is some redundancy here (the number of protons is already specified by the chemical symbol), the isotope is sometimes written ^{12}C and read "carbon-12."

The Alchemists' Dream

The discovery of radioactivity changed another long-held belief about atoms. If the particle emitted during radioactive decay is charged, the resulting nucleus (commonly called the **daughter nucleus**) does not have the same charge as the original (the **parent nucleus**). The parent and daughter are not the same element because the charge on the nucleus determines the number of electrons in the neutral atom and this number determines the atom's chemical properties.

The belief that atoms are permanent was wrong; they can change into other elements. We see that nature has succeeded where the alchemists failed.

Although we can't control the process of nuclear transformation, we can bombard materials as Rutherford did and change one element into another. But what kinds of reactions are possible? Can we, perhaps, change lead into gold? As scientists examined various processes that might change nuclei, they observed that the conservation laws are obeyed. These laws include the conservation of mass-energy, linear momentum, angular momentum, and charge, which we studied earlier. In addition, a new conservation law was discovered, the conservation of nucleons. Although the nucleons may be rearranged or a neutron changed into a proton, the total number of nucleons after the process is the same as before.

◀ conservation of nucleons

The reaction in which Rutherford produced oxygen (O) and a proton (p) by bombarding nitrogen (N) with alpha particles (α) can be written in the form

$$^{4}_{2}\alpha + ^{14}_{7}N \rightarrow ^{17}_{8}O + ^{1}_{1}p$$

where the arrow separates the initial nuclei on the left from the final nuclei on the right. Notice that the conservation of charge is obeyed; there are $2 + 7 = 9$ protons on the left-hand side of the arrow and $8 + 1 = 9$ protons on the right-hand side. The number of nucleons must also be conserved. There are $4 + 14 = 18$ nucleons on the left and $17 + 1 = 18$ nucleons on the right.

Neutrons are very effective at producing nuclear transformations because they do not have a positive charge and can penetrate to the nuclei, even with low energies. An example that occurs naturally in the atmosphere is the conversion of nitrogen to carbon via neutron (n) bombardment:

$$^{1}_{0}n + ^{14}_{7}N \rightarrow ^{14}_{6}C + ^{1}_{1}p$$

Let's now look at changes in the nucleus that occur naturally. For example, when a nucleus emits an alpha particle, it loses two neutrons and two protons. Therefore, the parent nucleus changes into a nucleus that is two elements lower in the periodic chart. This daughter nucleus has a nucleon number that is four less. For instance, when $^{226}_{88}Ra$ decays by alpha decay, the daughter nucleus has $88 - 2 = 86$ protons and $226 - 4 = 222$ nucleons. The periodic table printed on the inside front cover of this text tells us that the element with 86 protons is radon, which has the symbol Rn. Therefore, the daughter nucleus is $^{222}_{86}Rn$. This process can be written in symbolic form as

$$^{226}_{88}Ra \rightarrow ^{222}_{86}Rn + ^{4}_{2}\alpha$$

◀ alpha decay

Question What daughter results from the alpha decay of $^{232}_{90}Th$?

Answer The daughter will have $232 - 4 = 228$ nucleons and $90 - 2 = 88$ protons, so it must be $^{228}_{88}Ra$.

Flawed Reasoning

Two students are discussing alpha decay:

Corey: "Our teacher claims that the alchemists were right; atoms can be changed from one element to another. She said that radium becomes radon when it emits an alpha particle, but that doesn't make sense. When radium gives up two protons and two neutrons, it still has the same number of electrons. Therefore, it should still react chemically like radium."

Grant: "Well, maybe the radium atom loses two electrons after the alpha decay, so it becomes electrically neutral. I think the atom is still radium until the electrons leave."

What important principle are these students misunderstanding?

continues on next page

> **Answer** It is the number of protons, not electrons, in an atom that defines its chemical identity. A chlorine atom with an extra electron is a chlorine ion, not an argon atom. As soon as the alpha particle is ejected from the nucleus of the radium atom, the energy levels change to those of the radon atom.

The electron that is emitted in beta (β) decay is not one of those around the nucleus, nor is it one that already existed in the nucleus. The electron is ejected from the nucleus when a neutron decays into a proton. (We will study the details of this process in Chapter 27.) A new element is produced because the number of protons increases by one. The number of neutrons decreases by one, but it is the change in the proton number that causes the change in element. The daughter nucleus is one element higher in the periodic chart and has the same number of nucleons. For example,

beta minus decay ▶

$$^{24}_{11}\text{Na} \rightarrow {}^{24}_{12}\text{Mg} + {}^{0}_{-1}\beta^{-}$$

> **Question** What is the daughter of $^{228}_{89}$Ac if it undergoes beta decay?

> **Answer** The daughter must have the same number of nucleons and one more proton. Therefore, it is $^{228}_{90}$Th.

An inverse beta-decay process has also been observed. In this process an atomic electron (e) in an inner shell is captured by one of the protons in the nucleus to become a neutron. This **electron capture** decreases the number of protons by one and increases the number of neutrons by one. Therefore, the daughter nucleus belongs to the element that is one lower in the periodic table and has the same number of nucleons as the parent. For example,

electron capture ▶

$$^{7}_{4}\text{Be} + {}^{0}_{-1}\text{e} \rightarrow {}^{7}_{3}\text{Li}$$

This process is detected by observing the atomic X rays given off when the outer electrons drop down to fill the energy level vacated by the captured electron. The evidence for this transformation is very convincing; the spectral lines are characteristic of the daughter nucleus and not the parent nucleus.

> **Question** What daughter is produced when $^{234}_{93}$Np undergoes electron capture?

> **Answer** Once again, the number of nucleons does not change, only this time the number of protons decreases by one to yield $^{234}_{92}$U.

A third type of beta decay was observed in 1932. In this case the emitted particle has one unit of positive charge and a mass equal to that of an electron. This positive electron **(positron)** is the antiparticle of the electron, a concept that will be described further in Chapter 27. The emission of a positron results from the decay of a proton into a neutron. This process is called *beta plus decay* to distinguish it from *beta minus decay*, the process that produces a negative electron. Like electron capture, this decreases the number of protons by one and increases the number of neutrons by one. For example,

beta plus decay ▶

$$^{15}_{8}\text{O} \rightarrow {}^{15}_{7}\text{N} + {}^{0}_{1}\beta^{+}$$

> **Question** What is the result of $^{11}_{6}$C decaying via beta plus decay?

> **Answer** Because a proton turns into a neutron, we have $^{11}_{5}$B.

Table 25-2 | Changes in the Number of Protons, Neutrons, and Nucleons for Each Type of Radioactive Decay

Decay	Changes in the Number of		
	Protons	Neutrons	Nucleons
α	-2	-2	-4
β^-	$+1$	-1	0
Electron capture	-1	$+1$	0
β^+	-1	$+1$	0
γ	0	0	0

Nuclei that emit gamma rays do not change their identities, because gamma rays are high-energy photons and do not carry charge. The nucleus has discrete energy levels analogous to those in the atom. If the nucleus is not in the ground state, it may change to a lower energy state with the emission of the gamma ray. This process often happens after a nucleus has undergone one of the other types of decay to become an excited state of the daughter nucleus. Table 25-2 summarizes the changes produced by each type of decay.

Radioactive Decay

✓ MATH

Equal amounts of different radioactive materials do not give off radiation at the same rate. For example, 1 gram of radium emits 20 million times as much radiation per unit time as 1 gram of uranium. The **activity** of a radioactive sample is a measure of the number of decays that take place in a certain time. The unit of activity is named the **curie** (Ci) after Marie Curie and has a value of 3.7×10^{10} decays per second, which is the approximate activity of 1 gram of radium.

◄ curie

✓ Extended presentation available in the *Problem Solving* supplement

Two factors determine the activity of a sample of material. First, the activity is directly proportional to the number of radioactive atoms in the sample. Two grams of radium will have twice the activity of 1 gram. Second, the activity varies with the type of nucleus. Some nuclei are quite stable, whereas others decay in a matter of seconds and still others in millionths of a second.

Both factors can be illustrated with an analogy that emphasizes the random nature of radioactive decay. Imagine that you have 36 dice. Throwing the dice can simulate the radioactive decay process. Each throw represents a certain elapsed time. Assume that any die whose number 1 is face up represents an atom that has decayed during the last period.

Let's look at the "activity" of the sample of dice. How many dice would you expect to have 1s up after the first throw? Because each die has six faces that are equally likely to appear as the top face, there is a 1-in-6 chance of having 1s up. Because there are 36 dice in the sample, we expect that, on the average, 6 (that is, $\frac{1}{6}$ of 36) will "decay" on the first throw. This number represents the activity of the sample.

On the average, 6 of the 36 dice will have 1s up.

If you double the number of dice in your sample, how many of the 72 dice should show 1s up after the first throw? On average, 12 ($\frac{1}{6}$ of 72) will have 1s up. From this analogy we see that the level of radioactivity is directly proportional to the number of radioactive atoms in a sample.

In a sample of radioactive material, the number of nuclei of the original isotope continually decreases, which decreases its activity. To illustrate this, imagine that we once again have 36 dice in our sample. After each throw of the dice, we remove the dice with 1s up. They are no longer part of the sample because they have changed to a new element. After the first throw there will be, on the average, 6 dice that decay. Removing them reduces the sample size to 30.

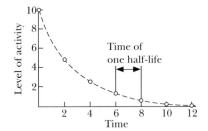

Figure 25-5 A graph of the activity of a radioactive material versus time. Notice that the time needed for the activity to decrease by a factor of 2 is always the same.

Question   On average, how many of these 30 dice do you expect to have 1s up on the next throw?

Answer   The number of 1s up will probably be $\frac{1}{6} \times 30 = 5$.

Removing the 5 dice that are expected to have 1s up on the second throw leaves a sample size of 25. Each successive throw of the dice yields a smaller sample size and consequently a lower level of activity on the next throw. The same is true of the radioactive sample. As with the dice, when a nucleus decays, it is no longer part of the sample. The activity of the radioactive sample decreases with time.

This change in the activity with time has a simple behavior due to the probabilistic nature of radioactivity. The graph in Figure 25-5 shows this behavior. Notice that the time it takes the activity to drop to one-half its value is constant throughout the decay process. This time is called the **half-life** of the sample. In 1 half-life the activity of the sample decreases by a factor of 2. During the next half-life, it decreases by another factor of 2, to $\frac{1}{4}$ of its initial value. After each additional half-life, the activity would be $\frac{1}{8}, \frac{1}{16}, \frac{1}{32}, \ldots$ that of the original value. It does not matter when you begin counting. If you wait 1 half-life, your counting rate drops by one-half.

Question   If a sample of material has an activity of 20 millicuries, what activity do you expect after waiting 2 half-lives?

Answer   After waiting 1 half-life, the activity will be half as much, namely, 10 millicuries. After waiting the other half-life, the activity will drop to half of this value. Therefore, the activity will be 5 millicuries.

This decay law also applies to the number of radioactive nuclei remaining in the sample. The time it takes to reduce the population of radioactive nuclei by a factor of 2 is also equal to the half-life. The important point is that you do not get rid of the radioactive material in 2 half-lives. Let's say that you initially have 1 kilogram of a radioactive element. One half-life later you will have $\frac{1}{2}$ kilogram. After 2 half-lives, you will have $\frac{1}{4}$ kilogram, and so on. In other words, one-half of the remaining radioactive nuclei decays during the next half-life.

The second factor that determines the activity of a sample depends on the character of the nuclei. As we saw in the previous chapter, quantum mechanics determines the probability that something will occur. Therefore, the dice analogy is especially appropriate for nuclear decays. The probability in our dice analogy depends on the number of faces on each die. Six faces means there is a probability of 1 in 6 that a given face will be up on each throw of the dice. Imagine that instead of using 6-sided dice, we use Arabian dice, which have 12 sides. In this case there are more alternatives, and the probability of 1s up is lower. Each die has a 1-in-12 chance of decaying. On the first throw of 36 dice, we would expect an average of 3 dice with 1s up. The half-life of the Arabian dice is twice as long. This variation also occurs with radioactive nuclei, but with a much larger range of probabilities. Nuclear half-lives range from microseconds to trillions of years.

Twelve-sided Arabian dice have a smaller probability for a given number to be on top.

This process is all based on the probability of random events. With dice it is the randomness of the throwing process; with radioactive nuclei it is the quantum-mechanical randomness of the behavior of the nucleons within the nucleus. As with all probabilities, any prediction is based on the mathematics of statistics. Each time we throw 36 dice, we should not expect exactly 6 dice to have 1s up. Sometimes there will be only 5; other times there might be 7 or 8. In fact, the number can vary from 0 to 36! However, statistics tells us that as the total number of dice increases, our ability to predict also increases. Because there are more than 10^{15} nuclei in even a small sample of radioactive material, our predictions of the number of nuclei that will decay in a half-life are very reliable.

Although our predictions get better with large numbers, we must remember the probabilistic nature of dice and radioactive nuclei; we cannot predict the behavior of an individual nucleus. One particular nucleus might last a million years, whereas an "identical" one might last only a millionth of a second.

PHYSICS | ON YOUR OWN

Experiment with the radioactivity analogy. To get good results, you will need to make many throws of many dice. About 100 dice thrown six to eight times gives acceptable results. Because it is difficult to obtain 100 dice, use 100 sugar cubes with one side marked with a dot. If you get 16 "dots up" on the first throw of the 100 cubes, you would use only $100 - 16 = 84$ cubes for the second "time interval." Determine the half-life of your sample by graphing the number of cubes remaining after successive intervals.

Radioactive Clocks

The decay rate of a radioactive sample is unaffected by physical and chemical conditions ordinarily found on Earth (excluding the extreme conditions found in nuclear reactors or bombs). Normal conditions involve energies that can rearrange electrons around atoms but cannot cause changes in nuclei. The nucleus is protected by its electron cloud and the large energies required to change the nuclear structure, which means that radioactive samples are good time probes into our history. Knowing the half-life of a particular isotope and the products into which it decays, we can determine the relative amount of parent and daughter atoms and calculate how long the isotope has been decaying. In effect, we have a radioactive "clock" for dating events in the past.

One of the first such clocks gave us the age of Earth. Before this, many different estimates were made by different groups. A 17th-century Irish bishop traced the family histories in the Old Testament and obtained a figure of 6000 years. Another estimate was obtained by calculating the length of time required for all the rivers of the world to bring initially freshwater oceans up to their present salinity. Another value was obtained by assuming the Sun's energy output was due to its slow collapse under the influence of the gravitational force. These latter two methods yielded estimates on the order of 100,000 years. Charles Darwin suggested that Earth was much older, as his theory of evolution indicated more time was required to produce the observed biodiversity.

The best value was obtained through the radioactivity of uranium. Uranium and the other heavy elements are formed only during supernova explosions that occur near the end of some stars' lives. Therefore, any uranium found on Earth was present in the gas and dust from which the Solar System formed. Uranium-238 decays with a half-life of 4.5 billion years into a stable isotope of lead ($^{206}_{82}$Pb) through a long chain of decays. Suppose we find a piece of igneous rock and determine that one-half of the uranium-238 has decayed into lead-206. We then know that the rock was formed at a time in the past equal to 1 half-life. The oldest rocks are found in Greenland and have a ratio of lead-206 to uranium-238 of a little less than 1, meaning that a little less than one-half of the original uranium-238 has decayed. This indicates that the rocks are about 4 billion years old. Incidentally, the rocks brought back from the Moon have similar ratios, establishing that Earth and Moon were formed about the same time.

Another radioactive clock is used to date organic materials. Living organ-

This is one of the rocks brought back from the Moon by the Apollo astronauts.

Smoke Detectors

Many of the uses of radioactive isotopes are not often apparent to most of us. For example, some smoke detectors (Figure A) commonly found in our homes use radioactive americium to detect the smoke particles. As shown in the schematic diagram of Figure B, the weak radioactive source ionizes the air, allowing a small current to flow between the electrodes. When smoke enters the detector, the ions are attracted by the smoke particles and stick to them. The larger mass of the smoke particles causes them to move slower than the ions, reducing the current in the circuit and setting off the alarm.

Americium-241 is an alpha emitter with a half-life of 433 years, so the level of ionization does not decrease significantly over the lifetime of the detector. This isotope is obtained as a by-product of the processing of fuel rods from nuclear reactors.

Figure A A household smoke detector.

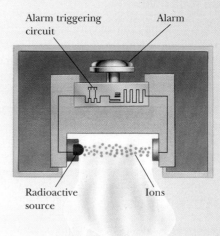

Figure B Diagram of a smoke detector.

isms contain carbon, which has two isotopes of interest—the stable isotope $^{12}_{6}C$ and the radioactive isotope $^{14}_{6}C$, which has a half-life of 5700 years. Cosmic rays bombarding our atmosphere continually produce carbon-14 to replace those that decay. Because of this replacement, the ratio of carbon-12 to carbon-14 in the atmosphere is relatively constant. While the plant or animal is living, it continually exchanges carbon with its environment and therefore maintains the same ratio of the two isotopes in the tissues that exists in the atmosphere. As soon as it dies, however, the exchange ceases and the amount of carbon-14 decreases due to the radioactive decay. By examining the ratio of the amount of carbon-14 to that of carbon-12 in the plant or animal, we can learn how long ago death occurred.

WORKING IT OUT | Radioactive Dating

As an example of radioactive dating, suppose a piece of charred wood is found in a primitive campsite. To find out how long ago the campsite was occupied, one examines the ratio of carbon-14 to carbon-12 in the wood. Suppose that the ratio is only one-eighth of the atmospheric value. Because $\frac{1}{2} \times \frac{1}{2} \times \frac{1}{2} = \frac{1}{8}$, the tree died 3 half-lives ago. This gives an age of 3×5700 years = 17,100 years.

Radiocarbon dating of this skeleton found in the Tyrolean Alps has helped estimate that the man lived around 5000 years ago.

Because the Dead Sea Scrolls were written on parchment, their age was determined by this technique. Similar studies indicate that the first human beings may have appeared on the North American continent about 27,000 years ago.

Radioactive dating with carbon-14 is not useful beyond 40,000 years (7 half-lives), as the amount of carbon-14 becomes very small. Other radioactive clocks can be used to go beyond this limit. These clocks indicate that humans, or at least prehumans, may have been around for more than 3.5 million years, mammals about 200 million years, and life for 3–4 billion years.

PHYSICS | **ON YOUR OWN**

> Find out more about carbon-14 dating at the website www.howstuffworks.com/carbon-14.htm.

Radiation and Matter   ✓ MATH

How would you know if a chunk of material was radioactive? Radiation is usually invisible. However, if the material is extremely radioactive, such as radium, the enormous amount of energy being deposited in the material by the radiation can make it hotter than its surroundings. In extreme cases it may even glow. A less radioactive sample might leave evidence of its presence over a longer time. Pierre Curie developed a radioactive burn on his side because of his habit of carrying a piece of radium in his vest pocket.

We learn about the various types of radiation through their interactions with matter. In the majority of cases, these interactions are outside the range of direct human observation. One exception is light. Although we don't see photons flying across the room, we do see the interaction of these photons with our retinas. Our eyes are especially tuned to a small range of photon energy. Other radiations are detected by extending our senses with instruments. For example, the electromagnetic radiation from radio and television stations is not detected by our bodies but interacts with the electrons in antennas.

A TV antenna is a radiation detector.

We also observe nuclear radiations through their interactions with matter. Alpha and beta particles interact with matter through their electric charge. For example, in Becquerel's discovery, the charged particles interacted with the photographic chemicals to expose the film. For the most part, these charged particles interact with the electrons in the material. As we saw in Rutherford's scattering experiment, they rarely collide with nuclei. As the alpha and beta particles pass through matter, they interact with atomic electrons, causing them to jump to higher atomic levels or to leave the atom entirely. This latter process is called **ionization**.

The relatively large mass of alpha particles compared with electrons means that the alpha particles travel through the material in essentially straight lines. This is like a battleship passing through a flotilla of canoes. Because beta particles are electrons, they are more like canoes hitting canoes and have paths that are more ragged. In both cases the colliding particles deposit energy in the material in a more or less continuous fashion. Therefore, the distance they travel in a material is a measure of their initial energy. We can determine this energy experimentally by measuring the distance the particle travels—that is, its range. Some sample ranges are given in Table 25-3.

The interaction of gamma rays with matter is quite different. Gamma rays do not lose their energy in bits and pieces. (There is no such thing as half a pho-

Table 25-3 | Ranges of Alpha Particles, Protons, and Electrons in Air and Aluminum

Energy (MeV*)	Range in Air (cm)			Range in Aluminum (cm)		
	α	p	e⁻	α	p	e⁻
1	0.5	2.3	314	0.0003	0.0014	0.15
5	3.5	34	2000	0.0025	0.019	0.96
10	10.7	117	4100	0.0064	0.063	1.96

*MeV (million electron volts) is an energy unit equal to 1.6×10^{-13} joules.

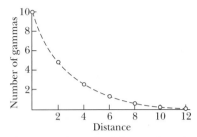

Figure 25-6 A graph of the number of gamma rays surviving at various distances through a material. Notice the similarity with the graph of the radioactivity in Figure 25-5.

Table 25-4 | Half Distances for the Absorption of Gamma Rays by Aluminum

Energy (MeV*)	Half Distance (cm)
1	4.2
5	9.1
10	11.1

*MeV (million electron volts) is an energy unit equal to 1.6×10^{-13} joules.

The 200,000 individual electronic components in this very small crystalline chip are especially sensitive to radiation damage.

ton.) Whenever a photon interacts with matter, it is completely annihilated in one of three possible ways. The photon's energy can ionize an atom, be transferred to a free electron with the creation of a new photon, or be converted into a pair of particles according to Einstein's famous equation $E = mc^2$. If we count the number of gamma rays surviving at various distances in a material and make a graph, we would obtain a curve like the one in Figure 25-6. This curve has the same shape as the one for radioactive decay in Figure 25-5, which means that gamma rays do not have a definite range but are removed from the beam with a characteristic half distance. That is, one-half of the original gamma rays are removed in a certain length, half of the remaining are removed in the next region of this length, and so on. From Table 25-4 you can see that gamma rays are much more penetrating than alpha or beta particles of the same energy.

Question How far must a beam of 10-million-electron-volt gamma rays travel in aluminum before it is all gone?

Answer Theoretically, this never happens; some of the gamma rays will travel very large distances. However, by the time the beam has traveled 7 half-distances, less than 1% of the gamma rays remain.

Any radiation passing through matter deposits energy along its path, resulting in temperature increases, atomic excitations, and ionization. The ionization caused by radiation can damage a substance if its properties depend strongly on the detailed molecular or atomic structure. For example, radiation can damage the complex molecules in living tissues and cause cancer or genetic defects. Damage can also occur in nonliving objects, such as transistors and integrated circuits. These modern electronic devices are made of crystals in which the atoms have very definite geometric arrangements. In integrated circuits, such as those found in pocket calculators and computers, there are literally thousands of individual electronic components in very small crystalline chips. A disruption of these atomic structures can change the electronic properties of these components. If the radiation damage is severe, the device could fail. This hazard is obviously serious if the device is a control component for a nuclear reactor or part of the guidance system for a missile. There is, however, not enough radiation around the university computer for you to expect all of your grades to change into A's.

Biological Effects of Radiation

The ionization caused by radiation passing through living tissue can destroy organic molecules if the electrons are involved in molecular binding. If too many molecules are destroyed in this fashion or if DNA molecules are destroyed, cells may die or become cancerous.

The effects of radiation on our health depend on the amount of radiation absorbed by living tissue and the biological effects associated with this absorption. A **rad** (the acronym from *r*adiation *a*bsorbed *d*ose) is the unit used to designate the amount of energy deposited in a material. One rad of radiation deposits 0.01 joule per kilogram of material. Another unit, the **rem** (radiation *e*quivalent in *m*ammals), was developed to reflect the biological effects caused by the radiation. Different radiations have different effects on our bodies. Alpha particles, for example, deposit more energy per centimeter of path than electrons with the same energy; therefore, the alpha particles cause more biological damage. One rad of beta particles can result in 1–1.7 rem of exposure, 1 rad of fast neutrons or protons may result in 10 rem, and 1 rad of alpha particles produces between 10 and 20 rem. The rad and the rem are very nearly equal for photons. Because most human exposure involves photons and electrons, the two units are roughly interchangeable when talking about typical doses.

Knowing the amount of energy deposited or even the potential damage to cells from exposure to radiation is still not a prediction of the future health of an individual. The biological effects vary considerably among individuals depending on their age, health, the length of time over which the exposure occurs, and the parts of the body exposed. Large doses, such as those resulting from the nuclear bombs exploded over Japan or the accident at the nuclear reactor at Chernobyl in the former Soviet Union (Figure 25-7), cause radiation sickness, which can result in vomiting, diarrhea, internal bleeding, loss of hair, and even death. It is virtually impossible to survive a dose of more than 600 rem to the whole body over a period of days. An exposure of 100 rem in the same period causes radiation sickness, but most people recover. For comparison, the maximum recommended occupational dose is 5 rem per year, the maximum rate recommended for the general public by the U.S. Environmental Protection Agency is 0.5 rem per year, and the natural background radiation is about 0.3 rem per year.

Obtaining data on the health effects of exposures to low levels of radiation is very difficult, primarily because the effects are masked by effects not related to radiation and the effects do not show up for a long time. Although cancers such as leukemia may show up in 2 to 4 years, the more prevalent tumorous cancers take much longer. Imagine conducting a free-fall experiment with an apple in which the apple doesn't begin to move noticeably for 3 years. Even after you spot its motion, you are compelled to ask if it occurred because of your actions or something else. When cause and effect are so widely separated in time, making connections is difficult. For example, because there is already a high incidence of cancer, it is hard to determine if any increase is due to an increase in radiation or to some other factor. On the other hand, it is known that large doses of radiation do cause significant increases in the incidence of cancer. Marie Curie died of leukemia, most likely the result of her long-term exposure to radiation.

Genetic effects due to low-level radiation are also difficult to determine. These effects can be passed on to future generations through mutation of the reproductive cells. Although mutations are important in the evolution of species, most of them are recessive. It is generally agreed that an increase in the rate of mutations is detrimental.

Any additional nuclear radiation hitting our bodies is detrimental. However, we live in a virtual sea of radiation during our entire lives. Exposure to some of this radiation is beyond our control, but other exposure is within our control. Indeed, many of the "nuclear age" debates center on this issue of assessing the effects of *additional* radiation on the health of the population and evaluating its risks and benefits. As an example, the increased radiation exposure during air travel due to the reduced shielding of the atmosphere is about 0.5 millirem per hour. This increase is not much for an occasional traveler, but may result in an additional 0.5 rem per year for an airplane pilot, which equals the maximum recommended dose.

The average dosage rates for a variety of sources are given in Table 25-5. A word of caution is appropriate here. Although the concept of averaging helps make some comparisons, there are situations where the value of the average is quite meaningless. Consider, for example, the amount of radiation our population receives from the nuclear power industry. The bulk of this radiation exposure occurs for the workers in the industry. Most of the rest of us get very little radiation from this source. This situation is similar to putting one foot in ice water, the other foot in boiling water, and claiming that on the average you are doing fine.

Radiation around Us

It is impossible to get away from all radiation. The radiation that is beyond our control comes from the cosmic rays arriving from outer space, from reactions in our atmosphere initiated by these cosmic rays, from naturally occurring radio-

Figure 25-7 One of the world's major nuclear reactor accidents happened in April 1986 at Chernobyl in the former Soviet Union. This photograph shows the structure built to contain the leaking radiation.

Air travelers implicitly accept the risk of the increased radiation for the benefits of quicker travel.

Table 25-5 | Average Annual Radiation Doses*

Source	Dose (mrem/yr)	
Cosmic rays	30	
Surroundings	30	
Internal	40	
Radon	200	
Naturally occurring		300
Medical	50	
Consumer products	5	
Nuclear power	5	
Human-made		60
Total		360

*Frank E. Gallagher III, California Campus Radiation Safety Officers Conference, September 1993.

Radon

A source of radiation that is partly under our control is the radioactive gas radon. Much attention has been focused by the media on the presence of indoor radon, especially in the air of some household basements. This awareness came about in part because of energy conservation efforts. In attempts to reduce heating bills, many homeowners improved their insulation and sealed many air leaks. This resulted in less air circulation within homes and allowed the concentration of radon in the air to increase.

The source of the radon is uranium ore that exists in the ground; the ore's quantities vary geographically. Through a long chain of decays shown in the table, the uranium eventually becomes a stable isotope of lead. All the radioactive daughters are solids except one, the noble gas radon (Rn). This gas can travel through the ground and enter basements of homes through small (or large) cracks in the basement floors or walls. Because radon is an inert gas, it doesn't deposit on walls and leave the air. Thus, it can be inhaled into a person's lungs and become trapped there. In this case, each of the eight subsequent decays causes the lungs to experience additional radiation and possible damage. The maximum "safe" level of radon has been established by the Environmental Protection Agency at 4×10^{-12} curie per liter. Kits to test for excessive concentrations are available through retail outlets. If high levels are found, the homeowner can take measures to ventilate the radon and to block its entry into the house by sealing cracks around the foundation and in the basement.

Radioactive Decay

$^{238}_{92}U \rightarrow$	$^{234}_{90}Th$	$+\alpha$	(9.5×10^9 years)
	$\downarrow$		
	$^{234}_{91}Pa$	$+\beta^-$	(24 days)
	$\downarrow$		
	$^{234}_{92}U$	$+\beta^-$	(6.7 hours)
	$\downarrow$		
	$^{230}_{90}Th$	$+\alpha$	(2.5×10^5 years)
	$\downarrow$		
	$^{226}_{88}Ra$	$+\alpha$	(7.5×10^4 years)
	$\downarrow$		
	$^{222}_{86}Rn$	$+\alpha$	(1622 years)
	$\downarrow$		
	$^{218}_{84}Po$	$+\alpha$	(3.85 days)
	$\downarrow$		
	$^{214}_{82}Pb$	$+\alpha$	(3 minutes)
	$\downarrow$		
	$^{214}_{83}Bi$	$+\beta^-$	(27 minutes)
	$\downarrow$		
	$^{214}_{84}Po$	$+\beta^-$	(19.7 minutes)
	$\downarrow$		
	$^{210}_{82}Pb$	$+\alpha$	(10^{-4} seconds)
	$\downarrow$		
	$^{210}_{83}Bi$	$+\beta^-$	(19 years)
	$\downarrow$		
	$^{210}_{84}Po$	$+\beta^-$	(5 days)
	$\downarrow$		
	$^{206}_{82}Pb$	$+\alpha$	(138 days)

A commercially available kit for testing radon gas levels in homes.

active decays in the material around us, and from the decay of radioactive isotopes (mostly potassium-40) within our bodies. In fact, even table salts that substitute potassium for sodium are radioactive. Radiation is part of the world in which we live. The radiations within our control range from testing of nuclear weapons to nuclear power stations to medical radiations used for diagnosing and treating diseases.

Medical sources of radiation *on average* contribute less to our yearly dose than the background radiation. The risks of diagnostic X rays have decreased over the years with new improvements in technology. Modern machines and more sensitive photographic films have greatly reduced the exposure necessary to get the needed information. The exposure from a typical chest X ray is 0.01 rem.

We saw in Chapter 23 that X rays can be used to examine internal organs and other structures for abnormalities. During this process, however, some cells are damaged or killed. We need to ask, is the possible damage worth the benefits of a better treatment based on the information gained? In most cases the answer is yes. In other cases the answer ranges from a vague response to a definite no. If your physician suspects that your ankle is broken, the benefits of having an X ray are definitely worth the risks. On the other hand, having an X ray of your lungs as part of an annual physical exam is not believed to be worth the risks. If, on the other hand, there is good reason to suspect lesions or a tumor, the odds are in favor of having the X ray.

The largest doses of medically oriented radiation occur during radiation treatment of cancer. In these cases radiation is used to deliberately kill some cells to benefit the patient. Because a beam of radiation causes damage along its entire path, the beam is rotated around the body to minimize damage to normal cells. This procedure is shown in Figure 25-8. Only the cells in the overlap region (the cancerous tumor) get the maximum dosage. Alternative techniques use beams of charged particles. Because charged particles deposit energy at the highest rate near the end of their range, the damage to normal cells along the entrance path is minimized.

Sometimes radioactive materials are ingested for diagnostic or treatment purposes. Our bodies do not have the ability to differentiate between various isotopes of a particular chemical element. If a tumor or organ is known to accumulate a certain chemical element, a radioactive isotope of that element can be put into the body to treat the tumor or to produce a picture of the tumor or organ. For instance, because iodine collects in the thyroid gland, iodine-131, a beta emitter, can be used to kill cells in the thyroid with little effect on more distant cells.

With each source of radiation the debate returns to the question of risk versus benefit; does the potential benefit of the exposure exceed the potential risk?

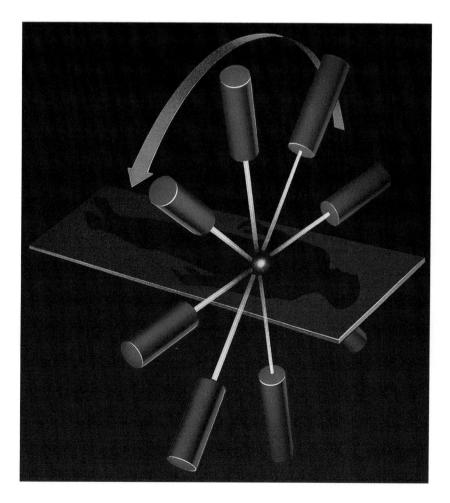

Figure 25-8 The direction of the radiation beam is changed to minimize the damage to normal cells surrounding a tumor.

Radiation Detectors

Many devices have been developed to detect nuclear radiation. The earliest was the fluorescent screen Rutherford used in his alpha-particle experiments. Alpha particles striking the screen ionized some of the atoms on the screen. When these atomic electrons returned to their ground states, visible photons were emitted. Rutherford and his students used a microscope to count the individual flashes.

The fluorescent screen is just one example of a *scintillation detector*, a detector that emits visible light upon being struck by radiation. The most familiar device of this type is the television screen. Scintillation detectors can be combined with electronic circuitry to automatically count the radiation. When a photon enters the device, it strikes a photosensitive surface and ejects an electron via the photoelectric effect (Chapter 23). A voltage then accelerates this electron so that it strikes a surface with sufficient energy to free two or more electrons. These electrons are accelerated and release additional electrons when they strike the next electrode. A typical photomultiplier, like the one in Figure 25-9, may have 10 stages, with a million electrons released at the last stage. This signal can then be counted electronically. Furthermore, the signal is proportional to the energy deposited in the scintillation material, and hence it gives the energy of the incident radiation.

The Geiger counter, popularized in old-time movies about prospecting for uranium, uses the ionization of a gas to detect radiation. The essential parts of a Geiger counter are shown in Figure 25-10. The cylinder contains a gas and has a wire running along its length. As the radiation interacts with the gas, it ionizes gas atoms. The electrons gain kinetic energy from the electric field in the cylinder, which they then lose in collisions with other gas atoms. These collisions release more electrons, and the process continues. This quickly results in an ava-

(a)

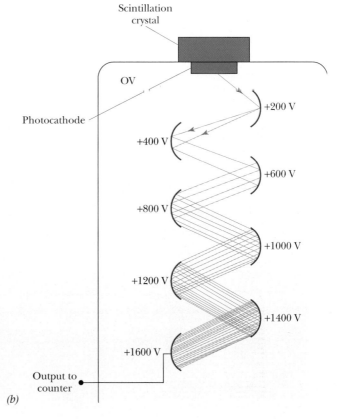
(b)

Figure 25-9 (a) Photograph and (b) schematic drawing of a photomultiplier tube used to detect radiation striking a scintillation counter.

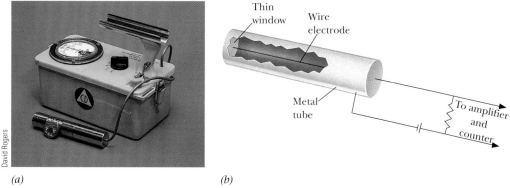

Figure 25-10 (a) A civil defense Geiger counter. (b) Schematic drawing of a Geiger counter.

lanche of electrons that produces a short electric current pulse that is detected by the electronics in the Geiger counter. In the movies the counter produces audible clicks. Many clicks per second means that the count rate is high and our hero is close to fame and fortune.

Another type of detector produces continuous tracks showing the paths of charged particles. These tracks are like those left by wild animals in snow. The big difference, however, is that although we may eventually see a deer, we cannot ever hope to see the charged particles. One track chamber is a bubble chamber in which a liquid is pressurized so that its temperature is just below the boiling point. When the pressure is suddenly released, boiling does not begin unless there is some disturbance in the liquid. When the disturbance is caused by the ionization of a charged particle, tiny bubbles form along its path. These bubbles are allowed to grow so that they can be photographed, and then the liquid is compressed to remove them and get ready for the next picture.

The paths of the charged particles are bent by placing the bubble chamber in a magnetic field; positively charged particles bend one way, and negatively charged particles the other way. Examination and measurement of the tracks in bubble chamber photographs, such as the one in Figure 25-11, yield information about the charged particles and their interactions.

Modern particle detectors come in a variety of forms and sizes. In many of these, the particles cannot be seen; their locations and identities are all deter-

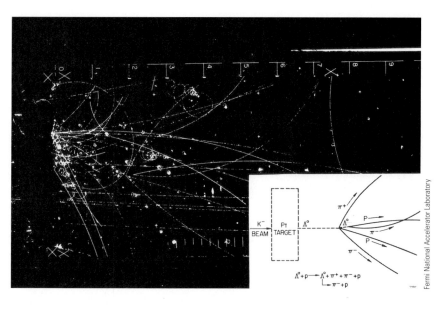

Figure 25-11 Photograph of tracks in a hydrogen bubble chamber shows the production of some strange particles that we will discuss in Chapter 27. The spiral track near the top of the photograph was produced by an electron.

Figure 25-12 This very large particle detector is used at the Fermi National Accelerator Laboratory to study collisions of particles at very high energies.

PhysicsNow™ Assess your understanding of this chapter's topics with sample tests and other resources found by logging into PhysicsNow at **http://physics.brookscole.com/kf6e**.

mined by large arrays of electronics connected to large computers like the one shown in Figure 25-12.

Summary

Radioactivity is a nuclear effect rather than an atomic one. It is virtually unaffected by outside physical conditions, such as strong electric and magnetic fields or extreme pressures and temperatures. Three types of radiation can occur: alpha (α) particles, or helium nuclei; beta (β) particles, which are electrons; and high-energy photons, or gamma (γ) rays. These different radiations have different penetrating properties, ranging from the gamma ray that can travel through many meters of air or even through thick walls to the alpha ray that won't penetrate a piece of paper.

Nuclei are composed of protons and neutrons. Nuclei of the same element have the same number of protons, but isotopes of this element have different numbers of neutrons. Isotopes of a particular element have their own nuclear properties but nearly identical chemical characteristics, because adding a neutron to a nucleus does not change the atom's electronic structure appreciably.

Radioactive decays are spontaneous, obey the known conservation laws (mass energy, linear momentum, angular momentum, electric charge, and nucleon number), and most often change the chemical nature of their atoms. There are five main naturally occurring nuclear processes: a nucleus (1) emits an alpha particle, producing a nucleus two elements lower in the periodic chart, (2) beta decays, emitting an electron and producing the next higher element, (3) inverse beta decays via electron capture, producing a daughter nucleus one lower in the periodic table, (4) beta decays with a positron, an antiparticle of the electron, decreasing the number of protons by one and increasing the number of neutrons by one, or (5) emits gamma rays, leaving its identity unchanged.

Equal amounts of different radioactive materials do not give off radiation at the same rate. The unit of activity, a curie, has a value of 3.7×10^{10} decays per second. Two factors determine the activity of a sample of material—the number of radioactive atoms in the sample and the characteristic half-life of the isotope. The quantum-mechanical nature of nucleons results in a wide range of decay rates and prohibits us from predicting the behavior of individual nuclei. Radioactive samples are good clocks for dating events in the past.

Radiation interacts with matter, often outside the range of direct human observation. Alpha and beta particles interact with matter through their electric charge, whereas gamma rays ionize atoms or convert into a pair of particles. All radiation interactions with matter deposit energy, resulting in temperature increases, atomic excitations, and ionization. The ionization caused by radiation passing through living tissue can kill or damage cells. It is impossible to get completely away from radiation.

Chapter 25 Revisited

If the artifact is made from material that was once living, its organic material had the same ratio of carbon-14 to carbon-12 as the atmosphere at the time the artifact was made. Over time, the ratio of these carbon isotopes decreases because the carbon-14 is radioactive while the carbon-12 is stable. The change in this isotopic ratio serves as a clock, dating the artifact.

KEY TERMS

activity: The number of radioactive decays that take place in a unit of time; measured in curies.

alpha particle: The nucleus of helium consisting of two protons and two neutrons.

alpha (α) radiation: The type of radioactive decay in which nuclei emit alpha particles (helium nuclei).

beta particle: An electron emitted by a radioactive nucleus.

beta (β) radiation: The type of radioactive decay in which nuclei emit electrons or positrons (antielectrons).

curie: A unit of radioactivity; 3.7×10^{10} decays per second.

daughter nucleus: The nucleus resulting from the radioactive decay of a parent nucleus.

electron capture: A decay process in which an inner atomic electron is captured by the nucleus. The daughter nucleus has the same number of nucleons as the parent but one less proton.

gamma (γ) radiation: The type of radioactive decay in which nuclei emit high-energy photons. The daughter nucleus is the same as the parent.

half-life: The time during which one-half of a sample of a radioactive substance decays.

ionization: The removal of one or more electrons from an atom.

isotope: An element containing a specific number of neutrons in its nuclei. Examples are $^{12}_{6}C$ and $^{14}_{6}C$, carbon atoms with six and eight neutrons, respectively.

neutron: The neutral nucleon; one of the constituents of nuclei.

nucleon: Either a proton or a neutron.

nucleus: The central part of an atom that contains the protons and neutrons.

parent nucleus: A nucleus that decays into a daughter nucleus.

positron: The antiparticle of the electron.

proton: The positively charged nucleon; one of the constituents of nuclei.

rad: The acronym for *r*adiation *a*bsorbed *d*ose; a rad of radiation deposits 0.01 joule per kilogram of material.

rem: The acronym for *r*adiation *e*quivalent in *m*ammals, a measure of the biological effects caused by radiation.

CONCEPTUAL QUESTIONS

The periodic table of the elements on the inside front cover is useful for answering some of these questions ▲.

1. What was the biggest difference between Becquerel's radiation and Roentgen's X rays?
2. What makes modeling the behavior of neutrons and protons inside the atomic nucleus more difficult than modeling the behavior of electrons in an atom?
3. Why can't we use visible light to look at nuclei?
4. What first led scientists to conclude that radioactivity is a nuclear phenomenon?
5. Which of the three types of radiation will interact with an electric field?
6. Why do beta rays and alpha rays deflect in opposite directions when moving through a magnetic field?
7. Which of the following names do *not* refer to the same thing: beta particles, cathode rays, alpha particles, and electrons? Explain.
8. Why are X rays not a fourth category of radiation?
9. Is the chemical identity of an atom determined by the number of neutrons, protons, or nucleons in its nucleus? Why?
10. How would the activity of 1 gram of uranium oxide compare with that of 1 gram of pure uranium?
11. What is a nucleon?
12. What observation told scientists that nuclei had to contain more than just protons?
13. What is the name of the element represented by the X in each of the following?
    a. $^{89}_{39}X$ b. $^{140}_{58}X$ c. $^{81}_{35}X$
14. What is the name of the element represented by the X in each of the following?
    a. $^{27}_{13}X$ b. $^{245}_{96}X$ c. $^{197}_{79}X$
15. How many neutrons, protons, and electrons are in each of the following atoms?
    a. $^{24}_{12}Mg$ b. $^{59}_{27}Co$ c. $^{208}_{82}Pb$
16. How many neutrons, protons, and electrons are in each of the following atoms?
    a. $^{11}_{5}B$ b. $^{64}_{29}Cu$ c. $^{239}_{93}Np$
17. What is the approximate mass of $^{90}_{40}Zr$ in atomic mass units?
18. What is the approximate mass of $^{64}_{30}Zn$ in atomic mass units?
19. The three naturally occurring isotopes of neon are $^{20}_{10}Ne$, $^{21}_{10}Ne$, and $^{22}_{10}Ne$. Given that the atomic mass of natural neon is 20.18 atomic mass units, which of these three isotopes must be the most common?
20. The two naturally occurring isotopes of chlorine are $^{35}_{17}Cl$ and $^{37}_{17}Cl$. Which of these isotopes must be the most common?
21. What happens to the charge of the nucleus when it decays via electron capture?
22. What happens to the charge of the nucleus when it decays via beta plus decay?
23. Each of the following isotopes decays by alpha decay. What is the daughter isotope in each case?
    a. $^{224}_{92}U$ b. $^{197}_{83}Bi$
24. Each of the following isotopes can be produced by alpha decay. What is the parent isotope in each case?
    a. $^{250}_{97}Bk$ b. $^{206}_{82}Pb$
25. What daughter is formed when each of the following undergoes beta minus decay?
    a. $^{18}_{7}N$ b. $^{90}_{38}Sr$
26. Each of the following isotopes can be produced by beta minus decay. What is the parent isotope in each case?
    a. $^{14}_{7}N$ b. $^{64}_{30}Zn$
27. Each of the following isotopes decays by electron capture. What is the daughter isotope in each case?
    a. $^{181}_{77}Ir$ b. $^{237}_{94}Pu$
28. Each of the following isotopes can be produced by electron capture. What is the parent isotope in each case?
    a. $^{65}_{28}Ni$ b. $^{253}_{98}Cf$
29. Each of the following isotopes decays by beta plus decay. What is the daughter isotope in each case?
    a. $^{28}_{15}P$ b. $^{22}_{11}Na$
30. Each of the following isotopes can be produced by beta plus decay. What is the parent isotope in each case?
    a. $^{50}_{23}V$ b. $^{52}_{24}Cr$
31. The isotope $^{40}_{15}P$ can beta decay to the isotope $^{39}_{16}S$. What else must have happened during this reaction?
32. Beta plus decay and electron capture both result in one additional neutron and one less proton in the nucleus. Why does beta minus decay not have a "mirror" reaction that produces the same change in the nucleus?
33. The isotope $^{64}_{29}Cu$ can decay via beta plus and via beta minus decay. What are the daughters in each case?
34. The isotope $^{153}_{66}Dy$ decays by alpha decay or by electron capture. What are the daughters in each case?
35. What type of decay process is involved in each of the following?
    a. $^{228}_{92}U \rightarrow ^{228}_{91}Pa + (?)$ b. $^{254}_{100}Fm \rightarrow ^{250}_{98}Cf + (?)$
36. What does the (?) stand for in each of the following decays?
    a. $^{233}_{90}Th \rightarrow ^{233}_{91}Pa + (?)$ b. $^{18}_{9}F \rightarrow ^{17}_{8}O + (?)$
▲ 37. Uranium-238 decays to a stable isotope of lead through a series of alpha and beta minus decays. Which of the following is a possible daughter of ^{238}U: ^{206}Pb, ^{207}Pb, ^{208}Pb, or ^{209}Pb?
▲ 38. What isotopes of lead (element 82) could be the decay products of $^{234}_{90}Th$ if the decays are all alpha and beta minus decays?
39. A proton strikes a nucleus of $^{20}_{10}Ne$. Assuming an alpha particle comes out, what isotope is produced?
40. What isotope is produced when a neutron strikes a nucleus of $^{10}_{5}B$ and an alpha particle is emitted?
41. What changes in the numbers of neutrons and protons occur when a nucleus is bombarded with a deuteron (the nucleus of deuterium containing one neutron and one proton) and an alpha particle is emitted?
42. A nucleus of $^{238}_{92}U$ captures a neutron to form an unstable nucleus that undergoes two successive beta minus decays. What is the resulting nucleus?
43. You place a chunk of radioactive material on a scale and find that it has a mass of 4 kilograms. The half-life of the material is 10 days. What will the scale read after 10 days?
44. You place a chunk of radioactive material on a scale and find that it has a mass of 10 kilograms. The half-life of the material is 20 days. How long will you have to wait for the scale to read 5 kilograms?

45. A radioactive material has a half-life of 50 days. How long would you have to watch a particular nucleus before you would see it decay?
46. The isotope $^{239}_{93}$Np has a half-life of about $2\frac{1}{2}$ days. Is it possible for a nucleus of this isotope to last for more than 1 year? Explain.
47. How do radioactive clocks determine the age of things that were once living?
▲ 48. Assume that the atmospheric ratio of ^{14}C to ^{12}C has been increasing by a small amount every 1000 years. Would determinations of age by radiocarbon dating be too short or too long? Why?
49. Can carbon-14 dating be used to date the stone pillars at Stonehenge? Why?

50. Why can carbon-14 dating not be used to determine the age of old-growth forests?
51. In decreasing order, rank the following in terms of range: a 10-million-electron-volt alpha particle, a 10-million-electron-volt electron, and a 10-million-electron-volt gamma ray.
52. A beam containing electrons, protons, and neutrons is aimed at a concrete wall. If all particles have the same kinetic energy, which particle will penetrate the wall the least?
53. Why do we often not worry about the distinction between rads and rems?
54. Which of the following causes the most biological damage if they all have the same energy: photons, electrons, neutrons, protons, or alpha particles?
55. Which of the following sources of radiation contributes the least to the average yearly dose received by humans: surroundings, medical, internal, cosmic rays, or nuclear power?
56. Which external source would most likely cause the most damage to internal organs, X rays or alpha particles? Why?
57. Why is it difficult to determine the biological effects of low levels of exposure to radiation?
58. What are the methods for using radiation to treat cancerous tumors?
59. Why are radioactive substances that emit alpha particles more dangerous inside the body than outside? Explain.
60. Would you expect more radiation damage to occur with a rad of X rays or a rad of alpha particles? Explain.

EXERCISES

1. Your younger brother plans to build a model of the gold atom for his science fair project. He plans to use a baseball (radius = 6 cm) for the nucleus. How far away should he put his outermost electrons?
2. The radius of the Sun is 7×10^8 m, and the average distance to Pluto is 6×10^{12} m. Are these proportions approximately correct for a scaled-up model of an atom?
3. What is the mass of the neutral carbon-12 atom in kilograms?
4. The atomic mass of neutral gold-197 is 196.97 amu. What is its mass in kilograms?
5. The neutral atoms of an unknown sample have an average atomic mass of 3.19×10^{-25} kg. What is the element?
6. What element has an average atomic mass of 4.48×10^{-26} kg?
7. A hydrogen atom consists of a proton and an electron. What is the atomic mass of the hydrogen atom?
▲ 8. A deuterium atom consists of a single electron orbiting a nucleus consisting of a proton and a neutron bound together. Compare the sum of the masses of three constituents of the deuterium atom to its measured atomic mass of 2.014 101 8 amu. How do you account for any difference?
9. The activity of a radioactive sample is initially 64 μCi (microcuries). What will the activity be after 3 half-lives have elapsed?
10. If the activity of a particular radioactive sample is 40 Ci and its half-life is 200 years, what would you expect for its activity after 400 years?
11. A radioactive material has a half-life of 10 min. If you begin with 512 trillion radioactive atoms, approximately how many would you expect to have after 40 min?
12. You have a sample of material whose activity is 64 Ci. After 60 minutes the material's activity has decreased to 8 Ci. What is the material's half-life?
13. If the ratio of carbon-14 to carbon-12 in a piece of bone is only one-eighth of the atmospheric ratio, how old is the bone?
14. If the ratio of carbon-14 to carbon-12 in a piece of parchment is only one-sixteenth of the atmospheric ratio, how old is the parchment?
15. A beam of 1-MeV (million electron volts) gamma rays is incident on a block of aluminum that is 12.6 cm thick. What fraction of the gamma rays penetrates the block?
16. How thick an aluminum plate would be required to reduce the intensity of a 5-MeV beam of gamma rays by a factor of 4?
17. How many chest X rays are required to equal the maximum recommended radiation dose for the general public?
18. Cosmic rays result in an average annual radiation dose of 41 mrem at sea level and 160 mrem in Leadville, Colorado (elevation 10,500 ft). How many chest X rays would a person living at sea level have to have each year to make up for this difference in natural exposure?

26 Nuclear Energy

In a nuclear power plant, heavy nuclei are split, releasing energy that is used to generate electricity.

Our world and its economy run on energy. Without an abundant supply of energy, people in third-world countries will not be able to obtain the standard of living of industrialized nations. With the depletion of fossil fuels—coal, oil, and natural gas—the need arises to explore alternatives. What options are available for using nuclear energy, and what are the primary issues in extracting this nuclear energy?

(See page 567 for the answer to this question.)

ALTHOUGH radioactivity indicates that nuclei are unstable and raises questions about why they are unstable, the more fundamental question is just the opposite: why do nuclei stay together? The electromagnetic and gravitational forces do not provide the answer. The gravitational force is attractive and tries to hold the nucleus together, but it is extremely weak compared with the repulsive electric force pushing the protons apart. Nuclei should fly apart. The stability of the majority of known nuclei, however, is a fact of nature. There must be another force—a force never imagined before the 20th century—that is strong enough to hold the nucleons together.

The existence of this force also means that there is a new source of energy to be harnessed. We have seen in previous chapters that an energy is associated with each force. For example, an object has gravitational potential energy because of the gravitational force. Hanging weights can run a clock, and falling water can turn a grinding wheel. The electromagnetic force shows up in household electric energy and in chemical reactions such as burning. When wood burns, the carbon atoms combine with oxygen atoms from the air to form carbon dioxide molecules. These molecules have less energy than the atoms from which they are made. The excess energy is given off as heat and light when the wood burns.

Similarly, the rearrangement of nucleons results in different energy states. If the energy of the final arrangement is less than that in the initial one, energy is given off, usually in the form of fast-moving particles and electromagnetic radiation. To understand the nuclear force, the possible types of nuclear reactions, and the potential new energy source, early experimenters needed to take a closer look at the subatomic world. This was accomplished with nuclear probes.

Physics Now™ Test your understanding of this chapter by logging into PhysicsNow at http://physics.brookscole.com/kf6e, selecting the chapter, and clicking on the "Take a Pre-Test" link.

The chemical energy released by the burning of the forests in Yellowstone National Park was a result of the electromagnetic forces between atoms.

Nuclear Probes

Information about nuclei initially came from the particles ejected during radioactive decays. This information is limited by the available radioactive nuclei. How can we study stable nuclei? Clearly, we can't just pick one up and examine it. Nuclei are 100,000 times smaller than their host atoms.

The situation is much more difficult than the study of atomic structure. Although the size of atoms is also beyond our sensory range, we can gain clues about atoms from the macroscopic world because it is governed by atomic characteristics. That is, the chemical and physical properties of matter depend on the atomic properties, and some of the atomic properties can be deduced from these everyday properties. On the other hand, there is no everyday phenomenon that reflects the character of stable nuclei.

The initial information about the structure of stable nuclei came from experiments using particles from radioactive decays as probes. These probes—initially limited to alpha and beta particles—failed to penetrate the nucleus. The relatively light beta particle is quickly deflected by the atom's electron cloud. As the Rutherford experiments showed, alpha particles could penetrate the electron cloud, but they were repelled by the positive charge on the nucleus. Higher-energy alpha particles had enough energy to penetrate the smaller nuclei, but they could not penetrate the larger ones. Even when penetration was possible, the matter-wave aspect of the alpha particles hindered the exploration. In Chapter 24 we learned that the wavelength of a particle gets larger as its momentum gets smaller. Typical alpha particles from radioactive processes have low momenta, which places a limit on the details that can be observed because diffraction effects hide nuclear structures that are smaller than the alpha particle's wavelength.

The nuclear energy released in the explosion of a nuclear bomb is a result of the strong force between nucleons.

Accelerators

This limit was lowered by constructing machines to produce beams of charged particles with much higher momenta and, consequently, much smaller wavelengths. **Particle accelerators**, such as the Van de Graaff generator shown in

Figure 26-1 This Van de Graaff generator is used to accelerate protons to high velocities to probe the structure of nuclei.

Figure 26-1, use electric fields to accelerate charged particles to very high velocities. In the process the charged particle acquires a kinetic energy equal to the product of its charge and the potential difference through which it moves (Chapter 20). The paths of the charged particles are controlled by magnetic fields.

There are two main types of particle accelerator: those in which the particles travel in straight lines and those in which they travel in circles. The simplest linear accelerator consists of a region in which there is a strong electric field and a way of injecting charged particles into this region. The maximum energy that can be given to a particle is limited by the maximum voltage that can be maintained in the accelerating region without electric discharges. The largest of these linear machines can produce beams of particles with energies up to 10 million electron volts (10 MeV), where 1 electron volt is equal to 1.6×10^{-19} joule. This energy is somewhat larger than the fastest alpha particles obtainable from radioactive decay.

Passing the charged particles through several consecutive accelerating regions can increase the maximum energy. The largest linear accelerator is the Stanford Linear Accelerator, shown in Figure 26-2. This machine is about 3 kilometers (2 miles) long and produces a beam of electrons with energies of 50 billion electron volts (or 50 GeV, where the G stands for *giga*, a prefix that means 10^9). Electrons with this energy have wavelengths approximately one-hundredth the size of the carbon nucleus, greatly reducing the diffraction effects.

An alternative technology, the circular accelerator, allows the particles to pass through a single accelerating region many times. The maximum energy of these machines is limited by the overall size (and cost), the strengths of the magnets required to bend the particles along the circular path, and the resulting radiation losses. Because particles traveling in circles are continuously being accelerated, they give off some of their energy as electromagnetic radiation. These radiation losses are more severe for particles with smaller masses, so circular accelerators are not as suitable for electrons. The most energetic accelerator of this type is at the Fermi National Accelerator Laboratory outside Chicago, shown in Figure 26-3. It has a diameter of 2 kilometers and can accelerate a beam of protons to energies of nearly 1 trillion electron volts (1 TeV, where T stands for *tera*, a prefix that means 10^{12}) with a corresponding wavelength thousands of times smaller than nuclei.

The Nuclear Glue

Experiments with particle accelerators provide a great deal of information about the structure of nuclei and the forces that hold them together. The nuclear force between protons is studied by shooting protons at protons and analyzing the angles at which they emerge after the collisions. Because it is not possible to make a target of bare protons, experimenters use the next closest thing, a hydrogen target. They also use energies that are high enough that the deflections due to the atomic electrons can be neglected.

At low energies the incident protons are repelled in the fashion expected by the electric charge on the proton in the nucleus. As the incident protons are given more and more kinetic energy, some of them pass closer and closer to the nuclear protons. When an incident proton gets within a certain distance of the target proton, it feels the nuclear force in addition to the electric force. This nuclear force changes the scattering. Some of the details of this new force can be deduced by comparing the scattering data with the known effects of the electrical interaction.

We know that the nuclear force has a very short range; it has no effect beyond a distance of about 3×10^{-15} meter (3 fermis). Inside this distance it is strongly attractive, approximately 100 times the strength of the electrical repulsion. At even closer distances, the force becomes repulsive, which means that the

Figure 26-2 An aerial view of the Stanford Linear Accelerator Center in California.

Figure 26-3 (a) An aerial view of the Fermi National Accelerator Laboratory. (b) This interior view shows magnets surrounding the two vacuum tubes in the main tunnel. The conventional magnets around the upper tube are red and blue, whereas the newer superconducting magnets are yellow.

protons cannot overlap but act like billiard balls when they get this close. The nuclear force is not a simple force that can be described by giving its strength as a function of the distance between the two protons. The force depends on the orientation of the two protons and on some purely quantum-mechanical effects.

The nuclear force also acts between a neutron and a proton (the n–p force) and between two neutrons (n–n force). The n–p force can be studied by scattering neutrons from hydrogen. The force between two neutrons, however, cannot be studied directly because there is no nucleus composed entirely of neutrons. Instead, this force is studied by indirect means—for example, by scattering neutrons from deuterons (the hydrogen nucleus that has one proton and one neutron) and subtracting the effects of the n–p scattering. These experiments indicate that when the effects of the proton's charge are ignored, the n–n, n–p, and p–p forces are very nearly the same. Therefore, the nuclear force is independent of charge.

Question Because neutrons are neutral, they cannot be accelerated in the same manner as protons. How might one produce a beam of fast neutrons?

Answer A beam of fast neutrons can be produced by accelerating protons and letting them collide with a foil. Head-on collisions of these protons with neutrons produce fast neutrons. The extra protons can be bent out of the way by a magnet.

There is another force in the nucleus. Investigations into the beta-decay process led to the discovery of a fourth force. This nuclear force is also short ranged but very weak. Although it is not nearly as weak as the gravitational force at the nuclear level, it is only about one-billionth the strength of the other nuclear force. The force involved in beta decay is thus called the **weak force**, and the force between nucleons is called the **strong force**. We concentrate on the strong force in this chapter because it is this force that is involved in the release of nuclear energy.

Nuclear Binding Energy

✓ MATH

Because nucleons are attracted to each other, a force is needed to pull them apart. This force acts through a distance and does work on the nucleons. Because it takes work to take a stable nucleus apart, the nucleus must have a lower energy than its separated nucleons. Therefore, we would expect energy to be released when we allow nucleons to combine to form one of these stable nuclei. This phenomenon is observed: a 2.2-million-electron-volt gamma ray is emitted when a neutron and a proton combine to form a deuteron (Figure 26-4).

Question How much energy would it require to separate the neutron and proton in a deuteron?

Answer Conservation of energy requires that the same amount of energy be supplied to separate them as was released when they combined—in the case above, it would be 2.2 million electron volts.

As an analogy, imagine baseballs falling into a hole. The balls lose gravitational potential energy while falling. This energy is given off in various forms such as heat, light, and sound. To remove a ball from the hole, we must supply energy equal to the change in gravitational potential energy that occurred when the ball fell in. The same occurs with nucleons. In fact, it is common for nuclear scientists to talk of nucleons falling into a "hole"—a nuclear potential well.

If we add the energies necessary to remove all the baseballs, we obtain the **binding energy** of the collection. This binding energy is a simple addition of the energy required to remove the individual balls because the removal of one ball has no measurable effect on the removal of the others. In the nuclear case, however, each removal is different because the removal of previous nucleons changes the attracting force. In any event, we can define a meaningful quantity called the *average binding energy per nucleon*, which is equal to the total amount of energy required to completely disassemble a nucleus divided by the number of nucleons.

✓ Extended presentation available in the *Problem Solving* supplement

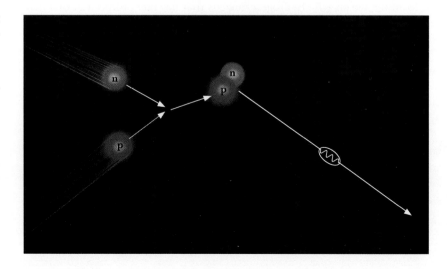

Figure 26-4 A 2.2-million-electron-volt gamma ray is emitted when a neutron and a proton combine to form a deuteron, indicating that the deuteron has a lower mass than the sum of the individual proton and neutron masses.

We don't have to actually disassemble the nucleus to obtain this number. The energy difference between two nuclear combinations is large enough to be detected as a mass difference. The assembled nucleus has less mass than the sum of the masses of its individual nucleons. The difference in mass is related to the energy difference by Einstein's famous mass-energy equation, $E = mc^2$. Therefore, we can determine the binding energy of a particular nucleus by comparing the mass of the nucleus with the total mass of its parts.

WORKING IT OUT | Nuclear Binding Energies ✓ MATH

As an example, we calculate the total binding energy and the average binding energy per nucleon for the helium nucleus. We add the masses of the component parts and subtract the mass of the helium nucleus:

2 protons	= 2 × 1.007 28 amu	=	2.014 56 amu
2 neutrons	= 2 × 1.008 67 amu	=	+2.017 34 amu
mass of parts			4.031 90 amu
1 helium nucleus		=	−4.001 50 amu
mass difference			0.030 40 amu

Therefore, the helium nucleus has a mass that is 0.030 40 amu less than its parts. Using Einstein's relationship, we can calculate the energy equivalent of this mass. A useful conversion factor is that 1 amu is equivalent to 931 MeV. Therefore, the total binding energy is

$$(0.030\ 4\ \text{amu}) \left[\frac{931\ \text{MeV}}{1\ \text{amu}} \right] = 28.3\ \text{MeV}$$

Dividing by 4, the number of nucleons, yields 7.08 MeV per nucleon.

Figure 26-5 shows a graph of the average binding energy per nucleon for the stable nuclei. The graph is a result of experimental work; it is simply a reflection of a pattern in nature. The graph shows that some nuclei are more tightly bound together than others.

Question   Which nuclei are the most tightly bound?

Answer   According to the graph in Figure 26-5, the most tightly bound nuclei have about 60 nucleons.

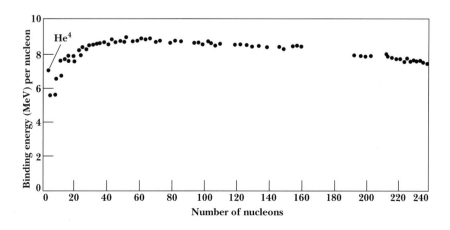

Figure 26-5 A graph of the average binding energy per nucleon versus nucleon number. Notice that the graph peaks in the range of 50 to 80 nucleons.

These differences mean that energy can be released if we can find a way of rearranging nucleons. For example, combining light nuclei or splitting heavier nuclei would release energy. But all of these nuclei are stable. To either split or join them we need to know more about their stability and the likelihood of various reactions.

> ### Flawed Reasoning
>
> Two students are discussing the conservation of mass:
>
> **Tyler:** "I have always been taught that mass is conserved, but if we put a chunk of radioactive material in a sealed bottle, half of it will be gone in 1 half-life."
>
> **Madison:** "*Gone* is the wrong word to use. Half of the radioactive material will have changed into something else, but the mass of the bottle will remain the same."
>
> Madison has cleared up a misconception held by Tyler, but **is Madison's claim entirely correct**?
>
> **Answer** Einstein taught us that mass can be converted to energy and energy to mass through his relationship $E = mc^2$. When a nucleus decays, the products will always have less mass than the parent. The missing mass is converted to energy. If some of the energy leaves the bottle (for example, as heat, light, or gamma rays), a careful measurement will show that the mass of the bottle does not remain the same. Of course, if the bottle is open and some of the daughters are gases, the change in mass will be easily observed.

Stability

Not all nuclei are stable. As we saw in the previous chapter, some are radioactive, whereas others seemingly last forever. A look at the stable nuclei shows a definite pattern concerning the relative numbers of protons and neutrons. Figure 26-6 is a graph of the stable isotopes plotted according to their numbers of neutrons and protons. The curve, or **line of stability** as it is sometimes called, shows that the light nuclei have equal, or nearly equal, numbers of protons and neutrons. As we follow this line into the region of heavier nuclei, it bends upward, meaning that these nuclei have more neutrons than protons.

To explain this pattern we try to "build" the nuclear collection that exists naturally in nature. That is, assuming we have a particular light nucleus, can we decide which particle—a proton or neutron—is the best choice for the next nucleon? Of course, the best choice is the one that occurs in nature.

Looking at the line of stability tells us that in the heavier nuclei adding a proton is not usually as stable an option as adding a neutron. Why is this so? Consider the characteristics of the two strongest forces involved. (We assume that gravity and the weak force play no role.) If you add a proton to a nucleus, it feels two forces—the electrical repulsion of the other protons and the strong nuclear attraction of the nearby nucleons. Although the electrical repulsion is weaker than the nuclear attraction, it has a longer range. The new proton feels a repulsion from *each* of the other protons in the nucleus but feels a nuclear attraction only from its nearest neighbors. With even heavier nuclei, the situation gets worse; the nuclear attraction stays roughly constant—there are only so many nearest neighbors it can have—but the total repulsion grows with the number of protons.

An equally valid way of explaining the upward curve in the line of stability is to recall that quantum mechanics is valid in the nuclear realm. Imagine the nuclear potential-energy well described in the previous section. Quantum me-

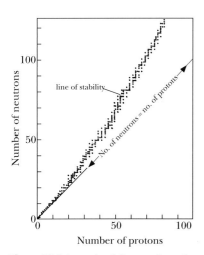

Figure 26-6 A graph of the number of neutrons versus the number of protons for the stable nuclei.

chanics tells us that the well contains a number of *discrete* energy levels for the nucleons. The Pauli exclusion principle that governs the maximum number of electrons in an atomic shell (Chapter 24) has the same effect here, but there is a slight difference. Because we have two different types of particles, there are two discrete sets of levels. Each level can contain only two neutrons or two protons.

In the absence of electric charge, the neutron and proton levels would be side-by-side (Figure 26-7) because the strong force is nearly independent of the type of nucleon. If this were true—that is, if level 4 for neutrons were the same height above the ground state as level 4 for protons—we would predict that there would be no preferential treatment when adding a nucleon. We would simply fill the proton level with two protons and the neutron level with two neutrons and then move to the next pair of levels. It would be unstable to fill proton level 5 without filling neutron level 4 because the nucleus could reach a lower energy state by turning one of its protons in level 5 into a neutron in level 4.

However, the electrical interaction between protons means that the separations of the energy levels are not identical for protons and neutrons. The additional force means that proton levels will be higher (the protons are less bound) than the neutron levels, as shown schematically in Figure 26-8. This structure indicates that at some point it becomes energetically more favorable to add neutrons rather than protons. Thus, the number of neutrons would exceed the number of protons for the heavier nuclei. Of course, too many neutrons will also result in an unstable situation.

Understanding the line of stability adds to our understanding of instability as well. We can now make predictions about the kinds of radioactive decay that should occur for various nuclei. If the nucleus in question is above the line of stability, it has extra neutrons. It would be more stable with fewer neutrons. In rare cases the neutron is expelled from the nucleus, but usually a neutron decays into a proton and an electron via beta minus decay. If the isotope is below the line of stability, we expect alpha or beta plus decay or electron capture to take place to increase the number of neutrons relative to the number of protons. In practice alpha decay is rare for nucleon numbers less than 140, and beta plus decay is rare for nucleon numbers greater than 200.

When the nuclei get too large, there is no stable arrangement. None of the elements beyond uranium (element 92) occur naturally on Earth. Some of them existed on Earth much earlier but have long since decayed. They can be artificially produced by bombarding lighter elements with a variety of nuclei. The list of elements (and isotopes) continues to grow through the use of this technique.

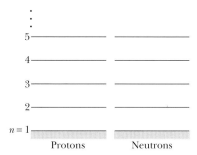

Figure 26-7 The proton and neutron energy levels in a hypothetical nucleus without electric charge occur at the same values.

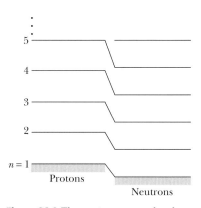

Figure 26-8 The proton energy levels in the hypothetical nucleus are raised due to the mutual repulsion of their positive charges.

Question What kind of decay would you expect to occur for $^{214}_{82}$Pb?

Answer Because this isotope is above the line of stability, it should undergo beta minus decay.

Nuclear Fission

The discovery of the neutron added a new probe for studying nuclei and initiating new nuclear reactions. Enrico Fermi, an Italian physicist, quickly realized that neutrons made excellent nuclear probes. Neutrons, being uncharged, have a better chance of probing deep into the nucleus. Working in Rome in the mid-1930s, Fermi was able to produce new isotopes by bombarding uranium with neutrons. These new isotopes were later shown to be heavier than uranium. Although this discovery is exciting, it perhaps isn't surprising; it seems reasonable that adding a nucleon to a nucleus results in a heavier nucleus.

GOEPPERT-MAYER | Magic Numbers

Like Marie Curie, Maria Goeppert-Mayer (1906–1972) was born in Poland but of a German family. When she was 4 years old, her pediatrician father moved the family to Göttingen, where she enrolled as a university student in mathematics in 1924. Excitement in the new field of quantum mechanics led her to physics, and in 1930 she earned her doctorate under Max Born, James Franck, and Adolf Windaus, world-renowned leaders in the field. While a graduate student, she met, fell in love with, and married a postdoctoral fellow from America, Joseph E. Mayer, who was working in physical chemistry.

Goeppert-Mayer accompanied her husband to the United States, where he taught at Johns Hopkins University in Baltimore. They coauthored an important book on statistical mechanics in 1940. In 1939 they relocated to Columbia University in New York City. She taught there, as she had at Johns Hopkins, only as a volunteer because of the nepotism clauses in her husband's contract. She joined Harold Urey in the early work on the separation of uranium isotopes that led to the use of atomic energy.

After the war, Goeppert-Mayer accepted a position at the University of Chicago, which had been the home base for so many refugees from Hitler's Europe. Her correspondence is a gold mine of fascinating information concerning these colleagues. Her work on the structure of nuclei was conducted at the Argonne National Laboratory in the Chicago area.

Maria Goeppert-Mayer

"For a long time I have considered even the craziest ideas about the atomic nucleus, and suddenly I discovered the truth," she said. She was trying to find a way of arranging the neutrons and protons into shells within nuclei to explain why nuclei with so-called magic numbers of neutrons and protons were particularly stable. One day, her old friend Enrico Fermi asked her if there was evidence of spin coupling. "When he said that, it all fell into place. In ten minutes I knew." The magic numbers made sense.

J. Hans Daniel Jensen independently arrived at the same theory, and they collaborated on the important book on nuclear shell structure that was published in 1955. In 1963 Goeppert-Mayer and Jensen were awarded the Nobel Prize for their work in nuclear structure. One hundred and seventy-eight Nobel Prizes have been awarded in physics, but only two have gone to women. Women in science faced an uphill battle until some obstacles were removed so that they could participate more fully in the great work of the new academy.

—*Pierce C. Mullen, historian and author*

Sources: Sharon Bertsch McGrayne, *Nobel Prize Women in Science: Their Lives, Struggles and Momentous Discoveries*, rev. ed. (New York: Carol Publishing Group, 1998); J. Dash, *Maria Goeppert-Mayer: A Life of One's Own* (New York: Paragon, 1973).

The real surprise came later when scientists found that their samples contained nuclei that were much less massive than uranium. Were these nuclei products of a nuclear reaction with uranium, or were they contaminants in the sample? It seemed unbelievable that they could be products. Fermi, for example, didn't even think about the possibility that a slow-moving neutron could split a big uranium nucleus. But that was what was happening.

The details of this process are now known. The capture of a low-energy neutron by uranium-235 results in another uranium isotope, uranium-236. This nucleus is unstable and can decay in several possible ways. It might give up the excitation energy by emitting one or more gamma rays, it might beta decay, or it might split into two smaller nuclei. This last alternative is called **fission**. The fissioning of uranium-235 is shown in Figure 26-9. A typical fission reaction is

$$n + {}^{235}_{92}U \rightarrow {}^{236}_{92}U \rightarrow {}^{142}_{56}Ba + {}^{91}_{36}Kr + 3n$$

The fission process releases a large amount of energy. This energy comes from the fact that the product nuclei have larger binding energies than the uranium nucleus. As the nucleons fall deeper into the nuclear potential wells, energy is released. The approximate amount of energy released can be seen from the graph in Figure 26-5. A nucleus with 236 nucleons has an average binding energy of 7.6 million electron volts per nucleon. Assume for simplicity that the nucleus splits into two nuclei of equal masses. A nucleus with 118 nucleons has an average binding energy of 8.5 million electron volts per nucleon. Therefore, each nucleon is more tightly bound by about 0.9 million electron volts, and the 236 nucleons must release about 210 million electron volts. This is a *tremendously* large energy for a single reaction. Typical energies from chemical reactions are only a few electron volts per atom, 100 million times less.

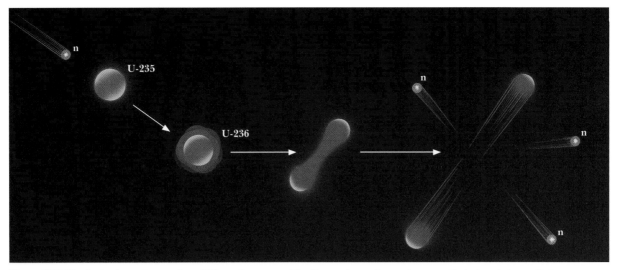

Figure 26-9 The fissioning of a uranium-235 nucleus caused by the capture of a neutron results in two intermediate mass nuclei, two or more neutrons, and some energy as gamma rays and kinetic energy.

Question How many joules are there in 210 million electron volts?

Answer $(2.1 \times 10^8 \text{ electron volts})(1.6 \times 10^{-19} \text{ joule per electron volt}) = 3.36 \times 10^{-11}$ joule. Although this is indeed a very small amount of energy, it is huge relative to other single-reaction energies.

Even though these energies are much larger than chemical energies, they are small on an absolute scale. The energy released by a single nuclear reaction as heat or light would not be large enough for us to detect without instruments. Knowledge of this led Rutherford to announce in the late 1930s that it was idle foolishness to even contemplate that the newly found process could be put to any practical use.

Chain Reactions

Rutherford's cynicism about the practicality of nuclear power seems silly in hindsight. But he was right about the amount of energy released from a single reaction. If you were holding a piece of uranium ore in your hand, you would not notice anything unusual; it would look and feel like ordinary rock. And yet, nuclei are continuously fissioning.

What is the difference between the ore in your hand and a nuclear power plant? The key to releasing nuclear energy on a large scale is the realization that a single reaction has the potential to start additional reactions. The accumulation of energy from many, many such reactions gets to levels that are not only detectable but can become tremendous. This occurs because the fission fragments have too many neutrons to be stable. The line of stability in Figure 26-6 shows that the ratio of protons to neutrons for nuclei in the 100-nucleon range is different from that of uranium-235. These fragments must get rid of the extra neutrons. In practice they usually emit two or three neutrons within 10^{-14} second. Even then the resulting nuclei are neutron-rich and need to become more stable by emitting beta particles.

The release of two or three neutrons in the fissioning of uranium-235 means that the fissioning of one nucleus could trigger the fissioning of others; these could trigger the fissioning of still others, and so on. Because a single reaction can cause a chain of events as illustrated in Figure 26-10, this sequence is called a **chain reaction**.

FERMI | A Man for All Seasons

The Italian Navigator has landed in the New World.

—Enrico Fermi

On the bitterly cold afternoon of December 2, 1942, Italian American physicist Enrico Fermi (1901–1954) and his team removed neutron-absorbing control rods from a massive ellipsoidal stack of graphite blocks and uranium oxide to achieve the first sustained nuclear chain reaction. The United States was fighting World War II at that time, and this nuclear experiment led to the development of weapons that ended the world conflict. A coded message informing American scientific authorities of this magnificent success began with the quote above. It was a new world and a new age, and the Italian Navigator illuminated it with beams of neutrons.

Enrico Fermi

Fermi was the son of an Italian railroad communications inspector and a mother who also had achieved a good education. The young Fermi was intellectually quick and curious. His application for a full scholarship to the selective Scuola Normale Superiore at the University of Pisa was a highly advanced study of sound characteristics. Gifted with mathematical ability, Fermi used differential equations and Fourier transforms to analyze sound vibrations. He began his studies at a level beyond most of the faculty, and when he graduated, he was an acknowledged authority on relativity and quantum mechanics—the hot new areas in physics.

At age 26, Fermi was appointed professor of theoretical physics at the University of Rome. He moved easily into the mainstream of European science and was recognized as one of an outstanding crop of brilliant movers and shakers in science. Early in his career, he became interested in the newly discovered neutron. Through a series of serendipitous discoveries, Fermi and his team noticed that neutrons that were slowed in passing through paraffin blocks were better at producing artificially radioactive isotopes. The Rome group produced a plethora of such isotopes, and the biomedical uses of some of these were important and highly publicized.

Going outside regular channels, Bohr told Fermi that he would receive the 1938 Nobel Prize in physics. This warning allowed Fermi to plan for his escape from Mussolini's increasingly anti-Semitic Italy—his wife, Laura, was Jewish. He accepted the prize in Stockholm, came directly to New York, and ultimately to Chicago. At the University of Chicago, Fermi worked on a great many problems associated with nuclear fission and weapons. Although he was adamantly opposed to the construction of thermonuclear weapons, his work led to the development of the plutonium bomb that devastated Nagasaki in August 1945.

He died in 1954 of stomach cancer. He had lived vigorously, in harmony with his peers and family, and displayed a wonderful balance of good humor and serious thinking. Element 100, fermium, is named in his honor. He may well have been the last physicist who could comprehend both experimental and theoretical physics. He was a man for all seasons.

—Pierce C. Mullen, historian and author

Sources: Laura Fermi, *Atoms in the Family* (Chicago: University of Chicago Press, 1954); Emilio Segrè, *Enrico Fermi: Physicist* (Chicago: University of Chicago Press, 1970); Daniel J. Kevles, *The Physicists: The History of a Scientific Community in Modern America* (New York: Knopf, 1978).

The world's first nuclear reactor was assembled in Chicago in 1942. There were no photographs taken of this reactor due to wartime secrecy.

Question How many neutrons must be emitted when $^{236}_{92}U$ fissions to become $^{141}_{55}Cs$ and $^{92}_{37}Rb$?

Answer Because the total number of nucleons remains the same, we expect to have $236 - 141 - 92 = 3$ neutrons emitted.

Imagine the floor of your room completely covered with mousetraps. Each mousetrap is loaded with three marbles, as shown in Figure 26-11, so that when a trap is tripped, the three marbles fly into the air. What happens if you throw a marble into the room? It will strike a trap, releasing three marbles. Each of these will trigger another trap, releasing its marbles. In the beginning the number of marbles released will grow geometrically. The number of marbles will be 1, 3, 9, 27, 81, 243, ... In a very short time the air will be swarming with marbles. Because the number of mousetraps is limited, the process dies out. However, the number of atoms in a small sample of uranium-235 is very large (on the order of 10^{20}), and the number of fissionings taking place can grow to be extremely large, releasing a lot of energy. If the chain reaction were to continue in the fashion we have described, a sample of uranium-235 would blow up.

But our piece of uranium ore doesn't blow up; it doesn't even get warm because most of the neutrons do not go on to initiate further fission reactions. Several factors affect this dampening of the chain reaction. One is size. Most of the neutrons leave a small sample of uranium before they encounter another uranium nucleus—a situation analogous to putting only a dozen mousetraps on the

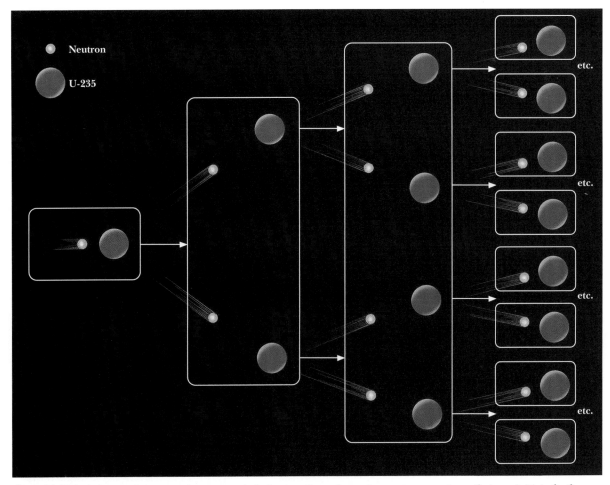

Figure 26-10 A chain reaction occurs because each fission reaction releases two or more neutrons that can initiate further fission reactions.

Figure 26-11 A chain reaction of mousetraps is initiated when the first marble is thrown into the array. Each mousetrap releases three marbles, which can trip additional mousetraps.

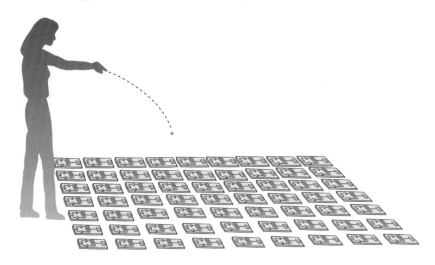

floor. If you trigger one trap, it is unlikely that one of its marbles will trigger another trap, but occasionally it happens. As the sample of material gets bigger, fewer and fewer of the neutrons escape the material.

But even a large piece of uranium ore does not blow up. Naturally occurring uranium consists of two isotopes. Only 0.7% of the nuclei are uranium-235; the remaining 99.3% are uranium-238. These uranium-238 nuclei will occasionally fission, but usually they capture the neutrons and decay by beta minus or alpha emission. Captured neutrons cannot go on to initiate other fission reactions. The chain reaction is **subcritical**, and the process dies out. To make our analogy correspond to a piece of naturally occurring uranium, we would need 140 unloaded mousetraps for each loaded one.

Extracting useful energy from fission is much easier if the percentage of uranium-235 is increased through what is known as enrichment. Any enrichment scheme is very difficult because the atoms of uranium-235 and uranium-238 are chemically the same, making it difficult to devise a process that preferentially interacts with uranium-235. Their masses differ by only a little more than 1%. The various enrichment processes take advantage of the slight differences in the charge-to-mass ratios, the rates of diffusion of their gases through membranes, resonance characteristics, or their densities. The enrichment processes must be repeated many, many times because only a small gain is made each time.

If enough enriched uranium is quickly assembled, the chain reaction can become **supercritical**. On average more than one neutron from each fission reaction initiates another reaction, and the number of reactions taking place grows rapidly. This process was utilized in the nuclear bombs exploded near the end of World War II, showing that Rutherford vastly underestimated the power of the fission reaction.

PHYSICS | ON YOUR OWN

> Experience a chain reaction by gathering a large group of friends in the center of a room. Give each friend three Styrofoam balls and instructions to throw the balls into the air whenever a ball strikes them. Watch what happens as you throw a ball to initiate the process. How would you create a subcritical situation?

Nuclear Reactors

Harnessing the tremendous energy locked up in nuclei requires controlling the chain reaction. The process must not become either subcritical or supercritical because the energy must be released steadily at manageable rates. In a nuclear re-

MEITNER | A Physicist Who Never Lost Her Humanity

When the name Lise Meitner (1878–1968) is mentioned, scientists most often think of a person who should have shared in the Nobel Prize but did not for both gender and political reasons. Before World War II, Meitner was at the top of her field in experimental physics and directed the department of nuclear physics at the Kaiser Wilhelm Institute in Berlin. Her associate, Otto Hahn, directed the radiochemistry department. The two had discovered protactinium in 1917, and they had been at the forefront of research on radioactivity ever since. German admirers referred to her as "Our Madam Curie." Before Meitner was driven, penniless and alone, from Germany, she had been investigating many of the phenomena for which Enrico Fermi received the Nobel Prize in 1938.

Lise Meitner

She was in exile in Sweden when she received news from her coworkers in Berlin that inexplicable chemical results occurred when uranium was bombarded with neutrons. After a traditional Swedish Christmas dinner, Meitner and her young nephew, Otto Frisch, went out into the evening snow to ponder the Berlin findings. They sat on a log calculating possibilities and suddenly discovered that, in Frisch's words, "Whenever mass disappears energy is created, according to Einstein's formula.... So here was the source of that energy; it all fitted." It was hard to believe because no serious scientist expected atoms to fission. Frisch returned to Copenhagen to tell Bohr, and Bohr in turn brought the terrible news to New York shortly after the New Year in 1939.

Hahn received the Nobel Prize for the discovery, which he attributed solely to his discipline, chemistry. After the war, Heisenberg, Hahn, and other non-Nazi Germans who had remained in the Third Reich were credited with work for which Meitner should have received her due. But the woman who had, against all odds, moved into the male domain of science, who had a Jewish grandparent, and who was not physically present in Germany during the war, received only minimum credit lest it reflect darkly upon the scientific establishment in wartime Germany. In a strange way, she was also a casualty of the cold war. The Western alliance needed to rebuild a strong Germany to shore up defenses against the Soviet Union. Therefore, Lise Meitner would not share in the credit for opening the field of nuclear energy.

Only later did Meitner receive due recognition. The United States honored Meitner as the first female recipient of the Fermi Award from its Atomic Energy Commission; Glenn Seaborg presented the medal to her in Vienna in 1966. Twenty years earlier she had been named Woman of the Year by the Women's National Press Club at a banquet presided over by President Harry Truman. The name meitnerium (Mt) has now been accepted for element 109. She is buried in Bramley, Hampshire, England, under a stone with the inscription that heads this biographical sketch.

—*Pierce C. Mullen, historian and author*

Sources: Ruth Lewin Sime, *Lise Meitner: A Life in Physics* (Berkeley: University of California Press, 1996); O. R. Frisch et al., *Trends in Atomic Physics: Essays Dedicated to Lise Meitner, Otto Hahn, and Max von Vaue on the Occasion of their 80th Birthday* (New York: Interscience Publishers, 1959).

---

actor, the conditions are adjusted so that an average of one neutron per fission initiates further fission reactions. Under these conditions the chain reaction is **critical**; it is self-sustaining. Energy is released at a steady rate and extracted from the reactor to generate electricity.

Several factors can be adjusted to ensure the criticality of the reactor. We have seen that the amount of uranium fuel (the core) must be large enough so that the fraction of neutrons escaping from it is small. It is also important that the nonfissionable uranium-238 nuclei not capture too large a fraction of the neutrons, which can be accomplished by enriching the fuel or by reducing the speed of the neutrons or both.

The likelihood that a neutron will cause a uranium-235 nucleus to fission varies with the speed of the incoming neutron. Initially, you might think that faster neutrons would be more likely to split the uranium-235 nucleus because they would impart more energy to the nucleus. This is not the case, because the splitting of the nucleus is a quantum-mechanical effect. Slow neutrons are much more likely to initiate the fission process. An added benefit is that the probability of a uranium-238 nucleus capturing a neutron decreases as the neutron speed decreases.

The neutrons are primarily slowed by elastic collisions with nuclei. A material (called the **moderator**) is added to the core of the reactor to slow the neutrons without capturing too many of them. Neutrons can transfer the most energy to another particle when their masses are the same. (In fact, a head-on collision will leave the neutron with little or no kinetic energy.) Hydrogen would seem to be the ideal moderator, but its probability of capturing the neutron is large. The material with the next lightest nucleus, deuterium, is fine but costly,

The Vallecitos boiling-water reactor and surrounding facilities.

and because it is a gas, it is hard to get enough mass into the core to do the job. Canadian reactors use deuterium as the moderator but in the form of water, called heavy water because of the presence of the heavy hydrogen. The world's first reactor used graphite (a form of carbon) as the moderator. Most current U.S. reactors use ordinary water as the moderator, requiring the use of enriched fuel.

Reactors require control mechanisms for fine-tuning and for adjusting to varying conditions as the fuel is used. Inserting rods of a material that is highly absorbent of neutrons controls the chain reaction. Boron is quite often used. The control rods are pushed into the core to decrease the number of neutrons available to initiate further fission reactions. The mechanical insertion and withdrawal of control rods could not control a reactor if all the neutrons were given off promptly in the usual 10^{-14} second. A small percentage of the neutrons are given off by the fission fragments and may take seconds to appear. These delayed neutrons allow time to adjust the reactor to fluctuations in the fission rate.

The last thing needed to make a reactor a practical energy source is a way of removing the heat from the core so that it can be used to run electric generators, which is accomplished in a variety of ways. In boiling-water reactors, water flows through the core and is turned to steam. Pressurized-water reactors use water under high pressure so that it doesn't boil in the reactor. Still other reactors use gases. A diagram of a nuclear reactor is shown in Figure 26-12.

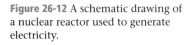

Play with the computer simulation found at www.ida.liu.se/~her/npp/demo.html. Try to keep the Kärnobyl fission reactor stable as it encounters three different failure sequences.

Breeding Fuel

We are exhausting many of our energy sources. This is as true for fission reactors as it is for coal- and oil-powered plants. It is estimated that there is enough uranium to run existing reactors for only the 30–40 years of expected operation of

Figure 26-12 A schematic drawing of a nuclear reactor used to generate electricity.

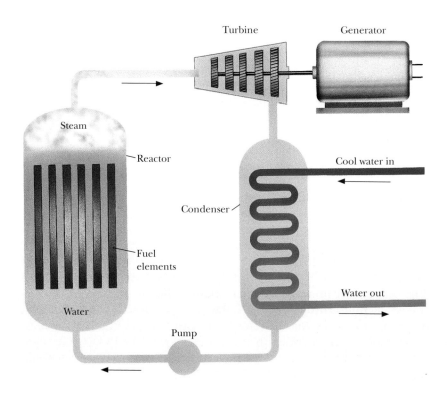

Natural Nuclear Reactors

In 1972 the remains of naturally occurring nuclear reactors were discovered in Gabon, a country in equatorial Africa. Using the long-lived radioactive products of the fission reactions as clocks, it is known that the reactors were active 1–2 billion years ago and operated for as long as 1 million years. These reactor sites are important resources because researchers can examine the migration of nuclear waste over a billion years, information that is useful for designing disposal facilities for human-manufactured nuclear waste.

Besides a supply of uranium with an enriched amount of uranium-235, the conditions required for a nuclear reactor to occur naturally are the same as those required for nuclear power plants: (1) a high concentration of uranium, (2) a critical size, (3) a moderator, and (4) few neutron absorbers. The 17 known natural reactors all occurred in uranium deposits with high concentrations of uranium ore. It is believed that water served as the moderator and as the control mechanism. If the reactors got too hot, the water vaporized, reducing the concentration of water molecules. This in turn slowed the reaction rate, reducing the temperature of the reactor and allowing water to flow into the region once again.

Could such reactors exist today? No; the ratio of uranium-235 to uranium-238 at the time the reactors were active was 3%, just sufficiently large to allow water to serve as a moderator. Since that time, the ratio of uranium-235 to uranium-238 has decreased to the current 0.7% because uranium-235 has a shorter half-life.

These ancient nuclear reactors were also the first breeder reactors. Besides breeding plutonium-239, the reactors operated for so long that part of the plutonium-239 decayed by alpha particle decay to form additional uranium-235.

This natural nuclear reactor is located in the Oklo Uranium Mine in Gabon, Africa. The yellowish rock is uranium oxide.

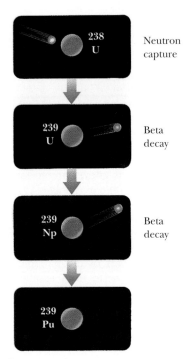

Figure 26-13 A scheme for converting uranium-238 to plutonium-239.

A breeder reactor at the Idaho National Laboratory.

each reactor. If uranium-235 were the only fuel, the nuclear-power age would turn out to be short-lived.

Imagine, however, that somebody suggests that it is possible to make new fuel for these reactors. Does this claim sound like a sham? Does it imply that we can get something for nothing? The claim is not a sham, but it is also not getting something for nothing. The laws of conservation of mass and energy still hold. The new process takes an isotope that does not fission and through a series of nuclear reactions transmutes that isotope into one that does.

When a $^{238}_{92}$U nucleus captures a neutron, it usually undergoes two beta minus decays to become plutonium, as diagrammed in Figure 26-13. The important point is that $^{239}_{94}$Pu is a fissionable nucleus that can be used as fuel in fission reactors. (A similar process transmutes $^{232}_{90}$Th into $^{232}_{92}$U, another fissionable nucleus.)

Some plutonium is produced in a normal reactor because there is uranium-238 in the core. However, not much is produced because the neutrons are slowed to optimize the fission reaction. Special reactors have been designed so that an average of more than one neutron from each fission reaction is captured by uranium-238. Such a reactor generates more fuel than it uses and is therefore known as a *breeder reactor*. Because most of the uranium is uranium-238 and because there is about the same amount of $^{232}_{90}$Th, these reactors greatly extend the amount of available fuel.

Like most other energy options, there are serious concerns about breeder reactors. Briefly, these concerns center on the technology of running these reactors, the assessment of the risks involved, and the security of the plutonium that is produced. Plutonium-239 is bomb-grade material that can be separated relatively inexpensively from uranium because it has different chemical properties.

Fusion Reactors ✓ MATH

There are other nuclear energy options. Returning to the binding energy curve in Figure 26-5 reveals another way to get energy from nuclear reactions. The increase in the curve for the average binding energy per nucleon for the light nuclei indicates that some light nuclei can be combined to form heavier ones with a release of energy. For instance, a deuteron ($^2_1$H) and a triton ($^3_1$H) can be fused to form helium with a release of a neutron and 17.8 million electron volts:

$$^2_1\text{H} + ^3_1\text{H} \rightarrow ^4_2\text{He} + n + 17.8 \text{ MeV}$$

Although this process releases a lot of energy per gram of fuel, it is much more difficult to initiate than fission. The interacting nuclei have to be close enough together so that the nuclear force dominates. (This wasn't a problem with fission because the incoming particle was an uncharged neutron.) The particles involved in the **fusion** reaction won't overcome the electrostatic repulsion of their charges to get close enough unless they have sufficiently high kinetic energies.

High kinetic energies mean very high temperatures. The required temperatures are on the order of millions of degrees, matching those found inside the Sun. In fact, the source of the Sun's energy is fusion. The first occurrence of fusion on Earth was in the explosion of hydrogen bombs in the 1950s. The extremely high temperatures needed for these bombs was obtained by exploding the older fission bombs (commonly called atomic bombs).

Making a successful fusion power plant involves harnessing the reactions of the hydrogen bomb and the Sun. Fusion requires not only high temperature but also the confinement of a sufficient density of material for long enough that the reactions can take place and return more energy than was necessary to initiate the process. At first this task might seem impossible and dangerous. Whether it is possible is still being determined. Most people in this field believe that it is a technological problem that can be solved. The characterization of fusion as dangerous is false and arises from confusion about the concepts of heat and temperature (Chapter 13). Heat is a flow of energy; temperature is a measure of the

Figure 26-14 The Tokamak at Princeton University uses a magnetic bottle to confine the plasma so that fusion reactions will occur.

average molecular kinetic energy. Something can have a very high temperature and be quite harmless. Imagine, for example, the vast differences in potential danger between a thimble and a swimming pool full of boiling water. The thimble of water has very little heat energy and thus is relatively harmless. The same is true of fusion reactors. Although the fuel is very hot, it is a rarefied gas at these temperatures. So the problem is not with it melting the container but rather the reverse—the container will cool the fuel.

Two schemes are being investigated for creating the conditions necessary for fusion. One method is to confine the plasma fuel in a magnetic "bottle" (Figure 26-14). The *magnetic field* interacts with the charged particles and keeps them in the bottle. The other, *inertial confinement*, uses tiny pellets of solid fuel. As these pellets fall through the reactor, they are bombarded from many directions by laser beams. This produces rapid heating of the pellet's outer surface, causing a compression of the pellet and even higher temperatures in the center.

To date, no one has been able to simultaneously produce all the conditions required to get more energy out of the process than was needed to initiate it. Successes have been limited to achieving some of these conditions but not all at the same time. Fusion reactors on a commercial scale seem many years away.

Some of the features of fusion, however, make it an attractive option. The risks are believed to be much lower than for fission reactors, and there is a lot of fuel. One possible reaction uses deuterium. Deuterium occurs naturally as one atom out of every 6000 hydrogen nuclei and thus is a constituent of water. This heavy water is relatively rare compared with ordinary water, but there is enough of it in a pail of water to provide the equivalent energy of 700 gallons of gasoline!

Solar Power

Throughout history people have puzzled about the source of the Sun's energy. What is the fuel? How long has it been burning? And how long will it continue to emit its life-supporting heat and light?

Many schemes have been suggested—some reasonable, some absurd. Early people thought of the Sun as an enormous "campfire" because fires were the only known source of heat and light. But calculations showed that if wood or coal were the Sun's fuel, it could not have been around for very long. Its lifetime would be much shorter than estimates of how long the Sun had already existed. Another idea involved a Sun heated by the constant bombardment of meteorites, which could account for the long lifetime of the Sun (the collisions continued indefi-

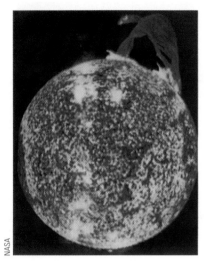

The source of the Sun's energy is the fusion of hydrogen into helium in its core.

nitely) but which also predicted that the mass of the Sun would increase. Earth would then spiral into the Sun due to the ever-increasing gravitational force. Another scheme had the Sun slowly collapsing under its own gravitational attraction. The loss in gravitational potential energy would be radiated into space. The flaw in this last scenario became apparent from geologic data suggesting that Earth was much older than the Sun. Also, if we mentally uncollapse the Sun, going back in time, we find that the Sun would extend beyond Earth's orbit at a time less than the assumed age of Earth.

The source of the Sun's power was a major conflict at the beginning of the 20th century. Astronomers were suggesting that the Sun was approximately 100,000 years old, but geologists and biologists were saying that Earth was very much older. Both sides couldn't be right. The discovery of radioactivity and other nuclear reactions showed that the geologists and biologists were right. Scientists now believe that our entire Solar System formed from interstellar debris left over from earlier stars. As the matter collapsed, it heated up due to the loss in gravitational potential energy. At some point the temperature in the interior of the Sun became high enough to initiate nuclear fusion. Now we have a Sun in which hydrogen is being converted into helium via the fusion reaction. The amount of energy released by the Sun is such that the mass of the Sun is decreasing at a rate of 4.3 billion kilograms per second! Yet this is such a small fraction of the Sun's mass that the change is hardly noticeable.

Knowing the mechanism and the mass of the Sun, we are able to calculate its lifetime. Our Sun is believed to be about 4.5 billion years old and will probably continue its present activity for another 4.5 billion years.

Flawed Reasoning

Your friend voices the following concern: "Scientists claim that a fusion reactor would never melt down like Chernobyl. Isn't the Sun a perfect example of a fusion reactor that is out of control?" **How might you respond?**

Answer Fusion reactors should be much safer than fission reactors. The two types of reactors have very different answers to the question, "What's the worst that could happen?" In a fission reactor, the core can go supercritical and produce much more energy than can be controlled, and a meltdown of the core could occur. In a fusion reactor, the fusion process stops, and very little additional energy is produced. The fuel in the Sun is held close together at high temperatures by the enormous gravitational pressure at its core. This mechanism is not possible on Earth.

PHYSICS | ON YOUR OWN

Learn more about the Sun by visiting the excellent website http://solar.physics.montana.edu/YPOP/. Start with the "Solar Tour" and then try some of the activities in "The Solar Classroom."

Physics♦Now™ Assess your understanding of this chapter's topics with sample tests and other resources found by logging into PhysicsNow at http://physics.brookscole.com/kf6e.

Summary

Nuclei stay together despite the electromagnetic repulsions between protons because of a nuclear force. This strong force between two nucleons has a very short range, is about 100 times as strong as the electric force, has a repulsive core, and is independent of charge. A second force in the nucleus, the weak force involved in the beta-decay process, is also short-ranged but very weak, only about one-billionth the strength of the strong force.

Information about nuclei initially came from particles ejected during radioactive decays. If a nucleus is above the line of stability, it has extra neutrons, which usually results in a neutron decaying into a proton and an electron via beta minus decay. If the isotope is below the line of stability, alpha or beta plus decay or electron capture increases the number of neutrons relative to the number of protons.

Later, the particles from radioactive decays were used as probes to study the structure of stable nuclei. Finally, particle accelerators produced beams of charged particles with much higher momenta and, consequently, much smaller wavelengths. The largest of these accelerators produce beams with energies exceeding 1 trillion electron volts.

The average binding energy per nucleon varies for the stable nuclei; some nuclei are more tightly bound than others, reaching a maximum near iron. Combining light nuclei or splitting heavier nuclei releases energy. The energy differences between nuclei are large enough to be detected as mass differences.

Bombarding uranium with neutrons splits the uranium nuclei, releasing large amounts of energy. Typical reaction energies are 100 million times as large as those of chemical reactions. The fission reaction is a practical energy source because a single reaction emits two or three neutrons that can trigger additional fission reactions.

Another way of releasing energy, fusion, combines light nuclei to form heavier ones. Fusion requires high temperatures and the confinement of a sufficient density of material for long enough so the reactions can take place and return more energy than was needed to initiate the process. This technology is still being developed. The Sun, however, is a working fusion reactor.

Chapter 26 Revisited

Nuclear energy can be used to heat water and make steam to turn turbines, just like the other options. The differences are at the front end—releasing the energy to make the steam. With coal, oil, and natural gas, we burn the fuel. In conventional nuclear power plants, we create a controlled chain reaction of nuclear disintegrations. Future fusion reactors will combine hydrogen nuclei to form heavier nuclei. With all nuclear reactors, the questions revolve around the use, production, and disposal of radioactive materials. Because radioactivity is unalterable by normal means, we need to isolate these materials from the biosphere. This is a formidable technological challenge.

KEY TERMS

binding energy: The amount of energy required to take a nucleus apart.

chain reaction: A process in which the fissioning of one nucleus initiates the fissioning of others.

critical chain reaction: A chain reaction in which an average of one neutron from each fission reaction initiates another reaction.

fission: The splitting of a heavy nucleus into two or more lighter nuclei.

fusion: The combining of light nuclei to form a heavier nucleus.

line of stability: The locations of the stable nuclei on a graph of the number of neutrons versus the number of protons.

moderator: A material used to slow the neutrons in a nuclear reactor.

particle accelerator: A device for accelerating charged particles to high velocities.

strong force: The force responsible for holding the nucleons together to form nuclei.

subcritical: A chain reaction that dies out because an average of less than one neutron from each fission reaction causes another fission reaction.

supercritical: A chain reaction that grows rapidly because an average of more than one neutron from each fission reaction causes another fission reaction, an extreme example of which is the explosion of a nuclear bomb.

weak force: The force responsible for beta decay.

CONCEPTUAL QUESTIONS

1. What electric potential difference is required to accelerate a proton to an energy of 10 million electron volts?
2. What electric potential difference is required to accelerate an alpha particle to an energy of 20 million electron volts?
3. An electron, a positron, and a proton are each accelerated through a potential difference of 1 million volts. Which of these, if any, acquires the largest kinetic energy? Explain your reasoning.
4. An electron, a positron, and a proton are each accelerated through a potential difference of 1 million volts. Which of these, if any, acquires the greatest final speed? Explain your reasoning.
5. Which acquires a larger kinetic energy when accelerated by the same potential difference, a proton or an alpha particle? Why?
6. Which acquires a greater final speed when accelerated by the same potential difference, a proton or an alpha particle? Why?
7. Why is a circular accelerator more suitable for protons than electrons?
8. If the Stanford Linear Accelerator were to be used to accelerate positrons instead of electrons, would the positron source be at the same end as the electron source or at the opposite end? Explain.
9. List the four fundamental forces in the order of decreasing strength.
10. Which of the forces in nature is associated with beta decay?
11. How do we know that there must be a strong force?
12. What evidence do we have to support the idea that the strong force is stronger than the electromagnetic force?
13. What are the similarities and the differences between the strong force and the electromagnetic force?
14. For the electric force, the charge of the particles determines whether it is attractive or repulsive. What determines whether the strong force is attractive or repulsive?
15. What is released when a neutron and a proton combine to form a deuteron?
16. Which nucleus would have the greater total binding energy, $^{56}_{26}$Fe or $^{112}_{48}$Cd? Explain.
17. Which of the following has the largest mass: 90 protons and 150 neutrons, $^{240}_{90}$Th, or $^{110}_{40}$Zr plus $^{130}_{50}$Sn? How do you know this?
18. Which of the following has the smallest mass: 90 protons and 150 neutrons, $^{240}_{90}$Th, or $^{110}_{40}$Zr plus $^{130}_{50}$Sn? Explain.
19. Both $^{12}_{7}$N and $^{12}_{5}$B decay to the stable nucleus $^{12}_{6}$C. Which of these three nuclei has the smallest mass? How do you know this?
20. $^{14}_{6}$C decays to $^{14}_{7}$N via beta minus decay. Which nucleus has the larger mass? Why?
21. $^{17}_{7}$N beta decays to $^{17}_{8}$O with a reaction energy of 8.68 MeV. $^{17}_{9}$F beta plus decays to $^{17}_{8}$O with a reaction energy of 2.76 MeV. Which parent nucleus has the greater mass? Explain.
22. $^{24}_{11}$Na with a mass of 23.991 atomic mass units beta decays to $^{24}_{12}$Mg. $^{24}_{13}$Al with a mass of 24.000 atomic mass units decays by electron capture to $^{24}_{12}$Mg. Which reaction has the greater decay energy? How do you know this?
23. The nuclear fusion process in stars much more massive than our Sun continues to fuse lighter elements together to form heavier ones with the release of energy. Use Figure 26-5 to explain why iron-56 is the heaviest element produced in this fashion.
▲ 24. Suppose the curves for the average binding energy were those shown in the figure instead of the actual binding energy curve shown in Figure 26-5. Would fission and fusion be possible in each case? If so, for approximately what range of nucleon number would they occur?

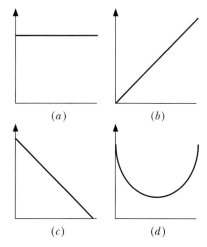

(a) (b) (c) (d)

25. How do the numbers of neutrons and protons compare for most stable nuclei with small atomic numbers? How do we account for this?
26. What general statement about the relative numbers of neutrons and protons can you make about nuclei with large atomic numbers? How do we account for this?
27. Why can't a stable nucleus contain only protons?
28. You may have learned the law "Matter can neither be created nor destroyed." Is this statement in agreement with modern-day physics?
29. Would you expect the nucleus $^{100}_{50}$Sn to be stable? If not, how would you expect it to decay?
30. Would you expect the nucleus $^{130}_{50}$Sn to be stable? If not, how would you expect it to decay?
31. What is the most likely decay mode for $^{20}_{11}$Na?
32. How would you expect an unstable nucleus of $^{80}_{30}$Zn to decay?
33. What is nuclear fission?
34. Would a uranium nucleus release more or less energy by splitting into three equal mass nuclei rather than two? Explain.
35. How many neutrons are released in the following fission reaction?

$$^{1}_{0}n + ^{235}_{92}U \rightarrow ^{140}_{54}Xe + ^{94}_{38}Sr + (?)^{1}_{0}n$$

36. Assume that a $^{235}_{92}$U nucleus absorbs a neutron and fissions with the release of three neutrons. If one of the fission fragments is $^{144}_{56}$Ba, what is the other?

37. Why can the fissioning of $^{235}_{92}$U produce a chain reaction?
38. What factors determine whether a piece of uranium undergoes a subcritical or supercritical reaction?
39. Why is it important for fission reactions to emit neutrons?
40. Why is it critical for nuclear fission chain reactions that heavier elements have a greater ratio of neutrons to protons than lighter elements?
41. In which device, a nuclear reactor or a nuclear bomb, does the greater number of neutrons per fission event go on to initiate another reaction?
42. Why don't chain reactions occur in naturally occurring deposits of uranium?
43. Why does a nuclear reactor have control rods?
44. What is the difference between a moderator and a control rod in a fission reactor?
45. The fissioning of plutonium-239 yields an average of 2.7 neutrons per reaction compared with 2.5 for uranium-239. Which substance would have the smaller critical mass?
46. Would it have been easier or harder to develop fission reactors if the average number of neutrons released per fission of uranium-235 were 2.0 instead of 2.5?
47. What is bred in a breeder reactor?
48. What problem is solved by a breeder reactor?
49. What is nuclear fusion?
50. Why are high temperatures required for nuclear fusion?
51. The temperature of the plasma in a typical household fluorescent light is 20,000°C. Why can you touch an operating light without being burned?
52. In a Tokamak fusion reactor, magnetic fields are used to hold plasma that must be heated to temperatures comparable to those in the Sun's core. If this "magnetic bottle" were to fail, would the reactor melt down? Why or why not?
53. What advantages would a fusion reactor have over a fission reactor?
54. What are the two basic approaches to developing fusion reactors?
55. Why do scientists believe that the Sun is 4.5 billion years old?
56. What are the conditions in the interiors of stars that make fusion possible?
57. What is the basic difference between fusion and fission?
58. Does $E = mc^2$ apply to both fusion and fission? What about an explosion of dynamite?

EXERCISES

1. A proton is accelerated through a potential difference of 10^6 V. Find the proton's momentum and its wavelength.
2. An alpha particle is accelerated through a potential difference of 10^6 V. Find the alpha particle's momentum and its wavelength.
3. A proton is accelerated through a potential difference of 4×10^{11} V. Show that a nonrelativistic calculation yields a final velocity greater than the speed of light.
4. A proton is accelerated through a potential difference of 4×10^{11} V. For energies much greater than the rest-mass energy ($E_0 = mc^2 = 938$ MeV, for a proton), a relativistic treatment yields a momentum of E/c, where c is the speed of light. What wavelength does this proton have?
5. Given that 1 amu equals 1.66×10^{-27} kg, show that 1 amu has an energy equivalent of 931 MeV.
6. On average, how many fission reactions of uranium-235 would it take to release 1 J of energy?
7. Given that the neutral nitrogen atom with seven neutrons has a mass of 14.003 074 amu, what is its total binding energy?
8. Calculate the total binding energy of the $^{12}_{6}$C nucleus.
9. The mass of the neutral lithium-7 atom is 7.016 004 amu. Find the mass of the bare lithium-7 nucleus.
10. The mass of the neutral tritium atom is 3.016 049 amu. Find the mass of the bare nucleus, called the triton.
11. How much energy is released when $^3_1$H decays to form $^3_2$He? The masses of the neutral hydrogen and helium atoms are 3.016 049 and 3.016 029 amu, respectively.
12. How much energy is released when $^{14}_{6}$C decays to form $^{14}_{7}$N? The masses of the neutral carbon and nitrogen atoms are 14.003 242 and 14.003 074 amu, respectively.
13. How much energy is released in the alpha decay of $^{239}_{94}$Pu? The masses of the neutral plutonium, uranium, and helium atoms are 239.052 158, 235.0439 25, and 4.002 603 amu, respectively.
14. How much energy is released in the alpha decay of $^{214}_{84}$Po? The masses of the neutral polonium, lead, and helium atoms are 213.995 190, 209.990 069, and 4.002 603 amu, respectively.
15. Use Figure 26-5 to estimate the energy released if $^{239}_{94}$Pu fissions to become $^{96}_{40}$Zr and $^{141}_{54}$Xe with the release of two neutrons.
16. Use Figure 26-5 to estimate the energy released if $^{236}_{92}$U fissions to become $^{142}_{45}$Ba and $^{91}_{36}$Kr with the release of three neutrons.
17. Show that the fusion reaction given below releases 17.6 MeV of energy. The masses of the deuteron and triton are 2.013 55 and 3.015 50 amu, respectively.

$$^2_1H + ^3_1H \rightarrow ^4_2He + ^1_0n$$

18. How much energy is released in the following fusion reaction? The masses of deuteron and triton are given in Exercise 17.

$$^2_1H + ^2_1H \rightarrow ^3_1H + ^1_0n$$

19. In the first cycle of a fission chain reaction, a single nucleus fissions and produces three neutrons. If every free neutron initiates a new fission event, then three nuclei fission in the second cycle, for a total of four. What is the total number of fission events after five cycles?
20. In the first cycle of a fission chain reaction, a single nucleus fissions and produces two neutrons. If every free neutron initiates a new fission event, then two nuclei

fission in the second cycle, for a total of three. If each fission event releases 210 MeV on average, what is the total energy released after eight cycles?

21. Given that the mass of the Sun is decreasing at the rate of 4.3×10^9 kg/s, what is the present energy radiated by the Sun each second?

22. Approximately how much mass has the Sun lost during its lifetime? How does this compare with its current mass of 2×10^{30} kg?

27 Elementary Particles

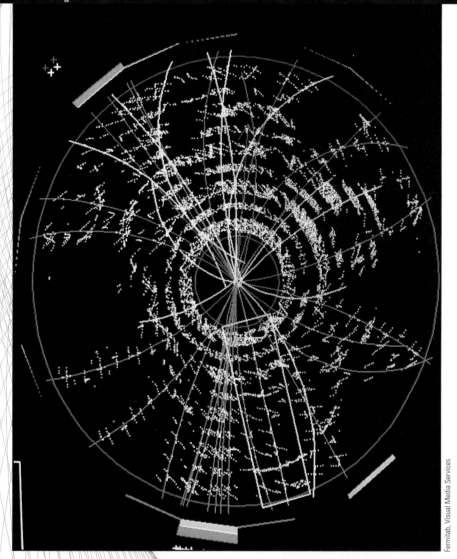

Computer-generated reconstruction of a collision in the Collider Detector at Fermilab. It is possible that the collision of a proton and an antiproton produced a top quark and antiquark.

Throughout recorded history we have searched for the primary building blocks of nature. Aristotle's four elements became the chemical elements, which gave way to electrons, protons, and neutrons. As we delved deeper, we found new, fascinating layers. Have we found the ultimate building blocks? If not, where will our search end?

(See page 587 for an answer to this question.)

THE idea that all of the diverse materials in the world around us are composed of a few simple building blocks is very appealing; because of its appeal, the idea has existed for more than 20 centuries. The search for these elementary components of matter is fueled by the desire to simplify our understanding of nature.

The search began with the Aristotelian world view, which assumed that everything was made of four basic elements: earth, fire, air, and water. During the 18th and 19th centuries, these four were eventually replaced by the modern chemical elements. Although the initial list numbered only a few dozen elements, it grew to more than 100 entries. A hundred different building blocks are not as appealing as four. Things improved, however, when atoms were discovered to be divisible and composed of three even more basic building blocks. The elementary particles described in 1932 consisted of the three constituents of atoms—the electron, proton, and neutron—and the quantum of light—the photon. These four building blocks restored an elegant simplicity to the physics world view.

This beautiful picture did not survive for long. As experimenters probed deeper and deeper into the subatomic realm, new particles were discovered. The experimenters were constantly confronted with new, seemingly bizarre observations. The first, and perhaps most bizarre, observation occurred during the same year the neutron was discovered and resulted in the discovery of an entirely new kind of matter.

Antimatter

A new type of particle was discovered in 1932. When high-energy photons (gamma rays) collided with nuclei, two particles were produced: one was an electron, the other an unknown. Double spirals in bubble-chamber photographs showed tracks of the pairs of particles that were created (Figure 27-1). This **pair production** is a dramatic example of Einstein's equation $E = mc^2$; energy (the massless gamma ray) is converted into mass (the pair of particles). The curvature of the new particle's path in the bubble chamber's magnetic field revealed that it had the same charge-to-mass ratio as an electron, but because it curved in the opposite direction, it had a positive charge. All the properties of this new particle were of the same magnitude as the electron's, although some properties had the opposite sign. For example, the masses were identical, and the electric charges

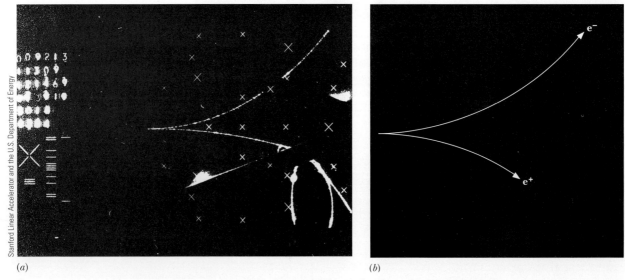

Figure 27-1 (a) A bubble-chamber photograph and (b) drawing of the tracks of a positron–electron pair.

The antinucleons were discovered at the Bevatron, a particle accelerator in Berkeley, California.

were the same size but opposite in sign. This positively charged electron was named the **positron** and is the **antiparticle** of the electron.

The prediction of the existence of the positron was contained in a quantum-mechanical theory for the electron developed by the English physicist P. A. M. Dirac in 1928. Until its discovery, however, Dirac's "other electron" seemed more like a mathematical oddity in the theory than a physical possibility. After its discovery, the full significance of Dirac's ideas was recognized.

Dirac's theory contained similar predictions for the antiparticles of the proton and neutron, and after the discovery of the positron they were assumed to exist. Finding these heavier antiparticles took two decades. A primary reason for the delay was the large amount of energy needed. To create a pair of particles, the available energy has to be larger than the energy needed to create the rest masses of the two particles. (The special theory of relativity tells us that mass and energy are equivalent. Therefore, it takes energy to create the masses of the particles.) Protons and neutrons are so much more massive than electrons that the production of their antiparticles had to await the development of huge accelerators to achieve these great energies. The antiproton was observed in 1955 and the antineutron in 1956.

Antiparticles are usually designated by putting a bar over the symbol of the corresponding particle. Thus, an antiproton becomes $\bar{p}$ and an antineutron $\bar{n}$. However, the positron is usually written as e^+.

Antiparticles don't survive long around matter. When particles and antiparticles come into contact, they annihilate, converting their combined mass into energy. This process is the reverse of pair production. Because the world we know is assumed to be "regular" matter, antiparticles have little chance of surviving. Once they are created, their lifetimes are determined by how long it takes them to meet their corresponding particle. The positron is slowed by collisions with particles and is eventually captured by an electron. They orbit each other briefly to form an "atom." In a time typically much less than a millionth of a second, the two annihilate, converting their combined mass back into photons.

If an antiparticle did not meet its counterpart, annihilation would not occur, and the antiparticle would exist for a long time. This means that antiatoms could be formed from antielectrons, antiprotons, and antineutrons. The only reason that this doesn't usually happen is the extremely low probability of these antiparticles finding each other in a world so predominantly composed of ordi-

Could any of these stars be composed of antimatter?

nary particles. However, under special circumstances, this can be achieved. Antideuterium—an antinucleus consisting of one antiproton and one antineutron—has been observed, and in 1995 antihydrogen atoms were produced for very brief periods of time.

The existence of antiatoms leads to the fascinating question of whether antiworlds might exist somewhere in the Universe. There doesn't seem to be any reason to believe they don't exist. Antiatoms should behave the same as atoms. In particular, they would display the same spectral lines. And, because photons and antiphotons are identical, looking at a distant star won't reveal whether it is composed of matter or antimatter. However, evidence shows that each cluster of galaxies must be either matter or antimatter. If there were matter and antimatter galaxies in a single cluster, the intergalactic dust particles would annihilate, giving off characteristic photons. These photons have not been observed.

Because radio waves are composed of photons, we would have no trouble communicating via radio with antihumans on an antiworld. Although distances make the possibility extremely unlikely, any attempt to communicate by visiting would result in an explosion larger than any bomb that we could build. The entire mass of the spaceship and an equal mass of the antiworld would annihilate each other.

The discovery of antiparticles provided a reassuring demonstration that the conservation laws of momentum and energy hold in the subatomic world. Every annihilation yields at least two photons. Suppose, for example, a positron–electron pair were orbiting each other as shown in Figure 27-2(a). If we are at rest relative to the pair, they have equal energies and equal but oppositely directed momenta. Before the annihilation they have a total energy that is twice the mass of one of them, but their total momentum is zero due to their traveling in opposite directions. If only one photon were produced by the annihilation, the total momentum would not be zero but would be equal to that of the photon—an obvious violation of the conservation of momentum. This situation never occurs; there are always at least two photons produced in the annihilation. The photons' total momentum is zero, as illustrated in Figure 27-2(b). The conservation rules have been extremely useful in helping us understand the details of the elementary particles' interactions.

Question   Would a decay into three photons be forbidden by the laws of conservation of linear momentum and energy?

Answer   No. The momenta of three photons can be arranged in many ways to get a total momentum of zero and a given total energy. Therefore, these two laws do not forbid this process from occurring.

Figure 27-2 The two photons produced in an electron–positron annihilation have equal but oppositely directed momenta to match the zero momentum of the electron–positron pair.

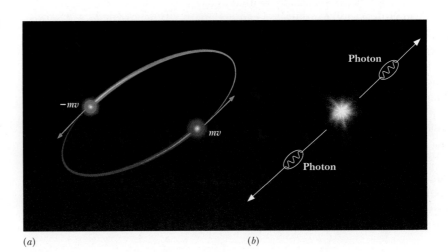

(a)   (b)

The Puzzle of Beta Decay

In Chapter 25 we saw that one element can spontaneously change into another element by undergoing beta decay. On the nucleon level, this means that a neutron turns into a proton, or vice versa. This process led to an interesting puzzle because beta decay did not appear to satisfy the very basic conservation laws for energy and linear momentum.

Imagine you are sitting in a reference system in which a neutron is at rest. The linear momentum of the neutron is zero, and its energy is that associated with its rest mass. If we assume that beta decay changes the neutron into a proton by emitting an electron, momentum can be conserved only if the electron goes off in one direction and the newly created proton recoils in the opposite direction with the same size momentum (Figure 27-3). The value of these momenta is determined by the requirement that energy also be conserved. Because this can happen only in one way, it was expected that the electron must always emerge with the same kinetic energy.

This was the expected result, but experiments showed that this doesn't happen. The ejected electrons do not have a single kinetic energy. The graph of experimental data in Figure 27-4 shows a continuous range of kinetic energies from zero up to a maximum value that is equal to the value predicted above.

Scientists were in a dilemma. There seemed to be two choices: they could abandon the conservation laws or assume that one or more additional particles were emitted along with the electron. In 1930 Pauli proposed that a third particle, the **neutrino**, was involved. (*Neutrino* means "little neutral one.") Using the conservation laws, he even predicted its properties. The neutrino has to be neutral because charge is already conserved. The fact that the electron sometimes emerges with all the kinetic energy predicted for the decay without the neutrino means that the neutrino sometimes carries away little or no energy. For this to be possible, the neutrino's rest mass must be very, very small because it would require some energy to produce its rest mass.

Even though the neutrino was not observed, faith in its existence continued to grow. It was such a nice solution to the beta-decay puzzle that experimental verification of the neutrino's existence seemed like it would only be a matter of time. "Only a matter of time" eventually became 26 years. In 1956 Clyde Cowan and Frederick Reines finally detected neutrinos, using an intense beam of radiation from a nuclear reactor (Figure 27-5). The observed reaction had an antineutrino $\bar{\nu}$ strike a proton, yielding a neutron and a positron.

$$\bar{\nu} + p \rightarrow n + e^+$$

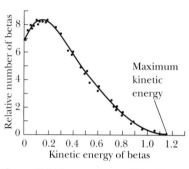

Figure 27-3 If beta decay of a neutron produced only a proton and an electron, they would have to emerge with equal but oppositely directed momenta.

Figure 27-4 The spectrum of kinetic energies of electrons emitted in the beta decay of neutrons.

◄ discovery of the neutrino

Figure 27-5 This apparatus was used by Cowan and Reines at the nuclear reactor in Savannah River, South Carolina, to detect the neutrino.

The properties of the neutrino were confirmed by the study of the dynamics of this interaction. Its rest mass had been shown to be very small; it was often assumed to be zero. However, in 1980 some experimental results indicated that the mass of the neutrino may not be exactly zero, and recent experiments have indicated that the neutrino mass is not zero (Chapter 28).

These results could affect our understanding of the evolution of stars and the Universe. Neutrinos are so abundant in the Universe that they may contribute enough mass to the Universe that it may eventually quit expanding and collapse under its own gravitational attraction (Chapter 28).

The long delay in detecting neutrinos was due to the extremely weak interaction of neutrinos with other particles; neutrinos do not participate in the electromagnetic or the strong interactions, only in the weak interactions. In fact, the neutrinos' interaction with other particles is so weak that only one of a trillion neutrinos passing through Earth is stopped.

Exchange Forces

During the development of Newton's law of universal gravitation, a nagging question arose about forces: what is it that reaches through empty space and pulls on the objects? Newton's idea of an action at a distance seemed unsatisfactory. A couple of hundred years later, the concept of a field provided an alternative (Chapter 20). One object creates a change in space (the field), and a second object responds to this field. At least empty space was filled with something, but it was still somewhat unsatisfying.

With the discovery of the quantum of energy, a new problem arose. An object moving through a field would gain energy continuously. However, the fact that the energy was quantized (Chapter 24) meant that the object should receive energy only in discrete lumps. This conclusion led to a picture of elementary particles interacting with each other through the exchange of still other elementary particles. Electrons, for example, exchange photons.

Instead of an electron being repelled by another electron due to an action at a distance or by a field, each electron continuously emits and absorbs photons. Because we are talking about effects that are entirely quantum mechanical, it is risky (if not foolhardy) to rely too heavily on classical analogies. Uncomfortable as it seems, we should be content to say that the interaction properties can be explained by the assumption that photons are exchanged and, depending on the photon properties, the particles attract or repel each other. Richard Feynman created a way of showing these interactions graphically. Imagine the Feynman diagram in Figure 27-6 as a graph in which the vertical dimension is time. Two electrons approaching each other exchange a photon and repel each other.

We should rely on the quantum-mechanical explanation, and yet analogies sometimes make things plausible. Consider the following: Imagine two people standing on skateboards, as shown in Figure 27-7. One throws a basketball to the

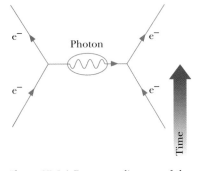

Figure 27-6 A Feynman diagram of the interaction between two electrons through the exchange of a photon.

Figure 27-7 A classical analogy of a repulsive exchange force is tossing a basketball back and forth.

FEYNMAN | Surely You're Joking, Mr. Feynman

The word *play* occurs often in scientists' writings. Newton said that he felt as if he were a child playing on the seashore as the great ocean of undiscovered law lay before him. A brilliant physicist, Richard Feynman (1918–1988), made play a central element in his work on what he called "strange particles" and in his conceptual breakthrough that led to the development of quantum electrodynamics (QED). Faith is an occupational hazard for physicists; doubt is a supreme virtue. In one of his many talks to students at the California Institute of Technology (Cal Tech), Feynman invited them to join him in studying nanotechnology—the study of very small things. The talk was titled "There's Plenty of Room at the Bottom." Most of us would like to head for the top, but Feynman played with words to invite them to explore the very small, where nature provides immense challenges. Playfulness provides us with the gusto and enthusiasm to explore our natural world.

Richard Feynman

Feynman was a playful prodigy as a child. He was the son of Jewish parents, a second-generation Russian father and a Polish mother. As a child in Far Rockaway, New York, he devoured science and mathematics. Afraid of being seen as a sissy, he developed a vigorous and almost combative playful style. In mathematics he moved ahead on his own at a furious pace, even inventing his own notation for trigonometric functions. He soon realized that no one else could use his notation and returned to the conventional notation. He was an undergraduate at MIT and even there began an unconventional course of study that led him to neglect required subjects such as history. When he petitioned to become a doctoral candidate, his mentor at MIT urged that he transfer to Princeton where he would be better challenged. Admission to Princeton in the late 1930s was difficult for a young Jewish student. Unspoken but real quotas existed in elite universities. When he presented his first seminar at Princeton, the audience included Einstein, Pauli, and John von Neumann, the inventor of game theory. Feynman received his Ph.D. in physics in 1942 and then joined the project to construct an atomic bomb at Los Alamos, New Mexico.

It was a taxing effort to restrain the boisterous young theorist. Feynman challenged the system by playing bongo drums at odd hours, sending coded letters to friends to frustrate the censors, and picking locks on safes containing classified documents and inserting little notes for the security staff. He described these high jinks in an autobiographical work he published 30 years later: *Surely You're Joking, Mr. Feynman* (New York: Norton, 1985).

After spending time on the faculty at Cornell University, in 1950 Feynman moved to Cal Tech in Pasadena. There he produced his magisterial works on the interaction of particles and atoms in radiation fields. His lifelong fascination with spacetime led him to develop *Feynman diagrams*, an innovative system that provided a visual representation of particle interactions. Along with Julian Schwinger and Shinichiro Tomonaga, he was awarded the Nobel Prize in 1965 for this work. He called his Nobel Prize "a pain in the neck" because of the increased demands on his time.

He was always interested in practical applications of science and later in life said that he wished he had taken up administration in large-scale enterprises like NASA because they offered unique new challenges. Feynman was called in to help investigate the tragic *Challenger* explosion. He demonstrated that when O rings (round rubber rings) are chilled, as they had been on that frosty night in Florida, they inevitably fail. He urged NASA engineers and managers to develop more rigorous procedures for testing all components under realistic conditions. In cases like this, it behooves scientists and engineers to beware their little bit of ignorance.

Stomach cancer afflicted him in later years. Before he died at the age of 69, he wrote, "The vastness of nature stretches my imagination. Stuck on this carousel my little eye can catch one-million-year-old light. . . . It does not harm the mystery to know a little about it."

—*Pierce C. Mullen, historian and author*

Sources: Jagdish Mehra, *The Beat of a Different Drum: The Life and Science of Richard Feynman* (New York: Oxford, 1994); James Gleick, *Genius: The Life and Science of Richard Feynman* (New York: Pantheon, 1992); Richard P. Feynman, *What Do You Care What Other People Think?* (New York: Norton, 1987).

---

other. The person throwing the basketball gives it some momentum and therefore must acquire some momentum in the backward direction. The person catching the basketball must absorb this momentum and therefore acquires momentum in the direction away from the thrower. The total interaction behaves like a repulsive force between the two people. Although this analogy illustrates that exchanging particles can affect particles at a distance and is easy to visualize, it fails for an attractive force. We must return to the reality that these are quantum-mechanical effects and that our commonsense world view does not serve us very well in the subatomic world of elementary particles.

We can ask if the idea of exchanging photons accounts for the observations. It does; however, it also poses a new problem. Imagine an electron at rest in space. If we explain its effect on other particles by saying it emits photons, it must recoil after the emission due to the conservation of momentum. But if it recoils, it has kinetic energy. In fact, the emitted photon also has some energy. These energies are over and above the rest mass of the electron and thus in violation of the law of conservation of energy (Figure 27-8). The same argument can be made for the electron that feels the interaction by absorbing the photon.

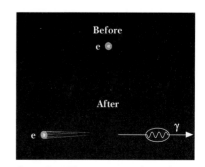

Figure 27-8 The emission of a real photon by an isolated particle violates the laws of conservation of energy and momentum.

Once again we are confronted with a situation in which energy and momentum are not conserved. Only in this case, the solution is different; we are not bailed out by the discovery of a new particle. This violation exists. However, it does not mean the abandonment of the conservation rules. Overall, momentum and energy are conserved. The violations created by the emission of the photon are canceled by the absorption of the photon by the other particle. It is only during the time that the photon travels from one particle to the other that the violation exists.

We invoke the uncertainty principle (Chapter 24) to understand this situation. One form of the uncertainty principle says that we can determine the energy of a system only to within an uncertainty ΔE that is determined by the time Δt taken to measure the energy. The product of these two uncertainties is in fact always greater than Planck's constant. This is our escape from the dilemma. If a violation of energy conservation by an amount ΔE takes place for less than a time Δt, no physical measurement can verify the violation. We now believe that unmeasurable violations of the conservation of energy (and momentum) can and do take place. (We have to be a bit careful with this resolution of the problem because we are assigning a classical type of trajectory to the photon like we tried to do with electrons in atoms. Once again it is really a quantum-mechanical effect; energy and momentum are conserved for the interaction.)

Because photons have no rest mass, their energies can be very small. Therefore, the violation of energy conservation due to the emission of a photon can be very small. Very small energy violations can last for very long times, and these photons can travel large distances before they are reabsorbed, which explains the infinite range of the electromagnetic force. The fact that the electromagnetic force decreases with distance is explained by the observation that exchange photons that have long ranges have low energies (and momenta) and therefore produce smaller effects.

Exchange Particles

Although perhaps sounding bizarre, the idea of exchanging particles explains more than previous models. This new concept of a force satisfies the requirements of quantum mechanics and provides a way of understanding all forces. As an example, because the different forces have different characteristics, presumably they have different exchange particles.

In 1935 Japanese physicist Hideki Yukawa used this radical idea to show why the nuclear force abruptly "shuts off" after a very short distance. Yukawa reasoned that the short range of the strong force required the exchange particle to have a nonzero mass. The mere creation of its mass requires an energy violation. This minimum energy violation must be at least as large as the rest-mass energy, which means that there is a limit to the time that the violation can occur and therefore a maximum distance the particle can travel before it must be absorbed by another particle. In other words, the exchange of nonzero rest-mass particles means that the force has a limited range.

Finding the Yukawa exchange particle was difficult. It couldn't be detected in flight between two nucleons because of the consequences of the uncertainty principle. The hope was that it might show up in other interactions. We observe photons, for example, when they are created in jumps between atomic levels, not from being "caught" between two electrically charged particles.

During the 1930s cosmic rays were the only known source of particles with energy high enough to create the new exchange particles. Cosmic rays are continually bombarding Earth's atmosphere, creating many other particles that rain down on Earth in extensive showers. In 1938 a new particle was discovered in a cosmic-ray shower that had a mass 207 times that of the electron and a charge equal to the electron's charge. For a while it was thought that this was the particle predicted by Yukawa, but it did not interact strongly with protons and neutrons. Therefore, the new particle, now known as the **muon**, could not be the exchange particle for the strong force.

The search for Yukawa's particle continued, and it was finally discovered 10 years later in yet another cosmic-ray experiment. The **pion** (short for pi meson) has a mass between that of the electron and the proton and comes with three possible charges: +1, 0, and −1 times that on the electron. Although the pion is no longer considered to be an exchange particle, it played a pivotal role in the acceptance of the idea of exchange forces. (We return to the question of the exchange particle for the strong force in a later section.)

The success of this model for the interactions between particles led to the hypothesis that all forces are due to the exchange of particles. The gravitational force is presumably due to the exchange of **gravitons**. Because of the similarities between the gravitational force and the electromagnetic force, the graviton should have properties similar to the photon. It should have no rest mass and travel with the speed of light. Although the graviton has not been observed, most physicists believe it exists.

The exchange particles for the weak force, however, have been detected. The weak force occurs through the exchange of particles known as **intermediate vector bosons**. These three particles were discovered in 1983: the W comes in two charge states, +1 and −1, and the Z^0 is neutral. One of the reasons it was so difficult to discover these particles is that they are very massive, each one having more than 100 times the mass of the proton. Their discovery had to wait for the construction of new accelerators.

Flawed Reasoning

While discussing the difference between the ranges of the electric force and the weak force, your classmate asserts, "Particles with mass, like the intermediate vector bosons, cannot travel at the speed of light like photons can. This is what limits the range of the weak force." **How do you respond to this?**

Answer It is not the speed of the intermediate vector bosons that limit their range, but their rest mass. The Heisenberg uncertainty principle allows a particle to be created and exchanged in violation of energy conservation as long as the particle exists for a short enough time that the violation cannot be physically observed. The larger the energy required to create the particle, the shorter the time it can exist. Photons can have arbitrarily small energies because they have no rest mass. They can exist for arbitrarily long times and travel arbitrarily large distances. However, intermediate vector bosons have large rest masses, providing a lower limit on the amount of energy required to create them. They can exist for very short times and travel extremely short distances.

The Elementary Particle Zoo

By 1948 the discovery of antiparticles and exchange particles had nearly tripled the number of known elementary particles. And the situation got worse. During the next 7 years, four other particles were discovered. Because the behavior of these particles did not match that of the known particles, they became known as the **strange particles**. The existence of the neutrino was confirmed in 1956. A second type of neutrino was discovered in 1962, and even a third type exists.

The 1960s also witnessed the discovery of another new phenomenon. Particles were discovered that live for such a short time that they decay into other particles before they travel distances that are visible even under a microscope.

Typical lifetimes for these particles are 10^{-23} second (the time it takes light to travel across a nucleus!).

Question What does the uncertainty principle say about the mass of a particle that has such a short lifetime?

Answer The uncertainty in the energies (and consequently in their masses due to the equivalence of mass and energy) must be large because the product of the uncertainties in the energy and the lifetime must exceed Planck's constant. This has been confirmed by many experiments.

Before long the number and variety of particles became so large that physicists began calling the collection a zoo. The proliferation of new particles had once again destroyed the hope that the complex structures in nature could be built from a relatively small number of simple building blocks. The number of "elementary" particles exceeded a few hundred, and the number continued to grow. Particle physicists began to feel organizational problems similar to those of zookeepers.

Much as zookeepers build order into their zoos by grouping the animals into families, particle physicists began grouping the elementary particles into families. Making families helps organize information and may result in new discoveries. When confronted with many seemingly unrelated facts, scientists often begin by looking for patterns. Mendeleev developed the periodic table of chemical elements using this technique. A Swiss mathematics teacher, Johann Balmer, decoded the data on the spectral lines of the hydrogen atom by arranging and rearranging the wavelengths. In the process he discovered a formula that gave the correct results. Both of these discoveries were empirical relationships, not results derived from fundamental understandings of nature. They served, however, to classify the data and provide some guidance for further experimental work.

Elementary particle physics was in a similar condition. Large amounts of data had been accumulated. Scientists were looking for patterns that might provide clues for the development of a comprehensive theory. Just as any collection of buttons can be classified in many ways—size, color, shape, and so forth—the elementary particles can be classified in different ways. (Of course, macroscopic attributes like color and size don't apply here.) One fruitful way is to group them according to the types of interaction in which they participate.

All particles participate in the gravitational interaction—even the massless photon because of the equivalence of mass and energy. Therefore, this interaction doesn't yield any natural divisions for the particles. Furthermore, gravitation is so small at the nuclear level that it usually isn't included in discussions of particle behavior.

The particles that participate in the strong interaction are called **hadrons**. This family includes most of the elementary particles. The hadrons are further divided into two subgroups according to their spin quantum numbers. The **baryons** have spins equal to $\frac{1}{2}, \frac{3}{2}, \frac{5}{2}, \ldots$ of the quantum unit of spin, whereas the **mesons** have whole-number units of spin. The best-known baryons are the neutron and proton. Table 27-1 lists some common hadrons and their properties. There are others, but their lifetimes are extremely short—less than a billionth of most of those listed.

The **lepton** family includes the electron, the muon, the tau (discovered in 1977), and their associated neutrinos. The tau lepton is even more massive than the muon. The word *lepton* means "light particle" and refers to the observation that (with the exception of the tau) leptons are less massive than the hadrons. In fact, they appear to be pointlike, having no observable size and no evidence

Table 27-1 | Properties of Some of the Hadrons

Name	Symbol	Spin ($h/2\pi$)	Rest Mass (MeV/c^2)	Half-Life (s)	Strangeness
Baryons					
Proton	p	$\frac{1}{2}$	938.3	Stable	0
Neutron	n	$\frac{1}{2}$	939.6	614	0
Lambda	Λ^0	$\frac{1}{2}$	1116	1.82×10^{-10}	-1
Sigma	Σ^+	$\frac{1}{2}$	1189	0.56×10^{-10}	-1
	Σ^0	$\frac{1}{2}$	1193	5.1×10^{-20}	-1
	Σ^-	$\frac{1}{2}$	1197	1.03×10^{-10}	-1
Xi	Ξ^0	$\frac{1}{2}$	1315	2.01×10^{-10}	-2
	Ξ^-	$\frac{1}{2}$	1321	1.14×10^{-10}	-2
Omega	Ω^-	$\frac{3}{2}$	1672	0.57×10^{-10}	-3
Mesons					
Pion	π^+	0	139.6	1.80×10^{-8}	0
	π^0	0	135.0	5.8×10^{-17}	0
	π^-	0	139.6	1.80×10^{-8}	0
Kaon	K^+	0	493.7	8.85×10^{-9}	$+1$
	K^0*	0	497.6	6.21×10^{-11}	$+1$
				3.59×10^{-8}	

*The K^0 has two lifetimes, 50% decay via each mode.

of any internal structure. Table 27-2 lists the known leptons. The questions of the masses and stability of the neutrinos are rather complex issues and are discussed in the next chapter.

All leptons and hadrons participate in the weak interaction. The only particles that fail to get listed in the hadron or lepton families are exchange particles.

Conservation Laws

Conservation laws provide insight into the puzzles of the elementary particles. We have already seen how the conservation laws were used to unravel beta decay. In some situations new conservation laws have been invented as a result of the experiences of viewing many, many particle collisions. The success of the conservation laws has been responsible for a guiding philosophy: *if it can happen, it will*. That is, any process not forbidden by the conservation laws will occur.

The classical laws of conserving energy (mass-energy), linear momentum, angular momentum (including spin), and electric charge are valid in the elementary particle realm. Any reaction or decay that occurs satisfies these laws. For instance, the neutron decays into a proton, an electron, and an antielectron neutrino via beta decay,

$$n \to p + e^- + \bar{\nu}_e$$ ◀ allowed

Table 27-2 | Properties of the Leptons

Name	Symbol	Spin ($h/2\pi$)	Rest Mass (MeV/c^2)	Half-Life (s)
Electron	e^-	$\frac{1}{2}$	0.511	Stable
Electron neutrino	ν_e	$\frac{1}{2}$	$\neq 0$	—
Muon	μ^-	$\frac{1}{2}$	105.7	1.52×10^{-6}
Mu neutrino	ν_μ	$\frac{1}{2}$	$\neq 0$	—
Tau	τ^-	$\frac{1}{2}$	1777	2.01×10^{-13}
Tau neutrino	ν_τ	$\frac{1}{2}$	$\neq 0$	—

but has never been observed to decay via

forbidden ▶

$$n \to p + e^+ + \nu_e$$

Question Why is this second alternative forbidden?

Answer The second alternative is forbidden by charge conservation because the initial state (the neutron) has zero charge and the final state (proton, positron, and neutrino) has a charge of +2.

A little less obvious are situations that are forbidden because they violate energy conservation. A particle cannot decay in a vacuum unless the total rest mass of the products is less than the decaying particle's rest mass. To see this, view the decay from a reference system at rest with respect to the original particle. The principle of relativity (Chapter 9) states that conclusions made in one reference system must hold in another. In the rest system, the total energy is the rest-mass energy of the particle. After the decay, however, the energy consists of the rest-mass energies of the products *plus* their kinetic energies. There has to be enough energy to create the decay particles even if they have no kinetic energy. Some of these decays that are forbidden by the conservation of energy can, however, occur in nuclei because the decaying particle can acquire kinetic energy from other nucleons to produce the extra mass. In reactions involving collisions of particles, kinetic energies must also be included in calculating conservation of energy.

Additional conservation laws were created as more information about the elementary particles became known. The total number of baryons is constant in all processes. So the concept of baryon number and its conservation was invented to reflect this discovery. Baryons are assigned a value of +1, antibaryons a value of −1, and all other particles a value of 0. In any reaction the sum of the baryon numbers before the reaction must equal the sum afterward. For example, suppose a negative pion collides with a proton. One result that could not happen is

forbidden ▶

$$\pi^- + p \to \pi^+ + \bar{p}$$
$$0 \;\; +1 \neq 0 \;\; -1 \quad \text{baryon numbers}$$

because the baryon number is +1 before and −1 after. On the other hand, we could expect to observe

allowed ▶

$$\pi^- + p \to p + \bar{p} + p + \pi^-$$
$$0 \;\; +1 = 1 \; -1 \; +1 \; +0 \quad \text{baryon numbers}$$

if the kinetic energy of the pion is sufficiently high.

Similarly, we don't expect the proton to decay by a process such as

forbidden ▶

$$p \to \pi^0 + e^-$$
$$1 \neq 0 \; +0 \quad \text{baryon numbers}$$

In fact, the proton cannot decay at all because it is the baryon with the smallest mass. Any decay that would be allowed by baryon conservation would require more energy than the rest mass of the proton. Some recent results have indicated that the conservation of baryon number might not be strictly obeyed. The proton might possibly decay, but its half-life is known to be at least a billion trillion times the age of the Universe!

There is no comparable conservation of meson number. Mesons can be created and destroyed, provided the other conservation rules are not violated. There are no conservation laws for the number of any of the exchange particles.

A conservation law for leptons has been discovered, but it is more complicated than that for baryons. There are separate lepton conservation laws for electrons, muons, and presumably taus. Furthermore, the neutrino associated with the electron is not the same as that associated with the muon or tau. The book-keeping procedure is slightly more detailed, but the procedure is essentially the same. The electron, muon, and tau lepton numbers must be separately conserved.

Some quantities are conserved by one type of interaction but not another. Each new particle is studied and grouped according to its properties: how fast it decays, its mass, its spin, and so on. One group of particles became known as the strange particles because their half-lives didn't seem to fit into the known interactions. If they decayed via the strong interaction, their half-lives should be about 10^{-23} second. If they decayed via the electromagnetic interaction, the predicted half-lives should be about 10^{-16} second. However, these particles are observed to live about 10^{-10} second, at least a million times as long as they "should" live. Something must be prohibiting these decays. A property called **strangeness** and an associated conservation law were invented. The various strange particles were given strangeness values, and the conservation law stated that any process that proceeds via the strong or electromagnetic interaction conserves strangeness, whereas those that proceed via the weak interaction can change the strangeness by a maximum of 1 unit.

The idea of a strangeness quantum number that is conserved seems quite foreign to our experiences. And it should. This attribute is clearly in the nuclear realm; we don't see its manifestation in our everyday world. In fact, we should probably be cautious about the feeling of comfort we have with other quantities. Consider electric charge. Most of us feel quite comfortable talking about the conservation of electric charge, perhaps because electricity is familiar to us. Imagine a world in which we had no experience with electricity. We would feel uneasy if someone suggested that if we assign a +1 to protons, a −1 to electrons, and 0 to neutrons, we might have conservation of something called "cirtcele" (electric spelled backward). It isn't unreasonable that unfamiliar quantities emerge as scientists explore the subatomic realm.

Question Given the values of strangeness in Table 27-1, would you expect the decay of the lambda particle to a proton and a pion to proceed via the strong or weak interaction? Does this agree with its lifetime?

Answer Because the strangeness numbers assigned to the lambda and the proton differ by 1, it should be a weak decay. The mean lifetime agrees with this conclusion.

Quarks

The continual rise in the number of elementary particles once again raised the question of whether the known particles were the simplest building blocks. At the moment the leptons appear to be elementary; there is no evidence that they have any size or internal structure. The exchange particles also appear to be truly elementary for the same reasons. On the other hand, there is experimental evidence that the hadrons have some internal structure.

Particle physicists asked themselves, "Is it possible to imagine a smaller set of particles with properties that could be combined to generate all the known hadrons?" The most successful of the many attempts to build the hadrons is the **quark** model proposed by two American physicists, Murray Gell-Mann and George Zweig, in 1964. Their original model hypothesized the existence of three **flavors** of quarks (and their corresponding antiquarks), now called the "up" (u), "down" (d), and "strange" (s) quarks. The strange quark has a strangeness number of −1, whereas the other quarks have no strangeness. Each quark has $\frac{1}{2}$ unit of spin and a baryon number of $\frac{1}{3}$.

Perhaps the boldest claim made in this model is the assignment of fractional electric charge to the quarks. There has never been any evidence for the exis-

Table 27-3 | Properties of the Quarks

Flavor	Symbol	Charge	Spin	Baryon No.	Strangeness	Charm	Bottomness	Topness
Down	d	$-\frac{1}{3}$	$\frac{1}{2}$	$\frac{1}{3}$	0	0	0	0
Up	u	$+\frac{2}{3}$	$\frac{1}{2}$	$\frac{1}{3}$	0	0	0	0
Strange	s	$-\frac{1}{3}$	$\frac{1}{2}$	$\frac{1}{3}$	-1	0	0	0
Charm	c	$+\frac{2}{3}$	$\frac{1}{2}$	$\frac{1}{3}$	0	$+1$	0	0
Bottom	b	$-\frac{1}{3}$	$\frac{1}{2}$	$\frac{1}{3}$	0	0	$+1$	0
Top	t	$+\frac{2}{3}$	$\frac{1}{2}$	$\frac{1}{3}$	0	0	0	$+1$

tence of anything other than whole-number multiples of the charge on the electron. These fractional charges should (but apparently don't) make the quarks easy to find. Yet the scheme works. The quark model has had remarkable successes describing the overall characteristics of the hadrons.

The properties assigned to the various flavors of quark are given in Table 27-3. The antiquarks have signs opposite to their related quarks for the baryon number, charge, and some other properties such as strangeness.

To see how this concept works, let's "build" a proton. We must first list the proton's properties: a proton has baryon number +1, strangeness 0, and electric charge +1. Examination of the quarks' properties confirms that the proton can be made from two up quarks and one down quark (uud), as shown in Figure 27-9. Other baryons can be created with different combinations of three quarks.

Question What three quarks form a neutron (baryon number +1, strangeness 0, and electric charge 0)?

Answer udd.

The mesons have 0 baryon number so they must be composed of equal numbers of quarks and antiquarks. The simplest assumption is one of each. For instance, the positive pion (Figure 27-10) is composed of an up quark and a down antiquark (u$\bar{\text{d}}$), giving it a charge of +1, a spin of zero, and a strangeness of 0.

Question What two different combinations of up and down quarks and antiquarks would yield a neutral pion?

Answer u$\bar{\text{u}}$ and d$\bar{\text{d}}$.

All of this was fine until 1974. In that year a neutral meson called J/ψ with a mass three times the mass of the proton was discovered. What made the particle unusual was its "long" lifetime. It was expected to decay in a typical time of 10^{-23} second, but it lived 1000 times as long. The quark model was able to account for this anomaly only with the addition of a fourth quark that possessed a property called **charm** (c). The J/ψ particle represents a bound state of a charmed quark and its antiquark (c$\bar{\text{c}}$). The next year a charmed baryon (udc) was observed, adding further support for the existence of this quark.

It is interesting to note that, based on symmetry arguments, the existence of a fourth quark had been proposed several years earlier. At that time, four leptons were known—the electron, the muon, and their two neutrinos. Why should there be four leptons and only three quarks? This discomfort with asymmetry led to the idea that there should be four quarks. The nice symmetry was quickly destroyed with the discovery of the tau lepton!

With the discovery of the tau and the presumed existence of its corresponding tau neutrino, two additional quarks were predicted so that there would be six leptons and six quarks. The discovery of the upsilon in 1977 was the first evi-

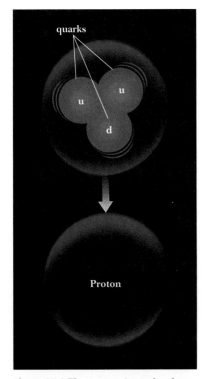

Figure 27-9 The proton is made of two up quarks and a down quark.

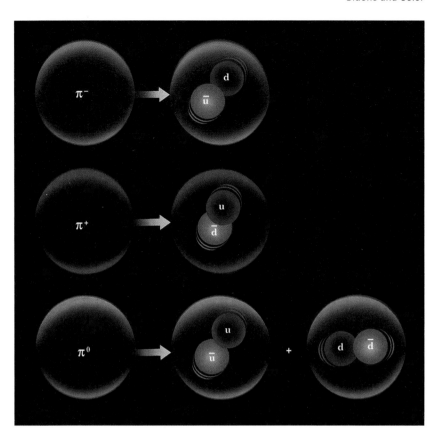

Figure 27-10 Each pion is composed of a quark and an antiquark.

dence for the fifth quark. The upsilon is a bound state of the **bottom** flavor of quark and its antiquark (b$\bar{\text{b}}$). There is also evidence for "bare bottom"—baryons and mesons that contain a bottom quark without a bottom antiquark. The sixth quark has the flavor called **top**, and its existence was confirmed in 1995.

PHYSICS | **ON YOUR OWN**

> Visit the online Lederman Science Center at http://www-ed.fnal.gov/ed_lsc.html and play the many educational games in the Fermilabyrinth. Be sure to try the game Baryon Bonanza.

Gluons and Color

Earlier we attributed the force between hadrons to the exchange of other hadrons such as pions. If the hadrons are actually composite particles made of quarks, we need to take our earlier idea one level further and ask what holds the quarks together in hadrons. The particles exchanged by quarks are known as **gluons**. (They "glue" the quarks together.)

Another problem of the simple quark theory is illustrated by the omega minus (Ω^-) particle, which has a strangeness of -3. It should be composed of three strange quarks with all three spins pointing in the same direction to account for its $\frac{3}{2}$ units of spin. However, the exclusion principle (Chapter 24) forbids identical particles with $\frac{1}{2}$ unit of spin from having the same set of quantum numbers. The exclusion principle can be satisfied if quarks have a new quantum number. This new quantum number has been named *color* and has three values: red, green, and blue. All observable particles must be "white" in color; that is, if the colors are considered to be lights, they must combine to form white light (Chapter 17). Therefore, baryons consist of three quarks, one of each color, and mesons con-

586 Chapter **27** Elementary Principles

Figure 27-11 The addition of the color quantum number increases the possible combinations of quarks (and antiquarks) that make up particles such as the proton and the positive pion.

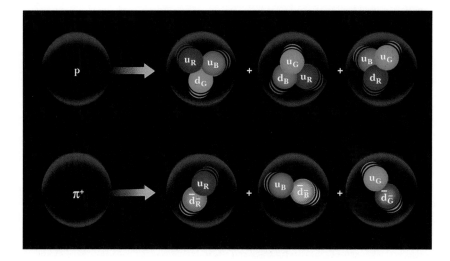

sist of a colored quark and an antiquark that has the complementary color (Figure 27-11).

Although the idea of the color quantum number began as an ad hoc way of accommodating the exclusion principle, it soon became a central feature of the quark model. Each quark is assumed to carry a *color charge*, similar to electric charge, and the force between quarks is called the *color force*. This theory requires that there be eight varieties of gluon, which differ only in their color properties. The quarks interact with each other through the exchange of gluons.

The strong force that holds the nucleons together to form nuclei is due to the color force between the quarks making up the nucleons. Therefore, gluons have replaced the mesons as the exchange particles of the strong force.

Flawed Reasoning

A classmate claims: "In Chapter 23 we learned that blue photons have more energy than red photons. Therefore, blue quarks must have more energy than red quarks, with green quarks somewhere in the middle." **How do you respond to this?**

Answer Your classmate is reading too much into a name. The Pauli exclusion principle demanded the existence of an additional quantum number to explain some of the quark combinations that were observed. Murray Gell-Mann decided to call this new quantum number *color* because it can take on three distinct values and only certain combinations of these values are allowed by nature. Just as white light can be made by combining red, green, and blue light or by combining any of these three colors with their complementary color, hadrons are composed of one quark of each of the three colors or a quark and an antiquark of the complementary color. Gell-Mann recognized this coincidence and used the term *color* to make the new concept more intuitive.

Physics⚛Now™ Assess your understanding of this chapter's topics with sample tests and other resources found by logging into PhysicsNow at **http://physics.brookscole.com/kf6e**.

Summary

The search for elementary components of matter began with Aristotle's basic elements—earth, fire, air, and water—and has led us through the chemical elements to the constituents of atoms—electrons, protons, and neutrons—to quarks.

Antimatter, an entirely new kind of matter, is produced by pair production or by the collisions of very energetic particles. Antiparticles of all the fundamental particles exist but don't survive long because particles and antiparticles annihilate each other when they come into contact.

The apparent failure of the conservation laws in beta decay led Pauli to propose the existence of the neutrino. The neutrino is electrically neutral, has a very small rest mass, and interacts extremely weakly with other particles.

Newton's idea of an action at a distance was replaced by the concept of a field, which in turn has been replaced by a picture of elementary particles interacting with each other through the exchange of still other elementary particles. The electromagnetic force occurs via the exchange of photons, the strong (or color) force via the eight gluons, the weak force via the intermediate vector bosons (W and Z^0), and the gravitational force via gravitons.

The discovery of the antiparticles, the muon, the tau, their neutrinos, and the very large collection of mesons and baryons quickly destroyed the concept of the elementary particles as being elementary. Some sense was brought to the particle zoo through the use of the conservation laws and classifying them according to their interactions. Further simplification came with the quark model.

The quark model has had remarkable success describing the overall characteristics of the elementary particles. The Universe appears to be made of six leptons and six quarks (each in three colors). The baryons, for example, consist of three quarks, one in each of the three colors, whereas the mesons consist of a quark and an antiquark of the complementary color.

Chapter 27 Revisited

Nobody knows where the search for the fundamental building blocks will end. The current theory of quarks and leptons is appealing in its completeness, and no experiments have contradicted our belief that these particles do not have internal structures. Maybe the search has ended. Maybe not.

KEY TERMS

antiparticle: A subatomic particle with the same-size properties as those of the particle, although some may have the opposite sign. The positron is the antiparticle of the electron.

baryon: A type of hadron having a spin of $\frac{1}{2}, \frac{3}{2}, \frac{5}{2}, \ldots$ times the smallest unit. The most common baryons are the proton and the neutron.

bottom: The flavor of the fifth quark.

charm: The flavor of the fourth quark.

flavor: The types of quark: up, down, strange, charm, bottom, or top.

gluon: An exchange particle responsible for the force between quarks. The eight gluons differ only in their color quantum numbers.

graviton: The exchange particle responsible for the gravitational force.

hadron: The family of particles that participates in the strong interaction. Baryons and mesons are the two subfamilies.

intermediate vector boson: The exchange particle of the weak nuclear interaction: the W^+, W^-, and Z^0 particles.

lepton: The family of elementary particles that includes the electron, muon, tau, and their associated neutrinos.

meson: A type of hadron with whole-number units of spin. This family includes the pion, kaon, and eta.

muon: A type of lepton; often called a heavy electron.

neutrino: A neutral lepton; one exists for each of the charged leptons—the electron, the muon, and the tau.

pair production: The conversion of energy into matter in which a particle and its antiparticle are produced. This usually refers to the production of an electron and a positron (antielectron).

pion: The least massive meson. The pion has three charge states: $+1$, 0, and -1.

positron: The antiparticle of the electron.

quark: A constituent of hadrons. Quarks come in six flavors of three colors each. Three quarks make up the baryons, whereas a quark and an antiquark make up the mesons.

strangeness: The flavor of the third quark.

strange particle: A particle with a nonzero value of strangeness. In the quark model, it is made up of one or more quarks carrying the quantum property of strangeness.

top: The flavor of the sixth quark.

CONCEPTUAL QUESTIONS

1. What particles were on the list of "elementary particles" in 1932?
2. Which particles on the list of "elementary particles" in 1932 are not on the current list?
3. What is the antiparticle of an electron? How does the charge of the antielectron compare to the charge of the electron?
4. What is the antiparticle of the photon? How does the charge of the antiphoton compare with the charge of the photon?
5. Do antiparticles have negative mass? Explain.
6. How much energy would be given off if an antiproton and an antineutron combined to form an antideuteron?
7. What is the ultimate fate of an antiparticle here on Earth?
8. The rest mass of the positron is 0.511 million electron volt. Why is more than 0.511 million electron volt released when a positron is annihilated?
9. The bubble-chamber tracks of an electron and a positron are clearly distinguishable. Why could you not use a bubble chamber to identify the pair production of a neutron–antineutron pair?
10. Particles and antiparticles annihilate when they come in contact. However, objects that are orbiting one another do not touch. Use the concept of probability clouds to explain why an electron and a positron "orbiting" each other can annihilate.
11. Explain why momentum conservation requires the emission of at least two photons when a positron and an electron annihilate. (The photon has a momentum equal to E/c.) For simplicity, assume that the electron and positron are at rest in your reference system.
12. Which elementary particles could be used to communicate with an antiworld?
13. The initial observations of beta decay indicated that some of the classical conservation laws might be violated. Which (if any) were obeyed without the invention of the neutrino?
14. Ten identical bombs are designed to each explode into two equal fragments—one red and one blue. If all ten bombs are exploded and the energy of the red fragment is measured, it will be the same in each trial because of the laws of conservation of energy and the conservation of momentum. However, if the ten identical bombs were instead designed to explode into three equal fragments—one red and two blue—the energy of the red fragment could now be different in each explosion. Use these results to interpret Figure 27-4 as requiring the existence of the neutrino.
15. Through which force does the neutrino interact with the rest of the world?
16. Why did it take so long for experimentalists to detect the neutrino?
17. Why would the whole theory of exchange particles have been considered absurd before the development of quantum mechanics?
▲ 18. One attempt at creating an analogy of an attractive exchange force utilizes boomerangs. The person on the right throws a boomerang toward the right, gaining momentum toward the left. The boomerang travels along a semicircle and is caught coming in from the left. When he catches the boomerang, the person on the left gains some momentum toward the right. Why is this not a good analogy?

19. The existence of intermediate vector bosons was predicted many years before they were detected in an accelerator. Why did this discovery take so long when scientists knew what they were looking for?
20. If the Heisenberg uncertainty principle allows energy conservation to be violated, why is energy conservation still a useful principle?
21. How does the uncertainty principle explain the infinite range of the electromagnetic interaction?
22. How does the uncertainty principle account for the decrease in the strength of the coulomb (electrostatic) force with increasing distance? How would this be different if photons had a rest mass?
23. What feature of the electromagnetic interaction requires the exchange particles to be massless?
24. What argument can be used to support the idea that the graviton has zero rest mass?
25. What are the differences between baryons and mesons?
26. Which particles do not participate in the strong interaction?
27. What are the important differences between the hadrons and the leptons?
28. What particles belong to the lepton family?
29. Do neutrons interact via the weak force? Explain.
30. Particles and antiparticles have the same properties except for the sign of some of them. Which particles in Table 27-1 could be particle-antiparticle pairs?
31. Roughly what would you expect for the lifetime of the decay $\Omega^- \to \Xi^0 + \pi^-$?
32. Roughly what would you expect for the lifetime of the decay $K^0 \to \pi^+ + \pi^-$?
33. Why can't a proton beta decay outside the nucleus?
34. If free protons can decay, the lifetime is greater than a billion trillion times the age of the Universe. Does this necessarily mean that there has not yet been a free proton decay in the universe?
35. What does the X stand for in the pion decay $\pi^+ \to \mu^+ + X$?
36. What does the X stand for in the antimuon decay $\mu^+ \to \nu_e + \bar{\nu}_\mu + X$?

37. Name at least one conservation law that prohibits each of the following:
    a. $\mu^- \rightarrow e^- + \nu_e + \bar{\nu}_\mu$ b. $p \rightarrow \pi^+ + \pi^+ + \pi^-$
    c. $\Omega^- \rightarrow \Lambda^0 + \pi^-$
38. Name at least one conservation law that prohibits each of the following:
    a. $\pi^- + p \rightarrow \Sigma^+ + \pi^0$ b. $\mu^- \rightarrow \pi^- + \nu_\mu$
    c. $\Sigma^0 \rightarrow \Lambda^0 + \pi^0$
39. If you observe the decay $X \rightarrow \Lambda^0 + \gamma$ with a lifetime of approximately 10^{-20} second, what can you say about the (a) baryon number, (b) strangeness, and (c) charge of X?
40. A particle X is observed to decay by $X \rightarrow \pi^+ + \pi^-$ with a lifetime of 10^{-10} second. What are possible values for the (a) baryon number, (b) strangeness number, and (c) charge of X?
41. That the lifetime of the π^0 is roughly 10^{-16} second indicates that the pion decays via the electromagnetic interaction. What would you guess would be the products of this decay?
42. The lifetime of the π^- is roughly 10^{-8} second. What decay products would you expect?
43. What combinations of quarks correspond to the antiproton and the antineutron?
44. What combination of quarks makes up the negative pion?
45. The K^- particle is the antiparticle of the K^+ particle. What combination of quarks makes up the K^- particle? What value does it have for strangeness?
46. Which hadron corresponds to the combination of a strange antiquark and an up quark?
47. The Δ^{++} particle has charge +2 and strangeness 0. What combination of quarks makes up the Δ^{++} particle?
48. Which hadron is composed of two up quarks and a strange quark?
49. What quarks make up a Ξ^0?
50. What combination of quarks corresponds to the Σ^-?
51. What quark and antiquark make up a K^0? Would this particle be its own antiparticle?
52. Why is the π^0 its own antiparticle?
53. In the quark model, is it possible to have a baryon with strangeness -1 and charge +2? Explain.
54. Why is it impossible to make a meson of charge +1 and strangeness -1?
55. A particle consists of a top quark and its corresponding antiquark. Is this particle a meson or a baryon? What is its charge?
56. What charge would a baryon have if it was composed of an up quark, a strange quark, and a top quark?
57. In the original quark model, the Ω^- was believed to be composed of three strange quarks. This assumption causes problems because quarks are expected to obey the Pauli exclusion principle. How did physicists get around this problem?

28 | Frontiers

When Albert Einstein conceived the general theory of relativity, he believed that the Universe was static; the Universe was neither expanding nor contracting. To account for this, he added a term to his equations known as the *cosmological constant* to prevent the Universe from collapsing under the influence of its own gravity. Later, when astronomers showed that the Universe was expanding rather rapidly, Einstein felt that the introduction of the cosmological constant was his biggest blunder. Will the Universe continue to expand forever, or will the expansion slow and the Universe collapse upon itself?

(See page 602 for the answer to this question.)

The imaginary sea serpent's head in this Hubble Space Telescope photograph of the Eagle Nebula is actually a cloud of molecular hydrogen and dust in which new stars are forming.

THE physics world view is a dynamic one. Ideas are constantly being proposed, debated, and tested against the material world. Some survive the scrutiny of the community of physicists; some don't. The inclusion of new ideas often forces the modification or outright rejection of previously accepted ones. Some firmly accepted ideas in the world view are very difficult to discard; in the long run, however, experimentation wins out over personal biases.

In this chapter we look at a few, selected areas on the frontiers of physics research. The ideas presented in this chapter are not as firmly established as those presented in the earlier chapters. Some of the ideas discussed here will survive and others will not. But that is the nature of an evolving science.

In their search for new physics, researchers do not have carte blanche to make up any theory they please. New theories must agree with the increasingly large body of experimental results and be compatible with the well-established theories. This puts very stringent constraints on what can be proposed.

In the first chapter we quoted Newton: "I do not know what I may appear to the world, but to myself I seem to have been only like a boy playing on the sea-shore, and diverting myself in now and then finding a smoother pebble or a prettier shell than ordinary, whilst the great ocean of truth lay all undiscovered before me." With the help of some of our colleagues, let's go looking for prettier shells.

Gravitational Waves

When Einstein proposed his general theory of relativity in 1916, he also postulated the existence of gravitational waves. In many ways gravitational waves are analogous to electromagnetic waves, which we discussed in Chapter 22. The acceleration of electric charges produces electromagnetic waves that we experience as light, radio, radar, TV, and X rays. Gravitational waves result from the acceleration of masses. Both types of waves carry energy through space, travel at the speed of light (3×10^8 m/s), and decrease in intensity as the inverse square of the distance from the source.

Because electromagnetic waves are so easily detected by a $10 radio and our eyes, why have gravitational waves not been detected? The first reason is that the gravitational force is 10^{43} times weaker than the electromagnetic force. Except in the most favorable cases, gravitational waves are weaker by this same factor. In addition, the detectors are less sensitive in intercepting gravitational waves by at least another factor of 10^{43}.

The detection of gravitational waves is one of the most fundamental challenges in modern physics. Joseph Weber, a physicist at the University of Maryland, pioneered the efforts to detect gravitational waves in the 1960s. He used a large, solid cylinder, 2 meters long and weighing several tons, as a detector. A passing gravitational wave causes vibrations in the length of the cylinder due to the differences in the forces on atoms in various parts of the cylinder. The detection system is amazingly sensitive; it can detect changes in the length of the cylinder of 2×10^{-16} meter, about one-fifth the radius of a proton. However, a cylinder can vibrate for many other reasons, such as passing trucks. To eliminate these vibrations, Weber used two cylinders placed 1000 miles apart and required that both vibrate together.

Although there may be many sources of gravitational waves, only a few should emit strong enough signals and be close enough to be within the range of current instruments. The details of the various processes are not well understood, so the calculations are rough estimates. When a supernova occurs, a very large burst of gravitational waves should be given off. This process may give off a large enough pulse to be detected but is expected to happen only once every 15 years within the Galaxy. Gravitational waves should also be given off strongly from a pair of neutron stars or black holes orbiting each other at close range.

Figure 28-1 Aerial view of the end station and one arm of the LIGO facility in Louisiana.

In 1991 the U.S. Congress approved the construction of the Laser Interferometer Gravitational-Wave Observatory (LIGO). LIGO consists of identical facilities in the states of Washington and Louisiana (Figure 28-1), widely separated locations to eliminate extraneous vibrations. Each facility is looking for changes in the distance between pairs of mirrors 4 kilometers apart. The distances are measured by splitting a laser beam into two beams that travel at right angles to each other, bounce off mirrors, and are then recombined. When the beams recombine, very slight changes in the distances to the distant mirrors alter the brightness of the light. This observatory should ultimately be a million times more sensitive than Weber's experiment.

The next generation of gravitational wave detectors will be built in space. The Laser Interferometer Space Antenna (LISA) will consist of three satellites at the corners of an equilateral triangle that is 5 million kilometers on a side. The group of satellites will lag 20 degrees behind Earth in its orbit around the Sun, as shown in Figure 28-2. The experimental technique is the same as for LIGO, but the longer arms will provide sensitivity in the low-frequency gravitational wave band from 0.0001 to 1 hertz, compared with LIGO's high-frequency band from 10 to 1000 hertz. In the low-frequency band, LISA will observe gravitational waves generated by binary systems or by the very massive black holes at the centers of many galaxies, including our own Milky Way Galaxy. The schedule as of fall 2005 called for LISA to be operational in 2013.

Although we do not have direct evidence of gravitational waves, there is very strong indirect evidence for the existence of gravitational waves. In 1974 Joseph Taylor and Russell Hulse discovered a pair of neutron stars orbiting each other at close range. Neutron stars have masses slightly larger than the mass of our Sun but are only about 10 kilometers in diameter. At these very great densities, electrons and protons combine to form neutrons, so the star consists almost entirely of neutrons. The two neutron stars orbit each other every 8 hours and reach orbital speeds of 0.13% of the speed of light. One of the neutron stars is a pulsar that emits a pulse of radiation every 59 milliseconds. This acts like a clock that is as good as any atomic clock that we have on Earth. This clock allows very, very precise timings of the orbits, and 30 years of measurements indicate that

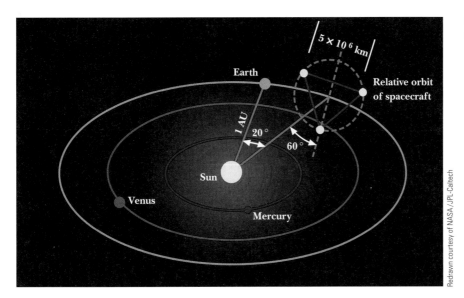

Figure 28-2 The orbit of the LISA satellites.

the orbital period is decreasing. In fact, it is decreasing at precisely the rate expected from the loss of energy due to gravitational radiation according to Einstein's theory of general relativity. The 1993 Nobel Prize in physics was awarded to Taylor and Hulse for this work.

Physicists expect that gravitational waves will soon be detected directly; it is only a matter of continuing to develop the technology for improving the sensitivity of the detectors.

Unified Theories

The view that the elementary building blocks in nature are the 6 leptons, the 6 quarks, and the 13 exchange particles is a great simplification of the hundreds of subnuclear particles that have been discovered since 1932.* The list of elementary particles totals 25, a reasonably small number for building the many diverse materials in the world. But can we reduce the complexity even more?

The present scheme contains two very distinct classes of particles—the leptons and the quarks. Leptons have whole-number units of the electron charge, whereas quarks have charges that are multiples of one-third of the basic unit. Quarks participate in the strong interaction through their color, whereas leptons are colorless and do not participate. Leptons are observed as free particles, but quarks have yet to be isolated—in fact, current theory predicts they cannot be isolated. There have been no observations that indicate that leptons can turn into quarks, and vice versa. Why should there be two classes? Why not only one?

Similarly, we have four different forces—gravitational, electromagnetic, weak, and strong (or color). Each appears to have its own strength, and there are three different dependencies on distance and 13 different exchange particles. Why should there be four forces? Why not only one?

Physicists have asked similar questions for a long time and have sought to reduce the number of particles and the number of forces as far as possible, ideally to one class of particles and one force. Theorists don't aim to eliminate (or overlook) the differences that are so apparent but rather to show that these are all manifestations of something much more basic and therefore more elementary.

Much progress has been made. Before the beginning of the 20th century, Maxwell was able to show that electricity and magnetism were really two different aspects of a common force, the electromagnetic force. During the 1960s,

---

*Written with the assistance of William Hiscock, Department of Physics, Montana State University.

Sheldon Glashow, Steven Weinberg, and Abdus Salam developed the electroweak theory that unified the weak and electromagnetic interactions.

The electroweak theory and the color theory of the strong interaction are the main components of the *standard model* of elementary particles that evolved in the early 1970s. This model seems to be able to explain every experiment that can be done. However, even though the model seems to be complete, its mathematical complexity is such that many calculations cannot be done with presently available techniques.

The unification efforts continue; some success has been achieved in combining the electroweak interaction and the color interaction in what are called *grand unified theories*. These theories predict that baryon number is not strictly conserved, and as a consequence the proton should decay, although the predicted lifetime far exceeds the present age of the Universe. So far, experiments have not seen any hint of proton decay, and this has ruled out some of the simpler grand unified theories.

If efforts to produce a grand unified theory succeed, there will still be the separate existence of the gravitational force. To reach the ultimate goal, gravity must also be included in an ultimate "theory of everything." Since 1984 many physicists have been working on a promising candidate for such a theory, *superstring* theory. In superstring theory, the fundamental objects from which all matter and forces are built are not point particles but tiny loops of material that cannot be further subdivided. The size of these loops defines the smallest distance in nature, which in most superstring theories is the *Planck length*, about 10^{-35} meter, more than a billion billion times smaller than a proton. The Planck length comes from combining gravity with Planck's constant h from quantum mechanics. Although experiment has yet to confirm any of these proposed further simplifications or unifications, theoretical physicists continue to explore what sorts of simple fundamental models are compatible with what is already known.

Cosmology

A cosmic connection exists between all the forces and particles that we have studied. The connection is made in the Big Bang model of the creation of the Universe. Experimental evidence indicates that the Universe had a beginning and that this beginning involved such incredibly high densities and temperatures that everything was a primordial soup beyond which it is impossible to look. According to this model, the Universe erupted in a big bang about 14 billion years ago.

Theorists divide the development of the Universe into seven stages (Figure 28-3). Immediately after the Big Bang, up until something like 10^{-44} second, the conditions were so extreme that the laws of physics as we now know them did not exist. As the Universe expanded, it cooled, the four forces developed their individual characteristics, and the particles that we currently observe were formed. The particular details of what happened early in the process are speculative but rest on firmer ground as time unfolds.

During the second stage—between 10^{-44} and 10^{-37} second—the gravitational force emerges as a separate force, leaving the electromagnetic and the two nuclear forces unified, as described by the grand unification theories (Figure 28-4). During this time the energies of particles and photons are so great that there is continual creation and annihilation of massive particle–antiparticle pairs.

During the third stage—between 10^{-37} and 10^{-10} second—temperatures drop from approximately 10^{28} K to approximately 10^{15} K, and the strong force splits off from the electroweak force, resulting in three forces.

Near the beginning of the fourth stage—which ends only a microsecond after the Big Bang—further cooling results in the electroweak force splitting into

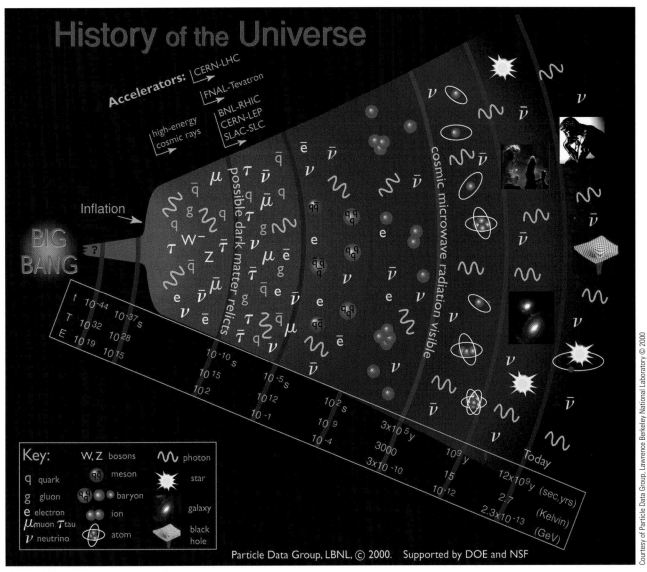

Figure 28-3 The history of the Universe.

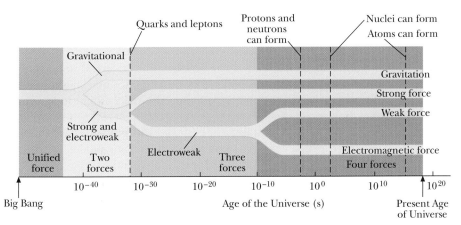

Figure 28-4 As the Universe expands and cools, the single original force splits into new forces that eventually become the four forces that we currently observe in nature.

the weak force and the electromagnetic force, giving us the four forces we can now detect. The Universe is filled with radiation (photons), quarks, leptons, and their antiparticles. The temperatures are still too hot for quarks to stick together to form protons and neutrons.

The fifth stage takes us up to 3 minutes after the Big Bang. By this time the temperature has dropped to 1 billion K, which is still very much hotter than the interior of our Sun. At the end of this stage, typical photons can no longer form electron–positron pairs, and some neutrons and protons have formed helium nuclei, with about 25% of the Universe's mass as helium and the rest as hydrogen. The model predicts, and the observations agree, that there should be one neutron for every seven protons. (Due to the extra mass of the neutron, it is harder to convert protons into neutrons than it is to convert neutrons into protons.)

The sixth stage ends when the temperature drops to approximately 3000 K, about 300,000 years after the Big Bang. Suddenly, matter becomes transparent to photons, and the photons can travel throughout the Universe. This moment defines the boundaries of the observable Universe: looking out into space is just like looking backward in time to this beginning. The radiation we see from objects at a distance of 1 billion light-years left there 1 billion years ago. Therefore, we are seeing the object as it was in the past, not as it is at the present.

The final stage begins with the formation of atoms and includes the formation of the galaxies, stars, and planets, such as Earth.

Cosmic Background Radiation

In 1965 Bell Laboratory scientists Arno Penzias and Robert Wilson made an accidental discovery that has had a tremendous impact on our view of the Universe. While testing a sensitive microwave receiver (Figure 28-5), Penzias and Wilson detected a faint background hiss that caused problems with their satellite communications. They tried thoroughly cleaning the receiver after evicting a flock of pigeons, and they tried cooling the electronics. However, the background persisted. In addition, the intensity of the background was the same, regardless of the direction they pointed their microwave detector. This suggested that the source was outside the Solar System or even outside the Galaxy.

Luckily, Penzias and Wilson learned that a group of scientists at Princeton University had predicted the existence of a *cosmic background radiation*. This radiation would have been emitted at the time the Universe became transparent to visible radiation some 300,000 years after the Big Bang. Although the radiation had a temperature of 3000 K at that time, the radiation has cooled as the Universe expanded and should now appear with a temperature of approximately 3 K!

Penzias and Wilson caused quite a stir in the scientific community when they announced that they had already detected the cosmic background radiation. Because their measurement was at a single wavelength, they could not determine the temperature. However, measurements by other groups quickly showed that the radiation is an excellent match for a temperature of 2.725 K, in agreement with the prediction.

The cosmic background radiation's existence and properties are the most compelling experimental evidence supporting the Big Bang model. Penzias and Wilson received the Nobel Prize in 1978 for their discovery.

This discovery, however, led to a new problem. The temperature was very accurately the same in all directions. How could a Universe full of galaxies, stars, and planets form from something that was so uniform? More recent data from satellite measurements show variations in temperature on the order of 0.0002 K, which may be sufficient to account for the structure of the Universe.

Figure 28-5 Arno Penzias (right) and Robert Wilson stand in front of their microwave receiver.

Dark Matter and Dark Energy

It has to be one of the most discomfiting results of modern science: only about 5% of the Universe is made of *ordinary* matter—the kind of matter that makes up stars, planets, rocks, and people.* Besides permitting the possibility that space is expanding, Einstein's general theory of relativity opened the door to the possibility that the idea of a flat space—one in which parallel lines never converge or diverge—might be a local illusion. For more than 90 years, scientists have been dealing with the possibility of a curved space. The overall density of matter in the Universe determines the shape of space (loosely speaking, whether it is concave, flat, or convex).

Recent results measuring minute ripples in the cosmic background radiation—the microwave radiation that is the remaining signature from the time the cooling Universe first became transparent to its own radiation—are confirming that we appear to live in a universe with exactly the critical matter density to make space flat on cosmic scales. Because this would seem far too unlikely to occur by chance, scientists have been searching for mechanisms that would favor a flat universe. They have generally settled on a mechanism called *inflation*—a period in which the Universe grew by up to 10^{30} in as little as 10^{-36} second early in our Universe's history—to explain why it developed exactly the critical density to be flat. The conundrum is that the density of all ordinary matter that we can find accounts for only about 5% of this critical density. So, what is the other 95%?

The first major clue that there is something out there we cannot see came from careful studies of the motions of stars within galaxies. A galaxy is a collection of billions of stars (like our own Milky Way Galaxy) in which individual stars orbit about the galactic center. For instance, the Sun takes about 250 million years to complete one orbit around our Galaxy. The speed at which a star moves about its galactic center depends upon the distribution of matter within the galaxy; it is the gravitational pull of the matter that holds the star in its orbit. By measuring the distribution of ordinary visible matter within a galaxy, it is possible to predict how fast the stars near the outside should be moving. These stars consistently move faster than predicted, indicating that there must be some unseen matter within the galaxies. This unseen matter has been dubbed *dark matter*.

The first question that scientists asked about dark matter is whether it is simply ordinary matter that is too cool to be seen by conventional methods—so-called *Massive Astrophysical Compact Halo Objects* (MACHOs)—or some form of exotic matter made of something other than our familiar protons, neutrons, and electrons—so-called *Weakly Interacting Massive Particles* (WIMPs). Yes, this was the battle between the MACHOs and the WIMPs! Although the less exotic MACHOs do compose some of the dark matter, most scientists agree that some kind of exotic matter of a form not yet identified must necessarily constitute the bulk of the dark matter. But, this is still not enough to get us to the critical mass density required for a flat universe. The most recent estimates are that about 25% of the Universe's matter is in the form of exotic dark matter. But along with the 5% from ordinary matter, this leaves 70% unaccounted for.

Recently, scientists studying extremely distant supernovae—extremely bright stellar explosions—have discovered that the most distant supernovae are actually somewhat dimmer than would be expected if the Universe's expansion were slowing at a rate expected due to the pull of gravity. The explanation for these dimmer-than-expected supernovae is that the Universe's expansion rate is actually increasing—placing the supernovae farther away. This implies that there must be some large-scale pressure in the Universe that is counteracting the

*Essay by Jeff Adams, Physics Department, Montana State University.

mutual pull of gravity caused by both ordinary and dark matter. This pressure force has been attributed to a *dark energy*, which must be uniformly distributed throughout the Universe. Furthermore, the density of dark energy could be exactly what is needed to make up the missing 70% to achieve the critical density for a flat universe. (Einstein told us that energy has an equivalent mass.) For now, scientists can say little about this dark energy. And yet, evidence is converging that dark energy may well be the most pervasive matter in the Universe.

Neutrinos

Neutrinos are unique among elementary particles in that much of what we know about them comes from sources beyond our control.* We learned a lot in the early days of neutrino physics from reactor and accelerator experiments, but in the last few years our most precise information has come from observing neutrinos from the Sun, cosmic rays, and a single supernova. These results have been hard won. The experiments themselves are long and difficult; they have to be performed in mines and tunnels far from the comforts and infrastructure of great laboratories to escape cosmic radiation that would swamp the tiny neutrino signals. It has also been necessary to fine-tune our theoretical understanding of the sources to an extraordinary degree to unravel fundamental neutrino physics from the data.

The 2002 Nobel Prize committee recognized two early pioneers in this field. Ray Davis of the University of Pennsylvania was the first to attempt the detection of solar neutrinos. He used hundreds of tons of cleaning fluid placed deep in a mine in South Dakota. The cleaning fluid was a cheap and safe form of chlorine, an efficient absorber of neutrinos. His first results obtained in the early 1970s showed the first hint that neutrinos were more complicated than the mathematical models suggested. Masatoshi Koshiba initiated a series of water-tank experiments in a mine in northern Japan that showed similar hints in the flux of neutrinos produced by cosmic rays striking Earth's atmosphere. The team he assembled went on to make the first real-time, directional measurement of solar neutrinos and to observe a small neutrino burst from supernova SN1987A—the first time ever that nonelectromagnetic radiation had been observed from an identifiable source beyond our Solar System.

The puzzle unearthed by Davis, Koshiba, and their collaborators showed up in the unexpectedly low rate of electron neutrinos coming from the Sun and a strange ratio of electron neutrinos to muon neutrinos seen in the cosmic ray signals. The results all seemed to point to an arcane feature of quantum mechanics, which allowed the possibility that the three neutrino types (called *flavors*)—electron, muon, and tau—could mutate one into another if their masses differed by a tiny amount. This "neutrino oscillation" was first proposed in the 1960s by the Italian Soviet physicist Bruno Pontecorvo. This in itself was a challenge to the prevailing notion that neutrinos were massless.

These strong experimental hints and weighty theoretical ideas required a major assault on the neutrino problem. Two types of experiment were designed and built to acquire crucial neutrino data. Radiochemical experiments, similar to Davis's but based on gallium, were mounted underneath the Apennines and the Caucasus Mountains. These provided a first look at the neutrinos arising directly from proton–proton fusion in the Sun. In Canada a unique supply of heavy water was used to construct the Sudbury Neutrino Observatory (SNO), which became operational in 1999. The heavy water allowed us, for the very first time, to unravel the neutrino flavors. Besides this new experimental evidence, theoretical astrophysicists have tightened the models of the solar interior and of cosmic

Artist's concept of the SNO detector. The acrylic vessel holds the heavy water. The framework holds the light sensors. The entire chamber is flooded with ordinary water to support the acrylic chamber and avoid problems of light transmission from the chamber.

*Essay by Chris Waltham, Department of Physics and Astronomy, University of British Columbia, and Sudbury Neutrino Observatory.

The SNO detector during construction.

rays. The gains have been impressive, especially in the case of the Sun; we can now truly claim to have a precision understanding of the solar interior and its thermonuclear furnace. As a result, the peculiarities of the neutrinos themselves have been thrown into sharp relief.

The picture that has emerged is that the Sun is powered by thermonuclear reactions that produce electron neutrinos in the numbers expected. This mechanism of energy production has been conjectured since the 1930s, but now we have hard proof. However, the neutrinos we see at Earth are only 34% electron neutrino; the other 66% are of the muon and tau types.

Cosmic rays produce muon and electron neutrinos in Earth's atmosphere, as expected, but about half of the muon neutrinos reach us as tau neutrinos.

The quantum mechanics of Pontecorvo has the following explanation. Neutrinos are born as distinct flavors—electron, muon, and tau—but propagate as distinct mass types—say, 1, 2, and 3. Unlike their charged electron, muon, and tau counterparts, the flavors and the masses do not correspond. Each is a thorough, distinct mix of the other. Because the masses 1, 2, and 3 are slightly different, they travel at different speeds and get out of phase with one another after a while. Hence, an electron neutrino produced in a nuclear reaction will soon start to look a bit "muonish" or "tauish" after traveling for a while. In a vacuum, neutrino 1 seems to be mostly electron, with a dash of muon and tau; neutrino 2 is a more or less equal mix of all three; and neutrino 3 is approximately equal parts muon and tau. In the dense core of the Sun (and to a lesser extent, Earth's core), matter effects change these mixtures. A neutrino created with an electron flavor in the Sun propagates as almost pure neutrino 2. Once outside the Sun, this neutrino finds itself a mix of all three flavors, and this results in the electron fraction of 34% that we actually measure.

A detailed look at the neutrino flavors actually observed tells us a lot about the tiny mass differences between neutrinos 1, 2, and 3. Neutrinos 1 and 2 are of the order 0.01 electron volt divided by the speed of light squared (eV/c^2), while 2 and 3 are of the order 0.1 eV/c^2 apart. These mass differences are tiny; the next lightest particle, the electron, is a whopping 0.5 million eV/c^2. At present we do not know much more about the neutrino mass differences than their orders of magnitude (the electron mass is known to six significant figures). In addition, we do not know the overall mass scale, although recent evidence from observing ripples in the cosmic microwave background radiation suggests that the sum of all three neutrino masses cannot be larger than 0.7 eV/c^2. This result from the

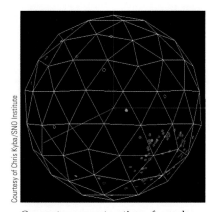

Computer reconstruction of a probable SNO solar-neutrino event. When a neutrino strikes the heavy water of the SNO detector, a faint cone of light spreads out from that point to SNO's light sensors. The green dots indicate which light sensors detected light.

Wilkinson Microwave Anisotropy Probe in 2003 demonstrates how properties of the smallest units of matter determine the largest-scale structure of the Universe.

What does it all mean? It's not clear at the moment. We are at one of those junctures in fundamental physics where a temporary increase in complexity hints at a deeper understanding just around the corner. Many new experiments are being designed for underground labs in Canada, the United States, Japan, and Europe to improve the measurements. Meanwhile, theorists are working hard to understand the fundamental mechanism of mass generation, which gives rise to these seemingly bizarre results.

We should not forget supernovae, which are still a bit of a mystery. The best calculations on the fastest computers still cannot produce a realistic bang. It is clear, however, that real supernovae cannot go bang without neutrinos. We need more data, and the world's neutrino detectors stand ready for the next little burst of supernova neutrinos, whenever that may come.

Quarks, the Universe, and Love

How does physics offer answers to questions about the workings of the world or, more broadly, the Universe?* Albert Einstein, *Time* magazine's Person of the Twentieth Century, expressed both puzzlement and hope when he wrote, "The most incomprehensible thing about the universe is that it is comprehensible." One can interpret Einstein's remark as saying that we cannot comprehend why the Universe is the way it is, but despite that, it happens that we can describe the Universe in terms of physical laws that apply not only on Earth but also everywhere in the Universe.

How is it that we are able to describe the universe with physical laws discovered by earthbound scientists? Richard Feynman, arguably the most admired and influential physicist of the second half of the 20th century, gave us many clues. One of his most memorable insights is that the single most important finding of science is that everything is made of atoms. Once everything is seen as being made of atoms, one may infer that understanding the laws that govern the behavior of individual atoms and their constituents, including the quarks, leads with sufficient effort to understanding complex systems containing huge numbers of atoms. This is called the *reductionist view*. It assumes that any physical system, no matter how complex, may be understood in terms of its component parts.

The study of the origin and fate of the Universe is called cosmology. Using the laws of physics, we can reconstruct the history of the Universe, which began about 14 billion years ago with the Big Bang. A chart of the history of the Universe (Figure 28-3) shows that before the Universe was a few millionths of a second old, it was an incredibly dense and hot mixture of matter and antimatter in the form of elementary particles, including quarks and antiquarks. As time progressed, the Universe expanded and cooled. By the time the Universe was about 3 minutes old, quarks had combined to form protons and neutrons, antiquarks had combined to form antiprotons and antineutrons, quarks had combined with antiquarks to form mesons, and protons and neutrons had combined to form elements. As the cooling process continued even further, atoms were formed as electrons bound to atomic nuclei. The force of gravity played a more dominant role as the Universe approached 1 billion years old. Gravity attracted atoms to each other, forming stars and clusters of stars. Stars then became sources of energy as atoms were drawn inward by each star's enormous gravitational attraction. As nuclei collided at the center of stars, they underwent nuclear fusion and released energy in the form of electromagnetic radiation that we see as sunlight.

*Essay by Robert S. Panvini, Department of Physics, Vanderbilt University.

Planets were formed by smaller clusters of atoms, too few to have sufficient gravitational pull to initiate the nuclear fusion that characterizes a star.

The most amazing thing in the history of the Universe, at least from our perspective, happened when one special planet—Earth—was formed. We may discover other planets like Earth someday, but what is special about Earth is that some of the atoms collected to form highly complex molecules and combinations of molecules, and enough of these complex molecules, in combination with other less complex molecules, formed living creatures, including dogs, cats, fish, mosquitoes, alligators, and people.

Remember how we got to this point. Fourteen billion years ago everything was in the form of elementary particles. The quarks of an atom's nucleus were part of the primordial soup, as were electrons that eventually combined with the nuclei to form atoms. Atoms collected to form stars and planets, and then, on at least one special planet, atoms really got fancy and life was formed.

We finish this essay with questions, not answers. If atoms are mostly made of quarks (and electrons) and if everything is made of atoms, including people, what is it that causes very large numbers of quarks, in the form of atoms, to exhibit the complex properties of living things? The characteristic of human beings that seems most different from inanimate matter includes consciousness and the realization of emotion, including love. Perhaps the reductionist view does not apply to systems as complex as living creatures. If not, why not? Whatever the ultimate answer, it will be fascinating to discover.

The Search Goes On

In Chapter 1 we set ourselves the task of expanding your world view. We began with a commonsense world view and carefully added bits and pieces of the physics world view. It is impossible for us to know which pieces have become parts of your own world view. Experience has shown us that expansion of a world view enables a person to see new connections between events. Some "see" sound waves, others feel the pull of gravity in a new way, and still others report experiencing new beauties in a rainbow or a red sunset.

This expansion of the world view produces different kinds of wonderment in different people. Some of us look at individual phenomena and marvel at the connections that can be made between seemingly unrelated things. Others of us wonder about the possibility of making sense out of the Universe.

We hope that you have gained some insight into the ways the physics world view evolves. We hope, for example, that you know that there is no single, static physics world view. Rather, a central core of relatively stable components is surrounded by a fuzzy, very fluid boundary. Metaphorically, it is an organism with spurts of growth and regions of maturation, decay, death, and even rebirth.

Part of our purpose was to show you that this growth proceeds within very definite constraints; it is certainly not the case that anything goes. Although intuitive feelings motivate new paths, they often result in dead ends. There were many more dead-end paths than we mentioned in this book. We picked up new ideas and discarded old ones. But within the space of this book, we couldn't follow the many, many blind paths that occurred in building the current physics world view. For the most part we had to follow the main paths.

Even the main paths, however, demonstrate why scientists believe what they believe. As one journeys from Aristotle's motions to Newton's inventions of concepts like gravity and force, it is tempting to look back in amusement at the more primitive ideas. Perhaps you had these feelings. This superiority should have quickly vanished as we adopted the spacetime ideas of Einstein and the wave–particle duality of the submicroscopic world. All these notions are creations of the human mind. Our task in building a world view is to create, but to create in a way different from the artist or poet.

The creations are different because we have a different answer to the question of why we believe what we believe. Our ideas must agree with nature's results. For a new idea to replace an old one, it must do all that the old idea did and more. Old ideas fail because experiments give results that don't agree with these ideas. The new ideas must meet the challenge: they must account for the observations that were in agreement with the old theory and also encompass the anomalous data. Einstein's idea of a warped spacetime includes Newton's concepts of gravity but does much more.

People might claim that a particular idea is a lot of malarkey. For example, they might claim that the speed of light is not the maximum speed limit in the Universe. For these critics to make a contribution, they will need to determine ways of testing their ideas. Their theories must be able to make predictions that can be tested by experiments.

So, science has an interesting split personality. During the birth of an idea, its personality is strongly dependent on the very personal, intuitive feeling of the scientist. But science also has a strong, cold, and very impersonal style of ruthlessly discarding ideas that have failed. There is no idea that escapes this threat of abandonment. No single idea is so appealing that it can circumvent the tests of nature.

It is very appealing to think that perpetual motion machines could exist and would solve our energy crisis, that the visual positions of the celestial objects give insight into the future, or that there is some magic potion that can free us from the entropy of old age. These ideas are appealing to most people. However, the appeal of an idea is not the only criterion for inclusion into the physics world view.

Now we arrive at our final point: the search goes on and the world view continually evolves. The locations of future growth spurts, however, are virtually impossible to predict. A survey of physicists would yield various possibilities. The differences depend on the responder's area of expertise and, perhaps, a few most cherished ideas.

And so, the search goes on.

Chapter 28 Revisited

The most recent experimental evidence indicates that the Universe's rate of expansion may actually be increasing and will therefore continue to expand forever. This and other experimental evidence has led to the idea that the Universe is filled with "dark energy" that causes the increase in the rate of expansion.

CONCEPTUAL QUESTIONS

1. What are three possible sources of gravity waves?
2. Why are gravity waves so hard to detect?
3. The Bohr model failed to account for the stability of atoms. Classical orbiting electrons would be constantly accelerating, would be constantly radiating energy in the form of electromagnetic waves, and would spiral into the nucleus. If orbiting planets are constantly radiating energy in the form of gravitational waves, why don't the planets spiral into the Sun?
4. The Planck model accounts for the stability of atoms by asserting that atoms radiate only when an electron jumps from one allowed orbit to another allowed orbit. Do planets radiate gravitational waves continually or just when they change orbits? Explain.
5. What indirect evidence has been found to support the existence of gravitational waves?
6. Neutron stars are a little more massive than our Sun. Why would a binary system of neutron stars be a strong source of gravitational waves?
7. If the predicted lifetime for the proton far exceeds the present age of the Universe, is it foolish to design an experiment to observe proton decay? Explain.
8. Current theories predict that the lifetime of the proton far exceeds the present age of the Universe. Does this mean that a proton has never decayed? Explain.
9. Which of the fundamental forces of nature is not included in the grand unification theories?
10. What does superstring theory try to do that other grand unified theories do not?

11. What are the fundamental building blocks of matter and force in the current superstring theories?
12. How big are the fundamental building blocks in current superstring theories?
13. How old is the Universe?
14. There was a period of time right after the Big Bang when the laws of physics as we know them did not exist and it was not necessary to take college physics classes. How long did this period last?
15. Which of the fundamental forces was the first to become distinct from the others after the Big Bang?
16. How long did it take after the Big Bang before the four fundamental forces of nature were distinct from each other? Can you blink your eyes this fast?
17. Why do neutrino experiments have to be performed in mineshafts deep below Earth's surface?
18. The fusion reactions in the Sun are now well understood. These reactions predict a much higher ratio of electron neutrinos compared with the other two flavors than is observed here on Earth. How do scientists account for this discrepancy?
19. What evidence led scientists to believe that the Universe is "flat"?
20. What observation led scientists to conclude that the rate of expansion of the Universe is increasing?
21. Describe the reductionist world view in your own words.
22. A friend claims, "My Uncle Fred wrote a book that explains everything. In his theory, shooting stars are angels traveling faster than light and dinosaurs were killed in the Great Flood. I don't know why Einstein is more famous than my uncle. After all, relativity is *just a theory*." Should Einstein be more famous than Uncle Fred? Explain.

Appendix

Nobel Laureates in Physics

The Nobel Prizes are awarded under the will of Alfred Nobel (1833–1896), the Swedish chemist and engineer who invented dynamite and other explosives. The annual distribution of prizes began on December 10, 1901, the anniversary of Nobel's death. No Nobel Prizes were awarded in 1916, 1931, 1934, 1940, 1941, or 1942. You can learn a lot more about the Nobel Prizes in all categories at the Nobel Prize Internet Archive, http://nobelprizes.com.

1901 **Wilhelm Roentgen*** (Germany), for the discovery of X rays.

1902 **Hendrik Lorentz** and **Pieter Zeeman** (both Netherlands), for investigation of the influence of magnetism on radiation.

1903 **Henri Becquerel*** (France), for the discovery of radioactivity, and **Pierre*** and **Marie Curie***[†] (France), for the study of radioactivity.

1904 **Lord Rayleigh** (Great Britain), for the discovery of argon.

1905 **Philipp Lenard** (Germany), for research on cathode rays.

1906 **Sir Joseph Thomson*** (Great Britain), for research on the electrical conductivity of gases.

1907 **Albert A. Michelson*** (U.S.), for spectroscopic and metrologic investigations.

1908 **Gabriel Lippmann** (France), for photographic reproduction of colors.

1909 **Guglielmo Marconi** (Italy) and **Karl Braun** (Germany), for the development of wireless telegraphy.

1910 **Johannes van der Waals** (Netherlands), for research concerning the equation of the state of gases and liquids.

1911 **Wilhelm Wien** (Germany), for laws governing the radiation of heat.

1912 **Gustaf Dalén** (Sweden), for the invention of automatic regulators used in lighting lighthouses and light buoys.

1913 **Heike Kamerlingh Onnes*** (Netherlands), for investigations into the properties of matter at low temperatures and the production of liquid helium.

1914 **Max von Laue** (Germany), for the discovery of the diffraction of X rays by crystals.

1915 **Sir William Bragg** and **Sir Lawrence Bragg** (both Great Britain), for the analysis of crystal structure using X rays.

1916 No award.

*These winners are mentioned in the text.

[†]Marie Curie also received a Nobel Prize in chemistry in 1911.

1917 **Charles Barkla** (Great Britain), for the discovery of the characteristic X rays of the elements.

1918 **Max Planck*** (Germany), for the discovery of the quantum theory of energy.

1919 **Johannes Stark** (Germany), for the discovery of the Doppler effect in canal rays and the splitting of spectral lines by electric fields.

1920 **Charles Guillaume** (Switzerland), for the discovery of anomalies in nickel–steel alloys.

1921 **Albert Einstein*** (Germany), for the explanation of the photoelectric effect.

1922 **Niels Bohr*** (Denmark), for the investigation of atomic structure and radiation.

1923 **Robert A. Millikan*** (U.S.), for work on the elementary electric charge and the photoelectric effect.

1924 **Karl Siegbahn** (Sweden), for investigations in X-ray spectroscopy.

1925 **James Franck** and **Gustav Hertz** (both Germany), for the discovery of laws governing the impact of electrons on atoms.

1926 **Jean B. Perrin** (France), for work on the discontinuous structure of matter and the discovery of equilibrium in sedimentation.

1927 **Arthur H. Compton** (U.S.), for the discovery of the Compton effect, and **Charles Wilson** (Great Britain), for a method of making the paths of electrically charged particles visible by vapor condensation (cloud chamber).

1928 **Sir Owen Richardson** (Great Britain), for the discovery of the Richardson law of thermionic emission.

1929 **Prince Louis de Broglie*** (France), for the discovery of the wave nature of electrons.

1930 **Sir Chandrasekhara Raman** (India), for work on light diffusion and discovery of the Raman effect.

1931 No award.

1932 **Werner Heisenberg*** (Germany), for the development of quantum mechanics.

1933 **Erwin Schrödinger*** (Austria) and **Paul A. M. Dirac*** (Great Britain), for the discovery of new forms of atomic theory.

1934 No award.

- 1935 **Sir James Chadwick*** (Great Britain), for the discovery of the neutron.
- 1936 **Victor Hess** (Austria), for the discovery of cosmic radiation, and **Carl D. Anderson** (U.S.), for the discovery of the positron.
- 1937 **Clinton J. Davisson*** (U.S.) and **Sir George P. Thomson** (Great Britain), for the discovery of the diffraction of electrons by crystals.
- 1938 **Enrico Fermi*** (Italy), for identification of new radioactive elements and the discovery of nuclear reactions effected by slow neutrons.
- 1939 **Ernest Lawrence** (U.S.), for development of the cyclotron.
- 1940–1942 No awards.
- 1943 **Otto Stern** (U.S.), for the discovery of the magnetic moment of the proton.
- 1944 **Isidor I. Rabi** (U.S.), for work on nuclear magnetic resonance.
- 1945 **Wolfgang Pauli*** (Austria), for discovery of the Pauli exclusion principle.
- 1946 **Percy Bridgman** (U.S.), for studies and inventions in high-pressure physics.
- 1947 **Sir Edward Appleton** (Great Britain), for discovery of the Appleton layer in the ionosphere.
- 1948 **Lord Patrick Blackett** (Great Britain), for discoveries in nuclear physics and cosmic radiation using an improved Wilson cloud chamber.
- 1949 **Hideki Yukawa*** (Japan), for prediction of the existence of mesons.
- 1950 **Cecil Powell** (Great Britain), for the photographic method of studying nuclear process and discoveries about mesons.
- 1951 **Sir John Cockroft** (Great Britain) and **Ernest Walton** (Ireland), for work on the transmutation of atomic nuclei.
- 1952 **Edward Purcell** and **Felix Bloch** (both U.S.), for the discovery of nuclear magnetic resonance in solids.
- 1953 **Frits Zernike** (Netherlands), for the development of the phase contrast microscope.
- 1954 **Max Born** (Great Britain), for work in quantum mechanics, and **Walter Bothe** (Germany), for work in cosmic radiation.
- 1955 **Polykarp Kusch** (U.S.), for measurement of the magnetic moment of the electron, and **Willis E. Lamb, Jr.** (U.S.), for discoveries concerning the hydrogen spectrum.
- 1956 **William Shockley**, **Walter Brattain**, and **John Bardeen** (all U.S.), for development of the transistor.
- 1957 **Tsung-Dao Lee** and **Chen Ning Yang** (both China), for discovering violations of the principle of parity.
- 1958 **Pavel Cerěnkov**, **Ilya Frank**, and **Igor Tamm** (all U.S.S.R.), for the discovery and interpretation of the Cerěnkov effect.
- 1959 **Emilio Segrè** and **Owen Chamberlain** (both U.S.), for confirmation of the existence of the antiproton.
- 1960 **Donald Glaser** (U.S.), for the invention of the bubble chamber.
- 1961 **Robert Hofstadter** (U.S.), for determination of the size and shape of nuclei, and **Rudolf Mössbauer** (Germany), for the discovery of the Mössbauer effect of gamma-ray absorption.
- 1962 **Lev D. Landau** (U.S.S.R.), for theories about condensed matter (superfluidity in liquid helium).
- 1963 **Eugene Wigner**, **Maria Goeppert-Mayer*** (both U.S.), and **J. Hans D. Jensen*** (Germany), for research on the structure of nuclei.
- 1964 **Charles Townes** (U.S.), **Nikolai Basov**, and **Alexandr Prokhorov** (both U.S.S.R.), for work in quantum electronics leading to the construction of instruments based on maser-laser principles.
- 1965 **Richard Feynman,*** **Julian Schwinger*** (both U.S.), and **Shinichiro Tomonaga*** (Japan), for research in quantum electrodynamics.
- 1966 **Alfred Kastler** (France), for work on atomic energy levels.
- 1967 **Hans Bethe** (U.S.), for work on the energy production of stars.
- 1968 **Luis Alvarez** (U.S.), for the study of subatomic particles.
- 1969 **Murray Gell-Mann*** (U.S.), for the study of subatomic particles.
- 1970 **Hannes Alfvén** (Sweden), for theories in plasma physics, and **Louis Néel** (France), for discoveries in antiferromagnetism and ferrimagnetism.
- 1971 **Dennis Gabor*** (Great Britain), for the invention of the hologram.
- 1972 **John Bardeen,*** **Leon Cooper,*** and **John Schrieffer*** (all U.S.), for the theory of superconductivity.
- 1973 **Ivar Giaever** (U.S.), **Leo Esaki** (Japan), and **Brian Josephson** (Great Britain), for theories and advances in the field of electronics.
- 1974 **Anthony Hewish** (Great Britain), for the discovery of pulsars, and **Martin Ryle** (Great Britain), for radiotelescope probes of outer space.
- 1975 **James Rainwater** (U.S.), **Ben Mottelson**, and **Aage Bohr** (both Denmark), for the development of the theory of the structure of nuclei.
- 1976 **Burton Richter** and **Samuel Ting** (both U.S.), for discovery of the subatomic J/ψ particles.
- 1977 **Philip Anderson**, **John Van Vleck** (both U.S.), and **Nevill Mott** (Great Britain), for work underlying computer memories and electronic devices.
- 1978 **Arno Penzias*** and **Robert Wilson*** (both U.S.), for the discovery of the cosmic microwave background radiation, and **Pyotr Kapitsa** (U.S.S.R.), for research in low temperature physics.
- 1979 **Steven Weinberg,*** **Sheldon Glashow*** (both U.S.), and **Abdus Salam*** (Pakistan), for developing the theory that the electromagnetic force and the weak force are facets of the same phenomenon.
- 1980 **James Cronin** and **Val Fitch** (both U.S.), for work concerning the asymmetry of subatomic particles.
- 1981 **Nicolaas Bloembergen** and **Arthur Schawlow** (both U.S.), for contributions to the development of laser spectroscopy, and **Kai Siegbahn** (Sweden), for contributions to the development of high-resolution electron spectroscopy.
- 1982 **Kenneth Wilson** (U.S.), for the study of phase transitions in matter.
- 1983 **Subramanyan Chandrasekhar** (U.S.), for theoretical studies of the processes important in the evolution of stars, and **William Fowler** (U.S.), for studies of the formation of the chemical elements in the Universe.
- 1984 **Carlo Rubbia** (Italy) and **Simon van der Meer** (Netherlands), for work in the discovery of intermediate vector bosons.
- 1985 **Klaus von Klitzing** (Germany), for the discovery of the quantized Hall effect.
- 1986 **Ernst Ruska** (Germany), for the design of the electron microscope, and **Gerd Binning** (Germany) and **Heinrich Rohrer** (Switzerland), for the design of the scanning tunneling microscope.

1987	**K. Alex Müller** (Switzerland) and **J. Georg Bednorz** (Germany), for development of a "high-temperature" superconducting material.
1988	**Leon Lederman,* Melvin Schwartz**, and **Jack Steinberger** (all U.S.), for the development of a new tool for studying the weak nuclear force and the discovery of the muon neutrino.
1989	**Norman Ramsay** (U.S.), for various techniques in atomic physics, and **Hans Dehmelt** (U.S.) and **Wolfgang Paul** (Germany), for the development of techniques for trapping single charged particles.
1990	**Jerome Friedman, Henry Kendall** (both U.S.), and **Richard Taylor** (Canada), for experiments important to the development of the quark model.
1991	**Pierre de Gennes** (France), for discovering methods for studying order phenomena in complex forms of matter.
1992	**Georges Charpak** (France), for his invention and development of particle detectors.
1993	**Russell Hulse*** and **Joseph Taylor*** (both U.S.), for discovering evidence of gravity waves.
1994	**Bertram Brockhouse** (Canada) and **Clifford Shull** (U.S.), for pioneering contributions to the development of neutron-scattering techniques for studies of condensed matter.
1995	**Martin Perl** and **Frederick Reines*** (both U.S.), for pioneering experimental contributions to lepton physics.
1996	**David Lee, Douglas Osheroff**, and **Robert Richardson** (all U.S.), for their discovery of superfluidity in helium-3.
1997	**Steven Chu** (U.S.), **Claude Cohen-Tannoudji** (Algeria), and **William Phillips** (U.S.), for the development of methods to cool and trap atoms with laser light.
1998	**Robert Laughlin** (U.S.), **Horst Stormer** (Germany), and **Daniel Tsui** (China), for their discovery of a new form of quantum fluid with fractionally charged excitations.
1999	**Gerardus 't Hooft** and **Martinus Veltman** (both Netherlands), for elucidating the quantum structure of electroweak interactions.
2000	**Zhores I. Alferov** (Russia) and **Herbert Kroemer** (U.S.), for developing semiconductor heterostructures used in high-speed- and opto-electronics, and **Jack St. Clair Kilby** (U.S.), for his part in the invention of the integrated circuit.
2001	**Eric A. Cornell, Wolfgang Ketterle**, and **Carl E. Wieman** (all U.S.), for the achievement of Bose–Einstein condensation in dilute gases of alkali atoms and for early fundamental studies of the properties of the condensates.
2002	**Raymond Davis, Jr.,* Riccardo Giacconi** (both U.S.), and **Masatoshi Koshiba*** (Japan), for pioneering contributions to astrophysics.
2003	**Alexei A. Abrikosov, Vitaly L. Ginzburg** (both Russia), and **Anthony J. Leggett** (Great Britain and U.S.), for pioneering contributions to the theory of superconductors and superfluids.
2004	**David J. Gross, H. David Politzer**, and **Frank Wilczek** (all U.S.), for the discovery of asymptotic freedom in the theory of the strong interaction.
2005	**Roy J. Glauber** (U.S.), for contributions to the quantum theory of optical coherence, and **John L. Hall** (U.S.) and **Theodor W. Hänsch** (Germany), for contributions to the development of laser-based precision spectroscopy.

Answers to Most Odd-Numbered Questions and Exercises

Chapter 1

Questions

1. A physics world view incorporates data from outside the range of human sensations.
3. It does not have any scientific basis.
5. It must (a) account for known data, (b) make testable predictions, and (c) have a scientific basis.
7. The more prestigious the scientist who proposes the theory, the more likely the scientific community will commit resources to test the theory.
9. United States
11. About 170 cm
13. About 2.5 m
15. About 85 kg
17. Sentinel (centi-nel)
19. 10^{10} people (6.5 billion)

Exercises

1. 86,400 s
3. 109 yd
5. 39.4 in.
7. (a) 6.82×10^4 m (b) 4.56×10^{-10} g
9. (a) 3480 s (b) 0.000 011 1 kg
11. (a) 6.21×10^1 (b) 3.2×10^9
13. 10^4 times

Chapter 2

Questions

1. The puck speeds up and then slows down.
3. In the middle
5. ● ● ●   ●   ●   ●   ●   ● ● ● ● ●●
7. ●● ● ●   ●   ●   ●   ●   ●   ●   ● ●●
9. The car
11. Less than 45 mph
13. They had the same average speed, but Chris had higher instantaneous speeds.
15. The average speed is greater than the instantaneous speed at C and less than the instantaneous speed at D.
17. Determine the time it takes to travel between mile markers.
19. No, the average speed doesn't tell us any instantaneous speeds.
21. A stopwatch and an odometer give average speed; a speedometer gives instantaneous speed.
23. Velocity has a direction.
25. They both have zero acceleration.
27. Anything that changes the speed (or direction) is an accelerator.
29. The bicycle has the greatest acceleration.
31. The motorcycle
33. Carlos could be going 60 mph and slowing while Andrea is going 5 mph and speeding up.
35. They are falling in a vacuum.
37. 40 m/s
39. It stays the same.
41. You will rise to the same height.
43. They hit at the same time.
45. Galileo concluded that the object falls with a constant acceleration when air resistance is ignored. Aristotle hypothesized that the object quickly reaches a constant speed.
47. The heavier ball hits first.
49. The marble has the greater acceleration.
51. The accelerations are the same.
53. The increasing upward force due to the air resistance causes the downward acceleration to continually decrease.
55. Greater than
57. Half the time

Exercises

1. 3530 km/h
3. 66 mph
5. 6.05 mph
7. 560 miles
9. 2.4 mph
11. 3.33 s, compared with 10 s for humans
13. 25 h; no
15. 12.5 mph/s
17. 5 mph/s
19. 27 m/s
21. 13 m/s
23. 20 m
25.

Time (s)	Height (m)	Velocity (m/s)
0	45	0
1	40	10
2	25	20
3	0	30

27. 30 m/s; 3 s

Chapter 3

Questions

1. The snacks appear to fall straight down.
3. An unbalanced force (friction) opposes the motion of the car.
5. The net force on each car is zero.
7. Because of its inertia
9. Because of your inertia
11. The inertia of the anvil keeps it from moving.
13. No, inertia refers to how hard it is to speed up or slow down.
15. No, the net force will not normally point in the direction of any of the individual forces.
17. 50 N; 130 N
19. The wagon now accelerates (speeds up indefinitely) because the net force is not zero.
21. East; still east
23. The net force is up.
25. It doubles.
27. 4 m/s^2
29. Her mass doesn't change.
31. The weight is six times as large.
33. For a skier slowing down and moving to the right:

35. If there is no air resistance
37. Upward force of resistance balances downward force of gravity, so acceleration is zero.
39. The acceleration of each is zero, so the net force on each must be zero.
41. Newton's first law is always valid. There must be other forces opposing the friction so that the net force is zero.
43. The frictional force is equal to 400 N because the net force is zero.
45. 250 N
47. They are equal and opposite by Newton's third law.
49. They are equal and opposite by Newton's third law.
51. Zero
53. According to Newton's third law, there is a reaction force on the cannon.
55. The frictional force of the floor on your feet.
57. 40 N
59. The forces act on different objects. The frictional force of the ground on the horse's hooves allows the horse to move the cart.
61. By Newton's second law because the net force is zero.

Exercises

1. 14 N; 2 N; 10 N
3. 250 N backward
5. 3 m/s^2
7. 900,000 m/s^2
9. 180 N
11. 1.67 m/s^2
13. 75 kg
15. 5 N down
17. 4 m/s^2
19. 162 N
21. 10 N
23. 3 m/s^2

Chapter 4

Questions

1. At each point the instantaneous velocity is in the direction of motion (tangent to the circle), and the net force is toward the center of the circle.
3. Gravity provides the required centripetal force.
5. The velocity is tangent to the path; all the others point radially inward.
7. They point in opposite directions; they have the same length.
9. The speed changes.
11. (a) (b)

13. (a) Speeding up in a straight line (b) Speeding up and turning left
15. (a) Any example that is moving down and speeding up would work, such as a falling rock. (b) Any example that is moving up and slowing down would work, such as an elevator slowing down as it reaches the 10th floor from the lobby.
17. The friction force exerted by the ground on the bicycle
19. The tension force must be greater than the monkey's weight.
21. At the bottom
23. It must be nearly uniform.
25. Doubling speed would quadruple acceleration; reducing radius to half would only double acceleration.
27. Only the force of gravity
29. The ball's velocity is horizontal toward home plate; the net force and acceleration are vertically downward.
31. The accelerations would be the same. Acceleration is independent of horizontal velocity.
33. They hit at the same time. They start with the same vertical velocity.
35. The demonstration works.
37. Kicks the ball at a larger angle
39. 5 N, by Newton's third law
41. The diagram contains only the gravitational force exerted by the Sun.

43. Point C is most likely nearest the object's center of mass.

Exercises

1. (a) 5 m/s west (b) 5 m/s east (c) 15 m/s east
3. 50 m/s at 37° north of west
5. 5 m/s^2 (53° north of east)
7. (a) 2 m/s^2 (b) 240 N
9. (a) 3.15×10^7 s (b) 9.42×10^{11} m (c) 5.96×10^{-3} m/s^2 (d) 3.56×10^{22} N
11. 45 m/s; 8 m/s
13. 125 m; 200 m
15. 2 s; 60 m
17. 9.83 m/s^2

Chapter 5

Questions

1. None
3. Same, by Newton's third law
5. Same
7. The surface area quadruples for cube and sphere.
9. 200 N
11. Because the Moon's acceleration does not depend on its mass
13. Twice as great
15. Because it is also moving sideways fast enough to match Earth's curvature
17. He is in free fall.
19. They are in free fall.
21. No
23. The gravitational forces are very small.
25. Atmospheric friction, though weak, still acts.
27. Send a satellite to orbit Venus.
29. Larger
31. A slow decrease in size of the planetary orbits
33. No, Paris is not on the equator.
35. The satellite would appear to move westward.
37. Because the mass of Earth is so much larger than that of the Moon, it has a much smaller orbit about their common center of mass.
39. Nearer low tide
41. Half the rotational period, 4 h 55 min
43. d; the effects of the Moon and Sun act together.
45. The effects reinforce, producing a larger bulge.
47. The extra tidal force is minuscule.
49. Decreases with the square of the distance
51. Not possible

Exercises

1. 0.006 m/s^2
3. $g/4 = 2.5$ m/s^2
5. 60 N
7. $2F_{old}$; no
9. 0.25 $\mathbf{F}_{Earth}$
11. 12,660 km; 6290 km
13. 1.07×10^{-7} N compared with 200 N
15. 200 N
17. 3.8 m/s^2
19. 264,000 km; 3060 m/s
21. 1.93
23. 40 N

Chapter 6

Questions

1. Its numerical value does not change.
3. Toby is correct.
5. They have large inertia.
7. The net force on an object is equal to the change in its momentum divided by the time required to make the change.
9. They lengthen the time for your leg to stop.
11. Same
13. The impulses are equal.
15. 8 kg·m/s directed up
17. Jeff feels the greater force because his flowerpot experienced the greater change in momentum.
19. They lengthen the interaction time and decrease the forces.
21. At each interaction there are equal and opposite momentum changes. The total momentum is conserved at all times.
23. 4 N acting for 4 s
25. Conservation of linear momentum requires the bullet to go one way and the rifle the other.
27. The net force exerted by the exhaust gases on the rocket-plate system is nearly zero.
29. Earth's large mass allows it to acquire an equal and opposite momentum to the ball without a measurable velocity.
31. At rest
33. Moving to the left
35. 250 kg·m/s
37. Zero
39. They give the rowboat momentum away from the dock.
41. It is conserved.
43. Not without outside help unless the astronaut has something to throw
45. Path B
47. Path D

Exercises

1. 36,000 kg·m/s
3. 62 m/s
5. 5625 N
7. 35,000 kg·m/s (= 35,000 N·s)

9. Impulse = 45,000 N·s; F_{av} = 5625 N
11. 240 N upward
13. 900 N
15. 1.5 m/s
17. Zero
19. 66,800 kg·m/s (north)
21. Zero

Chapter 7

Questions

1. The initial momentum of the system is zero.
3. Conservation of momentum
5. Yes
7. Motorboat pulling a water skier
9. Minivan
11. Same kinetic energies
13. Two balls leave the other side with the same speed as the incoming balls.
15. How high she lifted the ball
17. Decrease
19. Kinetic energies are the same; momentum is greater for the heavier sled.
21. Frictional forces and air resistance do an equal amount of negative work.
23. Less than
25. Both cars stop in the same distance.
27. No, the kinetic energy lost by one object during a collision could be converted to many different forms of energy.
29. The first
31. The choice for the zero of potential energy is arbitrary. Only differences in potential energy are important.
33. The gain in the kinetic energy equals the loss in the gravitational potential energy, keeping the mechanical energy the same.
35. Potential energy is a maximum at either end, and kinetic energy is a maximum at the bottom.
37. It will decrease.
39. As the satellite moves closer to Earth, its gravitational potential energy decreases, and its kinetic energy increases. The reverse happens as the satellite moves farther from Earth.
45. The work done in plowing through the dirt or sand reduces the kinetic energy of the truck.
47. The chemical potential energy of the battery is converted to thermal energy in the socks.
49. We cannot recover the energy lost due to frictional effects.
51. If the winch will run for more than 1 s, it can do more than 600 J of work.
53. A unit of energy
55. A kilowatt-hour is a unit of energy.

Exercises

1. 360,000 J
3. Momentum is not conserved; therefore, the collision could not have taken place.
5. 10 J
7. 2.1 J
9. 11 N
11. Negative for the next 6 months; zero for the year
13. 396 kJ
15. (a) −8.7 J (b) 0 (c) 0
17. 30 J
19. 144 kJ
21. 2.73 horsepower
23. 120 Wh = 0.12 kWh

Chapter 8

Questions

1. Rotational speed is the same for all points on a rotating body.
3. The same
5. Toward the North Star
7. Rotational inertia
11. F_2 because it acts farther from the pivot
13. The force on the nail is larger because it acts through a smaller radius than that of the force applied to the handle.
15. Your torque about the base of the ladder is balanced by the torque exerted by the window—the higher you climb, the greater your torque.
17. Because Sam has to exert the greater force, Sam must be closer to the center of mass.
19. The flywheel of larger radius has the greater rotational inertia and is harder to stop.
21. Increases; decreases
23. Sphere
25. There is no appreciable external torque acting on Earth.
27. The front–back location can be determined by separately weighing the front and back tires on a truck scale. The ratio of those two readings will give the ratio of the distances of the center of mass from the two axles. The same procedure works for the left–right position.
29. Below the foot of the figure
31. One diagram should show that the force acting at the center of mass lies inside the base for the upside-down cone even when it is tipped.
33. Stable equilibrium
35. The wall prevents you from rocking backward, to keep your center of mass over your feet.
37. The jumper's body is curved as it passes over the bar, causing the center of mass to lie outside the jumper's body below the bar.
39. The cylinder wins! Both have the same kinetic energy, but the hoop has more of this energy in the

form of rotational kinetic energy because of its larger rotational inertia.
41. To counteract the torque of the main rotor acting on the helicopter. This torque is a reaction to the torque applied to the main rotor.
43. Tuck
45. The lower rotational inertia requires a larger rotational speed to conserve angular momentum.
47. No, one part of its body has angular momentum in one direction while the rest of its body has an equal angular momentum in the opposite direction.
49. The grooves give the bullet angular momentum so that it doesn't tumble in flight.
51. The push exerted by the exhaust gases produces a torque on the merry-go-round.
53. In the same (horizontal) direction
55. The car would not go around corners because of the large angular momentum of the flywheel. Use two flywheels rotating in opposite directions.
57. Clockwise
59. Polaris appears stationary because Earth's spin axis is aligned with it.
61. (a) toward the ground (b) toward the sky (c) toward the ground (d) zero
63. Reducing the diameter of the wheels would reduce their angular momenta.

Exercises

1. 200 rpm, or 3.33 rev/s
3. 2.78×10^{-4} rev/s
5. -2500 rpm/s
7. $210 \text{ N} \cdot \text{m}$
9. $1800 \text{ N} \cdot \text{m}$
11. 2 m
13. Left-hand side will fall because it has larger torque.
15. $300 \text{ kg} \cdot \text{m}^2/\text{s}$
17. $L_m = 0.132 L_e$; Earth has the larger angular momentum.

Chapter 9

Questions

1. No, Newton's first law is only true in inertial reference frames.
3. Alice would not see any motion of the jar relative to her. Someone sitting on a shelf would see the jar fall freely.
5. No experiment can distinguish between the two.
7. The ball lands on the white spot.
9. Both would agree that there are no horizontal forces.
11. The woman would claim that the ball began with zero kinetic energy, whereas the observer on the ground would calculate a nonzero initial kinetic energy. Both observers would agree on the change in kinetic energy.
13. They would not agree on the value of the kinetic energy at an instant, but they would both agree that the change in kinetic energy is zero.
15. Forward because of its inertia
17. If the velocity is to the right, it is speeding up; if the velocity is to the left, it is slowing down.
19. The ball will land farther from her feet than the white spot. The inertial force will be in the direction opposite to the acceleration, independent of the velocity.
21. It would initially be zero and increase in the direction away from her.
23. The woman would claim that there is a constant horizontal force directed away from her. The observer on the ground would claim that there is no horizontal force.
25. Greater because the inertial force acts in the same direction as the gravitational force
27. Less because the inertial force acts in the opposite direction to the gravitational force
29. Same
31. Inertial forces act backward in the noninertial reference system of the accelerating plane.
33. The same because you are comparing masses. The weights increase by the same amount during the acceleration.
35. If the train's velocity is to the west, it must be speeding up; if its velocity is to the east, it must be slowing down.
37. (a) Up (b) Up (c) Down
39. Up is not defined.
41. (a) Constant velocity (b) Slowing down
43. The "up" direction is toward the ground.
45. Your figure should have the plants leaning toward the center.
47. The centrifugal force pulls the mud off.
49. Parallax would be easier to observe.
51. No, because there is no Coriolis force on the equator.
53. There is no Coriolis force on the equator.
55. The equator is pulled outward by the noninertial centrifugal force.

Exercises

1. 40 m/s; 10 m/s
3. (a) 70 mph (b) 30 mph backward
5. (a) 5 m/s^2 downward (b) 5 m/s^2 upward
7. 0; 25 J; 25 J
9. 1500 N
11. 360 N
13. 100 N
15. 240 N
17. 10 m/s^2

Chapter 10

Questions

1. Because the physical laws are the same in all inertial systems, it is impossible to determine your speed.
3. They contained the speed of light.
5. 70 mph
7. He will obtain a speed equal to c.
9. No, the special theory of relativity does not allow one to travel backward in time.
11. Yes
13. The flash at the front of the train occurred first.
15. The one at the back of the skateboard. The third person must have a speed to the right that is larger than that of the skateboard.
17. A could not have triggered B, so the order could be reversed.
19. Less than 3 min
21. Same note
23. Shorter
25. Greater
27. Peter will be the younger.
29. It is not possible to travel backward in time.
31. The observers would agree because the clocks move toward or away from the light by the same amount.
33. Earth's reference system
35. Less than 100 m
37. Less than 400 m
39. Same
41. (a) The caboose entered first. (b) The tunnel is longer. (c) Yes.
43. Decrease
45. Matter and energy can be converted to each other.
47. The energy radiated away causes the mass to decrease.
49. Special relativity does not include noninertial reference systems, which are included in the general theory.
51. No
53. Blake is comparing the gravitational masses. Jordan is comparing the inertial masses.
55. The penny
57. Passengers in spaceship B
59. Over laboratory distances the bending of light is less than the diameter of an atom.
61. (a) Slower due to the stronger gravitational field (b) Slower due to the centripetal acceleration

Exercises

1. 500 s = 8 min 20 s; 1.28 s
3. 300 s = 5 min
5. 1.09
7. 0.379 ns
9. 28.4 h
11. 69 m
13. 200 m
15. 1.31×10^{13} m
17. 672 N
19. 0.866 c
21. 5 m/s² backward
23. 11 m/s² downward
25. 5 m
27. $3 \times 180° = 540°$. It is almost a great circle, but it has slight bends in three places; for instance, at one pole and both crossings of the equator.

Chapter 11

Questions

1. Lack of predictive power
3. Seal off the machine, thus taking away essentials such as food and water for a period of time.
5. The next brown can might sink. A model can never be proven true.
7. The atomistic nature of matter is not important for most day-to-day activities.
9. Water, salt, and granite are not elements.
11. It is a compound because the mercury combines with something in the air.
13. Compounds are new substances with their own properties that result from substances combining according to the law of definite proportions.
15. 65 amu
17. 1 g of hydrogen
19. They have the same number of molecules.
21. 32 g
23. Ideal gas model does not apply to liquids, but the particles must be much closer together in the liquid.
25. Ball
27. Increased pressure at lower elevation results in decreased volume.
29. More molecules will be hitting the walls per unit time.
31. The perfume molecules travel very crooked paths.
33. To amplify the rise in the narrow tube
35. Atmospheric pressure and the purity of the water must be maintained.
37. 39°C
39. 39°C
41. 273 K
43. Average kinetic energy of the molecules
45. The average kinetic energies are equal because they are at the same temperature.
47. The particles have more kinetic energy. Therefore, they are moving faster and strike the walls more frequently and with more momentum.
49. Volume

51. Temperature drops by one-half on the Kelvin scale.
53. The particles strike the walls more frequently, producing a larger average impulse with the walls of the container.
55. The more energetic molecules leave via evaporation, lowering the average kinetic energy and therefore the temperature of the remaining water.
57. From the cooling due to evaporation

Exercises

1. 3 g
3. 18 g
5. 20 sandwiches; 1200 g; 800 g of ham
7. 5.02×10^{22} atoms
9. 3.35×10^{25} molecules
11. 1 nitrogen and 3 hydrogens
13. 10^{16}
15. 382 kPa $\cong$ 3.7 atm
17. $\frac{1}{3}$ L
19. 160 balloons
21. 1.6 atm

Chapter 12

Questions

1. Solid, liquid, gas, and plasma
3. They have the same densities.
5. Magnesium
7. The gold atoms must be closer together.
9. They are not the same because the crystals have different shapes.
11. Diamond has strong bonds in all directions; graphite has stronger bonds between atoms in two-dimensional layers.
13. Solid oxygen has a lower melting temperature.
15. Spherical
17. The surface tension allows the surface of the water to rise without overflowing.
19. The water molecules are more strongly attracted to the glass than to each other.
21. Gas particles are neutral, whereas the particles in a plasma are electrons and charged ions.
23. The reduced surface area in the case of the gravel requires greater pressure.
25. The volume of liquid water is less than that of vapor, resulting in lower pressure in the can.
27. Denver has a lower atmospheric pressure, so fewer horses would be needed.
29. Denver would always be listed as a low-pressure region.
31. Your ears would hurt the same in both.
33. Same
35. Greater for the cold water
37. It is the atmospheric pressure that determines the maximum height.
39. With the pump at the bottom, you can apply much higher pressures.
41. The purple liquid has the larger density.
43. Higher in salt water
45. The buoyant forces are the same.
47. The reduction in volume reduces the buoyant force and causes you to sink.
49. Lead will float and gold will sink.
51. The volume does not change. Expelling water reduces weight.
53. The buoyant force is the same because they both displace the same volume.
55. It stays the same because the volume below the surface is the same as the volume of the ice when melted.
57. The fast-moving air creates lower pressure above the dime, which results in a net upward force.
59. So the balls will drop over the net
61. The wind blowing by the base reduces the outside pressure.

Exercises

1. 10.5 g/cm^3; silver
3. 0.3 g/cm^3
5. 1200 kg
7. 0.09 m^3
9. 1 cm^3
11. 100,000 N; 100,000 Pa; same
13. 7.5 in.
15. 15 lb/in.^{2}
17. 750 kg/m^3; float
19. 5 N
21. 0.9 g/cm^3
23. 79,300 N

Chapter 13

Questions

1. The gravitational potential energy is converted to thermal energy.
3. The heat seemed to be produced endlessly, implying that it was not a substance.
5. Both change the internal energy of a system and are measured in joules. Heat operates at a microscopic level, whereas work is a macroscopic concept.
7. If the thermometer is initially at a different temperature, heat must flow.
9. Yes, if they start with different temperatures.
11. Temperature is not a physical quantity and therefore cannot flow between objects.
13. Heat is a flow of thermal energy, whereas temperature is a measure of the average kinetic energy of

the atoms and molecules. Notice that temperature is not a form of energy.
15. Neither student is correct.
17. Under all conditions
19. Water
21. No
23. Less
25. It takes a tremendous amount of thermal energy to change the temperature of the water in the oceans.
27. It will freeze if heat is removed and melt if heat is added.
29. It requires a lot of thermal energy to melt the ice without changing its temperature. Also, ice is not a very good conductor of thermal energy.
31. Less than 40°C
33. More internal energy in steam
35. The internal energy must increase by the first law. However, if the increase in internal energy causes a change of state, the system can remain at a fixed temperature.
37. The molecules of the rod exchange kinetic energy via collisions.
39. The order from best to worst is polyurethane foam, static air, glass, and concrete.
41. It will not freeze.
43. Stainless steel has a thermal conductivity that is about one-sixth that of iron.
45. The ground beneath the road keeps the road warm for a while.
47. The air beneath the clouds is shielded from the Sun and is cooler. Being cooler, the air is denser and creates a downdraft.
49. Both cars are heated by radiation.
51. The vacuum reduces conduction and convection, and the silvering reduces radiation.
53. Expansion and contraction of the roof
55. The spacing would be smaller on the wide portion.

Exercises

1. 4000 cal
3. 1200 J; 286 cal
5. 10 Cal
7. 8 J
9. 330 J
11. 0.278 cal/g·°C
13. 124 cal
15. −1°C; +4°C
17. (a) 1600 cal (b) 1600 cal (c) It stays the same.
19. 6771 kJ
21. 1.73×10^6 J; 1188 kJ
23. 0.8 mm

Chapter 14

Questions

1. It converts heat to mechanical energy.
3. This would violate the second law of thermodynamics because heat will not naturally flow between two reservoirs at the same temperature.
5. A heat engine can never convert heat into work without exhausting some heat.
7. Yes
9. Yes, provided the water that the tube accesses is at a different temperature.
11. You cannot get more energy out than you put in.
13. First: You cannot get more energy out than you put it. Second: You cannot convert all the heat to mechanical work.
15. About 20% of the thermal energy available in the food that is consumed shows up as mechanical work.
17. The efficiency decreases.
19. Heat engine A is more efficient.
21. This would allow the hot region to operate at a higher temperature.
23. Because an efficiency of 1 or more means that the mechanical energy produced is as large or larger than the thermal input.
25. In the shade
27. No
29. Temperature rises
31. A refrigerator will always need a motor.
33. The statement implies that work must be expended to make the thermal energy flow from a cooler region to a hotter region.
35. Greater than 1
37. Both strategies are equally invalid.
39. Sums 2 and 12 have highest order; sum 7 has the lowest order.
41. There are more ways of getting this sum than any other.
43. There is only one way of getting this arrangement.
45. The Universe continually becomes more disordered.
47. The disorder increases as collisions cause all the particles to reach a common kinetic energy.
49. The entropy of the universe increases.
53. The order increases. Entropy must increase elsewhere.
55. It is highly unlikely for all the molecules to move in the same direction at the same time.
57. No, because atomic motions are symmetric with time. The direction of time in the release of air from a balloon, however, would be observable.
59. b
61. Electricity
63. If the oven heats the kitchen, the baseboard heaters will run less.

Exercises

1. 600 kJ
3. 3000 cal (each minute)
5. 0.25, or 25%

7. 2250 J
9. 9000 cal/min; 0.25
11. 0.0273, or 2.73%
13. 750 K = 477°C
15. 800 J
17. 700 J (each second)
19. 1600 J (each second)
21. HHHH, HHHT, HHTH, HTHH, THHH, HHTT, HTTH, TTHH, HTHT, THTH, THHT, HTTT, THTT, TTHT, TTTH, TTTT
23. $\frac{6}{36}$, or 16.7%
25. $\frac{6}{216}$ = 1/36, or 2.78%

Chapter 15

Questions

1. Its inertia carries it through the equilibrium point.
3. Down toward the equilibrium point in both cases
5. The period will increase, and the frequency will decrease.
7. A 3-kg block
9. The period stays the same.
11. The period is 60 s, and the frequency is 1/(60 s) = 0.017 Hz.
13. Move the mass down to lengthen the period.
15. The exhaust system has a natural frequency at 2000 rpm.
17. 2 Hz and 1 Hz
19. No
21. It is a longitudinal wave because the medium is a fluid.
23. No
25. Tension and mass per unit length
27. The reflected pulse will be on the opposite side and will cancel with the first one.
29.

31.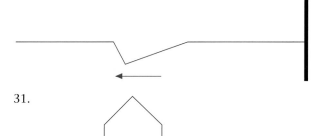

33. Frequency and wavelength
35. The wave with the lower frequency
37. The wavelength is doubled.
39. The frequency stays the same, but the spacing decreases.

41.

43. Five times as high
45.

47. One-half the rope's length
49. Twice the length of the rod
51. Antinode
53. The spacing decreases.
55. Increase
57. Decrease

Exercises

1. 1.5 s
3. 0.167 Hz
5. 3 cm
7. 0.628 s
9. 4
11. 4.45 s
13.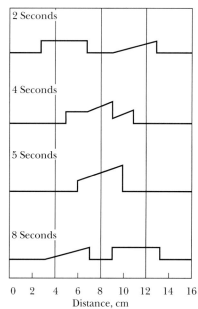

15. 4.17 m/s
17. 100 cm/s
19. 15 m
21. 343 Hz
23. 4 m; (4 m)/2 = 2 m; (4 m)/3 = 1.33 m; (4 m)/4 = 1 m; (4 m)/5 = 0.8 m
25. 2.5 Hz

Chapter 16

Questions

1. Interference and superposition
3. Temperature

5. Water
7. The speed of light is much faster than the speed of sound.
9. Sound waves of all frequencies have the same speed.
11. The intensity is reduced to one-hundredth.
13. Yes
15. Mainly frequency
17. Equal to
19. 400 Hz
21. Increasing the tension or decreasing its mass per unit length
23. Twice the length of the string
25. One quarter the distance from either end; no
27. No
29. The same
31. Nodes
33. Twice the length of the pipe
35. The wavelength doubles.
37. Open at both ends
39. Decreases
41. Equal to
43. The unique blend of higher harmonics is hard to reproduce electronically.
45. Increases
47. You cannot tell.
49. Increases
51. The frequency of the whistle decreases as the train passes by.
53. Speed
55. The frequency is decreasing, and the sound is getting louder.
57. Billy's has the higher frequency.
59. Both produce a sharp rise in pressure.
61. You hear nothing until the shock wave arrives. You will then hear the characteristic boom-boom followed by engine noise.

Exercises

1. 0.003 82 s
3. 0.327 m
5. 343 Hz
7. 17.2 m
9. 3.09 km
11. 600 m
13. 30 dB
15. 125 Hz, 250 Hz, 375 Hz, 500 Hz, ...
17. 588 Hz
19. $f_1 = 85.8$ Hz, $f_3 = 257$ Hz, $f_5 = 429$ Hz, ...
21. 17.4 cm
23. 75 Hz
25. 260 Hz or 264 Hz

Chapter 17

Questions

1. Unless there is dust in the air, none of the light gets scattered into the eyes until the light hits the wall.
3. No light scatters back toward your eyes.
5. Point source
7. Shrinking
9. There will be no umbra when the shadow is not long enough to reach the screen. This occurs when the Moon's shadow does not reach Earth.
11. The image gets brighter and fuzzier.
13. The image is upside down and left–right reversed.
15. C
17. A
19. 0.7 m
21. B and C
23. One; infinite
25. Half her height; no
27. The images are progressively farther from the observer. They are virtual.
29. To produce a beam of light
31. The concave mirror
33. Concave
35. The image is very close to the mirror on the back side and only slightly magnified. Moving away makes the image get larger and move away.
37. The light actually converges to form a real image; there is no light at the location of a virtual image.
39. At the focal point
41. The speed of light is much faster than the speed of sound.
43. The time delay to send a signal to Earth and get a response was on the order of 20 min.
45. Magenta
47. Red; green; black
49. The letter would blend into the background because both would appear red.
51. With light beams you are adding colors, whereas with pigments you are subtracting colors. Red paint absorbs all but red, and green paint will then absorb the red, leaving muddy brown.
53. Cyan
55. Magenta
57. The sky would appear green, and the Sun would appear magenta.

Exercises

1. 20-cm diameter
5. 5 m
7. 36.9°
9. 8 m
11. The image is real, inverted, and reduced in size.
13. The image is located 37.5 cm in front of the mirror and is 2 cm tall.

15. The image is located 20 cm behind the mirror.
17. The image is located 20 cm behind the mirror and is two-thirds as big.
19. The image is located 40 cm in front of the mirror.
21. 0.1 ms; it is 1/2000 as big.
23. 16 ms
25. 9.46×10^{15} m

Chapter 18

Questions

1. C
3. C
5. The bending of light at the surface
7. Greater at the water–air surface
9. The light rays hit the sides of the light pipe at angles greater than the critical angle.
11. 141 million miles
13. Decrease (remember that the Moon is shortened in the vertical direction)
15. The Sun appears to rise before it actually does. Therefore, sunrise is earlier.
17. They are shortened in the vertical direction, yielding oval shapes.
19. Assuming that there is no deflection of the spear as it enters the water, you would aim low, because the fish appears to be higher than it actually is.
21. No
23. The bend will not be a problem for the blue light, but red light will leak out.
25. Refer to Figure B in the "Mirages" feature.
27. Yellow and green
29. The angle formed by the line from the Sun to the raindrop and the line from your eye to the raindrop must have a fixed value for a given color.
31. In the west at about midmorning
33. Diverging lens
35. Send the two beams in parallel to the optic axis and find the point where they cross.
37. Send the two beams in parallel to the optic axis and trace their diverging paths. Extend the paths backward to find their crossing point.
39. Converging lens
41. Through the principal focal point
43. The image becomes dimmer.
45. The lens with the larger index of refraction
47. The light converges to form the image and then diverges again. Your eye must be to the right of the image to see this diverging light.
49. Real image
51. To regulate the amount of light entering the eye
53. Dispersion occurs when light is refracted, not reflected.
55. Converging
57. Converging
59. Larger
61. Consider the symmetry of the letters.

Exercises

1. 40°; 36°
3. From Figure 18-2, the fish is down 32° from normal, or 13° below image of the fish.
5. 39°–40°
7. 48°
11. The image is virtual, located at the other focal point, and the magnification is 2.
13. 26.7 cm on the near side of the lens
15. Farther than $2f$ from the lens
17. The diverging lens will spread the principal rays slightly, such that their convergence will be shifted to the right.
19. The image is virtual, erect, and reduced in size.
21. 5 diopters
23. 0.143 m
25. −0.5 m; diverging

Chapter 19

Questions

1. The particles must be moving very fast to not have noticeable deflections due to gravity.
3. Friction would change the component of the momentum parallel to the surface, and an inelastic collision would change the normal component. Either change would affect the angle of reflection.
5. Less than
7. The light particles experience forces directed into the material with the higher index of refraction.
9. Frequency
11. Red light
13. Amplitude
15. The wave equation ($v = \lambda f$) predicts that if the wavelength decreases while the frequency remains constant, the speed decreases.
17. From a slow medium to a fast medium
19. Red light has a longer wavelength.
21. All colors emerge at the same angle.
23. Blue light
25. Increase
27. Yellow light
29. Phase
31. Diffraction effects are more pronounced for sound because the wavelength is much longer.
33. Increase
35. 300 nm
37. Molecules are smaller than the wavelength of visible light.
39. 20
41. No
43. Thicker

45. Thicker
47. Yes
49. Rotate the lens of one pair 90° relative to the lens of another. If both lenses are polarized, no light will pass through.
51. Tilt your head by 90°.
53. Objects at different depths in the hologram move relative to one another when the hologram is viewed from different angles.
55. Monochromatic light with a constant phase relationship
57. A powerful laser can expose the film before the person can move less than the wavelength of the light.

Exercises

1. 1.76×10^8 m/s
3. 2.42
5. 1.5
7. 1.85
9. 211 m
11. 4.74×10^{14} Hz
13. 372 nm
15. 244 nm
17. 0.006 84
19. 1.38×10^{-2} arcseconds
21. 100 nm
23. 221 nm

Chapter 20

Questions

1. Insulator
3. The charge will flow to all points on the surface of the rod.
5. The moisture allows some of the charge to leave the balloon.
7. To prevent the buildup of charge that might cause sparks
9. Either the fur and plastic produced opposite charges on the rods or one rod is not charged.
11. Neutral objects have equal positive and negative charges.
13. The north pole of a magnet is attracted to positive objects and negative objects but not to neutral objects. It must also be electrically neutral.
15. Repel
17. Zero
19. The balloon induces a charge in the wall.
21. The induced charges on the near sides are always opposite the charged object, producing attraction.
23. Bring the charged rod next to the electroscope. If the leaves get closer together, the rod is negative. If the leaves get farther apart, the rod is positive.
25. Touch the rod to the electroscope.
27. The negative charges flow through the hand to the ground, neutralizing the electroscope.
29. Sphere A is negative, and sphere B is neutral.
31. Touching a charged sphere to a neutral one yields two spheres with one-half charge on each one, using two neutral spheres yields charges of one-third, and so on.
33. Diagram d indicates equal and opposite forces required by Newton's third law.
35. The force decreases by a factor of 9.
37. The force increases by a factor of 3.
39. 3
41. The gravitational force is always attractive, the elementary electric charge is not proportional to inertial mass, electric charge comes in one size, and gravity is much weaker.
43. You have very little net electric charge, and the gravitational force is too weak.
45. Because they have different masses
47. 30 N directed to the left
49. The same force at any point between the plates
51. Equal in size, opposite in direction, and force on proton is upward
53. b > a > c = d, and tangent to the field lines.
55. Greater
57. The speed at point B is greater than v_0.
59. Work required to bring 1 coulomb from reference zero to point in question
61. Moves toward lower electric potential energy and lower electric potential

Exercises

1. 1.6×10^{-19} C
3. 1.5×10^{-17} C
5. 4.05×10^{10} N
7. 2.13×10^{-6} N (attractive)
9. 1.23×10^{36}
11. 800 N/C north; 16 N south
13. 1.13×10^{10} N/C toward the charge
15. 2.16×10^{11} N/C outward
17. 3.6×10^{10} N/C toward the 2-C charge
19. 8×10^{-16} N; 4.79×10^{11} m/s^2
21. -10 J
23. 24 mJ
25. 39,000 V

Chapter 21

Questions

1. Lifetime
3. 18 V
5. Differences are voltage, maximum current available, and maximum charge available; both supply direct current.
7. c; a

9. b; d
11. Turn the battery end for end.
13. No difference
15.

```
B D F    B D F    B D F    B D F
| /|     | /|     | /|     | /|
A C E    A C—E    A C—E    A C—E
```

17. A volt is a measure of potential difference, and an ampere is a measure of current.
19. Water, volume per unit time, and pressure
21. All of them
23. Doubles
25. The increased movement of the atoms with increased temperature causes more collisions with the electrons.
27. Wire 22 bulbs in series.
29. Equal to
31. Increase; decrease
33. Increased; decreased
35. Greater
37. Less than
39. Batteries in series with the bulbs in parallel
41. A is the brightest; B and C are equally dim.
43. Stay the same
45. They will all be shorted.
47. They are equally bright.
49. They all go out.
51. A is shorted out; the others get brighter.
55. Parallel
57. The power is cut in half.
59. The same
61. 120-W bulb

Exercises

1. 138 Ω
3. 9.17 A
5. 4 V
7. 0.5 A
9. 6 Ω; 2A
11. 6 A; 2 Ω
13. 70 W
15. 0.55 A
17. 3 A; 36 W
19. 4400 W
21. 202 Ω
23. 0.16 kWh
25. 36¢

Chapter 22

Questions

1. Both ends of the unmagnetized rod will attract both ends of the other two rods.
3. The second experiment
5. No
7. 4
9. The north pole of a compass needle points in the direction of the field.
11. Clockwise
13. Same direction
15. Down at A; up at B
17. Into the page
19. The magnetic field from one wire cancels that from the other wire.
21. Out of the page
23. He could strike the needle while holding it in Earth's magnetic field.
25. Decrease
27. 4 N/m in the opposite direction, by Newton's third law
29. If two parallel wires each carry a current of 1 ampere, the force per unit length on each wire will be 2×10^{-7} newtons per meter.
31. Magnetic north pole
33. 20° west of north
35. The cosmic rays spiral along the magnetic field lines toward the South Pole.
37. No
39. They are bent in opposite directions. The electron will have the larger acceleration due to its smaller mass.
41. Up; counterclockwise
43. Away; push against it
45. Equal to
47. Inserting the south end into the coil or removing the north end
49. Larger
51. To reverse the direction of the current in the loop, ensuring that the torque is in the same direction
53. The electric and magnetic fields are perpendicular to the direction of travel and to each other. They oscillate in phase and travel at the speed of light.
55. Sound
57. They have different frequencies and wavelengths.
59. By modulating the amplitude of the carrier wave
61. 102.1 MHz

Exercises

1. 4.5×10^5 G
3. 4.3×10^{-4} T
5. 9.6×10^{-13} N; 1.05×10^{18} m/s^2
7. 1.2×10^{-3} N; 0.6 m/s^2
9. 16.7 μC
11. 1.5×10^6 m/s
13. 0.12 T
15. 30 turns
17. 0.025 A; 20
19. 20 A

21. 1.28 s
23. 12.2 cm; about one-fifth the size of the oven
25. 1×10^{15} Hz
27. 200–545 m
29. 275 m

Chapter 23

Questions

1. The structure of the atom was not understood in Mendeleev's time.
3. Carbon
5. Each gas has its own unique collection of spectral lines.
7. Absorption
9. Element A and element B
11. There are many fewer lines in the absorption spectrum.
13. The wavelength of lines can be determined by position alone.
15. B
17. Their deflections by electric and magnetic fields; no
19. Excess
21. Because the ratio of the electron's charge to mass was already known, allowing the mass to be determined
23. They were electrically charged and had a mass less than that of the hydrogen atom.
25. Accelerating charges radiate away energy.
27. The intensity curves are the same for all materials.
29. Decreasing
31. Object B
33. A star's temperature determines the color it appears to be.
35. That the atomic oscillators have quantized energies that are whole-number multiples of a lowest energy, $E = hf$
37. The number emitted from the surface per unit of time
39. Different electrons require different amounts of energy to reach the surface.
41. Each photon in red light has an energy below the minimum.
43. (1) Only angular momenta equal to whole-number multiples of a smallest angular momentum are allowed. (2) Electrons do not radiate when they are in allowed orbits. (3) A single photon is emitted or absorbed when an electron changes orbits.
45. They are equal.
47. The spectrum is produced by a very large number of hydrogen atoms.
49. The inner electrons shield part of the positive nucleus, resulting in one electron orbiting a hydrogen-like nucleus.
51. The first two shells are filled with two and eight electrons, and the outer shell has the remaining six electrons.
53. It easily gives up its single outer electron and forms a bond with other atoms.
55. X ray
57. The X-ray photon has higher energy, shorter wavelength, and higher frequency.
59. The electrons in the innermost orbit become more tightly bound as nucleus charge increases.

Exercises

1. 1.76×10^{11} C/kg
3. 2.7×10^3 kg/m^3; 2.7 times as much
5. 5×10^{-6} m
7. 3.32×10^{-19} J
9. 4.52×10^{14} Hz
11. 9.95×10^{-3} m
13. $h/2\pi = 1.06 \times 10^{-34}$ J·s
15. 4.77×10^{-10} m
17. 9.65×10^{14} Hz
19. 2.46×10^{15} Hz; above
21. 1:64
23. 8.29 keV

Chapter 24

Questions

1. *Successes*: Accounting for the stability of atoms, the numerical values for wavelengths of spectral lines in hydrogen and hydrogen-like atoms, and the general features of the periodic table. *Failures*: Could not account for why accelerating electrons didn't radiate; the spectral lines in nonhydrogen-like atoms; the splitting of spectral lines into two or more lines; the relative intensities of the spectral lines; details of the periodic table, including the capacity of each shell; and relativity.
3. 20 Hz
5. No, the electrons would still be accelerating.
7. Photoelectric effect
9. This implies that photons are massless.
11. Energy
13. Electrons and photons are neither particles nor waves; they simply exhibit behavior that we interpret as wavelike and particle-like.
15. No
17. The wavelengths associated with the electrons must form standing wave around the nucleus.
19. The wavelength is too small.
21. Both electrons and photons behave as waves when producing the pattern and as particles when detected.
23. Electrons are deflected by electric and magnetic fields, electrons have mass and charge, photons travel at the speed of light.
25. An interference pattern
27. Intensities
29. The probability of finding the "particle" at a particular location

31. In the middle of the box
33. You cannot meaningfully say that the electron was ever at one particular location in the box, and therefore it does not have to move.
35. Along the x axis, not too near the origin and not too far away
37. 10
39. $n = 1$, $\ell = 0$, $m_\ell = 0$, and $m_s = +\frac{1}{2}$; $n = 1$, $\ell = 0$, $m_\ell = 0$, and $m_s = -\frac{1}{2}$; $n = 2$, $\ell = 0$, $m_\ell = 0$, and $m_s = +\frac{1}{2}$; $n = 2$, $\ell = 0$, $m_\ell = 0$, and $m_s = -\frac{1}{2}$.
41. The Heisenberg uncertainty principle places very precise limits on our simultaneous knowledge of specific variables that are important for systems on the atomic scale. It should not be generalized to apply outside this domain.
43. Momentum and energy do not form a Heisenberg pair.
45. There must be uncertainty in the momentum and therefore in the wavelength. A range of wavelengths is used to describe a localized electron.
47. The electrons have precisely defined positions and velocities.
49. Einstein was troubled by the idea that the probability associated with quantum mechanics was all that can be known.
51. Time
53. Nothing
55. The electrons return to their lower energy levels through a series of jumps, emitting visible light.
57. The electrons must remain in the excited energy levels for a relatively long time.
59. A photon induces an electron in an excited state to return to a lower state, producing a second, identical photon.

Exercises

1. 1.75×10^{-38} m
3. 2.88×10^{-11} m
5. 3.11×10^{-7} m
7. 7.28×10^{6} m/s
9. 1.66×10^{-25} kg·m/s
11. 1.77×10^{-33} kg·m/s
13. 1.33×10^{-23} kg·m/s
15. 2.88×10^{-36} m
17. 0.132 nm
19. 1.46×10^{7} m/s; 607 eV
21. 2.07×10^{-7} eV
23. 2.07×10^{-11} s

Chapter 25

Questions

1. Becquerel's radiation was found to occur naturally.
3. The wavelengths of visible light are very much bigger than the size of nuclei.
5. Alpha and beta radiation
7. Alpha particles
9. Protons
11. A neutron or a proton
13. (a) Yttrium (b) Cerium (c) Bromine
15. (a) 12 neutrons, 12 protons, and 12 electrons
    (b) 32 neutrons, 27 protons, and 27 electrons
    (c) 126 neutrons, 82 protons, and 82 electrons
17. 90
19. $^{20}_{10}$Ne
21. Decreases by 1
23. (a) $^{220}_{90}$Th (b) $^{193}_{81}$Tl
25. (a) $^{18}_{8}$O (b) $^{90}_{39}$Y
27. (a) $^{181}_{76}$Os (b) $^{237}_{93}$Np
29. (a) $^{28}_{14}$Si (b) $^{22}_{10}$Ne
31. A neutron must leave the nucleus.
33. $^{64}_{28}$Ni or $^{64}_{30}$Zn
35. (a) Beta plus (b) Alpha
37. ^{206}Pb
39. $^{17}_{9}$F
41. There would be a decrease of one proton and one neutron.
43. 4 kg
45. It could happen at any time because the process is random.
47. By determining the fraction of the radioactive ^{14}C that has decayed to ^{12}C
49. No
51. Gamma ray, electron, and alpha particle
53. Most natural radiation exposure comes from photons and electrons, for which there is little difference between rads and rems.
55. Nuclear power
57. It takes a very long time for the effects to show up, and there are other causes.
59. Skin provides protection because alpha particles cannot penetrate very far.

Exercises

1. 6 km
3. 1.99×10^{-26} kg
5. 192.2 amu; Ir
7. 1.007 825 amu
9. 8 µCi
11. 32 trillion
13. 17,100 years
15. $\frac{1}{8}$
17. 50

Chapter 26

Questions

1. 10,000,000 V
3. They all acquire the same kinetic energy.

5. Alpha particle due to its larger charge
7. The energy loss due to the radiation of the accelerating charges is less for massive particles.
9. Strong, electromagnetic, weak, gravitational
11. There must be a strong force to counteract the electric repulsive force and hold the nucleons together.
13. Both are fundamental forces that can be attractive or repulsive. The strong force is stronger, has a finite range, and changes from attractive to repulsive at very short distances. The electromagnetic force has an infinite range and remains either attractive or repulsive.
15. A 2.2-MeV gamma ray
17. 90 protons and 150 neutrons
19. Carbon nucleus
21. Nitrogen-17
23. The process will continue until the nucleons are most tightly bound, which occurs in the region of the peak of the average-binding-energy curve at iron.
25. They are about equal because the energy spacing between proton states and neutron states is about the same.
27. It is energetically more favorable to add neutrons than it is to add protons.
29. It is unstable and would decay via beta plus decay.
31. Beta plus decay
33. The splitting of a heavy nucleus into two or more lighter ones
35. Two
37. Because it releases more than one neutron on the average.
39. To initiate additional fission reactions
41. Nuclear bomb
43. To absorb enough neutrons to ensure that an average of one neutron from each fission process initiates another
45. Plutonium-239
47. New fuel is bred by converting an isotope that does not readily fission into one that does.
49. The combining of two or more light nuclei to form a heavier one
51. The temperature is high, but the density is low, so there is little heat energy.
53. Fuel is much more available, and there is less risk.
55. Agrees with solar models and with the age of Earth, as determined by radioactive dating
57. Fusion is the combining of lighter elements to form heavier ones, whereas fission is the splitting of heavier elements to form lighter ones.

Exercises

1. 2.31×10^{-20} kg·m/s; 2.87×10^{-14} m
3. Use $E = \frac{1}{2}mv^2$ and solve for $v = 8.75 \times 10^9$ m/s $> c$.
5. Use $E = mc^2$ and convert units.
7. 105 MeV
9. 7.012 161 amu
11. 18.6 keV
13. 5.24 MeV
15. Approximately 200 MeV
17. Subtract the product and reactant masses and convert to million electron volts (MeV).
19. 121
21. 3.87×10^{26} W

Chapter 27

Questions

1. Electron, proton, neutron, and photon
3. Positron; same magnitude but opposite sign
5. No
7. It will mutually annihilate with its corresponding particle.
9. Because they are electrically neutral, they do not leave tracks.
11. The total linear momentum is zero. If a single photon were emitted, it would carry away momentum E/c, yielding nonzero total momentum.
13. Conservation of charge and nucleon number (or baryon number)
15. Weak
17. The uncertainty principle provides the mechanism to create exchange particles without violating conservation of energy.
19. The intermediate vector bosons are quite massive, and their discovery awaited more energetic particle accelerators.
21. Exchange photons with very little energy can exist for very long times and can therefore travel infinite distances.
23. Its infinite range
25. Mesons have spins that are whole numbers, whereas baryons have spins of $\frac{1}{2}, \frac{3}{2}, \ldots$
27. Hadrons are not elementary; they are composed of quarks. Hadrons also participate in the strong interaction.
29. Yes, as evidenced by its beta decay.
31. 10^{-10} s
33. It doesn't have enough rest-mass energy.
35. e^+
37. (a) Muon and electron lepton numbers (b) Spin and baryon number (c) Strangeness
39. (a) 1 (b) -1 (c) 0
41. Two photons
43. uud for the antiproton, and udd for the antineutron
45. $\bar{u}s$; -1
47. uuu
49. dss
51. ds; no
53. No

55. It would be a neutral meson.
57. By assigning the color quantum number to the quarks

Chapter 28

Questions

1. Supernovae, orbiting neutron stars, and orbiting black holes
3. The energy radiated in the form of gravitational waves is very small.
5. Taylor and Hulse found a pair of neutron stars orbiting each other and measured the orbital period to be decreasing, as predicted by theory.
7. Not if we observe a very large number of protons
9. Gravitational force
11. Tiny loops
13. 15 billion years
15. Gravitational
17. To escape cosmic radiation, which would swamp the tiny neutrino signals
19. Measurements of ripples in the cosmic-ray background
21. Any physical system, no matter how complex, may be understood in terms of its component parts.

Credits

This page constitutes an extension of the copyright page. We have made every effort to trace the ownership of all copyrighted material and to secure permission from copyright holders. In the event of any question arising as to the use of any material, we will be pleased to make the necessary corrections in future printings. Thanks are due to the following authors, publishers, and agents for permission to use the material indicated.

Text

Chapter 1
p. 3: Albert Einstein Archives. The Hebrew University of Jerusalem, Israel. p. 5: top, © 1992 by Nick Downes from Big Science.

Chapter 2
p. 17: top, © Sidney Harris. Used by permission. p. 23: bottom left, ©1992 by Nick Downes from Big Science.

Chapter 5
p. 88: top left, © 1992 by Sidney Harris. Used by permission.

Chapter 7
p. 113: R. P. Feynman, R. B. Leighton, and M. Sands, *The Feynman Lectures on Physics* (Glenview, IL: Addison-Wesley, 1963), 1: 4-1 and 4-2. Reprinted by permission.

Chapter 10
p. 192: top left, © Sidney Harris. Used by permission.

Chapter 16
p. 325: bottom center, © 1992 by Nick Downes from Big Science.

Chapter 17
p. 349: center right, © 1992 Sidney Harris. Used by permission.

Chapter 23
p. 491: bottom left, © Sidney Harris. Used by permission.

Chapter 24
p. 513: bottom center, © Sidney Harris. Used by permission.

Chapter 28
p. 595: top, Courtesy of Particle Data Group/Lawrence Berkeley National Lab © 2000.

Photographs

Pages 10, 74, 82, 83, 86, 89, 90, 92, 149, 161, 162, 164, 169, 170, 174, 356, 376, 391, 414, 464, 535, 566, 570, 574, 590: Courtesy of NASA.
Pages 28, 36, 41, 58, 63, 96, 102, 104, 105, 108, 137, 138, 139, 140, 141, 144, 148, 155, 256, 276, 293, 321, 339, 340, 343, 344, 345, 346, 347, 348, 350, 351, 359, 361, 362, 367, 380, 404, 409, 410, 447, 450, 451, 453, 474, 533, 534, 542, 543: David Rogers.
Pages 18, 30, 72, 224, 232, 242, 243, 244, 265, 280, 294, 309, 352, 356, 391, 393, 437, 457, 537: George Semple.
Pages 16, 196, 208, 225, 256, 342, 372, 388, 395, 438, 523: Gerald F. Wheeler.
Pages 34, 43, 168, 210, 231, 233, 243, 263, 279, 284, 287, 379, 390, 396, 427, 432, 482, 536, 540: Charles D. Winters.

Chapter 1
p. 1: Atlas Image Courtesy of 2Mass/IPAC–Caltech/NASA/NSF. p. 4: Gianni Tortoli/Photo Researchers; © Telegraph Colour Library/FPG/Getty Images. p. 7: Larry D. Kirkpatrick. p. 9: Jacques Descloitres, MODIS Land Rapid Response Team, NASA/GSFC. p. 10: © Ken Edward/Photo Researchers.

Chapter 2
p. 14: © Georgina Bowater/Corbis/Stock Market. p. 15: © Royalty-Free/Corbis; © Pat O'Hara/Corbis. p. 16: © Simon Bruty/Stone/Getty Images. p. 18: Harold and Esther Edgerton Foundation, 2002, Courtesy of Palm Press, Inc. p. 19: © Gary Hershorn/Reuters/Corbis. p. 21: Robin Smith/Stone/Getty Images. p. 24: North Wind Picture Archive. p. 26: © Kenneth Edward/Photo Researchers. p. 27: © Digital Vision/Getty Images.

Chapter 3
p. 33: Courtesy of U.S. Army Parachute Team, Golden Knights. p. 37: © Bettmann/Corbis. p. 47: Amoz Eckerson/Visuals Unlimited. p. 48: © TempSport/Corbis. p. 52: A. Copley/Visuals Unlimited. p. 55: © Jason Szenes/Corbis SYGMA.

Chapter 4
p. 56: © Brand-X Pictures/Getty Images. p. 60: © David Madison/Duomo. p. 62: © Duomo/Corbis. p. 63: © 1990 Richard Megna/Fundamental Photographs. p. 64: UPI/Corbis-Bettmann. p. 67: © 1990 Richard Megna/Fundamental Photographs. p. 68: © Marc Carter/Stone/Getty Images. p. 72: © Simon McComb/Stone/Getty Images. p. 73: © Kevin Schafer/Stone/Getty Images.

Chapter 5
p. 76: © Bettmann/Corbis. p. 77: © Jim Sugar/Corbis. p. 82: © Royalty-Free/Corbis. p. 83: Courtesy of DIRECTV, Inc. p. 86: Courtesy of Nova Scotia Tourism, Culture and Heritage. p. 87: California Institute of Technology/Palomar Observatory.

Chapter 6
p. 97: © Joel W. Rogers/Corbis; © Royalty-Free/Corbis. p. 98: © American Honda Motor Co., Inc.; Superstock, Inc. p. 103: © Lon C. Diehl/PhotoEdit. p. 104: © Photodisc Green/Getty Images. p. 105: Courtesy of the Special Collections Department, Bryn Mawr College Library.

Chapter 7
p. 111: © Thomas Del Brase/Stone/Getty Images. p. 114: Courtesy of Arbor Scientific, Inc. p. 117: © David Young-Wolff/PhotoEdit. p. 121: © Bob Torrez/Stone/Getty Images. p. 123: © VCG/FPG/Getty Images. p. 124: Larry D. Kirkpatrick; © First Light/Corbis; SOHO/NASA/ESA. p. 126: Bureau of Reclamation. p. 127: © Corbis. p. 130: Larry D. Kirkpatrick. p. 131: © Richard Hutchings/PhotoEdit; © David Young-Wolff/Stone/Getty Images. p. 132: David Frazier Photolibrary.

Chapter 8
p. 134: © Royalty-Free/Corbis. p. 144: © Gerard Lacz/NHPA. p. 148: Courtesy of Pacific Science Center, photograph by Dick Milligan; © Kathy Ferguson/PhotoEdit. p. 149: © Arnulf Husmo/Stone/Getty Images; © Neal Preston/Corbis.

Chapter 9
p. 154: David Malin/Anglo-Australian Observatory. p. 163: Cedar Point photos by Dan Feicht. p. 167: Courtesy of John Kielkopf, Department of Physics and Astronomy, University of Louisville. p. 173: Adam Jones/Dembinsky Photo Associates.

Chapter 10
p. 176: Telegraph Colour Library/FPG/Getty Images. p. 177: © Lucien Aigner/Corbis. p. 191: AIP Niels Bohr Library. p. 193: ® Registered U.S. Patent Office. © 2005 Hertz System, Inc. p. 201: Courtesy of Academy of Sciences, Washington, D.C. © 1978 by

C-1

Robert Berks. Photo by Alex Jamison. p. 202: Dr. Seth Shostak/Science Photo Library/Photo Researchers.

Chapter 11
p. 207: James Randklev/Stone/Getty Images. p. 221: Jeff Smith/FOTOSMITH. p. 224: Jeff Greenberg/Visuals Unlimited.

Chapter 12
p. 227: © Ted Kinsman/Photo Researchers. p. 229: University of California, Lawrence Livermore Laboratory and the U.S. Department of Energy. p. 231: Richard C. Walters/Visuals Unlimited; Courtesy of Leonard Fine. p. 232: Aaron Haupt/Photo Researchers. p. 233: Jack Finch/SPL/Photo Researchers; © Photodisc Green/Getty Images. p. 235: Courtesy of Sharp. p. 236: Rod Catanach/Woods Hole Oceanographic Institute. p. 237: T. Nakamura/Superstock. p. 239: © Taxi/Getty Images. p. 241: David J. Sams/Stone/Getty Images. p. 243: © Photodisc Green/Getty Images. p. 245: Amwell/Stone/Getty Images; Gary Bonner.

Chapter 13
p. 247: Mark Harmel/FPG/Getty Images. p. 251: The Royal Society, London/The Bridgeman Art Library. p. 257: Courtesy of Corning Glass Works. p. 258: A. A. Bartlett, University of Colorado, Boulder; © Janez Skok/Corbis; © Royalty-Free/Corbis. p. 259: © Tony McConnell/Photo Researchers. p. 262: Edward M. Wheeler; Courtesy of Honeywell, Inc. p. 264: © Cosmo Condina/Stone/Getty Images. p. 265: Ira Rubin/Dembinsky Photo Associates. p. 266: Richard Hamilton Smith/Dembinsky Photo Associates.

Chapter 14
p. 268: Stan Osolinski/Dembinsky Photo Associates. p. 269: Courtesy of VWR Corporation. p. 270: Rufus Cone, Physics Department, Montana State University; © American Honda Motor Co., Inc. p. 274: © Royalty-Free/Corbis; © Larry Lefever/Grant Heilman Photography. p. 275: Courtesy of Marc Sherman. p. 282: U.S. Environmental Science Services Administration. p. 284: Pearl Levi; Courtesy of Anne Sherman.

Chapter 15
p. 288: © Don Bonsey/Stone/Getty Images. p. 292: Richard Megna/Fundamental Photographs, NYC. p. 293: © Geoffrey Wheeler. p. 295: Ian O'Leary/Stone/Getty Images. p. 296: Special Collections Division, University of Washington Libraries, photos by Farquharson. pp.: 305 and 306: *PSSC Physics*, 2nd ed., 1965, D. C. Heath & Co. and Educational Development Center, Inc., Newton, Mass. p. 307: Courtesy of Sabrina Zigman/Benjamin Cardozo High School and by permission of *The Physics Teacher* 37 (1999): 55. p. 310: Courtesy of Philadelphia International Airport. p. 311: Courtesy of VWR Corporation.

Chapter 16
p. 313: © Jake Rajs/Stone/Getty Images. p. 317: Courtesy of SENSO Digital Hearing Aid by Widex. p. 318: Courtesy of Brooks/Cole Publishing Co. p. 319: Catgut Acoustical Society, Inc. p. 321: Corbis. p. 322: Erwin C. "Bud" Nielsen/Visuals Unlimited. p. 327: Courtesy of U.S. Navy. p. 328: © 1973 Kim Vandiver and Harold E. Edgerton/Courtesy of Palm Press, Inc.; © Jason Hawkes/Stone/Getty Images. p. 330: Courtesy of C. F. Martin & Co., Nazareth, Penn.; Courtesy of Henry Leap. p. 332: Courtesy of PASCO Scientific.

Chapter 17
p. 336: John C. Muegge/Visuals Unlimited. p. 338: Jim Anderson, Physics Department, Montana State University; Courtesy of Mike Murray. p. 344: © Ross Anania/Stone/Getty Images. p. 345: Mark E. Gibson/Visuals Unlimited. p. 351: Courtesy of American Association of Physics Teachers. Photo by Caitlin C. Morgan. p. 352: Courtesy of American Association of Physics Teachers. Photo by Rhian E. Vanderburg.

Chapter 18
p. 358: © 1996 Pekka Parviainen/Dembinsky Photo Associates. p. 362: Dennis O'Clair/Stone/Getty Images. p. 363: Paul Swenson. p. 364: David Parker/Science Photo Library/Photo Researchers. p. 365: David Cavagnaro/Visuals Unlimited; Patrick J. Endres/Visuals Unlimited. p. 366: Courtesy of Robert Greenler. p. 379: Thomas Hallstein/Outsight Environmental Photos.

Chapter 19
p. 382: Robert Mark, Princeton University. pp. 383 and 384: *PSSC Physics*, 2nd ed., 1965; D. C. Heath & Co. and Educational Development Center, Inc., Newton, Mass. pp. 386 and 387: Courtesy of Joshua Francis. p. 388: Gerald F. Wheeler and Henry Cruz. p. 389: M. Cagnet, M. Francon, and J. C. Thierr, *Atlas of Optical Phenomena*, Berlin, Springer-Verlag, 1962, plate 16. p. 391: Courtesy of Vincent Mallette. p. 396: Peter Aprahamian/Science Photo Library/Photo Researchers. p. 398: David Nunuk/SPL/Photo Researchers. p. 399: Phil Jude/SPL/Photo Researchers. p. 400: © Photodisc Red/Getty Images.

Chapter 20
p. 403: Jean-Loup Charmet/SPL/Photo Researchers. p. 407: © The Corcoran Gallery of Art/Corbis. p. 413: Courtesy of Prof. Clint Sprott, University of Wisconsin–Madison. p. 414: © Royalty-Free/Corbis. p. 419: © Photodisc Red/Getty Images.

Chapter 21
p. 426: © David H. Wells/Corbis. p. 428: Courtesy of Vector Products, Inc. p. 438: Courtesy of Marc Sherman.

Chapter 22
p. 446: © Taxi/Getty Images. pp. 448 and 449: *PSSC Physics*, 2nd ed., 1965, D. C. Heath & Co. and Educational Development Center, Inc., Newton, Mass. p. 451: Dembinsky Photo Associates. p. 452: Argonne National Laboratory and the U.S. Department of Energy. p. 455: © Kevin Schafer/Photographer's Choice/Getty Images. p. 459: Courtesy of the Gillette Company. p. 463: © Mark Kelley/Stone/Getty Images; North Wind Picture Archive. p. 464: Photo courtesy of DIRECTV, Inc. p. 465: Courtesy of Panasonic. p. 470: Martin Dohrn/Science Photo Library/Photo Researchers.

Chapter 23
p. 473: David Parker/SPL/Photo Researchers. p. 477: AIP Emilio Segrè Visual Archives, W. F. Meggers Collection. p. 481: U.K. Atomic Energy Authority, Courtesy AIP Emilio Segrè Visual Archives. p. 484: AIP Niels Bohr Library, W. F. Meggers Collection. p. 490: Niels Bohr Archives, Courtesy AIP Emilio Segrè Visual Archives. p. 492: Courtesy of Burndy Library, Electra Square, Conn. p. 496: Jerry Schad/Photo Researchers.

Chapter 24
p. 498: Super Stock. p. 499: AIP Niels Bohr Library, W. F. Meggers Collection. p. 500: Courtesy of PASCO Scientific. p. 501: Courtesy of VEECO Instruments and Purdue University. p. 502: ZYGO Corporation; Courtesy of RCA/General Electric Corporate Research and Development. p. 509: General Electric Company. p. 510: Oesper Collection in the History of Chemistry/University of Cincinnati. p. 511: Courtesy of University of Hamburg. p. 514: AIP Niels Bohr Library. p. 516: Alexander Tsiaras/Science Source/Photo Researchers; Rick Poley/Visuals Unlimited. p. 520: Aaron Haupt/Photo Researchers; © American Honda Motor Co., Inc.

Chapter 25
p. 524: Scott Camazine/Photo Researchers. p. 527: AIP Niels Bohr Library, W. F. Meggers Collection. p. 536: Paul Hanny/Liaison. p. 538: AT&T Bell Laboratories/Lucent Technologies. p. 539: Sovfoto/Eastfoto; © Royalty-Free/Corbis. pp. 543 and 544: Fermi National Accelerator Laboratory. p. 547: © Royalty-Free/Corbis.

Chapter 26
p. 548: © Brand X Pictures/Getty Images. p. 549: Don Collins, Biology Department, Montana State University; U.S. Army, White Sands Missile Range. p. 550: Courtesy of Gene Sprouse, SUNY at Stony Brook; Stanford Linear Accelerator and the U.S.

Department of Energy. p. 551: Fermi National Accelleratory Laboratory. p. 554: Courtesy of Louise Barker/AIP Niels Bohr Library. p. 558: AIP Emilio Segrè Visual Archives; Argonne National Laboratory and the U.S. Department of Energy. p. 561: Photograph by Francis Simon, courtesy of AIP Emilio Segrè Visual Archives. p. 562: © Royalty-Free/Corbis. p. 563: Courtesy of F. Gauthier-Lafaye/CNRS. p. 564: Idaho National Laboratory and the U.S. Department of Energy. p. 565: Princeton University Plasma Physics Laboratory.

Chapter 27
p. 571: Fermilab, Visual Media Services. p. 572: Stanford Linear Accelerator and the U.S. Department of Energy. p. 573: University of California Lawrence Livermore Laboratory and the U.S. Department of Energy. p. 575: Savannah River Laboratory. p. 577: © Shelley Gazin/Corbis.

Chapter 28
p. 592: Courtesy of Caltech/LIGO. p. 593: Redrawn courtesy of NASA/JPL-Caltech. p. 595: Courtesy of Particle Data Group, Lawrence Berkeley National Laboratory © 2000. p. 596: Photo by Robert Isear. Courtesy of AIP Emilio Segrè Visual Archives, Physics Today Collection. pp. 598 and 599: Courtesy of SNO Institute. p. 600: Courtesy of Chris Kyba/SNO Institute.

Interlude Images
p. 93: Mark J. Thomas/Dembinsky Photo Associates. p. 94: by Susan Schwartzenberg, © Exploratorium, www.exploratorium.edu. p. 152: © Doug Armand/Stone/Getty Images. p. 204: Lawrence Berkeley Laboratory/SPL/Photo Researchers. p. 205: Courtesy of Marc Sherman. p. 286: © Raymond Gendreau/Stone/Getty Images. p. 333: Jerry Irwin/Photo Researchers. p. 334: Bill Kamin/Visuals Unlimited. p. 401: Leif Skoogfors/The Franklin Institute Science Museum/Woodfin Camp & Associates. p. 471: © Floyd Dean/Taxi/Getty Images. p. 472: © Bettmann/Corbis. p. 522: ICRR, Institute of Cosmic Ray Research, University of Tokyo.

Glossary

aberration A defect in a mirror or lens causing light rays from a single point to fail to focus at a single point in space.

absolute zero The lowest possible temperature: 0 K, −273°C, or −459°F.

absorption spectrum The collection of wavelengths missing from a continuous distribution of wavelengths due to the absorption of certain wavelengths by the atoms or molecules in a gas.

activity The rate at which a collection of radioactive nuclei decay. One curie corresponds to 3.7×10^{10} decays per second; also called radioactivity.

alloy A metal produced by mixing other metals.

alpha particle The nucleus of helium, consisting of two protons and two neutrons.

alpha (α) radiation The type of radioactive decay in which nuclei emit alpha particles (helium nuclei).

ampere The SI unit of electric current; 1 coulomb per second. The current in each of two parallel wires when the magnetic force per unit length between them is 2×10^{-7} newton per meter.

amplitude The maximum distance from the equilibrium position that occurs in periodic motion.

angular momentum A vector quantity giving the rotational momentum. For an object orbiting a point, the magnitude of the angular momentum is the product of the linear momentum and the radius of the path, $L = mvr$. For a spinning object, it is the product of the rotational inertia and the rotational velocity, $L = I\omega$.

antinode One of the positions in a standing-wave or interference pattern where there is maximum movement; that is, the amplitude is a maximum.

antiparticle A subatomic particle with the same-size properties as those of the particle, although some may have the opposite sign. The positron is the antiparticle of the electron.

Archimedes' principle The buoyant force is equal to the weight of the displaced fluid.

astigmatism An aberration, or defect, in a mirror or lens that causes the image of a point to spread out into a line.

atom The smallest unit of an element that has the chemical and physical properties of that element. An atom consists of a nucleus surrounded by an electron cloud.

atomic mass The mass of an atom in atomic mass units. Sometimes this refers to the atomic mass number—the number of neutrons and protons in the nucleus.

atomic mass unit One-twelfth the mass of a neutral carbon atom containing six protons and six neutrons.

atomic number The number of protons in the nucleus or the number of electrons in the neutral atom of an element. This number also gives the order of the elements in the periodic table.

average acceleration The change in velocity divided by the time it takes to make the change, $\bar{a} = \Delta v/\Delta t$; measured in units such as (meters per second) per second. An acceleration can result from a change in speed, a change in direction, or both.

average speed The distance traveled divided by the time taken, $\bar{s} = d/t$; measured in units such as meters per second or miles per hour.

average velocity The change in position—displacement—divided by the time taken, $\bar{v} = \Delta x/\Delta t$.

Avogadro's number The number of molecules in 1 mole of any substance. Equal to 6.02×10^{23} molecules.

baryon A type of hadron having a spin of $\frac{1}{2}, \frac{3}{2}, \frac{5}{2}, \ldots$ times the smallest unit. The most common baryons are the proton and neutron.

beats A variation in the amplitude resulting from the superposition of two waves that have nearly the same frequencies. The frequency of the variation is equal to the difference in the two frequencies.

Bernoulli's principle The pressure in a fluid decreases as its velocity increases.

beta particle An electron emitted by a radioactive nucleus.

beta (β) radiation The type of radioactive decay in which nuclei emit electrons or positrons (antielectrons).

binding energy The amount of energy required to take a nucleus apart. The analogous amount of energy for other bound systems.

black hole A massive star that has collapsed to such a small size that its gravitational force becomes so strong that not even light can escape from its "surface."

bottom The flavor of the fifth quark.

British thermal unit The amount of heat required to raise the temperature of 1 pound of water by 1°F.

buoyant force The upward force exerted by a fluid on a submerged or floating object. *See* Archimedes' principle.

calorie The amount of heat required to raise the temperature of 1 gram of water by 1°C.

camera obscura A room with a small hole in one wall, used by artists to produce images.

cathode ray An electron emitted from the negative electrode in an evacuated tube.

center of mass The balance point of an object. The location in an object that has the same translational motion as the object if it were shrunk to a point.

centi A prefix meaning $\frac{1}{100}$. A centimeter is $\frac{1}{100}$ meter.

centrifugal An adjective meaning "center-fleeing."

centrifugal force A fictitious force arising in a rotating reference system. It points away from the center, in the direction opposite the centripetal acceleration.

centripetal An adjective meaning "center-seeking."

centripetal acceleration The acceleration of an object directed toward the center of its circular path. For uniform circular motion, it has a magnitude v^2/r.

centripetal force The force on an object directed toward the center of its circular path. For uniform circular motion, it has a magnitude mv^2/r.

chain reaction A process in which the fissioning of one nucleus initiates the fissioning of others.

change of state The change in a substance between solid and liquid or between liquid and gas.

charge A property of an elementary particle that determines the strength of its electric force with other particles possessing charge. Measured in coulombs, or in integral multiples of the charge on the proton.

charged Possessing a net negative or positive charge.

charm The flavor of the fourth quark.

chromatic aberration A property of lenses that causes different colors (wavelengths) of light to have different focal lengths.

coherent A property of two or more sources of waves that have the same wavelength and maintain constant phase differences.

complementarity principle The idea that a complete description of an atomic entity, such as an electron or a photon, requires both a particle description and a wave description, but not at the same time.

complementary color For lights, two colors that combine to form white.

complete circuit A continuous conducting path from one end of a battery (or other source of electric potential) to the other end of the battery.

compound A combination of chemical elements that forms a new substance with its own properties.

conduction, thermal The transfer of thermal energy by collisions of the atoms or molecules within a substance.

conductor A material that allows the passage of electric charge or the easy transfer of thermal energy. Metals are good conductors.

conservation of angular momentum If the net external torque on a system is zero, the total angular momentum of the system does not change.

conservation of charge In an isolated system, the total charge is conserved.

conservation of energy The total energy of an isolated system does not change.

conservation of linear momentum The total linear momentum of a system does not change if there is no net external force.

conservation of mass The total mass in a closed system does not change even when physical and chemical changes occur.

conserved This term is used in physics to mean that a number associated with a physical property does not change; it is invariant.

convection The transfer of thermal energy in fluids by means of currents such as the rising of hot air and the sinking of cold air.

Coriolis force A fictitious force that occurs in rotating reference frames. It is responsible for the direction of the winds in hurricanes.

coulomb The SI unit of electric charge. The amount of charge passing a given point in each section in a conductor carrying a current of 1 ampere. The charge of 6.24×10^{18} protons.

crest The peak of a wave disturbance.

critical angle The minimum angle of incidence for which total internal reflection occurs.

critical chain reaction A chain reaction in which an average of one neutron from each fission reaction initiates another reaction.

critical mass The minimum mass of a substance that will allow a chain reaction to continue without dying out.

crystal A material in which the atoms are arranged in a definite geometric pattern.

curie A unit of radioactivity; 3.7×10^{10} decays per second.

current A flow of electric charge; measured in amperes.

cycle One complete repetition of a periodic motion. It may start anywhere in the motion.

daughter nucleus The nucleus resulting from the radioactive decay of a parent nucleus.

definite proportions, law of When two or more elements combine to form a compound, the ratios of the masses of the combining elements have fixed values.

density A property of material equal to the mass of the material divided by its volume; measured in kilograms per cubic meter or grams per cubic centimeter.

diaphragm An opening that is used to limit the amount of light passing through a lens.

diffraction The spreading of waves passing through an opening or around a barrier.

diffuse reflection The reflection of rays from a rough surface. The reflected rays do not leave at fixed angles.

diopter A measure of the focal length of a mirror or lens, equal to the inverse of the focal length measured in meters.

disordered system A system with an arrangement equivalent to many other possible arrangements.

dispersion The spreading of light into a spectrum of color; the variation in the speed of a periodic wave due to its wavelength or frequency.

displacement A vector quantity giving the straight-line distance and direction from an initial position to a final position. In wave (or oscillatory) motion, the distance of the disturbance (or object) from its equilibrium position.

Doppler effect A change in the frequency of a periodic wave due to the motion of the observer, the source, or both.

efficiency The ratio of the work produced to the energy input. For an ideal heat engine, the Carnot efficiency is given by $1 - T_c/T_h$.

elastic A collision or interaction in which kinetic energy is conserved.

electric field The space surrounding a charged object, where each location is assigned a vector equal to the force experienced by one unit of positive charge placed at that location; measured in newtons per coulomb.

electric field lines A representation of the electric field in a region of space. The electric field is tangent to the field line at any point and its magnitude is proportional to the local density of field lines.

electric potential The electric potential energy divided by the object's charge; the work done in bringing a positive test charge of 1 coulomb from the zero reference location to a particular point in space; measured in joules per coulomb.

electric potential energy The work done in bringing a charged object from some zero reference location to a particular point in space; measured in joules.

electromagnet A magnet constructed by wrapping wire around an iron core. The electromagnet can be turned on and off by turning the current in the wire on and off.

electromagnetic wave A wave consisting of oscillating electric and magnetic fields. In a vacuum, electromagnetic waves travel at the speed of light.

electron A basic constituent of atoms; a lepton.

electron capture A decay process in which an inner atomic electron is captured by the nucleus. The daughter nucleus has the same number of nucleons as the parent but one fewer proton.

electron volt A unit of energy equal to the kinetic energy acquired by an electron or proton falling through an electric potential difference of 1 volt; equal to 1.6×10^{-19} joule.

element Any chemical species that cannot be broken up into other chemical species.

emission spectrum The collection of discrete wavelengths emitted by atoms that have been excited by heating or by electric currents.

entropy A measure of the order of a system. The second law of thermodynamics states that the entropy of an isolated system tends to increase.

equilibrium position A position where the net restoring force is zero.

equivalence principle Constant acceleration is completely equivalent to a uniform gravitational field.

ether The hypothesized medium through which light was believed to travel.

exclusion principle No two electrons can have the same set of quantum numbers. This statement also applies to protons, neutrons, and other baryons.

field A region of space where each location is assigned a value or a vector. *See* electric, gravitational, and magnetic fields.

fission The splitting of a heavy nucleus into two or more lighter nuclei.

flavor The types of quark: down, up, strange, charm, bottom, or top.

fluorescence The property of a material whereby it emits visible light when it is illuminated by ultraviolet light.

focal length The distance from a mirror or the center of a lens to its focal point.

focal point The location at which a mirror or a lens focuses rays parallel to the optic axis or from which such rays appear to diverge.

force A push or a pull, measured by the acceleration it produces on a standard, isolated object, $\mathbf{F}_{net} = m\mathbf{a}$; measured in newtons.

frequency The number of times a periodic motion repeats in a unit of time. It is equal to the inverse of the period, $f = 1/T$; measured in hertz.

fundamental frequency The lowest resonant frequency for an oscillating system.

fusion The combining of light nuclei to form a heavier nucleus.

gamma (γ) radiation The type of radioactive decay in which nuclei emit high-energy photons; the range of frequencies of the electromagnetic spectrum that lies beyond the X rays.

gas Matter with no definite shape or volume.

gauss A unit of magnetic field strength; 10^{-4} tesla.

geocentric model A model of the Universe with Earth at its center.

gluon An exchange particle responsible for the force between quarks. The eight gluons differ only in their color quantum numbers.

gravitation, law of universal $F = Gm_1m_2/r^2$, where F is the force between any two objects, G is a universal constant, m_1 and m_2 are the masses of the two objects, and r is the distance between their centers.

gravitational field The space surrounding an object where each location is assigned a vector equal to the gravitational force experienced by a 1-kilogram mass placed at that location.

gravitational mass The property of a particle that determines the strength of its gravitational interaction with other particles.

gravitational potential energy The work done by the force of gravity when an object falls from a particular point in space to the location assigned the value of zero, $GPE = mgh$.

graviton The exchange particle responsible for the gravitational force.

gravity wave A wave disturbance caused by the acceleration of masses.

grounding Establishing an electrical connection to Earth to neutralize an object.

ground state The lowest energy state of a system allowed by quantum mechanics.

hadron The family of particles that participates in the strong interaction. Baryons and mesons are the two subfamilies.

half-life The time during which one-half of a sample of a radioactive substance decays.

halo A ring of light that appears around the Sun or Moon. It is produced by refraction in ice crystals.

harmonic A frequency that is a whole-number multiple of the fundamental frequency.

heat Energy flowing due to a difference in temperature.

heat capacity The amount of heat required to raise the temperature of an object by 1°C.

heat engine A device for converting heat into mechanical work.

heat pump A reversible heat engine that acts as a furnace in winter and an air conditioner in summer.

heliocentric model A model of the Universe with the Sun at its center.

hologram A three-dimensional record of visual information.

holography The photographic process for producing three-dimensional images.

hyperopia Farsightedness. Images of distant objects are formed beyond the retina.

ideal gas An enormous number of very tiny particles separated by relatively large distances. The particles have no internal structure, are indestructible, and do not interact with each other except when they collide; all collisions are elastic.

ideal gas law $PV = nRT$, where P is the pressure, V is the volume, T is the absolute temperature, n is the number of moles, and R is the gas constant.

impulse The product of the force and the time during which it acts, $\mathbf{F} \Delta t$. This vector quantity is equal to the change in momentum.

index of refraction An optical property of a substance that determines how much light bends upon entering or leaving it. The index is equal to the ratio of the speed of light in a vacuum to that in the substance.

inelastic A collision or interaction in which kinetic energy is not conserved.

inertia An object's resistance to a change in its velocity. *See* inertial mass.

inertia, law of *See* motion, Newton's first law of.

inertial force A fictitious force that arises in accelerating (noninertial) reference systems. Examples are centrifugal and Coriolis forces.

inertial mass An object's resistance to a change in its velocity; measured in kilograms.

inertial reference system Any reference system in which the law of inertia (Newton's first law of motion) is valid.

in phase Two or more waves with the same wavelength and frequency that have their crests lined up.

instantaneous speed The limiting value of the average speed as the time interval becomes infinitesimally small. The magnitude of the instantaneous velocity.

insulator A material that does not allow the passage of electric charge or is a poor conductor of thermal energy. Ceramics are good electrical insulators; wood and stationary air are good thermal insulators.

interference The superposition of waves.

intermediate vector bosons The exchange particles of the weak nuclear interaction: the W^+, W^-, and Z^0 particles.

internal energy The total microscopic energy of an object, which includes its atomic and molecular translational and rotational kinetic energies, vibrational energy, and the energy stored in the molecular bonds.

inverse-square A relationship in which a quantity is related to the reciprocal of the square of a second quantity. Examples include the force laws for gravity and electricity; the force is proportional to the inverse square of the distance.

inversely proportional A relationship in which two quantities have a constant product. If one quantity increases by a certain factor, the other decreases by the same factor.

ion An atom with missing or extra electrons.

ionization The removal of one or more electrons from an atom.

isotope An element containing a specific number of neutrons in its nuclei. Examples are $^{12}_{6}C$ and $^{14}_{6}C$, carbon atoms with six and eight neutrons, respectively.

joule The SI unit of energy equal to 1 newton acting through a distance of 1 meter.

kilo A prefix meaning 1000. A kilometer is 1000 meters.

kilogram The SI unit of mass. A kilogram of material weighs about 2.2 pounds on Earth.

kilowatt-hour A unit of energy; 3,600,000 joules. One kilowatt-hour of energy is transformed to other forms when a machine runs at a power of 1000 watts for 1 hour.

kinetic energy The energy of motion, $\frac{1}{2}mv^2$; measured in joules.

kinetic friction The frictional force between two surfaces in relative motion. This force does not depend very much on the relative speed.

Kirchhoff's junction rule The sum of the currents entering any junction in a circuit must equal the sum of the currents leaving that junction.

Kirchhoff's loop rule Along any path from the positive terminal to the negative terminal of a battery, the voltage drops across the resistive elements encountered must add up to the battery voltage.

laser A device that uses stimulated emissions to produce a coherent beam of electromagnetic radiation. Laser is the acronym from *l*ight *a*mplification by *s*timulated *e*mission of *r*adiation.

latent heat The amount of heat required to melt (or vaporize) a unit mass of a substance. The same amount of heat is released when a unit mass of the same substance freezes (or condenses); measured in calories per gram or kilojoules per kilogram.

lepton A family of elementary particles that includes the electron, muon, tau, and their associated neutrinos.

light ray A line that represents the path of light in a given direction.

linear momentum A vector quantity equal to the product of an object's mass and its velocity, $\mathbf{p} = m\mathbf{v}$.

line of stability The locations of the stable nuclei on a graph of the number of neutrons versus the number of protons.

liquid Matter with a definite volume that takes the shape of its container.

liquid crystal A liquid that exhibits a rough geometrical ordering of its atoms.

longitudinal wave A wave in which the vibrations of the medium are parallel to the direction the wave is moving.

macroscopic The bulk properties of a substance such as mass, size, pressure, and temperature.

magnetic field The space surrounding a magnetic object, where each location is assigned a value determined by the torque on a compass placed at that location. The direction of the field is in the direction of the north pole of the compass.

magnetic monopole A hypothetical, isolated magnetic pole.

magnetic pole One end of a magnet; analogous to an electric charge.

magnitude The size of a vector quantity. For example, speed is the magnitude of a velocity.

mass *See* inertial mass, gravitational mass, critical mass, and center of mass.

matter-wave amplitude The wave solution to Schrödinger's equation for atomic and subatomic particles. The square of the matter-wave amplitude gives the probability of finding the particle at a particular location.

mechanical energy The sum of the kinetic energy and various potential energies, which may include the gravitational and the elastic potential energies.

meson A type of hadron with whole-number units of spin. This family includes the pion and kaon.

meter The SI unit of length equal to 39.37 inches, or 1.094 yards.

microscopic Properties not visible to the naked eye such as atomic speeds or the masses and sizes of atoms.

milli A prefix meaning $\frac{1}{1000}$. A millimeter is $\frac{1}{1000}$ meter.

mirage An optical effect that produces an image that looks like it has been reflected from the surface of a body of water.

moderator A material used to slow the neutrons in a nuclear reactor.

mole The amount of a substance that has a mass in grams numerically equal to the mass of its molecules in atomic mass units.

molecule A combination of two or more atoms.

momentum Usually refers to linear momentum. *See* angular momentum, linear momentum, and conservation of momentum.

motion, Newton's first law of The velocity of an object remains constant unless an unbalanced force acts on the object.

motion, Newton's second law of The net force on an object is equal to its mass times its acceleration: $\mathbf{F}_{net} = m\mathbf{a}$. The net force and the acceleration are vectors that always point in the same direction.

motion, Newton's third law of If an object exerts a force on a second object, the second object exerts an equal force back on the first object.

muon A type of lepton; often called a heavy electron.

myopia Nearsightedness. Images of distant objects are formed in front of the retina.

neutrino A neutral lepton; one exists for each of the charged leptons (electron, muon, and tau).

neutron The neutral nucleon; a member of the baryon and hadron families of elementary particles.

newton The SI unit of force. A net force of 1 newton accelerates a mass of 1 kilogram at a rate of 1 (meter per second) per second.

node One of the positions in a standing-wave or interference pattern where there is no movement; that is, the amplitude is zero.

noninertial reference system Any reference system in which the law of inertia (Newton's first law of motion) is not valid. An accelerating reference system is noninertial.

normal A line perpendicular to a surface or curve.

nucleon Either a proton or a neutron.

nucleus The central part of an atom that contains the protons and neutrons.

ohm The SI unit of electrical resistance. A current of 1 ampere flows through a resistance of 1 ohm under 1 volt of potential difference.

Ohm's law The resistance of an object is equal to the voltage across it divided by the current through it, $R = V/I$.

optic axis A line passing through the center of a curved mirror and the center of the sphere from which the mirror is made; a line passing through a lens and both focal points.

order of magnitude The value of a quantity rounded off to the nearest power of 10.

ordered system A system with an arrangement belonging to a group with the smallest number (possibly one) of equivalent arrangements.

oscillation A vibration about an equilibrium position or shape.

pair production The conversion of energy into matter in which a particle and its antiparticle are produced. This usually refers to the production of an electron and a positron (antielectron).

parallel Two circuit elements are wired in parallel when the current can flow through one or the other but not both. Elements that are wired parallel to each other are directly connected to each other at both terminals.

parent nucleus A nucleus that decays into a daughter nucleus.

particle accelerator A device for accelerating charged particles to high speeds.

penumbra The transition region between the darkest shadow and full brightness. Only part of the light from the source reaches this region.

period The shortest length of time it takes a periodic motion to repeat. It is equal to the inverse of the frequency, $T = 1/f$.

periodic wave A wave in which all the pulses have the same size and shape. The wave pattern repeats itself over a distance of one wavelength and over a time of one period.

phosphorescence The property of a material whereby it continues to emit visible light after it has been illuminated by ultraviolet light.

photoelectric effect The ejection of electrons from metallic surfaces by illuminating light.

photon A particle of light. The energy of a photon is given by the relationship $E = hf$, where f is the frequency of the light and h is Planck's constant. It is the exchange particle for the electromagnetic interaction.

pion The least massive meson. The pion has three charge states: +1, 0, and −1.

plasma The fourth state of matter, in which one or more electrons have been stripped from the atoms forming an ion gas.

polarized A property of a transverse wave when its vibrations are all in a single plane.

positron The antiparticle of the electron.

pound The unit of force in the U.S. customary system; the weight of 0.454 kilogram on Earth.

power The rate at which energy is converted from one form to another, $P = \Delta E/\Delta t$; measured in joules per second, or watts. In electric circuits the power is equal to the current times the voltage, $P = IV$.

powers-of-ten notation A method of writing numbers in which a number between 1 and 10 is multiplied or divided by 10 raised to a power.

pressure The force per unit area of surface; measured in newtons per square meter, or pascals.

projectile motion A type of motion that occurs near Earth's surface when the only force acting on the object is that of gravity.

proportional A relationship in which two quantities have a constant ratio. If one quantity increases by a certain factor, the other increases by the same factor.

proton The positively charged nucleon; a member of the baryon and hadron families of elementary particles.

quantum (pl., *quanta*) The smallest unit of a discrete property. For instance, the quantum of charge is the charge on the proton.

quantum mechanics The rules for the behavior of particles at the atomic and subatomic levels.

quantum number A number giving the value of a quantized quantity. For instance, a quantum number specifies the angular momentum of an electron in an atom.

quark A constituent of hadrons. Quarks come in six flavors of three colors each. Three quarks make up the baryons, whereas a quark and an antiquark make up the mesons.

rad The acronym from *r*adiation *a*bsorbed *d*ose. A rad of radiation deposits 0.01 joule per kilogram of material.

radiation The transport of energy via electromagnetic waves; particles emitted in radioactive decay.

real image An image formed by the convergence of light.

reference system A collection of objects not moving relative to each other that can be used to describe the motion of other objects. *See* inertial and noninertial reference systems.

reflecting telescope A type of telescope using a mirror as the objective.

reflection, law of The angle of reflection (measured relative to the normal to the surface) is equal to the angle of incidence. The incident ray, the reflected ray, and the normal all lie in the same plane.

refracting telescope A type of telescope using a lens as the objective.

refraction The bending of light that occurs at the interface between two transparent media. It occurs when the speed of light changes.

refrigerator A heat engine running backward.

relativity, Galilean principle of The laws of motion are the same in all inertial reference systems.

relativity, general theory of An extension of the special theory of relativity to include the concept of gravity.

relativity, special theory of A comprehensive theory of space and time that replaces Newtonian mechanics when velocities get very large.

rem The acronym from *radiation equivalent in mammals*, a measure of the biological effects caused by radiation.

resistance The impedance to the flow of electric current; measured in volts per ampere, or ohms. The resistance is equal to the voltage across the object divided by the current through it, $R = V/I$.

resonance A large increase in the amplitude of a vibration when a force is applied at a natural frequency of the medium or object.

rest-mass energy The energy associated with the mass of a particle; given by $E_o = mc^2$, where c is the speed of light.

retroreflectors Three flat mirrors at right angles to each other that reflect light back to its source.

rotational acceleration The change in rotational speed divided by the time it takes to make the change.

rotational inertia The property of an object that measures its resistance to a change in its rotational speed.

rotational kinetic energy Kinetic energy associated with the rotation of a body, $KE = \frac{1}{2}I\omega^2$; measured in joules.

rotational speed The angle of rotation or revolution divided by the time taken, $\omega = \Delta\theta/\Delta t$.

rotational velocity A vector quantity that includes the rotational speed and the direction of the axis of rotation.

series An arrangement of resistances (or batteries) on a single pathway so that the current flows through each element.

shell A collection of electrons in an atom that have approximately the same energy.

shock wave The characteristic cone-shaped wave front that is produced whenever an object travels faster than the speed of the waves in the surrounding medium.

short circuit A path in an electric circuit that has very little resistance.

solid Matter with a definite size and shape.

sonar Sound waves in water.

spacetime A combination of time and three-dimensional space that forms a four-dimensional geometry expressing the connections between space and time.

special relativity, first postulate of The laws of physics are the same for all inertial reference systems.

special relativity, second postulate of The speed of light in a vacuum is a constant regardless of the speed of the source or the speed of the observer.

specific heat The amount of heat required to raise the temperature of a unit mass of a substance by 1 degree; measured in calories per gram-degree Celsius or joules per kilogram-kelvin.

spherical aberration A property of lenses and mirrors caused by grinding the surface to a spherical rather than a parabolic shape.

spring constant The amount of force required to stretch a spring by 1 unit of length; measured in newtons per meter.

stable equilibrium An equilibrium position or orientation to which an object returns after being slightly displaced.

standing wave The interference pattern produced by two waves of equal amplitude and frequency traveling in opposite directions. The pattern is characterized by alternating nodal and antinodal regions.

static friction The frictional force between two surfaces at rest relative to each other. This force is equal and opposite to the net applied force if the force is not large enough to make the object accelerate.

stimulated emission The emission of a photon from an atom due to the presence of an incident photon. The emitted photon has the same energy, direction, and phase as the incident photon.

strange particle A particle with a nonzero value of strangeness. In the quark model, it is made up of one or more quarks carrying the quantum property of strangeness.

strangeness The flavor of the third quark.

strong force The force responsible for holding the nucleons together to form nuclei.

subcritical A chain reaction that dies out because an average of less than one neutron from each fission reaction causes another fission reaction.

supercritical A chain reaction that grows rapidly because an average of more than one neutron from each fission reaction causes another fission reaction. An extreme example of this is the explosion of a nuclear bomb.

superposition The combining of two or more waves at a location in space.

Système International d'Unités The French name for the metric system, or International System (SI), of units.

temperature, absolute The temperature scale with its zero point at absolute zero and degrees equal to those on the Celsius scale. Also known as the Kelvin temperature scale.

temperature, Celsius The temperature scale with the values of 0°C and 100°C for the temperatures of freezing and boiling water, respectively. Its degree is $\frac{9}{5}$ that of the Fahrenheit degree.

temperature, Fahrenheit The temperature scale with the values of 32°F and 212°F for the temperatures of freezing and boiling water, respectively.

temperature, Kelvin The temperature scale with its zero point at absolute zero and a degree equal to that on the Celsius scale. Also called the absolute temperature scale.

terminal speed The speed obtained in free fall when the upward force of air resistance is equal to the downward force of gravity.

tesla The SI unit of magnetic field.

thermal energy Internal energy.

thermal equilibrium A condition in which there is no net flow of thermal energy between two objects. This occurs when the two objects obtain the same temperature.

thermal expansion The increase in size of a material when heated.

thermodynamics The area of physics that deals with the connections between heat and other forms of energy.

thermodynamics, first law of The increase in internal energy of a system is equal to the heat added plus the work done on the system.

thermodynamics, second law of There are three equivalent forms: (1) It is impossible to build a heat engine to perform mechanical work that does not exhaust heat to the surroundings. (2) It is impossible to build a refrigerator that can transfer heat from a lower temperature region to a higher temperature region without expending mechanical work. (3) The entropy of an isolated system tends to increase.

thermodynamics, third law of Absolute zero may be approached experimentally but can never be reached.

thermodynamics, zeroth law of If objects A and B are each in thermodynamic equilibrium with object C, then A and B are in thermodynamic equilibrium with each other. All three objects are at the same temperature.

top The flavor of the sixth quark.

torque The rotational analog of force. It is equal to the radius multiplied by the force perpendicular to the radius, $\tau = rF$; measured in newton-meters. A net torque produces a change in an object's angular momentum.

total internal reflection A phenomenon that occurs when the angle of incidence of light traveling from a material with a higher index of refraction into one with a lower index of refraction exceeds the critical angle.

translational motion Motion along a path.

transverse wave A wave in which the vibrations of the medium are perpendicular to the direction the wave is moving.

trough A valley of a wave disturbance.

umbra The darkest part of a shadow where no light from the source reaches.

uncertainty principle The product of the uncertainty in the position of a particle along a certain direction and the uncertainty in the momentum along this same direction must be greater than Planck's constant, or $\Delta p_x \Delta x > h$. A similar relationship applies to the uncertainties in energy and time.

unstable equilibrium An equilibrium position or orientation from which an object leaves after being slightly displaced.

vector A quantity with a magnitude and a direction. Examples are displacement, velocity, acceleration, momentum, and force.

velocity A vector quantity that includes the speed and the direction of the object; the displacement divided by the time taken, $\bar{\mathbf{v}} = \Delta \mathbf{x}/\Delta t$.

vibration An oscillation about an equilibrium position or shape.

virtual image The image formed when light only appears to come from the location of the image.

viscosity A measure of the internal friction within a fluid.

volt The SI unit of electric potential, 1 joule per coulomb. One volt produces a current of 1 ampere through a resistance of 1 ohm.

watt The SI unit of power, 1 joule per second.

wave The movement of energy from one place to another without any accompanying matter.

wavelength The shortest repetition length for a periodic wave. For example, it is the distance from crest to crest or from trough to trough.

weak force The force responsible for beta decay. This force occurs through the exchange of the W and Z^0 particles. All leptons and hadrons interact via this force.

weight The support force needed to maintain an object at rest relative to a reference system. For inertial systems, the weight is the force of attraction of Earth for an object, $W = mg$.

work The product of the force along the direction of motion and the distance moved, $W = Fd$. Measured in energy units, joules.

X ray A high-energy photon, usually produced by cathode rays or emitted by electrons falling to lower energy states in atoms; the range of frequencies in the electronic spectrum lying between the ultraviolet and the gamma rays.

Index

e = equation
g = definition in glossary
t = table

A

Abbott, Edwin, 197
aberration, 371g
  chromatic, 371, 376
  spherical, 371, 376
absolute zero, 252g
acceleration
  absolute nature, 153
  average, 21g–21e–27, 42e, 58–58e–64
  Brownian motion, 215
  centripetal, 60–61g–61e–62
  of charges, 420, 462, 464–465, 482, 499, 501, 507, 542, 549–551, 591
  and classical relativity, 177
  constant, 26, 33, 41, 63, 155
  due to electric field, 432, 452
  and equivalence principle, 194
  free-body diagram, 45
  and general relativity, 192
  due to gravitational vs. electric field, 413
  due to gravity, 26, 82, 88, 160, 292–293
  in inertial reference systems, 157
  due to magnetic field, 455
  and net force, 37–38, 41, 46, 116
  in noninertial reference systems, 159
  and relativistic adjustment factor, 192
  rotational, 135g–135e, 137–138
  and special relativity, 188
  translational, 135, 137–138
  zero, 33, 47, 155
accelerator, 549g–550, 598
acoustics, 316
action at a distance, 523, 576
activity, 533g–534
  unit, 533
Aesop, 13
air conditioner, 262, 274–275, 437
air resistance. See resistance, air
alchemist, 530–531
alchemy, 209
Allen, Bryan, 127
alpha
  decay, 531, 555, 563
  emission, 560
  emitter, 536
  particle, 479g–480, 527g–532, 537–538, 549, 550
ammeter, 455, 457
amorphous, 235
ampere (unit), 431g, 451g
Ampère, André, 449
amplify, 316
amplitude, 289g–297, 300–305, 316, 324–325, 328, 392
angle
  critical, 362g, 364
  of incidence, 340, 359–362, 366, 383–384
  of reflection, 359, 383
  of refraction, 359–362, 384
  resolving, 389
angular momentum. See momentum, angular
angular separation, 389
  size, 374
annihilation, 573–574
  particle, 594
anode, 477
antiatom, 573–574
antibaryon, 582
antideuterium, 574
antielectron, 413, 573, 581
antihydrogen, 574
antimatter, 574, 600
antimeson, 600
antineutrino, 575
antineutron, 573–574, 600
antinodal line, 305–306, 324
antinodal region, 385
antinode, 303g–304, 320, 321–322
antinucleon, 573
antinucleus, 574
antiparticle, 532, 573g, 579, 594g–596
antiphoton, 574
antiproton, 571, 573–574, 600
antiquark, 571, 584–586, 600
Archimedes, 237–239
  principle, 237g–238
Aristotle, 23–25, 34–35, 46, 165, 205, 338, 601
  four elements, 34, 571–572
artifact, age determination, 524
astigmatism, 372–373
atom, 207–208g, 396, 402, 472–493, 498–502, 507–516, 523
  at absolute zero, 252
  anti-, 573–574
  Bohr, 493, 507
  crystal, 230–231
  and current, 432
  divisibility, 572
  dominant force, 414
  evidence for existence, 207–208
  formation, 596
  freezing water, 237
  ionization, 536–537g–538
  isotopes, 529–530, 555–556
  and lasers, 515–516
  and light, 482, 487
  liquid, 231
  and magnets, 449, 452
  mass, 529
  molecule formation, 600–601
  net electric charge, 407
  number in material, 213, 214
  periodicity, 491, 511
  photon interaction, 487, 538
  plasma, 228, 233
  properties, 491, 507
  radioactive, 533, 559–560
  seeing, 501
  size, 501, 549
    analogy, 525
  and spectral lines, 422, 488–489
  stable, 482, 492, 499, 507
  structure, 484
  unstable, 486
  uranium-235, 559–560
atomic
  behavior, 208
  bomb, 565
  energy (see energy, atomic)
  fingerprint, 473, 475
  level, 499, 505
  mass, 213g–214, 491, 493g, 529–530
  mass unit, 213g–214, 529–530
  model (see model, atomic)
  motion, 215
  number, 481, 493g
  ordering, 231
  oscillator, 483–485
  size, 213
  spacing, 213
  spectra, 474–476, 488–489, 530
  speeds, 207–208, 216–217, 220
  structure, 208, 228, 473, 490
  weight, 474
  world view, 208
attack (music), 317
attraction, 405–411, 414, 432, 447, 453, 554, 576
aurora australis, 455
aurora borealis, 233, 455
Avogadro, Amedeo, 212
  number, 214g
axis
  Earth's magnetic, 453
  optic, 344g–347, 367g–371
  polarization, 392–393
  rotation, 135–139, 144, 164–165, 453
  rotational, 453

B

balance, 67, 94, 157
Balmer, Johann, 580
Bardeen, John, 452
barometer, 234
baryon, 580g, 582, 585
  charmed, 584
  number, 582–583, 594
    conservation, 582
battery, 427–439, 455
  and bulb, 429–438
  dry cell, 428
  parallel, 428–429
  series, 428
  storage, 428
  symbol, 439
  voltaic, 251
beat, 324g
Becquerel, Henri, 472, 481, 525–527, 537
Bernoulli, Daniel, 239
  principle, 238–239g–241
beta
  decay, 532, 551–552, 555, 564, 575, 581
  puzzle, 575
  emitter, 541
  minus decay, 532–533, 555, 560, 564
  plus decay, 532–533, 555
Betelgeuse, 259
binocular, 375, 391
black hole, 2, 196, 591–592
Bode, William, 5
  law, 5–6
Bohr, Margrethe, 490
Bohr, Niels, 191, 481, 486–492, 499–500, 513–514, 558, 561
  biography, 490
  See model, atomic
boiling point, 255t, 543
bomb
  atomic, 564
  fission, 564
  hydrogen, 564
  nuclear, 549, 560
  plutonium, 558
Bondi, Sir Hermann, 4, 472
Born, Max, 556
bottom quark, 584t–585g
Boyle's law, 220–221
Brahe, Tycho, 76
British thermal unit (Btu), 249g
Brown, Robert, 206, 215

bubble chamber, 543, 572
bulk properties, 208

C

caloric, 248
calorie (unit), 249g–250
Calorie (unit), 250
calorimeter, 251
camera, 371
    obscura, 339
    pinhole, 339, 371
carbon-14 dating, 536
Carnot, Sadi, 271
cathode, 477
    ray, 477g–479, 492–493
Cavendish, Henry, 80
Celsius, Anders, 219
center of mass, 67g–68, 139g–145
centi, 7g
centigrade (unit), 219
ceramic, 452
Chadwick, James, 528–529
chain reaction, 557g–560, 562
    critical, 561g
    subcritical, 560g
    supercritical, 560g, 566
change of state, 255g
charge, 401, 404g–420, 427–435, 447–455, 459–462
    accelerating, 507
    alpha particle, 479–480, 527–528
    and cathode rays, 477–478
    color, 586
    conservation, 406g–407, 430, 434–435, 531, 575, 581–582
    and current, 432–435, 451
    effects on proton and neutron levels, 555
    and electric field, 414–418, 448
    and electric force, 411–412
    and electromagnetic waves, 462, 479, 482, 591
    electroscope, 409–411, 427, 447
    elementary, 412, 478, 527
    fractional, 583–584
    induced, 408, 413
    by induction, 407–408, 411
    intermediate vector boson, 579
    vs. mass, 413–414
    measuring, 478
    muon, 578
    neutron, 529
    photon, 533
    pion, 579
    positron, 532, 573–574
    quarks, 584
    radiation frequency, 487
    smallest unit, 478
    test, 415–418
    to mass ratio, 478–479, 527, 560, 572
    two kinds, 404–406, 413
    unit, 45g, 412g
    charged, 404g–420, 477–483, 492, 527, 530, 537, 541–543, 549, 573, 578
charm quark, 584g–584t
chemical reaction, 248, 526, 549–550, 556
chemistry, 209
Churchill, Winston, 490
circuit, 412, 419, 427–440, 458, 465, 538
    breaker, 437
    complete, 429g–430
    household, 428–431, 437, 439
    parallel, 435–437
    series, 434–435
    short, 435g, 438
clock, 181–185, 187, 191, 293
    atomic, 293, 592
    light, 185–187
    pendulum, 293
    radioactive, 535–537
    synchronized, 181–184
    water, 293
collision, 95–96, 98, 101–104, 315, 550–551, 561, 565
    atomic, 250, 257
    elastic, 114g–115, 214, 561
    inelastic, 114g–115
    particle, 216
color, 208, 334–335, 350–353
    adding, 350, 352
    complementary, 350–351g, 353
    and dispersion, 363–364, 385
    frequency of light, 482, 493
    gas absorption or emission, 473–475
    and interference, 385–386, 390–391
    perception, 350, 352
    photoelectric effect, 485–486, 515
    primary, 352
    printing, 352
    psychedelic, 509
    quark, 593
    rainbow, 364–366
    subtracting, 352
    television, 352
common sense, 4–5, 34, 164, 209, 387, 418, 513
commutator, 458
compass, 448–453
complementarity principle, 513g
component electronic, 538
compound, 209g–212
compression, 289, 291, 299, 314
compressor, 275
conceptual leap, 74
Concorde, 19, 328
condensation, 327
condenser, 275
conduction, 256g–259
conductivity, thermal, 257t
conductor, 257g, 404g–405, 428–431, 452
conservation, law of, 105, 177, 530, 581
conserved, 94–95, 100g

constant
    cosmological, 590
    Coulomb's, 412
    gravitational, 79–80
    Planck's, 580, 594
contact, thermal, 250, 253, 270, 275
convection, 256, 258g–260
converge, 345
cooling, evaporative, 221
Cooper, Leon, 452
Copernicus, Nicholas, 75–76, 165
    planetary motion, 24
cosmic ray, 455, 536, 539, 578, 598–599
cosmology, 594, 600
coulomb (unit), 412g, 431g
Coulomb, Charles, 411
    constant, 412
    law, 412
Count Rumford, 248–249
Cowan, Clyde, 575
crest, 298g–300, 303, 321–327, 384–385, 389, 500
crystal, 227, 230g–231
crystalline structure, 227, 231
curie (unit), 533g
Curie, Eve, 527
Curie, Marie Sklodowska, 526–527, 533, 556
    biography, 527
Curie, Pierre, 526–527, 537
current, 403, 426–429g–430e–440, 472, 475, 477, 501
    alternating, 428, 457
    atomic, 449–450
    convection, 258
    direct, 429, 458
    direction, 431
    and Earth's magnetic field, 453–454
    electromagnet, 451
    induced, 455–459
    due to lightning, 419
    loops, 449–450
    and magnetic field, 449, 454–457, 460–461
    model, 433
    and power, 439e–440
    and resistance, 432e
    standard, 433
    super, 452
    unit, 431, 451
    and voltage model, 435
    in water model, 430
curve ball, 240
cycle, 289g, 304

D

Dalton, John, 211, 252
Darwin, Charles, 530
da Vinci, Leonardo, 371
Davis, Ray, 598
Davisson, C. J., 501
de Broglie, Louis, 499–502, 507
decay, 533, 481
    electron capture, 532–533
    gamma, 533t

    law, 534
    proton, 594
    radioactive, 187
    rate, 535
    See alpha, decay; beta, decay
decibel (unit), 322
    level for common sounds, 322t
degrees (unit), 135
Democritus, 205
density, 228g–228e–232, 565, 597
    air, 229
    atmosphere, 234, 362, 364
    body fat, 239
    common materials, 229t
    critical matter, 597–598
    dark energy, 598
    Earth, 230
    floating, 237
    ice, 237, 263
    mass, spring, 298
    muscle, 239
    neutron star, 229
    silica aerogel, 229
    space, 229
    Universe, 597–598
    white dwarf, 229
Descartes, René, 514
detector
    gravitational wave, 591–592
    radiation, 537, 541–543
    scintillation, 542
deuterium, 530, 561–562, 565
deuteron, 551–552
diffraction, 306g–307, 387–389, 511
    effect, 486, 502, 549
    grating, 475, 489
    limit, 389
    pattern, 387–389, 502
diopter, 372
Dirac, P. A. M., 573
dispersion, 363–364g–365, 371, 385g
displacement, 20g, 38, 300g, 505
distance, 116, 179–182, 185–188, 195
    Earth to Moon, 73, 344
    stopping, 117
diverge, 346
Doppler, Christian, 326
    effect, 325–326g–326
down quark, 583–584t–586
Dubouchet, Karine, 48
dynamo, 251

E

ear, 316–317
Earth
    acceleration, 169
    acceleration near surface, 44, 80–82, 413
    age of, 537
    air pressure at surface, 233
    angular momentum, 142
    atmosphere, 233–234, 258–260, 362, 389, 579, 598–599
    auroras, 455
    core, 299, 599

curvature, 66
density, 230
eclipses, 338
formation, 596
geocentric, 164g–165
gravity, 26–27
heliocentric, 24, 76, 165g
interior, 299
kinetic energy, 114, 118
magnetic axis, 453
magnetic field, 450, 453–454
magnetic poles, 453
mass, 80, 101
momentum change, 101
motion, 15, 164–170
motion relative to the surface, 153
nearly inertial system, 164–166, 170
noninertial effects, 166–169
orbit, 118, 164–166, 566
orbital radius, 6
orbital speed, 19, 164, 178
precession, 145
radiation received, 259
radius, 80
reference system, 155, 169
rotation, 15, 83–84, 154, 164–170, 453
rotational axis, 453
rotation direction, 165–168
seasons, 253
shadows, 338
tides, 84–86
torque on, 144
weighing, 80
work on, 118
earthquake, 299
eclipse, 338–339
efficiency, 273g–273e–274
  Carnot, ideal, 274e
Einstein, Albert
  belief in God, 34
  bending of light, 194
  biography, 191
  and Bohr, 490
  Brownian motion, 206, 215
  and Curie, 527
  and de Broglie, 501
  energy/mass equivalence, 598
  equivalence principle, 194
  and Feynman, 577
  length contraction, 187–188
  mass–energy equation, 192, 538, 553–554, 561, 572
  and Maxwell, 463
  and Meitner, 561
  and modern physics, 514–515
  nature of light, 396
  and Newton, 463
  photoelectric effect, 485–486
  and Planck, 484
  quantum ideas, 499
  quote, 515, 600
  relativity, 153, 499
    general, 105, 192, 195–196, 590–591, 593, 597
    special, 177–179

scientific process, 3–4
simultaneity, 179–180
spacetime, 190, 601–602
time dilation, 185
Eisenhower, Dwight, 490
elasticity, 206, 299
electric, 402
  charge (see charge)
  circuit (see circuit)
  conductor, 257g, 404g–405, 428–431, 452
  current (see current)
  deflection, 529
  energy (see energy, electric)
  field (see field, electric)
  force (see force, electric)
  generator, 452, 457–458, 562
  grounding, 404–405g
  insulator, 404g–407, 410
  motor, 452, 457–458
  potential, 420g–420e, 452
    units, 420
  potential difference, 420, 427–428, 431–432, 437, 550
  power, 439g–439e, 457
  resistance, 431–432g–432e–437, 440, 452
  spark, 403, 405, 420
  voltage (see voltage)
  work, 418, 431
electricity, 403–407, 413–414, 419, 481, 484, 583
  from battery, 430
  cost, 438
  danger, 433
  flow, 407, 430–432
  generating, 458, 548, 561–562
  and gravity, 412–414
  household, 427–429, 434, 438
  lightning, 419
  and magnetism, 446–449, 455, 460–463, 593
  static, 427
  usage, 427
electrode, 428
electrolysis, 210
electrolyte, 428
electromagnet, 451g, 461
electromagnetism, 446, 499
electron, 4, 404–407, 412–414, 418–420, 431–432, 439, 449, 452, 464–465, 479g, 523
  accelerating, 413, 482
  anti-, 413, 573, 581
  and atomic periodicity, 479, 491–493
  behavior, 472
  and Bohr model, 486–487
  bubble chamber interaction, 543, 572
  capture, 532g–533, 556
  charge, 412
  cloud, 535, 549
  determining chemical properties of atom, 530–532
  diffraction, 501
  discovery, 478
  duality, 499, 504, 511, 514

  and electric force, 409, 414, 432
  and electricity, 405, 413–414
  and electromagnetic waves, 465
  and exclusion principle, 555
  existence in nucleus, 528–529
  free, 537
  and gravitational force, 414
  interference, 501–502, 504, 513
  and laser emission, 515–516
  and leptons, 580
  and light, 426
  lightning, 419
  mass, 414, 479–480, 529t, 537
  nature
    particle, 504–507, 513
    wave, 499–500, 505, 513
  photoelectric effect, 484–486
  and positron, 572–574, 596
  shell, 491g–492, 499, 509–510, 526
  and spectral emission, 488–489
  spin, 507–508
  state, 508–512
  and superconductivity, 452
  wavelength, 500e, 550
  and X rays, 464
electron volt (unit), 488g, 550
electroscope, 409–411, 427, 447
electrostatic, 454, 472
element, 209g–211, 555
  absorption and emission spectra, 476, 479, 499
  Aristotelian, 209
  Aristotle's four, 571–572
  chemical, 228, 474
  formation, 600
  isotope, 530, 555–556
  Lavoisier's, 210t
  naturally occurring, 556
  noble, 492
  periodicity, 474, 482, 491–493, 509–510, 517
  radioactive, 479, 526–534
  relative mass, 212–213
emission
  spontaneous, 516
  stimulated, 515g
energy, 111–129
  atomic, 481, 490, 554–555
  atomic kinetic, 251
  binding, 552g–553, 564
    average per nucleon, 552–553, 556, 564
    total, 553
  chemical, 124, 274, 428, 549
  conservation, 105, 111–129, 206, 221, 256, 271, 275, 289, 391, 439, 487–488, 526, 540, 552, 574–578, 581–582
    circuits, 436
    fails, 248, 577–579
    first law of thermodynamics, 251–252, 273
    hoax?, 125

  inertial reference system, 157–158, 197
  kinetic, 114–115
  mass–energy, 192, 553
  mechanical, 119–122
  special relativity, 192
  thermal, 253
  convection, 258
  crisis, 281, 602
  dark, 597–598
  discrete, 483–484, 487
  elastic potential, 123
  electric, 111, 125–128, 280, 428, 436–439, 452, 549
  electric potential, 255, 418g, 420, 480, 487
  electromagnetic potential, 124
  electromagnetic wave, 591
  electron, 484–487
  equivalence to mass, 553–554, 572–573, 580, 598
  excitation, 556
  frictional potential, 124–125
  gravitational potential, 118g–118e–125, 142, 158, 239, 249, 255, 278, 418, 436, 549, 552, 566
  gravitational wave, 591, 593
  heat, 344
  internal, 249–251g–253, 255, 268–279
  invariant, 95
  kinetic, 208, 232, 239, 248–249, 257, 278–279, 418, 432, 480, 488, 557, 561, 577, 582
    average
      molecular, 221, 228, 252, 256–257, 565
      atomic, 251
    from beta decay, 575
    change in, 115–118e–129
    from electric field, 542, 550
    inertial reference system, 158, 177
    particle in a box, 506–507
    photoelectric effect, 484–486
    quantized, 487
    relativistic, 191–192e, 197
    rotational, 142g–142e, 251–253
    units, 142
    translational, 113g–113e–126, 142, 251, 253
    units, 113
  level, 500, 506–509, 530–533
    diagram, 488, 500, 506–509
    discrete, 555
    neutron, 555
    proton, 555
    quantized, 507, 515
  loss, 280
  macroscopic, 279
  mechanical, 119g–119e–125, 248, 250, 269–273, 290, 487

energy—mechanical (continued)
  total, 120–121
  microscopic, 251
  molecular bonds, 251
  nuclear, 233, 490, 523, 548–549, 552, 555–557, 561, 564
  nuclear potential, 124
  perpetual motion, 272
  photon, 485e–489, 515, 528, 533, 537–538
  potential, 248–249, 272
  power, 127
  purchasing, 438–439
  quality, 281
  quanta, 483e–484, 576
  quantum nature, 472
  rest-mass, 192g–192e, 578, 582
  sound, 322, 326
  state, 549
    lowest, 506, 510–511
  Sun's, 564–566, 599
  and temperature, 216, 220–221, 250, 252
  thermal, 125–126, 247–249g–250, 255–259, 269–275, 280, 290, 296, 344, 439, 452
  to reverse Earth's magnetic, 454
  uncertainty principle, 512
  units, 249–250
  usage, 124–125
  vector, 113
  vibrational, 251, 253
  wave, 287, 295, 314
enrichment uranium, 560
entropy, 278g–279
equation
  mass-energy, 553–554
  Maxwell's, 177–178
  Schrödinger, 505–507, 510
equilibrium
  position, 289g–292, 295–300
  stable, 141g
  thermal, 250g, 256, 259–260, 269
  unstable, 141g
equivalent states, 277–278
ether, 178g
evaporation, 260
evaporator, 275
events, simultaneous, 179–181
evolution, theory of, 535
excited state, 506, 516, 533
exclusion principle, 510g–511, 555, 585–586
expansion, thermal, 262g–262e–263
  coefficient, 262
exponent, 9, 11
exponential growth, 124–125
eye, 371–372, 389, 498, 516–517
  glasses, 373
  schematic, 372

F

Fahrenheit, Gabriel, 218
Faraday, Michael, 251, 407, 455–458, 462–463

Faraday's law, 456
farsightedness, 373
Fermi, Enrico, 555–556, 558, 561
  biography, 558
Fermi, Laura, 558
Feynman diagram, 576–577
Feynman, Richard, 112, 129, 502, 576–577, 600
  biography, 577
fiber optic, 516–517
field, 87g, 576
  electric, 413–414g–415e–420, 439, 448, 461–463, 477–478, 492, 501, 526, 528
    effect on a charged particle, 542, 550
    lines, 416g–418
    and magnetic, 526, 528
    around negative charge, 415
    around positive charge, 415–416
    units, 415
  gravitational, 87g–88, 194–195, 413, 415, 418
  magnetic, 415, 448g–465, 477–480, 492, 499, 526–528, 543, 550, 565, 572
    determining direction, 449
    of Earth, 450, 453–455
    reversal, 454
    effect on a charged particle, 527, 543, 550, 565, 572
    and electric, 526, 528
    induced, 455–457
    near bar magnet, 448
    near solenoid, 449
    near wire, 449
    theoretical limit, 451
    unit, 451g
  radiation, 577
filament, 429
fission, 555–556g–562, 564–566
Fizeau, Hippolyte, 349
flavor
  of neutrinos, 598–599
  of quarks, 583g
flawed reasoning
  absorption spectrum, 489
  alpha decay, 531–532
  average speed, 21
  Avogadro's number, 214
  bending light, 195
  bulbs in series, 434
  buoyant force, 238
  center of mass, 140
  centrifugal forces, 164
  change in bulb brightness, 436
  circular motion, 60
  collision, 104
  color quantum number, 586
  comet gravity, 81
  conservation of mass, 554
  conservation of mechanical energy, 122
  Doppler effect, 325–326
  electrical charge, 408
  electrical force, 412
  electromagnetic wave, 462
  electron interference, 504

Fahrenheit temperature scale, 220
fusion reactor, 566
gravitational potential energy, 118g–119
heat capacity, 254
heat index, 261
image location, 342
images from a lens, 370
inertial reference systems, 158
intermediate vector bosons, 579
Lenz's law, 457
mixing colors, 352
momentum conservation, 101
Moon gravity, 82
net force, 45
orbits, 67
periodic wave, 301–302
pressure, 236
probability, 278
radioactivity, 526
rainbow, 366
second law of thermodynamics, 273
simultaneity, 181
speed of sound, 315
thin films, 389
third law forces, 50
torque, 139
uncertainty principle, 513
wave speed, 298
float, 236–237
Fludd, Robert, 272
fluid, 232–233, 237–239, 241, 258
  and transverse waves, 299
fluorescence, 509, 525, 542
focal length
  lens, 367g–375
  mirror, 344g–348
focal point
  lens, 367g–369, 374–375
  mirror, 344g, 347
  "other," 367–368
  principal, 367–369
foot (unit), 6
force, 37g–50, 95, 289, 292, 449, 451, 454, 477, 507, 514, 591, 593–594, 601
  and air resistance, 240
  attractive, 576
  bonding, 228
  buoyant, 231, 236g–239, 249
  centrifugal, 57, 60, 67–68, 163g, 168
  centripetal, 57g, 60–62, 68, 164, 454
  color, 586, 593
  constant, 40
  Coriolis, 168g–169
  elastic, 289
  electric, 159, 228, 402, 407, 411–415e, 418, 432, 461, 481, 487, 504, 525, 549–550, 579
  electromagnetic, 75, 124, 523, 549, 578–579, 591–596
  electrostatic, 499
  electroweak, 594
  exchange, 576, 579

  fictitious, 159–160, 163
  four basic, 75
  free-body diagram, 44–45, 60
  friction, 57, 60, 62–63, 117–129, 163–164, 280
  gravitational, 42, 62–68, 75, 79–88, 99, 117, 144–145, 240, 249, 413–414, 535, 549, 552, 554, 566, 579, 591, 594–595, 600
    on atomic level, 228–229, 481
    and center of mass, 84
    and development with Universe, 595, 600
    near Earth, 81
    and electric force, 487, 523, 549, 591
    as energy source, 549
    exchange particle, 579
    and grand unified theory, 594
    on Moon, 83
    and noninertial reference systems, 159–162, 168, 192
    and orbits, 66, 566
    on other planets, 82
    projectile motion, 60
    as restoring, 292, 297
    and source of Sun's energy, 535
    and tides, 85
    between two objects, 79
  of inertia, 67
  inertial, 159g–163, 166–169, 188, 192, 194
  intermolecular, 221, 231, 235
  magnetic, 159, 402, 451–454e–455, 459
  net, 40–41e–50, 57, 59, 97e–98, 116–117, 121, 137, 143, 160–161, 233, 237–239
    change momentum, 134
    constant, 157
    zero, 38, 49, 137, 156, 160, 164
  n–n, 551
  normal, 44–45
  n–p, 551
  nuclear, 124, 288, 523, 549–552, 564, 578, 594
  periodic, 294
  p–p, 551
  and pressure, 215, 233–234, 598
  repulsive, 576
  restoring, 289, 292, 297–298, 323–324
  and rotations, 135–141
  separation during Big Bang, 594
  strong, 75, 549–551g, 555, 578–579, 593–595
  third law, 47–50, 164
  unbalanced, 38
  unified, 595
  vector, 41, 88
  weak, 75, 552g, 554, 579, 595–596
fossil fuel, 548
Foucault, J.B.L., 166

Foucault, Jean, 384–385
fourth dimension, 153
Franck, James, 556
Franklin, Benjamin, 406
  biography, 407
Frayn, Michael, 490
free-body diagram, 44–45, 60
free fall, 24–27, 45–46, 155, 157
frequency, 290g–290e–305, 314–326, 463–464, 511, 513, 592
  audio, 464
  beat, 324
  carrier, 464–465
  Doppler, 326
  electromagnetic wave, 493
  fundamental, 303g–304, 317, 319–323
  harmonic, 319–320
  infrasonic, 318
  in a medium, 384
  natural, 293–296
    animals, 318t
    human, 316–318, 322
    range, 316–318
    visible light, 387, 463
  and photon energy, 483e, 485–487
  resonant, 294, 303–304, 317, 323
  sound, 319–327
  standing wave, 303–304, 319–320, 324
  ultrasonic, 318
  units, 290
friction, 35, 46, 60, 62, 145, 178, 205, 232, 279, 289–290
  force (see force, friction)
  kinetic, 47g
  loss, 271, 273
  static, 47g
Frisch, Otto, 561
fuse, 437
fusion, 564g–566, 600–601

G

Gabor, Dennis, 394
galaxy, 1, 96, 144
  Andromeda, 86
  Milky Way, 10, 86, 169–170, 196
Galileo, 24–27, 35–37, 95, 402
  and absolutes, 153
  and acceleration due to gravity, 25
  and Aristotle, 23, 46
  biography, 24
  and clocks, 293
  and Copernicus, 166
  energy conservation, 120
  and Huygens, 95
  inertia, 166, 205
  and Kepler, 76
  measuring the speed of light, 348
  motion, 25
  and Newton, 36, 46
  and pendula, 291
  principle of relativity, 157g, 177
  relative motion, 156

scientific style, 24–27
  spirit of, 214
  telescope, 4, 374
  thermometer, 218
  thought experiment, 35
Galloping Gertie, 296
Galvani, Luigi, 427
gamma ray, 464, 528, 533, 552–557, 572
  absorption, 537–538
gas, 227–228, 232g–233
  constant, 220
  convection in, 258
  ideal, 214g–216, 219–221, 232, 262, 271
    assumptions, 214
    law, 207, 220g–220e
    macroscopic properties, 207–208
  noble, 511
  real, 214, 219–220, 253
  spectral lines, 474–475, 479
gauss (unit), 451g
Geiger counter, 542–543
Gell-Mann, Murray, 583, 586
generator, 452, 457–458, 562
geographic pole, 453
geothermal, 280
Germer, L. H., 501
Gilbert, William, 404, 406, 448
Glashow, Sheldon, 594
global positioning system, 193
gluon, 585g–586
God, 34, 76
Goeppert-Mayer, Maria, 556
  biography, 556
Goitschel, Philippe, 48
Goldberg, Rube, 4
grand jeté, 68
gravitation, 407
  constant, 80–82
  field (see field, gravitational)
  force (see force, gravitational)
  universal law of, 79g–79e–84, 153, 194, 576
graviton, 579g
gravity, 2, 38, 74, 77, 402, 590, 597–598
  acceleration (see acceleration, due to gravity)
  artificial, 164
  bending light, 194
  black hole, 2, 196, 591–592
  center of mass, 139, 143–145
  concept, 75
  effect on atmosphere, 233
  and Einstein, 191
  on Moon, 82
  pendulum period, 292
  on planets, 82
  projectile motion, 60, 62–68
  wave (see wave, gravitational)
  zero, 162–163
greenhouse effect, 259
grounding, 404–405g
ground state, 487–489, 533, 542, 555
Guerrouj, Hicham El, 19
gyroscope, 144

H

hadron, 580g–585
Hahn, Otto, 561
half-distance, 538
half-life, 534g–535, 554, 583
Halley, Edmund, 37
halo, 366–367
harmonic, 303g–304, 317–323
Harrison, John, 293
heat, 117, 249g, 549
  capacity, 253g–254
  from chemical reaction, 557
  engine, 269g–275, 278, 280
    Carnot's ideal, 271, 274
    efficiency, 274e
    Hero's, 269
    schematic, 271
  and fission, 562
  and fusion, 564–565
  index, 260–261t
  and Joule, 250–251
  latent, 255g–255t–256
    units, 255
  nature of, 248
  pump, 275g
  and radiation, 482
  specific, 252–253g–253t–254, 258
    units, 253
  and temperature, 251–252
  unit, 249
  and work, 249
Heisenberg, Werner, 490, 511–512, 561
Helms, Susan, 162
Hero of Alexandria, 269
hertz (unit), 290
Hertz, Heinrich, 462, 472
Hilbert, David, 105
Hitler, Adolf, 105
hologram, 394
holography, 394, 517
Hooke, Robert, 37, 374
horsepower (unit), 127
Hulse, Russell, 592–593
Huygens, Christian, 37, 95, 293, 334, 402
hyperopia, 373

I

image
  camera, 339
  curved mirror, 343–348
  diffracted, 388–389
  erect, 346–348, 369, 374
  eye, 367, 371–373
  flat mirror, 341–342
  holographic, 394–395
  inverted, 339, 345, 348, 369–371, 375
  lens, 368–370
  magnetic resonance, 524
  magnified, 346–348, 369–370, 374–376
  multiple, 342–343
  negative, 351
  positive, 351
  real, 346g–348, 369–372, 375
  refracted, 361
  telescope, 374–377
  virtual, 336, 346g–348, 361, 374

impulse, 98g–98e–99, 105
  and kinetic energy, 116–117
  and momentum, 192
  and resonance, 294, 296
  unit, 98
induction, charging by, 407–408, 411
Industrial Revolution, 269, 272
inertia, 35g–38, 97, 205
  confinement, 565
  force (see force, inertial)
  and Galileo, 166
  and Kepler, 76
  law of, 37g, 58, 159
  rotational, 138g, 142–143
  and simple harmonic motion, 289, 292, 303
  translational, 138
inflation, 597
instrument
  percussion, 323–325
  stringed, 319–321
  wind, 321–323
insulator
  electric, 404g–407, 410
  thermal, 257g–258, 263
interaction
  color, 594
  electromagnetic, 576, 583, 594
  electroweak, 594
  gravitational, 580
  strong, 576, 580, 583, 593–594
  weak, 576, 581, 583
interference, 304g–307, 385–391
  of bullets, 503
  constructive, 391
  destructive, 391
  effect, 486, 513
  of light, 385–387, 502
  pattern, 305–306, 385–387, 391, 394–395, 501–505
    for electrons, 501–504
    for light, 502
    two-slit, 385–387
  thin film, 389–391
  two-slit, 502–504, 513
  of water waves, 503–504
intermediate vector boson, 579g
invariant, 94–95, 112, 114, 122, 192
inversely proportional, 41g
inverse-square, 77g–78, 411g–413
ion, 228, 233, 420, 478g–479, 529, 531, 536
ionization, 536–537g–538
isotope, 529g–530, 533–536, 541, 555–558, 564
  distinguishing, 530
  of hydrogen, 530
  radioactive, 530, 533–536

J

Jefferson, Thomas, 6
Jensen, J. Hans Daniel, 556
Joliot-Curie, Irene, 527
Jordan, Michael, 63
joule (unit), 113g

Joule, James, 249–250, 269, 273
  biography, 251
Joyner, Florence Griffith, 19

K

Kepler, Johannes, 75–76, 78, 144
  biography, 76
  third law, 76
kilo, 7g
kilogram (unit), 6, 8g, 41g
kilometer (unit), 7
kilowatt-hour (unit), 128, 438–439
Kirchhoff, Gustav, 435
  junction rule, 435g
  loop rule, 436g–437
Kittinger, Joseph, 48
Koshiba, Masatoshi, 598

L

laser, 344, 373, 388, 394, 475, 498, 515g, 565, 592
  acronym, 515
  uses, 516–517
Lavoisier, Antoine, 95
  periodic table, 210
law
  conservation, 105, 177, 530, 581
  of definite proportions, 211g–212
  fundamental, of physics, 406
  nature, 94, 96, 191
  physical, 600
  See also specific law
Leibniz, Gottfried, 37
length, 177
  contraction, 187–190
  focal
    lens, 367g–375
    mirror, 344g–348
  pendulum, 291–292, 294, 297–298
lens
  binocular, 391
  camera, 371
  coating, 391
  contact, 373
  converging, 367–374
  diverging, 367–370
  eye, 371–373
  fisheye, 366
  focal length, 367g–375
  focal point, 367g–369, 374–375
  magnifying, 368–369
  microscope, 367
  optic axis, 367g–368, 370–371
  special rays, 368
  spherical, 371
  telescope, 391
  thin, 367–368
Lenz's law, 456, 457
lepton, 580g–584, 593, 596
  number, 583
    conservation, 582–583
  properties, 581

Leucippus, 205
light, 248, 334–353, 358–377, 382–396, 439
  analogy for understanding quark color, 585–586
  atomic spectra, 473–479, 482, 487
  bending, 194
  black, 509
  and black holes, 196
  from a cathode tube, 477, 479
  coherent, 515, 517
  color, 350–353
  combinations, 350
  diffraction (see diffraction)
  electromagnetic wave, 591
  as energy release, 549, 552, 554, 557
  extended source, 337
  fluorescent, 509
  from hot objects, 482–483
  interference (see interference, of light)
  lasers (see laser)
  linear momentum (see momentum, linear)
  nature
    particle, 383–384, 499, 513
    wave, 383–385, 389–391, 396, 501, 513
  particle, 383–384, 485–486
  and photoelectric effect, 484–486
  photon interaction with matter, 537
  pipe, 362
  point source, 337, 388–389
  polarization, 392–394
  quantized, 485
  quantum, 572
  ray, 337g–346, 360–368, 383, 390–391
    diagram, 341, 346–347, 367–370, 383–384
    incident, 340, 345
    reflected, 340, 344–347
  reflection (see reflection)
  refraction (see refraction)
  and relativity, 177–187
  sight, 373
  speed (see speed, of light)
  of Sun, 565
  ultraviolet, 472, 492, 509
  visible, 463–464, 482, 501, 509, 525, 528, 542
    range, 386, 463
  wave, 483–486
    theory, 485
  white, 350–351, 364, 386, 395, 476, 489
light bulb, 429–438
  in parallel, 434–435
  in series, 434–435, 438
  standard, 434–435, 437
lightning, 419
light-year, 8
line of stability, 554g–557
liquid, 208, 221, 227–228, 231g–235, 253, 255–257, 262–263
  crystal, 235
  wave propagation, 299

looming, 364
Lord Kelvin. See Thomson, William

M

MacCready, Paul, 127
MACHOs, 597
macroscopic, 206
  properties, 207–208g, 214–216, 220
magnet, 408, 447–465, 477, 528, 550–551
  bar, 408, 448–449, 455–457
  horseshoe, 451
  permanent, 447
  superconducting, 551
magnetic, 402
  deflection, 529
  electro-, 451g–461
  field (see field, magnetic)
  force (see force, magnetic)
  monopole, 448g
  pole, 447g–449, 453, 455
  variation, 453
magnetism, 446–463, 593
magnetize, 447–450, 454, 459
magnification, 343, 347, 369, 374–375
magnifying glass, 368–369, 374–375
magnitude, 20g
mass, 2, 8, 41g–44, 46, 49–50, 62, 77, 79–88, 94–95, 206, 418
  alpha particle, 480
  and angular momentum, 42, 486
  atomic, 213g–214, 491, 493, 529
    unit, 213g
  cause of gravitational waves, 591
  center of, 67g–68, 139g–145
  and classical relativity, 177
  conservation, 95g, 157, 531, 554, 562
  critical, 597
  and density, 228–229, 263
  Earth, 49
  electron, 385–387, 413–414, 500, 529t, 599
  and energy, 113, 117, 121–123, 191, 192
  equivalence to energy, 553–554, 572–573, 580–581, 598
  gamma ray, 572
  gravitational, 194g, 413
  graviton, 579
  hadrons, 581t
  and heat, 248, 252–254
  inertial, 41, 194g
  invariant, 192
  isotopes, 530
  and law of definite proportions, 211–212
  and law of universal gravitation, 79
  leptons, 581t

and linear momentum, 97–104, 512
Moon, 82
neutrino, 576, 598–599
neutron, 529t
neutron star, 592
in outer space, 162
and pair production, 572–573
positron, 532
proton, 413–414, 529t
ratio to charge, 478–479, 527, 560, 572
rest, 192, 575–582
stars, 196
Sun, 566
unit, 8, 41
Universe, 596
and waves, 314
vs. weight, 43
white dwarf, 229
mass on a spring, 289–294, 303
  period, 291e
matter, 94, 205, 478, 537–538, 545, 549, 566, 572, 594–597
  dark, 597–598
  wave, 549
    amplitude, 505g
  and waves, 288–289
Maxwell, James Clerk, 523, 593
  biography, 463
  and Einstein, 191
  electromagnetic waves, 472
  electromagnetism, 177
  equations, 177, 462, 463, 499
  laws, 484, 505
Mayer, Joseph E., 556
McKinney, Steve, 48
mechanics
  Newtonian, 499, 514
  quantum (see quantum, mechanics)
mechanistic view, 514
Meitner, Lise, 490
  biography, 561
melting, 230–233
  point, 255t–256
  temperature, 228
Mendeleev, Dmitri, 474, 491, 580
meson, 579–580g–582, 585–586, 600
  number, 582
    conservation, 582
metastable state, 516
meter (unit), 6–7g
Michelson, Albert A., 178, 385
microscope, 367, 374–375, 389, 501, 542
  compound, 374
  electron, 208, 501
  optical, 208
  scanning tunneling, 501
microscopic, 206, 231
  properties, 207–208g, 218, 220
microwave, 463
milli, 8g
Millikan, Robert, 478

mirage, 364
mirror, 362, 368, 376, 389, 391, 516
  concave, 343–348, 376
  convex, 343–348
  curved, 343, 368
  cylindrical, 344–345
  flat, 341–342
  focal length, 344g–348
  focal point, 344g, 347
  fun-house, 343
  multiple, 342–343
  optic axis, 344g–347
  spherical, 344–348
mixture, 212
model, 34
  Aristotelian, 211
  atomic, 3, 208–214, 228, 474, 482, 486, 509
    Bohr's, 486–493, 499, 507, 510
      first postulate, 486
      second postulate, 487
      third postulate, 487
    Dalton, 211
    plum-pudding, 479
    Rutherford's, 479–482, 486
    solar system, 481–482
    Thomson's, 479, 480, 482
  Big Bang, 594, 596
  continuum, 211
  developing, 208–209
  electric fluid, 404–406
  geocentric, 164g
  heliocentric, 165g
  ideal gas, 214–216, 221
  for light, 382–383
    particle, 383–384
    wave, 383–384, 388, 391, 396
  mathematical, 208–209
  physical, 4
  quantum-mechanical, 508, 523
  scientific, 2
  solar system, 228
  standard, 594
moderator, 561g, 563
modulation
  amplitude, 464
  frequency, 464
mole (unit), 214g
molecule, 4, 211g, 414, 420
  air in a room, 277
  and change of state, 255–256
  evaporative cooling, 221
  formation, 208, 211–212, 549, 601
  in gas, 232
  in liquid, 231–232
  number in a mole, 212–216
  and sound wave, 314–315, 322
momentum, 208, 216, 506, 514
  angular, 95, 142g–142e–144, 486
    conservation, 105, 143g–143, 531, 581
    quantized, 486e
  change in, 116
  and classical relativity, 177
  conservation, 157, 192
  linear, 95, 97g–106, 134, 144, 574–575
    change in, 98e–105
    conservation, 96, 97e, 99–100g–106, 114–115, 143, 531, 574–578
    relativistic, 191
    unit, 97
  not conserved, 574–578
  quantized, 507
  and uncertainty principle, 512e–513
  and waves, 549
monophonic, 465
Moon
  acceleration, 77–78
  atmosphere, 25
  cause of tides, 83
  determining month, 293
  distance from Earth, 73, 344
  eclipses, 338
  elliptical appearance, 363
  mass, 73, 82
  motion, 74
  orbit, 10, 75–77
  orbital path, 165
  orbital period, 73, 83
Morley, E. W., 178
Moseley, Henry G., 481, 493
motion
  absolute, 177
  atomic, 278
  Brownian, 215
  celestial, 75–80
  circular, 57–62, 68, 75
  classical, 191
  Earth, 15, 164–170
  football, 15
  Galileo, 26
  laws of, 37–50, 153, 157, 177, 191
  macroscopic, 278
  natural, 23–26
  one-dimensional, 15–27, 37–50, 57, 63
  particle, 215, 288
  periodic, 289, 293
  perpetual, 205
  planetary, 166
  projectile, 60g, 63–66, 135, 143, 157, 240
  relative, 155–157, 179, 181, 188–190, 194
  rotational, 66–68, 134–145
  simple harmonic, 500
  translational, 67g–68, 134–139
  two-dimensional, 57–68
  vibrational, 252, 289
  wave, 296
motor, 452, 457–458
muon, 187, 579g–584
music, 313, 318, 321, 324
Mussolini, 558
myopia, 373

N

Narang, 64–66
natural place, 34–35, 46–47, 75
  hierarchy, 34
nature
  dual, of light and electrons, 513
  law of, 94, 96, 191
  rule of, 159, 340
nearsightedness, 373
nebula, 590
neutrino, 522, 575g–576, 580–582, 585, 598–600
  electron, 581–582, 598–599
  mass, 575–576, 598–599
  mu, 581–582, 598–599
  tau, 580–582, 585, 598–599
neutron, 11, 406, 523, 529g–533, 538, 551–564, 575–576, 592–597, 600
  anti-, 573–574, 600
  baryon, 580
  and beta decay, 531–532, 575
  charge, 529
  deuteron formation, 552
  discovery, 529
  and electron capture, 532
  formation during Big Bang, 596, 600
  and fusion, 564
  and isotopes, 529–530
  mass, 529t
  nuclear force (see force, nuclear)
  nuclear reaction, 556–564
  and quarks, 584
  stability, 554–557
neutron star, 591–592
newton (unit), 42g
Newton, Sir Isaac, 36–50, 218, 334, 402, 407, 462–463, 601–602
  action at a distance, 523, 576
  and alchemy, 209
  biography, 37
  dispersion experiments, 363–364
  and Einstein, 191
  first law, 36–37g–38, 55–58, 120, 135, 157, 236, 238
    for rotation, 136
  and Franklin, 407
  and Galileo, 36, 46
  and gravity, 67–68, 74, 76–88
    law of universal gravitation, 79, 576
  and Halley, 37
  and Hooke, 37
  and Huygens, 37
  and invariants, 95
  and Kepler, 76
  laws, 78, 86–88, 95, 103, 153, 155, 159, 177, 191, 514–515
  and Leibniz, 37
  and Maxwell, 191
  mechanics, 484
  particle nature of light, 383–385
  and Planck, 505
  and play, 577
  and pressure, 234
  and Queen Anne, 37
  quote, 5, 591
  second law, 38–40g–50, 59–60, 88, 95, 97, 116, 121, 162, 191–194
    relativistic form, 192
    for rotation, 138
  thermometer, 218
  thin-film interference, 391
  third law, 47–48g–50, 99–100, 139, 148, 240, 412, 451
Nightingale, Florence, 527
nodal line, 305–306, 323, 385
node, 303g–304, 319, 322
Noether, Amalie Emmy, 105
  biography, 105
normal, 340g–341, 359–360, 384
  force (see force, normal)
  incidence, 368
northern lights, 233
nuclear
  decay, 534
  energy (see energy, nuclear)
  potential well, 552–556
  reaction, 549, 556–558, 564, 566
  reactor, 558, 560, 563, 575
    boiling-water, 562
    breeding, 563–564
    fission, 566
    fusion, 566
    naturally occurring, 563
    pressurized-water, 562
    schematic, 562
nucleon, 529g–534, 549–559, 564, 578, 582, 586
  conservation, 531
  number, 548–564
nucleus, 228, 480g, 523–525g–537, 548–564, 601
  daughter, 530g–532, 554
  magic numbers, 556
  parent, 530g–532, 554
  radioactive, 526, 534
  size, 480, 525
  stable, 549, 552, 554–555
  structure, 554–555
  and surrounding electrons, 499–500, 507
  unstable, 549, 555–556

O

octave, 318, 320
Oersted, Hans Christian, 449
ohm (unit), 432g
Ohm's law, 432g–433
Onnes, Heike Kamerlingh, 452
optic axis, 344g–347, 367g–371
optical fiber, 362
order of magnitude, 9g
Orion, 259
oscillation, 289g, 291, 293, 296

P

pair production, 572g
paradox, twin, 188

parallax, 166
parallel, 428g
particle, 314, 477–479, 499–514, 523, 527–530, 536–538, 541–543, 593–594
  accelerator, 549g–550
  alpha (see alpha, particle)
  in a box, 506–507
  classical, 525
  creation, 594
  detector, 542–543
  elementary, 523, 571–582, 593–594, 598–601
  exchange, 578–585, 593
  lambda, 583
  momentum, 506, 512
  nature of light, 334
  omega minus, 585
  strange, 577, 579g, 583
  subnuclear, 593
  Yukawa's, 578–579
pascal (unit), 215
Pauli, Wolfgang, 510, 575, 577
pendulum, 289, 291–294, 300, 303
  Foucault, 167–168
  period, 292e
penumbra, 337g–338, 388
Penzias, Arno, 596
period, 289g–293, 300, 304, 321
  and frequency, 290e–291
  mass on a spring, 291e–292
  orbital, 78, 83, 349
  pendulum, 291–292e–293
  of revolution, 349
  wave, 295, 300–306, 462, 466
periodic table, 474, 479, 491–492, 499, 509, 531–532, 545
  Lavoisier's, 210
  modern, 474, 491–492
perpetual motion, 272–273, 602
phase, 505, 515
  difference, 385
  in, 304g–305
  out of, 305, 389
phosphorescence, 509
photoelectric effect, 191, 481, 484g–485, 542
photograph, strobe, 16, 21
photomultiplier, 522, 542
  schematic, 542
photon, 485g, 523, 528, 533, 537–539, 572–580, 586, 594, 596
  absorption, 516
  anti-, 574
  coherent, 515
  color, 486
  and complementarity principle, 513
  duality, 504, 511, 513
  emission, 499, 515–516
  energy (see energy, photon)
  exchange particle, 576–578
  fluorescence and phosphorescence, 509
  frequency, 485–486
  gamma ray, 464, 528, 537–538, 552, 554, 556–557, 572
  interference, 504–505
  laser, 515–516
  particle, 485
  and uncertainty principle, 511–512
  X ray, 493
Physics on Your Own
  absorption and emission spectra, 491
  angular momentum, 145
  astigmatism, 372
  average speed, 16
  baryon, 585
  battery, 427
  beats, 324
  Bernoulli's effect, 240
  billiards, 104
  buoyancy, 237
  camera obscura, 339
  carbon-14 dating, 537
  cathode rays, 479
  center of mass, 67
  center of mass, 140
  chain reaction, 560
  change of state, 255
  clocks, 293
  color pictures, 352
  compass, 448
  complete circuits, 430
  conductivity, 257
  conservation of potential and kinetic energy, 114
  Coriolis forces, 168
  crystal, 230
  crystalline structure, 231
  density, 230
  diffraction, 307
  diffraction pattern, 388
  electric energy, 439
  electric potential, 420
  emission spectra, 476
  flat mirrors, 341
  focal length, 368
  free fall, 25
  friction, 47
  geosynchronous orbit, 84
  heat engine, 269
  image in a flat mirror, 341
  image of the retina, 372
  impulse, 98
  induced attraction, 408
  inertia, 36
  inertial forces, 163
  laser, 516
  law of definite proportions, 212
  longitudinal wave, 315
  magnetic materials, 447
  magnification, 375
  making a magnet, 450
  material elasticity, 114
  model rocket, 106
  noninertial reference system, 160
  noninertial reference system, 163
  nuclear reactor, 562
  order and disorder, 277
  pendulum, 292
  pendulum height, 121
  polarization, 396
  polarization of light sources, 393
  projectile motion, 63
  radioactivity, 535
  rainbow, 366
  reaction time, 27
  relative motion, 157
  resonant frequency, 294
  right-angle mirrors, 343
  rotational inertia, 143
  satellite orbits, 83
  skyhook, 141
  sparks, 405
  speed of sound, 316
  surface tension, 231
  temperature, 218
  the Sun, 566
  thermal energy, 256
  thermal equilibrium, 251
  thermometer calibration, 219
  three-way switches, 438
  torque, 136
  two-slit interference, 387
  two-slit interference, 504
  types of charge, 406
  vibration on a string, 320
  waves, 298
  weight on an elevator, 50
  wind instrument, 323
Piaget, Jean, 94
pion, 578g, 582–586
pitch, 314, 316, 326
Planck, Max, 472, 483–486, 499
  biography, 484
  constant, 483, 500, 512
  length, 594
plasma, 228, 233g
point of view, Newtonian, 506
Polaris, 145
polarization, 235
  axes of, 392–393
  plane of, 392–393
polarized, 392g–394
  horizontal, 393
  plane, 392–393
  vertical, 392
Pontecorvo, Bruno, 597, 599
population inversion, 516
position, 153, 155, 177
  of a particle in a box, 506
  and uncertainty principle, 511–512e–514
positron, 532g, 572–573g–575, 582, 596
potential difference, 420, 427–428, 431–432, 437, 550
pound (unit), 6, 42
Powell, Asafa, 19
power, 127g–128e, 269, 402
  electric, 439g–439e, 457
  human, 127
  nuclear, 91, 557
  solar, 565
  unit, 127, 439
powers-of-ten notation, 9g–11
precession, 145
pressure, 206, 215g–215e, 472, 526, 597
  Archimedes' principle, 237
  atmospheric, 233–236
  Bernoulli's effect, 238–241
  and bubble chamber, 543
  change of state, 255
  cone, 328
  dark energy, 597–598
  in fluid, 233–236
  ideal gas law, 220e–221
  macroscopic property, 207–208, 233
  sound, 314–316, 322
  and speed of sound, 315
  in stars, 229
  thermometer, 218
  units, 215, 234
  in water model, 431
primary coil, 457, 459
principle
  Archimedes', 237
  complementarity, 513g
  equivalence, 194g
  exclusion, 510g–511, 555, 585–586
  of relativity, 157g, 177
  superposition, 302
  uncertainty, 511–514
prism, 362–367, 375, 475, 489
probability, 277–278
  cloud, 508
  distribution, 506
  matter waves, 549
  radioactivity, 533–535
property
  atomic, 526, 538, 549
  chemical, 527, 530, 549, 564
  electronic, 538
  magnetic, 529
  nuclear, 530
  physical, 549
proportional, 40g
  inversely, 41g, 220
proton, 4, 11, 404, 406–407, 523, 528g–533, 538–539, 549–557, 571–586, 591–600
  acceleration, 413, 550–551
  anti-, 571–574, 600
  atomic magnet, 528–529
  baryon, 580, 582
  beta decay, 575, 581–582
  binding energy, 552–555
  building, 584–586, 595–596, 600
  charge, 406–407, 412–414, 451, 527, 550–551, 555–556, 583
  decay, 531–533, 582, 594
  deuteron formation, 552
  discovery, 528
  electric force, 413–414, 549
  and electron capture, 532, 596
  gravitational force, 413–414
  mass, 413–414, 529t, 552, 579
  neutron formation, 531–532, 575, 592
  nuclear force, 550–551
  properties, 581t
  and quarks, 584, 586, 596, 600
  range, 537t
  reactions, 531–532, 575, 583

size, 11, 594
stability, 554–558
Proxima Centauri, 188
Ptolemy, 165
pulsar, 592
pulse, 295–298, 300
   transverse, 298, 315

Q

quantization, 472, 506–507, 515
quantum, 478g, 483g–484
   angular momentum, 507, 510
   electrodynamics, 508
   leap, 488
   mechanics, 86, 471, 484, 504–505g–511, 558, 573, 594, 599
      atom, 507–508, 511
      development, 252, 334
      effects, 551, 561, 576, 578
      and Einstein, 515
      and the future, 514
      interactions, 252, 334
      laws, 514, 525
      and line of stability, 554–555
      model, 508, 523
      probability for occurrence, 534
      and quantum electrodynamics, 508
      and relativity, 509
   number, 486g, 507–508, 510, 585–586
      angular momentum direction, 507, 510
      color, 585–586
      energy, 507, 510
      ground state, 510t
      spin, 507–508, 580
      strangeness, 583
   theory, 481
quantum electrodynamics, 508, 577
quark, 2, 10, 523, 583g–586, 593–596, 600–601
   bottom, 584–585g
   charm, 584g
   down, 583–584
   flavors, 583–584
   fractional charge, 583–584
   properties, 584
   strange, 583–585
   top, 571, 585g
   up, 583–584
Queen Anne, 37

R

rad (unit), 538g
radar, 591
radians, 135
radiation, 247, 256, 515, 525–528, 533, 537–541, 550, 575, 592, 596
   alpha, 481, 526g–527
   average annual dose, 539
   background, 539–540
   beta, 481, 526g–527
   cancer treatment, 541
   cosmic background, 596–599
   detector, 537, 541–543
   electromagnetic, 259g, 463, 472, 482, 493, 537, 549–550, 600
   gamma, 528g
   gravitational, 593
   infrared, 464
   medical sources, 540
   microwave, 597
   nonelectromagnetic, 598
   nuclear, 541
   sickness, 539
   spectrum, heated objects, 472
   ultraviolet, 464
radio, 591
   AM, 464–465
   FM, 464–465
   operation, 464
   wave (see wave, radio)
radioactive, 525–530, 533–541
   dating, 536
   decay, 530, 533t, 536, 538, 549–550, 555
radioactivity, 479, 481, 549, 554–555, 558–563, 566
radius, 60–62, 136, 142
   allowed, quantized, 486
rainbow, 333, 359, 364–366, 601
   secondary, 365–366
range
   audible, 316–318, 322
   particles, 537t
rarefaction, 314
ray, 337g–346, 360–368, 383, 390–391
   diagram, 341, 346–347, 367–370, 383–384
   incident, 340, 345
   light, 337–347, 360–364, 368, 370
   reflected, 340, 344–347
reductionist view, 600–601
reflection, 298–299, 315, 324, 336, 340, 359–366, 375
   angle of, 340
   coating, 391
   and color, 350–353
   diffuse, 340g
   fixed end, 298
   free end, 298
   law of, 340g, 343, 345, 383
   multiple, 342
   and polarization, 392–393
   rainbow formation, 364–366
   retro-, 344, 516
   rule for, 359
   superposition, 385–388
   total internal, 361–362g–364
reflector, 376
   Cassegrain, 377
   Newtonian, 377
refraction, 299, 358–359g–366, 372–374, 384g–385
   angle, 359–360, 362
   atmospheric, 358, 362–363
   bent image, 361
   dispersion, 364
   eye, 372
   halos and sundogs, 366
   index of, 359g–364, 371, 384–385g, 390
   magnification, 374
   rainbows, 364
   total internal reflection, 361
refractor, 375
refrigerator, 274g–275
   schematic, 275
Reines, Frederick, 575
relativity, 481, 507, 516, 558, 573, 582
   centrifugal force, 57, 60, 67–68, 163g, 168
   classical, 156, 177
   Coriolis force, 168g–169
   Earth's rotation, 164–169
   Einstein's theory of, 499
   Galilean principle, 157g, 177
   general, 86, 176, 191–192g–196, 590–591, 597
      black hole, 196
      equivalence principle, 194g
      gravitational wave, 196
      and light, 196
      spacetime, 190g–191, 195
   inertial force, 159g–163, 166–169, 188, 192, 194
   special, 86, 153, 177g, 181, 186–195
      adjustment factor, 186e–191
      clock, 181–184
      energy, 192
      ether, 178g
      first postulate, 177g–180, 184–186, 190
      length contraction, 187–190
      mass, 192
      principle of, 177–178, 192, 455
      rest–mass energy (see energy, rest–mass)
      second postulate, 178g–179
      simultaneity, 179–183, 188
      synchronization, 181–185
      time as fourth dimension, 153
      time dilation, 185–188
      twin paradox, 188
rem (unit), 538g
reproducibility, 34
repulsion, 405–414, 419, 432, 447, 549–550, 554, 561, 576
resistance, 33
   air, 46–47, 64, 67–68, 105, 240
   bulbs in parallel, 435
   bulbs in series, 434
   electric, 431–432g–432e–437, 440, 452
   Ohm's law, 432g–433
   unit, 432
   various materials, 432
resistor, 432, 439
   symbol, 439
resolution
   diffraction, 389
   microscope, 501
   telescope, 389
resonance, 293–294g, 560
resonator, 316
retina, 371–374, 537
retroreflector, 344, 516
revolutions, 135
right-hand rule, 135, 449
Roemer, Ole, 349
Roentgen, Wilhelm, 472, 492, 525
Roosevelt, Franklin, 191, 490
ROY G. BIV, 364–365
rotation
   acceleration, 135g–138
   axis, 135–139, 144, 164–165, 453
   Earth's, 15, 83–84, 154, 164–170, 453
   inertia, 138g, 142–144
   kinetic energy (see energy, kinetic, rotational)
   motion, 66–67, 134–135, 138–139
   Newton's first law, 135–136, 139
   Newton's second law, 138–139
   reference system, 163
   speed, 135g–136, 143, 145, 163, 349
   torque (see torque)
   velocity, 135g, 142, 144
Rutherford, Ernest, 479–482, 486, 490, 527–531, 542, 549, 557, 560
   biography, 481

S

St. John the Evangelist, 334
Salam, Abdus, 594
scale, atomic, 513
scattering, 353
Schrieffer, J. Robert, 452
Schrödinger, Erwin, 505
   equation, 505–507, 510
Schwinger, Julian, 577
scientific
   acceptance, 4
   basis, 6
   creative leap, 2
   merit, 2
   model, 2
   process, 3–4
   proof, 4
Seaborg, Glenn, 561
second (unit), 6
secondary coil, 457, 459
series, 428g
shadow, 337–338
shell, 491g–492, 499, 509–510, 526
silica aerogel, 229
simultaneity, 179–183, 188
   relative, 180
sink, 236–237

Sklodowska, Bronya, 527
slug (unit), 42
smoke detector, 536
Snell's law, 359
Soddy, Frederick, 481
Solar System, 24, 165, 169
solenoid, 449, 459
solid, 227–228, 230g, 233, 235, 255
  structure, 230
Solovine, Maurice, 3
sonar, 316
sonic boom, 327
sound, 114, 178, 313–328, 348, 350, 439, 552, 558
  animal hearing, 318t
  barrier, 327
  Doppler effect, 325–326
  harmonics, 303g–304, 318–323
  infra-, 318
  intensity, 316–317, 322
  levels, 322t
  loudest, 322
  music, 313–314, 318–323
  perception, 316
  softest, 322
  speed (see speed, of sound)
  ultra-, 314, 318
  wave (see wave, sound)
source point, 370
space, 15
  warped, 4, 153
spacetime, 190g, 602
  four-dimensional flat, 195
  warped, 191, 195
spark, 403, 405, 420
spectral line, 491, 512, 532, 580
  absorption, 476
  of antiatoms, 574
  comparison, 476, 488
  emission, 475–476, 479
  explaining, 476, 479, 482, 486, 488
  explaining discrete orbits, 499
  width, 512
spectrum, 364–365
  absorption, 476g, 489
  continuous, 482–483, 489, 493
  electromagnetic, 463–464, 488–489, 509
  emission, 475g–476, 479, 488–489
  radiation, 483
  white light, 476
  X ray, 493
speed, 240, 458, 463
  10-m dash, 19
  absolute nature, 153
  and adjustment factor, 186–192
  of alpha particle, 480
  animal, 19
  Apollo, 19
  of atomic particles, 207–208, 216–217, 220
  average, 15g–15e–27, 64, 216, 221
    electron in wire, 432
  Concorde, 19
  constant, 21–23, 35, 57, 60–61, 67, 75, 155, 165
    relative to the ground, 168
  continental drift, 19
  Earth orbit, 19, 164
  electron, 499–500
  fastest, 19
  instantaneous, 18g–28
  and kinetic energy, 113, 121
  of light, 19, 158, 315, 384–387, 463, 579, 591–592, 599, 602
    constant, 182, 185–186
    dependence on frequency or wavelength, 483
    in material, 385e
    measuring, 348–349
    and relativity, 178–186
  machine, 19
  of planets, 19
  pulse, 297–298
  relative, 178–179, 186
  rotational, 135g–136, 142–143, 163, 349
    units, 135
  slowest, 19
  of sound, 186, 315, 327
  space shuttle, 19
  SR-71A Blackbird, 19
  and stopping distance, 117–118
  supersonic, 327
  terminal, 46g–48, 97
  unit, 17
  wave, 297–300e–304, 314, 319–323, 328, 463e
spin, 528, 580–585
  down, 508–511
  electron, 507–511
  up, 508–511
spontaneous emission, 479
spring, 289–295, 298
  constant, 291g, 292
  units, 291
stability, 141
  line of, 554g–557
  nuclear, 549, 554–557
state
  angular momentum, 501
  discrete, 510
  energy, 506, 510–512, 516
  ground, 506, 510, 512
  of matter, 228, 230
  metastable, 516
  spin, 510
steam engine, 270
stereophonic, 465
Stradivari, Antonio, 319
strangeness, 583g–584
  conservation, 583
  quantum number, 583
strange quark, 583, 585
stress pattern, 382, 396
stroboscope, 17
structure
  atomic, 538, 549
  electronic, 527, 530
  nuclear, 535
subsonic, 327
sundog, 366
superconductivity, 452
supercrest, 305

supernova, 591, 597
superposition, 299–300g–302, 304, 314, 317–319, 324, 327, 385–388, 450
  principle, 302
supersonic, 327
supertrough, 305
switch, 431, 438
  symbol, 438–439
  three-way, 438
symbol, chemical, 530
system
  absolute reference, 177
  accelerating, 158–159
  disordered, 276g–278
  global positioning (GPS), 193
  inertial reference, 157g–160, 166, 169, 177–193
  isolated, 100, 112, 126, 278, 406
  moving, 185–186
  noninertial reference, 159g–162
  optical, 373
  ordered, 276g–278
  reference, 155g–159, 177, 181–183, 189–192
  rotating, 163, 164, 168
  wide area augmentation (WAAS), 193

T

Tacoma Narrows Bridge, 296
tau
  lepton, 580–584
  neutrino, 598
Taylor, Joseph, 592–593
telescope, 367, 374–377, 389, 391
  Galileo, 4
  Hubble Space, 376, 389
  prime focus, 377
  radio, 446
  reflecting, 376g
  refracting, 375g
temperature, 206, 293, 472, 487, 526, 538, 543, 563–566, 594, 596
  absolute, 219–221
  absolute zero, 252g
  and atomic speed, 216
  body, 218–219
  boiling, 256
  critical, 452
  decomposition, 228
  effect on continuous spectra, 482
  and engines and refrigerators, 269–270
  equivalent, 260
  and ideal gas, 220e–221
  macroscopic property, 207–208
  melting, 228
  and number of molecules in a volume, 214
  of radiating objects, 482–483
  and resistance in a wire, 432, 434, 440

scale, 218–219
  absolute, 219g–220, 252
  Celsius, 218–219g–220, 274
  Fahrenheit, 218g–220
  Kelvin, 219g–220, 252, 274
  and speed of sound, 315
  and thermal expansion, 262
  of the Universe, 596
  vaporization, 228
tension, 45, 292, 298, 321
  surface, 231
Tereskova, Valentina, 83
tesla (unit), 451g
theory
  BCS, 452
  color, 594
  electroweak, 594
  of everything, 594
  of evolution, 535
  grand unified, 594
  Maxwell's, 177
  quantum, 191
  superstring, 594
  unified, 593–594
thermal, 258
  contact, 250, 253, 270, 275
  efficiency, 251
  energy (see energy, thermal)
thermodynamics, 249g–250, 463, 472, 484
  first law, 251–252g, 269, 271–273, 281
  laws, 250
  second law, 251g, 269, 273–277
    entropy form, 278–280
    heat engine form, 271g
    refrigerator form, 275
  third law, 252g
  zeroth law, 250g
thermometer, 218, 251
thermonuclear, 558
  reaction, 599
thermostat, 262
Thomson, J. J., 477–481, 490, 529
Thomson, William (Lord Kelvin), 219, 251
tides, 84–88
time, 15, 176–177, 189–190, 279, 512–513
  dilation, 185–188
  machine, 185
  reaction, 117
  relativistic, 187
  slowed down, 4
Titus of Wittenberg, 5
Tomonaga, Shinichiro, 577
top quark, 571, 584t–585g
torque, 135–136g–145, 408, 448, 459
  equation, 136
  net, 137–140
  units, 136
transformer, 440, 457–458
tritium, 530
trough, 298g–305, 321–326, 389
Truman, Harry, 561

U

ultrasound, 314
umbra, 337g–338, 388
uncertainty principle, 511–512g–514, 528, 578–580
 energy/time, 512e
 position/momentum, 512e
units
 conversion, 7–8
 customary system, 6–8
 metric system, 6–8
 SI, 7g–8
 See also specific quantity
up quark, 583–584t–586
upsilon, 584
Urey, Harold, 556
Usachev, Yury, 162

V

vacuum
 electromagnetic waves, 259
 and free fall, 24–25, 45
 and light, 178–179, 334, 396
 and particle decay, 582
 propulsion, 105
 pump, 477
 refractive index, 360
 and speed of light, 348–349, 385
 tube, 551
Van Allen radiation belts, 233
Van de Graaff generator, 549–550
vector, 20g, 25, 47–50, 58–59, 98–99, 135–136, 144, 157, 416–417
 acceleration, 22, 25, 39, 42
 addition, 38–40
 angular momentum, 144
 average velocity, 20
 displacement, 20, 39
 electric field, 415–418
 force, 39, 42, 58
 impulse, 98–99
 momentum, 99, 101, 104
 rotational, 135
 subtraction, 58
 torque, 145
 velocity, 39, 58–59
velocity, 20g–27, 38, 45, 57, 60, 95, 97–104, 116, 122, 208, 238–239, 550
 and acceleration, 21–22, 40, 45
 average, 20g–20e
 change in, 58–59
 circular motion, 57
 and classical relativity, 177
 and collisions, 95
 constant, 38, 64, 67, 157, 160, 177, 182
 and Doppler effect, 326
 Earth's orbital, 169
 inertial reference system, 157
 initial, 58–60
 instantaneous, 66
 and magnetic force, 454
 and momentum, 512
 noninertial reference system, 168, 177
 relative, 156–158, 179–180, 188
  constant, 156–157, 170, 180
 rotational, 135g, 142–146
vibration, 288–289g–303, 314, 317–319, 392
 periodic, 296
 resonant, 289
view, atomistic, 205
viscosity, 232g
volt (unit), 420g
Volta, Alessandro, 427–428
voltage, 403, 420, 427–440, 452, 457–459, 477, 542, 550
 model, 436–437
 unit, 420
volume, 206–207, 212, 214, 218–221, 228–232, 237–239, 255
von Guericke, Otto, 234
von Neumann, 577
Vulovic, Vesna, 99

W

water
 heavy, 562, 598, 600
 model, 430–432
  limitations, 431
 vapor, 260
watt (unit), 127g, 439g
watt-hour (unit), 128
Watt, James, 127, 270
 steam engine, 270
wave, 287g, 288–295g–307, 383–394, 402, 477, 485–486, 499–506, 513
 amplitude, 505–506
 compression, 315
 de Broglie's, 499–501, 506
 electromagnetic, 177, 259, 326, 461–462g–465, 472, 479, 481–482, 499, 515, 591
 equation, 301, 505
 function, 505
 fundamental standing, 506
 gravitational, 196, 591–593
 intensity, 503, 505
 light, 177–178, 289, 383–394, 462, 502, 505
 longitudinal, 296g–297, 299, 314, 321, 392
 matter, 289, 505–506, 513
 mechanical, 316, 505
 nature of light, 334
 particle dilemma, 502
 particle duality, 499, 504, 511, 513, 601
 periodic, 300g–306
 primary, 299
 probability, 505, 515
 radio, 177, 289, 300, 349, 463–464, 574
 secondary, 299
 shock, 328g
 sound, 178, 287, 289, 300, 314–316, 321, 324, 464, 505, 601
 speed, 287
 standing, 296, 302g–304, 319, 323, 499–500
 television, 289
 tidal, 295
 transverse, 296g–297, 299, 302, 315, 321, 392
 traveling, 299, 302–304, 500
 ultraviolet, 464
 water, 503–505
 wavelength, 300g–306, 319–327, 385–391, 395, 464, 482–484, 549
  alpha particle, 549
  de Broglie, 501–502
  diffraction, 306
  fundamental, 319, 320, 323
  harmonic, 303–304
  interference, 305–306
  laser, 515
  in material, 390e
  quantized, 500, 506–507
  resonant, 303
  spectral lines, 475–476, 488–489, 499
  standing wave, 499–500
  symbol, 300
  and uncertainty principle, 511–513
Weber, Joseph, 591–592
weight, 8, 40–43g–44e–50, 62, 124, 137, 139–141, 233, 237, 239, 269, 289
 atomic, 463
 elevator, 160–163
 free fall, 24
 gravity, 88
 and mass, 8, 40–43g–44e
 on other planets, 82
 weightlessness, 162–163
Weinberg, Steven, 594
Wide Area Augmentation System (WAAS), 193
Wilson, Robert, 596
WIMPs, 597
Windaus, Adolf, 556
wind chill, 260–261t
work, 116g–116e–119, 552
 and efficiency, 273
 electric, 418, 431
 and heat, 248–251
 heat engine, 269–271
 and internal energy, 248, 251–252, 268
 and kinetic energy, 116–118
 mechanical, 270–271, 280
 refrigerator, 275
 and temperature, 248
 unit, 116
Working It Out
 acceleration, 23
 average speed, 18
 centripetal acceleration, 61
 collision, 103
 conservation of energy, 115
 cost of electric energy, 440
 density, 230
 electric power, 439–440
 energy of a quantum, 483
 gravitational and electrical force, 414
 gravity, 81
 momentum, 100
 nuclear binding energies, 553
 Ohm's law, 432–433
 period of a mass on a spring, 291
 period of a pendulum, 292
 power, 128
 projectile motion, 64
 radioactive dating, 536
 relativistic lengths, 189
 relativistic times, 187
 second law, 42
 should you jump? 26
 specific heat, 254
 speed of a wave, 301
 thermal expansion, 262
 weight, 44
world
 atomic, 503, 511
 common sense, 287
 macroscopic, 502, 549
 material, 205–206, 228
 real, 2
world view, 1–4, 15, 37, 75, 97, 190, 404, 427, 474, 513–514, 601
 Aristotelian, 34, 205, 209, 572
 atomic, 209
 building, 27
 common sense, 4, 8, 15, 164, 506, 576, 601
 Einsteinian, 195
 Greek, 166
 modern, 34, 36
 Newtonian, 37, 49, 86, 159, 190, 195
 physics, 2, 4, 8, 20, 24, 34, 36, 38, 43, 46, 87, 95, 155, 177, 181, 192, 195, 205–206, 287, 335, 384, 402, 406, 415, 455, 472, 478, 486, 515, 523, 572, 591, 601–602
 scientific, 269
 your, 2

X

X ray, 196, 464, 481, 492g–493, 501, 525–528, 532, 540, 591

Y

Yeager, Chuck, 327
Young, Thomas, 385–387
Yukawa, Hideki, 578

Z

Zweig, George, 583